Selected Titles in This Series

85 **V. A. Kozlov, V. G. Maz'ya, and J. Rossmann,** Spectral problems associated with corner singularities of solutions to elliptic equations, 2001

84 **László Fuchs and Luigi Salce,** Modules over non-Noetherian domains, 2001

83 **Sigurdur Helgason,** Groups and geometric analysis: Integral geometry, invariant differential operators, and spherical functions, 2000

82 **Goro Shimura,** Arithmeticity in the theory of automorphic forms, 2000

81 **Michael E. Taylor,** Tools for PDE: Pseudodifferential operators, paradifferential operators, and layer potentials, 2000

80 **Lindsay N. Childs,** Taming wild extensions: Hopf algebras and local Galois module theory, 2000

79 **Joseph A. Cima and William T. Ross,** The backward shift on the Hardy space, 2000

78 **Boris A. Kupershmidt,** KP or mKP: Noncommutative mathematics of Lagrangian, Hamiltonian, and integrable systems, 2000

77 **Fumio Hiai and Dénes Petz,** The semicircle law, free random variables and entropy, 2000

76 **Frederick P. Gardiner and Nikola Lakic,** Quasiconformal Teichmüller theory, 2000

75 **Greg Hjorth,** Classification and orbit equivalence relations, 2000

74 **Daniel W. Stroock,** An introduction to the analysis of paths on a Riemannian manifold, 2000

73 **John Locker,** Spectral theory of non-self-adjoint two-point differential operators, 2000

72 **Gerald Teschl,** Jacobi operators and completely integrable nonlinear lattices, 1999

71 **Lajos Pukánszky,** Characters of connected Lie groups, 1999

70 **Carmen Chicone and Yuri Latushkin,** Evolution semigroups in dynamical systems and differential equations, 1999

69 **C. T. C. Wall (A. A. Ranicki, Editor),** Surgery on compact manifolds, second edition, 1999

68 **David A. Cox and Sheldon Katz,** Mirror symmetry and algebraic geometry, 1999

67 **A. Borel and N. Wallach,** Continuous cohomology, discrete subgroups, and representations of reductive groups, second edition, 2000

66 **Yu. Ilyashenko and Weigu Li,** Nonlocal bifurcations, 1999

65 **Carl Faith,** Rings and things and a fine array of twentieth century associative algebra, 1999

64 **Rene A. Carmona and Boris Rozovskii, Editors,** Stochastic partial differential equations: Six perspectives, 1999

63 **Mark Hovey,** Model categories, 1999

62 **Vladimir I. Bogachev,** Gaussian measures, 1998

61 **W. Norrie Everitt and Lawrence Markus,** Boundary value problems and symplectic algebra for ordinary differential and quasi-differential operators, 1999

60 **Iain Raeburn and Dana P. Williams,** Morita equivalence and continuous-trace C^*-algebras, 1998

59 **Paul Howard and Jean E. Rubin,** Consequences of the axiom of choice, 1998

58 **Pavel I. Etingof, Igor B. Frenkel, and Alexander A. Kirillov, Jr.,** Lectures on representation theory and Knizhnik-Zamolodchikov equations, 1998

57 **Marc Levine,** Mixed motives, 1998

56 **Leonid I. Korogodski and Yan S. Soibelman,** Algebras of functions on quantum groups: Part I, 1998

55 **J. Scott Carter and Masahico Saito,** Knotted surfaces and their diagrams, 1998

54 **Casper Goffman, Togo Nishiura, and Daniel Waterman,** Homeomorphisms in analysis, 1997

(Continued in the back of this publication)

Spectral Problems Associated with Corner Singularities of Solutions to Elliptic Equations

Mathematical
Surveys
and
Monographs

Volume 85

Spectral Problems Associated with Corner Singularities of Solutions to Elliptic Equations

V. A. Kozlov

V. G. Maz´ya

J. Rossmann

American Mathematical Society

Editorial Board

2000 *Mathematics Subject Classification.* Primary 31B30, 35J05, 35J40,
35J55, 35P15, 35Q30, 47A75, 74B05.

Library of Congress Cataloging-in-Publication Data

Kozlov, Vladimir, 1954–
    Spectral problems associated with corner singularities of solutions of elliptic equations / by
V. A. Kozlov, V. G. Mazya and J. Rossmann.
        p. cm. — (Mathematical surveys and monographs, ISSN 0076-5376 ; v. 85)
    Includes bibliographical references and index.
    ISBN 0-8218-2727-8 (alk. paper)
    1. Differential equations, Elliptic—Numerical solutions. 2. Boundary value problems—
Numerical solutions. 3. Singularities (Mathematics) 4. Mathematical physics. I. Maz'ia, V. G.
II. Rossmann, J. (Jürgen), 1954– III. Title. IV. Mathematical surveys and monographs ; no. 85.

QA377.K66 2000
515′.353—dc21   00-045110

Contents

Introduction   1

Part 1. Singularities of solutions to equations of mathematical physics   7

Chapter 1. Prerequisites on operator pencils   9
   1.1. Operator pencils   10
   1.2. Operator pencils corresponding to sesquilinear forms   15
   1.3. A variational principle for operator pencils   21
   1.4. Elliptic boundary value problems in domains with conic points: some basic results   26
   1.5. Notes   31

Chapter 2. Angle and conic singularities of harmonic functions   35
   2.1. Boundary value problems for the Laplace operator in an angle   36
   2.2. The Dirichlet problem for the Laplace operator in a cone   40
   2.3. The Neumann problem for the Laplace operator in a cone   45
   2.4. The problem with oblique derivative   49
   2.5. Further results   52
   2.6. Applications to boundary value problems for the Laplace equation   54
   2.7. Notes   57

Chapter 3. The Dirichlet problem for the Lamé system   61
   3.1. The Dirichlet problem for the Lamé system in a plane angle   64
   3.2. The operator pencil generated by the Dirichlet problem in a cone   74
   3.3. Properties of real eigenvalues   83
   3.4. The set functions Γ and F_ν 88
   3.5. A variational principle for real eigenvalues   91
   3.6. Estimates for the width of the energy strip   93
   3.7. Eigenvalues for circular cones   97
   3.8. Applications   100
   3.9. Notes   105

Chapter 4. Other boundary value problems for the Lamé system   107
   4.1. A mixed boundary value problem for the Lamé system   108
   4.2. The Neumann problem for the Lamé system in a plane angle   120
   4.3. The Neumann problem for the Lamé system in a cone   125
   4.4. Angular crack in an anisotropic elastic space   133
   4.5. Notes   138

Chapter 5. The Dirichlet problem for the Stokes system   139

5.1. The Dirichlet problem for the Stokes system in an angle   142
5.2. The operator pencil generated by the Dirichlet problem in a cone   148
5.3. Properties of real eigenvalues   155
5.4. The eigenvalues $\lambda=1$ and $\lambda=-2$ 159
5.5. A variational principle for real eigenvalues   168
5.6. Eigenvalues in the case of right circular cones   175
5.7. The Dirichlet problem for the Stokes system in a dihedron   178
5.8. Stokes and Navier–Stokes systems in domains with piecewise smooth boundaries   192
5.9. Notes   196

Chapter 6. Other boundary value problems for the Stokes system in a cone   199
6.1. A mixed boundary value problem for the Stokes system   200
6.2. Real eigenvalues of the pencil to the mixed problem   212
6.3. The Neumann problem for the Stokes system   223
6.4. Notes   225

Chapter 7. The Dirichlet problem for the biharmonic and polyharmonic equations   227
7.1. The Dirichlet problem for the biharmonic equation in an angle   229
7.2. The Dirichlet problem for the biharmonic equation in a cone   233
7.3. The polyharmonic operator   239
7.4. The Dirichlet problem for Δ^2 in domains with piecewise smooth boundaries   246
7.5. Notes   248

Part 2. Singularities of solutions to general elliptic equations and systems   251

Chapter 8. The Dirichlet problem for elliptic equations and systems in an angle   253
8.1. The operator pencil generated by the Dirichlet problem   254
8.2. An asymptotic formula for the eigenvalue close to m 263
8.3. Asymptotic formulas for the eigenvalues close to $m-1/2$ 265
8.4. The case of a convex angle   272
8.5. The case of a nonconvex angle   275
8.6. The Dirichlet problem for a second order system   283
8.7. Applications   286
8.8. Notes   291

Chapter 9. Asymptotics of the spectrum of operator pencils generated by general boundary value problems in an angle   293
9.1. The operator pencil generated by a regular boundary value problem   293
9.2. Distribution of the eigenvalues   299
9.3. Notes   305

Chapter 10. The Dirichlet problem for strongly elliptic systems in particular cones   307
10.1. Basic properties of the operator pencil generated by the Dirichlet problem   308

10.2. Elliptic systems in $\mathbb{R}^n$ 313
10.3. The Dirichlet problem in the half-space   319
10.4. The Sobolev problem in the exterior of a ray   321
10.5. The Dirichlet problem in a dihedron   332
10.6. Notes   344

Chapter 11. The Dirichlet problem in a cone   345
11.1. The case of a "smooth" cone   346
11.2. The case of a nonsmooth cone   350
11.3. Second order systems   353
11.4. Second order systems in a polyhedral cone   365
11.5. Exterior of a thin cone   368
11.6. A cone close to the half-space   376
11.7. Nonrealness of eigenvalues   383
11.8. Further results   384
11.9. The Dirichlet problem in domains with conic vertices   386
11.10. Notes   387

Chapter 12. The Neumann problem in a cone   389
12.1. The operator pencil generated by the Neumann problem   391
12.2. The energy line   396
12.3. The energy strip   398
12.4. Applications to the Neumann problem in a bounded domain   411
12.5. The Neumann problem for anisotropic elasticity in an angle 414
12.6. Notes   415

Bibliography   417

Index   429

List of Symbols   433

Introduction

*"Ce problème est, d'ailleurs, indissoluble-
ment lié à la recherche des points sin-
guliers de f, puisque ceux-ci constituent,
au point de vue de la théorie moderne
des fonctions, la plus importante des pro-
priétés de f."*
Jacques Hadamard
Notice sur les travaux scientifiques,
Gauthier-Villars, Paris, 1901, p.2

Roots of the theory. In the present book we study singularities of solutions
to classical problems of mathematical physics as well as to general elliptic equations
and systems. Solutions of many problems of elasticity, aero- and hydrodynamics,
electromagnetic field theory, acoustics etc., exhibit singular behavior inside the
domain and at the border, the last being caused, in particular, by irregularities of
the boundary. For example, fracture criteria and the modelling of a flow around
the wing are traditional applications exploiting properties of singular solutions.

The significance of mathematical analysis of solutions with singularities had
been understood long ago, and some relevant facts were obtained already in the
19th century. As an illustration, it suffices to mention the role of the Green and
Poisson kernels. Complex function theory and that of special functions were rich
sources of information about singularities of harmonic and biharmonic functions,
as well as solutions of the Lamé and Stokes systems.

In the 20th century and especially in its second half, a vast number of mathematical papers about particular and general elliptic boundary value problems in
domains with smooth and piecewise smooth boundaries appeared. The modern theory of such problems contains theorems on solvability in various function spaces,
estimates and regularity results, as well as asymptotic representations for solutions
near interior points, vertices, edges, polyhedral angles etc. For a factual and historical account of this development we refer to our recent book [**136**], where a detailed
exposition of a theory of linear boundary value problems for differential operators
in domains with smooth boundaries and with isolated vertices at the boundary is
given.

Motivation. The serious inherent drawback of the elliptic theory for nonsmooth domains is that most of its results are conditional. The reason is that
singularities of solutions are described in terms of spectral properties of certain

1

pencils[1] of boundary value problems on spherical domains. Hence, the answers to natural questions about continuity, summability and differentiability of solutions are given under a priori conditions on the eigenvalues, eigenvectors and generalized eigenvectors of these operator pencils.

The obvious need for the unconditional results concerning solvability and regularity properties of solutions to elliptic boundary value problems in domains with piecewise smooth boundaries makes spectral analysis of the operator pencils in question vitally important. Therefore, in this book, being interested in singularities of solutions, we fix our attention on such a spectral analysis. However, we also try to add another dimension to our text by presenting some applications to boundary value problems. We give a few examples of the questions which can be answered using the information about operator pencils obtained in the first part of the book:

• Are variational solutions of the Navier-Stokes system with zero Dirichlet data continuous up to the boundary of an arbitrary polyhedron?

• The same question for the Lamé system with zero Dirichlet data.

• Are the solutions just mentioned continuously differentiable up to the boundary if the polyhedron is convex?

One can easily continue this list, but we stop here, since even these simply stated questions are so obviously basic that the utility of the techniques leading to the answers is quite clear. (By the way, for the Lamé system with zero Neumann data these questions are still open, despite all physical evidence in favor of positive answers.)

Another impetus for the spectral analysis in question is the challenging program of establishing unconditional analogs of the results of the classical theory of general elliptic boundary value problems for domains with piecewise smooth boundaries. This program gives rise to many interesting questions, some of them being treated in the second part of the book.

Singularities and pencils. What kind of singularities are we dealing with, and how are they related to spectral theory of operator pencils? To give an idea, we consider a solution to an elliptic boundary value problem in a cone. By Kondrat'ev's theorem [**109**], this solution, under certain conditions, behaves asymptotically near the vertex $\mathcal{O}$ as

$$(1) \qquad |x|^{\lambda_0} \sum_{k=0}^{s} \frac{1}{k!} (\log|x|)^k u_{s-k}(x/|x|),$$

where λ_0 is an eigenvalue of a pencil of boundary value problems on a domain, the cone cuts out on the unit sphere. Here, the coefficients are: an eigenvector u_0, and generalized eigenvectors $u_1, \ldots, u_s$ corresponding to λ_0. In what follows, speaking about singularities of solutions we always mean the singularities of the form (1).

It is worth noting that these power-logarithmic terms describe not only point singularities. In fact, the singularities near edges and vertices of polyhedra can be characterized by similar expressions.

The above mentioned operator pencil is obtained (in the case of a scalar equation) by applying the principal parts of domain and boundary differential operators

---

[1] The operators polynomially depending on a spectral parameter are called *operator pencils*, for the definition of their eigenvalues, eigenvectors and generalized eigenvectors see Section 1.1

to the function $r^\lambda u(\omega)$, where $r = |x|$ and $\omega = x/|x|$. Also, this pencil appears under the Mellin transform of the same principal parts. For example, in the case of the n-dimensional Laplacian Δ, we arrive at the operator pencil $\delta + \lambda(\lambda + n - 2)$, where δ is the Laplace-Beltrami operator on the unit sphere. The pencil corresponding to the biharmonic operator Δ^2 has the form:

$$\delta^2 + 2\left(\lambda^2 + (n-5)\lambda - n + 4\right)\delta + \lambda(\lambda - 2)(\lambda + n - 2)(\lambda + n - 4).$$

Even less attractive is the pencil generated by the Stokes system

$$\begin{pmatrix} U \\ P \end{pmatrix} \to \begin{pmatrix} -\Delta U + \nabla P \\ \nabla \cdot U \end{pmatrix},$$

where U is the velocity vector and P is the pressure. Putting

$$U(x) = r^\lambda u(\omega) \quad \text{and} \quad P(x) = r^{\lambda - 1} p(\omega),$$

one can check that this pencil looks as follows in the spherical coordinates (r, θ, φ):

$$\begin{pmatrix} u_r \\ u_\theta \\ u_\varphi \\ p \end{pmatrix} \to \begin{pmatrix} -\delta u_r - (\lambda - 1)(\lambda + 2)u_r + 2\dfrac{\partial_\theta(\sin\theta\, u_\theta) + \partial_\varphi u_\varphi}{\sin\theta} + (\lambda - 1)p \\[2ex] -\delta u_\theta - \lambda(\lambda + 1)u_\theta + \dfrac{u_\theta + 2\cos\theta\, \partial_\varphi u_\varphi}{\sin^2\theta} - 2\partial_\theta u_r + \partial_\theta p \\[2ex] -\delta u_\varphi - \lambda(\lambda + 1)u_\varphi + \dfrac{u_\varphi - 2\cos\theta\, \partial_\varphi u_\theta}{\sin^2\theta} - \dfrac{2\partial_\varphi u_r - \partial_\theta p}{\sin\theta} \\[2ex] \dfrac{\partial_\theta(\sin\theta\, u_\theta) + \partial_\varphi u_\varphi}{\sin\theta} + (\lambda + 2)u_r \end{pmatrix}.$$

Here ∂_θ and ∂_φ denote partial derivatives.

In the two-dimensional case, when the pencil is formed by ordinary differential operators, its eigenvalues are roots of a transcendental equation for an entire function of a spectral parameter λ. In the higher-dimensional case and for a cone of a general form one has to deal with nothing better than a complicated pencil of boundary value problems on a subdomain of the unit sphere.

Fortunately, many applications do not require explicit knowledge of eigenvalues. For example, this is the case with the question whether solutions having a finite energy integral are continuous near the vertex. For $2m < n$ the affirmative answer results from the absence of nonconstant solutions (1) with $m - n/2 < \operatorname{Re}\lambda_0 \leq 0$.

Since the investigation of regularity properties of solutions with the finite energy integral is of special importance, we are concerned with the widest strip in the λ-plane, free of eigenvalues and containing the "energy line" $\operatorname{Re}\lambda = m - n/2$. Information on the width of this "energy strip" is obtained from lower estimates for real parts of the eigenvalues situated over the energy line. Sometimes, we are able to establish the monotonicity of the energy strip with respect to the opening of the cone. We are interested in the geometric, partial and algebraic multiplicities of eigenvalues, and find domains in the complex plane, where all eigenvalues are real or nonreal. Asymptotic formulae for large eigenvalues are also given.

The book is principally based on results of our work and the work of our collaborators during last twenty years. Needless to say, we followed our own taste in the choice of topics and we neither could nor wished to achieve completeness in description of the field of singularities which is currently in process of development.

We hope that the present book will promote further exploration of this field.

Organization of the subject. Nowadays, for arbitrary elliptic problems there exist no unified approaches to the question whether eigenvalues of the associated operator pencils are absent or present in particular domains on the complex plane. Therefore, our dominating principle, when dealing with these pencils, is to depart from boundary value problems, not from methods.

We move from special problems to more general ones. In particular, the two-dimensional case precedes the multi-dimensional one. By the way, this does not always lead to simplifications, since, as a rule, one is able to obtain much deeper information about singularities for $n = 2$ in comparison with $n > 2$.

Certainly, it is easy to describe singularities for particular boundary value problems of elasticity and hydrodynamics in an angle, because of the simplicity of the corresponding transcendental equations. (We include this material, since it was never collected before, is of value for applications, and of use in our subsequent exposition.) On the contrary, when we pass to an arbitrary elliptic operator of order $2m$ with two variables, the entire function in the transcendental equation depends on $2m + 1$ real parameters, which makes the task of investigating the roots quite nontrivial.

It turns out that our results on the singularities for three-dimensional problems of elasticity and hydrodynamics are not absorbed by the subsequent analysis of multi-dimensional higher order equations, because, on the one hand, we obtain a more detailed picture of the spectrum for concrete problems, and, on the other hand, we are not bound up in most cases with the Lipschitz graph assumption about the cone, which appears elsewhere. (The question can be raised if this geometric restriction can be avoided, but it has no answer yet.) Moreover, the methods used for treating the pencils generated by concrete three-dimensional problems and general higher order multi-dimensional equations are completely different. We mainly deal with only constant coefficient operators and only in cones, but these are not painful restrictions. In fact, it is well known that the study of variable coefficient operators on more general domains ultimately rests on the analysis of the model problems considered here.

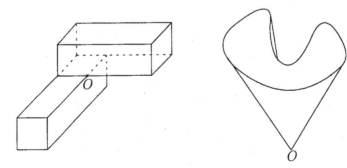

FIGURE 1. On the left: a polyhedron which is not Lipschitz in any neighborhood of O. On the right: a conic surface smooth outside the point O which is not Lipschitz in any neighborhood of O.

Briefly but systematically, we mention various applications of our spectral results to elliptic problems with variable coefficients in domains with nonsmooth boundaries. Here is a list of these topics: L_p- and Schauder estimates along with the corresponding Fredholm theory, asymptotics of solutions near the vertex, pointwise estimates for the Green and Poisson kernels, and the Miranda-Agmon maximum principle.

Structure of the book. According to what has been said, we divide the book into two parts, the first being devoted to the power-logarithmic singularities of solutions to classical boundary value problems of mathematical physics, and the second dealing with similar singularities for higher order elliptic equations and systems.

The first part consists of Chapters 1-7. In Chapter 1 we collect basic facts concerning operator pencils acting in a pair of Hilbert spaces. These facts are used later on various occasions. Related properties of ordinary differential equations with constant operator coefficients are discussed. Connections with the theory of general elliptic boundary value problems in domains with conic vertices are also outlined. Some of results in this chapter are new, such as, for example, a variational principle for real eigenvalues of operator pencils.

The Laplace operator, treated in Chapter 2, is a starting point and a model for the subsequent study of angular and conic singularities of solutions. The results vary from trivial, as for boundary value problems in an angle, to less straightforward, in the many-dimensional case. In the plane case it is possible to write all singular terms explicitly. For higher dimensions the singularities are represented by means of eigenvalues and eigenfunctions of the Beltrami operator on a subdomain of the unit sphere. We discuss spectral properties of this operator.

Our next theme is the Lamé system of linear homogeneous isotropic elasticity in an angle and a cone. In Chapter 3 we consider the Dirichlet boundary condition, beginning with the plane case and turning to the space problem. In Chapter 4, we investigate some mixed boundary conditions. Then by using a different approach, the Neumann problem with tractions prescribed on the boundary of a Lipschitz cone is studied. We deal with different questions concerning the spectral properties of the operator pencils generated by these problems. For example, we estimate the width of the energy strip. For the Dirichlet and mixed boundary value problems we show that the eigenvalues in a certain wider strip are real and establish a variational principle for these eigenvalues. In the case of the Dirichlet problem this variational principle implies the monotonicity of the eigenvalues with respect to the cone.

Parallel to our study of the Lamé system, in Chapters 5 and 6 we consider the Stokes system. Chapter 5 is devoted to the Dirichlet problem. In Chapter 6 we deal with mixed boundary data appearing in hydrodynamics of a viscous fluid with free surface. We conclude Chapter 6 with a short treatment of the Neumann problem. This topic is followed by the Dirichlet problem for the polyharmonic operator, which is the subject of Chapter 7.

The second part of the book includes Chapters 8-12. In Chapter 8, the Dirichlet problem for general elliptic differential equation of order $2m$ in an angle is studied. As we said above, the calculation of eigenvalues of the associated operator pencil leads to the determination of zeros of a certain transcendental equation. Its study is based upon some results on distributions of zeros of polynomials and meromorphic functions. We give a complete description of the spectrum in the strip $m - 2 \leq \operatorname{Re} \lambda \leq m$.

In Chapter 9 we obtain an asymptotic formula for the distribution of eigenvalues of operator pencils corresponding to general elliptic boundary value problems in an angle.

In Chapters 10 and 11 we are concerned with the Dirichlet problem for elliptic systems of differential equations of order $2m$ in a n-dimensional cone. For the cases when the cone coincides with $\mathbb{R}^n \setminus \{O\}$, the half-space $\mathbb{R}^n_+$, the exterior of a ray, or a dihedron, we find all eigenvalues and eigenfunctions of the corresponding operator pencil in Chapter 10. In the next chapter, under the assumptions that the differential operator is selfadjoint and the cone admits an explicit representation in Cartesian coordinates, we prove that the strip $|\operatorname{Re} \lambda - m + n/2| \leq 1/2$ contains no eigenvalues of the pencil generated by the Dirichlet problem. From the results in Chapter 11, concerning the Dirichlet problem in the exterior of a thin cone, it follows that the bound $1/2$ is sharp.

The Neumann problem for general elliptic systems is studied in Chapter 12, where we deal, in particular, with eigenvalues of the corresponding operator pencil in the strip $|\operatorname{Re} \lambda - m + n/2| \leq 1/2$. We show that only integer numbers contained in this strip are eigenvalues.

The applications listed above are placed, as a rule, in introductions to chapters and in special sections at the end of chapters. Each chapter is finished by bibliographical notes.

This is a short outline of the book. More details can be found in the introductions to chapters.

Readership. This volume is addressed to mathematicians who work in partial differential equations, spectral analysis, asymptotic methods and their applications. We hope that it will be of use also for those who are interested in numerical analysis, mathematical elasticity and hydrodynamics. Prerequisites for this book are undergraduate courses in partial differential equations and functional analysis.

Acknowledgements. V. Kozlov and V. Maz'ya acknowledge the support of the Royal Swedish Academy of Sciences, the Swedish Natural Science Research Council (NFR) and the Swedish Research Council for Engineering Sciences (TFR). V. Maz'ya is grateful to the Alexander von Humboldt Foundation for the sponsorship during the last stage of the work on this volume. J. Roßmann would like to thank the Department of Mathematics at Linköping University for hospitality.

Part 1

Singularities of solutions to equations of mathematical physics

CHAPTER 1

Prerequisites on operator pencils

In this chapter we describe the general operator theoretic means which are used in the subsequent analysis of singularities of solutions to boundary value problems. The chapter is auxiliary and mostly based upon known results from the theory of holomorphic operator functions. At the same time we have to include some new material concerning parameter-depending sesquilinear forms and variational principles for their eigenvalues.

Our main concern is with the spectral properties of operator pencils, i.e., operators polynomially depending on a complex parameter λ. We give an idea how the pencils appear in the theory of general elliptic boundary value problems in domains with conic vertices.

Let $\mathcal{G}$ be a domain in the Euclidean space $\mathbb{R}^n$ which coincides with the cone $\mathcal{K} = \{x \in \mathbb{R}^n : x/|x| \in \Omega\}$ in a neighborhood of the origin, where Ω is a subdomain of the unit sphere. We consider solutions of the differential equation

$$(1.0.1) \qquad LU = F \quad \text{in } \mathcal{G}$$

satisfying the boundary conditions

$$(1.0.2) \qquad B_k U = G_k, \quad k = 1, \dots, m,$$

outside the singular points of the boundary $\partial \mathcal{G}$. Here L is a $2m$ order elliptic differential operator and B_k are differential operators of orders m_k. We assume that the operators $L, B_1, \dots, B_m$ are subject to the ellipticity condition.

It is well known that the main results about elliptic boundary value problems in domains with smooth boundaries are deduced from the study of so-called model problems which involve the principal parts of the given differential operators with coefficients frozen at certain point. The same trick applied to the situation we are dealing with leads to the model problem

$$L^\circ U = \Phi \quad \text{in } \mathcal{K},$$
$$B_k^\circ U = \Psi_k \quad \text{on } \partial \mathcal{K} \backslash \{0\}, \ k = 1, \dots, m,$$

where L°, B_k° are the principal parts of L and B_k, respectively, with coefficients frozen at the origin. Passing to the spherical coordinates r, ω, where $r = |x|$ and $\omega = x/|x|$, we arrive at a problem of the form

$$\mathcal{L}(r\partial_r) U = r^{2m} \Phi \quad \text{in } \Omega \times (0, \infty),$$
$$\mathcal{B}_k(r\partial_r) U = r^{m_k} \Psi_k \quad \text{on } \partial\Omega \times (0, \infty), \ k = 1, \dots, m.$$

Now the application of the Mellin transform

$$\tilde{U}(\lambda) = (2\pi)^{-1/2} \int_0^\infty r^{-\lambda-1} U(r) \, dr$$

leads to the boundary value problem

$$\mathcal{L}(\lambda)\,u = f \quad \text{in } \Omega,$$
$$\mathcal{B}_k(\lambda)\,u = g_k \quad \text{on } \partial\Omega, \ k = 1, \dots, m,$$

with the complex parameter λ. Let us denote the polynomial operator (operator pencil) of this problem by $\mathfrak{A}(\lambda)$. Properties of the pencil $\mathfrak{A}$ are closely connected with those of the original boundary value problem (1.0.1), (1.0.2), in particular, with its solvability in various function spaces and the asymptotics of its solutions near the vertex of $\mathcal{K}$ (see Section 1.4). One can show, for example, that the solutions U behave asymptotically like a linear combination of the terms

$$r^\lambda \sum_{k=0}^{s} \frac{1}{k!} \left(\log r\right)^k u_{s-k}(\omega)$$

where λ is an eigenvalue of the pencil $\mathfrak{A}$, u_0 is an eigenfunctions and $u_1, \dots, u_s$ are generalized eigenfunctions corresponding to the eigenvalue λ. Thus, one has been naturally led to the study of spectral properties of polynomial operator pencils.

1.1. Operator pencils

1.1.1. Basic definitions. Let $\mathcal{X}, \mathcal{Y}$ be Hilbert spaces with the inner products $(\cdot, \cdot)_{\mathcal{X}}, (\cdot, \cdot)_{\mathcal{Y}}$ and the norms $\|\cdot\|_{\mathcal{X}}, \|\cdot\|_{\mathcal{Y}}$, respectively. We denote by $L(\mathcal{X}, \mathcal{Y})$ the set of the linear and bounded operators from $\mathcal{X}$ into $\mathcal{Y}$. If $A \in L(\mathcal{X}, \mathcal{Y})$, then by $\ker A$ and $\mathcal{R}(A)$ we denote the kernel and the range of the operator A. The operator A is said to be Fredholm if $\mathcal{R}(A)$ is closed and the dimensions of $\ker A$ and the orthogonal complement to $\mathcal{R}(A)$ are finite. The space of all Fredholm operators is denoted by $\Phi(\mathcal{X}, \mathcal{Y})$.

The operator polynomial

$$(1.1.1) \qquad \mathfrak{A}(\lambda) = \sum_{k=0}^{l} A_k\, \lambda^k, \quad \lambda \in \mathbb{C},$$

where $A_k \in L(\mathcal{X}, Y)$, is called operator pencil.

The point $\lambda_0 \in \mathbb{C}$ is said to be *regular* if the operator $\mathfrak{A}(\lambda_0)$ is invertible. The set of all nonregular points is called the *spectrum* of the operator pencil $\mathfrak{A}$.

DEFINITION 1.1.1. The number $\lambda_0 \in G$ is called an *eigenvalue* of the operator pencil $\mathfrak{A}$ if the equation

$$(1.1.2) \qquad \mathfrak{A}(\lambda_0)\,\varphi_0 = 0$$

has a non-trivial solution $\varphi_0 \in \mathcal{X}$. Every such $\varphi_0 \in \mathcal{X}$ of (1.1.2) is called an *eigenvector* of the operator pencil $\mathfrak{A}$ corresponding to the eigenvalue λ_0. The dimension of $\ker \mathfrak{A}(\lambda)$ is called the *geometric multiplicity* of the eigenvalue λ_0.

DEFINITION 1.1.2. Let λ_0 be an eigenvalue of the operator pencil $\mathfrak{A}$ and let φ_0 be an eigenvector corresponding to λ_0. If the elements $\varphi_1, \dots, \varphi_{s-1} \in \mathcal{X}$ satisfy the equations

$$(1.1.3) \qquad \sum_{k=0}^{j} \frac{1}{k!}\, \mathfrak{A}^{(k)}(\lambda_0)\,\varphi_{j-k} = 0 \qquad \text{for } j = 1, \dots, s-1,$$

where $\mathfrak{A}^{(k)}(\lambda) = d^k\mathfrak{A}(\lambda)/d\lambda^k$, then the ordered collection $\varphi_0, \varphi_1, \ldots, \varphi_{s-1}$ is said to be a *Jordan chain* of $\mathfrak{A}$ corresponding to the eigenvalue λ_0. The vectors $\varphi_1, \ldots, \varphi_{s-1}$ are said to be *generalized eigenvectors* corresponding to φ_0.

The maximal length of all Jordan chains formed by the eigenvector φ_0 and corresponding generalized eigenvectors will be denoted by $m(\varphi_0)$.

DEFINITION 1.1.3. Suppose that the geometric multiplicity of the eigenvalue λ_0 is finite and denote it by I. Assume also that

$$\max_{\varphi\in\ker\mathfrak{A}(\lambda_0)\backslash\{O\}} m(\varphi) < \infty .$$

Then a set of Jordan chains

$$\varphi_{j,0}, \varphi_{j,1}, \ldots, \varphi_{j,\kappa_j-1}, \quad j = 1, \ldots, I,$$

is called *canonical system of eigenvectors and generalized eigenvectors* if

1. the eigenvectors $\{\varphi_{j,0}\}_{j=1,\ldots,I}$ form a basis in $\ker\mathfrak{A}(\lambda_0)$,
2. Let $\mathfrak{M}_j$ be the space spanned by the vectors $\varphi_{1,0}, \ldots, \varphi_{j-1,0}$. Then

$$m(\varphi_{j,0}) = \max_{\varphi\in\ker\mathfrak{A}(\lambda_0)\backslash\mathfrak{M}_j} m(\varphi), \quad j = 1, \ldots, I.$$

The numbers $\kappa_j = m(\varphi_j, 0)$ are called the *partial multiplicities* of the eigenvalue λ_0. The number κ_1 is also called the *index* of λ_0. The sum $\kappa = \kappa_1 + \ldots + \kappa_I$ is called the *algebraic multiplicity* of the eigenvalue λ_0.

1.1.2. Basic properties of operator pencils. The following well-known assertion (see, for example, the book of Kozlov and Maz'ya [**135**, Appendix]) describes an important for applications class of operator pencils whose spectrum consists of isolated eigenvalues with finite algebraic multiplicities.

THEOREM 1.1.1. *Let G be a domain in the complex plane $\mathbb{C}$. Suppose that the operator pencil $\mathfrak{A}$ satisfies the following conditions:*

(i) *$\mathfrak{A}(\lambda) \in \Phi(\mathcal{X}, \mathcal{Y})$ for all $\lambda \in G$.*
(ii) *There exists a number $\lambda \in G$ such that the operator $\mathfrak{A}(\lambda)$ has a bounded inverse.*

Then the spectrum of the operator pencil $\mathfrak{A}$ consists of isolated eigenvalues with finite algebraic multiplicities which do not have accumulation points in G.

The next direct consequence of Theorem 1.1.1 is useful in applications.

COROLLARY 1.1.1. *Let the operator pencil (1.1.1) satisfy the conditions:*

a) *The operators $A_j : \mathcal{X} \to \mathcal{Y}$, $j = 1, \ldots, l$, are compact.*
b) *There exists at least one regular point of the pencil $\mathfrak{A}$.*

Then the result of Theorem 1.1.1 with $G = \mathbb{C}$ is valid for the pencil $\mathfrak{A}$.

The following remark shows that sometimes one can change the domain (of definition) of operator pencils without changing their spectral properties.

REMARK 1.1.1. Let $\mathcal{X}_0$, $\mathcal{Y}_0$ be Hilbert spaces imbedded into $\mathcal{X}$ and $\mathcal{Y}$, respectively. We assume that the operator $\mathfrak{A}(\lambda)$ continuously maps $\mathcal{X}_0$ into $\mathcal{Y}_0$ for arbitrary $\lambda \in \mathbb{C}$ and that every solution $u \in \mathcal{X}$ of the equation

$$\mathfrak{A}(\lambda)\, u = f$$

belongs to $\mathcal{X}_0$ if $f \in \mathcal{Y}_0$. Then the spectrum of the operator pencil (1.1.1) coincides with the spectrum of the restriction

$$\mathfrak{A}(\lambda) \,:\, \mathcal{X}_0 \to \mathcal{Y}_0.$$

The last pencil has the same eigenvectors and generalized eigenvectors as the pencil (1.1.1).

In order to describe the structure of the inverse to the pencil $\mathfrak{A}$ near an eigenvalue, we need the notion of *holomorphic operator functions*.

Let $G \subset \mathbb{C}$ be a domain. An operator function

$$\Gamma \,:\, G \to L(\mathcal{X}, Y)$$

is called holomorphic if in a neighborhood of every point λ_0 it can be represented as a convergent in $L(\mathcal{X}, Y)$ series

$$\Gamma(\lambda) = \sum_{j=0}^{\infty} \Gamma_j (\lambda - \lambda_0)^j,$$

where $\Gamma_j \in L(\mathcal{X}, Y)$ can depend on λ_0.

THEOREM 1.1.2. *Let the operator pencil $\mathfrak{A}$ satisfy the conditions in Theorem 1.1.1. If $\lambda_0 \in G$ is an eigenvalue of $\mathfrak{A}$, then the inverse operator to $\mathfrak{A}(\lambda)$ has the representation*

$$\mathfrak{A}(\lambda)^{-1} = \sum_{j=1}^{\sigma} \frac{T_j}{(\lambda - \lambda_0)^j} + \Gamma(\lambda)$$

in a neighborhood of the point λ_0, where σ is the index of the eigenvalue λ_0, $T_1, \dots, T_\sigma$ are linear bounded finite-dimensional operators and $\Gamma(\lambda)$ is a holomorphic function in a neighborhood of λ_0 with values in $L(\mathcal{X}, \mathcal{Y})$.

The following two theorems help to calculate the total algebraic multiplicity of eigenvalues situated in a certain domain. Their proofs can be found, for example, in the book by Gohberg, Goldberg and Kaashoek [**71**, Sect.XI.9].

THEOREM 1.1.3. *Let the conditions of Theorem 1.1.1 be satisfied. Furthermore, let G be a simply connected domain in $\mathbb{C}$ which is bounded by a piecewise smooth closed curve ∂G and let $\mathfrak{A} : \overline{G} \to L(\mathcal{X}, \mathcal{Y})$ be invertible on ∂G. Then*

$$(1.1.4) \qquad \frac{1}{2\pi i} \operatorname{tr} \int_{\partial G} \mathfrak{A}^{(1)}(\lambda) \, (\mathfrak{A}(\lambda))^{-1} \, d\lambda = \kappa(\mathfrak{A}, G),$$

where $\kappa(\mathfrak{A}, G)$ denotes the sum of the algebraic multiplicities of all eigenvalues of the operator pencil $\mathfrak{A}$ which are situated in the domain G.

Note that, by Theorem 1.1.2, the integral on the left-hand side of (1.1.4) is a finite-dimensional operator. Therefore, the trace of this integral is well defined.

THEOREM 1.1.4. *Let G be a simply connected domain in $\mathbb{C}$ which is bounded by a piecewise smooth closed curve ∂G and let $\mathfrak{A}$, $\mathfrak{B}$ be operator pencils satisfying the conditions of Theorem 1.1.1. Furthermore, we assume that $\mathfrak{A}(\lambda)$ is invertible for $\lambda \in \partial G$ and*

$$\left\| \mathfrak{A}(\lambda)^{-1} \, (\mathfrak{A}(\lambda) - \mathfrak{B}(\lambda)) \right\|_{L(\mathcal{X}, \mathcal{X})} < 1 \qquad for \ \lambda \in \partial G.$$

Then $\mathfrak{B}(\lambda)$ is invertible for $\lambda \in \partial G$ and $\kappa(\mathfrak{A}, G) = \kappa(\mathfrak{B}, G)$.

As a consequence of the last result which is a generalization of Rouché's theorem, we obtain the following assertion.

COROLLARY 1.1.2. *Let $\mathfrak{A}(t,\lambda)$ be an operator pencil with values in $L(\mathcal{X},Y)$ whose coefficients are continuous with respect to $t \in [a,b]$. Furthermore, we suppose that the pencil $\mathfrak{A}(t,\cdot)$ satisfies the conditions of Theorem 1.1.1 for every $t \in [a,b]$. If $\mathfrak{A}(t,\lambda)$ is invertible for $t \in [a,b]$ and $\lambda \in \partial \mathcal{G}$, then $\kappa\big(\mathfrak{A}(t,\cdot),G\big)$ is independent of t.*

REMARK 1.1.2. All definitions and properties of this and the preceding subsections can be obviously extended to holomorphic operator functions. For details we refer the reader to the books by Gohberg, Goldberg and Kaashoek [71], Kozlov and Maz'ya [135].

1.1.3. Ordinary differential equations with operator coefficients. Let $\mathfrak{A}(\lambda)$ be the operator pencil (1.1.1). We are interested in solutions of the ordinary differential equation

$$(1.1.5) \qquad \mathfrak{A}(r\partial_r)\, U(r) = 0 \qquad \text{for } r > 0$$

which have the form

$$(1.1.6) \qquad U(r) = r^{\lambda_0} \sum_{k=0}^{s} \frac{(\log r)^k}{k!}\, u_{s-k}\,,$$

where $\lambda_0 \in \mathbb{C}$ and $u_k \in \mathcal{X}$ $(k = 0, \dots, s)$. Here and elsewhere ∂_r denotes the derivative d/dr.

THEOREM 1.1.5. *The function (1.1.6) is a solution of (1.1.5) if and only if λ_0 is an eigenvalue of the pencil $\mathfrak{A}$ and $u_0, u_1, \dots, u_s$ is a Jordan chain corresponding to the eigenvalue λ_0.*

Proof: We have

$$
\begin{aligned}
(1.1.7) \quad \mathfrak{A}(r\partial_r)\, U(r) &= r^{\lambda_0}\, \mathfrak{A}(\lambda_0 + r\partial_r) \sum_{k=0}^{s} \frac{1}{k!}\, (\log r)^k\, u_{s-k} \\
&= r^{\lambda_0} \sum_{j=0}^{l} \frac{1}{j!}\, \mathfrak{A}^{(j)}(\lambda_0)\, (r\partial_r)^j \sum_{k=0}^{s} \frac{1}{k!}\, (\log r)^k\, u_{s-k} \\
&= r^{\lambda_0} \sum_{k=0}^{s} \frac{1}{k!}\, (\log r)^k \sum_{j=0}^{s-k} \frac{1}{j!}\, \mathfrak{A}^{(j)}(\lambda_0)\, u_{s-k-j}\,.
\end{aligned}
$$

Hence $U(r)$ is a solution of (1.1.5) if and only if the coefficients of $(\log r)^k$ on the right-hand side of the last formula are equal to zero. This proves the theorem. ∎

Let λ_0 be an eigenvalue of the operator pencil $\mathfrak{A}(\lambda)$. We denote by $\mathcal{N}(\mathfrak{A},\lambda_0)$ the space of all solutions of (1.1.5) which have the form (1.1.6). As a consequence of Theorem 1.1.5 we get the following assertion.

COROLLARY 1.1.3. *The dimension of $\mathcal{N}(\mathfrak{A},\lambda_0)$ is equal to the algebraic multiplicity of the eigenvalue λ_0. The maximal power of $\log r$ of the vector functions of $\mathcal{N}(\mathfrak{A},\lambda_0)$ is equal to $m-1$, where m denotes the index of the eigenvalue λ_0.*

Now we consider the inhomogeneous differential equation

(1.1.8) $$\mathfrak{A}(r\partial_r)\, U(r) = F(r).$$

THEOREM 1.1.6. *Suppose that the operator pencil $\mathfrak{A}$ satisfies the conditions of Theorem 1.1.1 and F is a function of the form*

$$F(r) = r^{\lambda_0} \sum_{k=0}^{s} \frac{(\log r)^k}{k!}\, f_{s-k},$$

where $\lambda_0 \in \mathbb{C}$ and $f_k \in \mathcal{Y}$ for $k = 0, \dots, s$. Then equation (1.1.8) has a solution of the form

(1.1.9) $$U(r) = r^{\lambda_0} \sum_{k=0}^{s+\sigma} \frac{(\log r)^k}{k!}\, u_{s+\sigma-k},$$

where $u_0, u_1, \dots, u_{s+\sigma}$ are elements of the space $\mathcal{X}$, σ is the index of λ_0 if λ_0 is an eigenvalue of $\mathfrak{A}$, while $\sigma = 0$ if λ_0 is a regular point.

Proof: By Theorem 1.1.2, the inverse of $\mathfrak{A}(\lambda)$ admits the representation

$$\mathfrak{A}^{-1}(\lambda) = \sum_{k=-\infty}^{\sigma} T_k\, (\lambda - \lambda_0)^k.$$

Here, by the identity $\mathfrak{A}(\lambda)\,\mathfrak{A}^{-1}(\lambda) = I$, the operators T_j satisfy the equalities

(1.1.10) $$\sum_{j=0}^{\sigma+k} \frac{1}{j!}\, \mathfrak{A}^{(j)}(\lambda_0)\, T_{j-k} = \begin{cases} I & \text{for } k = 0, \\ 0 & \text{for } k = -\sigma, \dots, -1, +1, +2, \dots. \end{cases}$$

Let U be the function (1.1.9). Then, analogously to (1.1.7), we get

$$r^{-\lambda_0}\, \mathfrak{A}(r\partial_r)\, U(r) = \sum_{k=0}^{s+\sigma} \frac{1}{k!}\, (\log r)^k \sum_{j=0}^{s+\sigma-k} \frac{1}{j!}\, \mathfrak{A}^{(j)}(\lambda_0)\, u_{s+\sigma-k-j}\, .$$

Setting

$$u_k = \sum_{\nu=0}^{\min(k,s)} T_{\sigma-k+\nu}\, f_n u, \quad k = 0, 1, \dots, s+\sigma,$$

we obtain

$$\begin{aligned}
r^{-\lambda_0}\, \mathfrak{A}(r\partial_r)\, U(r) &= \sum_{k=0}^{s+\sigma} \frac{1}{k!}\, (\log r)^k \sum_{\nu=0}^{\min(s+\sigma-k,s)} \sum_{j=0}^{\sigma+s-k-\nu} \frac{1}{j!}\, \mathfrak{A}^{(j)}(\lambda_0)\, T_{j-s+k+\nu}\, f_\nu \\
&= \sum_{\nu=0}^{s} \sum_{k=0}^{s+\sigma-\nu} \frac{1}{k!}\, (\log r)^k \Big(\sum_{j=0}^{\sigma+s-k-\nu} \cdot \frac{1}{j!}\, \mathfrak{A}^{(j)}(\lambda_0)\, T_{j-s+k+\nu} \Big) f_\nu\, .
\end{aligned}$$

According to (1.1.10), the right side of the last equality is equal to

$$\sum_{\nu=0}^{s} \sum_{k=0}^{s+\sigma-\nu} \frac{1}{k!}\, (\log r)^k \delta_{s-k-\nu,0}\, f_\nu = \sum_{\nu=0}^{s} \frac{1}{(s-\nu)!}\, (\log r)^{s-\nu}\, f_\nu\, .$$

This proves the lemma. ∎

REMARK 1.1.3. If λ_0 is a regular point of the pencil $\mathfrak{A}$, then the solution (1.1.9) is uniquely determined. In the case of an eigenvalue the solution (1.1.9) is uniquely determined up to elements of $\mathcal{N}(\mathfrak{A}, \lambda_0)$.

1.1.4. The adjoint operator pencil. Let $\mathfrak{A}(\lambda)$ be the operator (1.1.1). We set

$$\mathfrak{A}^*(\lambda) = \sum_{j=0}^{l} A_j^* \, \lambda^j \, ,$$

where $A_j^* : \mathcal{Y}^* \to \mathcal{X}^*$ are the adjoint operators to A_j. This means that the operator $\mathfrak{A}^*(\lambda)$ is adjoint to $\mathfrak{A}(\bar{\lambda})$ for every fixed λ. A proof of the following well-known assertions can be found, e.g., in the book by Kozlov and Maz'ya [**135**, Appendix].

THEOREM 1.1.7. *Suppose that the conditions of Theorem 1.1.1 are satisfied for the pencil $\mathfrak{A}$. Then the spectrum of $\mathfrak{A}^*$ consists of isolated eigenvalues with finite algebraic multiplicities.*

If λ_0 is an eigenvalue of $\mathfrak{A}$, then $\bar{\lambda}_0$ is an eigenvalue of the pencil $\mathfrak{A}^*$. The geometric, partial, and algebraic multiplicities of these eigenvalues coincide.

1.2. Operator pencils corresponding to sesquilinear forms

1.2.1. Parameter-depending sesquilinear forms. Let $\mathcal{H}_+$ be a Hilbert space which is compactly imbedded into and dense in the Hilbert space $\mathcal{H}$, and let $\mathcal{H}_-$ be its dual with respect to the inner product in $\mathcal{H}$. We consider the pencil of sesquilinear forms

$$(1.2.1) \qquad a(u,v;\lambda) = \sum_{j=0}^{l} a_j(u,v) \, \lambda^j \, ,$$

where $a_j(\cdot,\cdot)$ are bounded sesquilinear forms on $\mathcal{H}_+ \times \mathcal{H}_+$ which define linear and continuous operators $A_j : \mathcal{H}_+ \to \mathcal{H}_-$ by the equalities

$$(1.2.2) \qquad (A_j u, v)_{\mathcal{H}} = a_j(u,v) \, , \quad u,v \in \mathcal{H}_+ \, .$$

Then the operator

$$(1.2.3) \qquad \mathfrak{A}(\lambda) = \sum_{j=0}^{l} A_j \, \lambda^j$$

satisfies the equality

$$(1.2.4) \qquad (\mathfrak{A}(\lambda)u, v)_{\mathcal{H}} = a(u,v;\lambda) \qquad \text{for all } u,v \in \mathcal{H}_+ \, .$$

It can be easily verified that a number λ_0 is an eigenvalue of the operator pencil $\mathfrak{A}$ and $\varphi_0, \varphi_1, \dots, \varphi_{s-1}$ is a Jordan chain of $\mathfrak{A}$ corresponding to λ_0 if and only if

$$(1.2.5) \qquad \sum_{k=0}^{j} \frac{1}{k!} \, a^{(k)}(\varphi_{j-k}, v; \lambda_0) = 0 \qquad \text{for all } v \in \mathcal{H}_+, \ j = 0, 1, \dots, s-1,$$

where $a^{(k)}(u,v;\lambda) = d^k a(u,v;\lambda)/d\lambda^k$.

We suppose that the following conditions are satisfied:

(i) There exist a constant c_0 and a real-valued function c_1 such that

$$|a(u,u;\lambda)| \geq c_0 \, \|u\|_{\mathcal{H}_+}^2 - c_1(\lambda) \, \|u\|_{\mathcal{H}}^2 \qquad \text{for all } u \in \mathcal{H}_+$$

and for every $\lambda \in \mathbb{C}$.

(ii) There exists a real number γ such that the quadratic form $a(u,u;\lambda)$ has real values for $\operatorname{Re} \lambda = \gamma/2$, $u \in \mathcal{H}_+$.

REMARK 1.2.1. Suppose that the operators $A_1, \ldots, A_l$ are compact and there exists a number λ_0 such that

$$(1.2.6) \qquad |a(u, u; \lambda_0)| \geq c_0 \|u\|_{\mathcal{H}_+}^2 - c_1 \|u\|_{\mathcal{H}}^2 \qquad \text{for all } u \in \mathcal{H}_+,$$

where c_0 is a positive constant. Then condition (i) is satisfied.

Indeed, by the compactness of A_j, for every positive ε there exists a constant c_ε such that

$$|(A_j u, u)_{\mathcal{H}}| \leq \varepsilon \|u\|_{\mathcal{H}_+}^2 + c_\varepsilon \|u\|_{\mathcal{H}}^2 \qquad \text{for all } u \in \mathcal{H}_+.$$

Hence

$$
\begin{aligned}
|a(u, u; \lambda)| &\geq |a(u, u; \lambda_0)| - |a(u, u; \lambda) - a(u, u; \lambda_0)| \\
&\geq c_0 \|u\|_{\mathcal{H}_+}^2 - \sum_{j=1}^{l} |\lambda^j - \lambda_0^j| \left(\varepsilon \|u\|_{\mathcal{H}_+}^2 + c_\varepsilon \|u\|_{\mathcal{H}}^2 \right) - c_1 \|u\|_{\mathcal{H}}^2.
\end{aligned}
$$

Setting $\varepsilon = \dfrac{1}{2} c_0 \left(\displaystyle\sum_{j=1}^{l} |\lambda^j - \lambda_0^j| \right)^{-1}$, we get (1.2.6).

THEOREM 1.2.1. *Suppose that condition* (i) *is satisfied and there exists a complex number* λ_0 *such that*

$$(1.2.7) \qquad |a(u, u; \lambda_0)| > 0 \qquad \text{for all } u \in \mathcal{H}_+ \backslash \{0\}.$$

Then the operator $\mathfrak{A}(\lambda)$ *is Fredholm for every* $\lambda \in \mathbb{C}$ *and the spectrum of the pencil* $\mathfrak{A}$ *consists of isolated eigenvalues with finite algebraic multiplicities.*

Proof: First we prove that the kernel of $\mathfrak{A}(\lambda)$ has a finite dimension for arbitrary $\lambda \in \mathbb{C}$. By condition (i), we have

$$(1.2.8) \qquad \|u\|_{\mathcal{H}_+}^2 \leq \frac{c_1(\lambda)}{c_0(\lambda)} \|u\|_{\mathcal{H}}^2 \qquad \text{for all } u \in \ker \mathfrak{A}(\lambda).$$

Since the operator of the imbedding $\mathcal{H}_+ \subset \mathcal{H}$ is compact, the inequality (1.2.8) can be only valid on a finite-dimensional subspace. Consequently, $\dim \ker \mathfrak{A}(\lambda) < \infty$.

Now we prove that the range of $\mathfrak{A}(\lambda)$ is closed in $\mathcal{H}_+^*$ for arbitrary λ. We assume that $\mathfrak{A}(\lambda) u_k = f_k$ for $k = 1, 2, \ldots$ and the sequence $\{f_k\}_{k \geq 1}$ converges in $\mathcal{H}_+^*$ to a certain element f. Using (1.2.4), we get

$$c_0(\lambda) \|u_k\|_{\mathcal{H}_+}^2 - c_1(\lambda) \|u_k\|_{\mathcal{H}}^2 \leq |(f_k, u_k)_{\mathcal{H}}| \leq \|f_k\|_{\mathcal{H}_+^*} \|u_k\|_{\mathcal{H}_+} \leq c + \frac{1}{2} c_0(\lambda) \|u_k\|_{\mathcal{H}_+}^2.$$

Hence

$$\frac{1}{2} c_0(\lambda) \|u_k\|_{\mathcal{H}}^2 - c_1(\lambda) \|u\|_{\mathcal{H}}^2 \leq c.$$

By compactness of the imbedding $\mathcal{H}_+ \subset \mathcal{H}$, it follows from the last inequality that the sequence $\{u_k\}_{k \geq 1}$ is bounded in $\mathcal{H}_+$. Consequently, there exists a weakly convergent subsequence $\{u_{k_j}\}_{j \geq 1}$. Let u be the weak limit of this subsequence. Then for every $v \in \mathcal{H}_+$ we have

$$
\begin{aligned}
(\mathfrak{A}(\lambda) u, v)_{\mathcal{H}} &= (u, \mathfrak{A}(\lambda)^* v)_{\mathcal{H}} = \lim_{j \to \infty} (u_{k_j}, \mathfrak{A}(\lambda)^* v)_{\mathcal{H}} \\
&= \lim_{j \to \infty} (\mathfrak{A}(\lambda) u_{k_j}, v)_{\mathcal{H}} = \lim_{j \to \infty} (f_{k_j}, v)_{\mathcal{H}} = (f, v)_{\mathcal{H}},
\end{aligned}
$$

i.e., $\mathfrak{A}(\lambda) u = f$. Thus, we have proved that the range of $\mathfrak{A}(\lambda)$ is closed.

We show that the cokernel of $\mathfrak{A}(\lambda)$ has a finite dimension. For this it suffices to prove that $\ker \mathfrak{A}(\lambda)^*$ is finite-dimensional. The equality

$$(1.2.9) \qquad (\mathfrak{A}(\lambda)^* u, u)_{\mathcal{H}} = (u, \mathfrak{A}(\lambda) u)_{\mathcal{H}} = \overline{a(u, u; \lambda)}$$

yields

$$\left| (\mathfrak{A}(\lambda)^* u, u)_{\mathcal{H}} \right| \geq c_0(\lambda) \|u\|_{\mathcal{H}_+}^2 - c_1(\lambda) \|u\|_{\mathcal{H}}^2 .$$

Hence, arguing as in the proof of $\dim \ker \mathfrak{A}(\lambda) < \infty$, we obtain $\dim \ker \mathfrak{A}(\lambda)^* < \infty$. Consequently, the operator $\mathfrak{A}(\lambda)$ is Fredholm for every $\lambda \in \mathbb{C}$.

Furthermore, by (1.2.7), the kernel of $\mathfrak{A}(\lambda_0)$ is trivial and from (1.2.9) it follows that $\ker \mathfrak{A}(\lambda_0)^* = \{0\}$. Therefore, the operator $\mathfrak{A}(\lambda_0)$ has a bounded inverse. Using Theorem 1.1.1, we get the above assertion on the spectrum of the pencil $\mathfrak{A}$. ∎

THEOREM 1.2.2. *Let condition* (ii) *be satisfied.*
1) *Then the equality*

$$(1.2.10) \qquad \mathfrak{A}(\lambda)^* = \mathfrak{A}(\gamma - \overline{\lambda})$$

is valid for all $\lambda \in \mathbb{C}$,

2) *If* λ_0 *is an eigenvalue of the pencil* $\mathfrak{A}$, *then* $\gamma - \overline{\lambda}_0$ *is also an eigenvalue. The geometric, algebraic, and partial multiplicities of the eigenvalues* λ_0 *and* $\gamma - \overline{\lambda}_0$ *coincide.*

Proof: In order to prove (1.2.10), we have to show that

$$(1.2.11) \qquad a(u, v; \lambda) = \overline{a(v, u; \gamma - \overline{\lambda})} \qquad \text{for all } u, v \in \mathcal{H}_+ .$$

We set

$$b(u, v; \lambda) = a(u, v; \lambda) - \overline{a(v, u; \gamma - \overline{\lambda})}.$$

By condition (ii), the polynomial (in λ) $b(u, u; \lambda)$ vanishes on the line $\operatorname{Re} \lambda = \gamma/2$ and, therefore, on the whole complex plane. Thus, we have

$$2\, b(u, v; \lambda) = b(u + v, u + v; \lambda) + i\, b(u + iv, u + iv; \lambda) = 0$$

for all $u, v \in \mathcal{H}_+$. This implies (1.2.11).

The second assertion is a consequence of Theorem 1.1.7. ∎

DEFINITION 1.2.1. Let condition (ii) be satisfied. Then the line $\operatorname{Re} \lambda = \gamma/2$ is called *energy line*. The strip

$$|\operatorname{Re} \lambda - \gamma/2| < c$$

is called *energy strip* if there are no eigenvalues of the pencil $\mathfrak{A}$ in the set $0 < |\operatorname{Re} \lambda - \gamma/2| < c$.

The following lemma contains a sufficient condition for the absence of eigenvalues on the energy line.

LEMMA 1.2.1. 1) *Suppose the inequality*

$$(1.2.12) \qquad a(u, u, \gamma/2 + it) > 0$$

is satisfied for all $u \neq 0$ *and all real* t. *Then the line* $\operatorname{Re} \lambda = \gamma/2$ *does not contain eigenvalues of the pencil* $\mathfrak{A}$.

2) *If condition* (i) *is satisfied, inequality* (1.2.12) *is valid for all* $u \neq 0$ *and large* t, *and there are no eigenvalues of the pencil* $\mathfrak{A}$ *on the line* $\operatorname{Re} \lambda = \gamma/2$, *then* (1.2.12) *is valid for all real* t.

Proof: The first assertion is obvious. We prove the second one. Suppose that $a(u_0, u_0, \gamma/2 + it_0) \leq 0$ for some $u_0 \in \mathcal{H}_+\backslash\{0\}$, $t_0 \in \mathbb{R}$. From condition (i) and (1.2.10) it follows that the operator $\mathfrak{A}(\gamma/2 + it)$ is selfadjoint, semibounded from below, and has a discrete spectrum $\mu_1(t) \leq \mu_2(t) \leq \cdots$. Since the function $\mu_1(t)$ is continuous, positive for large $|t|$ and nonpositive for $t = t_0$, it vanishes for some $t = t_1$. Then the number $\lambda = \gamma/2 + it_1$ is an eigenvalue of the pencil $\mathfrak{A}$ on the line $\text{Re}\,\lambda = \gamma/2$. ∎

LEMMA 1.2.2. *Suppose that condition* (i) *is satisfied and that the quadratic form* $a(u, u; \lambda)$ *is nonnegative for* $\text{Re}\,\lambda = \gamma/2$. *If the form* $a(u, u; \gamma/2)$ *vanishes on the subspace* $\mathcal{H}_0$, *then* $\mathcal{H}_0$ *is the space of the eigenvectors of the pencil* $\mathfrak{A}$ *corresponding to the eigenvalue* $\lambda_0 = \gamma/2$. *Furthermore, every eigenvector corresponding to this eigenvalue has at least one generalized eigenvector.*

Proof: Since the form $a(u, u; \gamma/2)$ is nonnegative, we get

$$|a(u, v; \gamma/2)|^2 \leq a(u, u; \gamma/2) \cdot a(v, v; \gamma/2) = 0$$

for $u \in \mathcal{H}_0$, $v \in \mathcal{H}_+$. This implies $\mathfrak{A}(\gamma/2)u = 0$ for $u \in \mathcal{H}_0$. Conversely, every eigenvector u of the pencil $\mathfrak{A}$ corresponding to the eigenvalue $\gamma/2$ satisfies the equation $a(u, u; \gamma/2) = 0$. Thus, according to (1.2.10), we obtain $\ker \mathfrak{A}(\gamma/2) = \ker \mathfrak{A}(\gamma/2)^* = \mathcal{H}_0$.

We show that every eigenvector u_0 corresponding to the eigenvalue $\lambda = \gamma/2$ has at least one generalized eigenvector, i.e., there exists a vector u_1 satisfying the equation

$$a(u_1, v; \gamma/2) + a^{(1)}(u_0, v; \gamma/2) = 0 \qquad \text{for all } v \in \mathcal{H}_+$$

or, what is the same,

$$\mathfrak{A}(\gamma/2)\, u_1 = -f_1 \,,$$

where $f_1 \in \mathcal{H}_+^*$ denotes the functional $\mathcal{H}_+ \ni v \to a^{(1)}(u_0, v; \gamma/2)$. The last equation is solvable if

$$(1.2.13) \qquad (f_1, v)_{\mathcal{H}} = a^{(1)}(u_0, v; \gamma/2) = 0 \qquad \text{for all } v \in \mathcal{H}_0 \,.$$

We consider the function $t \to a(v, v; \gamma/2 + it)$, $v \in \mathcal{H}_0$, which is nonnegative for all real t and equal to zero for $t = 0$. Consequently, we have $a^{(1)}(v, v; \gamma/2) = 0$ for all $v \in \mathcal{H}_0$. This implies

$$a^{(1)}(u, v; \gamma/2) = \frac{1}{2}\left(a^{(1)}(u + v, u + v; \gamma/2) + i\, a^{(1)}(u + iv, u + iv; \gamma/2)\right) = 0$$

for all $u, v \in \mathcal{H}_0$, i.e., condition (1.2.13) is satisfied. The proof is complete. ∎

1.2.2. Ordinary differential equations in the variational form. Let $\mathcal{H}_+$, $\mathcal{H}$ be the same Hilbert spaces as in the foregoing subsection. Furthermore, let $a_{j,k}(\cdot, \cdot)$, $j, k = 1, \ldots, m$, be bounded sesquilinear forms on $\mathcal{H}_+ \times \mathcal{H}_+$.

We seek functions $U = U(r)$ of the form

$$(1.2.14) \qquad U(r) = r^{\lambda_0} \sum_{k=0}^{s} \frac{1}{k!} (\log r)^k u_{s-k}, \qquad u_k \in \mathcal{H}_+ \,,$$

which satisfy the integral identity

$$(1.2.15) \qquad \int_0^\infty \sum_{j,k=0}^m a_{j,k}\big((r\partial_r)^k U(r),\, (r\partial_r)^j V(r)\big)\, r^{-\gamma}\, \frac{dr}{r} = 0$$

for all $V \in C_0^\infty((0,\infty); \mathcal{H}_+)$. Here γ is a real number.

For $u, v \in \mathcal{H}_+$ we set

$$a(u,v;\lambda) = \sum_{j,k=0}^m a_{j,k}\big(\lambda^k u,\, (\gamma - \bar\lambda)^j v\big) = \sum_{j,k=0}^m a_{j,k}(u,v)\, \lambda^k\, (\gamma - \lambda)^j .$$

Furthermore, let $\mathfrak{A}(\lambda) : \mathcal{H}_+ \to \mathcal{H}_+^*$ be the operator defined by (1.2.4).

It can be easily verified that

$$a(u,v;\lambda) = \frac{1}{2|\log \varepsilon|} \int_\varepsilon^{1/\varepsilon} \sum_{j,k=0}^m a_{j,k}\big((r\partial_r)^k U(r),\, (r\partial_r)^j V(r)\big)\, r^{-\gamma}\, \frac{dr}{r} ,$$

where ε is a positive real number less than one, $U(r) = r^\lambda u$, $V(r) = r^{\gamma - \bar\lambda} v$.

THEOREM 1.2.3. *The function* (1.2.14) *is a solution of* (1.2.15) *if and only if λ_0 is an eigenvalue of the operator pencil $\mathfrak{A}$ and $u_0, \ldots, u_s$ is a Jordan chain corresponding to this eigenvalue.*

Proof: Integrating by parts in (1.2.15), we get

$$\int_0^\infty \sum_{j,k=0}^m a_{j,k}\big((-r\partial_r + \gamma)^j (r\partial_r)^k U(r),\, V(r)\big)\, r^{-\gamma}\, \frac{dr}{r} = 0.$$

This equality is valid for all $V \in C_0^\infty((0,\infty); \mathcal{H}_+)$ if and only if

$$\sum_{j,k=0}^m a_{j,k}\big((-r\partial_r + \gamma)^j (r\partial_r)^k U(r),\, v\big) = 0 \qquad \text{for } r > 0,\ v \in \mathcal{H}_+ .$$

The last equation can be rewritten in the form

$$\mathfrak{A}(r\partial_r)\, U(r) = 0 \qquad \text{for } r > 0.$$

Now our assertion is an immediate consequence of Theorem 1.1.5. ∎

THEOREM 1.2.4. *Suppose that there exists a number $\delta \in (0,1)$ such that*

$$(1.2.16) \qquad \Big| \int_0^\infty \sum_{j,k=0}^m a_{j,k}\big((r\partial_r)^k U(r),\, (r\partial r)^j U(r)\big)\, r^{-\gamma}\, \frac{dr}{r} \Big|$$

$$\geq c_0 \int_0^\infty \big(\|U(r)\|_{\mathcal{H}_+}^2 + \|\partial_r^m U\|_{\mathcal{H}}^2 - c_1 \|U\|_{\mathcal{H}}^2 \big)\, dr$$

for all $U \in C_0^\infty((0,\infty); \mathcal{H}_+)$ with support in $(1-\delta, 1+\delta)$. Then there exists a number T such that

$$(1.2.17) \qquad |a(u,u; it + \tfrac{\gamma}{2})| \geq c_0 (\|u\|_{\mathcal{H}_+}^2 + t^{2m} \|u\|_{\mathcal{H}}^2)$$

for all real t, $|t| > T$, and all $u \in \mathcal{H}_+$.

Proof: Let $\zeta = \zeta(r)$ be a smooth real-valued function on $(0, \infty)$ with support in $(1 - \delta, 1 + \delta)$ equal to one for $r = 1$. We set $U(r) = r^{it + \gamma/2} u$, where u is an element of $\mathcal{H}_+$. Then

$$(1.2.18) \qquad \Big| \int_0^\infty \sum_{j,k=0}^m a_{j,k}\big((r\partial_r)^k \zeta U(r), (r\partial_r)^j \zeta U(r)\big) \, r^{-\gamma} \frac{dr}{r}$$

$$- \int_0^\infty \sum_{j,k=0}^m \zeta^2 a_{j,k}\big((r\partial_r)^k U(r), (r\partial_r)^j U(r)\big) \, r^{-\gamma} \frac{dr}{r} \Big|$$

$$\leq c \sum_{\substack{0 \leq j,k \leq m \\ j+k \geq 1}} |a_{j,k}(u,u)| \, (1 + |t|)^{j+k-1} \, .$$

Furthermore,

$$(1.2.19) \qquad \int_0^\infty \sum_{j,k=0}^m \zeta^2 \, a_{j,k}\big((r\partial_r)^k U(r), (r\partial_r)^j U(r)\big) \, r^{-\gamma} \frac{dr}{r}$$

$$= a(u, u; it + \frac{\gamma}{2}) \int_0^\infty \zeta^2(r) \frac{dr}{r} \, .$$

Using (1.2.16), we get the estimate

$$(1.2.20) \qquad \Big| \int_0^\infty \sum_{j,k=0}^m a_{j,k}\big((r\partial_r)^k \zeta U(r), (r\partial_r)^j \zeta U(r)\big) \, r^{-\gamma} \frac{dr}{r} \Big|$$

$$\geq c(\|u\|_{\mathcal{H}_+}^2 + t^{2m}\|u\|_{\mathcal{H}}^2) - c_2(1 + |t|)^{2m-1}\|u\|_{\mathcal{H}}^2 \, .$$

From (1.2.18)–(1.2.20) we conclude that

$$|a(u, u; it + \frac{\gamma}{2})| \geq c(\|u\|_{\mathcal{H}_+}^2 + t^{2m}\|u\|_{\mathcal{H}}^2) - c_2(1 + |t|)^{2m-1}\|u\|_{\mathcal{H}_+}^2 \, .$$

This proves our assertion. ∎

THEOREM 1.2.5. *Suppose that the condition of* Theorem 1.2.4 *is satisfied. Furthermore, we assume that the operator*

$$(1.2.21) \qquad \qquad \mathfrak{A}(\lambda) - \mathfrak{A}(0) \,:\, \mathcal{H}_+ \to \mathcal{H}_+^*$$

is compact for all $\lambda \in \mathbb{C}$. *Then* $\mathfrak{A}(\lambda)$ *is Fredholm for all* $\lambda \in \mathbb{C}$ *and the spectrum of the pencil* $\mathfrak{A}$ *consists of isolated eigenvalues with finite algebraic multiplicities.*

Proof: From our assumption it follows that $\mathfrak{A}(\lambda) - \mathfrak{A}(\mu)$ is a compact operator from $\mathcal{H}_+$ into $\mathcal{H}_+^*$ for all $\lambda, \mu \in \mathbb{C}$. Hence for arbitrary $\varepsilon > 0$ there exists a constant c_ε depending on λ and μ such that

$$\big|\big((\mathfrak{A}(\lambda) - \mathfrak{A}(\mu))u, u\big)_{\mathcal{H}}\big| \leq \varepsilon \|u\|_{\mathcal{H}_+}^2 + c_\varepsilon \|u\|_{\mathcal{H}}^2$$

for $u \in \mathcal{H}_+$. Therefore, we obtain

$$|\big(\mathfrak{A}(\lambda)u, u\big)_{\mathcal{H}}| \;\geq\; \big|\big(\mathfrak{A}(it + \frac{\gamma}{2})u, u\big)_{\mathcal{H}}\big| - \big|\big((\mathfrak{A}(it + \frac{\gamma}{2}) - \mathfrak{A}(\lambda))u, u\big)_{\mathcal{H}}\big|$$

$$\geq\; c_0\big(\|u\|_{\mathcal{H}_+}^2 + t^{2m}\|u\|_{\mathcal{H}}^2\big) - \varepsilon\|u\|_{\mathcal{H}_+}^2 - c_\varepsilon\|u\|_{\mathcal{H}}^2 \, .$$

Now the result follows from Theorem 1.2.1. ∎

THEOREM 1.2.6. 1) *Suppose that*

$$(1.2.22) \qquad \int\limits_0^\infty \sum_{j,k=0}^m a_{j,k}\big((r\partial_r)^k U(r), (r\partial_r)^j U(r)\big)\, r^\gamma\, \frac{dr}{r}$$

is real for arbitrary $U \in C_0^\infty((0,\infty); \mathcal{H}_+)$. *Then* $a(u,u,it+\gamma/2)$ *is real for all* $u \in \mathcal{H}_+$, $t \in \mathbb{R}$.

2) *If* (1.2.22) *is nonnegative for all* $U \in C_0^\infty((0,\infty); \mathcal{H}_+)$, *then* $a(u,u,it+\gamma/2)$ *is also nonnegative for* $u \in \mathcal{H}_+$, $t \in \mathbb{R}$.

Proof: Let $u \in \mathcal{H}_+$, $\zeta \in C_0^\infty(\mathbb{R})$, and $\varepsilon > 0$. We set

$$U_\varepsilon(r) = \varepsilon^{1/2}\, r^{it+\gamma/2}\, \zeta(\varepsilon \log r)\, u.$$

Then

$$\int\limits_0^\infty \sum_{j,k=0}^m a_{j,k}\big((r\partial_r)^k U_\varepsilon(r), (r\partial_r)^j U_\varepsilon(r)\big)\, r^{-\gamma}\, \frac{dr}{r}$$

$$= \int\limits_{-\infty}^{+\infty} |\zeta(s)|^2\, ds \cdot a(u,u,it+\gamma/2) + O(\varepsilon).$$

Hence $a(u,u,it+\gamma/2)$ is real (nonnegative) if the left-hand side of the last equality is real (nonnegative). This proves the theorem. ∎

1.3. A variational principle for operator pencils

1.3.1. Assumptions. Let $\mathcal{H}_+, \mathcal{H}$ be the same Hilbert spaces as in the previous section. We consider the sesquilinear form (1.2.1), where a_j are sesquilinear, Hermitian and bounded forms on $\mathcal{H}_+ \times \mathcal{H}_+$. Then $a(u,u;\lambda)$ is real for real λ and $u \in \mathcal{H}_+$. The sesquilinear forms $a_j(\cdot,\cdot)$ and $a(\cdot,\cdot;\lambda)$ generate the operators $A_j : \mathcal{H}_+ \to \mathcal{H}_+^*$ and $\mathfrak{A}(\lambda) : \mathcal{H}_+ \to \mathcal{H}_+^*$ by (1.2.2) and (1.2.3), respectively.

We suppose that α, β are real numbers such that $\alpha < \beta$ and the following conditions are satisfied.

(I) There exist a positive constant c_1 and a continuous function $c_0(\cdot)$ on the interval $[\alpha, \beta]$, $c_0(\lambda) > 0$ in $[\alpha, \beta)$, such that

$$a(u,u;\lambda) \geq c_0(\lambda)\, \|u\|_{\mathcal{H}_+}^2 - c_1 \|u\|_{\mathcal{H}}^2 \quad \text{for all } u \in \mathcal{H}_+,\ \lambda \in [\alpha, \beta].$$

(II) The operator $\mathfrak{A}(\alpha)$ is positive definite.

(III) If $\mathfrak{A}(\lambda_0)u = 0$ for a certain $\lambda_0 \in (\alpha, \beta)$, $u \in \mathcal{H}_+$, $u \neq 0$, then

$$\frac{d}{d\lambda} a(u,u;\lambda)\Big|_{\lambda=\lambda_0} < 0.$$

1.3.2. Properties of the pencil $\mathfrak{A}$.

THEOREM 1.3.1. *Let conditions* (I)–(III) *be satisfied. Then the following assertions are valid.*

1) *The spectrum of the pencil $\mathfrak{A}$ on the interval $[\alpha, \beta)$ consists of isolated eigenvalues with finite algebraic multiplicities and the eigenvectors have no generalized eigenvectors.*

2) *For every $\lambda_0 \in [\alpha, \beta)$ the operator $\mathfrak{A}(\lambda_0)$ is selfadjoint, bounded from below and has a discrete spectrum with the unique accumulation point at $+\infty$.*

Proof: 1) There exists a open set $\mathcal{U} \subset \mathbb{C}$ containing the interval $[\alpha, \beta]$ such that

$$\operatorname{Re} a(u, u, \lambda) \geq \frac{1}{2} c_0(\operatorname{Re}\lambda) \, \|u\|_{\mathcal{H}_+}^2 - c_1 \, \|u\|_{\mathcal{H}_+}^2$$

for $\lambda \in \mathcal{U}$. Using Theorem 1.2.1, we can conclude that the spectrum of $\mathfrak{A}$ in $\mathcal{U}$ consists of isolated eigenvalues with finite algebraic multiplicities. Let λ_0 be an eigenvalue, φ_0 an eigenvector to λ_0 and φ_1 a generalized eigenvector associated to φ_0. Then

$$a(\varphi_1, v; \lambda_0) + \frac{d}{d\lambda} a(\varphi_0, v; \lambda) \Big|_{\lambda=\lambda_0} = 0$$

for all $v \in \mathcal{H}_+$. Setting $v = \varphi_0$, we get the equality

$$\frac{d}{d\lambda} a(\varphi_0, \varphi_0, \lambda) \Big|_{\lambda=\lambda_0} = 0$$

which contradicts condition (III).

2) From our assumption that the forms a_j are Hermitian and from condition (I) it follows that $\mathfrak{A}(\lambda_0)$ is selfadjoint and bounded from below for every $\lambda_0 \in [\alpha, \beta]$. Since the imbedding $\mathcal{H}_+ \subset \mathcal{H}$ is compact, it further follows that the spectrum is discrete. The proof is complete. ∎

1.3.3. Eigenvalues of the operator $\mathfrak{A}(\lambda)$. We consider the eigenvalue problem

(1.3.1) $$\mathfrak{A}(\lambda)u = \mu(\lambda)\, u$$

for fixed $\lambda \in [\alpha, \beta]$. Let $\mu_1(\lambda) \leq \mu_2(\lambda) \leq \dots$ be the nondecreasing sequence of the eigenvalues counted with their multiplicities. By Theorem 1.3.1, $\mu_j(\lambda)$ tends to infinity as $j \to \infty$.

THEOREM 1.3.2. 1) *The functions μ_j are continuous on the interval $[\alpha, \beta)$ and $\mu_j(\alpha) > 0$.*

2) *If $c_0(\beta) > 0$, then the functions μ_j are also continuous at the point $\lambda = \beta$.*

3) *For every index $j \geq 1$ the equation $\mu_j(\lambda) = 0$ has at most one root in the interval (α, β). If $\mu_j(\lambda_0) = 0$ for a certain $\lambda_0 \in (\alpha, \beta)$, then $\mu_j(\lambda) > 0$ for $\lambda \in (\alpha, \lambda_0)$ and $\mu_j(\lambda_0) < 0$ for $\lambda \in (\lambda_0, \beta)$.*

Proof: For assertions 1), 2) we refer to Kato's book [**96**, Ch.7,Th.1.8], where in fact a stronger assertion has been proved. In particular, it has been shown there that the functions μ_j are real-analytic on each of the intervals $(\lambda - \varepsilon, \lambda]$, $[\lambda, \lambda + \varepsilon)$, where $\lambda \in (\alpha, \beta)$, ε is a small positive number, and on the interval $[\alpha, \alpha + \varepsilon)$. If $c_0(\beta) > 0$, then this is also true on the interval $(\beta - \varepsilon, \beta]$.

3) Let $\lambda_0 \in (\alpha, \beta)$ be a root of the equation $\mu_j(\lambda) = 0$ and let $I = \dim \ker \mathfrak{A}(\lambda_0)$. As it has been proved in [**96**, Ch.7,Th.1.8], there exist analytic functions $m_k(\lambda)$ and vector functions $u_k(\lambda)$, $k = 1, \dots, I$, such that

(1.3.2) $$\mathfrak{A}(\lambda)u_k(\lambda) = m_k(\lambda)\, u_k(\lambda)$$

in a neighborhood of λ_0 and the linear closure of the vectors $u_1(\lambda_0), \dots, u_I(\lambda_0)$ coincides with $\ker \mathfrak{A}(\lambda_0)$. Obviously, m_k coincides with one of the functions μ_j on the left of the point λ_0 and possibly with a different μ_j on the right of λ_0. Differentiating (1.3.2) with respect to λ and setting $\lambda = \lambda_0$, we get

$$\mathfrak{A}(\lambda_0)\, u_k'(\lambda_0) + \mathfrak{A}'(\lambda_0)\, u_k(\lambda_0) = m_k'(\lambda_0)\, u_k(\lambda_0).$$

Multiplying both parts of the last equation by $u_k(\lambda)$, we obtain

$$m'_k(\lambda_0) = \left(\mathfrak{A}'(\lambda_0)u_k(\lambda_0)\,,\,u_k(\lambda_0)\right)_{\mathcal{H}} \cdot \|u_k(\lambda_0)\|_{\mathcal{H}}^{-2}\,.$$

By condition (III), the right-hand side is negative. Hence $m'_k(\lambda_0) < 0$. This proves the third assertion of the theorem. ∎

1.3.4. On the number of eigenvalues of the pencil $\mathfrak{A}$ in the interval (α, β). Let $\{u_j(\lambda)\}$ be a sequence of linearly independent eigenvectors of problem (1.3.1) corresponding to the eigenvalues $\mu_j(\lambda)$.

THEOREM 1.3.3. 1) *The number $\lambda_0 \in [\alpha, \beta)$ is an eigenvalue of the operator pencil $\mathfrak{A}$ with the geometric multiplicity I if and only if there exists a number $k \geq 1$ such that*

$$\mu_j(\lambda_0) = 0 \qquad for\ j = k, k+1, \ldots, k+I-1.$$

The corresponding characteristic space coincides with the linear closure of the vectors $u_k(\lambda_0), u_{k+1}(\lambda_0), \ldots, u_{k+I-1}(\lambda_0)$.

2) *Let $\beta_0 \in (\alpha, \beta)$ and let N be the maximum of all integer numbers s for which there exists a subspace $\mathcal{H}_s \subset \mathcal{H}_+$ of dimension s such that*

$$(1.3.3) \qquad a(u,u;\beta_0) < 0 \qquad for\ u \in \mathcal{H}_s \backslash \{0\}.$$

Then the interval (α, β_0) contains exactly N eigenvalues (counting their multiplicities) of the operator pencil $\mathfrak{A}$.

Proof: The first assertion is obvious. We show that assertion 2) is true. The number N can be characterized as follows. Let n be a number such that $\mu_n(\beta_0) < 0$ and $\mu_{n+1}(\beta_0) \geq 0$. Then it can be easily seen that $n = N$ and the space $\mathcal{H}_N$ can be chosen as the linear closure of the vectors $u_1(\beta_0), \ldots, u_N(\beta_0)$. Since $\mu_j(\alpha) > 0, \mu_j(\beta_0) < 0$ for $j = 1, \ldots, N$ and $\mu_j(\beta_0) \geq 0$ for $j > N$, it follows from Theorem 1.3.2 that every of the functions $\mu_1, \ldots, \mu_N$ has exactly one zero in the interval (α, β_0) and the functions μ_j for $j > N$ have no zeros in this interval. Using the first assertion of the theorem, we arrive at 2). ∎

THEOREM 1.3.4. 1) *Let $\mathcal{H}_s$ be a subspace of $\mathcal{H}_+$ with dimension s such that*

$$(1.3.4) \qquad a(u,u;\beta) < 0 \qquad for\ u \in \mathcal{H}_s \backslash \{0\}$$

Then the pencil $\mathfrak{A}$ has at least s eigenvalues (counting their multiplicities) in the interval (α, β).

2) *Suppose that the following condition is satisfied: If $a(u,u;\beta) \geq 0$, then there exists a positive number ε such that*

$$(1.3.5) \qquad a(u,u;\lambda) \geq 0 \qquad for\ \lambda \in (\beta - \varepsilon, \beta).$$

Under this condition, the number of the eigenvalues of the pencil $\mathfrak{A}$ in the interval (α, β) is equal to the maximal dimension of the spaces $\mathcal{H}_s$ for which (1.3.4) is true.

Proof: 1) If (1.3.4) is valid, then there exists a positive constant c such that

$$(1.3.6) \qquad a(u,u;\beta) \leq -c\,\|u\|_{\mathcal{H}_+}^2$$

for all $u \in \mathcal{H}_s$. For small $|\lambda - \beta|$ we get

$$|a(u,u;\lambda) - a(u,u,\beta)| \leq \frac{c}{2}\,\|u\|_{\mathcal{H}_+}^2 \qquad if\ u \in \mathcal{H}_+\,.$$

Consequently,

$$a(u, u; \lambda) \geq -\frac{c}{2} \|u\|_{\mathcal{H}_+}^2$$

for all $u \in \mathcal{H}_s$ and from the second part of Theorem 1.3.3 we obtain assertion 1).

2) Without loss of generality, we may assume that N is equal to the greatest number s for which the inequality (1.3.4) is satisfied. By the selfadjointness of the operator $\mathfrak{A}(\beta)$, the space $\mathcal{H}_+$ can be decomposed into a direct sum $\mathcal{H}_+ = \mathcal{H}_0 \oplus \mathcal{H}_1$ such that $\dim \mathcal{H}_0 = N$,

$$(1.3.7) \qquad \left\{ \begin{array}{ll} a(u, u; \beta) < 0 & \text{for} \quad u \in \mathcal{H}_0 \backslash \{0\}, \\ a(u, u; \beta) \geq 0 & \text{for} \quad u \in \mathcal{H}_1. \end{array} \right.$$

Since $\mathcal{H}_0$ is finite-dimensional, the inequality (1.3.6) is satisfied with certain constant c for all $u \in \mathcal{H}_0$. This and condition (1.3.5) imply the validity of the inequalities

$$a(u, u; \lambda) \leq -\frac{c}{2} \|u\|_{\mathcal{H}_+}^2 \qquad \text{for } u \in \mathcal{H}_0,$$

$$a(u, u; \lambda) \geq 0 \qquad \text{for } u \in \mathcal{H}_1$$

if λ lies in a sufficiently small neighborhood of β. Now 2) follows immediately from the second part of Theorem 1.3.3. ∎

1.3.5. On the smallest eigenvalue of the pencil $\mathfrak{A}$ in the interval (α, β). Let

$$\lambda_* = \sup \Big\{ \lambda \in [\alpha, \beta] : \ a(u, u; \mu) > 0 \text{ for all } \mu \in [\alpha, \lambda), \ u \in \mathcal{H}_+ \backslash \{0\} \Big\}.$$

The following assertion is obvious.

LEMMA 1.3.1. *If $\lambda_* < \beta$, then λ_* is the smallest eigenvalue of the operator pencil $\mathfrak{A}$ in the interval $[\alpha, \beta)$.*

We give another characterization of the number λ_*. Let $R(u)$ be the smallest root of the polynomial $\lambda \to a(u, u; \lambda)$, $u \in \mathcal{H}_+ \backslash \{0\}$, in the interval $[\alpha, \beta)$. If $[\alpha, \beta)$ does not contain such roots, then we set $R(u) = \beta$.

LEMMA 1.3.2. *The number λ_* is given by*

$$(1.3.8) \qquad \lambda_* = \inf \Big\{ R(u) : u \in \mathcal{H}_+ \backslash \{0\} \Big\}.$$

Proof: It suffices to consider the case $\lambda_* < \beta$. We denote the right-hand side of (1.3.8) by λ_1. Then from the definition of the number λ_* it follows that $\lambda_* \leq \lambda_1$. On the other hand, by Lemma 1.3.1, the number λ_* is an eigenvalue of $\mathfrak{A}$. Let φ_0 be an eigenvector corresponding to λ_*. Since $a(\varphi_0, \varphi_0; \lambda_*) = (\mathfrak{A}(\lambda_*)\varphi_0, \varphi_0)_{\mathcal{H}} = 0$, we get $\lambda_1 \leq R(\varphi_0) \leq \lambda_*$. This proves the lemma. ∎

1.3.6. Monotonicity of the eigenvalues of the operator pencil with respect to its domain. Let H_+ be a subspace of $\mathcal{H}_+$ and let H be the closure of the set H_+ in $\mathcal{H}$. Furthermore, let $\hat{a}(\cdot, \cdot; \lambda)$ be the restriction of the form $a(\cdot, \cdot; \lambda)$ to $H_+ \times H_+$, i.e.,

$$\hat{a}(u, v; \lambda) = a(u, v; \lambda) \qquad \text{for all } u, v \in H_+$$

We define the operators $\hat{A}_j : H_+ \to H_+^*$ by

$$(\hat{A}_j u, v) = a_j(u, v), \qquad u, v \in H_+,$$

and set

$$\hat{\mathfrak{A}}(\lambda) = \sum_{j=0}^{l} \hat{A}_j \, \lambda^j \,.$$

Obviously,

$$(\hat{\mathfrak{A}}(\lambda)\, u, v) = \hat{a}(u, v; \lambda) \quad \text{for all } u, v \in H_+,$$

and conditions (I), (II) are also valid for the form $\hat{a}(\cdot, \cdot; \lambda)$ and the operator $\hat{\mathfrak{A}}(\lambda)$. Additionally, we suppose that $\hat{\mathfrak{A}}$ satisfies condition (III). Then all results of Theorems 1.3.1–1.3.4 hold for the operator pencil $\hat{\mathfrak{A}}$ and the form $\hat{a}$.

Let $\lambda_1, \dots, \lambda_p$ be the eigenvalues of $\mathfrak{A}$ in the interval (α, β) counted with their multiplicities. Analogously, we denote the eigenvalues of $\hat{\mathfrak{A}}$ in (α, β) by $\hat{\lambda}_1, \dots, \hat{\lambda}_q$. We assume that the eigenvalues are numerated such that

$$\alpha < \lambda_1 \le \dots \le \lambda_p < \beta \quad \text{and} \quad \alpha < \hat{\lambda}_1 \le \dots \le \hat{\lambda}_q < \beta.$$

THEOREM 1.3.5. 1) *Under the above assumptions, we have*

$$q \le p \qquad and \qquad \lambda_j \le \hat{\lambda}_j \quad for \ j = 1, \dots, q.$$

2) *Suppose that for arbitrary $\lambda \in (\alpha, \beta)$ the vector $u = 0$ is the only element in H_+ subject to*

$$a(u, v; \lambda) = 0 \quad for \ all \ v \in \mathcal{H}_+ \,.$$

Then $\lambda_j < \hat{\lambda}_j$ for $j = 1, \dots, q$.

Proof: We denote the eigenvalues of the problems

$$\mathfrak{A}(\lambda)\, u = \mu\, u \qquad \text{and} \qquad \hat{\mathfrak{A}}(\lambda) u = \mu\, u$$

by $\mu_j(\lambda)$ and $\hat{\mu}_j(\lambda)$, respectively. The eigenvalues $\mu_j(\lambda)$ are given for $\lambda \in (\alpha, \beta)$ by

$$(1.3.9) \qquad \mu_j(\lambda) = \max_{L} \min_{u \in L \setminus \{0\}} \frac{a(u, u; \lambda)}{(u, u)_{\mathcal{H}}} \,,$$

where the maximum is taken over the set of all subspaces $L \subset \mathcal{H}_+$ of codimension $\ge j - 1$. Analogously,

$$\hat{\mu}_j(\lambda) = \max_{L} \min_{u \in L \setminus \{0\}} \frac{\hat{a}(u, u; \lambda)}{(u, u)_{\mathcal{H}}} \,,$$

where the maximum is taken over all subspaces $L \subset H_+$ of codimension $\ge j - 1$. Using the imbedding $H_+ \subset \mathcal{H}_+$, we obtain

$$(1.3.10) \qquad \mu_j(\lambda) \le \hat{\mu}_j(\lambda) \qquad \text{for } \lambda \in (\alpha, \beta), \ j = 1, 2, \dots$$

This together with the first part of Theorem 1.3.3 implies assertion 1).

2) We show that the equality $\hat{\mu}_j(\lambda) = 0$, $\lambda \in (\alpha, \beta)$, implies $\mu_j(\lambda) < 0$. To this end, we suppose that $\mu_j(\lambda) = \hat{\mu}_j(\lambda) = 0$ and denote by $u_1, \dots, u_j \in H_+$ an orthogonal system of eigenvectors of the operator $\hat{\mathfrak{A}}(\lambda)$ corresponding to the eigenvalues $\hat{\mu}_1(\lambda), \dots, \hat{\mu}_j(\lambda)$. By (1.3.9), we have

$$\mu_j(\lambda) \ge \min_{u \in \mathcal{H}_+^{(j)} \setminus \{0\}} \frac{a(u, u; \lambda)}{\|u\|_{\mathcal{H}}^2} \,,$$

where

$$\mathcal{H}_+^{(j)} = \{u \in \mathcal{H}_+ : (u, u_k) = 0 \text{ for } k = 1, \dots, j - 1\}.$$

Since $a(u_j, u_j; \lambda) = \hat{\mu}_j \|u\|_{\mathcal{H}}^2$ and $\mu_j = \hat{\mu}_j$, we obtain

$$(1.3.11) \qquad 0 = \mu_j(\lambda) = \min_{u \in \mathcal{H}_+^{(j)} \backslash \{0\}} \frac{a(u, u; \lambda)}{\|u\|_{\mathcal{H}}^2},$$

where the minimum in the right-hand side is attained at the vector $u = u_j$. From (1.3.11) it follows that there exist constants $c_1, \ldots, c_{j-1}$ such that

$$(1.3.12) \qquad a(u_j, v; \lambda) = \sum_{k=1}^{j-1} c_k \, a(u_k, v; \lambda) \qquad \text{for all } v \in \mathcal{H}_+.$$

Furthermore, since u_j is an eigenvector of the operator $\hat{\mathfrak{A}}(\lambda)$ corresponding to the eigenvalue $\hat{\mu}_j(\lambda) = 0$, we have $a(u_j, v; \lambda) = 0$ for all $v \in H_+$. Consequently,

$$\sum_{k=1}^{j-1} c_k \, a(u_k, v; \lambda) = 0 \qquad \text{for all } v \in H_+.$$

From this we obtain $c_k = 0$ if $\hat{\mu}_k(\lambda) \neq 0$. Hence by (1.3.12), we get

$$a(w, v; \lambda) = 0 \qquad \text{for all } v \in \mathcal{H}_+,$$

where

$$w = u_j - \sum_{\substack{1 \leq k \leq j-1 \\ \hat{\mu}_k(\lambda) = 0}} c_k \, u_k$$

is an element of H_+. By the assumption of the theorem, this implies $w = 0$, i.e., $u_j = 0$ and $c_k = 0$ for $k = 1, \ldots, j-1$. Therefore, the equality $\mu_j(\lambda) = \hat{\mu}_j(\lambda) = 0$ cannot be valid. Thus, $\mu_j(\lambda) < 0$ if $\hat{\mu}_j(\lambda) = 0$. Hence the zero of the function $\mu_j(\cdot)$ is less than the zero of the function $\hat{\mu}_j(\cdot)$. Using Theorem 1.3.3, we get the second assertion. ∎

1.4. Elliptic boundary value problems in domains with conic points: some basic results

As we said in the introduction to this chapter, pencils of boundary value problems on a subdomain of the unit sphere S^{n-1} appear naturally in the theory of elliptic boundary value problems on n-dimensional domains with conic vertices. In this section we formulate fundamental analytic facts of this theory. They depend upon general spectral properties of the corresponding operator pencils and are valid for arbitrary elliptic systems. A drawback of this generality is that one can deduce from them no explicit information on the continuity and differentiability properties of solutions. However, using these basic facts along with concrete results on the operator pencils to be obtained in the sequel, we shall be able to derive information of such a kind in the next chapters.

1.4.1. The operator pencil generated by the boundary value problem. Let $\mathcal{G}$ be a n-dimensional domain with d singular boundary points $x^{(1)}, \ldots, x^{(d)}$. Outside the set $\mathcal{S}$ of these points, the boundary $\partial \mathcal{G}$ is assumed to be smooth. Furthermore, we suppose that the domain $\mathcal{G}$ coincides with a cone $\mathcal{K}_\tau = \{x : (x - x^{(\tau)})/|x - x^{(\tau)}| \in \Omega_\tau\}$ in a neighborhood of $x^{(\tau)}$, $\tau = 1, \ldots, d$. By $r_\tau(x)$ we denote the distance of the point x to $x^{(\tau)}$ and by $r(x)$ a positive infinitely differentiable function on $\overline{\mathcal{G}} \backslash \mathcal{S}$ which coincides with the distance to $\mathcal{S}$ in a neighborhood of $\mathcal{S}$.

We consider the boundary value problem

(1.4.1) $$\sum_{j=1}^{N} L_{i,j}(x, \partial_x) U_j = F_i \ \text{in} \ \mathcal{G}, \ i = 1, \dots, N,$$

(1.4.2) $$\sum_{j=1}^{N} B_{k,j}(x, \partial_x) U_j = G_k \ \text{on} \ \partial\mathcal{G}\backslash\mathcal{S}, \ k = 1, \dots, M,$$

where $L_{i,j}$, $B_{k,j}$ are linear differential operators, ord $L_{i,j} \leq s_i + t_j$, ord $B_{k,j} \leq \sigma_k + t_j$ (s_i, t_j are given integer numbers, $s_1 + t_1 + \dots + s_N + t_N = 2M$, the operators $L_{i,j}$, $B_{k,j}$ are assumed to be zero if $s_i + t_j < 0$ and $\sigma_k + t_j < 0$, respectively).

We suppose that the coefficients of $L_{i,j}$, $B_{k,j}$ are smooth outside $\mathcal{S}$ and that problem (1.4.1), (1.4.2) is elliptic , i.e., the system (1.4.1) is elliptic in $\overline{\mathcal{G}}\backslash\mathcal{S}$ in the sense of Douglis and Nirenberg and the Lopatinskiĭ condition is satisfied on $\partial\mathcal{G}\backslash\mathcal{S}$ (see [**2, 49**]).

Consider a differential operator

(1.4.3) $$P(x, \partial_x) = \sum_{|\alpha| \leq k} p_\alpha(x) \, \partial_x^\alpha$$

of order k (for an arbitrary multi-index $\alpha = (\alpha_1, \dots, \alpha_n)$ with length $|\alpha| = \alpha_1 + \dots + \alpha_n$ we denote by ∂_x^α the partial derivative $\partial_{x_1}^{\alpha_1} \cdots \partial_{x_n}^{\alpha_n}$). This operator is said to be *admissible* in a neighborhood of $x^{(\tau)}$ if the coefficients p_α have the form

$$p_\alpha(x) = r_\tau^{|\alpha|-k} \, p_\alpha^{(\tau)}(r_\tau, \omega)$$

in this neighborhood, where $\omega = (x - x^{(\tau)})/r_\tau$, the functions $p_\alpha^{(\tau)}$ are smooth in $(0, \infty) \times \overline{\Omega}_\tau$, continuous in $[0, \infty) \times \overline{\Omega}_\tau$ and satisfy the condition

$$(r\partial_r)^i \, \partial_\omega^\beta \big(p_\alpha^{(\tau)}(\omega, r) - p_\alpha^{(\tau)}(\omega, 0) \big) \to 0 \ \text{as} \ r \to 0$$

uniformly with respect to $\omega \in \overline{\Omega}_\tau$ for all i and β. The differential operator

$$P^{(\tau)}(x, \partial_x) = \sum_{|\alpha| \leq l} r_\tau^{|\alpha|-k} \, p_\alpha^{(\tau)}(0, \omega) \, \partial_x^\alpha$$

is called the *leading part* of the operator (1.4.3) at $x^{(\tau)}$. For example, every differential operator $P(x, \partial_x)$ with infinitely differentiable coefficients on $\overline{\mathcal{G}}$ is admissible. In this case, the differential operator $P^{(\tau)}(x, \partial_x)$ is equal to the principal part of P with coefficients frozen at $x^{(\tau)}$, i.e., to the operator

$$P^\circ(x^{(\tau)}, \partial_x) = \sum_{|\alpha|=k} p_\alpha(x^{(\tau)}) \, \partial_x^\alpha$$

We assume that $L_{i,j}$, $B_{k,j}$ are admissible operators of order $s_i + t_j$ and $\sigma_k + t_j$, respectively, in a neighborhood of each of the points $x^{(1)}, \dots, x^{(d)}$.

For every $\tau = 1, \dots, d$ we consider the operator $\mathfrak{A}_\tau(\lambda)$ of the parameter-depending boundary value problem

(1.4.4) $$\sum_{j=1}^{N} \mathcal{L}_{i,j}^{(\tau)}(\lambda + t_j) \, u_j = f_i \ \text{in} \ \Omega_\tau, \ i = 1, \dots, N,$$

(1.4.5) $$\sum_{j=1}^{N} \mathcal{B}_{k,j}^{(\tau)}(\lambda + t_j) \, u_j = g_k \ \text{on} \ \partial\Omega_\tau \ \text{on} \ \partial\Omega_\tau, \ k = 1, \dots, M,$$

where

$$\mathcal{L}_{i,j}^{(\tau)}(\lambda)u(\omega) = r_\tau^{s_i + t_j - \lambda}\, L_{i,j}^{(\tau)}(x, \partial_x)\, r_\tau^\lambda u(\omega),$$

$$\mathcal{B}_{k,j}^{(\tau)}(\lambda)u(\omega) = r_\tau^{\sigma_k + t_j - \lambda}\, B_{k,j}^{(\tau)}(x, \partial_x)\, r_\tau^\lambda u(\omega)$$

and $L_{i,j}^{(\tau)}$, $B_{k,j}^{(\tau)}$ are the leading parts of L and $B_{k,j}$, respectively, at $x^{(\tau)}$.

In this book we are mostly concerned with the systems elliptic in the sense of Petrovskiĭ, which means that

$$t_1 = \ldots = t_N = 0 \quad \text{and} \quad s_1 = \ldots = s_N = 2m\,.$$

In this case system (1.4.1) can be written in the form

$$(1.4.6) \qquad \sum_{|\alpha| \le 2m} A_\alpha(x)\, \partial_x^\alpha U = F \text{ in } \mathcal{G},$$

where $A_\alpha(x)$ are $N \times N$-matrices, and U, F denote the vectors with the components $U_1, \ldots, U_N$ and $F_1, \ldots, F_N$, respectively. If the matrices $A_\alpha(x)$ are infinitely differentiable with respect to x in a neighborhood of $x^{(\tau)}$, then system (1.4.4) has the form $\mathcal{L}^{(\tau)}(\lambda)\, u = f$ in Ω_τ, where

$$\mathcal{L}^{(\tau)}(\lambda)\, u = r_\tau^{2m - \lambda} \sum_{|\alpha| = 2m} A_\alpha(x^{(\tau)})\, \partial_x^\alpha \left(r_\tau^\lambda u \right)\,.$$

1.4.2. Solvability of the boundary value problem in weighted Sobolev spaces. Let l be a nonnegative integer, $p \in (1, \infty)$, and $\vec{\beta} = (\beta_1, \ldots, \beta_d) \in \mathbb{R}^d$. We define the weighted Sobolev space $V_{p,\vec{\beta}}^l(\mathcal{G})$ as the closure of the set $C_0^\infty(\overline{\mathcal{G}} \backslash S)$ with respect to the norm

$$\|U\|_{V_{p,\vec{\beta}}^l(\mathcal{G})} = \left(\int_{\mathcal{G}} \left(\prod_{\tau=1}^d r_\tau^{p\beta_\tau} \right) \sum_{|\alpha| \le l} r^{p(|\alpha| - l)} |\partial_x^\alpha U(x)|^p \, dx \right)^{1/p}.$$

Sometimes, we will use the notation $V_{p,\beta}^l(\mathcal{G})$, by that we mean the above defined space with $\vec{\beta} = (\beta, \ldots, \beta) \in \mathbb{R}^d$. The space of traces of functions from $V_{p,\vec{\beta}}^l(\mathcal{G})$, $l \ge 1$, on $\partial\mathcal{G} \backslash S$ is denoted by $V_{p,\vec{\beta}}^{l-1/p}(\partial\mathcal{G})$. It is equipped with the norm

$$\|U\|_{V_{p,\vec{\beta}}^{l-1/p}(\partial\mathcal{G})} = \inf \left\{ \|V\|_{V_{p,\vec{\beta}}^l(\mathcal{G})}\ :\ V \in V_{p,\vec{\beta}}^l(\mathcal{G}),\ V = U \text{ on } \partial\mathcal{G} \backslash S \right\}.$$

It follows from $U \in V_{p,\vec{\gamma}}^l(\mathcal{G})$ that

$$\prod_{\tau=1}^d r_\tau^{p\delta_\tau} U \in V_{p,\vec{\gamma} - \vec{\delta}}^l(\mathcal{G}), \quad \vec{\delta} = (\delta_1, \ldots, \delta_d)\,.$$

We consider the operator

$$U = (U_1, \ldots, U_N) \to \left\{ \sum_{j=1}^N L_{i,j} U_j\,,\ \sum_{j=1}^N B_{k,j} U_j \right\}_{1 \le i \le N,\, 1 \le k \le M}$$

of the boundary value problem (1.4.1), (1.4.2). Clearly, under the above assumptions, this operator realizes a continuous mapping

$$(1.4.7) \qquad \mathcal{A}_{p,\vec{\beta}}^l : \prod_{j=1}^N V_{p,\vec{\beta}}^{l + t_j}(\mathcal{G}) \to \prod_{i=1}^N V_{p,\vec{\beta}}^{l - s_i}(\mathcal{G}) \times \prod_{k=1}^M V_{p,\vec{\beta}}^{l - \sigma_k - 1/p}(\partial\mathcal{G}),$$

where

(1.4.8) l is integer and $l \geq \max s_i, \ l > \max \sigma_k$.

We shall assume in all theorems formulated in this section that l (and a similar index l') are subject to (1.4.8). *As it was proved by Maz'ya and Plamenevskiĭ* [182] *(see also the monograph of Nazarov and Plamenevskiĭ* [207] *and, for the case $p = 2$, Kondrat'ev's paper* [109] *and the monographs of Grisvard* [78], *Dauge* [41], *Kozlov, Maz'ya and Roßmann* [136]), *the following statement is valid.*

THEOREM 1.4.1. *Suppose that the line* $\operatorname{Re} \lambda = l - \beta_\tau - n/p$ *does not contain eigenvalues of the pencil* $\mathfrak{A}_\tau, \ \tau = 1, \ldots, d.$ *Then the operator* $\mathcal{A}^l_{p,\vec{\beta}}$ *is Fredholm, and for any solution*

$$U = (U_1, \ldots, U_N) \in \prod_{j=1}^N V^{l+t_j}_{p,\vec{\beta}}(\mathcal{G})$$

of problem (1.4.1), (1.4.2) *there is the estimate*

$$\sum_{j=1}^N \|U_j\|_{V^{l+t_j}_{p,\vec{\beta}}(\mathcal{G})} \leq c \Big(\sum_{i=1}^N \|F_i\|_{V^{l-s_i}_{p,\vec{\beta}}(\mathcal{G})} + \sum_{k=1}^M \|G_k\|_{V^{l-\sigma_k-1/p}_{p,\vec{\beta}}(\partial\mathcal{G})} + \sum_{j=1}^N \|U_j\|_{L_p(\mathcal{G}')} \Big),$$

where $\mathcal{G}'$ is an arbitrary nonempty open set, $\overline{\mathcal{G}'} \subset \overline{\mathcal{G}} \backslash \mathcal{S}.$

In [136] an analogous result in weighted Sobolev spaces with arbitrary integer order is obtained. Note that the conditions on the eigenvalues of the pencils $\mathfrak{A}_\tau$ in Theorem 1.4.1 are also necessary.

From Theorem 1.4.3 below it follows that the kernel of the operator $\mathcal{A}^l_{p,\beta}$ depends only on the numbers $l - \beta_\tau - n/p, \ \tau = 1, \ldots, d$, if there are no eigenvalues of the pencils $\mathfrak{A}_\tau$ on the lines $\operatorname{Re} \lambda = l - \beta_\tau - n/p$. The same is true for the kernel of the adjoint operator. In Maz'ya and Plamenevskiĭ's paper [180, Le.8.1] the following formula for the *index* of the operator $\mathcal{A}^l_{p,\beta}$ was proved under the assumptions that $l - \beta_\tau - n/p < l' - \beta'_\tau - n/p'$ and there are no eigenvalues of the pencils $\mathfrak{A}_\tau$ on the lines $\operatorname{Re} \lambda = l - \beta_\tau - n/p$ and $\operatorname{Re} \lambda = l' - \beta'_\tau - n/p', \ \tau = 1, \ldots, d$ (in the case $p = p' = 2$ see also the books [136, Th.6.6] and [207, Ch.4,Th.3.3]):

(1.4.9) $\operatorname{ind} \mathcal{A}^l_{p,\beta} = \operatorname{ind} \mathcal{A}^{l'}_{p',\beta'} + \sum_{\tau=1}^d \kappa_\tau,$

where κ_τ is the sum of the algebraic multiplicities of all eigenvalues of the pencil $\mathfrak{A}_\tau$ in the strip $l - \beta_\tau - n/p < \operatorname{Re} \lambda < l' - \beta'_\tau - n/p', \ \tau = 1, \ldots, d.$

1.4.3. Regularity and asymptotics of the solution. By means of classical local regularity results for solutions of elliptic boundary value problems, the following assertion can be proved (see, e.g., [136, Le.6.3.1]).

THEOREM 1.4.2. *Let* $U_j \in V^0_{p,\vec{\beta}-(l+t_j)\vec{1}}(\mathcal{G}), \ j = 1, \ldots, N$ *(here $\vec{1}$ denotes the vector* $(1, 1, \ldots, 1) \in \mathbb{R}^d$), *be functions which belong to* $W^{l+t_j}_p(\mathcal{G}')$ *for every open set $\mathcal{G}', \ \overline{\mathcal{G}'} \subset \overline{\mathcal{G}} \backslash \mathcal{S}$. If the vector function $U = (U_1, \ldots, U_N)$ is a solution of problem* (1.4.1), (1.4.2), *where* $F_i \in V^{l-s_i}_{p,\vec{\beta}}(\mathcal{G}), \ G_k \in V^{l-\sigma_k-1/p}_{p,\vec{\beta}}(\partial\mathcal{G})$, *then* $U_j \in V^{l+t_j}_{p,\vec{\beta}}(\mathcal{G}).$

Furthermore, the inequality

$$\sum_{j=1}^{N} \|U_j\|_{V_{p,\vec{\beta}}^{l+t_j}(\mathcal{G})} \le c \Big(\sum_{i=1}^{N} \|F_i\|_{V_{p,\vec{\beta}}^{l-s_i}(\mathcal{G})} + \sum_{k=1}^{M} \|G_k\|_{V_{p,\vec{\beta}}^{l-\sigma_k-1/p}(\partial\mathcal{G})}$$

$$+ \sum_{j=1}^{N} \|U_j\|_{V_{p,\vec{\beta}-(l+t_j)\vec{1}}^{0}(\mathcal{G})} \Big)$$

is satisfied.

In contrast to Theorem 1.4.2, the following statement holds only under additional assumptions on the eigenvalues of the pencils $\mathfrak{A}_\tau$. We refer again to [**182, 207**] and, in the case $p = 2$, to [**41, 78, 109, 136**].

THEOREM 1.4.3. *Let* $U \in \prod V_{p,\vec{\beta}}^{l+t_j}(\mathcal{G})$ *be a solution of problem* (1.4.1), (1.4.2) *with* $F_i \in V_{p',\vec{\beta}'}^{l'-s_i}(\mathcal{G})$, $G_k \in V_{p',\vec{\beta}'}^{l'-\sigma_k-1/p'}(\partial\mathcal{G})$. *If there are no eigenvalues of the pencils* $\mathfrak{A}_\tau$, $\tau = 1, \dots, d$, *in the closed strip between the lines* $\operatorname{Re}\lambda = l - \beta_\tau - n/p$ *and* $\operatorname{Re}\lambda = l' - \beta'_\tau - n/p'$, *then* $U \in \prod V_{p',\vec{\beta}'}^{l'+t_j}(\mathcal{G})$.

One of the central questions in the theory of elliptic boundary value problems for domains with conic points is the question on the asymptotic behavior of the solutions near the singular boundary points. We give here an asymptotic formula for the case when the operators $L_{i,j}$, $B_{k,j}$ are model operators, i.e., they coincide with their leading parts near $x^{(\tau)}$.

THEOREM 1.4.4. *Let* $U = (U_1, \dots, U_N) \in \prod V_{p,\vec{\beta}}^{l+t_j}(\mathcal{G})$ *be a solution of problem* (1.4.1), (1.4.2) *with*

$$F_i \in V_{p',\vec{\beta}'}^{l'-s_i}(\mathcal{G}), \ G_k \in V_{p',\vec{\beta}'}^{l'-\sigma_k-1/p'}(\partial\mathcal{G}) \,,$$

where $l' - \beta'_\tau - n/p' > l - \beta_\tau - n/p$. *We suppose that near* $x^{(\tau)}$ *the operators* $L_{i,j}$, $B_{k,j}$ *coincide with their principal parts at* $x^{(\tau)}$. *Furthermore, we assume that the lines* $\operatorname{Re}\lambda = l - \beta_\tau - n/p$ *and* $\operatorname{Re}\lambda = l' - \beta'_\tau - n/p'$ *are free of eigenvalues of the pencil* $\mathfrak{A}_\tau$. *Then the functions* U_j, $j = 1, \dots, N$, *admit, near* $x^{(\tau)}$, *the asymptotic representation*

$$U_j = \sum_{\mu=1}^{M} \sum_{\nu=1}^{I_\nu} \sum_{s=0}^{\kappa_{\mu,\nu}-1} c_{\mu,\nu,s} r_\tau^{\lambda_\mu + t_j} \sum_{\sigma=0}^{s} \frac{1}{s - \sigma!} (\log r)^{s-\sigma} u_{j;\mu,\nu,\sigma}(\omega) + V_j,$$

where $V_j \in V_{p',\vec{\beta}'}^{l'+t_j}(\mathcal{G})$, $\lambda_1, \dots, \lambda_M$ *are the eigenvalues of the pencil* $\mathfrak{A}_\tau$ *in the strip* $l - \beta_\tau - n/p < \operatorname{Re}\lambda < l' - \beta'_\tau - n/p'$, ω *are coordinates on the sphere* $|x - x^{(\tau)}| = 1$, *and* $u_{\mu,\nu,\sigma} = (u_{1;\mu,\nu,\sigma}, \dots, u_{N;\mu,\nu,\sigma})$ *are eigenvectors* $(\sigma = 0)$ *and generalized eigenvectors* $(\sigma \ge 1)$ *of the pencil* $\mathfrak{A}_\tau$ *corresponding to the eigenvalue* λ_μ.

1.4.4. Solvability in weighted Hölder spaces. Let again l be nonnegative, $\vec{\beta} = (\beta_1, \dots, \beta_d) \in \mathbb{R}^d$, and $0 < \sigma < 1$. We denote by $N_{\vec{\beta}}^{l,\sigma}(\mathcal{G})$ the weighted Hölder space with the norm

$$\|U\|_{N_{\vec{\beta}}^{l,\sigma}(\mathcal{G})} = \sup_{x \in \mathcal{G}} \Big(\prod_{\tau=1}^{d} r_\tau^{\beta_\tau} \sum_{|\alpha| \le l} r^{|\alpha|-l-\sigma} |\partial_x^\alpha U(x)| \Big)$$

$$+ \sum_{|\alpha|=l} \sup_{x,y \in \mathcal{G}} |x-y|^{-\sigma} \Big| \prod_{\tau=1}^{d} r_\tau(x)^{\beta_\tau} \partial_x^\alpha U(x) - \prod_{\tau=1}^{d} r_\tau(y)^{\beta_\tau} \partial_y^\alpha U(y) \Big|$$

Furthermore, let $N_{\vec{\beta}}^{l,\sigma}(\partial\mathcal{G})$ be the trace space on $\partial\mathcal{G}$ for the space $N_{\vec{\beta}}^{l,\sigma}(\mathcal{G})$. If β is a real number, then by $N_\beta^{l,\sigma}(\mathcal{G})$ and $N_\beta^{l,\sigma}(\partial\mathcal{G})$ we mean the spaces just introduced with the weight parameter $\vec{\beta} = (\beta, \dots, \beta) \in \mathbb{R}^d$. Obviously, the space $N_{\beta+1}^{l+1,\sigma}(\mathcal{G})$ is continuously imbedded into $N_\beta^{l,\sigma}(\mathcal{G})$ for arbitrary l, β, and σ.

The operator of the boundary value problem (1.4.1), (1.4.2) realizes a continuous mapping

$$(1.4.10) \qquad \mathcal{A}_{\vec{\beta}}^{l,\sigma} : \prod_{j=1}^{N} N_{\vec{\beta}}^{l+t_j,\sigma}(\mathcal{G}) \to \prod_{i=1}^{N} N_{\vec{\beta}}^{l-s_i,\sigma}(\mathcal{G}) \times \prod_{k=1}^{m} N_{\vec{\beta}}^{l-\sigma_k,\sigma}(\partial\mathcal{G}) .$$

We state a result for the space $N_{\vec{\beta}}^{l,\sigma}(\mathcal{G})$ similar to Theorems 1.4.1 and 1.4.3 (cf. [**182**, Th.6.3,Th.6.4]).

THEOREM 1.4.5. 1) *Suppose that there are no eigenvalues of the pencils $\mathfrak{A}_\tau$ on the line* $\operatorname{Re}\lambda = l + \sigma - \beta_\tau$, $\tau = 1, \dots, d$. *Then operator* (1.4.10) *is Fredholm and every solution* $U \in \prod N_{\vec{\beta}}^{l+t_j,\sigma}(\mathcal{G})$ *of problem* (1.4.1), (1.4.2) *satisfies the estimate*

$$\sum_{j=1}^{N} \|U_j\|_{N_{\vec{\beta}}^{l+t_j,\sigma}(\mathcal{G})} \le c \Big(\sum_{i=1}^{N} \|F_i\|_{N_{\vec{\beta}}^{l-s_i,\sigma}(\mathcal{G})} + \sum_{k=1}^{M} \|G_k\|_{N_{\vec{\beta}}^{l-\sigma_k,\sigma}(\partial\mathcal{G})} + \|U\|_{L(\mathcal{G}')^N} \Big),$$

where $\mathcal{G}'$ is an arbitrary nonempty open set, $\overline{\mathcal{G}'} \subset \overline{\mathcal{G}} \backslash \mathcal{S}$.

2) *Let $U \in \prod V_{p,\vec{\beta}}^{l+t_j}(\mathcal{G})$ be a solution of problem* (1.4.1), (1.4.2), *where*

$$F_i \in N_{\vec{\beta}'}^{l'-s_i,\sigma}(\mathcal{G}) , \quad G_k \in N_{\vec{\beta}'}^{l'-\sigma_k,\sigma}(\partial\mathcal{G}) .$$

If there are no eigenvalues of the pencils $\mathfrak{A}_\tau$, $\tau = 1, \dots, d$, in the closed strip between the lines $\operatorname{Re}\lambda = l - \beta_\tau - n/p$ *and* $\operatorname{Re}\lambda = l' + \sigma - \beta'_\tau$, *then* $U \in \prod N_{\vec{\beta}'}^{l'+t_j,\sigma}(\mathcal{G})$.

We have restricted ourselves to the minimum information which is frequently used in this book. However, many other results of a similar nature are known, in particular, those relating function spaces with "nonhomogeneous" norms such as classical Sobolev and Hölder spaces (see, for example, [**41, 109, 136, 179**]). Combining such results with explicit facts about operator pencils to be obtained in the sequel, one can easily extend the scope of applications of these facts.

1.5. Notes

Pencils of general elliptic parameter dependent boundary value problems appeared first in the works Agranovich and Vishik [**4**], Agmon and Nirenberg [**3**] (cylindrical domains), and Kondrat'ev [**109**] (domain with conic vertices). Lopatinskiĭ [**156**] and Eskin [**51, 52**] arrived at holomorphic operator functions when studying boundary integral equations generated by elliptic boundary value problems in domains with corners.

Section 1.1. Basic facts from the theory of holomorphic operator functions in a pair of Banach spaces can be found in works of Gohberg, Goldberg and Kaashoek [**71**], Markus [**160**], Wendland [**265**], Mennicken and Möller [**200**], and Kozlov and Maz'ya [**135**]. The Laurent decomposition of the resolvent near a pole was constructed by Keldysh [**97, 98**] and extended to holomorphic operator functions by Markus and Sigal [**161**] and to meromorphic operator functions by Gohberg and Sigal [**73**] (for the proof see [**71**] and [**135**]). Operator versions of the logarithmic residual and Rouché's theorems were obtained in [**98**] for pencils, in [**161**] for holomorphic operator functions and in [**73**] for meromorphic operator functions. Theorems 1.1.7 and 1.1.9 are classical results in the theory of ordinary differential equations.

Section 1.2. The material is borrowed from the paper [**129**] by Kozlov and Maz'ya.

Section 1.3. All results can be found in the papers [**140, 141**] by Kozlov, Maz'ya and Schwab.

Section 1.4. There exists extensive bibliography concerning elliptic boundary value problems in domains with angle and conic vertices. A theory of general elliptic problems for these domains was initiated in the above mentioned works by Lopatinskiĭ, Eskin and Kondrat'ev. The first two authors dealt with boundary value problems in plane domains with angular points. They reduced the problem to an integral equation on the boundary and investigated this equation by using Mellin's transform. Kondrat'ev [**109**] studied boundary value problems for scalar differential operators in domains of arbitrary dimension with conic points by applying Mellin's transform directly to the differential operators. He established the Fredholm property in weighted and usual L_2-Sobolev spaces, and also found asymptotic representations of solutions near vertices. Maz'ya and Plamenevskiĭ extended these results to other function spaces (L_p-Sobolev spaces, Hölder spaces, spaces with inhomogeneous norms). They calculated the coefficients in asymptotics and described singularities of Green's kernels (see [**179, 180, 182, 186**]).

Formula (1.4.9) which describes dependence of the index on function spaces is proved in [**180**] by Maz'ya and Plamenevskiĭ. Eskin [**54**] obtained an index formula for elliptic boundary value problem in a plane domain with corners. In connection with the index formulas, we mention also the works of Gromov and Shubin [**87, 88**], Shubin [**245**], where the classical Riemann-Roch theorem was generalized to solutions of general elliptic equations with isolated singularities on a compact manifold.

A theory of pseudodifferential operators on manifolds with conic points was developed in works of Plamenevskiĭ [**228**], Schulze [**236, 238, 240, 241**], Melrose [**198**] and others.

The modern state of the theory of elliptic problems in domains with angular or conic points is discussed in the books of Dauge [**41**], Maz'ya, Nazarov and Plamenevskiĭ [**174**], Schulze [**237, 239**], Nazarov and Plamenevskiĭ [**207**], Kozlov, Maz'ya and Roßmann [**136**]. In these books and in the review of Kondrat'ev and Oleĭnik [**112**], many additional references concerning the subject can be found.

In a number of works, results of the same type as in Section 1.4 were extended to transmission problems. We mention here the papers by Kellogg [**100**], Ben M'Barek and Merigot [**17**], Lemrabet [**152**], Meister, Penzel, Speck and Teixeira [**197**] (Laplace, Helmholtz equations), the books of Leguillon and Sanchez-Palencia [**151**] (second order equations, elasticity), Nicaise [**211**] (Laplace, biharmonic equations), and the papers of Nicaise and Sändig [**212**] (general equations).

CHAPTER 2

Angle and conic singularities of harmonic functions

This chapter is an introduction to the theme of singularities. The Laplacian is a suitable object to begin with because of the simplicity of the corresponding operator pencil $\delta + \lambda(\lambda + n - 2)$, where δ is the Laplace-Beltrami operator on the unit sphere. Thus, we deal with a standard spectral problem for the operator δ (at least, in the case of the Dirichlet and Neumann boundary conditions).

We start with an example of singularities generated by the Laplace operator and say a few words about their applications. Let $\mathcal{G}$ be a bounded plane domain whose boundary $\partial\mathcal{G}$ contains the origin. We suppose that the arc $\partial\mathcal{G}\backslash\{0\}$ is smooth and that near the point $x = 0$ the domain coincides with the angle

$$\{x = r\,e^{i\varphi}\ :\ 0 < r < 1,\ 0 < \varphi < \alpha\},$$

where $\alpha \in (0, 2\pi]$. Consider the Dirichlet problem

(2.0.1) $-\Delta U = F$ on $\mathcal{G}$, $U = 0$ on $\partial\mathcal{G}\backslash\{0\}$

where F is a given function in $L_2(\mathcal{G})$. By the Riesz representation theorem there exists a unique (variational) solution U in the Sobolev space

$$W_2^1(\mathcal{G}) = \{U \in L_2(\mathcal{G})\ :\ \int_{\mathcal{G}} |\nabla U|^2\, dx < \infty\}$$

(see the book [**155**, Ch.2,Sect.9] by Lions and Magenes). If we assume additionally that $F = 0$ in a neighborhood of 0, then a direct application of the Fourier method leads to the representation of U near 0 in the form of the convergent series:

(2.0.2) $$U(x) = \sum_{k=1}^{\infty} c_k\, r^{k\pi/\alpha} \sin(k\pi\varphi/\alpha).$$

When applying the Fourier method, we see that the exponents $\lambda_k = k\pi/\alpha$ are eigenvalues of the spectral problem

(2.0.3) $-u''(\varphi) - \lambda^2 u(\varphi) = 0,$ $0 < \varphi < \alpha,$

(2.0.4) $u(0) = u(\alpha) = 0.$

and that $\sin(k\pi\varphi/\alpha)$ is an eigenfunction corresponding to λ_k.

In (2.0.2) we meet infinitely many singular, i.e., nonsmooth terms. In particular, all terms in this series are singular if π/α is irrational. Clearly, (2.0.2) contains all information about differentiability properties of u in a neighborhood of the vertex $x = 0$. We see, for example, that U and its derivatives up to order m are continuous if π/α is integer or $\alpha < \pi/m$.

We state a well-known regularity result, where the first term in (2.0.2) is important (see, for example, our book [**136**, Sect.6.6]). Consider the Dirichlet problem

(2.0.1) with an arbitrary $F \in L_2(\mathcal{G})$. The second derivatives of the solution U to (2.0.1) belong to $L_2(\mathcal{G})$ if and only if the same is true for the function $r^{\pi/\alpha} \sin(\pi\varphi/\alpha)$ or, equivalently, if and only if $\alpha \leq \pi$.

As in the two-dimensional case, the Fourier method can be used to describe singularities of solutions to problem (2.0.1) for the n-dimensional domain $\mathcal{G}$ coinciding with the cone $\mathcal{K} = \{x \in \mathbb{R}^n : 0 < r < 1, \omega \in \Omega\}$ near the origin. Here $r = |x|, \omega = x/|x|$ and Ω is a domain on the unit sphere. If $F = 0$ near the origin, the solution is given, for sufficiently small r, by the series

$$(2.0.5) \qquad U(x) = \sum_{k=1}^{\infty} c_k \, r^{\lambda_k} \, u_k(\omega),$$

where c_k=const, $\{u_k\}$ is a sequence of eigenfunctions and $\{\lambda_k\}$ is the sequence of positive eigenvalues of the spectral problem

$$-\delta u - \lambda(\lambda + n - 2)\, u = 0 \quad \text{in } \Omega, \quad u = 0 \quad \text{on } \partial\Omega,$$

As in the two-dimensional case, the decomposition (2.0.5) gives rise to theorems on regularity of solutions to problem (2.0.1), formulated in terms of λ_k. However, one can find these eigenvalues very seldom, which makes description of the regularity properties of solutions in a particular domain difficult even in the case of the Laplace operator.

In Section 2.1 we consider classical boundary value problems for harmonic functions in an angle for which the question of singularities is trivial. The Dirichlet problem for the Laplace operator in an n-dimensional cone is shortly discussed in Section 2.2, which is a collection of known results, with simple facts proved and deeper ones only formulated. Sections 2.3 and 2.4 deal with the multi-dimensional Neumann and oblique derivative problems in a cone. In Section 2.5 we present some qualitative information about eigenvalues obtained by asymptotic methods, and we conclude the chapter with historical notes.

2.1. Boundary value problems for the Laplace operator in an angle

2.1.1. The Dirichlet problem. Let $\mathcal{K} = \{(x_1, x_2) \in \mathbb{R}^2 : r > 0, 0 < \varphi < \alpha\}$ be a plane angle with vertex at the origin. Here r, φ are the polar coordinates of the point (x_1, x_2) and $0 < \alpha \leq 2\pi$. We are interested in harmonic functions in $\mathcal{K}$ which are positively homogeneous of degree λ, $\lambda \in \mathbb{C}$, and equal to zero on the boundary. This means, we seek the solutions of the problem

$$(2.1.1) \qquad -\Delta U = 0 \quad \text{on } \mathcal{K}, \quad U(r, 0) = U(r, \alpha) = 0$$

which have the form

$$(2.1.2) \qquad U(r, \varphi) = r^{\lambda} u(\varphi).$$

Since the Laplace operator has the representation

$$\Delta U = r^{-1}\partial_r \left(r\partial_r U\right) + r^{-2}\partial_{\varphi}^2 U$$

in polar coordinates, we get the Dirichlet problem (2.0.3), (2.0.4) for the function u. Let $\mathfrak{A}(\lambda) = -d^2/d\varphi^2 - \lambda^2$ be the operator on the left-hand side of (2.0.3) which is defined on the set of the functions equal to zero at the ends of the interval $(0, \alpha)$. The operator $\mathfrak{A}(\lambda)$ can be considered, e.g., as a mapping from the space $W_2^2((0, \alpha)) \cap \overset{\circ}{W}$

$\frac{1}{2}((0,\alpha))$ into $L_2((0,\alpha))$. Here $L_2((0,\alpha))$ is the space of square summable functions in $(0,\alpha)$, $W_2^l((0,\alpha))$ denotes the Sobolev space with the norm

$$\|u\|_{W_2^l((0,\alpha))} = \Big(\sum_{j=0}^{l} |\partial_\varphi u(\varphi)|^2 \, d\varphi \Big)^{1/2},$$

and $\overset{\circ}{W}{}_2^l((0,\alpha))$ denotes the closure of the set of all infinitely differentiable functions with support in $(0,\alpha)$ with respect to the $W_2^l((0,\alpha))$-norm.

By what has been shown above, the function (2.1.2) is a solution of problem (2.1.1) if and only if λ is an eigenvalue of the operator pencil $\mathfrak{A}$ and u is an eigenfunction corresponding to this eigenvalue. It can be easily verified that the eigenvalues of the pencil $\mathfrak{A}$ are the numbers $\lambda_{\pm j} = \pm j\pi/\alpha$, $j = 1,2,\ldots$ and that

$$v_{\pm j}(\varphi) = \sin \frac{j\pi\varphi}{\alpha}.$$

are the eigenfunctions corresponding to the eigenvalues $\lambda_{\pm j}$.

Now we seek solutions of (2.1.1) which have the form

$$(2.1.3) \qquad U = r^\lambda \sum_{k=0}^{s} \frac{1}{k!} (\log r)^k \, u_{s-k}(\varphi),$$

where $u_k \in W_2^2((0,\alpha)) \cap \overset{\circ}{W}{}_2^1((0,\alpha))$, $k = 0,1,\ldots,s$, $u_0 \neq 0$, $s \geq 1$. Inserting (2.1.3) into (2.1.1), we obtain

$$\begin{aligned}
0 \;=\; & r^\lambda \frac{(\log r)^s}{s!} \big(\lambda^2 u_0 + u_0''\big) + r^\lambda \frac{(\log r)^{s-1}}{(s-1)!} \big(\lambda^2 u_1 + u_1'' + 2\lambda u_0\big) \\
& + r^\lambda \sum_{k=0}^{s-2} \frac{(\log r)^k}{k!} \big(\lambda^2 u_{s-k} + u_{s-k}'' + 2\lambda u_{s-1-k} + u_{s-2-k}\big).
\end{aligned}$$

Consequently, the functions $u_0, u_1, \ldots, u_s$ satisfy the equalities

$$(2.1.4) \qquad -u_0'' - \lambda^2 u_0 \;=\; 0,$$
$$(2.1.5) \qquad -u_1'' - \lambda^2 u_1 \;=\; 2\lambda u_0,$$
$$(2.1.6) \qquad -u_k'' - \lambda^2 u_k \;=\; 2\lambda u_{k-1} + u_{k-2}, \quad k = 2,\ldots,s.$$

This means that λ is an eigenvalue of the pencil $\mathfrak{A}$ and the functions $u_0, u_1, \ldots, u_s$ form a Jordan chain of the pencil $\mathfrak{A}$ corresponding to the eigenvalue λ. In particular, we have $\lambda = \lambda_{\pm j}$ and $u_0 = c\,v_{\pm j}$, $c \neq 0$. Multiplying (2.1.5) by $v_{\pm j}$ and integrating over the interval $(0,\alpha)$, we obtain

$$2c\lambda_{\pm j} \int_0^\alpha |v_{\pm j}(\varphi)|^2 \, d\varphi = \int_0^\alpha \mathfrak{A}(\lambda_{\pm j}) u_1 \cdot v_{\pm j} \, d\varphi = \int_0^\alpha u_1 \cdot \mathfrak{A}(\lambda_{\pm j}) v_{\pm j} \, d\varphi = 0.$$

Since this contradicts the assumption $c \neq 0$, we conclude that the operator pencil $\mathfrak{A}$ has no generalized eigenfunctions. Thus, solutions of the form (2.1.3) with $u_0 \neq 0$, $s \geq 1$, do not exist.

The same results hold if the operator $\mathfrak{A}(\lambda)$ is considered as a mapping from $\overset{\circ}{W}{}_2^1((0,\alpha))$ into the dual space $W_2^{-1}((0,\alpha))$. Then the operator $\mathfrak{A}(\lambda)$ is given by the equality

$$\big(\mathfrak{A}(\lambda)\, u, \, v\big) = a(u,v;\lambda), \qquad u, v \in \overset{\circ}{W}{}_2^1((0,\alpha)),$$

where $(\cdot, \cdot)$ denotes the extension of the inner product in $L_2((0, \alpha))$ to the Cartesian product $W_2^{-1}((0, \alpha)) \times \overset{\circ}{W}{}_2^1((0, \alpha))$ and

$$(2.1.7) \qquad a(u, v; \lambda) = \int_0^\alpha \left(u'(\varphi) \, \overline{v'(\varphi)} - \lambda^2 \, u(\varphi) \, \overline{v(\varphi)} \right) d\varphi,$$

2.1.2. The Neumann problem. Next we consider the problem
$$-\Delta U = 0 \quad \text{in } \mathcal{K}, \quad \partial_\varphi U(r, 0) = \partial_\varphi U(r, \alpha) = 0.$$
As above, we seek solutions of the form (2.1.2). This leads to the problem

$$(2.1.8) \qquad\qquad -u''(\varphi) - \lambda^2 u(\varphi) = 0, \quad 0 < \varphi < \alpha,$$
$$(2.1.9) \qquad\qquad u'(0) = u'(\alpha) = 0.$$

The corresponding operator $\mathfrak{A}(\lambda)$ can be defined, e.g., as a mapping from the space $\{u \in W_2^2((0, \alpha)) : u'(0) = u'(\alpha) = 0\}$ into $L_2((0, \alpha))$.

The spectrum of the pencil $\mathfrak{A}$ consists of the eigenvalues $\lambda_{\pm j} = \pm j\pi/\alpha$, $j = 0, 1, 2, \dots$, and the corresponding eigenfunctions are

$$v_{\pm j}(\varphi) = \cos \frac{j\pi\varphi}{\alpha} \,.$$

It can be easily verified that all eigenvalues, except $\lambda_0 = 0$, are *simple*, i.e., (up to a scalar factor) there exists only one eigenfunction to every of these eigenvalues, while generalized eigenfunctions do not exist. The set of the eigenfunctions and generalized eigenfunctions corresponding to the eigenvalue $\lambda_0 = 0$ consists of the constant functions $u_0 = c_0$, $u_1 = c_1$.

Let $a(\cdot, \cdot, \lambda)$ be the sesquilinear form (2.1.7) on $W_2^1((0, \alpha)) \times W_2^1((0, \alpha))$. Then the Neumann problem (2.1.8), (2.1.9) can be understood in the sense of the integral identity

$$a(u, v; \lambda) = 0 \quad \text{for all } v \in W_2^1((0, \alpha)),$$

The form $a(\cdot, \cdot, \lambda)$ generates a continuous operator $W_2^1((0, \alpha)) \to W_2^1((0, \alpha))^*$ which quadratically depends on the parameter λ. This operator is an extension of the above considered operator $\mathfrak{A}(\lambda)$. It can be easily shown that the pencil defined by means of the form a has the same eigenvalues, eigenfunctions, and generalized eigenfunctions as $\mathfrak{A}$.

Both for the Dirichlet and the Neumann problem the quadratic form $a(u, u; \lambda)$ is nonnegative for $\operatorname{Re} \lambda = 0$. The line $\operatorname{Re} \lambda = 0$ is free of eigenvalues in the case of the Dirichlet problem, while it contains the eigenvalue $\lambda = 0$ of the pencil generated by the Neumann problem. The widest strip in the complex plane which contains the line $\operatorname{Re} \lambda = 0$ and which is free of eigenvalues outside this line is the strip $|\operatorname{Re} \lambda| < \pi/\alpha$.

2.1.3. The mixed problem. Now we are interested in solutions of the boundary value problem
$$-\Delta U = 0 \quad \text{in } \mathcal{K}, \quad U(r, 0) = \partial_\varphi U(r, \alpha) = 0$$
which are positively homogeneous of degree λ. This problem is connected with the parameter-depending boundary value problem

$$(2.1.10) \qquad\qquad -u''(\varphi) - \lambda^2 u(\varphi) = 0, \quad 0 < \varphi < \alpha,$$
$$(2.1.11) \qquad\qquad u(0) = 0, \ u'(\alpha) = 0.$$

The spectrum of the operator pencil corresponding to the problem (2.1.10), (2.1.11) consists of the eigenvalues $\lambda_{\pm j} = \pm j \frac{\pi}{2\alpha}$, $j = 1, 2, \ldots$. Here

$$v_{\pm j}(\varphi) = \sin \frac{j\pi\varphi}{2\alpha}$$

are the eigenfunctions corresponding to $\lambda_{\pm j}$. Generalized eigenfunctions do not exist.

Note that the strip $|\operatorname{Re} \lambda| < \frac{\pi}{2\alpha}$ does not contain eigenvalues.

2.1.4. The problem with oblique derivative. Let $\mathcal{K}$ denote the angle

$$\{(x_1, x_2) \in \mathbb{R}^2 : 0 < r < \infty, \, |\varphi| < \alpha/2\},$$

where $\alpha \in (0, 2\pi]$. We consider the boundary value problem

(2.1.12) $$-\Delta U = 0 \quad \text{in } \mathcal{K},$$

(2.1.13) $$a^\pm \, \partial_{\tau^\pm} U + b^\pm \, \partial_{\nu^\pm} U = 0 \quad \text{for } \varphi = \pm\alpha/2.$$

Here $a^\pm$, $b^\pm$ are real numbers, $b^\pm > 0$, $\tau^\pm$ are the directions of the rays $\varphi = \pm\alpha/2$, and $\nu^\pm$ are the exterior normals to the sides $\varphi = \pm\alpha/2$ of the angle $\mathcal{K}$. In polar coordinates the boundary conditions (2.1.13) have the form

$$(a^\pm \, \partial_r U \pm b^\pm r^{-1} \partial_\varphi U)\Big|_{\varphi = \pm\alpha/2} = 0.$$

Let $\mathfrak{A}(\lambda)$ be the operator

$$W_2^2(\Omega) \ni u \to \{-(\partial_\varphi^2 + \lambda^2)u, \, (\pm b^\pm \partial_\varphi u + \lambda a^\pm u)|_{\varphi=\pm\alpha/2}\} \in L_2(\Omega) \times \mathbb{C}^2,$$

where Ω denotes the interval $(-\alpha/2, +\alpha/2)$. Then the function $U = r^\lambda u(\varphi)$ is a solution of problem (2.1.12), (2.1.13) if and only if λ is an eigenvalue of the pencil $\mathfrak{A}$ and u is an eigenfunction corresponding to this eigenvalue, i.e., u is a nontrivial solution of the equation $\mathfrak{A}(\lambda) \, u = 0$. Furthermore, the function

$$U = r^\lambda \sum_{k=0}^{s} \frac{1}{k!} (\log r)^k u_{s-k}(\varphi)$$

with $u_0 \neq 0$ is a solution of (2.1.12), (2.1.13) if and only if the functions $u_0, u_1, \ldots, u_s$ form a Jordan chain of the pencil $\mathfrak{A}$ corresponding to the eigenvalue λ, i.e., if

(2.1.14) $$\mathfrak{A}(\lambda) \, u_0 = 0,$$

(2.1.15) $$\mathfrak{A}(\lambda) \, u_1 = -\mathfrak{A}'(\lambda) \, u_0,$$

(2.1.16) $$\mathfrak{A}(\lambda) \, u_k = -\mathfrak{A}'(\lambda) \, u_{k-1} - \frac{1}{2} \mathfrak{A}''(\lambda) u_{k-2}, \quad k = 2, \ldots, s.$$

Here $\mathfrak{A}'(\lambda) \, u = \{-2\lambda u, \, a^\pm u|_{\varphi=\pm\alpha/2}\}$ and $\mathfrak{A}''(\lambda) \, u = \{-2u, 0, 0\}$.

It can be directly verified that the set of the eigenvalues of the pencil $\mathfrak{A}$ consists of the numbers $\lambda_* = 0$ and

$$\lambda_{\pm j} = (k^+ + k^- \pm j\pi)/\alpha, \quad j = 0, 1, 2, \ldots,$$

where $k^\pm = \arctan(a^\pm/b^\pm)$. The eigenfunctions corresponding to the eigenvalues $\lambda_{\pm j}$ are

$$v_{\pm j}(\varphi) = \cos\big((k^+ + k^- + j\pi)\varphi/\alpha + (k^+ + k^- - j\pi)/2\big).$$

Generalized eigenfunctions corresponding to these eigenvalues do not exist.

The eigenvalue $\lambda_* = 0$ has the eigenfunction $u_0 = 1$. Equation (2.1.15) with $\lambda = 0$, $u_0 = 1$ is solvable if and only if $a^+ b^- + a^- b^+ = 0$. In this case there is the generalized eigenfunction $u_1 = a^+\varphi/b^+$ which is uniquely determined up to multiples of the eigenfunction $u_0 = 1$. Other generalized eigenfunctions do not exist.

Finally, let us mention that the strip determined by the inequalities

$$
\begin{array}{ll}
(k^+ + k^-)/\alpha > \operatorname{Re}\lambda > (k^+ + k^- - \pi)/\alpha & \text{if } a^+b^- + a^-b^+ > 0, \\
(k^+ + k^- + \pi)/\alpha > \operatorname{Re}\lambda > (k^+ + k^-)/\alpha & \text{if } a^+b^- + a^-b^+ < 0, \\
|\operatorname{Re}\lambda| < \pi/\alpha & \text{if } a^+b^- + a^-b^+ = 0
\end{array}
$$

contains only the eigenvalue $\lambda_* = 0$.

2.2. The Dirichlet problem for the Laplace operator in a cone

2.2.1. The operator pencil generated by the Dirichlet problem. Let
Ω be an open subset of the unit sphere S^{n-1}, $n \geq 3$, with the boundary $\partial\Omega$. Furthermore, let $\mathcal{K}$ be the cone

$$
\{x = (x_1, \dots, x_n) \in \mathbb{R}^n \; : \; r > 0, \; \omega \in \Omega\},
$$

where $r = |x|$ and $\omega = x/|x|$. We are interested in solutions of the Dirichlet problem

$$
(2.2.1) \qquad -\Delta U = 0 \text{ in } \mathcal{K}, \qquad U = 0 \text{ on } \partial\mathcal{K}\backslash\{0\}
$$

which have the form

$$
(2.2.2) \qquad U(x) = r^\lambda u(\omega),
$$

where λ is a complex number. If $\mathcal{K} = \mathbb{R}^n\backslash\{0\}$, i.e., $\Omega = S^{n-1}$, then (2.2.1) is considered without the Dirichlet boundary condition.

It is useful to write the Laplace operator in spherical coordinates $r, \theta_1, \dots, \theta_{n-1}$ which are connected with the Cartesian coordinates by the following formulas:

$$
\begin{array}{rcl}
x_1 & = & r\cos\theta_1 \\
x_2 & = & r\sin\theta_1\cos\theta_2\,, \\
& \vdots & \\
x_{n-1} & = & r\sin\theta_1\sin\theta_2\cdots\sin\theta_{n-2}\cos\theta_{n-1}\,, \\
x_n & = & r\sin\theta_1\sin\theta_2\cdots\sin\theta_{n-2}\sin\theta_{n-1}\,,
\end{array}
$$

$\theta_1, \dots, \theta_{n-2} \in [0, \pi]$, $\theta_{n-1} \in [0, 2\pi]$. Then we have

$$
(2.2.3) \qquad \Delta = \partial^2/\partial r^2 + (n-1)\, r^{-1}\, \partial/\partial r + r^{-2}\delta,
$$

where δ is the so-called *Beltrami operator*:

$$
(2.2.4) \qquad \delta = \sum_{j=1}^{n-1} \frac{1}{q_j(\sin\theta_j)^{n-j-1}} \frac{\partial}{\partial\theta_j} \left((\sin\theta_j)^{n-j-1}\frac{\partial}{\partial\theta_j}\right),
$$

$$
(2.2.5) \qquad q_1 = 1, \quad q_j = (\sin\theta_1\sin\theta_2\cdots\sin\theta_{j-1})^2 \quad \text{for } j = 2, \dots, n-1
$$

(see, e.g., Mikhlin's book [**201**, Ch.12,§2]). By means of (2.2.3), it can be easily shown that U is a solution of (2.2.1) if and only if the function $r^{2-n}\, U(x/r^2)$ is a solution of (2.2.1). The transformation $U(x) \to r^{2-n}\, U(x/r^2)$ is called *Kelvin transformation*.

Inserting (2.2.2) into (2.2.1), we get

$$
(2.2.6) \qquad -\delta u - \lambda(\lambda + n - 2)\, u = 0 \text{ in } \Omega, \quad u = 0 \text{ on } \partial\Omega.
$$

We introduce the norm

(2.2.7) $$\|u\|_{\overset{\circ}{W}_2^1(\Omega)} = \Big(\int\limits_{\Omega} \sum_{j=1}^{n-1} \frac{1}{q_j} \Big| \frac{\partial u}{\partial \theta_j} \Big|^2 d\omega \Big)^{1/2},$$

where

$$d\omega = \sin^{n-2}\theta_1 \, \sin^{n-3}\theta_2 \cdots \sin\theta_{n-2} \, d\theta_1 \cdots d\theta_{n-1}$$

denotes the measure on Ω. It can be easily seen that

$$-\int\limits_{\Omega} \delta u \cdot \overline{u} \, d\omega = \|u\|^2_{\overset{\circ}{W}_2^1(\Omega)}$$

for all $u \in C_0^\infty(\Omega)$. Let $W_2^1(\Omega)$ be the Sobolev space with the norm

$$\|u\|_{W_2^1(\Omega)} = \Big(\|u\|^2_{L_2(\Omega)} + \|u\|^2_{\overset{\circ}{W}_2^1(\Omega)} \Big)^{1/2}$$

and let $\overset{\circ}{W}_2^1(\Omega)$ be the closure of the set $C_0^\infty(\Omega)$ of all infinitely differentiable functions with support in Ω with respect to the norm (2.2.7). The space $W_2^{-1}(\Omega)$ is defined as the dual space of $\overset{\circ}{W}_2^1(\Omega)$ with respect to the inner product in $L_2(\Omega)$.

We denote the operator

$$-\delta - \lambda(\lambda + n - 2) : \overset{\circ}{W}_2^1(\Omega) \to W_2^{-1}(\Omega)$$

of problem (2.2.6) by $\mathfrak{A}(\lambda)$. Here δ is understood as Friedrichs' extension of the differential operator (2.2.4) given on $C_0^\infty(\Omega)$. The operator $-\delta$ is selfadjoint, nonnegative and has a discrete spectrum. It is evident that λ is an eigenvalue of the operator pencil $\mathfrak{A}$ if and only if $\lambda(\lambda + n - 2)$ is an eigenvalue of the operator $-\delta$.

Let $\{\Lambda_j\}_{j\geq 1}$ be the nondecreasing sequence of the eigenvalues of the operator $-\delta$ counted with their multiplicities. Furthermore, let $\{\varphi_j\}_{j\geq 1}$ be an orthonormal (in $L_2(\Omega)$) sequence of eigenfunctions corresponding to the eigenvalues Λ_j. Then

(2.2.8) $$\lambda_{\pm j} = 1 - n/2 \pm \sqrt{(1 - n/2)^2 + \Lambda_j}, \qquad j = 1, 2, \dots,$$

are the eigenvalues of the pencil $\mathfrak{A}$ and $v_{\pm j} = \varphi_j$ are the corresponding eigenfunctions. Since $\Lambda_j \geq 0$, we have $\lambda_j \geq 0$ and $\lambda_{-j} \leq 2 - n$ for $j = 1, 2, \dots$. In particular, the strip

$$|\mathrm{Re}\,\lambda - 1 + n/2| < \sqrt{(1 - n/2)^2 + \Lambda_1}$$

does not contain eigenvalues.

Equation (2.1.15) for the first generalized eigenfunction has the form

(2.2.9) $$\mathfrak{A}(\lambda_{\pm j})\,u = -(2\lambda_{\pm j} + n - 2)\,v_{\pm j}.$$

From (2.2.8) it follows that $\lambda_{\pm j} \neq 1 - n/2$. Hence the right-hand side of (2.2.9) is not orthogonal to the function $v_{\pm j}$ in $L_2(\Omega)$ and equation (2.2.9) has no solutions. Consequently, the pencil $\mathfrak{A}$ has no generalized eigenfunctions. From this it follows that there are no solutions of problem (2.2.1) which have the form

(2.2.10) $$U(x) = r^\lambda \sum_{k=0}^{s} \frac{1}{k!} (\log r)^k u_{s-k}(\omega),$$

where $u_k \in \overset{\circ}{W}_2^1(\Omega)$ for $k = 0, 1, \dots, s$, $s \geq 1$, $u_0 \neq 0$.

Using the positivity of Green's function for the Beltrami operator and the classical Jentzsch theorem on integral operators with a positive kernel (see, e.g., the books of Krasnosel'skiĭ [**146**, Ch.2] and Zeidler [**274**, §7.8]), we obtain the following result.

THEOREM 2.2.1. *The eigenvalue* Λ_1 *is simple and the corresponding eigenfunction is positive in* Ω.

2.2.2. The cases of the space and the half-space. The following theorem allows one to obtain a complete description of the spectrum of the pencil $\mathfrak{A}$ in the cases $\mathcal{K} = \mathbb{R}^n \backslash \{0\}$ and $\mathcal{K} = \mathbb{R}^n_+ = \{(x_1, \dots, x_n) \in \mathbb{R}^n : x_n > 0\}$.

THEOREM 2.2.2. 1) *Let* $\mathcal{K} = \mathbb{R}^n \backslash \{0\}$. *Then all solution to problem* (2.2.1) *of the form* (2.2.2), *where* $\lambda \geq 0$, *are homogeneous polynomials.*

2) *If* $\mathcal{K} = \mathbb{R}^n_+$, *then all solutions to the Dirichlet problem* (2.2.1) *of the form* (2.2.2) *with* $\lambda \geq 0$ *have the representation* $U(x) = x_n p(x)$, *where* p *is a homogeneous polynomial.*

Proof: 1) Since U is a harmonic function in $\mathbb{R}^n \backslash \{0\}$, it is smooth outside the origin. Using the mean value theorem, we conclude from the local boundedness of the solution that U is smooth everywhere. From the positive homogeneity of U it follows that U is a harmonic homogeneous polynomial.

2) Using classical regularity assertions for solutions of elliptic boundary value problems, we obtain $U \in C^\infty(\overline{\mathbb{R}^n_+})$. From this it follows again that U is a harmonic homogeneous polynomial. By the Dirichlet boundary condition, this polynomial has the form $x_n p(x)$. The theorem is proved. ∎

COROLLARY 2.2.1. 1) *Let* $\mathcal{K} = \mathbb{R}^n \backslash \{0\}$. *Then the spectrum of the pencil* $\mathfrak{A}$ *consists of the numbers* $\lambda_j = j - 1$ *and* $\lambda_{-j} = 3 - n - j$, $j = 1, 2, \dots$. *The eigenfunctions corresponding to the eigenvalues* $\lambda_{\pm j}$ *are the restrictions of harmonic homogeneous polynomials of degree* $j - 1$ *to the sphere* S^{n-1}.

2) *In the case* $\mathcal{K} = \mathbb{R}^n_+$ *the spectrum of the pencil* $\mathfrak{A}$ *consists of the numbers* $\lambda_j = j$ *and* $\lambda_{-j} = 2 - n - j$, $j = 1, 2, \dots$. *The eigenfunctions corresponding to* $\lambda_{\pm j}$ *are the restrictions of harmonic functions of the form* $x_n p(x)$, *where* p *is a homogeneous polynomial of degree* $j - 1$.

As a consequence of the last result, we obtain that the eigenvalues of the operator $-\delta$ are

$$\Lambda_j = (j-1)(j+n-3), \qquad j = 1, 2, \dots,$$

if Ω coincides with the sphere S^{n-1}. If Ω is the half-sphere $S^{n-1}_+ = S^{n-1} \cap \mathbb{R}^n_+$, then the spectrum of $-\delta$ consists of the eigenvalues

$$\Lambda_j = j(j+n-2), \qquad j = 1, 2, \dots.$$

2.2.3. Monotonicity of the eigenvalues. Now we study the dependence of the eigenvalues Λ_j on the domain Ω. The proofs of the following two theorems are based on a variational principle for the eigenvalues $\Lambda_j(\Omega)$ (see, e.g., Courant and Hilbert's book [**32**]):

$$(2.2.11) \qquad \Lambda_j(\Omega) = \max_{\{L\}} \min_{u \in L \backslash \{0\}} \frac{-\int_\Omega \delta u \cdot \overline{u} \, d\omega}{\int_\Omega |u|^2 d\omega},$$

where the maximum is taken over all subspaces $L \subset \overset{\circ}{W}{}_2^1(\Omega)$ with codimension $\geq j - 1$.

For the formulation and the proof of the following theorem we need the notion of the *harmonic capacity* and some elementary results connected with this notion which can be found in the book of Landkof [**149**]. Here the capacity of a closed set $K \subset S^2$ is defined as

$$(2.2.12) \qquad \mathrm{cap}\,(K) = \inf \left\{ \|u\|_{W_2^1(S^2)}^2 \, : \, u \in C^\infty(S^2), u \geq 1 \text{ on } K \right\}.$$

The reader who is not familiar with the theory of capacity can skip the following theorem without loss of understanding for the further results.

THEOREM 2.2.3. *Let K be contained in the open connected set Ω of the sphere S^{n-1}. Then $\Lambda_1(\Omega \backslash K) > \Lambda_1(\Omega)$ if and only if K has positive capacity.*

Proof: If the capacity of K is equal to zero, then $\overset{\circ}{W}{}_2^1(\Omega \backslash K) = \overset{\circ}{W}{}_2^1(\Omega)$ (see Maz'ya [**164**, Ch.9]). Hence the equality $\Lambda_1(\Omega \backslash K) = \Lambda_1(\Omega)$ follows from the variational principle (2.2.11).

We assume now that K has positive capacity. It suffices to show that the eigenfunction of $-\delta$ corresponding to the eigenvalue $\Lambda_1(\Omega)$ does not belong to the space $\overset{\circ}{W}{}_2^1(\Omega \backslash K)$. Indeed, in the opposite case this function vanishes on the set K except for a subset of zero capacity. This contradicts Theorem 2.2.1. ∎

Since the eigenfunctions φ_j, $j > 1$, have zeros in the interior of Ω, the assertion of Theorem 2.2.3 is not true, in general, for the eigenvalues $\Lambda_2, \Lambda_3, \dots$. However, the following statement holds for all eigenvalues Λ_j.

THEOREM 2.2.4. *If $\Omega_1 \subset \Omega_2$, then $\Lambda_j(\Omega_1) \geq \Lambda_j(\Omega_2)$. Moreover, $\Lambda_j(\Omega_1) > \Lambda_j(\Omega_2)$ if Ω_1 is a subset of the open connected set Ω_2 and $\Omega_2 \backslash \overline{\Omega}_1 \neq \emptyset$.*

Proof: If $\Omega_1 \subset \Omega_2$, then $\overset{\circ}{W}{}_2^1(\Omega_1)$ can be considered as a subspace of $\overset{\circ}{W}{}_2^1(\Omega_2)$. Using the variational principle (2.2.11), we conclude that $\Lambda_j(\Omega_1) \geq \Lambda_j(\Omega_2)$.

Now let Ω_2 be a connected open set and $\Omega_2 \backslash \overline{\Omega}_1 \neq \emptyset$. Suppose that for a certain index j the eigenvalue $\Lambda_j(\Omega_1)$ is equal to $\Lambda_j(\Omega_2)$. Furthermore, let $\varphi_1, \dots, \varphi_j$ be the eigenfunctions of the Dirichlet problem for the operator $-\delta$ in the domain Ω_1 corresponding to the eigenvalues $\Lambda_1(\Omega_1), \dots, \Lambda_j(\Omega_1)$. We introduce the subspaces

$$L_j^{(k)} \overset{def}{=} \left\{ u \in \overset{\circ}{W}{}_2^1(\Omega_k) : \int_{\Omega_k} u \cdot \overline{\varphi}_s \, d\omega = 0 \text{ for } s = 1, \dots, j-1 \right\}, \quad k = 1, 2.$$

Then

$$\Lambda_j(\Omega_2) = \Lambda_j(\Omega_1) = \min_{L_j^{(1)} \backslash \{0\}} \frac{-\int_{\Omega_1} \delta u \cdot \overline{u} \, d\omega}{\int_{\Omega_1} |u|^2 \, d\omega} = \min_{L_j^{(2)} \backslash \{0\}} \frac{-\int_{\Omega_2} \delta u \cdot \overline{u} \, d\omega}{\int_{\Omega_2} |u|^2 \, d\omega},$$

where both minima are realized for the function $u = \varphi_j$. From this it follows that

$$-\int_{\Omega_2} \delta \varphi_j \cdot \overline{u} \, d\omega - \Lambda_j(\Omega_2) \int_{\Omega_2} \varphi_j \overline{u} \, d\omega = 0$$

for all $u \in L_j^{(2)}$. Hence

$$-\delta\varphi_j - \Lambda_j(\Omega_2)\varphi_j = \sum_{k=1}^{j-1} c_k\,\varphi_k \quad \text{in } \Omega_2$$

with certain complex numbers $c_1,\ldots,c_{j-1}$. Since $-\delta\varphi_j = \Lambda_j(\Omega_1)\varphi_j$ in Ω_2 and the functions $\varphi_1,\ldots,\varphi_{j-1}$ are linearly independent, we get $c_k = 0$ for $k = 1,\ldots,j-1$. Consequently, $-\delta\,\varphi_j = \Lambda_j(\Omega_2)\,\varphi_j$ on Ω_2. From this we conclude that φ_j is real-analytic in Ω_2. On the other hand, the function φ_j vanishes on the open set $\Omega_2 \backslash \overline{\Omega}_1$. Therefore, $\varphi_j = 0$ in Ω_2 which contradicts our assumption on φ_j. The theorem is proved. ∎

Without proof we give a known *isoperimetric property* of the eigenvalue $\Lambda_1(\Omega)$ which was proved by Sperner [**248**], Friedland and Hayman [**63**].

THEOREM 2.2.5. *Among all sets Ω with fixed $(n-1)$-dimensional measure, the geodesic ball on S^{n-1} has the smallest eigenvalue $\Lambda_1(\Omega)$ of the operator $-\delta$. This ball is the only extremal domain.*

One can easily prove the following assertion.

LEMMA 2.2.1. *Let B_α be the geodesic ball $\{\omega \in S^{n-1} : 0 \leq \theta_1 < \alpha\}$ and let $\lambda_1 = \lambda_1(\alpha)$ be the smallest positive solution of the equation $C_\lambda^{(n-2)/2}(\cos\alpha) = 0$, where $C_\lambda^{(n-2)/2}(\cdot)$ denotes the Gegenbauer function (see, e.g., Magnus, Oberhettinger and Soni [**158**, §5.3]). Then*

$$\Lambda_1(B_\alpha) = \lambda_1(\alpha)\,(\lambda_1(\alpha) + n - 2)$$

is the first eigenvalue of the operator $-\delta$ (with the Dirichlet boundary conditions) in B_α.

Proof: It suffices to show that there exists a positive solution $u = u(\theta_1)$ of the Dirichlet problem for the equation $-\delta u = \Lambda_1(B_\alpha)u$ in B_α or, what is the same, of the problem

$$u''(\theta_1) + \frac{(n-2)\cos\theta_1}{\sin\theta_1}\,u'(\theta_1) + \lambda_1(\lambda_1 + n - 2)\,u(\theta_1) = 0 \quad \text{for } 0 \leq \theta_1 < \alpha,$$

$$u(\alpha) = 0.$$

The substitution $\cos\theta_1 = t$ leads to the problem

$$v''(t) + \frac{(n-1)t}{t^2 - 1}\,v'(t) - \frac{\lambda_1(\lambda_1 + n - 2)}{t^2 - 1}\,v(t) = 0, \qquad v(\cos\alpha) = 0$$

for the function $v(t) = u(\arccos t)$. This problem has the positive solution $v(t) = C_{\lambda_1}^{(n-2)/2}(t)$. The lemma is proved. ∎

Theorem 2.2.5 and Lemma 2.2.1 yield the following sharp lower estimate for the eigenvalue $\lambda_1 = \lambda_1(\Omega)$ of the pencil $\mathfrak{A}$ in terms of the measure of Ω. Let α be given by the equality

$$\text{mes}_{n-1}\Omega = \text{mes}_{n-2}S^{n-2} \int_0^\alpha (\sin\theta)^{n-2}\,d\theta$$

and let $\lambda_1 = \lambda_1(\alpha)$ be the smallest positive root of the equation $C_\lambda^{(n-2)/2}(\cos\alpha) = 0$. Then $\lambda_1(\Omega) \geq \lambda_1(B_\alpha) = \lambda_1(\alpha)$.

2.2.4. On the multiplicity of eigenvalues. We close this section with an upper estimate for the multiplicity of the eigenvalue Λ_j. For this we need the following lemma which was proved by Nadirashvili [**202**].

LEMMA 2.2.2. *Let G be a bounded domain in $\mathbb{R}^2$ with smooth boundary and let*

$$L = \sum_{i,j=1}^{2} \frac{\partial}{\partial x_i}\left(a_{i,j}(x)\frac{\partial}{\partial x_j}\right) + a(x) + \lambda\, b(x)$$

be a uniformly elliptic operator with coefficients $a, b, a_{i,j} \in C^\infty(\overline{G})$, $b > 0$, such that the matrix $(a_{i,j})_{1 \leq i,j \leq 2}$ is symmetric and positive definite in $\overline{G}$. We consider the spectral problem

$$Lu = 0 \quad in\ G, \quad Bu = 0 \quad on\ \partial G,$$

where B is the operator of the Dirichlet or Neumann boundary condition. By $\{\mu_j\}_{j \geq 1}$ we denote the sequence of the eigenvalues to this problem. Then the multiplicity of μ_j does not exceed $2j - 1$.

THEOREM 2.2.6. *If Ω is a domain on the sphere S^2 with smooth boundary, then the multiplicity of the eigenvalue $\Lambda_j(\Omega)$ does not exceed $2j - 1$.*

Proof: If $\Omega = S^2$, then, by Corollary 2.2.1, we have $\Lambda_j = j(j-1)$. The eigenfunctions corresponding to these eigenvalues are restrictions of harmonic homogeneous polynomials of degree $j - 1$ to the sphere S^2. Since there are exactly $2j - 1$ linearly independent polynomials of that kind, the theorem is true in the case $\Omega = S^2$.

Now let $\Omega \neq S^2$. Then, without loss of generality, we may assume that the south pole P of the sphere is a point of the set $S^2 \backslash \overline{\Omega}$. We introduce the coordinates $y = (y_1, y_2)$ on $S^2 \backslash P$, where

(2.2.13) $y_1 = \tan(\theta_1/2)\cos\theta_2\,, \quad y_2 = \tan(\theta_1/2)\sin\theta_2\,.$

The coordinate change $\omega \to y$ maps the domain Ω onto a bounded subdomain of the plane with smooth boundary. The spectral problem associated with (2.2.1) in the coordinates y has the form

$$\Delta_y v(y) + 4\lambda(1 + |y|^2)^{-2}\, v(y) = 0 \quad in\ G, \quad v(y) = 0 \quad on\ \partial G.$$

It remains to apply Lemma 2.2.2. ∎

2.3. The Neumann problem for the Laplace operator in a cone

2.3.1. The operator pencil generated by the Neumann problem. As in the preceding section, let $\mathcal{K}$ be the cone $\{x \in \mathbb{R}^n : x/|x| \in \Omega\}$, where Ω is an open and connected set on the unit sphere S^{n-1}, $n \geq 3$. We suppose in this section that the operator of the imbedding $W_2^1(\Omega) \subset L_2(\Omega)$ is compact. Necessary and sufficient and also more easy verifiable sufficient conditions for the compactness of this imbedding are given in the book of Maz'ya [**164**]. For example, the condition $\Omega \in C$ (i.e., the boundary of the domain can be locally given in a system of Cartesian coordinates by means of continuous functions) is sufficient for the compactness of the imbedding $W_2^1(\Omega) \subset L_2(\Omega)$ (see Courant and Hilbert's book [**32**] or [**164**]).

Let ν denote the exterior normal to $\partial\mathcal{K}\backslash\{0\}$. We consider the Neumann problem

$$(2.3.1) \qquad -\Delta U = 0 \ \text{in} \ \mathcal{K}, \quad \frac{\partial U}{\partial \nu} = 0 \ \text{on} \ \partial\mathcal{K}\backslash\{0\}$$

which can be understood in the sense of the integral identity

$$(2.3.2) \qquad \int_{\mathcal{K}} \nabla U \cdot \nabla \overline{V} \, dx = 0,$$

where $\eta U \in W_2^1(\mathcal{K})$ for all $\eta \in C_0^\infty(\overline{\mathcal{K}}\backslash\{0\})$, and V is an arbitrary function in $W_2^1(\mathcal{K})$ with compact support equal to zero in a neighborhood of the origin. Here ∇U denotes the gradient of U. As before, we seek solutions which are positively homogeneous of degree λ, i.e., which have the form $U(x) = r^\lambda u(\omega)$. If we insert $V(r,\omega) = \zeta(r)\,v(\omega)$ into (2.3.2), where $\zeta \in C_0^\infty((0,\infty))$, $v \in W_2^1(\Omega)$, we obtain the integral identity

$$(2.3.3) \qquad \int_\Omega \nabla_\omega u \cdot \nabla_\omega \overline{v} \, d\omega - \lambda(\lambda + n - 2) \int_\Omega u \cdot \overline{v} \, d\omega = 0.$$

Here

$$\nabla_\omega u \cdot \nabla_\omega \overline{v} = \sum_{j=1}^{n-1} \frac{1}{q_j} \partial_{\theta_j} u \cdot \partial_{\theta_j} \overline{v}$$

and the quantities q_j are given by (2.2.5). The sesquilinear form

$$W_2^1(\Omega) \times W_2^1(\Omega) \ni (u,v) \to \int_\Omega \nabla_\omega u \cdot \nabla_\omega \overline{v} \, d\omega$$

generates a selfadjoint operator $A : W_2^1(\Omega) \to W_2^1(\Omega)^*$. Obviously, the solution u of (2.3.3) satisfies the equation $\mathfrak{A}(\lambda)u = 0$, where $\mathfrak{A}(\lambda) = A - \lambda(\lambda + n - 2)$. By the compactness of the imbedding $W_2^1(\Omega) \subset L_2(\Omega)$, the spectrum of A is discrete. We denote by $\{N_j\}_{j\geq 1}$ the nondecreasing sequence of eigenvalues (which are counted with their multiplicities) of A. Let $\{\psi_j\}_{j\geq 1}$ be an orthonormal in $L_2(\Omega)$ sequence of eigenfunctions to N_j. Then

$$(2.3.4) \qquad \lambda_{\pm j} = 1 - n/2 \pm \sqrt{(1 - n/2)^2 + N_j}\,, \quad j = \pm 1, \pm 2, \ldots,$$

are the eigenvalues of the operator pencil $\mathfrak{A}$ with the corresponding eigenfunctions $v_{\pm j} = \psi_j$.

It can be easily seen that the operator $\mathfrak{A}(\lambda) = A - \lambda(\lambda + n - 2)$ is positive for λ on the line $\operatorname{Re}\lambda = 1 - n/2$. The widest strip containing this line which is free of eigenvalues of the pencil $\mathfrak{A}$ is the strip $2 - n < \operatorname{Re}\lambda < 0$. The boundary of this strip contains the simple eigenvalues $\lambda_1 = 0$ and $\lambda_{-1} = 2 - n$ with the corresponding eigenfunctions $v_{\pm 1} = const$.

Note that the equation (2.2.9) with $\mathfrak{A}(\lambda) = A - \lambda(\lambda + n - 2)$ has no solutions. Hence the eigenfunctions $v_{\pm j}$ do not have generalized eigenfunctions. From this it follows that there are no solutions of the Neumann problem (2.3.1) which have the form (2.2.10) with $u_k \in W_2^1(\Omega)$, $s \geq 1$, $u_0 \neq 0$.

Repeating the proof of Theorem 2.2.6, we arrive at the following result.

THEOREM 2.3.1. *If Ω is a domain on the sphere S^2 with smooth boundary, then the multiplicity of the eigenvalue $N_j(\Omega)$ does not exceed $2j - 1$.*

2.3.2. On the monotonicity of the eigenvalues. In general, the eigenvalues λ_j are not monotone functions of the domain Ω. To see this it suffices to observe the dependence of the eigenvalue λ_2 on the angle $\alpha \in [\pi/2, \pi]$ for the circle $\Omega = \{\omega \in S^2, 0 \le \theta_1 < \alpha, 0 \le \theta_2 < 2\pi\}$. In this case the eigenvalues $\lambda_2 = \lambda_3$ are the smallest positive solutions of the equation

$$\partial_\theta P_\lambda^1(\cos\theta)\Big|_{\theta=\alpha} = 0.$$

The corresponding eigenfunctions are $P_\lambda^1(\cos\theta_1)\, e^{\pm i\theta_2}$, where $\lambda = \lambda_2 = \lambda_3$ and P_λ^m denotes the associated Legendre function of first kind.

The graph of the function $\lambda = \lambda(\alpha)$ is represented in the figure below.

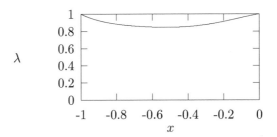

FIGURE 2. Dependence of the eigenvalue $\lambda = \lambda_2 = \lambda_3$ on $x = \cos\alpha$, $\alpha \in [\pi/2, \pi]$

However, in a particular but important case (a domain with a crack) the monotonicity of the functions $\lambda_j = \lambda_j(\Omega)$ holds.

THEOREM 2.3.2. *Let $\Omega_1 \subset \Omega_2$ and $\mathrm{mes}_{n-1}(\Omega_2 \backslash \Omega_1) = 0$. Then $\lambda_j(\Omega_1) \le \lambda_j(\Omega_2)$ for $j = 1, 2, \dots$.*

Proof: It suffices to show the inequality $N_j(\Omega_1) \le N_j(\Omega_2)$, $j = 1, 2, \dots$. Let $\psi_j \in W_2^1(\Omega_1)$ be an eigenfunction corresponding to the eigenvalue λ_j for the domain Ω_1. We introduce the spaces

$$H_j(\Omega_i) \stackrel{def}{=} \left\{ h \in W_2^1(\Omega_i) : \int_{\Omega_i} h\,\overline{\psi_k}\, d\omega = 0 \ \text{ for } k = 1, \dots, j-1 \right\}, \quad i = 1, 2.$$

Then, by the variational principle, we have

$$(2.3.5) \qquad N_j(\Omega_1) = \inf_{h \in H_j(\Omega_1)} \int_{\Omega_1} |\nabla_\omega h|^2\, d\omega \cdot \left(\int_{\Omega_1} |h|^2\, d\omega \right)^{-1}.$$

Clearly, for $h \in H_j(\Omega_2)$ we have $h|_{\Omega_1} \in H_j(\Omega_1)$,

$$\int_{\Omega_2} |\nabla_\omega h|^2\, d\omega = \int_{\Omega_1} \left| \nabla_\omega h|_{\Omega_1} \right|^2 d\omega \quad \text{and} \quad \int_{\Omega_2} |h|^2\, d\omega = \int_{\Omega_1} \left| h|_{\Omega_1} \right|^2 d\omega.$$

Hence the right-hand side of (2.3.5) does not exceed the value

$$\inf_{h \in H_j(\Omega_2)} \int_{\Omega_2} |\nabla_\omega h|^2\, d\omega \cdot \left(\int_{\Omega_2} |h|^2\, d\omega \right)^{-1}$$

which is not greater than

$$\max_{\{L\}} \inf_{h\in L} \int_{\Omega_2} |\nabla_\omega h|^2 \, d\omega \cdot \left(\int_{\Omega_2} |h|^2 \, d\omega \right)^{-1} = N_j(\Omega_2),$$

where the maximum is taken over all subspace $L \subset W_2^1(\Omega_2)$ of codimension $j-1$. This proves the theorem. ∎

It is worth mentioning that, under the conditions of Theorem 2.3.2, the eigenvalues of the operator pencil corresponding to the Neumann problem do not decrease with growing Ω, whereas in the case of the Dirichlet problem they do not increase. Furthermore, there is the following difference between the properties of the eigenvalues of the operator pencils corresponding to the Dirichlet and Neumann problems. If $\{\Omega_k\}_{k\geq 1}$ is an increasing sequence of subdomains $\Omega_k \subset \Omega$ exhausting the domain Ω, then $\Lambda_j(\Omega_k) \downarrow \Lambda_j(\Omega)$. The following example shows that all eigenvalues $N_j(\Omega_k)$ may converge to zero for $k \to \infty$, whereas $N_2(\Omega) > 0$.

Example. Let G be the square $(0,1) \times (0,1)$ and let R_k be the rectangle $(0,1) \times (2^{-k}, 1)$. For every integer $k \geq 1$ we consider the rectangles

$$P_{k,l} = \big((l-1/2)/k \,, l/k \big) \times \big(3 \cdot 2^{-k-2}, 7 \cdot 2^{-k-3} \big), \quad l = 1, \dots, k,$$

and the squares

$$Q_{k,l} = \big(l/k - 2^{-k-3}, l/k \big) \times [7 \cdot 2^{-k-3}, 2^{-k}], \quad l = 1, \dots, k.$$

The union of the rectangles R_k, $P_{k,l}$ and the squares $Q_{k,l}$, $l = 1, \dots, k$, is denoted by G_k (see Figure 3). Obviously, $\{G_k\}_{k\geq 1}$ is an increasing and exhausting sequence for the domain G.

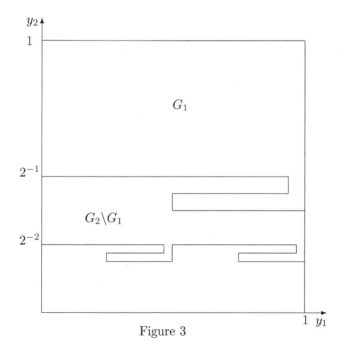

Figure 3

We introduce the functions

$$u_{k,l} = \begin{cases} 1 & \text{in } P_{k,l}, \\ 2^{k+3}(2^{-k} - y_2) & \text{in } Q_{k,l}, \\ 0 & \text{elsewhere.} \end{cases}$$

These functions belong to the Sobolev space $W_2^1(G_k)$ and the supports of the functions $u_{k,1}, \ldots, u_{k,k}$ do not intersect. Moreover,

$$\int_{G_k} |\nabla u_{k,l}|^2 \, dy \cdot \left(\int_{G_k} |u_{k,l}|^2 \, dy \right)^{-1} < \frac{\text{mes}_2 Q_{k,l}}{\text{mes}_2 P_{k,l}} = k \cdot 2^{-k-2}.$$

Applying Theorem 1.3.3 with

$$a(u, u; \lambda) = \int_{G_k} |\nabla u|^2 \, dx - \lambda \int_{G_k} |u|^2 \, dx,$$

$\alpha = -\varepsilon$, where ε is an arbitrarily small positive number, $\beta = k \, 2^{-k-2}$, $\mathcal{H} = L_2(G_k)$, $\mathcal{H}_+ = W_2^1(G_k)$, we obtain that the Laplace operator in the domain G_k with the Neumann boundary condition has at least k eigenvalues in the interval $[0, k \cdot 2^{-k-2})$. From this it follows that all eigenvalues of the Neumann problem for the Laplace operator converge to zero. Obviously, this is also true for the eigenvalues N_j to the Neumann problem for the equation

$$\Delta_y v(y) + 4 N_j (1 + |y|^2)^{-2} v(y) = 0 \quad \text{in } G_k.$$

Using the stereographic projection (i.e., the mapping $\omega \to y$ defined in the proof of Theorem 2.2.6), we obtain a sequence $\{\Omega_k\}_{k \geq 1}$ such that $N_j(\Omega_k) \to 0$ as $k \to \infty$ and $N_2(\Omega) > 0$.

Finally, we formulate an isoperimetric property of the first nonzero eigenvalue N_2 which was proved by Ashbaugh and Benguria [**8**].

THEOREM 2.3.3. *If Ω is contained in a hemisphere, then*

$$N_2(\Omega) \leq N_2(B),$$

where B is a geodesic ball in S^{n-1} of the same measure as Ω.

Note that for the first eigenvalue of the Beltrami operator with Dirichlet boundary condition the opposite inequality $\Lambda_1(\Omega) \geq \Lambda_1(B)$ is valid (see Theorem 2.2.5).

2.4. The problem with oblique derivative

Now let Ω be a subdomain of the sphere S^{n-1} with smooth boundary and let $\mathcal{K}$ be the cone $\{x \in \mathbb{R}^n : 0 < r < \infty, \omega \in \Omega\}$. We are interested in solutions

$$U(x) = r^\lambda u(\omega), \quad u \in W_2^2(\Omega),$$

of the boundary value problem

(2.4.1) $-\Delta U = 0 \quad \text{in } \mathcal{K}, \qquad BU = 0 \quad \text{on } \partial\mathcal{K} \backslash \{0\},$

where

$$BU = \partial_\nu U + a \, \partial_r U + \frac{1}{r} Q \, U.$$

Here ν denotes the exterior normal to $\partial\mathcal{K} \backslash \{0\}$, a is a smooth real-valued function on $\partial\Omega$, and Q is a differential operator of first order on $\partial\Omega$ with smooth real-valued coefficients such that $Q\,1 = 0$. Using the representation (2.2.3) for the Laplace operator, we get the problem $\mathfrak{A}(\lambda)u = 0$, where

$$\mathfrak{A}(\lambda)\,u = \big(-\delta u - \lambda(\lambda + n - 2)u\,,\,(\partial_\nu u + Qu + \lambda a u)|_{\partial\Omega}\big).$$

Here the operator $\mathfrak{A}(\lambda)$ is considered as a mapping from $W_2^2(\Omega)$ into $L_2(\Omega) \times W_2^{1/2}(\partial\Omega)$. From the ellipticity of problem (2.4.1) it follows, by a result of Agranovich and Vishik [**4**] (see also [**136**, Th.3.6.1,Le.6.1.5]) that the operator $\mathfrak{A}(\lambda)$ is Fredholm for all $\lambda \in \mathbb{C}$ and invertible for large λ situated near the imaginary axis. Consequently, the spectrum of the operator pencil $\mathfrak{A}$ consists of isolated eigenvalues with finite algebraic multiplicities (see Theorem 1.1.1). These eigenvalues, with the possible exception of finitely many, lie outside a double sector which contains the imaginary axis.

THEOREM 2.4.1. *The line* $\operatorname{Re}\lambda = 0$ *contains only the eigenvalue* $\lambda = 0$ *with the unique eigenfunction* $u = const.$

Proof: Let $\lambda = i\beta$ $(\beta \in \mathbb{R})$ be an eigenvalue of $\mathfrak{A}$ and let $u = \varphi + i\psi$, where φ, ψ are real-valued functions, be an eigenfunction corresponding to this eigenvalue. Then the function

$$U(x) = \operatorname{Re} r^{i\beta} u(\omega) = \cos(\beta\,\log r)\,\varphi(\omega) - \sin(\beta\,\log r)\,\psi(\omega)$$

is a solution of problem (2.4.1). If $\beta \neq 0$, then the coordinate change $r = e^t$ transforms U into the $2\pi/\beta$-periodic (with respect to the variable t) function

$$\cos(\beta t)\,\varphi(\omega) - \sin(\beta t)\,\psi(\omega).$$

Hence, both in the cases $\beta = 0$ and $\beta \neq 0$, the function U attains its maximum at a certain point $x \in \overline{\mathcal{K}}\backslash\{0\}$. By the maximum principle and the Giraud theorem on the sign of nontangential derivatives of harmonic functions (see, e.g., Protter and Weinberger [**230**, Ch.2,Th.7]), we get $U = const.$ Consequently, $\beta = 0$ and $u = \varphi = const.$ The theorem is proved. ∎

In order to find a necessary and sufficient condition for the simplicity of the eigenvalue $\lambda = 0$, we need the operator

$$\mathfrak{A}^+(0) = (-\delta\,,\,\partial_\nu + Q^+),$$

where Q^+ is the formally adjoint operator to Q on $\partial\Omega$. By Theorem 2.4.1, the kernel of $\mathfrak{A}^+(0)$ is one-dimensional. It is determined by a real-valued function v.

THEOREM 2.4.2. *The eigenvalue* $\lambda = 0$ *is simple if and only if*

$$(2.4.2) \qquad\qquad (n-2)\int\limits_\Omega v\,d\omega \neq \int\limits_{\partial\Omega} a\,v\,d\sigma$$

for $v \in \ker\mathfrak{A}^+(0).$

Proof: By Theorem 2.4.1, the function $u_0 = 1$ is the eigenfunction corresponding to the eigenvalue $\lambda = 0$. The boundary value problem for the generalized eigenfunction u_1 has the form

$$-\delta u_1 = n - 2 \quad \text{in } \Omega, \qquad \partial_\nu u_1 + Q\,u_1 = -a \quad \text{on } \partial\Omega.$$

Therefore,

$$(n-2)\int\limits_\Omega v\,d\omega = -\int\limits_\Omega v \cdot \delta u_1\,d\omega = \int\limits_{\partial\Omega}(\partial_\nu v \cdot u_1 - v \cdot \partial_\nu u_1)\,d\omega.$$

By the boundary conditions on the functions v and u_1, the last integral is equal to

$$\int_{\partial\Omega} \left(v(Qu_1 + a) - Q^+ v \cdot u_1 \right) d\sigma = \int_{\partial\Omega} av \, d\sigma.$$

Hence the condition (2.4.2) is equivalent to the nonexistence of a generalized eigenfunction. ∎

We close this section with the following property of a nontrivial function $v \in \ker \mathfrak{A}^+(0)$.

THEOREM 2.4.3. *The function v does not vanish in Ω.*

Proof: We assume that the function v takes values with different signs and introduce the sets

$$\Omega_\varepsilon = \{\omega \in \Omega : \ v(\omega) > \varepsilon\}, \qquad (\partial\Omega)_\varepsilon = \{\omega \in \partial\Omega : \ v(\omega) > \varepsilon\},$$

and

$$\Gamma_\varepsilon = \{\omega \in \Omega : \ v(\omega) = \varepsilon\},$$

where $\varepsilon > 0$. Furthermore, for an arbitrary function u in Ω, we set $u_+(x) = u(x)$ if $u(x) > 0$, $u_+(x) = 0$ if $u(x) < 0$.

Since v is a smooth function, the manifold Γ_ε is smooth for almost all ε. We denote the exterior (with respect to Ω_ε) normal to the manifold Γ_ε by ν_ε. Obviously, $\partial v/\partial\nu_\varepsilon \leq 0$ on Γ_ε and

$$0 = \int_{\Omega_\varepsilon} \delta v \, d\omega = \int_{\Gamma_\varepsilon} \partial_{\nu_\varepsilon} v \, d\sigma + \int_{(\partial\Omega)_\varepsilon} \partial_\nu v \, d\sigma = \int_{\Gamma_\varepsilon} \partial_{\nu_\varepsilon} v \, d\sigma - \int_{(\partial\Omega)_\varepsilon} Q^+ v \, d\sigma$$

for almost all ε. Here the last integral is equal to

$$\int_{(\partial\Omega)_\varepsilon} Q^+ (v - \varepsilon) \, d\sigma + \varepsilon \int_{(\partial\Omega)_\varepsilon} Q^+ 1 \, d\sigma = \int_{\partial\Omega} Q^+ \big((v - \varepsilon)_+ \big) \, d\sigma + \varepsilon \int_{(\partial\Omega)_\varepsilon} Q^+ 1 \, d\sigma.$$

Since $Q1 = 0$ on $\partial\Omega$, the first integral on the right-hand side is equal to zero. Consequently,

$$\int_{\Gamma_\varepsilon} \partial_{\nu_\varepsilon} v \, d\sigma = O(\varepsilon) \qquad \text{as } \varepsilon \to 0.$$

Let φ be an arbitrary function from $C_0^\infty(\Omega)$. Since $\partial v/\partial\nu_\varepsilon \leq 0$ on Γ_ε, we get

$$\int_{\Gamma_\varepsilon} \varphi \, \partial_{\nu_\varepsilon} v \, d\sigma = O(\varepsilon) \qquad \text{as } \varepsilon \to 0.$$

This implies

$$\int_\Omega \nabla_\omega v_+ \cdot \nabla_\omega \varphi \, d\omega = \lim_{\varepsilon\to 0} \int_{\Omega_\varepsilon} \nabla_\omega v \cdot \nabla_\omega \varphi \, d\omega = \lim_{\varepsilon\to 0} \int_{\Gamma_\varepsilon} \varphi \cdot \partial_{\nu_\varepsilon} v \, d\sigma = 0.$$

Consequently, $\delta v_+ = 0$ in Ω. Since $v_+ = 0$ on the nonempty open set $\{\omega \in \Omega : v(\omega) < 0\}$, we conclude that $v \equiv 0$. This proves the theorem. ∎

2.5. Further results

Additional information about the eigenvalues $\lambda_j(\Omega)$ can be obtained by means of asymptotic methods. Here we give asymptotic formulas for the eigenvalues in the cases when Ω is the unit sphere with a small hole and when Ω is a small domain on the sphere.

2.5.1. Asymptotic formulas for the eigenvalues of the pencil generated by the Dirichlet problem. Let $\lambda_j = \lambda_j(\Omega)$ be the eigenvalues of the pencil $\mathfrak{A}$ introduced in Section 2.2. First we consider the case when Ω is the sphere with a small hole. As before, let $\omega = x/|x|$ and $r = |x|$. Moreover, we set $x' = (x_1, \ldots, x_{n-1})$ and $\omega' = x'/r$. Let G be a given bounded domain in $\mathbb{R}^{n-1}$ and let ε be a small positive parameter. We introduce the following subset of the unit sphere S^{n-1}:

$$\Omega_\varepsilon = \{\omega \in S^{n-1} : \omega'/\varepsilon \in \overline{G}\}.$$

Samarskiĭ [232], Maz'ya, Nazarov and Plamenevskiĭ [171] showed that in the case $n > 3$ the asymptotic formula

$$\lambda_1(S^{n-1} \backslash \overline{\Omega}_\varepsilon) \sim \varepsilon^{n-3} \frac{(n-3)\, \mathrm{mes}_{n-2} S^{n-2}}{(n-2)\, \mathrm{mes}_{n-1} S^{n-1}}\, \mathrm{cap}\, G$$

is valid, where $\mathrm{cap}\, G$ is the classical Wiener capacity in $\mathbb{R}^{n-1}$. Furthermore, by [171], for $n = 3$ we have

$$(2.5.1) \qquad \lambda_1(S^2 \backslash \overline{\Omega}_\varepsilon) = h(|\log \varepsilon|^{-1}) + O(\varepsilon^{1-\sigma}),$$

where σ is an arbitrary positive number and h is an analytic function in a neighborhood of zero satisfying the condition $h(0) = 0$, $h'(0) = 1/2$.

For the Dirichlet problem in the small domain Ω_ε the main term in the asymptotics of $\Lambda_j(\Omega_\varepsilon)$ can be easily found by means of the perturbation theory. We have

$$\Lambda_j(\Omega_\varepsilon) = \frac{\mu_j(G)}{\varepsilon^2} + O(1),$$

where $\mu_j(G)$ is the j-th eigenvalue of the Dirichlet problem for the Laplace operator in the domain G. Consequently, the asymptotic formula

$$\lambda_j(\Omega_\varepsilon) = \frac{\sqrt{\mu_j(G)}}{\varepsilon} + O(1)$$

for the eigenvalues of the pencil $\mathfrak{A}$ holds.

2.5.2. Asymptotic formulas for the eigenvalues of the pencil generated by the Neumann problem. Now let $\lambda_{\pm j} = \lambda_{\pm j}(\Omega)$ be the numbers (2.3.4), where N_j are the eigenvalues of the operator $-\delta$ on Ω with the Neumann boundary conditions.

1. We start with the case when Ω is the sphere with a small hole which was considered by Maz'ya and Nazarov in [170]. Let G be a domain in the plane $\mathbb{R}^2$ with compact closure and smooth boundary. We introduce the domain

$$G_\varepsilon = \{y \in \mathbb{R}^2 : \varepsilon^{-1} y \in G\}$$

depending on a small positive parameter ε. Here $y = (y_1, y_2)$ are the coordinates given by (2.2.13). Then the domain

$$\Omega_\varepsilon = \{\omega \in S^2 : \tan(\theta_1/2)(\cos\theta_2, \sin\theta_2) \in G_\varepsilon\}$$

plays the role of a small hole on the sphere S^2. In [170] explicit formulas for $\lambda_j(S^2\backslash\overline{\Omega}_\varepsilon)$, $j = 2, 3, 4$, as $\varepsilon \to 0$ were found. These eigenvalues exhaust the spectrum in the interval $(0, 1)$ and in a neighborhood of the point $\lambda = 1$. The interval $(0, 1)$ contains exactly two numbers λ_2 and λ_3, where

$$\lambda_j(S^2\backslash\overline{\Omega}_\varepsilon) = 1 - \pi^{-1}\mu_j\,\varepsilon^2 + O(\varepsilon^3\,|\log\varepsilon|)\,, \quad j = 2,\,3,$$

and μ_2, μ_3 are the eigenvalues of the positive definite matrix $M = (m_{k,l})_{k,l=1,2}$ with the elements

$$m_{k,l} = \delta_{k,l}\operatorname{mes}_2 G + \int_{\mathbb{R}^2\backslash G} \nabla_y W_k(y) \cdot \nabla_y W_l(y)\,dy\,.$$

Here W_k are the harmonic functions in $\mathbb{R}^2\backslash\overline{G}$ satisfying the boundary condition $\partial W_k/\partial\nu_y = -\partial y_k/\partial\nu_y$ on ∂G. The eigenvalue $\lambda_4(S^2\backslash\overline{\Omega}_\varepsilon)$ which is greater than 1 has the asymptotics

$$\lambda_4(S^2\backslash\overline{\Omega}_\varepsilon) = 1 + 4\pi^{-1}\operatorname{mes}_2 G\,\varepsilon^2 + O(\varepsilon^3\,|\log\varepsilon|).$$

In particular, if G is the unit circle, then $M = 2\pi I$ and, therefore, the formulas

$$\lambda_{2,3}(S^2\backslash\overline{\Omega}_\varepsilon) \sim 1 - 2\varepsilon^2\,, \qquad \lambda_4(S^2\backslash\overline{\Omega}_\varepsilon) \sim 1 + 4\varepsilon^2$$

are valid for a circled hole Ω_ε.

2. Now let the domain Ω be the complement in S^2 to the segment

$$E_\alpha = \{\omega \in S^2 : \theta_1 = \pi/2,\ 0 \le \theta_2 \le \alpha\}$$

of the equator, where $\alpha \in (0, 2\pi)$. We are interested in the eigenvalues lying in the interval $(0, 1)$. In [170, Le.2.1] it was shown that for small α exactly three eigenvalues (counted with their multiplicities) are situated near the point $\lambda = 1$. Two of them, λ_3 and λ_4, are equal to 1 and the eigenfunctions corresponding to these eigenvalues are $\sin\theta_1\cos\theta_2$ and $\sin\theta_1\sin\theta_2$. For the eigenvalue λ_2 the asymptotic formulas

$$\lambda_2(S^2\backslash E_\alpha) = \frac{1}{2} + \frac{1}{\pi}\cos\frac{\alpha}{2} + \frac{1}{\pi^2}\cos^2\frac{\alpha}{2} + \left(\frac{1}{\pi^3} + \frac{1}{6\pi}\right)\cos^3\frac{\alpha}{2} + O(\cos^4\frac{\alpha}{2})$$
$$\text{as } \alpha \to \pi,$$

$$\lambda_2(S^2\backslash E_\alpha) = 1 - \frac{1}{2}(\sec\frac{\alpha}{2} - 1) + \frac{1}{4}(\sec\frac{\alpha}{2} - 1)^2 \log(\sec\frac{\alpha}{2} - 1)$$
$$+ \frac{3}{4}(1 - \log 2)(\sec\frac{\alpha}{2} - 1)^2 + O\left((\sec\frac{\alpha}{2} - 1)^3 \log(\sec\frac{\alpha}{2} - 1))^2\right)$$
$$\text{as } \alpha \to 0$$

were obtained by Brown and Stewartson [24]. The last formula yields

(2.5.2) $\lambda_2(S^2\backslash E_\alpha) < 1$ for all $\alpha \in (0, 2\pi)$.

We consider the spectral Neumann problem for the set $S^2\backslash E$, where E is the equator. Obviously, the eigenvalues $\lambda_{1,2}(S^2\backslash E)$ are equal to zero and the corresponding eigenfunctions are 1 and $\operatorname{sign}(\theta_1 - \pi/2)$. Furthermore, $\lambda_j(S^2\backslash E) = 1$ for $j = 3, 4, 5, 6$

and the corresponding eigenfunctions are the functions $\sin\theta_1\cos\theta_2$, $\sin\theta_1\sin\theta_2$, $\sin\theta_1\cos\theta_2\operatorname{sign}(\theta_1-\pi/2)$, and $\sin\theta_1\sin\theta_2\operatorname{sign}(\theta_1-\pi/2)$. Theorem 2.3.2 implies

$$(2.5.3) \qquad 1 = \lambda_3(S^2\backslash E) \le \lambda_3(S^2\backslash E_\alpha) \quad \text{for } \alpha \in (0, 2\pi).$$

Since the set $S^2\backslash E_\alpha$ is connected, the eigenvalue $\lambda_1(S^2\backslash E_\alpha) = 0$ is simple. Hence $\lambda_2(S^2\backslash E_\alpha) > 0$. From this and from the inequalities (2.5.2), (2.5.3) it follows that $\lambda_2(S^2\backslash E_\alpha)$ is the only eigenvalue in the interval $(0, 1)$.

We derive the asymptotics of the function $\lambda_2(S^2\backslash E_\alpha)$ for $\alpha \to 2\pi$. Let $u(\theta_1, \theta_2)$ be an eigenfunction to this eigenvalue. Then $u(\pi-\theta_1, \theta_2)$ is also an eigenfunction to this eigenvalue. Consequently, $u(\theta_1, \theta_2) = u(\pi-\theta_1, \theta_2)$ or $u(\theta_1, \theta_2) = -u(\pi-\theta_1, \theta_2)$. In the first case we obtain $\partial_{\theta_1}u(\theta_1, \theta_2) = 0$ for $\theta_1 = \pi/2$, $\theta_2 \in [0, 2\pi)$, i.e., $\lambda_2(S^2\backslash E_\alpha)$ is an eigenvalue of the operator pencil to the Neumann problem for the half-sphere S_+^2 and u is an eigenfunction corresponding to this eigenvalue. Moreover,

$$\int_{S_+^2} u\, d\omega = 0.$$

This implies $\lambda_2(S^2\backslash E_\alpha) = \lambda_j(S_+^2)$ for a certain index $j \ge 2$. Therefore, we get the inequality $\lambda_2(S^2\backslash E_\alpha) \ge 1$ which contradicts (2.5.2). Consequently, we have $u(\theta_1, \theta_2) = -u(\pi - \theta_1, \theta_2)$ and, in particular,

$$(2.5.4) \qquad u(\pi/2, \theta_2) = 0 \quad \text{for } \alpha < \theta_2 < 2\pi.$$

The function

$$v(\theta_1, \theta_2) = \begin{cases} u(\theta_1, \theta_2) & \text{for} \quad \theta_1 \in [0, \pi/2], \\ u(\pi - \theta_1, \theta_2) & \text{for} \quad \theta_1 \in [\pi/2, \pi] \end{cases}$$

satisfies the equation

$$-\delta v - \lambda_2(\lambda_2 + 1)\, v = 0 \quad \text{in } S^2\backslash\{\omega \in S^2 \,:\, \alpha \le \theta_2 \le 2\pi,\ \theta = \pi/2\},$$

where $\lambda_2 = \lambda_2(S^2\backslash E_\alpha)$, and the boundary condition (2.5.4). Using the asymptotic formula (2.5.1), we obtain

$$\lambda_2(S^2\backslash E_\alpha) \sim \frac{1}{2\,|\log(2\pi - \alpha)|} \qquad \text{as } \alpha \to 2\pi.$$

3. Finally, we consider the eigenvalues λ_j for a small domain on the unit sphere. Let G be a bounded domain in $\mathbb{R}^{n-1}$ and $\Omega_\varepsilon = \{\omega \in S^{n-1} : \omega'/\varepsilon \in G\}$. Analogously to the case of the Dirichlet problem, there is the asymptotic representation

$$\lambda_j(\Omega_\varepsilon) = \frac{\sqrt{\mu_j(G)}}{\varepsilon} + O(1) \qquad \text{for } j \ge 2,$$

where $\mu_j(G)$ is the j-th eigenvalue of the Neumann problem for the Laplace operator in the domain G.

2.6. Applications to boundary value problems for the Laplace equation

The results of this chapter lead to various assertions concerning regularity properties of solutions and solvability of boundary value problems for the Laplace equation in domains with singular boundary points, such as angular and conic vertices, edges. We restrict ourselves to a few examples.

2.6.1. The Dirichlet problem for the Laplace equation in plane domains with angular points. Let $\mathcal{G}$ be a plane domain with d singular boundary points $x^{(1)}, \ldots, x^{(d)}$. We suppose that for each point $x^{(\tau)}$, $\tau = 1, \ldots, d$, there exists a neighborhood $\mathcal{U}_\tau$ such that $\mathcal{U}_\tau \cap \mathcal{G}$ is the intersection of a plane angle $\mathcal{K}_\tau = \{x : (x - x^{(\tau)})/|x - x^{(\tau)}| \in \Omega_\tau\}$ with the unit disk centered at $x^{(\tau)}$. The opening of the interior angle at $x^{(\tau)}$ is denoted by α_τ.

Let F be an arbitrary function in $L_2(\mathcal{G}) = V_{2,0}^0(\mathcal{G})$ Then, by the Riesz representation theorem, there exists a unique variational solution $U \in \mathring{W}_2^1(\mathcal{G})$ of the Dirichlet problem

(2.6.1) $$-\Delta U = F \quad \text{in } \mathcal{G}, \qquad u = 0 \quad \text{on } \partial\mathcal{G},$$

i.e., a function satisfying the equality

$$\int_{\mathcal{G}} \nabla U \cdot \nabla V \, dx = \int_{\mathcal{G}} F V \, dx \quad \text{for all } V \in \mathring{W}_2^1(\mathcal{G}).$$

Since $\mathring{W}_2^1(\mathcal{G})$ is continuously imbedded into $V_{2,0}^1(\mathcal{G})$, it follows from Theorem 1.4.2 that this solution belongs to $V_{2,1}^2(\mathcal{G})$ and satisfies the estimate

$$\|U\|_{V_{2,1}^2(\mathcal{G})} \le c \, \|F\|_{V_{2,1}^0(\mathcal{G})}.$$

Applying Theorems 1.4.1 and 1.4.3 together with the results of Section 2.1, we obtain

THEOREM 2.6.1. *The Dirichlet problem* (2.6.1) *is uniquely solvable in* $V_{p,\vec{\beta}}^l(\mathcal{G})$ *for all* $F \in V_{p,\vec{\beta}}^{l-2}(\mathcal{G})$, $l \ge 2$, $p \in (1, \infty)$, $\vec{\beta} = (\beta_1, \ldots, \beta_d) \in \mathbb{R}^d$, *if and only if* $|l - \beta_\tau - 2/p| < \pi/\alpha_\tau$ *for* $\tau = 1, \ldots, d$.

If $(l - \beta_\tau - 2/p)\alpha_\tau/\pi$ is a nonzero integer for at least one τ, then the operator

(2.6.2) $$V_{p,\beta}^l(\mathcal{G}) \ni u \to \big(-\Delta u, u|_{\partial\mathcal{G} \setminus \mathcal{S}} \big) \in V_{p,\beta}^{l-2}(\mathcal{G}) \times V_{p,\beta}^{l-1/2}(\partial\mathcal{G})$$

is *not* Fredholm. We assume that

(2.6.3) $$k_\tau \pi/\alpha_\tau < l - \beta_\tau - 2/p < (k_\tau + 1)\pi/\alpha_\tau \quad \text{for } \tau = 1, \ldots, d,$$

where $k_1, \ldots, k_d$ are positive integers. One can deduce the following assertion from Section 2.1.1.

THEOREM 2.6.2. *Let inequalities* (2.6.3) *be satisfied with certain positive integers* k_τ. *Then there exist* $k = k_1 + \cdots + k_d$ *linearly independent and linear continuous functionals* Φ_j, $j = 1, \ldots, k$, *on* $V_{p,\beta}^{l-2}(\mathcal{G})$ *such that problem* (2.6.1) *is solvable in* $V_{p,\vec{\beta}}^l(\mathcal{G})$ *if and only if* $F \in V_{p,\beta}^{l-2}(\mathcal{G})$ *satisfies the conditions* $\Phi_j(F) = 0$ *for* $j = 1, \ldots, k$. *Under these conditions, the solution is unique.*

The functionals Φ_j can be explicitly described. Suppose, for simplicity, that $d = 1$, the angular point $x^{(1)}$ lies in the origin, and $\mathcal{G}$ coincides with the angle $\mathcal{K} = \{(x_1, x_2) \in \mathbb{R}^2 : 0 < \varphi = \arg(x_1 + ix_2) < \alpha\}$ in a neighborhood of $x^{(1)}$. Then for every $j = 1, 2, \ldots$ there exists a harmonic function ζ_j in $\mathcal{G}$ which vanishes on $\partial\mathcal{G} \setminus \mathcal{S}$ and has the asymptotics

(2.6.4) $$\zeta_j(x) = (j\pi)^{-1/2} r^{-j\pi/\alpha} \sin\frac{j\pi\varphi}{\alpha} + O(1)$$

as $r \to 0$. The functions $\zeta_1, \dots, \zeta_k$ belong to the kernel of the adjoint operator to (2.6.2). Consequently, one can set

$$\Phi_j(F) = \int_{\mathcal{G}} F(x)\,\zeta_j(x)\,dx \quad j = 1, \dots, k.$$

Analogous results are valid in the class of the weighted Hölder spaces $N_\beta^{l,\sigma}(\mathcal{G})$. In particular, Theorem 1.4.5 leads to the following statement.

THEOREM 2.6.3. *Problem* (2.6.1) *is uniquely solvable in* $N_\beta^{l,\sigma}(\mathcal{G})$ *for every* $F \in N_\beta^{l-2,\sigma}(\mathcal{G})$ *if and only if* $|l + \sigma - \beta_\tau| < \pi/\alpha_\tau$ *for* $\tau = 1, \dots, d$.

We consider again the solution $U \in \overset{\circ}{W}{}_2^1(\mathcal{G})$ of problem (2.6.1), where $F \in W_2^{-1}(\mathcal{G}) \cap V_{p,\beta}^{l-2}(\mathcal{G})$. Suppose that $l - \beta_\tau - 2/p \in \left(k\pi/\alpha_\tau,\, (k+1)\pi/\alpha_\tau\right)$ for a certain $\tau \in \{1, \dots, d\}$ and for some integer $k > 0$. For simplicity, we assume that $d = 1$, $x^{(1)}$ lies in the origin, and $\mathcal{G}$ coincides with the plane angle $\mathcal{K} = \{(x_1, x_2) \in \mathbb{R}^2 : 0 < \varphi = \arg(x_1 + ix_2) < \alpha\}$ in a neighborhood of $x^{(1)}$. Then, according to Theorem 1.4.4, the solution $U \in \overset{\circ}{W}{}_2^1(\mathcal{G})$ admits the representation

$$U = \sum_{j=1}^{k} c_j\,(j\pi)^{-1/2}\,r^{j\pi/\alpha_1}\,\sin\frac{j\pi\varphi}{\alpha_1} + V,$$

where c_j are constants and $V \in V_{p,\beta}^l(\mathcal{G})$. By [**180**, Th.9.1] (see also [**174**, Th.1.3.8]), the coefficients c_j are given by the formula

$$c_j = -\int_{\mathcal{G}} F(x)\,\zeta_j(x)\,dx,$$

where ζ_j is the function (2.6.4).

2.6.2. The mixed boundary value problem for the Laplace equation in domains with angular points. Let $\mathcal{G}$ be the same domain as in the preceding subsection. We consider the boundary value problem

(2.6.5) $-\Delta U = F$ in $\mathcal{G}$, $B_\tau U = 0$ on the side $x^{(\tau)}x^{(\tau+1)} \subset \partial\mathcal{G}$, $\tau = 1, \dots, d$,

where $x^{(d+1)} = x^{(1)}$ and $B_\tau = 1$ or $B_\tau = \partial/\partial\nu$ and suppose that the Dirichlet condition is given on at least one of the adjacent sides of the boundary. Then every function $U \in W_2^1(\mathcal{G})$ satisfying the boundary conditions in (2.6.5) belongs also to the space $V_{2,0}^1(\mathcal{G})$. Using the information on the spectrum of the operator pencil corresponding to the parameter-depending problem (2.1.10), (2.1.11) obtained in Section 2.1, we can deduce from Theorems 1.4.1–1.4.5 analogous results as in the case of the Dirichlet problem. In particular, the following statement holds.

THEOREM 2.6.4. *Let* $\lambda_1^{(\tau)} = \pi/\alpha_\tau$ *if the Dirichlet condition is given on both sides of* $\partial\mathcal{G}$ *adjacent to* $x^{(\tau)}$ *and* $\lambda_1^{(\tau)} = \pi/(2\alpha_\tau)$ *if on one side the Dirichlet condition and on the other side the Neumann condition are given. Then problem* (2.6.5) *is uniquely solvable in* $V_{p,\vec{\beta}}^l(\mathcal{G})$ *for all* $F \in V_{p,\vec{\beta}}^{l-2}(\mathcal{G})$ *if and only if*

$$|l - \beta_\tau - 2/p| < \lambda_1^{(\tau)} \quad \text{for } \tau = 1, \dots, d.$$

For the unique solvability in $N_{\vec{\beta}}^{l,\sigma}(\mathcal{G})$ *it is necessary and sufficient that* $|l + \sigma - \beta_\tau| < \lambda_1^{(\tau)}$ *for* $\tau = 1, \dots, d$.

2.6.3. The Dirichlet problem for the Laplace equation in a domain with conic points. Now we consider problem (2.6.1) in the same n-dimensional domain $\mathcal{G}$ with the conic vertices $x^{(1)}, \ldots, x^{(d)}$, as in Section 1.4. Theorems 1.4.1–1.4.3 together with the information obtained in Section 2.2 lead to the following assertion.

THEOREM 2.6.5. *Suppose that*

$$(2.6.6) \qquad \left| l - 1 - \beta_\tau + \frac{n}{2} - \frac{n}{p} \right|^2 < \left(1 - \frac{n}{2} \right)^2 + \Lambda_1^{(\tau)} \quad \text{for } \tau = 1, \ldots, d,$$

where $\Lambda_1^{(\tau)}$ denotes the first eigenvalue of the Dirichlet problem for the operator $-\delta$ on the intersection Ω_τ of the sphere $|x - x^{(\tau)}| = 1$ with the cone $\mathcal{K}_\tau$. Then the Dirichlet problem (2.6.1) is uniquely solvable in $V^l_{p,\vec{\beta}}(\mathcal{G})$ for every $F \in V^{l-2}_{p,\vec{\beta}}(\mathcal{G})$. Analogously, the condition

$$(2.6.7) \qquad \left| l + \sigma - 1 - \beta_\tau + \frac{n}{2} \right|^2 < \left(1 - \frac{n}{2} \right)^2 + \Lambda_1^{(\tau)} \quad \text{for } \tau = 1, \ldots, d$$

ensures the unique solvability of the Dirichlet problem in $N^{l,\sigma}_{\vec{\beta}}(\mathcal{G})$ for arbitrary $f \in N^{l-2,\sigma}_{\vec{\beta}}(\mathcal{G})$. The variational solution $U \in \overset{\circ}{W}{}^1_2(\mathcal{G})$ belongs to $V^l_{p,\vec{\beta}}(\mathcal{G})$ if $F \in V^{l-2}_{p,\vec{\beta}}(\mathcal{G})$ and (2.6.6) is satisfied. It belongs to $N^{l,\sigma}_{\vec{\beta}}(\mathcal{G})$ if $F \in N^{l-2,\sigma}_{\vec{\beta}}(\mathcal{G})$ and condition (2.6.7) is satisfied.

Since the eigenvalues $\Lambda_1^{(\tau)}$ are positive, we conclude that the variational solution U belongs to the Sobolev space $W^l_p(\mathcal{G})$ if $F \in W^{l-2}_p(\mathcal{G})$ and $l \leq n/p$. Stronger regularity results can be obtained using the monotonicity of $\Lambda_1^{(\tau)}$ with respect to Ω_τ. If, for example, $\mathcal{K}_\tau$ is contained in a half-space, then $\Lambda_1^{(\tau)} \geq n - 1$. Thus, in the case when all cones $\mathcal{K}_\tau$ are contained in half-spaces, we get $U \in W^l_p(\mathcal{G})$ if $F \in W^{l-2}_p(\mathcal{G})$ and $l \leq 1 + n/p$.

2.7. Notes

Section 2.1-2.4. Most of the results are standard. The isoperimetric property of the first eigenvalue of the Dirichlet problem for the Beltrami operator (Theorem 2.2.5) was first proved by Sperner [**248**]. The proof of the corresponding result for domains in the Euclidean space was given by Faber [**55**] and Krahn [**144, 145**] in the 1920s (see also the book of Pólya and Szegö [**229**]). In the book of Chavel [**26**] a modern proof is given which works for any of the model spaces including sphere. Another isoperimetric inequality was obtained by Shen and Shieh [**244**] who proved that among all spherical bands on S^2 with given area the band which is symmetric to the equator has the largest first eigenvalue of the Dirichlet problem.

The isoperimetric property for the first nonzero eigenvalue of the Neumann problem (Theorem 2.3.3) was first proved by Chavel [**25, 26**] under more restrictive hypotheses. Ashbaugh and Benguria [**8**] extended Chavel's result to arbitrary spherical domains contained in a hemisphere. The corresponding result for the Euclidean case was obtained by Szegö [**252**] and Weinberger [**264**].

Theorems 2.2.6 and 2.3.1 on the multiplicity of the eigenvalues of the Beltrami operator were essentially proved by Nadirashvili [**202**].

Figure 2 is taken from the paper [**167**] of Maz'ya and Levin. Other material in Sections 2.3.2 and 2.4 was unpublished.

Boersma [**22**] and Keller [**99**] studied singularities of solutions to the Dirichlet and Neumann problems for the Laplace equation in $\mathbb{R}^3 \backslash K_\alpha$, where K_α is a plane angular sector of opening angle α.

Section 2.5. Here we collected asymptotic formulas for eigenvalues which were obtained in papers by Samarskiĭ [**232**], Brown and Stewartson [**24**], Maz'ya, Nazarov and Plamenevskiĭ [**171**], Maz'ya and Nazarov [**170**].

Section 2.6. Treatments of boundary value problems for the Laplacian in domains with piecewise smooth boundaries can be found in the monographs of Grisvard [**78, 86**], Kufner and Sändig [**147**], Dauge [**41**], Maz'ya, Nazarov and Plamenevskiĭ [**174**], Nazarov and Plamenevskiĭ [**207**], Kozlov, Maz'ya and Roß-mann [**136**]. Regularity results for the solution of the Dirichlet problem in plane domains of polygonal type were obtained in 1956 by Nikol'skiĭ [**214, 215**] and in the 1960s by Volkov [**258, 259, 260**] and Fufaev [**64**]. They gave (necessary and sufficient) conditions ensuring that the solution belongs to the Nikolskiĭ space H_p^s and to a Hölder space, respectively. Birman and Skvortsov [**20**] got a W_2^2 regularity result for the case when the angles at the corners are less than π. For the case of a three-dimensional convex polyhedron an analogous result was proved by Hanna and Smith [**89**]. Wigley [**269**]–[**272**] studied the smoothness of the solution to the mixed problem and its asymptotic behavior near the angular points. Distribution solutions of the mixed problem in a polygon were considered by Aziz and Kellogg [**9**].

In [**255, 256**] Maz'ya and Verzhbinskiĭ developed an asymptotic theory of the Dirichlet problem for the Laplacian in n-dimensional domains with singular boundary points including conic and cuspidal vertices. As a consequence of their estimates for the Green function and harmonic measure, they obtained a L_p coercivity result for the case of a conic point (a proof of this result is given in [**257**]).

Grisvard [**74**]–[**86**] dealt with various boundary value problems for the Laplace equation in polygonal or polyhedral domains. He studied, in particular, the solvability in Sobolev spaces (without weight) and the singularities of the solutions. Fichera [**59**]–[**62**] dealt with the Dirichlet problem in three-dimensional polyhedral domains. In [**59**] he proved, for example, that the solution belongs to C^1 if the polyhedron is convex.

Theorems on the solvability of boundary value problems for the Laplace equation in domains with corners and edges as well as regularity assertions for the solutions can be also found in papers of Kondrat'ev [**110, 111**], Komech [**107**], Zajaczkowski and Solonnikov [**273**], Dauge [**39, 40, 43**], Dauge and Nicaise [**45**]. Solutions of mixed screen boundary value problems for the Helmholtz equation in the lifting surface theory were studied, e.g., by Hinder, Meister [**90**] and Nazarov [**205**].

Besides, there are many works which are concerned with the Laplace equation in domains of special classes (convex domains, domains with C^1 or Lipschitz boundaries). For example, Grisvard [**78**, Ch.3] proved that the Dirichlet problem (2.6.1) in an arbitrary bounded convex domain $\mathcal{G}$ of $\mathbb{R}^n$ is uniquely solvable in $W_2^2(\mathcal{G})$ for every $F \in L_2(\mathcal{G})$. Also for convex domains Adolfsson and Jerison [**1**] proved that the Neumann problem has a unique solution in $W_p^2(\mathcal{G})$, $1 < p \leq 2$, if the function on the right-hand side of the differential equation belongs to $L_p(\mathcal{G})$.

A deep theory of second order equations in Lipschitz domains including solvability of boundary value problems for L_p boundary data and regularity results for the solutions was developed in the works of Dahlberg [33], Dahlberg and Kenig [34], Kenig and Pipher [105, 106], Kenig [103], Jerison and Kenig [92, 93]. A survey of this development is given in Kenig's monograph [104]. Higher order regularity results for certain classes of Lipschitz domains can be found in the book of Maz'ya and Shaposhnikova [196].

Numerical methods for the determination of eigenvalues of operator pencils are not a subject of this book. Let us give here only a few references concerning this direction. The first result we mention here is that of Fichera [59] who considered the Dirichlet problem in the exterior of a cube and obtained two–sided estimates for the smallest positive eigenvalue of the corresponding pencil. Furthermore, we refer to the papers of Walden and Kellogg [263], Stephan and Whiteman [250], Beagles and Whiteman [15, 16], Babuška, von Petersdorff and Andersson [10], Schmitz, Volk and Wendland [235] which are also concerned with the numerical calculation of singularities of solutions to the Laplace equation near singular boundary points.

CHAPTER 3

The Dirichlet problem for the Lamé system

In this chapter we consider the Dirichlet problem for the Lamé system of isotropic elasticity

$$(3.0.1) \qquad \Delta U + (1 - 2\nu)^{-1} \nabla \nabla \cdot U = 0$$

in an angle and a cone with vertex at the origin. Here U is the displacement vector and ν denotes the Poisson ratio, $\nu < 1/2$.

Solutions to this problem may exhibit infinite stresses near singularities of the boundary, such as edges and vertices. Their occurrence is related to displacement field U which, near an isolated vertex, have the form $U = r^\lambda u(\omega)$, where r denotes the distance to the vertex and ω are spherical coordinates in the base of the cone. The quantity λ and the vector function u depend only on the geometry, the type of the boundary condition and the Poisson ratio, but are independent of the particular right-hand side and the boundary data. The knowledge of λ and u enables one to determine not only the asymptotics of stresses near conic points, but also the regularity of a weak solution in scales of Sobolev spaces. Moreover, λ and u are needed for the computation of coefficients in the asymptotic expansion of the stresses near the conic points, the so-called stress intensity factors. Knowledge of λ and u is crucial for the design of proper numerical approximation schemes (mesh grading in finite element schemes, for example).

The pairs (λ, u) can be characterized as eigenvalues and eigenvectors of a rather complicated operator pencil in a domain on the unit sphere. This pencil is Fredholm. By considering general singularities

$$(3.0.2) \qquad r^\lambda \sum_{k=0}^{s} \frac{1}{k!} (\log r)^k \, u^{(s-k)}(\omega)$$

one arrives at generalized eigenfunctions of the pencil.

We begin with the case of a plane angle. As for the Laplace equation (see Chapter 2), the singularities can be described by means of the eigenvalues, eigenvectors and generalized eigenvectors of certain pencils of ordinary differential operators on an interval. In Section 3.1 we derive transcendental equations for the eigenvalues and study the solvability of these equations in different strips of the complex plane. Furthermore, we give representations for the eigenfunctions and generalized eigenfunctions.

We give a simple example showing the usefulness of information obtained in Section 3.1. Let $\mathcal{G}$ be the same two-dimensional domain as in the introduction to Chapter 2. Recall that α denotes the opening of the angle. We consider the Dirichlet problem

$$\Delta U + (1 - 2\nu)^{-1} \nabla \nabla \cdot U = F \ \text{ on } \mathcal{G}, \quad U = 0 \ \text{ on } \partial \mathcal{G} \setminus \{0\},$$

where F is a given function equal zero near the angular boundary point. By combining Theorem 3.1.2 and Theorem 1.4.4, we can explicitly describe the asymptotic behavior of the solution U with a finite Dirichlet integral. According to Theorem 3.1.2, we must separate the cases $\alpha \in (0, \pi)$, $\alpha \in (\pi, 2\pi)$, and $\alpha = 2\pi$. Here we restrict ourselves to the case $\alpha \in (\pi, 2\pi)$. Let $x_-(\alpha)$ and $x_-^{(1)}(\alpha)$ be real roots of the equation

$$\frac{\sin z}{z} - \frac{\sin \alpha}{(3 - 4\nu)\alpha} = 0$$

and $x_+(\alpha)$ be a real root of

$$\frac{\sin z}{z} + \frac{\sin \alpha}{(3 - 4\nu)\alpha} = 0.$$

The inequalities

$$0 < x_+(\alpha) < \pi < x_-(\alpha) < x_-^{(1)}(\alpha) < 2\pi$$

should be valid. (The above roots are uniquely defined by these inequalities.) Now the asymptotics of U can be written in the form

$$U(x) = C_+ \, r^{\lambda_+} \, u_+(\varphi) + C_- \, r^{\lambda_-} \, u_-(\varphi) + C_+^{(1)} \, r^{\lambda_+^{(1)}} \, u_+^{(1)}(\varphi) + o(r^{2\pi/\alpha}),$$

where the constant coefficients $C_\pm$ and $C_-^{(1)}$ depend on the vector function F, whereas $u_\pm$ and $u_-^{(1)}$ are linear combinations of trigonometric functions (see (3.1.28) and 3.1.29)). The exponents $\lambda_\pm$ and $\lambda_-^{(1)}$ are introduced by

$$\lambda_+ = \frac{x_+(\alpha)}{\alpha}, \quad \lambda_- = \frac{x_-(\alpha)}{\alpha}, \quad \lambda_-^{(1)} = \frac{x_-^{(1)}(\alpha)}{\alpha} \, .$$

Note that, by Theorem 3.1.2, the functions $\lambda_\pm$ and $\lambda_-^{(1)}$ are strictly decreasing and subject to the inequalities $1/2 < \lambda_+ < \lambda_- < 1 < \lambda_-^{(1)}$ on $(\pi, 2\pi)$.

In Section 3.2 we turn to the incomparably more complicated three dimensional Dirichlet problem for the system (3.0.1) in a cone $\mathcal{K}$. By $U = (U_1, U_2, U_3)$ we denote the displacement vector. We are interested in particular solutions of the form (3.0.2). The exponents λ are (in general complex) eigenvalues of the operator pencil

$$\mathcal{L}(\lambda) : \overset{\circ}{W}_2^1(\Omega)^3 \to W_2^{-1}(\Omega)^3$$

on the intersection Ω of $\mathcal{K}$ with the unit sphere S^2, which is defined by

$$\mathcal{L}(\lambda) \, u = r^{2-\lambda} \left(\Delta r^\lambda u + (1 - 2\nu)^{-1} \nabla \nabla \cdot r^\lambda u \right).$$

We now outline some results of this chapter which concern the three-dimensional case. In Subsection 3.2.4, starting with Jordan chains of $\mathcal{L}$ generated by an eigenvalue λ_0, we describe Jordan chains of the same pencil corresponding to the eigenvalue $-1 - \lambda_0$. In passing, in Subsection 3.2.5 we obtain an analog of the Kelvin transform for solutions of the Lamé system:

If U satisfies the Lamé system (3.0.1), then the same is true for

$$V(x) = |x|^{-1} \left\{ (3 - 4\nu)I + \Xi(x|x|^{-1}) + (I - 2\Xi(x|x|^{-1})) \, x \cdot \partial_x \right\} U\left(x|x|^{-2}\right),$$

where I is the identity matrix and $\Xi(\omega) = \{\omega_i \omega_j\}_{i,j=1}^3$.

Our next result is that only real eigenvalues of the pencil $\mathcal{L}$ may occur in the domain

$$(\operatorname{Re}\lambda + 1/2)^2 - (\operatorname{Im}\lambda)^2 < F_\nu(\Omega)^2,$$

where $F_\nu(\Omega)$ is a decreasing function of Ω, which, in particular, admits the estimate

$$F_\nu(\Omega) \geq \left(\left(\frac{5}{2} - 2\nu\right)^2 + (1 - 2\nu)^2\right)^{1/2}$$

(Section 3.3). We prove that these eigenvalues do not admit generalized eigenvectors. Furthermore, we establish a variational principle for real eigenvalues of $\mathcal{L}$ in the interval $(-\frac{1}{2}, \beta)$, where

$$\beta = \min\left\{F_\nu(\Omega) - \frac{1}{2}, 3 - 4\nu\right\}.$$

In particular, this principle allows one to show that all eigenvalues of $\mathcal{L}$ in $(-\frac{1}{2}, \beta)$ depend monotonically on the domain Ω cut out by the cone $\mathcal{K}$ on the unit sphere S^2.

As a corollary we obtain that the strip $-1 \leq \operatorname{Re}\lambda \leq 0$ is free of the eigenvalues and that, moreover, the same holds for the strip $-2 \leq \operatorname{Re}\lambda \leq 1$ if $\Omega \subset S_+^2$ and $S_+^2 \backslash \overline{\Omega}$ contains a nonempty open set. Another corollary is that, if $S_+^2 \subset \Omega$ and $S^2 \backslash \overline{\Omega}$ contains a nonempty open set, then there are exactly three eigenvalues $\lambda_j \in (0, 1]$ which may coincide, but always have total multiplicity 3.

These results imply that solutions of the Dirichlet problem are either Hölder continuous near the vertex of a cone or have a singularity stronger than r^{-1}. If the cone is situated in a half space, then the stresses are Hölder continuous. Besides, in a neighborhood of the reentrant vertex the solution of the Dirichlet problem with finite energy integral has the asymptotics

$$U(x) = \sum_{j=1}^{3} r^{\lambda_j} u_j(\omega) + o(r^{1+\varepsilon}), \quad \varepsilon > 0.$$

where $0 < \lambda_1 \leq \lambda_2 \leq \lambda_3 < 1$.

The above results on the spectrum of $\mathcal{L}$ together with Theorems 1.4.1–1.4.5 enable one to draw conclusions about the regularity of the solution of the inhomogeneous Dirichlet problem

$$\Delta U + (1 - 2\nu)^{-1}\nabla \nabla \cdot U = F \quad \text{in } \mathcal{G}, \quad U = 0 \quad \text{on } \partial\mathcal{G}$$

where $\mathcal{G}$ is a domain in $\mathbb{R}^3$ with conic vertices. For example,

(i) $U \in C^{1,\alpha}(\mathcal{G})^3$ for some positive α if $F \in L_\infty(\mathcal{G})^3$ and all tangent cones at vertices are contained in a half-space.

(ii) $U \in W_2^2(\mathcal{G})^3$ if $F \in L^2(\mathcal{G})^3$ and the tangent cones are contained in a circular cone for which the smallest real part of the eigenvalues of $\mathcal{L}$ is greater than $1/2$.

Section 3.4 is dedicated to estimates of the width of a strip on the complex plane which contains no eigenvalues of $\mathcal{L}$. For example, it is shown that this is true for the strip

$$\left|\operatorname{Re}\lambda + \frac{1}{2}\right| \leq \frac{(2\gamma + 1)M}{M + 2\gamma + 4} + \frac{1}{2},$$

where M is a positive number such that $M(M + 1)$ is the first eigenvalue of the Laplace-Beltrami operator in Ω with zero Dirichlet data.

In the next section for circular cones we obtain an explicit transcendental equation for all eigenvalues of the pencil $\mathcal{L}$ and compute the dependence of the first eigenvalues on the opening angle of the cone. It is found that the dominant stress fields near a rotationally symmetric rigid inclusion are not rotationally symmetric.

We conclude this chapter by stating some applications of the above mentioned results to the Dirichlet problem for the Lamé system in a two- or three-dimensional domain with piecewise smooth boundary.

3.1. The Dirichlet problem for the Lamé system in a plane angle

We start with the Dirichlet problem for the Lamé system in a plane angle. In contrast to the boundary value problems for the Laplace operator, it is not possible to give explicit representations for the eigenvalues of the operator pencil generated by this problem. However, one can find entire functions whose zeros are the mentioned eigenvalues. By means of these functions, we study the distribution of the eigenvalues in the complex plane.

3.1.1. The operator pencil generated by the Dirichlet problem. Let

$$\mathcal{K} = \big\{(x_1, x_2) \in \mathbb{R}^2 \,:\, 0 < r < \infty, \, -\alpha/2 < \varphi < \alpha/2\big\}, \qquad \alpha \in (0, 2\pi],$$

be a plane angle. We consider the Lamé system

(3.1.1) $$\Delta U + (1 - 2\nu)^{-1} \nabla \nabla \cdot U = 0 \qquad \text{in } \mathcal{K}$$

for the displacement vector $U = (U_1, U_2)$ with the Dirichlet condition

(3.1.2) $$U = 0 \qquad \text{on } \partial\mathcal{K}\backslash\{0\}.$$

In polar coordinates r, φ the Lamé system has the form

$$\Delta U_r - \frac{1}{r^2} U_r - \frac{2}{r^2} \partial_\varphi U_\varphi + (1 - 2\nu)^{-1} \partial_r \Big(\frac{1}{r} \partial_r (rU_r) + \frac{1}{r} \partial_\varphi U_\varphi\Big) = 0,$$

$$\Delta U_\varphi - \frac{1}{r^2} U_\varphi + \frac{2}{r^2} \partial_\varphi U_r + (1 - 2\nu)^{-1} \frac{1}{r^2} \partial_\varphi \Big(\partial_r (rU_r) + \partial_\varphi U_\varphi\Big) = 0,$$

where

(3.1.3) $$\begin{pmatrix} U_r \\ U_\varphi \end{pmatrix} = \begin{pmatrix} \cos\varphi & \sin\varphi \\ -\sin\varphi & \cos\varphi \end{pmatrix} \cdot \begin{pmatrix} U_1 \\ U_2 \end{pmatrix}.$$

We seek solutions

(3.1.4) $$U_r(r, \varphi) = r^\lambda u_r(\varphi), \qquad U_\varphi(r, \varphi) = r^\lambda u_\varphi(\varphi)$$

which are positively homogeneous of degree λ. For the functions u_r, u_φ the following system of ordinary differential equations holds:

(3.1.5) $$\frac{1 - 2\nu}{2 - 2\nu} u_r'' + (\lambda^2 - 1) u_r + \Big(\lambda - 1 - \frac{1 - 2\nu}{2 - 2\nu}(\lambda + 1)\Big) u_\varphi' = 0,$$

(3.1.6) $$u_\varphi'' + \frac{1 - 2\nu}{2 - 2\nu}(\lambda^2 - 1) u_\varphi + \Big(\lambda + 1 - \frac{1.- 2\nu}{2 - 2\nu}(\lambda - 1)\Big) u_r' = 0.$$

Moreover, u_r and u_φ satisfy the boundary conditions

(3.1.7) $$u_r(\pm\alpha/2) = u_\varphi(\pm\alpha/2) = 0.$$

Let $\mathcal{L}(\lambda)$ denote the differential operator on the left-hand side of the system (3.1.5), (3.1.6). Furthermore, let $\mathcal{B}$ be the operator

$$(W_2^2(-\alpha/2, +\alpha/2))^2 \ni \begin{pmatrix} u_r \\ u_\varphi \end{pmatrix} \to \begin{pmatrix} u_r(\pm\alpha/2) \\ u_\varphi(\pm\alpha/2) \end{pmatrix} \in \mathbb{C}^4.$$

We introduce the operator

$$(W_2^2(-\alpha/2,+\alpha/2))^2 \ni \begin{pmatrix} u_r \\ u_\varphi \end{pmatrix} \to \mathfrak{A}(\lambda) \begin{pmatrix} u_r \\ u_\varphi \end{pmatrix} \stackrel{def}{=} \big(\mathcal{L}(\lambda), \mathcal{B}\big) \begin{pmatrix} u_r \\ u_\varphi \end{pmatrix}$$
$$\in (L_2(-\alpha/2,+\alpha/2))^2 \times \mathbb{C}^4.$$

Clearly, the spectrum of the operator pencil $\mathfrak{A}$ is discrete and consists of eigenvalues with finite algebraic multiplicities.

3.1.2. Two lemmas on operator functions. Let the space $\mathcal{Y}$ be the Cartesian product $\mathcal{Y}_0 \times \mathcal{Y}_1$ of two Hilbert spaces and let the operator pencil

$$(3.1.8) \qquad \mathfrak{A}(\lambda) = \big(\mathcal{L}(\lambda), \mathcal{B}(\lambda)\big) : \mathcal{X} \to \mathcal{Y}$$

consist of two pencils $\mathcal{L}(\lambda) \in L(\mathcal{X}, \mathcal{Y}_0)$, $\mathcal{B}(\lambda) \in L(\mathcal{X}, \mathcal{Y}_1)$. We suppose that there exist a Hilbert space $\mathcal{Z}$ and a holomorphic operator-function $\mathcal{P}(\lambda) \in L(\mathcal{Z}, \mathcal{X})$ defined for all complex λ in a domain $G \subset \mathbb{C}$ such that

$$(3.1.9) \quad \ker \mathcal{P}(\lambda) = \{0\}, \quad \mathcal{R}(\mathcal{P}(\lambda)) = \ker \mathcal{L}(\lambda), \quad \mathcal{L}(\lambda) \text{ is surjective for all } \lambda \in G,$$

where $\mathcal{R}(\mathcal{P}(\lambda))$ denotes the range of the operator $\mathcal{P}(\lambda)$.

The next lemma gives a connection between the spectral properties of the pencil (3.1.8) and the operator function

$$(3.1.10) \qquad \mathcal{B}(\lambda)\,\mathcal{P}(\lambda) : \mathcal{Z} \to \mathcal{Y}_1, \quad \lambda \in G.$$

LEMMA 3.1.1. *Suppose that condition* (3.1.9) *is satisfied. Then*
1) *the operator* $\mathfrak{A}(\lambda)$ *is Fredholm if and only if* $\mathcal{B}(\lambda)\,\mathcal{P}(\lambda)$ *is Fredholm,*
2) *the spectra in* G *of the operator functions* $\mathfrak{A}(\lambda)$ *and* $\mathcal{B}(\lambda)\,\mathcal{P}(\lambda)$ *coincide,*
3) *the number* $\lambda_0 \in G$ *is an eigenvalue of the operator-function* $\mathfrak{A}(\lambda)$ *if and only if* λ_0 *is an eigenvalue of the operator-function* $\mathcal{B}(\lambda)\,\mathcal{P}(\lambda)$. *The geometric, partial, and algebraic multiplicities of the eigenvalue* λ_0 *for the operator-functions* $\mathfrak{A}(\lambda)$ *and* $\mathfrak{B}(\lambda)\,\mathcal{P}(\lambda)$ *coincide.*
4) *If* $u_0, \ldots, u_{s-1} \in \mathcal{Z}$ *is a Jordan chain of* (3.1.10) *corresponding to the eigenvalue* λ_0, *then*

$$(3.1.11) \qquad v_j = \sum_{k=0}^{j} \frac{1}{(j-k)!} \mathcal{P}^{(j-k)}(\lambda_0)\, u_k, \quad j = 0, \ldots, s-1,$$

is a Jordan chain of (3.1.8) *corresponding to* λ_0. *Here* $\mathcal{P}^{(k)}(\lambda_0) = d^k \mathcal{P}(\lambda)/d\lambda^k|_{\lambda=\lambda_0}$.
5) *If* $v_0, \ldots, v_{s-1}$ *is a Jordan chain of* (3.1.8) *corresponding to* λ_0, *then there exists a Jordan chain* $u_0, \ldots, u_{s-1}$ *of* (3.1.10), *such that* (3.1.11) *is valid.*

Proof: a) *Assertions 1) and 2).* By (3.1.9), there exists a subspace $\mathcal{Z}_\lambda \subset \mathcal{Z}$ with the same dimension as $\ker \mathfrak{A}(\lambda)$ such that $\mathcal{P}(\lambda)\,\mathcal{Z}_\lambda = \ker \mathfrak{A}(\lambda)$. Consequently, $\ker \mathcal{B}(\lambda)\,\mathcal{P}(\lambda) = \mathcal{Z}_\lambda$ and

$$\dim \ker \mathfrak{A}(\lambda) = \dim \ker \mathcal{B}(\lambda)\,\mathcal{P}(\lambda).$$

Since $\mathcal{L}(\lambda)$ is surjective, we have

$$\mathcal{R}\big(\mathfrak{A}(\lambda)\big) = \mathcal{Y}_0 \times \mathcal{Y}_1',$$

where $\mathcal{Y}_1'$ is a linear subset of $\mathcal{Y}_1$. We show that the range of the operator $\mathcal{B}(\lambda)\,\mathcal{P}(\lambda)$ coincides with $\mathcal{Y}_1'$. Obviously, the range of $\mathcal{B}(\lambda)\,\mathcal{P}(\lambda)$ is contained in $\mathcal{Y}_1'$. Let ψ_1 be an arbitrary element of $\mathcal{Y}_1'$. Then there exists an element $\varphi \in \mathcal{X}$ such that

$$\mathcal{L}(\lambda)\,\varphi = 0, \quad \mathcal{B}(\lambda)\,\varphi = \psi_1.$$

Furthermore, by (3.1.9), there exists an element $\chi \in \mathcal{Z}$ such that $\mathcal{P}(\lambda)\chi = \varphi$ and, consequently, $\mathcal{B}(\lambda)\mathcal{P}(\lambda)\chi = \psi_1$. Thus, the range of $\mathcal{B}(\lambda)\mathcal{P}(\lambda)$ coincides with $\mathcal{Y}_1'$. From this we conclude, that the operators $\mathfrak{A}(\lambda)$ and $\mathcal{B}(\lambda)\mathcal{P}(\lambda)$ are simultaneously Fredholm. Moreover, they are simultaneously surjective. This proves 1) and 2).

b) *Generating functions.* We consider a general holomorphic operator function $\mathcal{C}(\lambda) : \mathcal{X} \to \mathcal{Y}$. Let λ_0 be an eigenvalue. A holomorphic function $\Phi(\lambda)$ with values in $\mathcal{X}$ is called generating function of rank s in λ_0 for $\mathcal{C}$ if

$$\Phi(\lambda_0) \neq 0 \quad \text{and} \quad \mathcal{C}(\lambda)\Phi(\lambda) = O(|\lambda - \lambda_0|^s).$$

Direct calculation shows that $\Phi(\lambda)$ is a generating function of rank s if and only if the vectors

$$\Phi_j = \frac{1}{j!}\Phi^{(j)}(\lambda_0), \quad j = 0, \ldots, s-1,$$

form a Jordan chain corresponding to λ_0.

c) *Proof of 4).* Let $\Psi(\lambda)$ be a generating function of rank s for $\mathcal{B}(\lambda)\mathcal{P}(\lambda)$, i.e.,

$$\Psi(\lambda_0) \neq 0 \quad \text{and} \quad \mathcal{B}(\lambda)\mathcal{P}(\lambda)\Psi(\lambda) = O(|\lambda - \lambda_0|^s).$$

Since $\mathcal{L}(\lambda)\mathcal{P}(\lambda) = 0$ and $\ker \mathcal{P}(\lambda_0) = \{0\}$, we get

$$\mathcal{P}(\lambda_0)\Psi(\lambda_0) \neq 0 \quad \text{and} \quad \mathfrak{A}(\lambda)\mathcal{P}(\lambda)\Psi(\lambda) = O(|\lambda - \lambda_0|^s).$$

Hence $\mathcal{P}(\lambda)\Psi(\lambda)$ is a generating function of rank s in λ_0 for $\mathfrak{A}(\lambda)$. Now the assertion follows from a).

d) *Proof of 5).* Let $\Phi(\lambda)$ be a generating function of rank s for $\mathfrak{A}(\lambda)$, i.e.,

$$\Phi(\lambda_0) \neq 0 \quad \text{and} \quad \mathfrak{A}(\lambda)\Phi(\lambda) = O(|\lambda - \lambda_0|^s).$$

We prove by induction that for every $k = 1, \ldots, s$ there exists a polynomial $\Psi(\lambda)$ of degree $k-1$ with coefficients in $\mathcal{Z}$ such that

$$(3.1.12) \qquad \mathcal{P}(\lambda)\Psi(\lambda) - \Phi(\lambda) = O(|\lambda - \lambda_0|^k)$$

In the case $k = 1$ we set $\Psi(\lambda) = \psi_0$, where ψ_0 is a solution of the equation $\mathcal{P}(\lambda_0)\Psi_0 = \Phi(\lambda_0)$. Since $\Phi(\lambda_0) \in \ker \mathcal{L}(\lambda_0)$, the existence of the solution Ψ_0 follows from (3.1.9).

Suppose that there exists a polynomial $\chi(\lambda)$ of degree $k-2 \leq s-2$ satisfying the condition

$$(3.1.13) \qquad \mathcal{P}(\lambda)\chi(\lambda) - \Phi(\lambda) = O(|\lambda - \lambda_0|^{k-1}).$$

Since $\Phi(\lambda)$ is a generating function of rank $s \geq k$ for $\mathfrak{A}(\lambda)$ and $\mathcal{L}(\lambda)\mathcal{P}(\lambda) = 0$, we have

$$\mathcal{L}(\lambda)\big(\mathcal{P}(\lambda)\chi(\lambda) - \Phi(\lambda)\big) = O(|\lambda - \lambda_0|^k).$$

This and (3.1.13) yield

$$\mathcal{L}(\lambda_0)\left(\frac{d^{k-1}}{d\lambda^{k-1}}\big(\mathcal{P}(\lambda)\chi(\lambda) - \Phi(\lambda)\big)\Big|_{\lambda=\lambda_0}\right) = 0.$$

Consequently, by (3.1.9), there exists an element $\Psi_{k-1} \in \mathcal{Z}$ such that

$$(3.1.14) \qquad \mathcal{P}(\lambda_0)\Psi_{k-1} + \frac{1}{(k-1)!}\frac{d^{k-1}}{d\lambda^{k-1}}\big(\mathcal{P}(\lambda)\chi(\lambda) - \Phi(\lambda)\big)\Big|_{\lambda=\lambda_0} = 0.$$

We set $\Psi(\lambda) = \chi(\lambda) + \Psi_{k-1}(\lambda - \lambda_0)^{k-1}$. Then, by (3.1.13) and (3.1.14), we obtain

$$\frac{d^{j-1}}{d\lambda^{j-1}}\left(\mathcal{P}(\lambda)\,\Psi(\lambda) - \Phi(\lambda)\right)\Big|_{\lambda=\lambda_0}$$

for $j = 0, 1, \ldots, k-1$, i.e., $\mathcal{P}(\lambda)\,\Psi(\lambda) - \Phi(\lambda) = O(|\lambda - \lambda_0|^k)$.

Furthermore, $\Psi(\lambda_0) \neq 0$, since the kernel of $\mathcal{P}(\lambda_0)$ is trivial. Therefore, $\Psi(\lambda)$ is a generating function of rank k for $\mathcal{B}(\lambda)\,\mathcal{P}(\lambda)$. Thus the existence of a function $\Phi(\lambda)$ satisfying (3.1.12) with $k = s$ is proved. This gives, in particular, the relations (3.1.11). To complete the proof of (v), it suffices to apply b).

e) The validity of 3) follows from 4) and 5). ∎

In the case $\dim \mathcal{X} = \dim \mathcal{Y} < \infty$ the algebraic multiplicities of the eigenvalues can be calculated by means of the following lemma (see, e.g., the book of Gohberg, Lancaster and Rodman [**72**]).

LEMMA 3.1.2. *Let $\mathcal{X}$ and $\mathcal{Y}$ be finite-dimensional spaces with the same dimension. If λ_0 is an eigenvalue of the operator-function $\mathfrak{A}$, then its algebraic multiplicity is equal to the multiplicity of the zero $\lambda = \lambda_0$ of the function $\det \mathfrak{A}$.*

3.1.3. A transcendental equation for the eigenvalues. The general solution of the system (3.1.5), (3.1.6) is a linear combination of the vectors

(3.1.15)
$$\begin{pmatrix} \cos(1+\lambda)\varphi \\ -\sin(1+\lambda)\varphi \end{pmatrix}, \begin{pmatrix} \sin(1+\lambda)\varphi \\ \cos(1+\lambda)\varphi \end{pmatrix}, \begin{pmatrix} A\cos(1-\lambda)\varphi \\ -B\sin(1-\lambda)\varphi \end{pmatrix}, \begin{pmatrix} A\sin(1-\lambda)\varphi \\ B\cos(1-\lambda)\varphi \end{pmatrix},$$

where $A = 3 - 4\nu - \lambda$ and $B = 3 - 4\nu + \lambda$. The vectors (3.1.15) are linearly dependent for $\lambda = 0$. Therefore, it is more convenient to represent the general solution in the form

(3.1.16)
$$u = \sum_{k=1}^{4} c_k\, u^{(k)},$$

where

$$u^{(1)} = \begin{pmatrix} u_r^{(1)} \\ u_\varphi^{(1)} \end{pmatrix} = \begin{pmatrix} \cos(1+\lambda)\varphi \\ -\sin(1+\lambda)\varphi \end{pmatrix},$$

$$u^{(2)} = \begin{pmatrix} u_r^{(2)} \\ u_\varphi^{(2)} \end{pmatrix} = \begin{pmatrix} \sin(1+\lambda)\varphi \\ \cos(1+\lambda)\varphi \end{pmatrix},$$

$$u^{(3)} = \begin{pmatrix} u_r^{(3)} \\ u_\varphi^{(3)} \end{pmatrix} = \frac{1}{\lambda}\begin{pmatrix} A\cos(1-\lambda)\varphi - (3-4\nu)\cos(1+\lambda)\varphi \\ -B\sin(1-\lambda)\varphi + (3-4\nu)\sin(1+\lambda)\varphi \end{pmatrix},$$

$$u^{(4)} = \begin{pmatrix} u_r^{(4)} \\ u_\varphi^{(4)} \end{pmatrix} = \frac{1}{\lambda}\begin{pmatrix} A\sin(1-\lambda)\varphi - (3-4\nu)\sin(1+\lambda)\varphi \\ B\cos(1-\lambda)\varphi - (3-4\nu)\cos(1+\lambda)\varphi \end{pmatrix}.$$

For $\lambda = 0$ we have

$$u^{(3)} = \begin{pmatrix} u_r^{(3)} \\ u_\varphi^{(3)} \end{pmatrix} = \begin{pmatrix} -\cos\varphi + 2(3-4\nu)\,\varphi\sin\varphi \\ -\sin\varphi + 2(3-4\nu)\,\varphi\cos\varphi \end{pmatrix},$$

$$u^{(4)} = \begin{pmatrix} u_r^{(4)} \\ u_\varphi^{(4)} \end{pmatrix} = \begin{pmatrix} -\sin\varphi - 2(3-4\nu)\,\varphi\cos\varphi \\ \cos\varphi + 2(3-4\nu)\,\varphi\sin\varphi \end{pmatrix}.$$

The boundary condition (3.1.2) yields the following algebraic system for the constants $c_1, \ldots, c_4$:

$$(3.1.17) \qquad \sum_{k=1}^{4} c_k\, u^{(k)}(\lambda, \pm\alpha/2) = 0$$

Since the components $u_r^{(1)}$, $u_\varphi^{(2)}$, $u_r^{(3)}$, $u_\varphi^{(4)}$ are even functions of the variable φ and the components $u_\varphi^{(1)}$, $u_r^{(2)}$, $u_\varphi^{(3)}$, $u_r^{(4)}$ are odd functions, this algebraic system is equivalent to the following two algebraic systems

$$(3.1.18) \qquad c_1 u^{(1)}(\lambda, \alpha/2) + c_3\, u^{(3)}(\lambda, \alpha/2) = 0,$$

$$(3.1.19) \qquad c_2 u^{(2)}(\lambda, \alpha/2) + c_4\, u^{(4)}(\lambda, \alpha/2) = 0.$$

This means,

$$(3.1.20) \quad \left\{ \begin{array}{rcl} (c_1 - \frac{3-4\nu}{\lambda} c_3)\cos(1+\lambda)\alpha/2 + \frac{1}{\lambda} c_3 A \cos(1-\lambda)\alpha/2 &=& 0, \\[2mm] (c_1 - \frac{3-4\nu}{\lambda} c_3)\sin(1+\lambda)\alpha/2 + \frac{1}{\lambda} c_3 B \sin(1-\lambda)\alpha/2 &=& 0, \end{array} \right.$$

and

$$(3.1.21) \quad \left\{ \begin{array}{rcl} (c_2 - \frac{3-4\nu}{\lambda} c_4)\sin(1+\lambda)\alpha/2 + \frac{1}{\lambda} c_4 A \sin(1-\lambda)\alpha/2 &=& 0, \\[2mm] (c_2 - \frac{3-4\nu}{\lambda} c_4)\cos(1+\lambda)\alpha/2 + \frac{1}{\lambda} c_4 B \cos(1-\lambda)\alpha/2 &=& 0. \end{array} \right.$$

The determinants of these systems are equal to

$$d_-(\lambda) = \lambda^{-1}\Big((3-4\nu)\sin\lambda\alpha - \lambda\sin\alpha\Big),$$

$$d_+(\lambda) = \lambda^{-1}\Big((3-4\nu)\sin\lambda\alpha + \lambda\sin\alpha\Big).$$

For $\lambda = 0$ these determinants are equal to $(3-4\nu)\alpha \mp \sin\alpha$ and do not vanish. Therefore, every eigenvalue of the pencil $\mathfrak{A}$ is a solution of one of the following equations:

$$(3.1.22) \qquad (3-4\nu)\sin\lambda\alpha - \lambda\sin\alpha = 0,$$

$$(3.1.23) \qquad (3-4\nu)\sin\lambda\alpha + \lambda\sin\alpha = 0.$$

The graphs of the real parts of the roots of equations (3.1.22), (3.1.23) are represented in Figures 4 and 5 below, the thick lines correspond to real eigenvalues and the thin lines to nonreal eigenvalues, multiple eigenvalues are indicated by $\otimes$).

It can be easily seen that the numbers λ, $-\lambda$, $\overline{\lambda}$, and $-\overline{\lambda}$ are simultaneously eigenvalues.

Let λ be a solution of the equation $d_-(\lambda) = 0$. Then every solution of (3.1.20) has the form

$$(3.1.24) \quad c_1 = A_1\,\lambda^{-1}\Big((3-4\nu)\cos(1+\lambda)\alpha/2 - A\cos(1-\lambda)\alpha/2\Big)$$
$$+ A_2\,\lambda^{-1}\Big((3-4\nu)\sin(1+\lambda)\alpha/2 - B\sin(1-\lambda)\alpha/2\Big)$$

$$(3.1.25) \quad c_3 = A_1\cos(1+\lambda)\alpha/2 + A_2\sin(1+\lambda)\alpha/2.$$

If we set

$$(3.1.26) \qquad A_1 = -\lambda\cos(1+\lambda)\alpha/2, \qquad A_2 = -\lambda\sin(1+\lambda)\alpha/2,$$

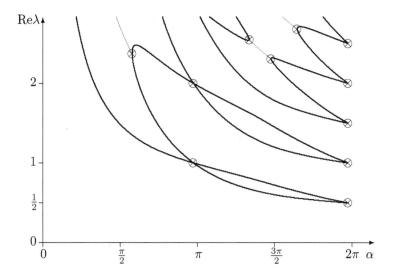

FIGURE 4. Real parts of eigenvalues of the pencil generated by the Dirichlet problem for $\nu = 0.17$

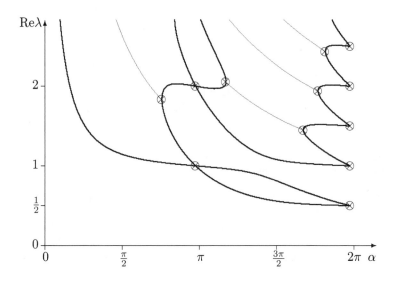

FIGURE 5. Real parts of eigenvalues in the case $\nu = 0.45$

we obtain

$$(3.1.27) \qquad c_1 = (3 - 4\nu)(\cos \lambda \alpha - 1) - \lambda \cos \alpha, \qquad c_3 = -\lambda.$$

Thus, for an eigenvalue of the operator pencil $\mathfrak{A}$ satisfying the equality $d_-(\lambda) = 0$ we get the eigenvector

$$(3.1.28) \begin{pmatrix} u_r^- \\ u_\varphi^- \end{pmatrix} = c_1 u^{(1)} + c_3 u^{(3)}$$

$$= \begin{pmatrix} -A \cos(1 - \lambda)\varphi \\ B \sin(1 - \lambda)\varphi \end{pmatrix} + ((3 - 4\nu) \cos \lambda \alpha - \lambda \cos \alpha) \begin{pmatrix} \cos(1 + \lambda)\varphi \\ -\sin(1 + \lambda)\varphi \end{pmatrix}.$$

If λ satisfies the equality $d_+(\lambda) = 0$, then, in a similar way, we obtain the eigenvector

$$(3.1.29) \quad \begin{pmatrix} u_r^+ \\ u_\varphi^+ \end{pmatrix} = -\begin{pmatrix} A\sin(1-\lambda)\varphi \\ B\cos(1-\lambda)\varphi \end{pmatrix}$$
$$+ \left((3-4\nu)\cos\lambda\alpha + \lambda\cos\alpha\right)\begin{pmatrix} \sin(1+\lambda)\varphi \\ \cos(1+\lambda)\varphi \end{pmatrix}.$$

Since each of the systems (3.1.18) and (3.1.19) has at most one linearly independent solution, we have obtained all eigenvectors in this way.

3.1.4. Existence of generalized eigenvectors. Obviously, the operator

$$\mathcal{P}(\lambda) : \mathbb{C}^4 \ni (c_1, c_2, c_3, c_4) \to \sum_{k=1}^{4} c_k\, u^{(k)}(\lambda, \varphi) \in \ker \mathcal{L}(\lambda).$$

satisfies condition (3.1.9) for all $\lambda \in \mathbb{C}$. Consequently, by Lemma 3.1.1, every eigenvalue of the pencil $\mathfrak{A}$ is also an eigenvalue of the pencil $\mathcal{BP}$ with the same geometric and algebraic multiplicities. Here $\mathcal{BP}(\lambda)$ is a linear mapping $\mathbb{C}^4 \to \mathbb{C}^4$. Hence the algebraic multiplicity of any eigenvalue λ_0 coincides with the multiplicity of the zero $\lambda = \lambda_0$ of the function $\det \mathcal{BP}(\lambda)$ (see Lemma 3.1.2) and, therefore, also with the multiplicity of the zero $\lambda = \lambda_0$ of the function $d_+(\lambda)\, d_-(\lambda)$. Thus, generalized eigenvectors corresponding to the eigenvalue λ_0 may only exist if

$$d_+(\lambda_0)\, d_-(\lambda_0) = 0 \quad \text{and} \quad d'_+(\lambda_0)\, d_-(\lambda_0) + d_+(\lambda_0)\, d'_-(\lambda_0),$$

i.e., if one of the conditions $d_+(\lambda_0) = d_-(\lambda_0) = 0$, $d_+(\lambda_0) = d'_+(\lambda_0) = 0$, and $d_-(\lambda_0) = d'_-(\lambda_0) = 0$ is satisfied. Obviously, only in the cases $\alpha = \pi$ and $\alpha = 2\pi$ the number λ_0 can be a common root of the equations $d_+(\lambda) = 0$ and $d_-(\lambda) = 0$. If $\alpha = \pi$, then the spectrum of the pencil $\mathfrak{A}$ consists of the eigenvalues $\pm 1, \pm 2, \dots$, and in the case $\alpha = 2\pi$ the eigenvalues are the numbers $\pm k/2$ $(k = 1, 2, \dots)$. In both cases two eigenvectors (3.1.28), (3.1.29) correspond to every eigenvalue, while generalized eigenvectors do not exist.

LEMMA 3.1.3. 1) *There are no generalized eigenvectors which correspond to nonreal eigenvalues.*

2) *The algebraic multiplicity of an arbitrary eigenvalue is not greater than two.*

Proof: 1) Let $d_-(\lambda_0) = d'_-(\lambda_0) = 0$ or $d_+(\lambda_0) = d'_+(\lambda_0) = 0$. Then one of the equalities

$$(3.1.30) \qquad\qquad (3 - 4\nu)\,\alpha\,\cos\lambda_0\alpha = \pm\sin\alpha$$

is satisfied. Hence, $\cos\lambda_0\alpha$ is real. Thus, either $\sin(\operatorname{Re}\lambda_0\alpha) = 0$ or $\sinh(\operatorname{Im}\lambda_0\alpha) = 0$. The equality $\sin(\operatorname{Re}\lambda_0\alpha) = 0$ can be excluded. Otherwise, we get $\cos\lambda_0\alpha = \pm\cosh(\operatorname{Im}\lambda_0\alpha)$ and, consequently, $(3-4\nu)\alpha\,|\cos\lambda_0\alpha| > \sin\alpha$. The last inequality contradicts (3.1.30). Consequently, we obtain $\sinh(\operatorname{Im}\lambda_0\alpha) = 0$, i.e., $\operatorname{Im}\lambda_0 = 0$.

2) Suppose that $d_-(\lambda_0) = d'_-(\lambda_0) = d''_-(\lambda_0) = 0$. Since

$$d''_-(\lambda_0) = -(3-4\nu)\alpha^2\,\sin\lambda_0\alpha,$$

we get $\sin\lambda_0\alpha = 0$ and the equality $(3-4\nu)\sin\lambda_0\alpha = \lambda_0\sin\alpha$ yields $\sin\alpha = 0$. This contradicts the equality $d'_-(\lambda_0) = 0$. In the same way, it can be shown that λ_0 cannot satisfy the equalities $d_+(\lambda_0) = d'_+(\lambda_0) = d''_+(\lambda_0) = 0$. This proves the lemma. ∎

LEMMA 3.1.4. *If λ_0 is a multiple root of one of the equations $d_-(\lambda) = 0$ or $d_+(\lambda) = 0$, then*

$$(3.1.31) \qquad \begin{pmatrix} v_r^{(\mp)}(\lambda_0, \varphi) \\ v_\varphi^{(\mp)}(\lambda_0, \varphi) \end{pmatrix} = \partial_\lambda \begin{pmatrix} u_r^{(\mp)}(\lambda, \varphi) \\ u_\varphi^{(\mp)}(\lambda, \varphi) \end{pmatrix}\Bigg|_{\lambda=\lambda_0}$$

is a generalized eigenvector associated with the eigenvectors (3.1.28) and (3.1.29), respectively.

Proof: Let $c_1 = c_1(\lambda)$, $c_3 = c_3(\lambda)$ be the functions (3.1.24) and (3.1.25), respectively. We show that

$$(3.1.32) \quad c_1(\lambda)\, u^{(1)}(\lambda, \alpha/2) + c_3(\lambda)\, u^{(3)}(\lambda, \alpha/2) = \lambda\, d_-(\lambda) \begin{pmatrix} \cos(1 + \lambda)\alpha/2 \\ -\sin(1 + \lambda)\alpha/2 \end{pmatrix}$$

for all $\lambda \in \mathbb{C}$. Indeed, it can be easily seen that

$$u_\varphi^{(3)}(\lambda, \alpha/2)\, u_r^{(1)}(\lambda, \alpha/2) - u_\varphi^{(1)}(\lambda, \alpha/2)\, u_r^{(3)}(\lambda, \alpha/2) = \begin{pmatrix} 0 \\ d_-(\lambda) \end{pmatrix},$$

$$u_r^{(3)}(\lambda, \alpha/2)\, u_\varphi^{(1)}(\lambda, \alpha/2) - u_r^{(1)}(\lambda, \alpha/2)\, u_\varphi^{(3)}(\lambda, \alpha/2) = \begin{pmatrix} d_-(\lambda) \\ 0 \end{pmatrix}.$$

This together with (3.1.24)–(3.1.26) implies (3.1.32).

Since $u_r^{(1)}$, $u_r^{(3)}$ are even and $u_\varphi^{(1)}$, $u_\varphi^{(3)}$ are odd functions, it holds

$$(3.1.33)$$

$$c_1(\lambda)\, u^{(1)}(\lambda, -\alpha/2) + c_3(\lambda)\, u^{(3)}(\lambda, -\alpha/2) = \lambda\, d_-(\lambda) \begin{pmatrix} \cos(1 + \lambda)\alpha/2 \\ \sin(1 + \lambda)\alpha/2 \end{pmatrix}.$$

Using (3.1.5), (3.1.6), (3.1.32), and (3.1.33), we arrive at

$$(3.1.34) \qquad \mathfrak{A}(\lambda) \begin{pmatrix} u_r^-(\lambda, \varphi) \\ u_\varphi^-(\lambda, \varphi) \end{pmatrix} = \lambda\, d_-(\lambda) \left\{ 0, \begin{pmatrix} \cos(1 + \lambda)\alpha/2 \\ \mp\sin(1 + \lambda)\alpha/2 \end{pmatrix} \right\}.$$

Now let $\lambda = \lambda_0$ be a multiple root of the equation $d_-(\lambda) = 0$. Differentiating (3.1.34) with respect to λ and setting $\lambda = \lambda_0$, we obtain

$$\mathfrak{A}'(\lambda_0) \begin{pmatrix} u_r^-(\lambda_0, \varphi) \\ u_\varphi^-(\lambda_0, \varphi) \end{pmatrix} + \mathfrak{A}(\lambda_0) \begin{pmatrix} v_r^-(\lambda_0, \varphi) \\ v_\varphi^-(\lambda_0, \varphi) \end{pmatrix} = 0.$$

This proves the assertion in the case "$-$". The proof for the case "$+$" proceeds analogously. ∎

Obviously, a multiple root of one of the equations $d_\mp(\lambda) = 0$ satisfies the equalities

$$(3 - 4\nu)\sin\lambda\alpha \mp \lambda\sin\alpha = 0,$$

$$(3 - 4\nu)\alpha\cos\lambda\alpha \mp \sin\alpha = 0.$$

Consequently, $\tan\lambda\alpha = \lambda\alpha$ and $\dfrac{\sin\alpha}{\alpha} = \pm(3 - 4\nu)(\lambda^2\alpha^2 + 1)^{-1/2}$. Therefore, the sequence of the roots x_n of the equation $\tan x = x$ generates a countable set of angles α satisfying the equality $\alpha^{-1}\sin\alpha = \pm(3 - 4\nu)(x_n^2 + 1)^{-1/2}$. This set has the accumulation points π and 2π. The corresponding eigenvalues are determined by the formula

$$(3.1.35) \qquad \lambda = \pm\left(\frac{(3 - 4\nu)^2}{\sin^2\alpha} - \frac{1}{\alpha^2} \right)^{1/2}.$$

3.1.5. Distribution of the eigenvalues in the half-plane $\operatorname{Re}\lambda > 0$.

THEOREM 3.1.1. *Let $\alpha \neq \pi$, $\alpha \neq 2\pi$.*

1) *The lines $\operatorname{Re}\lambda = k\pi/\alpha$ $(k = 0, 1, \ldots)$ contain no points of the spectrum of the pencil $\mathfrak{A}$.*

2) *If k is an odd number, $k \geq 1$, then the strip*

$$(3.1.36) \qquad k\frac{\pi}{\alpha} < \operatorname{Re}\lambda < (k+1)\frac{\pi}{\alpha}$$

contains two eigenvalues satisfying (3.1.23) for $\alpha < \pi$ and (3.1.22) for $\alpha > \pi$.

3) *If k is an even number, $k \geq 2$, then the strip (3.1.36) contains two eigenvalues of the pencil $\mathfrak{A}$ satisfying (3.1.22) for $\alpha < \pi$ and (3.1.23) for $\alpha > \pi$.*

4) *The strip*

$$(3.1.37) \qquad 0 < \operatorname{Re}\lambda < \frac{\pi}{\alpha}$$

contains exactly one eigenvalue which satisfies (3.1.22) for $\alpha < \pi$ and (3.1.23) for $\alpha > \pi$.

Proof: We consider the roots z of the equations

$$(3.1.38) \qquad \frac{\sin z}{z} - \frac{\sin \alpha}{(3 - 4\nu)\alpha} = 0,$$

$$(3.1.39) \qquad \frac{\sin z}{z} + \frac{\sin \alpha}{(3 - 4\nu)\alpha} = 0$$

which correspond to the equations $d_\mp(\lambda) = 0$ for $\lambda = z/\alpha$. Suppose that $z = k\pi + iy$, where $y \in \mathbb{R}$, $k = 1, 2, \ldots$, is a root of one of the equations (3.1.38), (3.1.39). Then we obtain

$$(-1)^k i\,(3 - 4\nu)\,\alpha \sinh y \mp (k\pi + iy) \sin \alpha = 0.$$

This is impossible, since $\alpha \neq \pi$ and $\alpha \neq 2\pi$. If $z = iy$, then we get the equation

$$\frac{\sinh y}{y} = \pm\frac{\sin \alpha}{(3 - 4\nu)\alpha}$$

which is also unsolvable, since the left-hand side is greater or equal to one and the right-hand side is less than one. Consequently, the lines $\operatorname{Re}\lambda = k\pi/\alpha$, $k = 0, 1, \ldots$, do not contain points of the spectrum.

From the obvious inequality

$$\left|\frac{\sin z}{z}\right| \geq \frac{|\sinh(\operatorname{Im} z)|}{|z|}$$

it follows that each of the equations (3.1.38) and (3.1.39) has no solutions on the set $\{z : |\sinh(\operatorname{Im} z)| > |z|\}$. Hence in an arbitrary strip $|\operatorname{Re} z| < a$ the modulus of the roots of (3.1.38), (3.1.39) is uniformly bounded (with respect to $\alpha \in (0, 2\pi]$) by a constant only depending on a.

Furthermore, in a standard way, it can be verified that for angles α sufficiently near to π every of the equations (3.1.37), (3.1.38) has a unique real-analytic solution

$$z_\mp(\alpha) = k\pi \mp \frac{(-1)^k k}{3 - 4\nu}(\alpha - \pi) + \cdots$$

in a small neighborhood of the point $z = k\pi$, $k = 1, 2, \ldots$. Therefore, for angles α near to π, the roots of the equations (3.1.38), (3.1.39) are distributed as follows: If k is an odd number, $k \geq 1$, then the strip $k\pi < \operatorname{Re} z < (k+1)\pi$ contains two roots of (3.1.39) for $\alpha < \pi$ and two roots of (3.1.38) for $\alpha > \pi$. If k is an even number, $k \geq 2$, then this strip contains two roots of (3.1.38) for $\alpha < \pi$ and two roots of (3.1.39) for $\alpha > \pi$. Moreover, one root of the equation (3.1.38) is contained in the strip $0 < \operatorname{Re} z < \pi$ if $\alpha < \pi$ and one root of the equation (3.1.39) is contained in this strip if $\alpha > \pi$. Since there are no roots of the equations (3.1.38), (3.1.39) on the line $\operatorname{Re} z = k\pi$, it follows from Rouché's theorem that the number of the roots of every of these equations in the strips (3.1.36) and (3.1.37) is independent of the angle α. This proves the theorem. ∎

Comparing the graphs of the functions $y = x^{-1}\sin x$ and $y = (3-4\nu)^{-1}x^{-1}\sin x$, we can conclude that the equation (3.1.38) has no real solutions in the interval $(0, \pi]$ for $\alpha \in (\pi, 2\pi)$ and that it has a unique solution $x_-(\alpha)$ in this interval for every $\alpha \in (0, \pi]$. Obviously,

$$x_-(\pi) = \pi, \qquad x_-(0) = c_0,$$

where $c_0 \in (0, \pi)$ denotes the root of the equation $\sin c = c/(3 - 4\nu)$. Furthermore, it can be easily seen that the root $z = x_-(\alpha)$ is simple. Consequently, the function $\alpha \to x_-(\alpha)$ is real-analytic in the interval $(0, \pi]$. This function admits an analytic extension to a neighborhood of the point π.

Analogously, the equation (3.1.39) has no real solutions in the interval $(0, \pi]$ if $\alpha \in (0, \pi)$ and a unique solution $x_+(\alpha)$ in this interval for every $\alpha \in [\pi, 2\pi]$, where

$$x_+(\pi) = x_+(2\pi) = \pi.$$

The function $\alpha \to x_+(\alpha)$ is real-analytic in $[\pi, 2\pi]$ and admits an analytic extension to a neighborhood of the point π.

We consider the real roots of equation (3.1.38) which are situated in the interval $(\pi, 2\pi)$ for $\alpha \in (\pi, 2\pi)$. It can be easily verified that this equation has exactly two roots $x_-^{(1)}(\alpha)$, $x_-^{(2)}(\alpha)$, $x_-^{(1)}(\alpha) > x_-^{(2)}(\alpha)$, in the interval $(\pi, 2\pi)$. These roots are simple and have analytic extensions to certain neighborhoods of the points $\alpha = \pi$ and $\alpha = 2\pi$. Moreover,

$$x_-^{(1)}(\pi) = x_-^{(1)}(2\pi) = 2\pi, \qquad x_-^{(2)}(\pi) = x_-^{(2)}(2\pi) = \pi.$$

The function $x_-^{(2)}$ is the analytic extension of the function x_- to the interval $(\pi, 2\pi)$. Therefore, in the sequel, we denote the function $x_-^{(2)}$ by x_-. Note that for every $\alpha \in (\pi, 2\pi)$ the inequalities

$$0 < x_+(\alpha) < \pi < x_-(\alpha) < x_-^{(1)}(\alpha) < 2\pi$$

are valid.

The above introduced functions $x_\mp$ and $x_-^{(1)}$ are used in the following theorem for the description of the spectrum of the operator pencil $\mathfrak{A}$ near the line $\operatorname{Re} \lambda = 0$.

THEOREM 3.1.2. 1) If $\alpha \in (0, \pi)$, then the spectrum of the pencil $\mathfrak{A}$ in the strip $0 \leq \operatorname{Re} \lambda \leq \pi/\alpha$ consists of the unique simple eigenvalue $\lambda_-(\alpha) = x_-(\alpha)/\alpha$.

2) Let $\alpha \in (\pi, 2\pi)$. Then the spectrum of $\mathfrak{A}$ in the strip $0 \leq \operatorname{Re} \lambda \leq 2\pi/\alpha$ consists of the following three real eigenvalues :

$$\lambda_+(\alpha) = \frac{x_+(\alpha)}{\alpha}, \quad \lambda_-(\alpha) = \frac{x_-(\alpha)}{\alpha}, \quad \lambda_-^{(1)}(\alpha) = \frac{x_-^{(1)}(\alpha)}{\alpha}.$$

3) *The function λ_- is strictly decreasing on the interval $(0, 2\pi]$, while λ_+ and $\lambda_-^{(1)}$ are strictly decreasing functions on $(\pi, 2\pi]$. Furthermore, the inequalities*

(3.1.40)
$$\begin{cases} \lambda_-(\alpha) > 1 \ \text{ for } \alpha \in (0, \pi), \\ 1/2 < \lambda_+(\alpha) < \lambda_-(\alpha) < 1 < \lambda_-^{(1)}(\alpha) \ \text{ for } \alpha \in (\pi, 2\pi) \end{cases}$$

are satisfied.

Proof: Assertions 1) and 2) are immediate consequences of Theorem 3.1.1 and of the definition of the functions x_-, x_+, $x_-^{(1)}$. We show that λ_-, λ_+, $\lambda_-^{(1)}$ are strictly decreasing functions. From the definition of λ_- it follows that

$$\lambda_-'(\alpha) = -\lambda_- \frac{(3 - 4\nu) \cos \lambda_- \alpha - \cos \alpha}{(3 - 4\nu) \alpha \cos \lambda_- \alpha - \sin \alpha}.$$

Since $\lambda_-(\pi) = 1$, $\lambda_-(2\pi) = 1/2$, it suffices to verify the inequality $\lambda_-'(\alpha) \neq 0$ for $\alpha \in (0, 2\pi)$. Suppose that $\lambda_-'(\alpha_0) = 0$. Then for $\lambda = \lambda_-(\alpha_0)$ we get

(3.1.41) $(3 - 4\nu) \sin \lambda\alpha_0 - \lambda \sin \alpha_0 = 0$, $(3 - 4\nu) \cos \lambda\alpha_0 - \cos \alpha_0 = 0$.

Hence

(3.1.42)
$$\frac{\tan \lambda\alpha_0}{\lambda\alpha_0} = \frac{\tan \alpha_0}{\alpha_0}.$$

By the definition of the function λ_-, we have $0 < \lambda\alpha_0 < \pi$ for $0 < \alpha_0 < \pi$ and $\pi < \lambda\alpha_0 < 2\pi$ for $\pi < \alpha_0 < 2\pi$. In both cases equation (3.1.42) has the unique solution $\lambda = 1$ which does not satisfy (3.1.41).

The inequality $(\lambda_-^{(1)})'(\alpha) \neq 0$ for $\alpha \in (\pi, 2\pi)$ can be proved in a similar manner.

We show that $\lambda_+'(\alpha) \neq 0$ for $\alpha \in (\pi, 2\pi)$. If $\lambda_+'(\alpha_0) = 0$, then for $\lambda = \lambda_+(\alpha_0)$ we obtain

(3.1.43) $(3 - 4\nu) \sin \lambda\alpha_0 + \lambda \sin \alpha_0 = 0$, $(3 - 4\nu) \cos \lambda\alpha_0 + \cos \alpha_0 = 0$.

As above, this implies (3.1.42). Since $\lambda\alpha_0 \leq \pi$, $\pi < \alpha_0 < 2\pi$, we have $\lambda \leq 1$. However, from (3.1.43) it follows that $(3-4\nu)^2 = \lambda^2 \sin^2 \alpha_0 + \cos^2 \alpha_0$ and, therefore, $\lambda \geq 3 - 4\nu > 1$. Consequently, the function λ_+ is strictly decreasing on $(\pi, 2\pi)$.

The inequalities (3.1.40) follow from the corresponding estimates for $x_\pm(\alpha)$ and $x_-^{(1)}(\alpha)$. ∎

3.2. The operator pencil generated by the Dirichlet problem in a cone

Let $\mathcal{K}$ be the cone $\{x \in \mathbb{R}^3 : 0 < |x| < \infty, \ x/|x| \in \Omega\}$, where Ω is a domain on the unit sphere. We consider the Lamé system for the displacement vector $U = (U_1, U_2, U_3)$

(3.2.1) $\Delta U + (1 - 2\nu)^{-1} \nabla \nabla \cdot U = 0$ in $\mathcal{K}$

with the Dirichlet boundary condition

(3.2.2) $U = 0$ on $\partial \mathcal{K} \backslash \{0\}$.

The description of all solutions of problem (3.2.1), (3.2.2) which have the form (3.0.2) with $u^{(k)} \in \overset{\circ}{W}_2^1(\Omega)^3$ requires the study of the spectral properties of a certain operator pencil $\mathcal{L}$ which is defined below. We prove in this section that the spectrum of this pencil consists of isolated points with finite algebraic multiplicities and that the numbers λ, $\overline{\lambda}$, $1 - \lambda$, and $1 - \overline{\lambda}$ are simultaneously eigenvalues or not.

Furthermore, we give a relation between the Jordan chains corresponding to the eigenvalues λ and $1 - \lambda$.

3.2.1. Sobolev spaces on a spherical domain.

In what follows, we will systematically employ the spherical coordinates r and $\omega = (\theta, \varphi)$ which are related to the Cartesian coordinates $x = (x_1, x_2, x_3)$ by

$$x_1 = r \sin\theta \cos\varphi, \qquad x_2 = r \sin\theta \sin\varphi, \qquad x_3 = r \cos\theta.$$

If (u_1, u_2, u_3) is a vector function in Cartesian coordinates, then its *spherical components* $(u_r, u_\theta, u_\varphi)$ are given by

$$\begin{pmatrix} u_r \\ u_\theta \\ u_\varphi \end{pmatrix} = J \begin{pmatrix} u_1 \\ u_2 \\ u_3 \end{pmatrix},$$

where

$$(3.2.3) \qquad J = J(\theta, \varphi) = \begin{pmatrix} \sin\theta \cos\varphi & \sin\theta \sin\varphi & \cos\theta \\ \cos\theta \cos\varphi & \cos\theta \sin\varphi & -\sin\theta \\ -\sin\varphi & \cos\varphi & 0 \end{pmatrix}.$$

Let us express the fact that $u = (u_1, u_2, u_3) \in W_2^1(\Omega)^3$ in terms of the spherical components of u.

We set $u_\omega = (u_\theta, u_\varphi)$ and denote by

$$\nabla_\omega \cdot u_\omega = \frac{1}{\sin\theta} \partial_\theta(\sin\theta\, u_\theta) + \frac{1}{\sin\theta} \partial_\varphi u_\varphi$$

the *spherical divergence* of u_ω and by

$$\nabla_\omega v = \begin{pmatrix} \partial_\theta v \\ (\sin\theta)^{-1} \partial_\varphi v \end{pmatrix}$$

the *spherical gradient* of the function v. Furthermore, we introduce the sesquilinear form

$$(3.2.4) \quad Q(u_\omega, v_\omega) = \int_\Omega \Big(\partial_\theta u_\theta \cdot \partial_\theta \overline{v}_\theta + \partial_\theta u_\varphi \cdot \partial_\theta \overline{v}_\varphi$$

$$+ \Big(\frac{1}{\sin\theta} \partial_\varphi u_\theta - \cot\theta\, u_\varphi \Big) \cdot \Big(\frac{1}{\sin\theta} \partial_\varphi \overline{v}_\theta - \cot\theta\, \overline{v}_\varphi \Big)$$

$$+ \Big(\frac{1}{\sin\theta} \partial_\varphi u_\varphi + \cot\theta\, u_\theta \Big) \cdot \Big(\frac{1}{\sin\theta} \partial_\varphi \overline{v}_\varphi + \cot\theta\, \overline{v}_\theta \Big) \Big)\, d\omega,$$

where $d\omega = \sin\theta\, d\theta\, d\varphi$. The space $h_2^1(\Omega)$ is defined as the set of all vector functions u_ω on Ω for which the quantity

$$\|u_\omega\|_{h_2^1(\Omega)} = \Big(Q(u_\omega, u_\omega) + \int_\Omega |u_\omega|^2\, d\omega \Big)^{1/2}$$

is finite.

LEMMA 3.2.1. *The Cartesian components of the vector function u belong to the space $W_2^1(\Omega)$ if and only if $u_r \in W_2^1(\Omega)$ and $u_\omega \in h_2^1(\Omega)$.*

Proof: The norm in $W_2^1(\Omega)$ is defined as

$$\left(\int_\Omega \left(|v|^2 + |\nabla_\omega v|^2 \right) d\omega \right)^{1/2}$$

(see Section 2.2). By a straightforward calculation, the following relations between the Cartesian components u_1, u_2, u_3 and the spherical components u_r, u_θ, u_φ of the vector function u can be verified:

$$\sum_{j=1}^{3} |\partial_\theta u_j|^2 = |\partial_\theta u_r|^2 + |\partial_\theta u_\theta|^2 + |\partial_\theta u_\varphi|^2 + |u_r|^2 + |u_\theta|^2 + 2\,\mathrm{Re}\,(u_r \partial_\theta \overline{u}_\theta - u_\theta \partial_\theta \overline{u}_r),$$

$$\sum_{j=1}^{3} |\partial_\varphi u_j|^2 = |\partial_\varphi u_r|^2 + |\partial_\varphi u_\theta|^2 + |\partial_\varphi u_\varphi|^2 + \sin^2\theta\,|u_r|^2 + \cos^2\theta\,|u_\theta|^2 + |u_\varphi|^2$$

$$+ 2\,\mathrm{Re}\,\Big(\sin\theta\,\cos\theta\,u_r\,\overline{u}_\theta + \sin\theta\,(u_r \partial_\varphi \overline{u}_\varphi - u_\varphi \partial_\varphi \overline{u}_r) \Big)$$

$$+ 2\cos\theta\,\mathrm{Re}\,\big(u_\theta \partial_\varphi \overline{u}_\varphi - u_\varphi \partial_\varphi \overline{u}_\theta \big).$$

From these formulas we obtain

$$(3.2.5) \qquad \sum_{j=1}^{3} |\nabla_\omega u_j|^2 = |\nabla_\omega u_r|^2 + |\partial_\theta u_\theta|^2 + |\partial_\theta u_\varphi|^2 + 2\,|u_r|^2 + |u_\omega|^2$$

$$+ \left| \frac{1}{\sin\theta}\,\partial_\varphi u_\theta - \cot\theta\,u_\varphi \right|^2 + \left| \frac{1}{\sin\theta}\,\partial_\varphi u_\varphi + \cot\theta\,u_\theta \right|^2$$

$$+ 2\,\mathrm{Re}\,(u_r\,\nabla_\omega \cdot \overline{u}_\omega - u_\omega \cdot \nabla_\omega \overline{u}_r),$$

where $|\cdot|$ is the Euclidean norm and $u \cdot v$ denotes the sum of products of the corresponding components of the vectors u and v. One checks directly that

$$(3.2.6) \qquad |\partial_\theta u_\theta|^2 + \left| \frac{1}{\sin\theta}\,\partial_\varphi u_\varphi + \cot\theta\,u_\theta \right|^2 \geq \frac{1}{2}\,|\nabla_\omega \cdot u_\omega|^2.$$

This and (3.2.5) imply the assertion of the lemma. ∎

We denote by $\overset{\circ}{h}{}_2^1(\Omega)$ the closure of $C_0^\infty(\Omega)^2$ in $h_2^1(\Omega)$. Then, as a consequence of Lemma 3.2.1, the following assertion holds.

COROLLARY 3.2.1. *The Cartesian components of the vector function u are in $\overset{\circ}{W}{}_2^1(\Omega)$ if and only if $u_r \in \overset{\circ}{W}{}_2^1(\Omega)$ and $u_\omega \in \overset{\circ}{h}{}_2^1(\Omega)$.*

Furthermore, due to the compactness of the imbedding $\overset{\circ}{W}{}_2^1(\Omega) \subset L_2(\Omega)$, we get the following result.

COROLLARY 3.2.2. *The imbedding $\overset{\circ}{h}{}_2^1(\Omega) \subset L_2(\Omega)^2$ is compact.*

3.2.2. Properties of the form Q.

LEMMA 3.2.2. *For all $u_\omega \in \overset{\circ}{h}{}_2^1(\Omega)$ there is the equality*

$$(3.2.7) \qquad Q(u_\omega, u_\omega) + \int_\Omega |u_\omega|^2\,d\omega$$

$$= \int_\Omega \left(|\nabla_\omega \cdot u_\omega|^2 + \left| \frac{1}{\sin\theta}\partial_\varphi u_\theta - \cot\theta\,u_\varphi - \partial_\theta u_\varphi \right|^2 \right) d\omega.$$

Proof: Applying the equalities

$$\int_{\Omega} |\nabla_{\omega} \cdot u_{\omega}|^2 \, d\omega \;=\; \int_{\Omega} \left(|\partial_{\theta} u_{\theta}|^2 + \left| \cot\theta\, u_{\theta} + \frac{1}{\sin\theta} \partial_{\varphi} u_{\varphi} \right|^2 \right) d\omega$$

$$+\, 2\,\mathrm{Re} \int_{\Omega} \partial_{\theta} u_{\theta} \left(\cot\theta\, \overline{u}_{\theta} + \frac{1}{\sin\theta} \partial_{\varphi} \overline{u}_{\varphi} \right) d\omega,$$

$$\int_{\Omega} \left| \frac{1}{\sin\theta} \partial_{\varphi} u_{\theta} - \cot\theta\, u_{\varphi} - \partial_{\theta} u_{\varphi} \right|^2 d\omega = \int_{\Omega} \left(|\partial_{\theta} u_{\varphi}|^2 + \left| \frac{1}{\sin\theta} \partial_{\varphi} u_{\theta} - \cot\theta\, u_{\varphi} \right|^2 \right) d\omega$$

$$-\, 2\,\mathrm{Re} \int_{\Omega} \partial_{\theta} u_{\varphi} \left(\frac{1}{\sin\theta} \partial_{\varphi} \overline{u}_{\theta} - \cot\theta\, \overline{u}_{\varphi} \right) d\omega$$

and making use of the identities

$$(3.2.8) \qquad 2\,\mathrm{Re} \int_{\Omega} \left(\partial_{\theta} u_{\theta} \cdot \overline{u}_{\theta} + \partial_{\theta} u_{\varphi} \cdot \overline{u}_{\varphi} \right) \cot\theta \, d\omega = \int_{\Omega} |u_{\omega}|^2 \, d\omega,$$

$$(3.2.9) \qquad \mathrm{Re} \int_{\Omega} \left(\partial_{\theta} u_{\theta} \cdot \partial_{\varphi} \overline{u}_{\varphi} - \partial_{\theta} u_{\varphi} \cdot \partial_{\varphi} \overline{u}_{\theta} \right) (\sin\theta)^{-1} \, d\omega = 0$$

which are verified by integration by parts, we arrive at (3.2.7). ∎

As a consequence of (3.2.7), we have

$$(3.2.10) \qquad Q(u_{\omega}, u_{\omega}) + \int_{\Omega} |u_{\omega}|^2 \, d\omega \geq \int_{\Omega} |\nabla_{\omega} \cdot u_{\omega}|^2 \, d\omega$$

for all $u_{\omega} \in \overset{\circ}{h}{}^1_2(\Omega)$.

LEMMA 3.2.3. *There is the inequality*

$$(3.2.11) \qquad Q(u_{\omega}, u_{\omega}) \geq \int_{\Omega} |u_{\omega}|^2 \, d\omega \qquad \text{for all } u_{\omega} \in \overset{\circ}{h}{}^1_2(\Omega).$$

Proof: First we establish the identity

$$(3.2.12) \quad Q(u_{\omega}, u_{\omega}) - \int_{\Omega} |u_{\omega}|^2 \, d\omega = \int_{\Omega} \left(2|\partial_{\theta} u_{\theta}|^2 + 2 \left| \frac{1}{\sin\theta} \partial_{\varphi} u_{\varphi} + \cot\theta\, u_{\theta} \right|^2 \right.$$

$$\left. +\, \left| \partial_{\theta} u_{\varphi} + \frac{1}{\sin\theta} \partial_{\varphi} u_{\varphi} - \cot\theta\, u_{\varphi} \right|^2 - |\nabla_{\omega} \cdot u_{\omega}|^2 \right) d\omega$$

for any vector function $u_\omega \in \overset{\circ}{h}{}_2^1(\Omega)$. In order to show this, we write the right-hand side of (3.2.12) in the form

$$Q(u_\omega, u_\omega) + \int_\Omega \left(|\partial_\theta u_\theta|^2 + \left| \frac{1}{\sin\theta} \partial_\varphi u_\varphi + \cot\theta\, u_\theta \right|^2 \right.$$

$$+ 2\,\mathrm{Re}\, \partial_\theta u_\varphi \left(\frac{1}{\sin\theta} \partial_\varphi \bar{u}_\theta - \cot\theta\, \bar{u}_\varphi \right) - |\nabla_\omega \cdot u_\omega|^2 \bigg)\, d\omega$$

$$= Q(u_\omega, u_\omega) + 2\,\mathrm{Re} \int_\Omega \left(\partial_\theta u_\varphi \left(\frac{1}{\sin\theta} \partial_\varphi \bar{u}_\theta - \cot\theta\, \bar{u}_\varphi \right) \right.$$

$$- \partial_\theta u_\theta \left(\frac{1}{\sin\theta} \partial_\varphi \bar{u}_\varphi + \cot\theta\, \bar{u}_\theta \right) \bigg)\, d\omega.$$

Transforming the last integral by means of (3.2.8) and (3.2.9), we arrive at (3.2.12). Since

$$2|\partial_\theta u_\theta|^2 + 2\left| \frac{1}{\sin\theta} \partial_\varphi u_\varphi + \cot\theta\, u_\theta \right|^2 - |\nabla_\omega \cdot u_\omega|^2 = \left| \partial_\theta u_\theta - \frac{1}{\sin\theta} \partial_\varphi u_\varphi - \cot\theta\, u_\theta \right|^2,$$

we obtain from (3.2.12) that

$$(3.2.13) \quad Q(u_\omega, u_\omega) - \int_\Omega |u_\omega|^2\, d\omega = \int_\Omega \left(\left| \partial_\theta u_\varphi + \frac{1}{\sin\theta} \partial_\varphi u_\theta - \cot\theta\, u_\varphi \right|^2 \right.$$

$$+ \left| \partial_\theta u_\theta - \frac{1}{\sin\theta} \partial_\varphi u_\varphi - \cot\theta\, u_\theta \right|^2 \bigg)\, d\omega.$$

This implies (3.2.11). ∎

3.2.3. Definition of the pencil $\mathcal{L}$.

The Lamé system in spherical coordinates has the following form (cf. Malvern's book [159, p.672]):

$$(3.2.14) \quad \Delta U_r - \frac{2}{r^2}(U_r + \nabla_\omega \cdot U_\omega) + \frac{1}{1-2\nu} \partial_r \left(\frac{1}{r^2} \partial_r (r^2 U_r) + \frac{1}{r} \nabla_\omega \cdot U_\omega \right) = 0,$$

$$(3.2.15) \quad \Delta U_\theta + \frac{2}{r^2} \left(\partial_\theta U_r - \frac{1}{2\sin^2\theta} U_\theta - \frac{\cot\theta}{\sin\theta} \partial_\varphi U_\varphi \right)$$

$$+ \frac{1}{(1-2\nu)r^2} \partial_\theta \left(\frac{1}{r} \partial_r (r^2 U_r) + \nabla_\omega \cdot U_\omega \right) = 0,$$

$$(3.2.16) \quad \Delta U_\varphi + \frac{2}{r^2 \sin\theta} \left(\partial_\varphi U_r + \cot\theta\, \partial_\varphi U_\theta - \frac{U_\varphi}{2\sin\theta} \right)$$

$$+ \frac{1}{(1-2\nu)r^2 \sin\theta} \partial_\varphi \left(\frac{1}{r} \partial_r (r^2 U_r) + \nabla_\omega \cdot U_\omega \right) = 0.$$

Taking $(U_r, U_\theta, U_\varphi) = r^\lambda (u_r, u_\theta, u_\varphi)$ and then multiplying (3.2.14)–(3.2.16) by $r^{2-\lambda}$, we arrive at the matrix differential operator

$$(3.2.17) \quad \mathcal{L}(\lambda) \begin{pmatrix} u_r \\ u_\omega \end{pmatrix} = \begin{pmatrix} -\delta u_r + \frac{2-2\nu}{1-2\nu}(1-\lambda)(\lambda+2)\, u_r + \frac{3-4\nu-\lambda}{1-2\nu} \nabla_\omega \cdot u_\omega \\ \\ -H_\nu u_\omega + (1-\lambda)(\lambda+2)\, u_\omega - \frac{4-4\nu+\lambda}{1-2\nu} \nabla_\omega u_r \end{pmatrix}.$$

Here δ denotes the Beltrami operator on the sphere S^2 and

$$H_\nu u_\omega = H u_\omega + (1-2\nu)^{-1} \nabla_\omega (\nabla_\omega \cdot u_\omega),$$

where

$$H = \begin{pmatrix} \delta + 2 - (\sin\theta)^{-2} & -2(\sin\theta)^{-1}\cot\theta\,\partial_\varphi \\ 2(\sin\theta)^{-1}\cot\theta\,\partial_\varphi & \delta + 2 - (\sin\theta)^{-2} \end{pmatrix}.$$

Note that $\delta v = \nabla_\omega \cdot (\nabla_\omega v)$ and that the matrix differential operator H satisfies the equalities

$$\begin{aligned} H\,\nabla_\omega v &= \nabla_\omega\,(\delta+2)\,v, \\ \nabla_\omega \cdot H\,u_\omega &= (\delta+2)\,\nabla_\omega \cdot u_\omega. \end{aligned}$$

LEMMA 3.2.4. *The operator*

$$(3.2.18) \qquad \mathcal{L}(\lambda) : \overset{\circ}{W}{}^1_2(\Omega) \times \overset{\circ}{h}{}^1_2(\Omega) \to \left(\overset{\circ}{W}{}^1_2(\Omega) \times \overset{\circ}{h}{}^1_2(\Omega) \right)^*$$

is continuous for every $\lambda \in \mathbb{C}$.

Proof: Let $u_r,\, v_r \in \overset{\circ}{W}{}^1_2(\Omega)$ and $u_\omega,\, v_\omega \in \overset{\circ}{h}{}^1_2(\Omega)$. Then

$$-\int_\Omega H u_\omega \cdot \overline{v}_\omega\, d\omega = \int_\Omega \left(\nabla_\omega u_\theta \cdot \nabla_\omega \overline{v}_\theta + \nabla_\omega u_\varphi \cdot \nabla_\omega \overline{v}_\varphi + (\cot^2\theta - 1)\, u_\omega \cdot \overline{v}_\omega \right.$$

$$\left. +2\,\frac{\cot\theta}{\sin\theta}\,(\partial_\varphi u_\varphi \cdot \overline{v}_\theta - \partial_\varphi u_\theta \cdot \overline{v}_\varphi) \right) d\omega$$

$$= Q(u_\omega, v_\omega) - \int_\Omega u_\omega \cdot \overline{v}_\omega\, d\omega.$$

Consequently,

$$(3.2.19) \qquad \int_\Omega \mathcal{L}(\lambda) \begin{pmatrix} u_r \\ u_\omega \end{pmatrix} \cdot \begin{pmatrix} \overline{v}_r \\ \overline{v}_\omega \end{pmatrix} d\omega = Q(u_\omega, u_\omega)$$

$$+ \int_\Omega \left(\nabla_\omega u_r \cdot \nabla_\omega \overline{v}_r + \frac{2 - 2\nu}{1 - 2\nu}\,(\lambda + 2)\,(1 - \lambda)\, u_r \overline{v}_r \right.$$

$$+ \left((\lambda+2)(1-\lambda) - 1 \right) u_\omega \cdot \overline{v}_\omega + \frac{1}{1 - 2\nu}(\nabla_\omega \cdot u_\omega)(\nabla_\omega \cdot \overline{v}_\omega)$$

$$\left. + \frac{3 - 4\nu - \lambda}{1 - 2\nu}(\nabla_\omega \cdot u_\omega)\,\overline{v}_r + \frac{4 - 4\nu + \lambda}{1 - 2\nu}\,u_r\,\nabla_\omega \cdot \overline{v}_\omega \right) d\omega.$$

This and (3.2.6) imply

$$\left| \int_\Omega \mathcal{L}(\lambda) \begin{pmatrix} u_r \\ u_\omega \end{pmatrix} \cdot \begin{pmatrix} \overline{v}_r \\ \overline{v}_\omega \end{pmatrix} d\omega \right| \leq c \left\| \begin{pmatrix} u_r \\ u_\omega \end{pmatrix} \right\|_{W^1_2(\Omega)\times h^1_2(\Omega)} \cdot \left\| \begin{pmatrix} u_r \\ u_\omega \end{pmatrix} \right\|_{W^1_2(\Omega)\times h^1_2(\Omega)}.$$

Thus, the operator $\mathcal{L}(\lambda)$ is continuous. ∎

From Theorem 1.1.5 it follows that the vector function (3.0.2) satisfies the system (3.2.1) or, equivalently, the equations (3.2.14)–(3.2.16) if and only if

$$(3.2.20) \qquad \sum_{k=0}^{j} \frac{1}{k!} \frac{\partial^k \mathcal{L}}{\partial \lambda^k}(\lambda_0) \begin{pmatrix} u_r^{(s-k)} \\ u_\omega^{(s-k)} \end{pmatrix} = 0 \quad \text{on } \Omega \qquad \text{for } j = 0, 1, \dots, s.$$

Here $u_r^{(k)}$ and $u_\omega^{(k)} = (u_\theta^{(k)}, u_\varphi^{(k)})$ are the spherical components of the vector function $u^{(k)}$. This means that the following assertion is true.

LEMMA 3.2.5. *The vector function* (3.0.2) *is a solution of problem* (3.2.1), (3.2.2) *if and only if* λ_0 *is an eigenvalue of the pencil* $\mathcal{L}$ *and* $u^{(0)}, \dots, u^{(s-1)}$ *is a Jordan chain of this pencil corresponding to the eigenvalue* λ_0.

3.2.4. Basic properties of the pencil $\mathcal{L}$. It can be easily seen that the operator $\mathcal{L}(\lambda)$ is selfadjoint for $\mathrm{Re}\,\lambda = -1/2$. In the following lemma we show, in particular, that $\mathcal{L}(\lambda)$ is positive for λ on the above line.

LEMMA 3.2.6. *Let* $\lambda = it - 1/2$, $t \in \mathbb{R}$. *Then*

$$(3.2.21) \qquad \int_\Omega \mathcal{L}(\lambda) \binom{u_r}{u_\omega} \cdot \binom{\overline{u}_r}{\overline{u}_\omega} \, d\omega$$

$$\geq c \left(Q(u_\omega, u_\omega) + \int_\Omega |\nabla_\omega u_r|^2 \, d\omega \right) + (c + t^2) \int_\Omega \left(|u_r|^2 + |u_\omega|^2 \right) d\omega$$

for every vector function $(u_r, u_\omega) \in \overset{\circ}{W}{}^1_2(\Omega) \times \overset{\circ}{h}{}^1_2(\Omega)$. *Here c is a positive constant which is independent of the domain Ω.*

Proof: Since the functions u_r and u_ω can be extended by zero outside Ω, it suffices to prove (3.2.21) for the case $\Omega = S^2$. By means of (3.2.19), we can rewrite the left-hand side of (3.2.21) in the form

$$Q(u_\omega, u_\omega) + \int_{S^2} \left(|\nabla_\omega u_r|^2 + \frac{2 - 2\nu}{1 - 2\nu} \left(\frac{9}{4} + t^2 \right) |u_r|^2 + \frac{1}{1 - 2\nu} |\nabla_\omega \cdot u_\omega|^2 \right.$$

$$\left. + \left(\frac{5}{4} + t^2 \right) |u_\omega|^2 + 2\,\mathrm{Re} \left(\frac{7/2 - 4\nu + it}{1 - 2\nu} u_r \nabla_\omega \cdot \overline{u}_\omega \right) \right) d\omega.$$

Using (3.2.7) and integrating by parts, we obtain that the last expression is equal to

$$\int_{S^2} \left(\left| \nabla_\omega u_r - \frac{1}{2} u_\omega \right|^2 + \left| \frac{3}{2} u_r + \nabla_\omega \cdot u_\omega \right|^2 + \frac{1}{1 - 2\nu} \left| \left(\frac{3}{2} + it \right) u_r + \nabla_\omega \cdot u_\omega \right|^2 \right.$$

$$\left. + t^2 \left(|u_r|^2 + |u_\omega|^2 \right) + \left| \frac{1}{\sin\theta} \partial_\varphi u_\theta - \cot\theta\, u_\varphi - \partial_\theta u_\varphi \right|^2 \right) d\omega.$$

Hence (3.2.21) will be proved if we show that

$$(3.2.22)$$

$$\int_{S^2} \left(\left| \nabla_\omega u_r - \frac{1}{2} u_\omega \right|^2 + \left| \frac{3}{2} u_r + \nabla_\omega \cdot u_\omega \right|^2 + \left| \frac{1}{\sin\theta} \partial_\varphi u_\theta - \cot\theta\, u_\varphi - \partial_\theta u_\varphi \right|^2 \right) d\omega$$

$$\geq c \left(Q(u_\omega, u_\omega) + \int_\Omega \left(|\nabla_\omega u_r|^2 + |u_r|^2 + |u_\omega|^2 \right) d\omega \right).$$

To this end, we show first that the left-hand side of (3.2.22) is not equal to zero unless the vector function (u_r, u_ω) vanishes identically. Indeed, let us assume that

$$(3.2.23) \qquad\qquad \nabla_\omega u_r - \frac{1}{2} u_\omega = 0, \qquad \frac{3}{2} u_r + \nabla_\omega \cdot u_\omega = 0.$$

Then we have

$$(3.2.24) \qquad\qquad \delta u_r + \frac{3}{4} u_r = 0, \qquad \int_{S^2} u_r \, d\omega = 0.$$

Since the second eigenvalue of $-\delta$ on the sphere is equal to 2 and the first eigenvalue is zero with eigenfunction $u = const$ (see Section 2.2.2), the identity (3.2.24) cannot be true if $u_r \neq 0$. Therefore, it follows from (3.2.23) that $u_r = 0$, $u_\omega = 0$. Consequently, the left-hand side of (3.2.22) is positive for $(u_r, u_\omega) \neq 0$. Furthermore, by Lemma 3.2.2, the left-hand side of (3.2.22) is majorized by

$$c \left(\|u_r\|^2_{W_2^1(\Omega)} + \|u_\omega\|^2_{\overset{\circ}{h}{}_2^1(\Omega)} \right)$$

and bounded from below by

$$c_1 \left(\|u_r\|^2_{W_2^1(\Omega)} + \|u_\omega\|^2_{\overset{\circ}{h}{}_2^1(\Omega)} \right) - c_2 \left(\|u_r\|^2_{L_2(\Omega)} + \|u_\omega\|^2_{L_2(\Omega)^2} \right).$$

Hence the left-hand side of (3.2.22) is an equivalent norm in $\overset{\circ}{W}{}_2^1(\Omega) \times \overset{\circ}{h}{}_2^1(\Omega)$. This proves the inequalities (3.2.22) and (3.2.21). ■

The following theorem contains some basic properties of the pencil $\mathcal{L}$.

THEOREM 3.2.1. 1) *The operator $\mathcal{L}(\lambda)$ is Fredholm for every $\lambda \in \mathbb{C}$ and positive definite for* $\operatorname{Re} \lambda = -1/2$.

2) *The spectrum of the pencil $\mathcal{L}$ consists of isolated eigenvalues with finite algebraic multiplicities.*

3) *For all $\lambda \in \mathbb{C}$ there is the equality*

(3.2.25) $(\mathcal{L}(\lambda))^* = \mathcal{L}(-1 - \overline{\lambda}).$

This implies, in particular, that the number λ_0 is an eigenvalue of the pencil $\mathcal{L}$ if and only if $-1 - \overline{\lambda}_0$ is an eigenvalue of $\mathcal{L}$. Moreover, the geometric, partial and algebraic multiplicities of the eigenvalues λ_0 and $-1 - \overline{\lambda}_0$ coincide.

4) *The number λ_0 is an eigenvalue of the pencil $\mathcal{L}$ if and only if $\overline{\lambda}_0$ is an eigenvalue of the pencil $\mathcal{L}$. The geometric, partial and algebraic multiplicities of these eigenvalues coincide. If $u^{(0)}, \ldots, u^{(s)}$ is a Jordan chain of the pencil $\mathcal{L}$ corresponding to the eigenvalue λ_0, then $\overline{u^{(0)}}, \ldots, \overline{u^{(s)}}$ is a Jordan chain corresponding to the eigenvalue $\overline{\lambda}_0$.*

Proof: 1) By (3.2.21), the operator $\mathcal{L}(\lambda)$ is positive definite and invertible for $\operatorname{Re} \lambda = -1/2$. Furthermore, for every $\lambda \in \mathbb{C}$ the operator

$$\mathcal{L}(\lambda) - \mathcal{L}(-1/2) : \overset{\circ}{W}{}_2^1(\Omega) \times \overset{\circ}{h}{}_2^1(\Omega) \to \left(\overset{\circ}{W}{}_2^1(\Omega) \times \overset{\circ}{h}{}_2^1(\Omega) \right)^*$$

is compact and, since $\mathcal{L}(-1/2)$ is invertible, it follows that $\mathcal{L}(\lambda)$ is a Fredholm operator for all $\lambda \in \mathbb{C}$.

2) The second assertion follows from assertion 1) and Theorem 1.1.1.

3) Applying Theorem 1.2.2, where $\mathfrak{A} = \mathcal{L}$ and $\gamma = -1$, we get (3.2.25).

4) For the proof of the last assertion we take the complex conjugate of (3.2.20) and utilize the fact that the coefficients of the Lamé system are real. The proof is complete. ■

3.2.5. A connection between the Jordan chains corresponding to the eigenvalues λ_0 and $-1 - \lambda_0$. We introduce the diagonal matrix

(3.2.26) $\mathcal{T}(\lambda) = \begin{pmatrix} 4 - 4\nu + \lambda & 0 & 0 \\ 0 & 3 - 4\nu - \lambda & 0 \\ 0 & 0 & 3 - 4\nu - \lambda \end{pmatrix}.$

It can be immediately verified that

$$(3.2.27) \qquad\qquad \mathcal{T}(\lambda)\,\mathcal{L}(\lambda) = \mathcal{L}(-1-\lambda)\,\mathcal{T}(\lambda).$$

THEOREM 3.2.2. *Let λ_0 be an eigenvalue and $u^{(0)},\dots,u^{(s)}$ a Jordan chain corresponding to this eigenvalue.*

1) *If λ_0 and $u^{(0)}$ satisfy one of the conditions*

$$\text{(i)} \qquad \lambda_0 \neq 3 - 4\nu \qquad \text{and} \qquad \lambda_0 \neq 4\nu - 4,$$

$$\text{(ii)} \qquad \lambda_0 \in \{3 - 4\nu,\, 4\nu - 4\} \qquad \text{and} \qquad \mathcal{T}(\lambda_0)u^{(0)} \neq 0,$$

then the vector functions

$$\mathcal{T}(\lambda_0)\,u^{(0)}, \quad (-1)^k \left(\mathcal{T}(\lambda_0)\,u^{(k)} + \begin{pmatrix} 1 & 0 & 0 \\ 0 & -1 & 0 \\ 0 & 0 & -1 \end{pmatrix} u^{(k-1)} \right), \quad k = 1,\dots,s,$$

form a Jordan chain of eigenvectors and generalized eigenvectors of the pencil $\mathcal{L}$ corresponding to the eigenvalue $-1 - \lambda_0$.

2) *Let either $\lambda_0 = 3 - 4\nu$ or $\lambda_0 = 4\nu - 4$. Furthermore, let $\mathcal{T}(\lambda_0)u^{(0)} = 0$ and $s \geq 1$. Then the vector functions*

$$(-1)^{k-1} \left(\mathcal{T}(\lambda_0)\,u^{(k)} + \begin{pmatrix} 1 & 0 & 0 \\ 0 & -1 & 0 \\ 0 & 0 & -1 \end{pmatrix} u^{(k-1)} \right), \qquad k = 1,\dots,s,$$

form a Jordan chain corresponding to the eigenvalue $-1 - \lambda_0$.

Proof: Let

$$u(\lambda) = \sum_{k=0}^{s} (\lambda - \lambda_0)^k \, u^{(k)}.$$

Then

$$\mathcal{L}(\lambda)\,u(\lambda) = O(|\lambda - \lambda_0|^{s+1}).$$

From this and (3.2.27) it follows that

$$(3.2.28) \qquad \mathcal{L}(-1-\lambda)\,\mathcal{T}(\lambda)\,u(\lambda) = O(|\lambda - \lambda_0|^{s+1}).$$

We set $v^{(0)} = \mathcal{T}(\lambda_0)\,u^{(0)}$ and

$$(3.2.29) \qquad v^{(k)} = (-1)^k \left(\mathcal{T}(\lambda_0)\,u^{(k)} + \mathcal{T}'(\lambda_0)\,u^{(k-1)} \right), \qquad k = 1,\dots,s.$$

Then, by (3.2.28), we have

$$\sum_{j=0}^{k} \frac{1}{j!}\, \mathcal{L}^{(j)}(-1-\lambda)\, v^{(k-j)} = 0$$

for $k = 0,\dots,s$. Furthermore, we have $v^{(0)} \neq 0$ in case 1) and $v^{(1)} \neq 0$ in case 2). This proves the theorem. ∎

REMARK 3.2.1. If $\lambda_0 \neq 3 - 4\nu$ and $\lambda_0 \neq 4\nu - 4$, then the formulas given in the first part of Theorem 3.2.2 establish a one-to-one correspondence between the eigenvectors and generalized eigenvectors of the pencil $\mathcal{L}$ corresponding to the eigenvalues λ_0 and $-1 - \lambda_0$, respectively.

3.2.6. The Kelvin transform for the Lamé system. The identity (3.2.27) implies the following analogue of the Kelvin transform for solutions of the Lamé system.

Let $U(x)$ be a solution of the system (3.2.1) in the domain $\mathcal{G} \subset \mathbb{R}^3$. Then the vector function

$$V(x) = \frac{1}{|x|} \left((3 - 4\nu) I + \Xi + (I - 2\Xi) \sum_{k=1}^{3} x_k \, \partial_{x_k} \right) U\left(\frac{x}{|x|^2} \right),$$

where I denotes the identity matrix and Ξ is the matrix with the elements $x_i x_j |x|^{-2}$, is also a solution of the Lamé system in a domain which is obtained from $\mathcal{G}$ via the inversion $x \to x|x|^{-2}$.

Indeed, we have $\mathcal{L}(r\partial_r) J U = 0$, where $J = J(\theta, \varphi)$ is the transformation matrix between spherical and Cartesian coordinates introduced in the beginning of this section. Since (3.2.27) can be written as

$$\mathcal{T}(r\partial_r) \, \mathcal{L}(r\partial_r) = \mathcal{L}(-1 - r\partial_r) \, \mathcal{T}(r\partial_r),$$

we get

$$\mathcal{L}(-1 - r\partial_r) \, \mathcal{T}(r\partial_r) \left(J(\theta, \varphi) \, U(x) \right) = 0$$

or, equivalently,

$$\mathcal{L}(-r\partial_r) \left(r \, \mathcal{T}(r\partial_r) \left(J(\theta, \varphi) \, U(x) \right) \right) = 0.$$

Replacing r by r^{-1}, we obtain

$$\mathcal{L}(r\partial_r) \left(r^{-1} \, \mathcal{T}(-r\partial_r) \left(J(\theta, \varphi) \, U(x/|x|^2) \right) \right) = 0,$$

i.e., the vector function $J^* \, r^{-1} \, \mathcal{T}(-r\partial_r) \, J U(x/|x|^2)$ satisfies the system (3.2.1). Using the definition of the matrices J and $\mathcal{T}$, we verify that this function is equal to V.

The Kelvin transform obtained here is useful for the construction of particular solutions of the Lamé system. For example, replacing U by the matrices I and $x_3 I$, we arrive, up to scaling factors, at the Kelvin-Somigliana tensor and the Poisson kernel for the Dirichlet problem in the half-space $x_3 > 0$.

3.3. Properties of real eigenvalues

Here we describe a domain symmetric with respect to the line $\mathrm{Re}\,\lambda = -1/2$ which contains only real eigenvalues. We also show that there are no generalized eigenvectors associated to eigenvalues in this set.

3.3.1. A domain containing only real eigenvalues. We introduce the quantities

$$(3.3.1) \qquad \Gamma(\Omega) = \inf_{0 \neq u_\omega \in \overset{\circ}{h}_2^1(\Omega)} \frac{Q(u_\omega, u_\omega) + (2 - 2\nu)^{-1} \int_\Omega |\nabla_\omega \cdot u_\omega|^2 \, d\omega}{\int_\Omega |u_\omega|^2 \, d\omega}$$

and

$$(3.3.2) \qquad F_\nu(\Omega) = \left(\frac{1}{2} \Gamma(\Omega) + \frac{3}{4} + (2 - 2\nu)(3 - 4\nu) \right)^{1/2}.$$

THEOREM 3.3.1. *In the region*

$$(3.3.3) \qquad \left(\operatorname{Re}\lambda + \frac{1}{2}\right)^2 - (\operatorname{Im}\lambda)^2 < (F_\nu(\Omega))^2$$

the pencil $\mathcal{L}$ has only real eigenvalues.

Proof: Let λ be a complex number such that $\operatorname{Im}\lambda \neq 0$, $\operatorname{Re}\lambda \neq -1/2$. We consider the quadratic form

$$q(u,u;\lambda) = \int_\Omega \mathcal{L}(\lambda)\begin{pmatrix} u_r \\ u_\omega \end{pmatrix} \cdot \begin{pmatrix} c\,\overline{u}_r \\ \overline{u}_\omega \end{pmatrix} d\omega,$$

where

$$(3.3.4) \qquad c = \frac{4 - 4\nu + \overline{\lambda}}{3 - 4\nu - \lambda}.$$

After a straightforward calculation we arrive at

$$(3.3.5) \quad q(u,u;\lambda) = Q(u_\omega, u_\omega) + \int_\Omega \left(c\,|\nabla_\omega u_r|^2 + c\,\frac{2 - 2\nu}{1 - 2\nu}(1 - \lambda)(\lambda + 2)|u_r|^2 \right) d\omega$$

$$+ \left((1 - \lambda)(\lambda + 2) - 1 \right) \int_\Omega |u_\omega|^2\, d\omega + \frac{1}{1 - 2\nu} \int_\Omega |\nabla_\omega \cdot u_\omega|^2\, d\omega$$

$$+ 2\operatorname{Re}\left(\frac{4 - 4\nu + \lambda}{1 - 2\nu} \int_\Omega u_r \nabla_\omega \cdot \overline{u}_\omega\, d\omega \right).$$

Using the relations

$$(3.3.6) \qquad c\,(1 - \lambda)\,(\lambda + 2) = |4 - 4\nu + \lambda|^2 - 2\,(1 - 2\nu)\,(5 - 4\nu)\,c,$$

$$(3.3.7) \qquad c = \frac{(4 - 4\nu + \operatorname{Re}\lambda)(3 - 4\nu - \operatorname{Re}\lambda) + (\operatorname{Im}\lambda)^2 + i\operatorname{Im}\lambda\,(2\operatorname{Re}\lambda + 1)}{|3 - 4\nu - \lambda|^2}$$

and separating real and imaginary parts of the form $q(u,u;\lambda)$, we find

$$\operatorname{Im} q(u,u;\lambda) = \operatorname{Im}\lambda\,(2\operatorname{Re}\lambda + 1)\left(\frac{1}{|3 - 4\nu - \lambda|^2} \int_\Omega |\nabla_\omega u_r|^2\, d\omega \right.$$

$$\left. - \frac{4(1 - \nu)(5 - 4\nu)}{|3 - 4\nu - \lambda|^2} \int_\Omega |u_r|^2\, d\omega - \int_\Omega |u_\omega|^2\, d\omega \right),$$

$$\operatorname{Re} q(u,u;\lambda) = Q(u_\omega, u_\omega) + \operatorname{Re} c \int_\Omega |\nabla_\omega u_r|^2\, d\omega$$

$$+ \left(\frac{2 - 2\nu}{1 - 2\nu}|4 - 4\nu + \lambda|^2 - 4(1 - \nu)(5 - 4\nu)\operatorname{Re} c \right) \int_\Omega |u_r|^2\, d\omega$$

$$+ \left(1 - (\operatorname{Re}\lambda)^2 - \operatorname{Re}\lambda + (\operatorname{Im}\lambda)^2 \right) \int_\Omega |u_\omega|^2\, d\omega$$

$$+ \frac{1}{1 - 2\nu} \int_\Omega |\nabla_\omega \cdot u_\omega|^2\, d\omega + 2\operatorname{Re}\left(\frac{4 - 4\nu + \lambda}{1 - 2\nu} \int_\Omega u_r \nabla_\omega \cdot \overline{u}_\omega\, d\omega \right).$$

From this we obtain

$$(3.3.8) \quad \operatorname{Re} q(u, u; \lambda) - \frac{\operatorname{Re} c \, |3 - 4\nu - \lambda|^2}{\operatorname{Im} \lambda \, (2\operatorname{Re} \lambda + 1)} \operatorname{Im} q(u, u; \lambda)$$

$$= Q(u_\omega, u_\omega) + \frac{1}{2 - 2\nu} \int_\Omega |\nabla_\omega \cdot u_\omega|^2 \, d\omega$$

$$+ \frac{2 - 2\nu}{1 - 2\nu} \int_\Omega \left| (4 - 4\nu + \lambda) \, u_r + \frac{1}{2 - 2\nu} \nabla_\omega \cdot u_\omega \right|^2 d\omega$$

$$+ \left(1 + (3 - 4\nu)(4 - 4\nu) + 2(\operatorname{Im} \lambda)^2 - 2\operatorname{Re} \lambda \, (\operatorname{Re} \lambda + 1) \right) \int_\Omega |u_\omega|^2 \, d\omega.$$

Consequently, if

$$(3.3.9) \qquad \Gamma(\Omega) + 1 + (3 - 4\nu)(4 - 4\nu) + 2(\operatorname{Im} \lambda)^2 - 2 \operatorname{Re} \lambda \, (\operatorname{Re} \lambda + 1) > 0,$$

then (3.3.8) is positive for $u_\omega \neq 0$.

Suppose now that λ is an eigenvalue satisfying the inequality (3.3.9) and (u_r, u_ω) is an eigenvector corresponding to this eigenvalue. Then $q(u, u; \lambda) = 0$ and, necessarily, also $u_\omega = 0$. From this and from the equation $\mathcal{L}(\lambda) \, u = 0$ we find that $(4 - 4\nu + \lambda)\nabla_\omega u_r = 0$ on Ω. Due to our assumption that $\operatorname{Im} \lambda \neq 0$, the last equality yields $\nabla_\omega u_r = 0$ and, therefore, $u_r = const$. Since $u_r \in \overset{\circ}{W}{}^1_2(\Omega)$, the function u_r must vanish. This proves the assertion in the case $\operatorname{Re} \lambda \neq -1/2$. Since, by Theorem 3.2.1, there are no eigenvalues on the line $\operatorname{Re} \lambda = -1/2$, the theorem is completely proved. ∎

As a consequence of Theorem 3.3.1, the following assertion holds.

COROLLARY 3.3.1. *In the strip*

$$(3.3.10) \qquad\qquad \left| \operatorname{Re} \lambda + \frac{1}{2} \right| \le F_\nu(\Omega)$$

the pencil $\mathcal{L}$ has only real eigenvalues.

An estimate for the set function F_ν will be given later.

3.3.2. Absence of generalized eigenvectors. We consider the real eigenvalues of the pencil $\mathcal{L}$ in the interval $(-F_\nu(\Omega) - \frac{1}{2}, \, F_\nu(\Omega) - \frac{1}{2})$. Our goal is to show that there are no generalized eigenvectors associated to these eigenvalues.

LEMMA 3.3.1. *Let λ_0 be a real eigenvalue of the pencil $\mathcal{L}$ and $u^{(0)}$ a corresponding eigenvector. Furthermore, let $c(\lambda) = (4 - 4\nu + \lambda)/(3 - 4\nu - \lambda)$.*
   1) If $-1/2 < \lambda_0 < \min\{3 - 4\nu, \, F_\nu(\Omega) - 1/2\}$, then

$$\left. \frac{d}{d\lambda} \int_\Omega \mathcal{L}(\lambda) \, u^{(0)} \cdot \begin{pmatrix} c(\lambda) \, \bar{u}_r^{(0)} \\ \bar{u}_\omega^{(0)} \end{pmatrix} d\omega \right|_{\lambda = \lambda_0} < 0.$$

   2) If $3 - 4\nu < \lambda_0 < F_\nu(\Omega) - 1/2$, then

$$\left. \frac{d}{d\lambda} \int_\Omega \mathcal{L}(\lambda) \, u^{(0)} \cdot \begin{pmatrix} c(\lambda) \, \bar{u}_r^{(0)} \\ \bar{u}_\omega^{(0)} \end{pmatrix} d\omega \right|_{\lambda = \lambda_0} > 0.$$

Proof: Let q be the sesquilinear form (3.3.5). For real λ the function (3.3.4) coincides with $c(\lambda) = (4 - 4\nu + \lambda)/(3 - 4\nu - \lambda)$. Differentiating (3.3.5), we get

$$(3.3.11) \quad \frac{d}{d\lambda} \int_\Omega \mathcal{L}(\lambda) u^{(0)} \cdot \begin{pmatrix} c(\lambda) \bar{u}_r^{(0)} \\ \bar{u}_\omega^{(0)} \end{pmatrix} d\omega \bigg|_{\lambda = \lambda_0} = \frac{d}{d\lambda} q(u^{(0)}, u^{(0)}; \lambda) \bigg|_{\lambda = \lambda_0}$$

$$= \int_\Omega \left(c'(\lambda_0) |\nabla_\omega u_r^{(0)}|^2 - (2\lambda_0 + 1) |u_\omega^{(0)}|^2 + \frac{2}{1 - 2\nu} \operatorname{Re} u_r^{(0)} \nabla_\omega \cdot \bar{u}_\omega^{(0)} \right.$$

$$\left. + \frac{2 - 2\nu}{1 - 2\nu} \left((1 - \lambda_0)(\lambda_0 + 2) c'(\lambda_0) - (2\lambda_0 + 1) c(\lambda_0) \right) |u_r^{(0)}|^2 \right) d\omega,$$

where $c'(\lambda) = (7 - 8\nu)(3 - 4\nu - \lambda)^{-2}$. Multiplying both sides of the equation $\mathcal{L}(\lambda_0) u^{(0)} = 0$ by the vector function $(u_r^{(0)}, 0)$, integrating over Ω and taking the real part, we find that the integral

$$\int_\Omega \left(|\nabla_\omega u_r^{(0)}|^2 + \frac{2 - 2\nu}{1 - 2\nu}(1 - \lambda_0)(\lambda_0 + 2) |u_r^{(0)}|^2 + \frac{3 - 4\nu - \lambda_0}{1 - 2\nu} \operatorname{Re}(\nabla_\omega \cdot u_\omega^{(0)}) \bar{u}_r^{(0)} \right) d\omega$$

vanishes. From this and from (3.3.11) we obtain

$$\frac{1 + 2\lambda_0}{(3 - 4\nu - \lambda_0)^2} \int_\Omega |\nabla_\omega u_r^{(0)}|^2 \, d\omega - \frac{4(1 + 2\lambda_0)(1 - \nu)(5 - 4\nu)}{(3 - 4\nu - \lambda_0)^2} \int_\Omega |u_r^{(0)}|^2 \, d\omega$$

$$-(2\lambda_0 + 1) \int_\Omega |u_\omega^{(0)}|^2 \, d\omega = \frac{d}{d\lambda} q(u^{(0)}, u^{(0)}; \lambda) \bigg|_{\lambda = \lambda_0}.$$

Furthermore, by a calculation analogous to that in the proof of Theorem 3.3.1, we arrive at the identity

$$-\frac{(4 - 4\nu + \lambda_0)(3 - 4\nu - \lambda_0)}{1 + 2\lambda_0} \frac{d}{d\lambda} q(u^{(0)}, u^{(0)}; \lambda) \bigg|_{\lambda = \lambda_0}$$

$$= q(u^{(0)}, u^{(0)}; \lambda_0) - \frac{(4 - 4\nu + \lambda_0)(3 - 4\nu - \lambda_0)}{1 + 2\lambda_0} \frac{d}{d\lambda} q(u^{(0)}, u^{(0)}; \lambda) \bigg|_{\lambda = \lambda_0}$$

$$= Q(u_\omega^{(0)}, u_\omega^{(0)}) + \frac{1}{2 - 2\nu} \int_\Omega |\nabla_\omega \cdot u_\omega^{(0)}|^2 \, d\omega$$

$$+ \frac{2 - 2\nu}{1 - 2\nu} \int_\Omega \left| (4 - 4\nu + \lambda_0) u_r^{(0)} + \frac{1}{2 - 2\nu} \nabla_\omega \cdot u_\omega^{(0)} \right|^2 d\omega$$

$$+ \left(1 + (3 - 4\nu)(4 - 4\nu) - 2 \operatorname{Re} \lambda_0 (\operatorname{Re} \lambda_0 + 1) \right) \int_\Omega |u_\omega^{(0)}|^2 \, d\omega.$$

This implies the assertion of the lemma. ∎

THEOREM 3.3.2. *There are no generalized eigenvectors associated to eigenvalues in the interior of the strip* (3.3.10).

Proof: Due to assertions 1) and 3) of Theorem 3.2.1, it is sufficient to consider the case $-1/2 < \lambda_0 < F_\nu(\Omega) - 1/2$. Let $u^{(0)}$ be an eigenvector corresponding to λ_0 and let $u^{(1)}$ be a generalized eigenvector for $u^{(0)}$. Then

$$(3.3.12) \qquad\qquad \mathcal{L}(\lambda_0) u^{(0)} = 0,$$

$$(3.3.13) \qquad\qquad \mathcal{L}(\lambda_0) u^{(1)} = -\mathcal{L}'(\lambda_0) u^{(0)}.$$

Assume first that $\lambda_0 \neq 3 - 4\nu$. Multiplying both sides of (3.3.13) by $(\bar{c}u_r^{(0)}, u_\omega^{(0)})$, where the quantity c has been defined in (3.3.4), setting $\lambda = \lambda_0$ and integrating over Ω, we get

$$q(u^{(1)}, u^{(0)}; \lambda_0) = -\frac{d}{d\lambda} q(u^{(0)}, u^{(0)}; \lambda)\Big|_{\lambda=\lambda_0}.$$

Here $q(\cdot, \cdot; \lambda)$ is the sesquilinear form (3.3.5). Since this form is symmetric for real λ, we have

$$q(u^{(1)}, u^{(0)}; \lambda_0) = \overline{q(u^{(0)}, u^{(1)}; \lambda_0)} = 0.$$

Therefore,

$$\frac{d}{d\lambda} q(u^{(0)}, u^{(0)}; \lambda)\Big|_{\lambda=\lambda_0} = 0.$$

However, by Lemma 3.3.1, this is impossible. Thus, the theorem is proved for the case $\lambda_0 \neq 3 - 4\nu$.

Now let $3 - 4\nu < F_\nu(\Omega) - 1/2$ and $\lambda_0 = 3 - 4\nu$. The equations (3.3.12), (3.3.13) for the components $u_r^{(0)}$ and $u_r^{(1)}$ take the form

$$\delta u_r^{(0)} + (4 - 4\nu)(5 - 4\nu) u_r^{(0)} = 0,$$

$$\delta u_r^{(1)} + (4 - 4\nu)(5 - 4\nu) u_r^{(1)} = -\frac{2(1 - \nu)(7 - 8\nu)}{1 - 2\nu} u_r^{(0)} - \frac{1}{1 - 2\nu} \nabla_\omega \cdot u_\omega^{(0)}.$$

The second equation may have a nontrivial solution only if

$$(3.3.14) \qquad 2(1 - \nu)(7 - 8\nu) \int_\Omega |u_r^{(0)}|^2 \, d\omega + \mathrm{Re} \int_\Omega (\nabla_\omega \cdot u_\omega^{(0)}) \overline{u}_r^{(0)} \, d\omega = 0.$$

Multiplying (3.3.12) by the vector function $(0, u_\omega^{(0)})$, integrating over Ω and taking the real part, we find

$$(3.3.15) \qquad 0 = Q(u_\omega^{(0)}, u_\omega^{(0)}) - \left(1 + 2(1 - 2\nu)(5 - 4\nu)\right) \int_\Omega |u_\omega^{(0)}|^2 \, d\omega$$

$$+ \frac{1}{1 - 2\nu} \int_\Omega |\nabla_\omega \cdot u_\omega^{(0)}|^2 \, d\omega + \frac{7 - 8\nu}{1 - 2\nu} \mathrm{Re} \int_\Omega u_r^{(0)} \nabla_\omega \cdot \bar{u}_\omega^{(0)} \, d\omega.$$

From (3.3.14) it follows that

$$\frac{2 - 2\nu}{1 - 2\nu} \int_\Omega \left| (7 - 8\nu) u_r^{(0)} + \frac{1}{2 - 2\nu} \nabla_\omega \cdot u_\omega^{(0)} \right|^2 \, d\omega$$

$$- \frac{1}{(1 - 2\nu)(2 - 2\nu)} \int_\Omega |\nabla_\omega \cdot u_\omega^{(0)}|^2 \, d\omega = \frac{7 - 8\nu}{1 - 2\nu} \mathrm{Re} \int_\Omega u_r^{(0)} \nabla_\omega \cdot \overline{u_\omega^{(0)}} \, d\omega.$$

Using this equality, we can write (3.3.15) in the form

$$Q(u_\omega^{(0)}, u_\omega^{(0)}) + \frac{2 - 2\nu}{1 - 2\nu} \int_\Omega \left| (7 - 8\nu) u_r^{(0)} + \frac{1}{2 - 2\nu} \nabla_\omega \cdot u_\omega^{(0)} \right|^2 \, d\omega$$

$$+ \frac{1}{2 - 2\nu} \int_\Omega |\nabla_\omega \cdot u_\omega^{(0)}|^2 \, d\omega - \left(1 + 2(1 - 2\nu)(5 - 4\nu)\right) \int_\Omega |u_\omega^{(0)}|^2 \, d\omega = 0.$$

Now, since $F_\nu(\Omega) - 1/2 > 3 - 4\nu$, we have $\Gamma(\Omega) > 1 + 2(1 - 2\nu)(5 - 4\nu)$. Therefore, the last equality cannot hold. The theorem is proved. ∎

3.4. The set functions Γ and F_ν

We are interested in estimates for the quantity $F_\nu(\Omega)$ which appeared in Theorems 3.3.1 and 3.3.2. For this we need information about the function $\Gamma(\Omega)$ defined in (3.3.1).

In the sequel, we will write $\Omega_0 \subset\subset \Omega_1$, where Ω_0, Ω_1 are subdomains of the sphere, if $\Omega_0 \subset \Omega_1$ and $\Omega_1 \backslash \Omega_0$ contains a nonempty open set.

3.4.1. Monotonicity of Γ and F_ν.

LEMMA 3.4.1. 1) If $\Omega_0 \subset \Omega_1$, then $\Gamma(\Omega_0) \geq \Gamma(\Omega_1)$.
2) If $\Omega_0 \subset\subset \Omega_1$, then $\Gamma(\Omega_0) > \Gamma(\Omega_1)$.

Proof: If $\Omega_0 \subset \Omega_1$, then $\overset{\circ}{h}{}^1_2(\Omega_0)$ is contained in $\overset{\circ}{h}{}^1_2(\Omega_1)$. This proves the monotonicity of Γ.

Let $\Omega_0 \subset\subset \Omega_1$. We assume that $\Gamma(\Omega_0) = \Gamma(\Omega_1)$ and denote by $u^{(0)}_\omega \in \overset{\circ}{h}{}^1_2(\Omega_0)$ a nonzero vector on which the infimum in (3.3.1) is attained. The extension of this vector function to Ω_1 by zero will be also denoted by $u^{(0)}_\omega$. From our assumption it follows that the infimum in (3.3.1) for the space $\overset{\circ}{h}{}^1_2(\Omega_1)$ is also attained on $u^{(0)}_\omega$. Therefore,

$$-H_{\nu-1/2} u^{(0)}_\omega = \Gamma(\Omega_1)\, u^{(0)}_\omega \quad \text{on } \Omega_1.$$

From this we obtain that $u^{(0)}_\omega$ is analytic on Ω_1 and, since $u^{(0)}_\omega = 0$ on $\Omega_1 \backslash \Omega_0$, it follows that $u^{(0)}_\omega = 0$ on Ω_1. This completes the proof. ∎

REMARK 3.4.1. If we let ν in (3.3.1) tend to $-\infty$, then $\Gamma(\Omega)$ is equal to the quantity

$$(3.4.1) \qquad s(\Omega) = \inf_{0 \neq u_\omega \in \overset{\circ}{h}{}^1_2(\Omega)} \left(\int_\Omega |u_\omega|^2 \, d\omega \right)^{-1} Q(u_\omega, u_\omega).$$

Repeating the proof of Lemma 3.4.1 verbatim, we verify that the assertions 1) and 2) of Lemma 3.4.1 also hold for $s(\Omega)$. According to Lemma 3.2.3, we have $s(\Omega) \geq 1$.

From the stated properties of $\Gamma(\Omega)$ we immediately obtain the following assertions.

COROLLARY 3.4.1. *Let $F_\nu(\Omega)$ be the quantity defined in (3.3.2).*
1) *If $\Omega_0 \subset \Omega_1$, then $F_\nu(\Omega_0) \geq F_\nu(\Omega_1)$.*
2) *If $\Omega_0 \subset\subset \Omega_1$, then $F_\nu(\Omega_0) > F_\nu(\Omega_1)$.*
3) *For all $\Omega \subset S^2$ there is the inequality*

$$(3.4.2) \qquad F_\nu(\Omega) \geq \left(\left(\frac{5}{2} - 2\nu \right)^2 + (1 - 2\nu)^2 \right)^{1/2}.$$

3.4.2. Estimates of $\Gamma(\Omega)$, $F_\nu(\Omega)$ for circular cones. Now we estimate the quantities $\Gamma(\Omega)$ and $F_\nu(\Omega)$ for right circular cones, i.e., for

$$\Omega = \Omega_\alpha = \{ (\theta, \varphi) : \ 0 \leq \theta < \alpha, \ 0 \leq \varphi < 2\pi \},$$

where $\alpha \in (0, \pi]$.

LEMMA 3.4.2. *The quantity $\Gamma(\Omega_\alpha)$ satisfies the inequalities*

(3.4.3) $$\gamma_0(\alpha) - 1 \geq \Gamma(\alpha_0) \geq \min\left\{\gamma_0(\alpha), \gamma_1^+(\alpha), \gamma_1^-(\alpha)\right\} - 1,$$

where

(3.4.4) $$\gamma_0(\alpha) = \inf_{f \neq 0,\, f(\cos\alpha)=0} \frac{\int\limits_{\cos\alpha}^{1}\left((1-x^2)\,|f'(x)|^2 + (1-x^2)^{-1}\,|f(x)|^2\right) dx}{\int\limits_{\cos\alpha}^{1} |f(x)|^2\, dx},$$

(3.4.5) $$\gamma_1^\pm(\alpha) = \inf_{f \neq 0,\, f(\cos\alpha)=0} \frac{\int\limits_{\cos\alpha}^{1}\left((1-x^2)\,|f'(x)|^2 + 2\,(1\pm x)^{-1}\,|f(x)|^2\right) dx}{\int\limits_{\cos\alpha}^{1} |f(x)|^2\, dx}.$$

Proof: Let us first establish the upper bound for $\Gamma(\Omega_\alpha)$. Taking the infimum in the right-hand side of (3.3.1) only over vector functions of the form $(0, g(\theta))$, where $g \in \overset{\circ}{h}{}_2^1(\Omega_\alpha)$, we get

$$\Gamma(\Omega_\alpha) \leq \inf_{g \neq 0,\, g(\alpha)=0} \frac{\int\limits_{0}^{\alpha}(|g'(\theta)|^2 + \cot^2\theta\,|g(\theta)|^2)\,\sin\theta\,d\theta}{\int\limits_{0}^{\alpha} |g(\theta)|^2\,\sin\theta\,d\theta}.$$

The change of variables $\cos\theta = x$ leads immediately to the upper bound in (3.4.3).

We show the lower bound. Since $\Gamma(\Omega_\alpha) \geq s(\Omega_\alpha)$, it is sufficient to estimate $s(\Omega_\alpha)$ from below. Observe that, if the infimum $s(\Omega_\alpha)$ in (3.4.1) is attained on the pair (u_θ, u_φ), then it is also attained on the pair $(u_\varphi, -u_\theta)$, since the form $Q(u_\omega, u_\omega)$ is invariant under this transformation. Consequently, there exists a minimizer of the form (f, if) or $(f, -if)$ on which $s(\Omega_\alpha)$ is attained. If we represent $f(\theta, \varphi)$ as

$$f = \sum_k f^{(k)}(\theta)\,\frac{e^{ik\varphi}}{\sqrt{2\pi}},$$

where $f^{(k)}(\alpha) = 0$, we find that

$$s(\Omega_\alpha) = \min_k \mu_k^\pm,$$

where

$$\mu_k^\pm = \inf_f \left(\int\limits_0^\alpha |f|^2\,\sin\theta\,d\theta\right)^{-1} \cdot \int\limits_0^\alpha \left(|f'(\theta)|^2 + \left(\frac{k}{\sin\theta} \pm \cot\theta\right)^2 |f(\theta)|^2\right)\sin\theta\,d\theta$$

and the infimum is taken over all functions f on the interval $[0, \alpha]$ such that $f(\alpha) = 0$. Since for $k \geq 2$ and $\theta \in (0, \alpha]$ the inequality

$$\left(\frac{k}{\sin\theta} \pm \cot\theta\right)^2 > \left(\frac{1}{\sin\theta} \pm \cot\theta\right)^2$$

is satisfied, we have $\mu_1^\pm < \mu_k^\pm$ for $k \geq 2$ and, therefore,

$$s(\Omega_\alpha) = \min(\mu_0, \mu_1^+, \mu_1^-).$$

Substituting $x = \cos\theta$ into the expressions for μ_0 and $\mu_1^\pm$, we get the quantities $\gamma_0 - 1$ and $\gamma_1^\pm - 1$, respectively. This proves the lemma. ∎

The figure below shows the dependence of the bounds $\gamma_0(\alpha) - 1$, $\gamma_1^+(\alpha) - 1$ for $\Gamma(\Omega_\alpha)$, $\alpha \in (\pi/2, \pi)$.

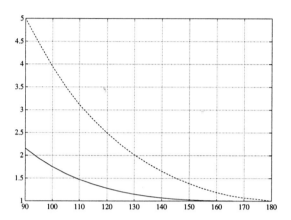

FIGURE 6. Upper bound $\gamma_0(\alpha) - 1$ and lower bound $\gamma_1^+(\alpha) - 1$ for $\Gamma(\Omega_\alpha)$

REMARK 3.4.2. Let $\alpha = \pi$. Then $\Gamma(S^2) = 1$ and the only extremal element (up to normalization) in (3.3.1) is the vector $(u_\theta, u_\varphi) = (0, \sin\theta)$.

Indeed, by Lemma 3.2.3, we have $\Gamma(S^2) \geq 1$. Since the right-hand side of (3.3.1) is equal to one for $\Omega = S^2$ and $(u_\theta, u_\varphi) = (0, \sin\theta)$, we conclude that $\Gamma(S^2) = 1$. In order to show that there are no other minimizers, it suffices to consider only those minimizers of the functional s for which $\nabla_\omega \cdot u_\omega = 0$. Since $\mu_k > \mu_1$ for $k \geq 2$, we need only consider functions which are either independent of φ or which are of the form $e^{i\varphi}v_\omega(\theta)$. Using formula (3.2.13), we obtain in the first case that

$$Q(u_\omega, u_\omega) = \int_{S^2} |u_\omega|^2 \, d\omega$$

and $u_\omega = (0, c\sin\theta)$ if $\nabla_\omega \cdot u_\omega = 0$. In the second case the right-hand side of (3.2.13) is always positive for nonzero $e^{i\varphi}v_\omega(\theta) \in \overset{\circ}{h}{}_2^1(S^2)$.

REMARK 3.4.3. Let $\gamma_0(\alpha) = \mu(\mu + 1)$, $\mu > 0$. As is easily seen, μ is the smallest positive root of the equation $P_\mu^1(\cos\alpha) = 0$, where P_μ^1 denotes the associated Legendre function. The corresponding minimizer is equal to $P_\mu^1(x)$.

In the next lemma we employ the notion of capacity, see (2.2.12). Note that

$$\mathrm{cap}\,(S^2\backslash\Omega) = 0 \quad \Leftrightarrow \quad h_2^1(S^2) = \overset{\circ}{h}{}_2^1(\Omega) \quad \Leftrightarrow \quad W_2^1(S^2) = \overset{\circ}{W}{}_2^1(\Omega).$$

As is well known, the capacity of any arc is positive.

LEMMA 3.4.3. Let $\Omega \subset S^2$ and $\mathrm{cap}\,(S^2\backslash\Omega) > 0$. Then $\Gamma(\Omega) > 1$.

Proof: It is known (see the article [**46**] by Deny and Lions or Landkof's book [**164**]) that every function $u \in \overset{\circ}{W}{}^1_2(\Omega)^3$, "precisée" in the sense of [**46**], is equal to zero quasi-everywhere on $S^2 \backslash \Omega$. By Lemma 3.2.1, the same holds for all $v \in \overset{\circ}{h}{}^1_2(\Omega)$.

Now the crucial observation is that the vector $(0, \sin\theta)$ for which, according to Remark 3.4.2, the infimum $\Gamma = 1$ is attained is nonzero quasi-everywhere on S^2 as it vanishes only at the poles. Hence $(0, \sin\theta) \notin \overset{\circ}{h}{}^1_2(\Omega)$ if $\operatorname{cap}(S^2 \backslash \Omega) > 0$. This implies $\Gamma > 1$ if $\operatorname{cap}(S^2 \backslash \Omega) > 0$. ∎

3.5. A variational principle for real eigenvalues

Let $\alpha = -1/2$, $\beta = \min\{F_\nu(\Omega) - 1/2, \, 3 - 4\nu\}$. By Theorem 3.3.1, the eigenvalues of the pencil $\mathcal{L}$ in the strip $\alpha \le \operatorname{Re}\lambda < \beta$ are real. By means of a variational principle, we show that these eigenvalues depend monotonically on the domain Ω.

We set

$$A(\lambda) \overset{def}{=} \mathcal{T}(\lambda)\,\mathcal{L}(\lambda),$$

where $\mathcal{T}(\lambda)$ is the matrix (3.2.26). Then every eigenvalue of the pencil $\mathcal{L}$ is also an eigenvalue of the pencil A. Furthermore, we have

$$\tilde{a}(u, u; \lambda) \overset{def}{=} \int\limits_\Omega A(\lambda) \begin{pmatrix} u_r \\ u_\omega \end{pmatrix} \cdot \begin{pmatrix} \overline{u}_r \\ \overline{u}_\omega \end{pmatrix} d\omega = (3 - 4\nu - \lambda)\, q(u, u; \lambda)$$

$$= (3 - 4\nu - \lambda) \left(Q(u_\omega, u_\omega) + (1 - \lambda - \lambda^2) \int\limits_\Omega |u_\omega|^2 \, d\omega \right)$$

$$+ (3 - 4\nu - \lambda)\left(\frac{1}{1 - 2\nu} \int\limits_\Omega |\nabla_\omega \cdot u_\omega|^2 \, d\omega + \frac{2(4 - 4\nu + \lambda)}{1 - 2\nu} \operatorname{Re} \int\limits_\Omega u_r \nabla_\omega \cdot \overline{u}_\omega \, d\omega \right)$$

$$+ (4 - 4\nu + \lambda)\left(\int\limits_\Omega |\nabla_\omega u_r|^2 \, d\omega + \frac{2 - 2\nu}{1 - 2\nu}(1 - \lambda)(\lambda + 2) \int\limits_\Omega |u_r|^2 \, d\omega \right)$$

for $\lambda \in \mathbb{R}$, $(u_r, u_\omega) \in \overset{\circ}{W}{}^1_2(\Omega) \times \overset{\circ}{h}{}^1_2(\Omega)$. The form $\tilde{a}$ and the pencil A satisfy conditions (I)–(III) in Section 1.3 if we set

$$\mathcal{H}_+ = \overset{\circ}{W}{}^1_2(\Omega) \times \overset{\circ}{h}{}^1_2(\Omega) \quad \text{and} \quad \mathcal{H} = L_2(\Omega)^3.$$

Condition (I) is a consequence of $3 - 4\nu - \lambda > 0$, while conditions (II) and (III) follow from Lemmas 3.2.6 and 3.3.1.

By Theorem 1.3.1, the operator $A(\lambda)$ is selfadjoint, bounded from below and has discrete spectrum for every fixed $\lambda \in [\alpha, \beta)$. We consider the eigenvalue problem

$$A(\lambda)\, u = \mu(\lambda)\, u$$

for fixed $\lambda \in [\alpha, \beta)$. By $\{\mu_j(\lambda)\}_{j \ge 1}$ we denote a nondecreasing sequence of eigenvalues of this problem counted with their multiplicities. As it was shown in Section 1.3 (see Theorem 1.3.2), the functions μ_j are continuous on the interval $[\alpha, \beta)$. Furthermore, as a consequence of Theorems 1.3.2, 1.3.3 and Corollary 3.3.1, we obtain the following results.

THEOREM 3.5.1. 1) *All eigenvalues of the pencil $\mathcal{L}$ in the strip $-1/2 \leq \operatorname{Re}\lambda < \beta$ are real and may be characterized by*

$$\{\lambda_j\}_{j=1,\dots,J} = \left\{\lambda \in [-\frac{1}{2}, \beta] \ : \ \mu_j(\lambda) = 0 \ for \ j = 1, \dots, J\right\},$$

where J is the largest index j for which the function μ_j has a zero in the interval $[-\frac{1}{2}, \beta)$. For every j the function μ_j has not more than one zero in $[-\frac{1}{2}, \beta)$.

2) *If $\lambda_0 \in [-\frac{1}{2}, \beta)$ is an eigenvalue of multiplicity I, then I is equal to the number of function μ_j which have a zero at λ_0.*

Now we study the dependence of the eigenvalues on the domain Ω. To this end, we will indicate the dependence of the quantities on the domain explicitly (for example $\beta(\Omega)$, $\mu_j(\lambda; \Omega)$, and so on). By the variational principle for eigenvalues of selfadjoint operators which are bounded from below, we have

$$\mu_j(\lambda; \Omega) = \max_L \ \min_{u \in L \setminus \{0\}} \frac{\tilde{a}(u, u; \lambda, \Omega)}{\|u\|^2_{L_2(\Omega)^3}} \quad \text{for } \lambda \in [-\frac{1}{2}, \beta(\Omega)),$$

where the maximum is taken over all subspaces $L \subset \overset{\circ}{W}{}^1_2(\Omega) \times \overset{\circ}{h}{}^1_2(\Omega)$ of codimension $\geq j - 1$.

We enumerate the eigenvalues of the pencil $\mathcal{L}(\lambda; \Omega)$ which are located in the interval $[-\frac{1}{2}, \beta(\Omega))$ in nondecreasing order:

$$-1/2 < \lambda_1(\Omega) \leq \dots \leq \lambda_J(\Omega) < \beta(\Omega).$$

Here the eigenvalues are counted with their geometric multiplicities, i.e., $J = J(\Omega)$ denotes the sum of the geometric multiplicities of all eigenvalues of $\mathcal{L}(\lambda; \Omega)$ in the interval $[-\frac{1}{2}, \beta)$.

THEOREM 3.5.2. *Let Ω_1, Ω_2 be open subsets of the unit sphere, $\Omega_1 \subset \Omega_1$. Then $J(\Omega_1) \leq J(\Omega_2)$ and*

(3.5.1) $\lambda_j(\Omega_2) \leq \lambda_j(\Omega_1) \quad for \ j = 1, \dots, J(\Omega_1).$

If, in addition, $\Omega_1 \subset\subset \Omega_2$, then in (3.5.1) even the strict inequality holds.

Proof: We apply Theorem 1.3.5. In our case

$$\mathcal{H}_+ = \overset{\circ}{W}{}^1_2(\Omega_2) \times \overset{\circ}{h}{}^1_2(\Omega_0) \quad \text{and} \quad \mathcal{H} = L_2(\Omega_2)^3.$$

The same spaces with Ω_1 instead of Ω_2 are denoted by H_+ and H. If we identify every function on Ω_1 with its extension by zero to Ω_2, we can consider the spaces H_+, H as subspaces of $\mathcal{H}_+$ and $\mathcal{H}$, respectively. Applying the first part of Theorem 1.3.5 to the form $\tilde{a}$, we obtain (3.5.1). In order to prove the strict inequality, we have to verify that the equality

$$\tilde{a}(u, v; \lambda) = 0 \quad \text{for all } v \in \mathcal{H}_+,$$

where $u \in H_+$, $-1/2 < \lambda < \beta(\Omega_1)$, implies $u = 0$. Indeed, from the above equality we obtain that $\mathcal{L}(\lambda)\,u = 0$ in Ω_2. Since the operator $\mathcal{L}(\lambda)$ is elliptic and has analytic coefficients, we obtain that u is analytic in Ω_2. Therefore, from the equality $u = 0$ in $\Omega_2 \setminus \Omega_1$ it follows that $u = 0$. Now it suffices to refer to part 2) of Theorem 1.3.5 to complete the proof. ∎

If $\mathcal{K} = \mathbb{R}^3$, then the only solutions of (3.2.1), (3.2.2) having the form (3.0.2) with $-\frac{1}{2} \leq \operatorname{Re}\lambda \leq 1$ are the vector functions $U = a$ and $U = b^{(1)}x_1 + b^{(2)}x_2 + b^{(3)}x_3$, where a, $b^{(1)}$, $b^{(2)}$, $b^{(3)} \in \mathbb{C}^3$. In the case $\mathcal{K} = \mathbb{R}^3_+ = \{x \in \mathbb{R}^3 : x_3 > 0\}$ the only solution of this form is the vector function $U = bx_3$, where $b \in \mathbb{C}^3$. Consequently, for $\mathcal{K} = \mathbb{R}^3$ the spectrum of the pencil $\mathcal{L}$ consists only of the eigenvalues $\lambda_0 = 0$ and $\lambda_1 = 1$ with geometric multiplicities 3 and 9, respectively. In the case $\mathcal{K} = \mathbb{R}^3_+$ the spectrum in this strip consists only of the eigenvalue $\lambda = 1$ with geometric multiplicity 3. In both cases there are no generalized eigenvectors.

Using Theorem 3.5.2 and this information on the spectrum of the pencil $\mathcal{L}$ for $\Omega = S^2_+$ and $\Omega = S^2$ we arrive at the following statement.

THEOREM 3.5.3. *If $\Omega \subset S^2_+$, then the strip $|\mathrm{Re}\,\lambda + \frac{1}{2}| \leq \frac{3}{2}$ contains no eigenvalues of the operator pencil $\mathcal{L}$. If $S^2_+ \subset \Omega \subset S^2$, then the strip $|\mathrm{Re}\,\lambda + \frac{1}{2}| \leq \frac{3}{2}$ contains exactly three eigenvalues of $\mathcal{L}$. They are real and the corresponding eigenfunctions do not have generalized eigenfunctions.*

3.6. Estimates for the width of the energy strip

We suppose that $\mathrm{cap}\,(S^2 \backslash \Omega) > 0$. Then the first eigenvalue of the Dirichlet problem for the operator $-\delta$ in Ω is positive. We represent this eigenvalue in the form $M(M + 1)$, $M > 0$.

Our goal is the estimation of the energy strip for the pencil $\mathcal{L}$ stated in terms of M. For arbitrary real t we set

$$\phi(t) = (2t + 1)(t + 1)(t + 2) - (3 - 4\nu - t)(M - t)(M + t + 1).$$

Furthermore, we denote by $t(M)$ the smallest solution of the equation

$$\phi(t) = 2(2t + 1)$$

lying in the interval $-1/2 < t < M$. Since $\phi(t) < 2(2t + 1)$ for $-1/2 < t \leq 0$ and $\phi(M) > 2(2M + 1)$, such a solution always exists and is positive.

THEOREM 3.6.1. *The strip determined by the inequality*

(3.6.1)
$$\left| \mathrm{Re}\,\lambda + \frac{1}{2} \right| \leq \min\{1, t(M)\} + \frac{1}{2}$$

does not contain eigenvalues of the pencil $\mathcal{L}$.

Proof: Since there are no eigenvalues of $\mathcal{L}(\lambda)$ on the line $\mathrm{Re}\,\lambda = -1/2$ and λ_0 is an eigenvalue if and only if $-1 - \bar{\lambda}_0$ is an eigenvalue (see Theorem 3.2.1), it suffices to prove the nonexistence of eigenvalues in the strip (3.6.1) for $\mathrm{Re}\,\lambda > -1/2$. Suppose λ_0 is such an eigenvalue and u is an eigenfunction corresponding to λ_0. Then $U(x) = r^{\lambda_0} u_0(\omega)$ is a solution of the boundary value problem (3.2.1), (3.2.2). Consequently, the pair (U, P), where $P = -(1 - 2\nu)^{-1}\,\mathrm{div}\,U$, satisfies the equations

(3.6.2) $$-\Delta U + \mathrm{grad}\,P = 0,$$
(3.6.3) $$\mathrm{div}\,U + (1 - 2\nu)\,P = 0$$

in $\mathcal{K}$. For arbitrary $\varepsilon > 0$ let $\mathcal{K}_\varepsilon$ be the domain $\{x \in \mathcal{K} : \varepsilon < |x| < 1/\varepsilon\}$. Then (3.6.2) implies

$$-\int_{\mathcal{K}_\varepsilon} \Delta U \cdot \overline{U}\,dx + \int_{\mathcal{K}_\varepsilon} \mathrm{grad}\,P \cdot \overline{U}\,dx = 0.$$

Integrating by parts and using (3.6.3), we get

$$\int_{\mathcal{K}_\varepsilon} |\nabla U|^2\, dx + \varepsilon^2 \int_{r=\varepsilon} \overline{U}\, \partial_r U\, d\omega - \varepsilon^{-2} \int_{r=1/\varepsilon} \overline{U}\, \partial_r U\, d\omega$$

$$+ (1-2\nu) \int_{\mathcal{K}_\varepsilon} |P|^2\, dx - \varepsilon^2 \int_{r=\varepsilon} P\cdot\overline{U}_r\, d\omega + \varepsilon^{-2} \int_{r=1/\varepsilon} P\cdot\overline{U}_r\, d\omega = 0.$$

Passing to the functions $u(\omega) = r^{-\lambda_0} U(x)$ and $p(\omega) = r^{-\lambda_0+1} P(x)$, we arrive at

$$\int_\varepsilon^{1/\varepsilon} r^{2\,\mathrm{Re}\,\lambda_0}\, dr \int_\Omega \left(|\nabla_\omega u|^2 + |\lambda_0|^2\, |u|^2 + (1-2\nu)\,|p|^2 \right) d\omega$$

$$+ \left(\varepsilon^{2\,\mathrm{Re}\,\lambda_0+1} - \varepsilon^{-2(\mathrm{Re}\,\lambda_0+1)} \right) \left(\lambda_0 \int_\Omega |u|^2\, d\omega - \int_\Omega p\,\overline{u}_r\, d\omega \right) = 0.$$

Here we have used the Cartesian components of the vector function u, and by $\nabla_\omega u$ we have denoted the vector function $(\nabla_\omega u_1, \nabla_\omega u_2, \nabla_\omega u_3)$. For $2\,\mathrm{Re}\,\lambda_0 > -1$ the last equality implies

(3.6.4)
$$\int_\Omega |\nabla_\omega u|^2\, d\omega + \left((\mathrm{Im}\,\lambda_0)^2 - \mathrm{Re}\,\lambda_0\,(\mathrm{Re}\,\lambda_0 + 1) \right) \int_\Omega |u|^2\, d\omega$$

$$+ (1-2\nu) \int_\Omega |p|^2\, d\omega + (2\,\mathrm{Re}\,\lambda_0 + 1)\,\mathrm{Re} \int_\Omega p\,\overline{u}_r\, d\omega = 0.$$

By (3.6.2) and (3.6.3), we have

$$r\partial_r P \;=\; x\cdot\mathrm{grad}\, P = x\cdot\Delta U = \Delta(x\cdot U) - 2\,\mathrm{div}\, U$$
$$=\; \Delta(rU_r) + 2(1-2\nu)P.$$

Taking $P = r^{\lambda_0-1} p(\omega)$, $U_r = r^{\lambda_0} u_r(\omega)$, we obtain

(3.6.5) $$p(\omega) = -(3-4\nu-\lambda_0)^{-1} \left(\delta u_r(\omega) + (\lambda_0+1)(\lambda_0+2) u_r(\omega) \right)$$

and, consequently,

$$\mathrm{Re} \int_\Omega p\,\overline{u}_r\, d\omega \;=\; \mathrm{Re}\,(3-4\nu-\lambda_0)^{-1} \int_\Omega |\nabla_\omega u_r|^2\, d\omega$$

$$-\;\mathrm{Re}\,\frac{(\lambda_0+1)(\lambda_0+2)}{3-4\nu-\lambda_0} \int_\Omega |u_r|^2\, d\omega.$$

From this identity and from (3.6.4) it follows that

(3.6.6) $$\int_\Omega |\nabla_\omega u|^2\, d\omega + \left((\mathrm{Im}\,\lambda_0)^2 - \mathrm{Re}\,\lambda_0\,(\mathrm{Re}\,\lambda_0 + 1) \right) \int_\Omega |u|^2\, d\omega$$

$$+ (1-2\nu) \int_\Omega |p|^2\, d\omega + (2\,\mathrm{Re}\,\lambda_0 + 1)\,\mathrm{Re}\,(3-4\nu-\lambda_0)^{-1} \int_\Omega |\nabla_\omega u_r|^2\, d\omega$$

$$-(2\,\mathrm{Re}\,\lambda_0 + 1)\,\mathrm{Re}\,\frac{(\lambda_0+1)(\lambda_0+2)}{3-4\nu-\lambda_0} \int_\Omega |u_r|^2\, d\omega = 0.$$

Obviously,

$$(3.6.7) \qquad \int_\Omega |\nabla_\omega u|^2 \, d\omega + \left((\operatorname{Im}\lambda_0)^2 - \operatorname{Re}\lambda_0 \, (\operatorname{Re}\lambda_0 + 1) \right) \int_\Omega |u|^2 \, d\omega$$

$$\geq \left(M(M+1) - \operatorname{Re}\lambda_0 \, (\operatorname{Re}\lambda_0 + 1) \right) \int_\Omega |u|^2 \, d\omega.$$

Furthermore, we have

$$\operatorname{Re}\left((\lambda_0 + 1)(\lambda_0 + 2)(3 - 4\nu - \overline{\lambda}_0) \right)$$
$$= (\operatorname{Re}\lambda_0 + 1)\,(\operatorname{Re}\lambda_0 + 2)\,(3 - 4\nu - \operatorname{Re}\lambda_0) - (\operatorname{Im}\lambda_0)^2\,(6 - 4\nu + \operatorname{Re}\lambda_0)$$
$$\leq (\operatorname{Re}\lambda_0 + 1)\,(\operatorname{Re}\lambda_0 + 2)\,(3 - 4\nu - \operatorname{Re}\lambda_0).$$

From this and from (3.6.6), (3.6.7) we obtain

$$(3.6.8) \qquad (3 - 4\nu - \operatorname{Re}\lambda_0)\,(1 - 2\nu) \int_\Omega |p|^2 \, d\omega + (2\operatorname{Re}\lambda_0 + 1) \int_\Omega |\nabla_\omega u_r|^2 \, d\omega$$

$$+ (3 - 4\nu - \operatorname{Re}\lambda_0) \left(M(M+1) - \operatorname{Re}\lambda_0 \, (\operatorname{Re}\lambda_0 + 1) \right) \int_\Omega |u_\omega|^2 \, d\omega$$

$$\leq \phi(\operatorname{Re}\lambda_0) \int_\Omega |u_r|^2 \, d\omega \, .$$

By (3.6.3), we have $\nabla_\omega \cdot u_\omega + (\lambda_0 + 2)\,u_r + (1 - 2\nu)p = 0$. Integrating this equality, we conclude that

$$\int_\Omega \left(u_r + \frac{1 - 2\nu}{\lambda_0 + 2}\,p \right) d\omega = 0.$$

We consider the minimization problem for the functional

$$(3.6.9) \qquad \frac{(3 - 4\nu - \operatorname{Re}\lambda_0)\,(1 - 2\nu)}{2\operatorname{Re}\lambda_0 + 1} \int_\Omega |q|^2 \, d\omega + \int_\Omega |\nabla_\omega v|^2 \, d\omega$$

on the set of pairs $(v, q) \in \overset{\circ}{W}{}_2^1(\Omega) \times L_2(\Omega)$ such that

$$(3.6.10) \qquad \int_\Omega \left(v + \frac{1 - 2\nu}{\lambda_0 + 2}\,q \right) d\omega = 0, \qquad \|v\|_{L_2(\Omega)} = 1.$$

Let $\Xi(\Omega, \lambda_0)$ be the minimum of this functional and let (v_0, q_0) be a pair realizing this minimum. Inserting $v = v_0$ and $q = q_0 + q_1$ into (3.6.9), where q_1 is an arbitrary function in $L_2(\Omega)$ orthogonal to 1, we obtain

$$\operatorname{Re} \int_\Omega q_0 \cdot \overline{q}_1 \, d\omega = 0.$$

Consequently, q_0 is constant and we can restrict ourselves to constants q in the above formulated variational problem. From this and from (3.6.10) it follows

$$\Xi(\Omega, \lambda_0) = \inf \left\{ \int_\Omega |\nabla_\omega v|^2 \, d\omega + \frac{(3 - 4\nu - \operatorname{Re}\lambda_0)\,|\lambda_0 + 2|^2}{(1 - 2\nu)\,(2\operatorname{Re}\lambda_0 + 1)\,|\Omega|} \left| \int_\Omega v \, d\omega \right|^2 \right\},$$

where the infimum is taken over all $v \in \overset{\circ}{W}^1_2(\Omega)$ with $L_2(\Omega)$-norm equal to one. This representation shows, in particular, that the function $\lambda \to \Xi(\Omega, \lambda)$ decreases on the interval $(-\frac{1}{2}, 3 - 4\nu]$ and that the function $\Omega \to \Xi(\Omega, \lambda)$ does not increase as Ω increases.

Representing v as a series of spherical harmonics, we obtain

$$\Xi(S^2, \lambda_0) \geq \min \left\{ 2, \, \frac{(3 - 4\nu - \operatorname{Re}\lambda_0)\,(\operatorname{Re}\lambda_0 + 2)^2}{(1 - 2\nu)\,(2\operatorname{Re}\lambda_0 + 1)} \right\}.$$

Since the function $t \to (3 - 4\nu - t)\,(t + 2)^2\,(1 - 2\nu)^{-1}\,(2t + 1)^{-1}$ decreases on the interval $(-\frac{1}{2}, 3 - 4\nu]$, there is the inequality

$$\frac{(3 - 4\nu - \operatorname{Re}\lambda_0)\,(\operatorname{Re}\lambda_0 + 2)^2}{(1 - 2\nu)\,(2\operatorname{Re}\lambda_0 + 1)} \geq 2$$

for $\operatorname{Re}\lambda_0 \leq 1$. Therefore, $\Xi(S^2, \lambda_0) \geq 2$ and hence, $\Xi(\Omega, \lambda_0) > 2$. By (3.6.8)

$$(3.6.11) \qquad (3 - 4\nu - \operatorname{Re}\lambda_0) \left(M(M + 1) - \operatorname{Re}\lambda_0\,(\operatorname{Re}\lambda_0 + 1) \right) \int_\Omega |u_\omega|^2 \, d\omega$$

$$\leq \left(\phi(\operatorname{Re}\lambda_0) - \Xi(\Omega, \lambda_0)(2\operatorname{Re}\lambda_0 + 1) \right) \int_\Omega |u_r|^2 \, d\omega.$$

Since $-1/2 < \operatorname{Re}\lambda_0 \leq \min\{1, t(M)\}$, we have

$$3 - 4\nu - \operatorname{Re}\lambda_0 > 0, \quad M(M + 1) - \operatorname{Re}\lambda_0\,(\operatorname{Re}\lambda_0 + 1) > 0$$

and $\phi(\operatorname{Re}\lambda_0) - \Xi(\Omega, \lambda_0)(2\operatorname{Re}\lambda_0 + 1) < 0$, from (3.6.11) we obtain $u_\omega = 0$ and $u_r = 0$. This proves the theorem. ∎

COROLLARY 3.6.1. *In the strip*

$$(3.6.12) \qquad |\operatorname{Re}\lambda + 1/2| \leq \min \left\{ 1, \frac{(3 - 4\nu)M}{M + 6 - 4\nu} \right\} + \frac{1}{2}$$

there are no eigenvalues of the pencil $\mathcal{L}$.

Proof: Let $0 < t(M + 6 - 4\nu) \leq (3 - 4\nu)M$. The last inequality can be rewritten in the form

$$t(t + 3) \leq (3 - 4\nu - t)\,(M - t).$$

Multiplying this estimate by $2t + 1$ and using the inequality $t < M$, we get

$$t(2t + 1)(t + 3) < (3 - 4\nu - t)(M - t)(M + t + 1)$$

or, what is the same, $\phi(t) < 2(2t + 1)$. Consequently,

$$t(M) > \frac{(3 - 4\nu)M}{M + 6 - 4\nu}$$

and, by Theorem 3.6.1, we obtain (3.6.12). ∎

REMARK 3.6.1. In Theorem 3.6.1 the guaranteed width of the strip not containing eigenvalues does not exceed 3. The following estimate shows that the width of the energy strip can be arbitrarily large if the domain Ω is sufficiently small: the spectrum of the pencil $\mathcal{L}$ lies outside the strip

$$(3.6.13) \qquad |\operatorname{Re}\lambda + 1/2| < \left(\frac{1 - 2\nu}{2 - 2\nu} \right)^{1/2} (M + 1/2).$$

Indeed, inequality (3.6.13) means that the quadratic function

$$\Big(M(M+1) - \operatorname{Re}\lambda\,(\operatorname{Re}\lambda + 1) + (\operatorname{Im}\lambda)^2\Big)\,t^2 + (2\operatorname{Re}\lambda + 1)\,t + 1 - 2\nu$$

is positive. On the other hand, it follows from (3.6.4) that

$$\Big(M(M+1) - \operatorname{Re}\lambda\,(\operatorname{Re}\lambda + 1) + (\operatorname{Im}\lambda)^2\Big) \int_\Omega |u|^2\,d\omega$$

$$+ (2\operatorname{Re}\lambda + 1)\operatorname{Re}\int_\Omega p\,\overline{u}_r\,d\omega + (1 - 2\nu)\int_\Omega |p|^2\,d\omega \le 0.$$

Hence $u = 0$, $p = 0$, i.e., there are no eigenvalues in the strip (3.6.13). Estimate (3.6.13) improves the result obtained in Theorem 3.6.1 for large M.

3.7. Eigenvalues for circular cones

3.7.1. The Boussinesq solution in spherical coordinates. Let $\Omega = \Omega_\alpha = \{\omega \in S^2 : 0 \le \theta < \alpha,\ 0 \le \varphi < 2\pi\}$, $\alpha \in (0, \pi]$. We are interested in nonnegative real eigenvalues of the pencil $\mathcal{L}$. For this we have to find special solutions of problem (3.2.1), (3.2.2) which have the form $U = r^\lambda u(\theta, \varphi)$ with nonnegative real λ.

To find such solutions, we use the following representation of the general solution of system (3.2.1) in terms of three harmonic functions Ψ, Θ, Λ which goes back to Boussinesq [**23**]:

$$(3.7.1) \qquad U = a\nabla\Psi + 2b\nabla \times (\Theta\vec{e}_3) + c\left(\nabla(x_3\Lambda) - 4(1-\nu)\Lambda\vec{e}_3\right),$$

where ∇ denotes the gradient in Cartesian coordinates, $\vec{e}_3$ is the unit vector in x_3 direction, and a, b, c are arbitrary constants. A general harmonic function W can be found by separation of variables: $W = R(r)\,T(\theta)\,S(\varphi)$. Since we are interested in nontrivial solutions of the form $U(x) = r^\lambda\,u(\omega)$, where λ is a real nonnegative number, let us consider a harmonic function of the form $W = r^\mu\,T_\mu^{(m)}(\theta)\,S_m(\varphi)$ in the cone $\mathcal{K}$. Then $T_\mu^{(m)}$ must satisfy the associated Legendre differential equation

$$(3.7.2) \qquad \frac{d^2 T_\mu^{(m)}}{d\theta^2} + \cot\theta\,\frac{dT_\mu^{(m)}}{d\theta} + \left(\mu(\mu+1) - \frac{m^2}{\sin^2\theta}\right) T_\mu^{(m)} = 0,$$

while S_m satisfies the equation

$$\frac{d^2 S_m}{d\varphi^2} + m^2\,S_m = 0 \quad \text{for } \varphi \in [0, 2\pi)$$

with the periodic boundary conditions $S_m(0) = S_m(2\pi)$, $S_m'(0) = S_m'(2\pi)$. This implies $m = 0, \pm 1, \pm 2, \ldots$ and $S_m(\varphi) = \cos(m\varphi)$ or $S_m(\varphi) = \sin(m\varphi)$. For integer m, the general solution of (3.7.2) has the form (see Bateman, Erdélyi et al. [**11**, Vol.1,Ch.3])

$$T_\mu^m(\theta) = c_1\,Q_\mu^m(\cos\theta) + c_2\,P_\mu^{-m}(\theta),$$

where P_μ^{-m}, Q_μ^m are the associated Legendre functions of the first and second kind, respectively. The requirement that W is bounded for finite r leads to $c_1 = 0$. Thus, an admissible harmonic function in $\mathcal{K}$ has the form

$$(3.7.3) \qquad W = r^\mu\,P_\mu^{-m}(\cos\theta)\,\sin(m\varphi) \quad \text{or} \quad W = r^\mu\,P_\mu^{-m}(\cos\theta)\,\cos(m\varphi).$$

The representation (3.7.1) has the form

$$
\begin{pmatrix} U_r \\ U_\theta \\ U_\varphi \end{pmatrix} = a \begin{pmatrix} \partial_r \Psi \\ r^{-1} \partial_\theta \Psi \\ (r \sin \theta)^{-1} \partial_\varphi \Psi \end{pmatrix} + 2 \frac{b}{r} \begin{pmatrix} \partial_\varphi \Theta \\ \cot \theta \, \partial_\varphi \Theta \\ -\sin \theta \, \partial_r (r\Theta) - \partial_\theta (\Theta \cos \theta) \end{pmatrix}
$$
$$
+ c \begin{pmatrix} \cos \theta \left(\partial_r (r\Lambda) - 4(1 - \nu)\Lambda \right) \\ \partial_\theta (\Lambda \cos \theta) + 4(1 - \nu)\Lambda \sin \theta \\ \cot \theta \, \partial_\varphi \Lambda \end{pmatrix}
$$

in spherical coordinates. We select Ψ, Θ and Λ in (3.7.1) of the form (3.7.3) such that the displacement field U has the form $U = r^\lambda u_\lambda^{(m)}(\theta, \varphi)$ and each of the components U_r, U_θ, U_φ contains only $\cos(m\varphi)$ or $\sin(m\varphi)$, i.e., the angular modes $\cos(m\varphi)$ or $\sin(m\varphi)$ in the displacement field separate. This leads to the selections

(3.7.4)
$$
\begin{cases} \Psi_m = r^{\lambda+1} P_{\lambda+1}^{-m}(\cos \theta) \cos(m\varphi), \\ \Theta_m = r^{\lambda+1} P_{\lambda+1}^{-m}(\cos \theta) \sin(m\varphi), \\ \Lambda_m = r^\lambda P_\lambda^{-m}(\cos \theta) \cos(m\varphi), \end{cases}
$$

and

(3.7.5)
$$
\begin{cases} \Psi_m = r^{\lambda+1} P_{\lambda+1}^{-m}(\cos \theta) \sin(m\varphi), \\ \Theta_m = r^{\lambda+1} P_{\lambda+1}^{-m}(\cos \theta) \cos(m\varphi), \\ \Lambda_m = r^\lambda P_\lambda^{-m}(\cos \theta) \sin(m\varphi), \end{cases}
$$

respectively. Let $M_m^+(\lambda, \theta)$, $M_m^-(\lambda, \theta)$ be the 3×3 matrices with the columns

$$
\begin{pmatrix} (\lambda + 1) P_{\lambda+1}^{-m} \\ (\lambda + 1) \cot \theta \, P_{\lambda+1}^{-m} - (\lambda + 1 - m)(\sin \theta)^{-1} P_\lambda^{-m} \\ \mp m (\sin \theta)^{-1} P_{\lambda+1}^{-m} \end{pmatrix},
$$

$$
\begin{pmatrix} \pm 2m P_{\lambda+1}^{-m} \\ \pm 2m \cot \theta \, P_{\lambda+1}^{-m} \\ 2(\lambda + 1) \cot \theta \, P_\lambda^{-m} - 2(\lambda + 1)(\sin \theta)^{-1} P_{\lambda+1}^{-m} \end{pmatrix}
$$

and

$$
\begin{pmatrix} (\lambda - 3 + 4\nu) \cos \theta \, P_\lambda^{-m} \\ (\lambda + m + 1) \cot \theta \, P_{\lambda+1}^{-m} + \left((3 - 4\nu) \sin \theta - (\lambda + 1) \cos \theta \cot \theta \right) P_\lambda^{-m} \\ \mp m \cot \theta \, P_\lambda^{-m} \end{pmatrix}.
$$

Here the upper sign corresponds to $M_m^+(\lambda, \theta)$ and the lower sign to $M_m^-(\lambda, \theta)$. Inserting (3.7.4), (3.7.5) into (3.7.1), then straightforward calculation yields the following result.

LEMMA 3.7.1. *Let m be a nonnegative integer and λ a real number. Then to the selections (3.7.4) and (3.7.5) for the harmonic functions Ψ, Θ and Λ in (3.7.1) correspond the displacement fields $r^\lambda u_\lambda^{(m)}(\theta, \varphi)$ with*

(3.7.6) $\quad u_\lambda^{(m)}(\theta, \varphi) = \begin{pmatrix} f_m^+(\theta) \cos(m\varphi) \\ g_m^+(\theta) \cos(m\varphi) \\ h_m^+(\theta) \sin(m\varphi) \end{pmatrix}$ *for selection (3.7.4)*

and

(3.7.7) $\quad u_\lambda^{(m)}(\theta, \varphi) = \begin{pmatrix} f_m^-(\theta) \sin(m\varphi) \\ g_m^-(\theta) \sin(m\varphi) \\ h_m^-(\theta) \cos(m\varphi) \end{pmatrix}$ *for selection (3.7.5), where*

$$(3.7.8) \qquad \begin{pmatrix} f_m^+ \\ g_m^+ \\ h_m^+ \end{pmatrix} = M_m^+(\lambda, \theta) \begin{pmatrix} a \\ b \\ c \end{pmatrix}, \qquad \begin{pmatrix} f_m^- \\ g_m^- \\ h_m^- \end{pmatrix} = M_m^-(\lambda, \theta) \begin{pmatrix} a \\ b \\ c \end{pmatrix}.$$

Note that $M_0^+(\lambda, \theta) = M_0^-(\lambda, \theta)$ and $\det M_m^+(\lambda, \theta) = \det M_m^-(\lambda, \theta)$ for $m = 1, 2, \ldots$.

3.7.2. The first eigenvalues of the pencil $\mathcal{L}$. There exist nonzero vector functions (3.7.8) satisfying the Dirichlet boundary conditions at $\theta = \alpha$ if

$$(3.7.9) \qquad\qquad\qquad \det M_m^+(\lambda, \alpha) = 0$$

for some integer m. The solutions of the transcendental equation (3.7.9) are eigenvalues of the pencil $\mathcal{L}$ in Ω_α. The vector functions (3.7.6), (3.7.7), where the vector (a, b, c) in (3.7.8) is an eigenvector of $\det M_m^+(\lambda, \alpha)$ and $\det M_m^-(\lambda, \alpha)$, respectively, are the corresponding eigenvectors.

The monotonicity of the eigenvalues of $\mathcal{L}$ with respect to the domain Ω established in Theorem 3.5.2 allows one to give sharp upper and lower bounds for the eigenvalues for an arbitrary domain Ω in terms of those for rotational cones.

THEOREM 3.7.1. *Let $\pi/2 \leq \alpha \leq \beta < \pi$ and $\Omega_\alpha \subset \Omega \subset \Omega_\beta$. Then in the strip $0 \leq \operatorname{Re}\lambda \leq 1$ the pencil $\mathcal{L}$ has exactly three eigenvalues which are real. The eigenvalues satisfy*

$$\lambda_j(\Omega_\beta) \leq \lambda_j(\Omega) \leq \lambda_j(\Omega_\alpha).$$

If $\Omega_\alpha \subset\subset \Omega \subset\subset \Omega_\beta$, then the strict inequalities hold here.

The figures below show the dependence of the eigenvalues $\lambda_j(\Omega_\alpha) \leq 1$ on $\alpha \in (\frac{\pi}{2}, \pi)$. In the first figure the roots of equation (3.7.9) for $m = 0$ are depicted for several values of ν including, in particular, the limiting case $\nu = 1/2$ which corresponds to the Dirichlet problem for the Stokes system. In this case $\lambda = 1$ is an eigenvalue for all angles α and it is simple except for $\alpha = \frac{2\pi}{3}$. For this angle the eigenvector corresponding to $\lambda = 1$ has exactly one generalized eigenvector (see Chapter 5). For fixed α the eigenvalues are monotonically increasing with respect to ν.

In the second figure the roots of (3.7.9) for $m = 1$ and the same values of ν are depicted. We point out that this eigenvalue lies below the corresponding eigenvalue for $m = 0$, has two eigenvectors and decreases monotonically for increasing ν and fixed α.

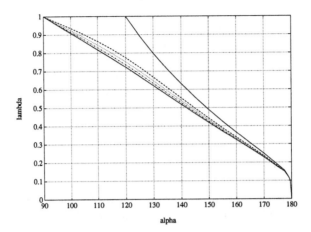

FIGURE 7. Solutions of (3.7.9) for $m = 0$, $\nu = 0.2, 0.25, 0.3, 0.35, 0.5$

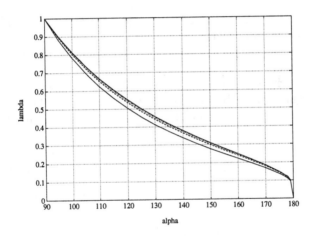

FIGURE 8. Solutions of (3.7.9) for $m = 1$, $\nu = 0.2, 0.25, 0.3, 0.35, 0.5$

3.8. Applications

The information about the spectral properties of the operator pencils obtained in Sections 3.1–3.7 enables one to answer various questions concerning the Dirichlet problem for the Lamé system

$$(3.8.1) \qquad \Delta U + (1 - 2\nu)^{-1} \nabla \nabla \cdot U = F \quad \text{in } \mathcal{G}, \quad U = 0 \quad \text{on } \partial\mathcal{G}$$

in a two- or three-dimensional domain with piecewise smooth boundary. In this section we discuss the solvability in weighted Sobolev and Hölder spaces, regularity properties of variational solutions, estimates for Green's tensor and the Miranda-Agmon maximum principle for the Lamé system.

3.8.1. The Lamé system in plane domains with corners. Let $\mathcal{G}$ be a plane bounded domain with d corners $x^{(1)}, \dots, x^{(d)}$ on the boundary. Outside the set $\mathcal{S} = \{x^{(1)}, \dots, x^{(d)}\}$, the boundary is assumed to be smooth. Let $\mathcal{G}$ coincide with a plane wedge in a neighborhood of each point $x^{(\tau)}$. The opening of the angle at $x^{(\tau)}$ is denoted by α_τ.

1. **Solvability in weighted Sobolev and Hölder spaces.** Let $\vec{\beta} = (\beta_1, \dots, \beta_d) \in \mathbb{R}^d$ and let $V^l_{p,\vec{\beta}}(\mathcal{G})$ and $N^{l,\sigma}_{\vec{\beta}}(\mathcal{G})$ be the weighted spaces introduced in Section 1.4.

For arbitrary $\alpha \in (0, 2\pi]$ we set

$$(3.8.2) \qquad \lambda_*(\alpha) = \frac{x_-(\alpha)}{\alpha} \text{ for } 0 < \alpha \le \pi, \quad \lambda_*(\alpha) = \frac{x_+(\alpha)}{\alpha} \text{ for } \pi < \alpha \le 2\pi,$$

where x_-, x_+ are the functions introduced in Section 3.1.4. Then the following assertion holds as a result of Theorems 3.1.2 and 1.4.1.

THEOREM 3.8.1. *Problem (3.8.1) is uniquely solvable in $V^l_{p,\vec{\beta}}(\mathcal{G})^2$ for arbitrary $F \in V^{l-2}_{p,\vec{\beta}}(\mathcal{G})^2$, $l \ge 2$, if and only if $|l - \beta_\tau - 2/p| < \lambda_*(\alpha_\tau)$ for $\tau = 1, \dots, d$. If $|l + \sigma - \beta_\tau| < \lambda_*(\alpha_\tau)$ for $\tau = 1, \dots, d$, then problem (3.8.1) is uniquely solvable in $N^{l,\sigma}_{\vec{\beta}}(\mathcal{G})^2$ for every $F \in N^{l-2,\sigma}_{\vec{\beta}}(\mathcal{G})^2$.*

2. **Regularity assertions and asymptotics of the solution.** Problem (3.8.1) has a unique solution $U \in \overset{\circ}{W}{}^1_2(\mathcal{G})^2$ for arbitrary $F \in W^{-1}_2(\mathcal{G})^2$. The following regularity result is another consequence of Theorems 3.1.2 and 1.4.3.

THEOREM 3.8.2. *The solution $U \in \overset{\circ}{W}{}^1_2(\mathcal{G})^2$ of (3.8.1) belongs to $V^l_{p,\vec{\beta}}(\mathcal{G})^2$ if $F \in V^{l-2}_{p,\vec{\beta}}(\mathcal{G})^2$, $l \ge 2$, and $l - \beta_\tau - 2/p < \lambda_*(\alpha_\tau)$. It belongs to $N^{l,\sigma}_{\vec{\beta}}(\mathcal{G})^2$ if $F \in N^{l,\sigma}_{\vec{\beta}}(\mathcal{G})^2$, $l \ge 2$, and $l + \sigma - \beta_\tau < \lambda_*(\alpha_\tau)$.*

Since $\lambda_*(\alpha) > 1$ for $\alpha < \pi$, we obtain, in particular, that under the assumption $\alpha_\tau < \pi$, $\tau = 1, \dots, d$,

$$U \in W^2_2(\mathcal{G})^2 \quad \text{if} \quad F \in L_2(\mathcal{G})^2$$

and

$$U \in C^{1,\sigma}(\overline{\mathcal{G}})^2 \quad \text{for some } \sigma > 0 \text{ if} \quad F \in L_p(\mathcal{G})^2, \, p > 2 \, .$$

The asymptotics of the solution $U \in (\overset{\circ}{W}{}^1_2(\mathcal{G}))^2$ near $x^{(\tau)}$ can be described by using Theorems 1.4.4 and 3.1.2. Suppose, for simplicity, that $x^{(\tau)}$ is the origin, the domain $\mathcal{G}$ coincides with the angle $\{x \in \mathbb{R}^2 : r > 0, |\varphi| < \alpha/2\}$ and $F = 0$ near $x^{(\tau)}$. Here, by r, φ we denote the polar coordinates. We have

$$(3.8.3) \qquad \begin{pmatrix} U_r \\ U_\varphi \end{pmatrix} = c\, r^{\lambda_-(\alpha_\tau)} \begin{pmatrix} u_r^-(\varphi) \\ u_\varphi^-(\varphi) \end{pmatrix} + o\big(r^{\pi/\alpha_\tau}\big)$$

if $\alpha_\tau \in (0, \pi)$. The function λ_- was introduced in Section 3.1, its values are contained in the interval $(1, \pi/\alpha_\tau)$. The vector function (u_r^-, u_φ^-) is defined by (3.1.28).

In the case $\alpha_\tau \in (\pi, 2\pi)$ the asymptotics

$$(3.8.4) \qquad \begin{pmatrix} U_r \\ U_\varphi \end{pmatrix} = c_1\, r^{\lambda_+(\alpha_\tau)} \begin{pmatrix} u_r^+(\varphi) \\ u_\varphi^+(\varphi) \end{pmatrix} + c_2\, r^{\lambda_-(\alpha_\tau)} \begin{pmatrix} u_r^-(\varphi) \\ u_\varphi^-(\varphi) \end{pmatrix}$$

$$+ c_3\, r^{\lambda_-^{(1)}(\alpha_\tau)} \begin{pmatrix} u_{r,1}^-(\varphi) \\ u_{\varphi,1}^-(\varphi) \end{pmatrix} + o\big(r^{2\pi/\alpha_\tau}\big)$$

holds. Here $1/2 < \lambda_+(\alpha) < \lambda_-(\alpha) < 1 < \lambda_-^{(1)}(\alpha) < 2\pi/\alpha$.

The term $o(r^{j\pi/\alpha_\tau})$ in (3.8.3), (3.8.4), where $j = 1$ for $\alpha_\tau \in (0, \pi)$ and $j = 2$ for $\alpha_\tau \in (\pi, 2\pi)$, can be replaced by a remainder

$$V \in V_{p,\vec\beta}^l(\mathcal{G})^2, \quad l \geq 2, \; l - \beta_\tau - 2/p = j\pi/\alpha_\tau + \varepsilon$$

(ε is a sufficiently small number), if $F \in V_{p,\vec\beta}^{l-2}(\mathcal{G})^2$, and by a remainder

$$V \in N_{\vec\beta}^{l,\sigma}(\mathcal{G})^2, \quad l \geq 2, \; l + \sigma - \beta_\tau = j\pi/\alpha_\tau + \varepsilon,$$

if $F \in N_{\vec\beta}^{l-2,\sigma}(\mathcal{G})^2$.

3.8.2. The Lamé system in a domain of polyhedral type. Now let $\mathcal{G}$ be a bounded domain in $\mathbb{R}^3$ with piecewise smooth boundary $\partial\mathcal{G}$ containing singularities of the types of conic vertices, edges and polyhedral vertices. This means that $\partial\mathcal{G}$ is the union of faces Γ_j, $j = 1, \ldots, T$, edges E_k, $k = 1, \ldots, q$, and vertices $x^{(\tau)}$, $\tau = 1, \ldots, d$. Outside the set $\mathcal{S}$ of the edges E_k and vertices $x^{(\tau)}$, the boundary is assumed to be smooth. Furthermore, we suppose that the domain $\mathcal{G}$ is diffeomorphic to a dihedron $\mathcal{D} = K \times \mathbb{R}$, where K is a plane angle, in a neighborhood of each edge point and that for each vertex $x^{(\tau)}$ there exist a neighborhood $\mathcal{U}_\tau$ and a diffeomorphism $\kappa : \mathcal{U}_\tau \to \mathbb{R}^3$ of class C^∞ such that $\kappa(\mathcal{U}_\tau \cap \mathcal{G})$ is the intersection of a cone $\mathcal{K}_\tau = \{x : x/|x| \in \Omega_\tau\}$ with the unit ball and $\kappa'(x^{(\tau)}) = I$. Here Ω_τ is a subdomain of the sphere S^2 the boundary of which is the union of smooth arcs intersecting transversally, and I denotes the unit matrix.

1. Solvability in weighted Sobolev and Hölder spaces. We denote by $\rho_\tau(x)$ the distance of x to $x^{(\tau)}$, by r_k the regularized distance to E_k (i.e. a smooth function on $\mathcal{G}$ which coincides with the distance to E_k in a neighborhood of E_k), by $r(x)$ the regularized distance to the set $\mathcal{S}$, and by $\rho(x)$ the distance to the set $\{x^{(1)}, \ldots, x^{(d)}\}$. Then the weighted Sobolev space $V_{\vec\beta,\vec\delta}^{l,p}(\mathcal{G})$ is defined for $1 < p < \infty$, $l = 0, 1, \ldots$, $\vec\beta = (\beta_1, \ldots, \beta_d) \in \mathbb{R}^d$ and $\vec\delta = (\delta_1, \ldots, \delta_q) \in \mathbb{R}^q$ as the closure of the set $C_0^\infty(\overline{\mathcal{G}}\backslash\mathcal{S})$ with respect to the norm

$$\|U\|_{V_{\vec\beta,\vec\delta}^{l,p}(\mathcal{G})} = \left(\int_{\mathcal{G}} \left(\prod_{\tau=1}^d \rho_\tau^{p\beta_\tau} \right) \left(\prod_{k=1}^q \left(r_k/\rho \right)^{p\delta_k} \right) \sum_{|\alpha|\leq l} r^{p(|\alpha|-l)} |D^\alpha U|^p \, dx \right)^{1/p}.$$

Note that $L_2(\mathcal{G}) = V_{0,0}^{0,2}(\mathcal{G})$ and $\mathring{W}_2^l(\mathcal{G}) \subset V_{0,0}^{l,2}(\mathcal{G}) \subset W_2^l(\mathcal{G})$ for $l = 1, 2, \ldots$.

Furthermore, let $N_{\vec\beta,\vec\delta}^{l,\sigma}(\mathcal{G})$ be the weighted Hölder space with the norm

$$\|U\|_{N_{\vec\beta,\vec\delta}^{l,\sigma}(\mathcal{G})} = \left\| \left(\prod_{\tau=1}^d \rho_\tau^{\beta_\tau} \right) \left(\prod_{k=1}^q \left(r_k/\rho \right)^{\delta_k} \right) U \right\|_{N^{l,\sigma}(\mathcal{G})},$$

where

$$\|U\|_{N^{l,\sigma}(\mathcal{G})} = \sum_{|\alpha|\leq l} \sup_{x\in\mathcal{G}} r^{|\alpha|-l-\sigma} |D^\alpha U(x)| + \sum_{|\alpha|=l} \sup_{x,y\in\mathcal{G}} \frac{|(D^\alpha U)(x) - (D^\alpha U)(y)|}{|x-y|^\sigma}.$$

If β and δ are real numbers, then by $V_{\beta,\delta}^{l,p}(\mathcal{G})$ and $N_{\beta,\delta}^{l,\sigma}(\mathcal{G})$ we mean the weighted spaces just defined, where the role of $\vec\beta$ and $\vec\delta$ is played by the tuples $(\beta, \ldots, \beta) \in$

$\mathbb{R}^d$ and $(\delta, \ldots, \delta) \in \mathbb{R}^q$. It can be easily seen that $V_{\beta,\delta}^{l,p}(\mathcal{G}) \subset V_{\beta-1,\delta-1}^{l-1,p}(\mathcal{G})$ and $N_{\beta,\delta}^{l,\sigma}(\mathcal{G}) \subset N_{\beta-1,\delta-1}^{l-1,p}(\mathcal{G})$ for $l \geq 1$.

The trace spaces for $V_{\vec{\beta},\vec{\delta}}^{l,p}(\mathcal{G})$, $l \geq 1$, and $N_{\vec{\beta},\vec{\delta}}^{l,\sigma}(\mathcal{G})$ on the boundary are denoted by $V_{\vec{\beta},\vec{\delta}}^{l-1/p,p}(\partial\mathcal{G})$ and $V_{\vec{\beta},\vec{\delta}}^{l,\sigma}(\partial\mathcal{G})$, respectively.

Finally, we introduce the following notation. For every $x \in \bigcup E_k$ let α_x be the angle at x between the tangent half-planes to $\partial\mathcal{G}\backslash\mathcal{S}$. The smallest positive solution of the equation

$$(3 - 4\nu)\sin(\lambda\alpha_x) \pm \lambda\sin\alpha_x = 0$$

is denoted by $\lambda_*(\alpha_x)$ (see (3.8.2) and Theorem 3.1.2). Furthermore, we set $\mu_\tau = \min_j \operatorname{Re} \lambda_j^{(\tau)}$, where $\lambda_j^{(\tau)}$ are the eigenvalues of the operator pencil $\mathcal{L}_\tau$ generated by the Dirichlet problem for the Lamé system in the cone $\mathcal{K}_\tau$ in the half-plane $\operatorname{Re}\lambda > 0$. (By Theorem 3.6.1, there are no eigenvalues in the strip $-1/2 \leq \operatorname{Re}\lambda \leq 0$.)

The following theorem was established in [189]. For the case of a three-dimensional domain with conic vertices we also refer to [188].

THEOREM 3.8.3. *Suppose that*

$$(3.8.5) \qquad |l - \delta_k - 2/p| < \inf_{x \in E_k} \lambda_*(\alpha_x) \quad and \quad |l - \beta_\tau - 3/p + 1/2| < \mu_\tau + 1/2$$

for $k = 1, \ldots, q$, $\tau = 1, \ldots, d$. Then the problem

$$(3.8.6) \qquad \Delta U + (1 - 2\nu)^{-1}\nabla\nabla \cdot U = F \quad in \ \mathcal{G}, \quad U = \Phi \quad on \ \partial\mathcal{G}$$

is uniquely solvable in $V_{\vec{\beta},\vec{\delta}}^{l,p}(\mathcal{G})^3$ for all $F \in V_{\vec{\beta},\vec{\delta}}^{l-2,p}(\mathcal{G})^3$, $\Phi \in V_{\vec{\beta},\vec{\delta}}^{l-1/p,p}(\partial\mathcal{G})^3$, $l \geq 2$. Analogously, the conditions

$$(3.8.7) \qquad |l + \sigma - \delta_k| < \inf_{x \in E_k} \lambda_*(\alpha_x) \quad and \quad |l + \sigma - \beta_\tau + 1/2| < \mu_\tau + 1/2$$

ensure the unique solvability of problem (3.8.6) in $N_{\vec{\beta},\vec{\delta}}^{l,\sigma}(\mathcal{G})^3$ for arbitrary $F \in N_{\vec{\beta},\vec{\delta}}^{l-2,\sigma}(\mathcal{G})^3$, $\Phi \in N_{\vec{\beta},\vec{\delta}}^{l,\sigma}(\partial\mathcal{G})^3$.

Lower estimates for the quantity μ_τ are given in Sections 3.5–3.7. For example, according to Corollary 3.6.1,

$$\mu_\tau > \min\left\{1, \frac{(3 - 4\nu)M_\tau}{M_\tau + 6 - 4\nu}\right\},$$

where M_τ is a positive number such that $M_\tau(M_\tau + 1)$ is the first eigenvalue of the Dirichlet problem for the operator $-\delta$ on Ω_τ.

2. Regularity of solutions. The following assertion for the solution $U \in W_2^1(\mathcal{G})^3$ of problem (3.8.6) can also be found in [189].

THEOREM 3.8.4. *Let $U \in W_2^1(\mathcal{G})^3$ be a solution of problem (3.8.6). If $F \in V_{\vec{\beta},\vec{\delta}}^{l-2,p}(\mathcal{G})^3$, $\Phi \in V_{\vec{\beta},\vec{\delta}}^{l-1/p,p}(\partial\mathcal{G})^3$ and (3.8.5) is satisfied, then $U \in V_{\vec{\beta},\vec{\delta}}^{l,p}(\mathcal{G})^3$. If $F \in N_{\vec{\beta},\vec{\delta}}^{l-2,\sigma}(\mathcal{G})^3$, $\Phi \in N_{\vec{\beta},\vec{\delta}}^{l,\sigma}(\partial\mathcal{G})^3$ and condition (3.8.7) is satisfied, then $U \in N_{\vec{\beta},\vec{\delta}}^{l,\sigma}(\mathcal{G})^3$.*

Note that $\lambda_*(\alpha_x) > 1/2$ for $\alpha_x \in (0, 2\pi)$ (see Theorem 3.1.2). Since $\lambda_*(\alpha_x) > 1$ for $\alpha_x < \pi$ and $\mu_\tau > 1$ for $\Omega_\tau \subset S_+^2$, we conclude from Theorem 3.8.4 that the solution $U \in \overset{\circ}{W}_2^1(\mathcal{G})^3$ of problem (3.8.1) belongs to $W_2^2(\mathcal{G})^3$ if $F \in L_2(\mathcal{G})^3$, all interior edge angles are less than π and the cones $\mathcal{K}_\tau$ are contained in half-spaces. The same assumptions on $\mathcal{G}$ guarantee that U belongs to the space $N^{2,\sigma}_{1,1}$ and, therefore, also to $C^{1,\sigma}(\mathcal{G})$ with a certain positive σ if $F \in N^{0,\sigma}_{1,1}(\mathcal{G})$.

3. **Point estimates of the Green matrix.** Explicit information about μ_τ is also of importance for the estimation of Green's tensor in a neighborhood of singular boundary points. Let the elements $G_{i,j}$, $i,j = 1,2,3$, of the matrix function $(x,y) \to G(x,y)$ satisfy the boundary value problem

$$\Delta_x G_{i,j}(x,y) + (1-2\nu)^{-1} \sum_{k=1}^{3} \partial_{x_i} \partial_{x_k} G_{k,j}(x,y) = \delta_{i,j}\,\delta(x-y), \quad x,y \in \mathcal{G},$$

$$G_{i,j}(x,y) = 0 \text{ for } x \in \partial\mathcal{G}\backslash\mathcal{S},\ y \in \mathcal{G}.$$

This problem is uniquely solvable in the class of all matrix functions G such that the functions $x \to G_{i,j}(x,y)$ belong to the space $W^1_2(\mathcal{G}\backslash\overline{\mathcal{U}})$ for every $y \in \mathcal{G}$ and every neighborhood $\mathcal{U}$ of y.

By [**189**, Th.11.9], the functions $G_{i,j}$ satisfy the following pointwise estimates in a neighborhood of the vertex $x^{(\tau)}$.

THEOREM 3.8.5. 1) *If $2\rho_\tau(x) < \rho_\tau(y)$, then*

$$(3.8.8) \qquad |G_{i,j}(x,y)| \le \frac{c}{\rho_\tau(y)} \left(\frac{\rho_\tau(x)}{\rho_\tau(y)}\right)^{\mu_\tau - \varepsilon} \prod_{\{k:x^{(\tau)}\in\overline{E}_k\}} \left(\frac{r_k(x)}{\rho_\tau(x)}\frac{r_k(y)}{\rho_\tau(y)}\right)^{\theta_k - \varepsilon},$$

where $\theta_k = \inf_{x\in E_k} \lambda_*(\alpha_x)$ and ε is an arbitrarily small positive number. In the case $2\rho_\tau(y) < \rho_\tau(x)$ one has to change the places of x and y in the above estimate.
2) *If $\rho_\tau(x)/2 < \rho_\tau(y) < 2\rho_\tau(x)$ and $|x-y| > \min(r(x),r(y))$, then*

$$(3.8.9) \qquad |G_{i,j}(x,y)| \le \frac{c}{|x-y|} \prod_{\{k:x^{(\tau)}\in\overline{E}_k\}} \left(\frac{r_k(x)r_k(y)}{|x-y|^2}\right)^{\theta_k - \varepsilon}.$$

3) *For $|x-y| < \min(r(x),r(y))$ the estimate $|G_{i,j}(x,y)| \le c\,|x-y|^{-1}$ holds.*

Analogous estimates are valid for the derivatives $D^\alpha_x D^\beta_y G(x,y)$ of Green's tensor.

4. **Weighted and nonweighted maximum principle.** The estimates in Theorem 3.8.5 lead to weighted L_∞ estimates for the solution. We denote by $L^\infty_{\vec{\beta},\vec{\delta}}(\mathcal{G})$ the set of all functions U on $\mathcal{G}$ such that

$$\left\| \left(\prod_{\tau=1}^{d} \rho^{\beta_\tau}_\tau\right) \left(\prod_{k=1}^{q} (r_k/\rho)^{\delta_k}\right) U \right\|_{L_\infty(\mathcal{G})} < \infty.$$

Analogously, the space $L^\infty_{\vec{\beta},\vec{\delta}}(\partial\mathcal{G})$ is defined.

We consider the boundary value problem

$$(3.8.10) \qquad \Delta U + (1-2\nu)^{-1}\nabla\nabla\cdot U = 0 \text{ in } \mathcal{G}, \quad U = \Phi \text{ on } \partial\mathcal{G}\backslash\mathcal{S},$$

where $\Phi \in L^\infty_{\vec{\beta},\vec{\delta}}(\partial\mathcal{G})^3$. The solution of this problem is defined as follows. Let $\{\Phi^{(k)}\}$ be a sequence of vector functions from $C^\infty_0(\partial\mathcal{G}\backslash\mathcal{S})^3$ such that

$$\Phi^{(k)} \to \Phi \text{ a.e. on } \partial\mathcal{G} \quad \text{and} \quad \|\Phi^{(k)}\|_{L^\infty_{\vec{\beta},\vec{\delta}}(\mathcal{G})^3} \le \text{const.}$$

Furthermore, let $U^{(k)} \in W_2^1(\mathcal{G})^3$ be the solution of the homogeneous Lamé system with Dirichlet data $\Phi^{(k)}$ on $\partial\mathcal{G}\backslash\mathcal{S}$. By means of Theorem 3.8.5, it can be shown that the sequence $U^{(k)}$ converges in every point of $\mathcal{G}$ if $\delta_k < \inf_{x\in E_k}\lambda_*(\alpha_x)$ for $k = 1, \ldots, q$ and $\beta_\tau < 1 + \mu_\tau$ for $\tau = 1, \ldots, d$. The limit U of the sequence $\{U^{(k)}\}$ does not depend on the choice of the approximating sequence $\{\Phi^{(k)}\}$ and is called solution of problem (3.8.10).

THEOREM 3.8.6. [**189**, Th.11.11] *Suppose that*

$$|\delta_k| < \inf_{x\in E_k}\lambda_*(\alpha_x) \quad and \quad |\beta_\tau - 1/2| < \mu_\tau + 1/2$$

for $k = 1, \ldots, q$ *and* $\tau = 1, \ldots, d$. *Then the solution* U *of problem* (3.8.10) *satisfies*

(3.8.11) $$\|U\|_{L^\infty_{\vec{\beta},\vec{\delta}}(\mathcal{G})^3} \leq c\,\|\Phi\|_{L^\infty_{\vec{\beta},\vec{\delta}}(\partial\mathcal{G})^3}$$

with a constant c *independent of* Φ.

Since $\mu_\tau > 0$, we can take $\beta_\tau = 0$ and $\delta_k = 0$ in order to obtain

$$\|U\|_{L_\infty(\mathcal{G})^3} \leq c\,\|\Phi\|_{L_\infty(\partial\mathcal{G})^3}\,.$$

Furthermore, we have $U \in C(\overline{\mathcal{G}})^3$ if $\Phi \in C(\partial\mathcal{G})^3$.

In other words, the maximum modulus of the displacement vector is majorized by that of its boundary values for all piecewise smooth domains without cusps.

3.9. Notes

Section 3.1. Singularities of solutions of two–dimensional elasticity at angular points were studied, among others, in papers by Vorovich [**261**], Grisvard [**80**]–[**82**], [**84, 85**], Sändig and Richter [**233**], Nicaise [**208, 209, 210**], Costabel and Dauge [**31**]. We refer also to the books of Parton and Perlin [**223**] and Grisvard [**86**]. In [**233**] graphs for the eigenvalues similar to Figures 4 and 5 are given. The case $|\theta| < \pi$ (the plane with a crack) is fundamental in fracture mechanics (see, for example, Liebowitz [**154**] and Cherepanov [**27, 28**]).

Sections 3.2–3.5. The results are taken from the paper [**140**] by Kozlov, Maz'ya and Schwab.

Section 3.6. The estimate for the width of the energy strip in Corollary 3.6.1 was obtained by Maz'ya and Plamenevskiĭ [**188**]. The other results were unpublished.

Section 3.7. Here we follow Kozlov, Maz'ya and Schwab [**139, 140**]. Equations for the eigenvalues of the pencil corresponding to the Dirichlet problem for the Lamé system in a circular cone were numerically solved by A.-M. and R. Sändig [**234**]. Rotationally symmetric solutions in a circular cone were studied by Bažant and Keer [**12**], Beagles and Sändig [**14**]. The papers [**12, 14, 139, 234**] include the Neumann problem.

Section 3.8. Using the estimate for the energy strip in Corollary 3.6.1, Maz'ya and Plamenevskiĭ [**188**] proved theorems on the solvability of the Dirichlet problem for the Lamé system in three-dimensional domains with conical points as well as regularity assertions for solutions, estimates for Green's matrix and a maximum principle of Miranda-Agmon type. In [**189**] they studied the more complicated situation when the domain is of polyhedral type. Section 3.8 contains formulations of the main results of the last paper. Regularity results for solutions of the Lamé system in three-dimensional polyhedral domains can be also found in the works by Grisvard [84, 86] and Nicaise [208, 210].

Energy estimates of Saint-Venant's principle type for solutions of boundary value problems for the system of elasticity in a neighborhood of singular boundary points were obtained by Oleĭnik and Yosifyan [217, 218], Kopachek and Oleĭnik [116, 117], Kondrat′ev, Oleĭnik and Kopachek [113, 115] (see also the review of Kondrat′ev and Oleĭnik [112]).

The Lamé system in Lipschitz domains with boundary data in L_p was studied by Fabes [56], Kenig [102], Dahlberg, Kenig and Verchota [37], Dahlberg and Kenig [35].

CHAPTER 4

Other boundary value problems for the Lamé system

The present chapter is concerned with the spectral properties of operator pencils generated by some boundary value problems for the Lamé system in a three-dimensional cone $\mathcal{K}$. In Section 4.1 we consider a mixed boundary value problem with the following three types of boundary conditions:

(i) $U = 0$,
(ii) $U_n = 0$ and $\sigma_{n,\tau}(U) = 0$,
(iii) $U_\tau = 0$ and $\sigma_{n,n}(U) = 0$.

We use the notation:

$U = (U_1, U_2, U_3)$ is the displacement vector,
$n = (n_1, n_2, n_3)$ is the exterior normal to $\partial\mathcal{K}\backslash\{0\}$,
$U_n = U \cdot n$ is the normal component of the vector U,
$U_\tau = U - U_n n$ is the tangential component of the vector U on the boundary,
$\sigma(U) = \{\sigma_{i,j}(U)\}$ is the stress tensor connected with the strain tensor

$$\left\{\varepsilon_{i,j}(U)\right\} = \left\{\frac{1}{2}\left(\partial_{x_j} U_i + \partial_{x_i} U_j\right)\right\}$$

by the Hooke law

$$\sigma_{i,j} = 2\mu\left(\frac{\nu}{1-2\nu}(\varepsilon_{1,1} + \varepsilon_{2,2} + \varepsilon_{3,3})\,\delta_{i,j} + \varepsilon_{i,j}\right)$$

(μ is the shear modulus, ν is the Poisson ratio, $\nu < 1/2$, and $\delta_{i,j}$ denotes the Kronecker symbol),
$\sigma_{n,n}(U)$ is the normal component of the vector $\sigma_n(U) = \sigma(U)\,n$, i. e.,

$$\sigma_{n,n}(U) = \sum_{i,j=1}^{3} \sigma_{i,j}(U)\,n_i\,n_j \,,$$

$\sigma_{n,\tau}(U)$ is the tangential component of the vector $\sigma_n(U)$.

We modify the method used in Chapter 3. As in the case of the Dirichlet problem, we show that the spectrum of the corresponding operator pencil in a certain strip centered about the line $\operatorname{Re}\lambda = -1/2$ consists only of real eigenvalues and that the eigenvectors corresponding to eigenvalues in the interior of this strip do not have generalized eigenvectors. This result combined with Theorem 1.4.4 implies that the solution U has the following asymptotics in a neighborhood of a conic vertex:

$$U = \sum_{j=1}^{J} c_j\, r^{\lambda_j} u^{(j)}(\omega) + O(r^\beta).$$

Here β is an arbitrary positive real number less than

$$\left(\left(\frac{5}{2} - 2\nu\right)^2 + (1 - 2\nu)^2\right)^{1/2} - \frac{1}{2},$$

c_j are constants, λ_j are the real eigenvalues of the above mentioned pencil in the interval $(-1/2, \beta)$, and $u^{(j)}$ are the corresponding eigenvectors. The feature of the above asymptotic representation is that the exponents λ_j are real and that it does not contain logarithmic terms. At the end of Section 4.1 we derive a variational principle for the eigenvalues $\lambda_1, \ldots, \lambda_J$.

Sections 4.2–4.4 are dedicated to the Neumann boundary conditions for the Lamé system. First we analyze the plane problem in the spirit of Section 3.1. For the three-dimensional case the method used previously does not work and we apply a different approach. We start with the situation when the cone $\mathcal{K}$ is given by the inequality $x_3 > \phi(x_1, x_2)$, where ϕ is a smooth positively homogeneous of degree 1 function on $\mathbb{R}^2 \backslash \{0\}$. Then we pass to a three-dimensional anisotropic medium with a crack which has the form of a plane angle. We prove that in both cases the operator pencil generated by the Neumann problem has only the eigenvalues 0 and -1 in the strip $-1 \leq \operatorname{Re} \lambda \leq 0$ and that both eigenvalues have geometric and algebraic multiplicity 3.

These assertions imply the Hölder continuity of the solution to the Neumann problem for the Lamé system in a neighborhood of the vertex of a polyhedral angle as well as the Hölder continuity of the displacement field in an anisotropic medium with a polygonal crack.

4.1. A mixed boundary value problem for the Lamé system

4.1.1. Formulation of the problem. Let $\mathcal{K}$ be the cone $\{x = (x_1, x_2, x_3) \in \mathbb{R}^3 : x/|x| \in \Omega\}$, where Ω is a domain on the unit sphere with Lipschitz boundary $\partial\Omega = \overline{\gamma}_1 \cup \cdots \cup \overline{\gamma}_N$ and $\gamma_1, \ldots, \gamma_N$ are pairwise disjoint open arcs. Then

$$\partial\mathcal{K} = \overline{\Gamma}_1 \cup \cdots \cup \overline{\Gamma}_N,$$

where $\Gamma_k = \{x : x/|x| \in \gamma_k\}$.

The homogeneous Lamé system

$$\Delta U + (1 - 2\nu)^{-1} \nabla \nabla \cdot U = 0$$

can be written in the form

$$(4.1.1) \qquad \sum_{j=1}^{3} \frac{\partial \sigma_{i,j}(U)}{\partial x_j} = 0 \quad \text{for } i = 1, 2, 3,$$

where $\{\sigma_{i,j}(U)\}$ denotes the stress tensor. Throughout this section, it is assumed that

$$-1 < \nu < 1/2.$$

We divide the set of the indices $1, \ldots, N$ into three subsets I_0, I_n, I_τ. Our goal is to find generalized solutions of the system (4.1.1) satisfying the boundary conditions

$$(4.1.2) \qquad \begin{cases} U = 0 & \text{on } \Gamma_k \quad \text{for } k \in I_0, \\ U_n = 0, \ \sigma_{n,\tau}(U) = 0 & \text{on } \Gamma_k \quad \text{for } k \in I_n, \\ U_\tau = 0, \ \sigma_{n,n}(U) = 0 & \text{on } \Gamma_k \quad \text{for } k \in I_\tau. \end{cases}$$

We introduce the notion of generalized solutions by means of the Green formula

(4.1.3)
$$\int_{\mathcal{K}} \sum_{i,j=1}^{3} \sigma_{ij}(U) \cdot \varepsilon_{ij}(\overline{V}) \, dx$$

$$= -\int_{\mathcal{K}} \sum_{i,j=1}^{3} \frac{\partial \sigma_{ij}(U)}{\partial x_j} \cdot \overline{V}_i \, dx + \int_{\partial \mathcal{K} \backslash \{0\}} \sum_{i,j=1}^{3} \sigma_{ij}(U) \, n_j \cdot \overline{V}_i \, d\sigma$$

which is satisfied for all $U, V \in C_0^\infty(\overline{\mathcal{K}} \backslash \{0\})^3$. If U is a formal solution of problem (4.1.1), (4.1.2) and $V \in W_2^1(\mathcal{K})^3$ is a vector function vanishing for small and large $|x|$ which satisfies the conditions

(4.1.4)
$$\begin{cases} V = 0 & \text{on } \Gamma_k & \text{for } k \in I_0, \\ V_n = 0 & \text{on } \Gamma_k & \text{for } k \in I_n, \\ V_\tau = 0 & \text{on } \Gamma_k & \text{for } k \in I_\tau, \end{cases}$$

then (4.1.3) implies

(4.1.5)
$$\sum_{i,j=1}^{3} \int_{\mathcal{K}} \sigma_{i,j}(U) \cdot \varepsilon_{i,j}(\overline{V}) \, dx = 0.$$

Let $\mathcal{H}$ be the space of all vector functions $u \in W_2^1(\Omega)^3$ such that

- $u = 0$ on γ_k for $k \in I_0$,
- $u_n = 0$ on γ_k for $k \in I_n$,
- $u_\tau = 0$ on γ_k for $k \in I_\tau$,

where $u_n = u \cdot n$ and u_τ is the projection of the vector u onto the tangent plane to $\partial \mathcal{K} \backslash \{0\}$.

DEFINITION 4.1.1. The function

(4.1.6)
$$U(x) = r^{\lambda_0} \sum_{k=0}^{s} \frac{1}{k!} (\log r)^k u^{(s-k)}(\omega), \quad u^{(s-k)} \in \mathcal{H},$$

is said to be a *generalized solution* of problem (4.1.1), (4.1.2) if the integral identity (4.1.5) is valid for all $V \in W_2^1(\mathcal{K})^3$ with compact support in $\overline{\mathcal{K}} \backslash \{0\}$ satisfying the boundary conditions (4.1.4).

4.1.2. The operator pencil generated by the boundary value problem. We use the same notations as in Section 3.2. In spherical coordinates, the components of the strain tensor have the form

(4.1.7)
$$\begin{cases} \varepsilon_{rr} = \partial_r U_r, \qquad \varepsilon_{\varphi\varphi} = \frac{1}{r \sin \theta} \partial_\varphi U_\varphi + \frac{U_r}{r} + \cot \theta \frac{U_\theta}{r} \\[2mm] \varepsilon_{\theta\theta} = \frac{1}{r} \partial_\theta U_\theta + \frac{U_r}{r}, \qquad \varepsilon_{r\varphi} = \frac{1}{2} \left(\frac{1}{r \sin \theta} \partial_\varphi U_r - \frac{U_\varphi}{r} + \partial_r U_\varphi \right), \\[2mm] \varepsilon_{r\theta} = \frac{1}{2} \left(\frac{1}{r} \partial_\theta U_r - \frac{U_\theta}{r} + \partial_r U_\theta \right), \\[2mm] \varepsilon_{\theta\varphi} = \frac{1}{2} \left(\frac{1}{r} \partial_\theta U_\varphi - \cot \theta \frac{U_\varphi}{r} + \frac{1}{r \sin \theta} \partial_\varphi U_\theta \right) \end{cases}$$

(see, for example, Malvern [**159**]). Furthermore, the Hooke law has the following representation:

$$\sigma_{rr} = 2\mu \left(\frac{\nu}{1-2\nu}\Theta + \varepsilon_{rr} \right), \quad \sigma_{\varphi\varphi} = 2\mu \left(\frac{\nu}{1-2\nu}\Theta + \varepsilon_{\varphi\varphi} \right),$$

$$\sigma_{\theta\theta} = 2\mu \left(\frac{\nu}{1-2\nu}\Theta + \varepsilon_{\theta\theta} \right), \quad \sigma_{r\varphi} = 2\mu\,\varepsilon_{r\varphi}, \quad \sigma_{\theta\varphi} = 2\mu\,\varepsilon_{\theta\varphi}, \quad \sigma_{r\theta} = 2\mu\,\varepsilon_{r\theta},$$

where $\Theta = \varepsilon_{rr}+\varepsilon_{\theta\theta}+\varepsilon_{\varphi\varphi}$. In the spherical coordinates, the integral identity (4.1.5) takes the form

$$(4.1.8) \quad \int_0^\infty \int_\Omega \Big(\sigma_{rr}(U)\,\varepsilon_{rr}(\overline{V}) + \sigma_{\theta\theta}(U)\,\varepsilon_{\theta\theta}(\overline{V}) + \sigma_{\varphi\varphi}(U)\,\varepsilon_{\varphi\varphi}(\overline{V}) $$
$$+ 2\sigma_{r\theta}(U)\,\varepsilon_{r\theta}(\overline{V}) + 2\sigma_{r\varphi}(U)\,\varepsilon_{r\varphi}(\overline{V}) + 2\sigma_{\theta\varphi}(U)\,\varepsilon_{\theta\varphi}(\overline{V}) \Big) r^2\,d\omega\,dr = 0.$$

We introduce the sesquilinear form

$$a(u,v;\lambda) \;=\; \frac{1}{2\mu\,|\log\varepsilon|} \int_\varepsilon^{1/\varepsilon} \int_\Omega \Big(\sigma_{rr}(U)\,\varepsilon_{rr}(\overline{V}) + \sigma_{\theta\theta}(U)\,\varepsilon_{\theta\theta}(\overline{V}) + \sigma_{\varphi\varphi}(U)\,\varepsilon_{\varphi\varphi}(\overline{V}) $$
$$+ 2\,\sigma_{r\theta}(U)\,\varepsilon_{r\theta}(\overline{V}) + 2\,\sigma_{r\varphi}(U)\,\varepsilon_{r\varphi}(\overline{V}) + 2\,\sigma_{\theta\varphi}(U)\,\varepsilon_{\theta\varphi}(\overline{V}) \Big) r^2\,d\omega\,dr$$

$$=\; \frac{1}{|\log\varepsilon|} \int_\varepsilon^{1/\varepsilon} \int_\Omega \Big(\varepsilon_{rr}(U)\,\varepsilon_{rr}(\overline{V}) + \varepsilon_{\theta\theta}(U)\,\varepsilon_{\theta\theta}(\overline{V}) + \varepsilon_{\varphi\varphi}(U)\,\varepsilon_{\varphi\varphi}(\overline{V})$$
$$+ 2\,\varepsilon_{r\theta}(U)\,\varepsilon_{r\theta}(\overline{V}) + 2\,\varepsilon_{r\varphi}(U)\,\varepsilon_{r\varphi}(\overline{V}) + 2\,\varepsilon_{\theta\varphi}(U)\,\varepsilon_{\theta\varphi}(\overline{V})$$
$$+ \frac{\nu}{1-2\nu}\,\Theta(U)\,\Theta(\overline{V}) \Big) r^2\,d\omega\,dr,$$

where $U = r^\lambda u(\omega)$, $V = r^{-1-\overline{\lambda}}\,v(\omega)$, and ε is a positive real number less than 1. It can be easily verified that the expression on the right side is independent of ε. Using the formulas for the stress and strain tensors in spherical coordinates given above, we obtain

$$(4.1.9) \quad a(u,v;\lambda) = [u_\omega, v_\omega] + \int_\Omega \Big(\nabla_\omega u_r \cdot \nabla_\omega \overline{v}_r + (\lambda+2)(1-\lambda)\frac{2-2\nu}{1-2\nu}\,u_r\,\overline{v}_r$$

$$+ (\lambda+2)(1-\lambda)\,u_\omega \cdot \overline{v}_\omega + \frac{2\nu}{1-2\nu}(\nabla_\omega \cdot u_\omega)\,\nabla_\omega \cdot \overline{v}_\omega$$

$$+ \Big(\frac{2\nu}{1-2\nu}(\lambda+2)+2 \Big) u_r\,\nabla_\omega \cdot \overline{v}_\omega + \Big(\frac{2\nu}{1-2\nu}(1-\lambda)+2 \Big)(\nabla_\omega \cdot u_\omega)\,\overline{v}_r$$

$$- (1-\lambda)\,u_\omega \cdot \nabla_\omega \overline{v}_r - (\lambda+2)\,\nabla_\omega u_r \cdot \overline{v}_\omega \Big)\,d\omega,$$

where

$$(4.1.10) \quad [u_\omega, v_\omega] = \int_\Omega \Big(2\partial_\theta u_\theta \cdot \partial_\theta \overline{v}_\theta + 2\Big(\frac{\partial_\varphi u_\varphi}{\sin\theta} + \cot\theta\,u_\theta \Big) \cdot \Big(\frac{\partial_\varphi \overline{v}_\varphi}{\sin\theta} + \cot\theta\,\overline{v}_\theta \Big)$$

$$+ \Big(\partial_\theta u_\varphi + \frac{\partial_\varphi u_\theta}{\sin\theta} - \cot\theta\,u_\varphi \Big) \cdot \Big(\partial_\theta \overline{v}_\varphi + \frac{\partial_\varphi \overline{v}_\theta}{\sin\theta} - \cot\theta\,\overline{v}_\varphi \Big) \Big)\,d\omega,$$

$$\nabla_\omega v = \begin{pmatrix} \partial_\theta v \\ (\sin\theta)^{-1}\partial_\varphi v \end{pmatrix},$$

$$u_\omega = \begin{pmatrix} u_\theta \\ u_\varphi \end{pmatrix}, \qquad \nabla_\omega \cdot u_\omega = \frac{1}{\sin\theta}\partial_\theta(\sin\theta\, u_\theta) + \frac{1}{\sin\theta}\partial_\varphi u_\varphi.$$

Note that the form (4.1.9) coincides with the right-hand side of (3.2.19) on the space $\overset{\circ}{W}{}^1_2(\Omega)\times \overset{\circ}{h}{}^1_2(\Omega)$.

We give some properties of the form $[\cdot,\cdot]$. Here we make use of the spaces $h^1_2(\Omega)$ and $\overset{\circ}{h}{}^1_2(\Omega)$ introduced in Section 3.2.

LEMMA 4.1.1. 1) *The inequalities*

(4.1.11) $$\qquad\qquad [u_\omega, u_\omega] \;\leq\; 2\,Q(u_\omega, u_\omega),$$

(4.1.12) $$\qquad\qquad [u_\omega, u_\omega] \;\geq\; \int_\Omega |\nabla_\omega \cdot u_\omega|^2\, d\omega.$$

are valid for arbitrary vector functions $u_\omega \in h^1_2(\Omega)$. *(For the definition of Q see (3.2.4).)*

2) *For all $u_\omega \in h^1_2(\Omega)$ the inequality*

(4.1.13) $$\qquad\qquad [u_\omega, u_\omega] + \int_\Omega |u_\omega|^2\, d\omega \geq c_0\, \|u_\omega\|^2_{h^1_2(\Omega)}$$

holds, where c_0 is a positive constant.

3) *Every vector function $u_\omega \in \overset{\circ}{h}{}^1_2(\Omega)$ satisfies the equality*

(4.1.14) $$\qquad [u_\omega, u_\omega] + \int_\Omega |u_\omega|^2\, d\omega = Q(u_\omega, u_\omega) + \int_\Omega |\nabla_\omega \cdot u_\omega|^2\, d\omega.$$

Proof: 1) The inequality (4.1.11) follows from the estimate

$$\left|\partial_\theta u_\varphi + \frac{1}{\sin\theta}\partial_\varphi u_\theta - \cot\theta\, u_\varphi\right|^2 \leq 2\left(|\partial_\theta u_\varphi|^2 + \left|\frac{1}{\sin\theta}\partial_\varphi u_\theta - \cot\theta\, u_\varphi\right|^2\right),$$

while (4.1.12) follows from the equality

(4.1.15) $$\quad [u_\omega, u_\omega] = \int_\Omega |\nabla_\omega \cdot u_\omega|^2\, d\omega + \int_\Omega \left|\partial_\theta u_\theta - \frac{1}{\sin\theta}\partial_\varphi u_\varphi - \cot\theta\, u_\theta\right|^2 d\omega$$

$$+ \int_\Omega \left|\partial_\theta u_\varphi + \frac{1}{\sin\theta}\partial_\varphi u_\theta - \cot\theta\, u_\varphi\right|^2 d\omega.$$

2) Let $\mathcal{K}_0$ be the set $\{x \in \mathcal{K} : \frac{1}{2} < |x| < 2\}$. Since the boundary of $\mathcal{K}_0$ is Lipschitz, the Korn inequality (see Gobert [**70**])

$$\int_{\mathcal{K}_0} \left(\sum_{i,j=1}^{3} |\varepsilon_{i,j}(U)|^2 + |U|^2\right) dx \geq c \int_{\mathcal{K}_0} \sum_{i,j=1}^{3} |\partial_{x_i} U_j|^2\, dx$$

is valid. Taking $U(x) = u(\omega)$ and writing the last inequality in the spherical components, we obtain

$$\int_{\Omega} \left(|\varepsilon_{rr}|^2 + |\varepsilon_{r\theta}|^2 + |\varepsilon_{r\varphi}|^2 + |\varepsilon_{\theta\theta}|^2 + |\varepsilon_{\varphi\varphi}|^2 + |\varepsilon_{\theta\varphi}|^2 \right) d\omega$$

$$+ \int_{\Omega} \left(|u_\omega|^2 + |u_r|^2 \right) d\omega \geq c \left(\|u_\omega\|^2_{h_2^1(\Omega)} + \|u_r\|^2_{W_2^1(\Omega)} \right).$$

Setting $u_r = 0$, we arrive at (4.1.13).

3) It can be easily verified that

$$[u_\omega, u_\omega] - \int_{\Omega} |\nabla_\omega \cdot u_\omega|^2 \, d\omega = Q(u_\omega, u_\omega)$$

$$+ 2 \operatorname{Re} \int_{\Omega} \left(\partial_\theta u_\varphi \left(\frac{1}{\sin\theta} \partial_\varphi \overline{u}_\theta - \cot\theta \, \overline{u}_\varphi \right) - \partial_\theta u_\theta \left(\frac{1}{\sin\theta} \partial_\varphi \overline{u}_\varphi + \cot\theta \, \overline{u}_\theta \right) \right) d\omega.$$

Integrating by parts in the last integral, we get the equality (4.1.14) for $u_\omega \in \overset{\circ}{h}{}_2^1(\Omega)$. The lemma is proved. ∎

We introduce the space $\mathcal{H}_s$ which consists of all vectors (u_r, u_ω) such that their Cartesian components belong to the space $\mathcal{H}$ defined in Section 4.1.1. It is evident that $\mathcal{H}_s$ is a closed subspace of $W_2^1(\Omega) \times h_2^1(\Omega)$ and

$$(4.1.16) \qquad \overset{\circ}{W}{}_2^1(\Omega) \times \overset{\circ}{h}{}_2^1(\Omega) \subset \mathcal{H}_s \subset W_2^1(\Omega) \times h_2^1(\Omega).$$

Using the estimates (4.1.11), (4.1.12), it can be shown that the form $a(\cdot, \cdot; \lambda)$ is continuous on $\mathcal{H}_s \times \mathcal{H}_s$ for every complex λ. Therefore, this form generates a continuous operator

$$(4.1.17) \qquad \mathfrak{A}(\lambda) : \mathcal{H}_s \to \mathcal{H}_s^*$$

which is defined by the equality

$$\left(\mathfrak{A}(\lambda) u, v \right)_{L_2(\Omega)^3} = a(u, v; \lambda), \qquad u, v \in \mathcal{H}_s.$$

Clearly, $\mathfrak{A}(\lambda)$ depends polynomially on λ. Furthermore, by Theorem 1.2.3, there is the following connection between the eigenvectors and generalized eigenvectors of the pencil $\mathfrak{A}$ and the solutions (4.1.6) of equation (4.1.5).

LEMMA 4.1.2. *The function* (4.1.6) *is a solution of problem* (4.1.1), (4.1.2) *if and only if λ_0 is an eigenvalue of the pencil $\mathfrak{A}$ and the functions $u^{(0)}, \ldots, u^{(s)}$ form a Jordan chain corresponding to this eigenvalue.*

4.1.3. Basic properties of the pencil $\mathfrak{A}$.

THEOREM 4.1.1. 1) *The operator $\mathfrak{A}(\lambda)$ is Fredholm for all $\lambda \in \mathbb{C}$.*

2) *The operator $\mathfrak{A}(-1/2 + it)$ is positive definite for all real t.*

3) *The spectrum of the pencil $\mathfrak{A}$ consists of isolated eigenvalues with finite algebraic multiplicities.*

4) *The number λ_0 is an eigenvalue of the pencil $\mathfrak{A}$ if and only if $-1 - \overline{\lambda}_0$ is an eigenvalue of this pencil. The geometric, algebraic, and partial multiplicities of the eigenvalues λ_0 and $-1 - \overline{\lambda}_0$ coincide.*

Proof: 1) We prove that there exists a constant $c_1(\lambda)$ such that

$$(4.1.18) \quad |a(u,u;\lambda)| \geq c_0 \left(\|u_r\|^2_{H^1(\Omega)} + \|u_\omega\|^2_{h^1(\Omega)} \right) - c_1(\lambda) \int_\Omega (|u_r|^2 + |u_\omega|^2)\, d\omega$$

for arbitrary $u \in \mathcal{H}_s$.

Using (4.1.12) and the fact that $\nu/(1-2\nu) > -1/3$ for $\nu > -1$, we get

$$\int_\Omega \frac{2\nu}{1-2\nu} |\nabla_\omega \cdot u_\omega|^2\, d\omega \geq -\frac{2}{3} [u_\omega, u_\omega].$$

Furthermore, we have

$$\left| \int_\Omega u_r \nabla_\omega \cdot \overline{u}_\omega\, d\omega \right| \leq \varepsilon \int_\Omega |\nabla_\omega \cdot u_\omega|^2\, d\omega + \frac{1}{4\varepsilon} \int_\Omega |u_r|^2\, d\omega$$

$$\leq 2\varepsilon\, [u_\omega, u_\omega] + \frac{1}{4\varepsilon} \int_\Omega |u_r|^2\, d\omega$$

and

$$\left| \int_\Omega u_\omega \cdot \nabla_\omega \overline{u}_r\, d\omega \right| \leq \varepsilon \int_\Omega |\nabla_\omega u_r|^2\, d\omega + \frac{1}{4\varepsilon} \int_\Omega |u_\omega|^2\, d\omega,$$

where ε is an arbitrary positive number. Let ε be sufficiently small. Then (4.1.9) yields

$$|a(u,u;\lambda)| \geq \frac{1}{4} \left([u_\omega, u_\omega] + \int_\Omega |\nabla_\omega u_r|^2\, d\omega \right) - c(\lambda) \int_\Omega (|u_r|^2 + |u_\omega|^2)\, d\omega$$

with a constant $c(\lambda)$ depending only on λ and ν. From this and from assertion 2) of Lemma 4.1.1, we obtain (4.1.18). This proves the Fredholm property of the operator $\mathfrak{A}(\lambda)$ for arbitrary $\lambda \in \mathbb{C}$.

2) Now let $\operatorname{Re}\lambda = -1/2$. Then the quadratic form $a(u,u;\lambda)$ is equal to

$$\frac{1}{|\log \varepsilon|} \int_\varepsilon^{1/\varepsilon} \int_\Omega \left(|\varepsilon_{rr}(U)|^2 + |\varepsilon_{\theta\theta}(U)|^2 + |\varepsilon_{\varphi\varphi}(U)|^2 + \frac{\nu}{1-2\nu} |\Theta(U)|^2 \right.$$

$$\left. + 2\,|\varepsilon_{r\theta}(U)|^2 + 2\,|\varepsilon_{r\varphi}(U)|^2 + 2\,|\varepsilon_{\theta\varphi}(U)|^2 \right) r^2\, d\omega\, dr,$$

where $U = r^\lambda u(\omega)$. Since $-1 < \nu < 1/2$, we have

$$|\varepsilon_{rr}(U)|^2 + |\varepsilon_{\theta\theta}(U)|^2 + |\varepsilon_{\varphi\varphi}(U)|^2 + \frac{\nu}{1-2\nu} |\Theta(U)|^2$$

$$\geq \min\left(1, \frac{\nu+1}{1-2\nu}\right) \cdot \left(|\varepsilon_{rr}(U)|^2 + |\varepsilon_{\theta\theta}(U)|^2 + |\varepsilon_{\varphi\varphi}(U)|^2 \right).$$

Hence the form $a(u,u;\lambda)$ is nonnegative for $\operatorname{Re}\lambda = -1/2$. Moreover, the equality $a(u,u;\lambda) = 0$ implies

$$\varepsilon_{rr}(U) = \varepsilon_{\theta\theta}(U) = \varepsilon_{\varphi\varphi}(U) = \varepsilon_{r\theta}(U) = \varepsilon_{r\varphi}(U) = \varepsilon_{\theta\varphi}(U) = 0.$$

Since $\varepsilon_{rr}(U) = \lambda r^{\lambda-1} u_r$, we conclude from this that $u_r = 0$. Analogously, by (4.1.7), we obtain $u_\varphi = u_\theta = 0$, i.e, $u_\omega = 0$. This proves 2).

Assertion 3) is a consequence of 1) and 2) combined with Theorem 1.1.1. Finally, 4) follows from 2) and Theorem 1.2.2. ∎

4.1.4. Properties of the space $\mathcal{H}_s$. In what follows, the properties of the space $\mathcal{H}_s$ given in the next lemma play a crucial role.

LEMMA 4.1.3. 1) *The subspace $\mathcal{H}_s$ admits the representation*

$$(4.1.19) \qquad\qquad \mathcal{H}_s = \mathcal{H}_s^r \times \mathcal{H}_s^\omega \,,$$

where $\mathcal{H}_s^r$, $\mathcal{H}_s^\omega$ are subspaces of $W_2^1(\Omega)$ and $h_2^1(\Omega)$, respectively, such that $\overset{\circ}{W}{}_2^1(\Omega) \subset \mathcal{H}_s^r$ and $\overset{\circ}{h}{}_2^1(\Omega) \subset \mathcal{H}_s^\omega$.

2) *The equality*

$$(4.1.20) \qquad\qquad \int_{\partial\Omega} u_n \bar{v}_r \, d\omega' = 0$$

or, equivalently,

$$(4.1.21) \qquad\qquad \int_{\Omega} \Big((\nabla_\omega \cdot u_\omega) \bar{v}_r + u_\omega \cdot \nabla_\omega \bar{v}_r \Big) d\omega = 0$$

is satisfied for all $u, v \in \mathcal{H}_s$.

Proof: 1) In order to prove (4.1.19) we have to show that $(u_r, 0) \in \mathcal{H}_s$ if $(u_r, u_\omega) \in \mathcal{H}_s$.

Let (u_r, u_ω) be an arbitrary element of $\mathcal{H}_s$. Then the Cartesian components of the vector function $(u_r, 0)$ are

$$w = \begin{pmatrix} w_1 \\ w_2 \\ w_3 \end{pmatrix} = J^* \begin{pmatrix} u_r \\ 0 \end{pmatrix} = \begin{pmatrix} \sin\theta\,\cos\varphi \\ \sin\theta\,\sin\varphi \\ \cos\theta \end{pmatrix} u_r \,.$$

Here J^* denotes the adjoint matrix to (3.2.3). If $k \in I_0 \cup I_\tau$, then $u_r = 0$ on γ_k and, therefore, $w = 0$ on γ_k. Furthermore, since the vector $(\sin\theta\,\cos\varphi, \sin\theta\,\sin\varphi, \cos\theta) = x/|x|$ is orthogonal to n, we have $w_n = 0$ on γ_k for every $k = 1, \ldots, N$. Thus, $w \in \mathcal{H}$ and, consequently, $(u_r, 0) \in \mathcal{H}_s$.

2) If $k \in I_0 \cup I_\tau$, then $v_r = 0$ on γ_k, while $u_n = 0$ on γ_k for $k \in I_0 \cup I_n$. Hence

$$\int_{\gamma_k} u_n \cdot \bar{v}_r \, d\omega' = 0$$

for $k = 1, \ldots, N$. This implies (4.1.20).

It remains to prove that the left-hand sides of (4.1.20) and (4.1.21) coincide. Using the representation of u_ω in terms of the Cartesian components of u, we get

$$(4.1.22) \qquad \int_{\Omega} \Big((\nabla_\omega \cdot u_\omega) \bar{v}_r + u_\omega \cdot \nabla_\omega \bar{v}_r \Big) d\omega$$

$$= \int_{\Omega} (\sin\theta)^{-1} \Big(\partial_\theta (\sin\theta\, u_\theta\, \bar{v}_r) + \partial_\varphi (u_\varphi \bar{v}_r) \Big) d\omega$$

$$= \int_{\Omega} \bigg(\cos\theta \left(\cos\varphi\, \partial_\theta(u_1 \bar{v}_r) + \sin\varphi\, \partial_\theta(u_2 \bar{v}_r) \right) - \sin\theta\, \partial_\theta(u_3 \bar{v}_r)$$

$$- \frac{\sin\varphi}{\sin\theta} \partial_\varphi(u_1 \bar{v}_r) + \frac{\cos\varphi}{\sin\theta} \partial_\varphi(u_2 \bar{v}_r) - 2\sin\theta\,\cos\varphi\, u_1 \bar{v}_r$$

$$- 2\sin\theta\,\sin\varphi\, u_2 \bar{v}_r - 2\cos\theta\, u_3 \bar{v}_r \bigg) d\omega.$$

Integrating by parts, we obtain

$$(4.1.23) \qquad \frac{1}{\log 2} \int_{\substack{\mathcal{K} \\ 1 < |x| < 2}} \nabla_x \cdot (r^{-2} u(\omega) \, \overline{v_r(\omega)}) \, dx$$

$$= \frac{1}{\log 2} \int_{\substack{\partial \mathcal{K} \\ 1 < |x| < 2}} n \cdot r^{-2} \, u(\omega) \, \overline{v_r(\omega)} \, dx' = \int_{\partial \Omega} u_n \, \overline{v}_r \, d\omega'.$$

The integrand $\nabla_x \cdot (r^{-2} u(\omega) \, \overline{v_r(\omega)})$ on the left-hand side of (4.1.23) is equal to

$$\frac{1}{r^3} \left(\cos \theta \Big(\cos \varphi \, \partial_\theta (u_1 \overline{v}_r) + \sin \varphi \, \partial_\theta (u_2 \overline{v}_r) \Big) - \sin \theta \, \partial_\theta (u_3 \overline{v}_r) - \frac{\sin \varphi}{\sin \theta} \, \partial_\varphi (u_1 \overline{v}_r) \right.$$

$$\left. + \frac{\cos \varphi}{\sin \theta} \partial_\varphi (u_2 \overline{v}_r) - 2 \sin \theta \, \cos \varphi \, u_1 \overline{v}_r - 2 \sin \theta \, \sin \varphi \, u_2 \overline{v}_r - 2 \cos \theta \, u_3 \overline{v}_r \right).$$

Hence (4.1.22) and (4.1.23) coincide. The lemma is proved. ∎

4.1.5. Connections between eigenvectors and generalized eigenvectors corresponding to the eigenvalues λ and $-1 - \lambda$. By (4.1.21), the sesquilinear form $a(\cdot, \cdot; \lambda)$ can be written as

$$(4.1.24) \quad a(u, v; \lambda) = [u_\omega, v_\omega] + \int_\Omega \left((\nabla_\omega u_r) \cdot \nabla_\omega \overline{v}_r + (\lambda + 2)(1 - \lambda) \frac{2 - 2\nu}{1 - 2\nu} u_r \overline{v}_r \right.$$

$$+ (\lambda + 2)(1 - \lambda) \, u_\omega \cdot \overline{v}_\omega + \frac{2\nu}{1 - 2\nu} (\nabla_\omega \cdot u_\omega) \nabla_\omega \cdot \overline{v}_\omega$$

$$\left. + \frac{4 - 4\nu + \lambda}{1 - 2\nu} u_r \nabla_\omega \cdot \overline{v}_\omega + \frac{3 - 4\nu - \lambda}{1 - 2\nu} (\nabla_\omega \cdot u_\omega) \overline{v}_r \right) d\omega$$

for $u, v \in \mathcal{H}_s$.

Let $\mathcal{T}(\lambda)$ be the matrix (3.2.26). Then for $u, v \in \mathcal{H}_s$ we have $\mathcal{T}(\lambda) u \in \mathcal{H}_s$, $\mathcal{T}(\overline{\lambda}) v \in \mathcal{H}_s$, and

$$a(\mathcal{T}(\lambda)u, v; -1 - \lambda) = a(u, \mathcal{T}(\overline{\lambda})v; \lambda).$$

Consequently, for all $\lambda \in \mathbb{C}$ there is the equality

$$(4.1.25) \qquad \mathcal{T}(\lambda) \, \mathfrak{A}(\lambda) = \mathfrak{A}(-1 - \lambda) \, \mathcal{T}(\lambda).$$

Repeating the proof of Theorem 3.2.2, we get the following result.

THEOREM 4.1.2. *Let λ_0 be an eigenvalue of the pencil $\mathfrak{A}$ and let $u^{(0)}, \dots, u^{(s)}$ be a Jordan chain corresponding to this eigenvalue. If λ_0 does not take the values $3 - 4\nu$ and $4\nu - 4$ or $\mathcal{T}(\lambda_0) \, u^{(0)} \neq 0$, then $-1 - \lambda_0$ is also an eigenvalue and the vector functions*

$$\mathcal{T}(\lambda_0) \, u^{(0)}, \quad (-1)^k \left(\mathcal{T}(\lambda_0) \, u^{(k)} + \begin{pmatrix} 1 & 0 & 0 \\ 0 & -1 & 0 \\ 0 & 0 & -1 \end{pmatrix} u^{(k-1)} \right), \quad k = 1, \dots, s,$$

form a Jordan chain corresponding to this eigenvalue.

REMARK 4.1.1. If $\lambda_0 \neq 3 - 4\nu$ and $\lambda_0 \neq 4\nu - 4$, then the formulas in the theorem determine a one-to-one relation between the eigenvectors and generalized eigenvectors of the pencil $\mathfrak{A}(\lambda)$ corresponding to the eigenvalues λ_0 and $-1 - \lambda_0$.

4.1.6. A strip containing only real eigenvalues of the pencil $\mathfrak{A}$.

THEOREM 4.1.3. *The strip*

$$(4.1.26) \qquad \left| \operatorname{Re} \lambda + \frac{1}{2} \right|^2 \le \left(\frac{5}{2} - 2\nu \right)^2 + (1 - 2\nu)^2$$

contains only real eigenvalues of the pencil $\mathfrak{A}$.

Proof: Let λ be a complex number such that $\operatorname{Re} \lambda \ne -1/2$, $\operatorname{Im} \lambda \ne 0$. We consider the sesquilinear form

$$(4.1.27) \qquad q(u, v; \lambda) = a\left(\begin{pmatrix} u_r \\ u_\omega \end{pmatrix}, \begin{pmatrix} \overline{c}v_r \\ v_\omega \end{pmatrix}; \lambda \right),$$

where

$$c = \frac{4 - 4\nu + \overline{\lambda}}{3 - 4\nu - \lambda}.$$

By (4.1.24), we have

$$(4.1.28) \quad q(u, u; \lambda) = [u_\omega, u_\omega] + \int_\Omega \left(c \left| \nabla_\omega u_r \right|^2 + (\lambda + 2)(1 - \lambda) \frac{2 - 2\nu}{1 - 2\nu} c \left| u_r \right|^2 \right.$$

$$+ (\lambda + 2)(1 - \lambda) \left| u_\omega \right|^2 + \frac{2\nu}{1 - 2\nu} \left| \nabla_\omega \cdot u_\omega \right|^2$$

$$\left. + 2 \operatorname{Re} \left(\frac{4 - 4\nu + \lambda}{1 - 2\nu} u_r \, \nabla_\omega \cdot \overline{u}_\omega \right) \right) d\omega.$$

Using the formulas

$$c(1 - \lambda)(\lambda + 2) = |4 - 4\nu + \lambda|^2 - 2(1 - 2\nu)(5 - 4\nu)\, c,$$

$$c = \frac{(4 - 4\nu + \operatorname{Re} \lambda)(3 - 4\nu - \operatorname{Re} \lambda) + (\operatorname{Im} \lambda)^2 + i \operatorname{Im} \lambda \, (2\operatorname{Re} \lambda + 1)}{|3 - 4\nu - \lambda|^2},$$

we get

$$\operatorname{Im} q(u, u; \lambda) \quad = \quad \operatorname{Im} \lambda \, (2 \operatorname{Re} \lambda + 1) \int_\Omega \left(\frac{1}{|3 - 4\nu - \lambda|^2} \left| \nabla_\omega u_r \right|^2 \right.$$

$$\left. - \frac{4(1 - \nu)(5 - 4\nu)}{|3 - 4\nu - \lambda|^2} \left| u_r \right|^2 - \left| u_\omega \right|^2 \right) d\omega,$$

while the real part of $q(u, u; \lambda)$ is equal to

$$[u_\omega, u_\omega] + \int_\Omega \left(\operatorname{Re} c \left| \nabla_\omega u_r \right|^2 + \left((1 - \operatorname{Re} \lambda)(\operatorname{Re} \lambda + 2) + (\operatorname{Im} \lambda)^2 \right) \left| u_\omega \right|^2 \right.$$

$$+ 2(1 - \nu) \left(\frac{|4 - 4\nu + \lambda|^2}{1 - 2\nu} - (10 - 8\nu) \operatorname{Re} c \right) \left| u_r \right|^2$$

$$\left. + \frac{2\nu}{1 - 2\nu} \left| \nabla_\omega \cdot u_\omega \right|^2 + 2 \operatorname{Re} \left(\frac{4 - 4\nu + \lambda}{1 - 2\nu} u_r \, \nabla_\omega \cdot \overline{u}_\omega \right) \right) d\omega.$$

This implies

$$
\operatorname{Re} q(u, u; \lambda) - \frac{\operatorname{Re} c \, |3 - 4\nu - \lambda|^2}{\operatorname{Im} \lambda \, (2 \operatorname{Re} \lambda + 1)} \operatorname{Im} q(u, u; \lambda)
$$

$$
= [u_\omega, u_\omega] + 2 \int_\Omega \Big((1 - \operatorname{Re} \lambda)(\operatorname{Re} \lambda + 2) + (1 - 2\nu)(5 - 4\nu) + (\operatorname{Im} \lambda)^2 \Big) |u_\omega|^2 \, d\omega
$$

$$
+ \int_\Omega \Big(\frac{2 - 2\nu}{1 - 2\nu} |4 - 4\nu + \lambda|^2 |u_r|^2 + \frac{2\nu}{1 - 2\nu} |\nabla_\omega \cdot u_\omega|^2 \Big) \, d\omega
$$

$$
+ 2 \int_\Omega \operatorname{Re} \Big(\frac{4 - 4\nu + \lambda}{1 - 2\nu} u_r \, \nabla_\omega \cdot \overline{u}_\omega \Big) \, d\omega
$$

$$
= [u_\omega, u_\omega] + 2 \int_\Omega \Big((1 - \operatorname{Re} \lambda)(\operatorname{Re} \lambda + 2) + (1 - 2\nu)(5 - 4\nu) + (\operatorname{Im} \lambda)^2 \Big) |u_\omega|^2 \, d\omega
$$

$$
+ \int_\Omega \Big(\frac{2 - 2\nu}{1 - 2\nu} \Big| (4 - 4\nu + \lambda) u_r + \frac{1}{2 - 2\nu} \nabla_\omega \cdot u_\omega \Big|^2 - \frac{1 - 2\nu}{2 - 2\nu} |\nabla_\omega \cdot u_\omega|^2 \Big) \, d\omega.
$$

If the inequality (4.1.26) is satisfied, then, by (4.1.12), the right side of the last equation is positive for $u \neq 0$. Hence $q(u, u; \lambda) \neq 0$ for all nonreal λ in the strip (4.1.26), $\operatorname{Re} \lambda \neq -1/2$, and all $u \in \mathcal{H}_s$. From this we conclude the assertion of the theorem in the case $\operatorname{Re} \lambda \neq -1/2$. Since the line $\operatorname{Re} \lambda = -1/2$ does not contain eigenvalues (see Theorem 4.1.1, assertion 2), the theorem is completely proved. ∎

4.1.7. Absence of generalized eigenvectors. Now we show that the eigenvectors corresponding to eigenvalues in the interior of the strip (4.1.26) have no generalized eigenvectors.

LEMMA 4.1.4. *Let λ_0 be a real eigenvalue of the pencil $\mathfrak{A}$ in the interval*

$$
(4.1.29) \qquad -\frac{1}{2} < \lambda < \Big(\Big(\frac{5}{2} - 2\nu \Big)^2 + (1 - 2\nu)^2 \Big)^{1/2} - \frac{1}{2}
$$

and let $u^{(0)}$ be an eigenvector corresponding to this eigenvalue. Then

$$
(4.1.30) \qquad \frac{d}{d\lambda} a\Big(\Big(\begin{matrix} u_r^{(0)} \\ u_\omega^{(0)} \end{matrix} \Big), \Big(\begin{matrix} \overline{c(\lambda)} u_r^{(0)} \\ u_\omega^{(0)} \end{matrix} \Big); \lambda \Big) \Big|_{\lambda = \lambda_0} < 0,
$$

where $c(\lambda) = (4 - 4\nu + \lambda)/(3 - 4\nu - \lambda)$.

Proof: Let $q(\cdot, \cdot; \lambda)$ be the sesquilinear form (4.1.27) with $c(\lambda)$ as in the formulation of the lemma. Then, according to (4.1.28), the left side of (4.1.30) is equal to

$$
\frac{d}{d\lambda} q(u^{(0)}, u^{(0)}; \lambda) \Big|_{\lambda = \lambda_0}
$$

$$
= \int_\Omega \Big(c'(\lambda_0) |\nabla_\omega u_r^{(0)}|^2 + \frac{2}{1 - 2\nu} \operatorname{Re} \big(u_r^{(0)} \, \nabla_\omega \cdot \overline{u_\omega^{(0)}} \big) - (2\lambda_0 + 1) |u_\omega^{(0)}|^2
$$

$$
+ \frac{2 - 2\nu}{1 - 2\nu} \Big((\lambda_0 + 2)(1 - \lambda_0) c'(\lambda_0) - (2\lambda_0 + 1) c(\lambda_0) \Big) |u_r^{(0)}|^2 \Big) \, d\omega,
$$

where $c'(\lambda_0) = (7 - 8\nu)(3 - 4\nu - \lambda_0)^{-2}$.

Furthermore, by the first part of Lemma 4.1.3, the vector function $(u_r^{(0)}, 0)$ belongs to the space $\mathcal{H}_s$. Consequently,

$$
0 = a\left(\begin{pmatrix} u_r^{(0)} \\ u_\omega^{(0)} \end{pmatrix}, \begin{pmatrix} u_r^{(0)} \\ 0 \end{pmatrix}; \lambda_0\right) = \int_\Omega \left(|\nabla_\omega u_r^{(0)}|^2 + \frac{2 - 2\nu}{1 - 2\nu}(\lambda_0 + 2)(1 - \lambda_0)|u_r^{(0)}|^2 \right.
$$
$$
\left. + \frac{3 - 4\nu - \lambda_0}{1 - 2\nu}(\nabla_\omega \cdot u_\omega^{(0)})\,\overline{u_r^{(0)}} \right)\,d\omega.
$$

From this we obtain

$$
\frac{d}{d\lambda} q(u^{(0)}, u^{(0)}; \lambda)\Big|_{\lambda = \lambda_0}
$$
$$
= \frac{d}{d\lambda} q(u^{(0)}, u^{(0)}; \lambda)\Big|_{\lambda = \lambda_0} - \frac{2}{3 - 4\nu - \lambda_0} \operatorname{Re} a\left(\begin{pmatrix} u_r^{(0)} \\ u_\omega^{(0)} \end{pmatrix}, \begin{pmatrix} u_r^{(0)} \\ 0 \end{pmatrix}; \lambda_0\right)
$$
$$
= (1 + 2\lambda_0) \int_\Omega \left(\frac{1}{(3 - 4\nu - \lambda_0)^2}|\nabla_\omega u_r^{(0)}|^2 - \frac{4(1 - \nu)(5 - 4\nu)}{(3 - 4\nu - \lambda_0)^2}|u_r^{(0)}|^2 - |u_\omega^{(0)}|^2 \right)\,d\omega.
$$

Since $q(u^{(0)}, u^{(0)}; \lambda_0) = 0$, we get, analogously to the proof of Lemma 3.3.1,

$$
(4.1.31) \quad -\frac{(4 - 4\nu + \lambda_0)(3 - 4\nu - \lambda_0)}{1 + 2\lambda_0} \frac{d}{d\lambda} q(u^{(0)}, u^{(0)}; \lambda)\Big|_{\lambda = \lambda_0}
$$
$$
= [u_\omega^{(0)}, u_\omega^{(0)}] + \frac{2 - 2\nu}{1 - 2\nu} \int_\Omega \left| (4 - 4\nu + \lambda_0)u_r^{(0)} + \frac{1}{2 - 2\nu}\nabla_\omega \cdot u_\omega^{(0)} \right|^2\,d\omega
$$
$$
+ 2\left((\tfrac{5}{2} - 2\nu)^2 + (1 - 2\nu)^2 - (\lambda_0 + \tfrac{1}{2})^2 \right) \int_\Omega |u_\omega^{(0)}|^2\,d\omega.
$$
$$
- \frac{1 - 2\nu}{2 - 2\nu} \int_\Omega |\nabla_\omega \cdot u_\omega^{(0)}|^2\,d\omega.
$$

Using (4.1.12), we obtain that the right side of (4.1.31) is positive for λ_0 from the interval (4.1.29). Since the upper bound of the interval (4.1.29) is less than $3 - 4\nu$, we get (4.1.30). ∎

THEOREM 4.1.4. *The eigenvectors corresponding to eigenvalues in the strip*

$$
\left| \operatorname{Re}\lambda + \frac{1}{2} \right|^2 < \left(\frac{5}{2} - 2\nu \right)^2 + (1 - 2\nu)^2
$$

do not have generalized eigenvectors.

Proof: By assertion 4) of Theorem 4.1.1, we can restrict ourselves to real eigenvalues in the interval (4.1.29).

Let λ_0 be an eigenvalue in this interval and let $u^{(0)}, u^{(1)}$ be a Jordan chain corresponding to this eigenvalue. Then for arbitrary $v \in \mathcal{H}_s$

$$
(4.1.32) \qquad a(u^{(0)}, v; \lambda_0) = 0,
$$

$$
(4.1.33) \qquad a(u^{(1)}, v; \lambda_0) + \frac{d}{d\lambda} a(u^{(0)}, v; \lambda)\Big|_{\lambda = \lambda_0} = 0.
$$

We denote by $q(\cdot, \cdot; \lambda)$ the same sesquilinear form as in the proof of Lemma 4.1.4. From (4.1.32) it follows that

$$
q(u^{(0)}, u^{(1)}; \lambda_0) = \frac{1}{3 - 4\nu - \lambda} a(u^{(0)}, \mathcal{T}(\bar{\lambda}_0) u^{(1)}; \lambda_0) = 0.
$$

Furthermore, since the form q is symmetric, we have

$$0 = q(u^{(1)}, u^{(0)}; \lambda_0) = \frac{1}{3 - 4\nu - \lambda} \, a(u^{(1)}, \mathcal{T}(\overline{\lambda}_0) \, u^{(0)}; \lambda_0).$$

Hence we get

$$\begin{aligned}
\frac{d}{d\lambda} q(u^{(0)}, u^{(0)}; \lambda)\big|_{\lambda=\lambda_0} &= \frac{d}{d\lambda} \frac{1}{3 - 4\nu - \lambda} \, a(u^{(0)}, \mathcal{T}(\overline{\lambda}) \, u^{(0)}; \lambda)\big|_{\lambda=\lambda_0} \\
&= \frac{1}{3 - 4\nu - \lambda} \frac{d}{d\lambda} a(u^{(0)}, \mathcal{T}(\overline{\lambda}) \, u^{(0)}; \lambda)\big|_{\lambda=\lambda_0} \\
&= \frac{1}{3 - 4\nu - \lambda} \left(a(u^{(0)}, \mathcal{T}'(\overline{\lambda}_0) \, u^{(0)}; \lambda_0) + \frac{d}{d\lambda} a(u^{(0)}, \mathcal{T}(\overline{\lambda}_0) \, u^{(0)}; \lambda)\big|_{\lambda=\lambda_0} \right) \\
&= \frac{1}{3 - 4\nu - \lambda} \left(a(u^{(0)}, \mathcal{T}'(\overline{\lambda}_0) \, u^{(0)}; \lambda_0) - a(u^{(1)}, \mathcal{T}(\overline{\lambda}_0) \, u^{(0)}; \lambda_0) \right) = 0.
\end{aligned}$$

This contradicts (4.1.30). The theorem is proved. ∎

REMARK 4.1.2. We have proved Theorems 4.1.2–4.1.4 for the operator pencil $\mathfrak{A}$ generated by the Lamé system (4.1.1) with the boundary conditions (4.1.2). These results hold also for other boundary conditions provided the assertions of Lemma 4.1.3 are true for the space $\mathcal{H}$ which determines these boundary conditions. Note that the second assertion of Lemma 4.1.3 is not valid for the space $\mathcal{H} = W_2^1(\Omega) \times h_2^1(\Omega)$, i.e., the Neumann boundary conditions are excluded.

4.1.8. A variational principle. We consider the operator

$$A(\lambda) = \mathcal{T}(\lambda) \, \mathfrak{A}(\lambda)$$

and the corresponding quadratic form

$$\begin{aligned}
\left(A(\lambda)u, u \right)_{L_2(\Omega)^3} &= \tilde{a}(u, u; \lambda) \overset{def}{=} (3 - 4\nu - \lambda) \, q(u, u; \lambda) = (3 - 4\nu - \lambda) \, [u_\omega, u_\omega] \\
&+ \int_\Omega \Bigg((4 - 4\nu + \lambda) |\nabla_\omega u_r|^2 + \frac{2 - 2\nu}{1 - 2\nu} (\lambda + 2)(1 - \lambda)(4 - 4\nu + \lambda) |u_r|^2 \\
&\quad + (\lambda + 2)(1 - \lambda)(3 - 4\nu - \lambda) |u_\omega|^2 + \frac{2\nu}{1 - 2\nu} (3 - 4\nu - \lambda) |\nabla_\omega \cdot u_\omega|^2 \\
&\quad + \frac{2(3 - 4\nu - \lambda)(4 - 4\nu + \lambda)}{1 - 2\nu} \operatorname{Re} u_r \, \nabla_\omega \cdot \overline{u}_\omega \Bigg) \, d\omega,
\end{aligned}$$

where $u \in \mathcal{H}_s$. We shall apply the results of Section 1.3 for the form $\tilde{a}$. In our case $\mathcal{H} = L_2(\Omega)^3$, $\mathcal{H}_+ = \mathcal{H}_s$, $\alpha = -\frac{1}{2}$ and $\beta = -\frac{1}{2} + \left((\frac{5}{2} - 2\nu)^2 + (1 - 2\nu)^2 \right)^{1/2}$. Note that $\beta < 3 - 4\nu$. Let us verify the validity of conditions (I)–(III) of Section 1.3. Indeed, for $\lambda \geq -\frac{1}{2}$ we have $c \geq 1$ and, therefore, $q(u, u; \lambda) \geq a(u, u; \lambda)$. Thus, condition (I) can be deduced from (4.1.18). Condition (II) follows from the identity $A(-\frac{1}{2}) = (\frac{7}{2} - 4\nu) \mathfrak{A}(-\frac{1}{2})$ and from the second assertion of Theorem 4.1.1. Finally, condition (III) is a consequence of Lemma 4.1.4.

We denote by $\{\mu_j(\lambda)\}$ a nondecreasing sequence of eigenvalues of the operator $A(\lambda)$ counting their multiplicities. From the positivity of the operator $A(-1/2)$ it follows that $\mu_j(-1/2) > 0$. The next theorem is a consequence of Theorems 1.3.2 and 1.3.3.

THEOREM 4.1.5. *The spectrum of the pencil $\mathfrak{A}$ has the following properties in the strip $-1/2 \leq \operatorname{Re} \lambda \leq \beta$.*

1) *All eigenvalues of the pencil $\mathfrak{A}$ are real and may be characterized by*

$$\{\lambda_j\}_{j=1,\dots,J} = \left\{ \lambda \in [-\frac{1}{2}, \beta] \; : \; \mu_j(\lambda) = 0 \; for \; some \; j = 1, \dots, J \right\},$$

where J is the largest index j for which the function μ_j has a zero in the interval $[-\frac{1}{2}, \beta)$. For every j the function μ_j has not more than one zero in $[-\frac{1}{2}, \beta)$.

2) *If $\lambda_0 \in [-\frac{1}{2}, \beta)$ is an eigenvalue of multiplicity I, then I is equal to the number of functions μ_j which have a zero at λ_0.*

If $\mathcal{H}_s \neq \overset{\circ}{W}{}_2^1(\Omega) \times \overset{\circ}{h}{}_2^1(\Omega)$, then the eigenvalues do not monotonically depend on the domain Ω. However, the following statement is a consequence of Theorem 1.3.5.

THEOREM 4.1.6. *Let $\mathcal{H}_{s,1}$, $\mathcal{H}_{s,2}$ be subspaces of $W_2^1(\Omega) \times h_2^1(\Omega)$ such that the assertions of Lemma 4.1.3 are valid. We denote by $\mathfrak{A}_1$, $\mathfrak{A}_2$ the operator pencils corresponding to these subspaces. Furthermore, let $\{\lambda_j^{(1)}\}$, $\{\lambda_j^{(2)}\}$ be the nondecreasing sequences of the eigenvalues of the pencils $\mathfrak{A}_1$ and $\mathfrak{A}_2$ in the interval $[-1/2, \beta)$ counted with their multiplicities. If $\mathcal{H}_{s,1} \subset \mathcal{H}_{s,2}$, then the number of the eigenvalues of the pencil $\mathfrak{A}_1$ in the interval $[-1/2, \beta)$ is not greater than the number of such eigenvalues of the pencil $\mathfrak{A}_2$. Furthermore, the inequality*

$$\lambda_j^{(2)} \leq \lambda_j^{(1)}$$

is valid.

4.2. The Neumann problem for the Lamé system in a plane angle

Now we study the eigenvalues, eigenfunctions and generalized eigenfunctions of the operator pencil generated by the Neumann problem for the Lamé system in a plane angle. As in the case of the Dirichlet problem, we obtain entire functions whose zeros are the mentioned eigenvalues. This enables us to obtain information on the distribution of these eigenvalues.

4.2.1. The operator pencil generated by the Neumann problem. Let $\mathcal{K} = \{(x_1, x_2) \in \mathbb{R}^2 : 0 < r < \infty, \; -\alpha/2 < \varphi < \alpha/2\}$ be a plane angle with opening $\alpha \in (0, 2\pi]$. We consider the Lamé system

$$\Delta U + (1 - 2\nu)^{-1} \nabla \nabla \cdot U = 0 \quad \text{in } \mathcal{K}$$

with the Neumann boundary condition $\sigma_\varphi = \tau_{r\varphi} = 0$ on $\partial\mathcal{K}\backslash\{0\}$, i.e.,

$$\partial_r U_r + \frac{1-\nu}{\nu}\left(\frac{1}{r}\partial_\varphi U_\varphi + \frac{1}{r}U_r\right) = 0, \quad \frac{1}{r}\partial_\varphi U_r + \partial_r U_\varphi - \frac{1}{r}U_\varphi = 0 \quad \text{for } \varphi = \pm\alpha/2.$$

Putting $U_r = r^\lambda u_r(\varphi)$ and $U_\varphi = r^\lambda u_\varphi(\varphi)$ into the Lamé system, we get the system

$$\mathcal{L}(\lambda)\left(\begin{array}{c} u_r \\ u_\varphi \end{array}\right) = 0$$

given by (3.1.5), (3.1.6), while the boundary conditions take the form

$$(4.2.1) \qquad \mathcal{B}_\pm(\lambda)\left(\begin{array}{c} u_r \\ u_\varphi \end{array}\right) = \left(\begin{array}{c} \lambda u_r + \frac{1-\nu}{\nu}(u_\varphi' + u_r) \\ u_r' + (\lambda - 1)u_\varphi \end{array}\right)\Bigg|_{\varphi=\pm\alpha/2} = 0.$$

We denote the operator

$$\left(\mathcal{L}(\lambda), \mathcal{B}_+, \mathcal{B}_-\right) \; : \; (W_2^2(-\alpha/2, +\alpha/2))^2 \rightarrow (L_2(-\alpha/2, +\alpha/2))^2 \times \mathbb{C}^2 \times \mathbb{C}^2$$

by $\mathfrak{A}(\lambda)$. As in the case of the Dirichlet problem, the operator $\mathfrak{A}(\lambda)$ is Fredholm for every $\lambda \in \mathbb{C}$ and the spectrum of the pencil $\mathfrak{A}$ consists of eigenvalues with finite algebraic multiplicities. If λ_0 is an eigenvalue, then $-\lambda_0$, $\overline{\lambda}_0$, $-\overline{\lambda}_0$ are also eigenvalues.

4.2.2. Equations for the eigenvalues of the operator pencil. Inserting the general solution (3.1.16) of system (3.1.5), (3.1.6) into the boundary conditions (4.2.1), we obtain the following algebraic system for the coefficients $c_1, \dots, c_4$:

$$-2\lambda \left(c_1 - \frac{3-4\nu}{\lambda} c_3\right) \begin{pmatrix} \cos(1+\lambda)\alpha/2 \\ \pm \sin(1+\lambda)\alpha/2 \end{pmatrix}$$

$$+2\lambda \left(c_2 - \frac{3-4\nu}{\lambda} c_4\right) \begin{pmatrix} \pm \sin(1+\lambda)\alpha/2 \\ \cos(1+\lambda)\alpha/2 \end{pmatrix}$$

$$+2c_3 \begin{pmatrix} (1+\lambda)\cos(1-\lambda)\alpha/2 \\ \pm(1-\lambda)\sin(1-\lambda)\alpha/2 \end{pmatrix} + 2c_4 \begin{pmatrix} \mp(1+\lambda\sin(1-\lambda)\alpha/2 \\ -(1-\lambda)\cos(1-\lambda)\alpha/2 \end{pmatrix} = 0.$$

Analogous to the case of the Dirichlet problem, this system is equivalent to two systems

$$(4.2.2) \quad \begin{cases} -\lambda \left(c_1 - \frac{3-4\nu}{\lambda} c_3\right) \cos(1+\lambda)\alpha/2 + (1+\lambda)c_3 \cos(1-\lambda)\alpha/2 = 0, \\[2mm] -\lambda \left(c_1 - \frac{3-4\nu}{\lambda} c_3\right) \sin(1+\lambda)\alpha/2 + (1-\lambda)c_3 \sin(1-\lambda)\alpha/2 = 0 \end{cases}$$

and

$$(4.2.3) \quad \begin{cases} \lambda \left(c_2 - \frac{3-4\nu}{\lambda} c_4\right) \sin(1+\lambda)\alpha/2 - (1+\lambda)c_4 \sin(1-\lambda)\alpha/2 = 0, \\[2mm] \lambda \left(c_2 - \frac{3-4\nu}{\lambda} c_4\right) \cos(1+\lambda)\alpha/2 + (1-\lambda)c_4 \cos(1-\lambda)\alpha/2 = 0 \end{cases}$$

with the determinants

$$d_+(\lambda) = \lambda \left(\sin \lambda\alpha + \lambda \sin \alpha\right),$$
$$d_-(\lambda) = \lambda \left(\sin \lambda\alpha - \lambda \sin \alpha\right).$$

Therefore, every eigenvalue of the pencil $\mathfrak{A}$ is a solution of one of the equations

$$(4.2.4) \qquad\qquad\qquad \sin \lambda\alpha + \lambda \sin \alpha = 0,$$
$$(4.2.5) \qquad\qquad\qquad \sin \lambda\alpha - \lambda \sin \alpha = 0.$$

The graphs of the real parts of the roots of the equations (4.2.4), (4.2.5) are represented in the figure below, the thick lines correspond to real and the thin lines to nonreal eigenvalues).

4.2.3. Eigenvectors and generalized eigenvectors of the pencil $\mathfrak{A}$. As in Section 3.1, let $A = 3 - 4\nu - \lambda$, $B = 3 - 4\nu + \lambda$. If λ is an eigenvalue satisfying the equality $d_+(\lambda) = 0$, then

$$(4.2.6) \quad \begin{pmatrix} u_r^+ \\ u_\varphi^+ \end{pmatrix} = (\cos \lambda\alpha + \lambda \cos \alpha) \begin{pmatrix} \cos(1+\lambda)\varphi \\ -\sin(1+\lambda)\varphi \end{pmatrix} + \begin{pmatrix} A\cos(1-\lambda)\varphi \\ -B\sin(1-\lambda)\varphi \end{pmatrix}$$

is an eigenvector corresponding to this eigenvalue. Analogously, the eigenvector

$$(4.2.7) \quad \begin{pmatrix} u_r^- \\ u_\varphi^- \end{pmatrix} = (\cos \lambda\alpha - \lambda \cos \alpha) \begin{pmatrix} \sin(1+\lambda)\varphi \\ \cos(1+\lambda)\varphi \end{pmatrix} + \begin{pmatrix} A\sin(1-\lambda)\varphi \\ B\cos(1-\lambda)\varphi \end{pmatrix}$$

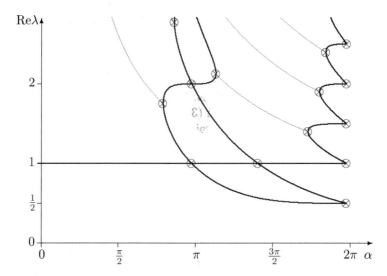

FIGURE 9. Real parts of eigenvalues of the pencil generated by the
Neumann problem in the strip $0 < \operatorname{Re}\lambda < 3$

corresponds to the eigenvalue λ satisfying the equality $d_-(\lambda) = 0$. Since each of the
systems (4.2.2), (4.2.3) has at most one nontrivial solution, we have obtained all
eigenvectors in this way.

The equations (4.2.4), (4.2.5) coincide with (3.1.22), (3.1.23) for $\nu = 1/2$.
Therefore, the spectra of the operator pencils corresponding to the Dirichlet and
Neumann problems have several common properties. In particular, there are no
generalized eigenvectors to nonreal eigenvalues and the length of the Jordan chains
corresponding to real eigenvalues is not greater than 2 (cf. Lemma 3.1.3). The
angles α for which generalized eigenvectors exist are determined by the relations
given after the proof of Lemma 3.1.4. Here, in the case of the Neumann problem,
we have to put $\nu = 1/2$.

Furthermore, formula (3.1.31) for the generalized eigenvectors is also valid for
the pencil of the Neumann problem. Here the functions $u_r^\pm$, $u_\varphi^\pm$ on the right-hand
side are determined by (4.2.6), (4.2.7). The proof of this fact is the same as for
Lemma 3.1.4.

In contrast to the Dirichlet problem, the spectrum of the pencil generated by
the Neumann problem contains the number $\lambda = 0$. This eigenvalue has the algebraic
multiplicity 4, since $\lambda = 0$ is a zero of multiplicity two both of $d_+(\lambda)$ and $d_-(\lambda)$.
This eigenvalue has the eigenvectors

$$\begin{pmatrix} u_r^+ \\ u_\varphi^+ \end{pmatrix} = \begin{pmatrix} \cos\varphi \\ -\sin\varphi \end{pmatrix}, \quad \begin{pmatrix} u_r^- \\ u_\varphi^- \end{pmatrix} = \begin{pmatrix} \sin\varphi \\ \cos\varphi \end{pmatrix}$$

and the corresponding generalized eigenvectors

$$\begin{pmatrix} v_r^+ \\ v_\varphi^+ \end{pmatrix} = \frac{1-2\nu}{2-2\nu}\varphi \begin{pmatrix} \sin\varphi \\ \cos\varphi \end{pmatrix} - \frac{1}{4(1-\nu)} \begin{pmatrix} \cos\varphi \\ \sin\varphi \end{pmatrix},$$

$$\begin{pmatrix} v_r^- \\ v_\varphi^- \end{pmatrix} = \frac{1-2\nu}{2-2\nu}\varphi \begin{pmatrix} -\cos\varphi \\ \sin\varphi \end{pmatrix} + \frac{1}{4(1-\nu)} \begin{pmatrix} -\sin\varphi \\ \cos\varphi \end{pmatrix}.$$

Obviously, the common roots $\lambda \neq 0$ of the equations $d_+(\lambda) = 0$ and $d_-(\lambda) = 0$ exist only in the cases $\alpha = \pi$ and $\alpha = 2\pi$. If $\alpha = \pi$, then the spectrum consists of the eigenvalues $0, \pm 1, \pm 2, \ldots$, and in the case $\alpha = 2\pi$ the numbers $\pm k/2$ ($k = 0, 1, \ldots$) are eigenvalues. In both cases there are two eigenvectors (4.2.6), (4.2.7) and generalized eigenvectors only exist for $\lambda = 0$.

The eigenvalue $\lambda = \frac{1}{2}$ of the operator pencil $\mathfrak{A}$ for $\alpha = 2\pi$ plays an important role in fracture mechanics (see Cherepanov [**27, 28**]). The eigenvectors corresponding to this eigenvalue are

$$\left(\begin{array}{c} (5 - 8\nu)\cos\varphi/2 - \cos 3\varphi/2 \\ -(7 - 8\nu)\sin\varphi/2 + \sin 3\varphi/2 \end{array} \right), \left(\begin{array}{c} (5 - 8\nu)\sin\varphi/2 - 3\sin 3\varphi/2 \\ (7 - 8\nu)\cos\varphi/2 - 3\cos 3\varphi/2 \end{array} \right).$$

Finally, we note that $\lambda = 1$ is a root of the equation $d_-(\lambda) = 0$ for every angle α. The eigenvector corresponding to this eigenvalue is

$$\left(\begin{array}{c} u_r^- \\ u_\varphi^- \end{array} \right) = \left(\begin{array}{c} 0 \\ 1 \end{array} \right).$$

A generalized eigenvector corresponding to this eigenvalue exists only for $\alpha = \alpha_*$, where α_* is the unique solution of the equation $\tan\alpha = \alpha$ in the interval $(0, 2\pi]$, $\alpha_* \approx 1.4303\,\pi$. From (3.1.31) it follows that the generalized eigenvector has the form

$$\left(\begin{array}{c} v_r^- \\ v_\varphi^- \end{array} \right) = -\frac{1}{4(1 - \nu)\cos\alpha_*} \left(\begin{array}{c} \sin 2\varphi \\ \cos 2\varphi \end{array} \right) - \frac{1 - 2\nu}{2 - 2\nu} \left(\begin{array}{c} \varphi \\ 0 \end{array} \right).$$

4.2.4. Distribution of the eigenvalues in the plane $\operatorname{Re}\lambda > 0$. For the proof of the following theorem it suffices to repeat the arguments used in the proof of Theorem 3.1.1. Here we have to put $\nu = 1/2$.

THEOREM 4.2.1. *Let $\alpha \neq \pi$, $\alpha \neq 2\pi$. Then the lines $\operatorname{Re}\lambda = k\pi/\alpha$, $k = 1, 2, \ldots$, do not contain eigenvalues of $\mathfrak{A}$, while the line $\operatorname{Re}\lambda = 0$ contains the unique eigenvalue $\lambda = 0$. Furthermore, the assertions 2)–4) of Theorem 3.1.1, where the constant ν in (3.1.22), (3.1.23) is replaced by $1/2$, are valid for the pencil $\mathfrak{A}$ generated by the Neumann problem.*

Before we formulate an analogous result to Theorem 3.1.2, we introduce two auxiliary functions $\xi_\pm$. Let $\xi_+(\alpha)$ be the smallest positive root of the equation

$$(4.2.8) \qquad \frac{\sin\xi}{\xi} + \frac{\sin\alpha}{\alpha} = 0 \qquad \alpha \in (0, 2\pi].$$

Positive roots exist only for $\alpha \geq \alpha_0$, where α_0 is the smallest positive root of the equation $\alpha^{-1}\sin\alpha = -\cos\alpha_*$, $\alpha_0 \approx 0.8128\pi$. For $\alpha < \alpha_0$ the equation (4.2.8) has only nonreal roots. The function ξ_+ is real-analytic for $\alpha \in (\alpha_0, 2\pi]$, assumes its maximum α_* at the point $\alpha = \alpha_0$ and its minimum α_0 at the point $\alpha = \alpha_*$. It decreases on the interval $[\alpha_0, \alpha_*]$ and increases on the interval $[\alpha_*, 2\pi]$. Moreover,

$$\xi_+(\pi) = \xi_+(2\pi) = \pi.$$

Furthermore, let $\xi_-(\alpha)$ be the smallest positive root of the equation

$$(4.2.9) \qquad \frac{1}{\xi - \alpha}\left(\frac{\sin\xi}{\xi} - \frac{\sin\alpha}{\alpha} \right) = 0.$$

The function ξ_- is defined for $\alpha \in [\alpha_1, 2\pi]$, where α_1 is the smallest positive root of the equation $\alpha^{-1}\sin\alpha = \cos\alpha_{**}$ and α_{**} is the second positive root of the equation

$\tan \alpha = \alpha$, $\alpha_{**} \approx 2.4590\,\pi$, $\alpha_1 \approx 0.88397\,\pi$. Note that ξ_- is real-analytic and strictly decreasing in $(\alpha_1, 2\pi]$. Furthermore,

$$\xi_-(\alpha_1) = \alpha_{**}, \quad \xi_-(\pi) = 2\pi, \quad \xi_-(\alpha_*) = \alpha_*, \quad \xi_-(2\pi) = \pi.$$

The following theorem describes the spectrum of the operator pencil $\mathfrak{A}$ near the line $\operatorname{Re}\lambda = 0$ and can be proved analogously to Theorem 3.1.2.

THEOREM 4.2.2. 1) *If* $\alpha \in (0, \pi)$, *then the spectrum of the operator pencil* $\mathfrak{A}$ *in the strip* $0 < \operatorname{Re}\lambda < \pi/\alpha$ *consists of the unique simple eigenvalue* $\lambda = 1$.

2) *Let* $\alpha \in (\pi, 2\pi)$. *Then the spectrum of* $\mathfrak{A}$ *in the strip* $0 < \operatorname{Re}\lambda < 2\pi/\alpha$ *consists of the eigenvalues*

$$\lambda_+(\alpha) = \frac{\xi_+(\alpha)}{\alpha}, \quad 1, \quad \lambda_-(\alpha) = \frac{\xi_-(\alpha)}{\alpha}.$$

3) *The functions* λ_+, λ_- *defined on the intervals* $[\alpha_0, 2\pi]$ *and* $[\alpha_1, 2\pi]$, *respectively, are strictly decreasing. Furthermore, the following inequalities are satisfied:*

$$\lambda_+(\alpha) < 1 < \lambda_-(\alpha) < \frac{2\pi}{\alpha} \quad \text{for } \alpha \in (\pi, \alpha_*),$$

$$\lambda_+(\alpha) < \lambda_-(\alpha) < 1 \quad \text{for } \alpha \in (\alpha_*, 2\pi).$$

Moreover, $\lambda_+(\pi) = 1$, $\lambda_-(\pi) = 2$, $\lambda_-(\alpha_*) = 1$, *and* $\lambda_+(2\pi) = \lambda_-(2\pi) = \frac{1}{2}$.

Now we deal with the spectrum of the pencil $\mathfrak{A}$ in the strip

(4.2.10)
$$\frac{\pi}{\alpha} < \operatorname{Re}\lambda < \frac{2\pi}{\alpha}$$

for $\alpha \in (0, \pi)$. If $\alpha \in (\alpha_0, \pi)$, then the equation (4.2.8) has two roots in the interval $(\pi, 2\pi)$. We denote the smallest of them by ξ_+, while the second is denoted by $\xi_+^{(1)}$. The function $\xi_+^{(1)}$ is strictly increasing and satisfies the equalities

$$\xi_+^{(1)}(\alpha_0) = \xi_+(\alpha_0) = \alpha_*, \quad \xi_+^{(1)}(\pi) = 2\pi.$$

Since (4.2.8) has no real roots for $\alpha \in (0, \alpha_0)$, according to Theorem 4.2.1, there are two nonreal eigenvalues $z(\alpha)/\alpha$ and $\overline{z(\alpha)}/\alpha$ in the strip (4.2.10), where $z(\alpha)$ is the root of (4.2.8) with positive imaginary part. The root $z(\alpha)$ is simple and real-analytic with respect to $\alpha \in (0, \alpha_0)$. Furthermore, the limits $z(+0)$ and $z(\alpha_0 - 0)$ exist. The first of these limits is the root of the equation $\sin z = -z$ with the smallest positive real part, $z(+0) \approx 4.2 + 2.3i$, while the second is equal to α_*. Moreover, $z(\alpha)$ has the asymptotics

$$z(\alpha) = \alpha_* + c\,i\,(\alpha_0 - \pi)^{1/2} + O(\alpha_0 - \alpha) \quad \text{as } \alpha \to \alpha_0 - 0,$$

where $c = \sqrt{2}\,(\alpha_0^{-1} - \cot \alpha_0)^{1/2} > 0$. Indeed, in a neighborhood of the point $\alpha = \alpha_0$, $z = \alpha_*$ the following decomposition holds:

(4.2.11) $\alpha \sin z + z \sin \alpha = (\sin \alpha_* + \alpha_* \cos \alpha_0)\,(\alpha - \alpha_0)$

$$- \alpha_0 \sin \alpha_* \frac{(z - \alpha_*)^2}{2} + \sum_{l+2k>2} a_{k,l}\,(\alpha - \alpha_0)^k\,(z - \alpha_*)^l.$$

Hence $z(\alpha)$ can be represented as a convergent series of powers of $(\alpha_0 - \alpha)^{1/2}$ (see, e.g., the book by Vainberg and Trenogin [254]). From (4.2.11) it follows immediately that

$$c = \left(2\,\frac{\sin \alpha_* + \alpha_* \cos \alpha_0}{\alpha_0 \sin \alpha_*} \right)^{1/2}.$$

Using the equalities $\tan\alpha_* = \alpha_*$, $\alpha_0^{-1}\sin\alpha_0 = -\cos\alpha_*$, we obtain the expression for the constant c given above.

THEOREM 4.2.3. 1) Let $\alpha \in (0, \alpha_0)$. Then the spectrum of the operator pencil $\mathfrak{A}$ in the strip (4.2.10) consists of two nonreal simple eigenvalues $z(\alpha)/\alpha$ and $\overline{z(\alpha)}/\alpha$. The function $\operatorname{Re} z(\alpha)$ is strictly increasing on $(0, \alpha_0)$.

2) If $\alpha = \alpha_0$, then the pencil $\mathfrak{A}$ has the only eigenvalue α_*/α in the strip (4.2.10). The algebraic multiplicity of this eigenvalue is equal to two.

3) If $\alpha \in (\alpha_0, \pi)$, then the spectrum of $\mathfrak{A}$ in the strip (4.2.10) consists of the simple real eigenvalues $\xi_+(\alpha)/\alpha$ and $\xi_+^{(1)}(\alpha)/\alpha$.

Proof: By Theorem 4.2.1, the sum of the algebraic multiplicities of all eigenvalues in the strip (4.2.10) is equal to two. A description of the eigenvalues has been given before the present theorem. Therefore, we have only to show the monotonicity of the function $\operatorname{Re} z(\alpha)$.

Obviously, the function $t(\alpha) = \alpha^{-1}\sin\alpha$ is positive and strictly decreasing in the interval in $(0, \alpha_0)$. We set $\zeta(t) = z(\alpha(t))$, where $\alpha = \alpha(t)$ is the inverse function to $t = t(\alpha)$. Differentiating the equation $\sin\zeta(t) + t\,\zeta(t) = 0$, we obtain $\zeta' = -(\cos\zeta + t)^{-1}\zeta$. Hence for $\zeta = a + bi$

$$(4.2.12) \qquad \operatorname{Re}\zeta' = \frac{-a(\cos a \cdot \cosh b + t) + b\,\sin a \cdot \sinh b}{|\cos\zeta + t|^2}.$$

Splitting the equation $\sin\zeta + t\zeta = 0$ into the real and imaginary parts, we get the system

$$(4.2.13) \qquad \sin a \cdot \cosh b + ta = 0, \qquad \cos a \cdot \sinh b + tb = 0.$$

Together with (4.2.12) this implies

$$\operatorname{Re}\zeta' = \frac{\sin 2a - 2at^2}{2t\,|\cos\zeta + t|^2}.$$

By (4.2.13), we have $\sin 2a = 4t^2\, ab/\sinh 2b > 0$ and, therefore,

$$\operatorname{Re}\zeta' = \frac{at(2b - \sinh 2b)}{\sinh 2b\,|\cos\zeta + t|^2} < 0.$$

Hence the function $\operatorname{Re} z(\alpha)$ is strictly increasing on $(0, \alpha_0)$. ∎

4.3. The Neumann problem for the Lamé system in a cone

In this section we deal with the pencil corresponding to the Neumann problem for the Lamé system in a three-dimensional cone. We describe its spectrum in the strip $-1 \le \operatorname{Re}\lambda \le 0$. We show that the spectrum in this strip consists of two eigenvalues $\lambda_0 = 0$ and $\lambda_1 = -1$ which have geometric and algebraic multiplicity 3 if the cone $\mathcal{K}$ is given by the inequality $x_3 > \phi(x_1, x_2)$ with a smooth positively homogeneous of degree 1 function ϕ.

4.3.1. The operator pencil generated by the Neumann problem for the Lamé system. We consider the Neumann problem for the Lamé system:

$$(4.3.1) \qquad \Delta U + (1 - 2\nu)^{-1}\nabla\nabla \cdot U = 0 \qquad \text{in } \mathcal{K},$$

$$(4.3.2) \qquad \sum_{j=1}^{3}\sigma_{i,j}(U)\,n_j = 0 \qquad \text{on } \partial\mathcal{K}\backslash\{0\}, \ i = 1, 2, 3.$$

Here again $\{\sigma_{i,j}(U)\}$ denotes the stress tensor. In what follows, it is assumed that the cone $\mathcal{K}$ is given by $x_3 > \phi(x_1, x_2)$, where ϕ is a positively homogeneous function of degree 1 which is smooth on $\mathbb{R}^2 \backslash \{0\}$.

We are interested in solutions of problem (4.3.1), (4.3.2) which have the form

$$(4.3.3) \qquad U = r^{\lambda_0} \sum_{k=0}^{s} \frac{1}{k!} (\log r)^k u^{(s-k)}(\omega),$$

where $u^{(k)} \in W_2^1(\Omega)^3$ for $k = 0, 1, \ldots, s$. This means that we look for vector functions (4.3.3) which satisfy the integral identity

$$(4.3.4) \qquad \sum_{i,j=1}^{3} \int_{\mathcal{K}} \sigma_{i,j}(U) \cdot \varepsilon_{i,j}(V) \, dx = 0$$

for all functions $V = V(r, \omega)$ of the class $C_0^\infty\big((0, \infty); W_2^1(\Omega)^3\big)$.

By $a(\cdot, \cdot; \lambda)$ we denote the parameter-depending sesquilinear form

$$(4.3.5) \qquad a(u, v; \lambda) = \frac{1}{2 \log 2} \int_{\substack{\mathcal{K} \\ 1/2 < |x| < 2}} \sum_{i,j=1}^{3} \sigma_{i,j}(U) \cdot \varepsilon_{i,j}(V) \, dx,$$

where $U(x) = r^\lambda u(\omega)$, $V(x) = r^{-1-\overline{\lambda}} v(\omega)$, and $u, v \in W_2^1(\Omega)^3$.

The sesquilinear form (4.3.5) generates a linear and continuous operator

$$(4.3.6) \qquad \mathfrak{A}(\lambda) : W_2^1(\Omega)^3 \to (W_2^1(\Omega)^3)^*$$

by the equality

$$(4.3.7) \qquad \big(\mathfrak{A}(\lambda)u, v\big)_{L_2(\Omega)^3} = a(u, v; \lambda), \qquad u, v \in W_2^1(\Omega)^3.$$

According to Lemma 1.2.3, the vector function (4.3.3) is a solution of the equation (4.3.4) if and only if λ_0 is an eigenvalue of the operator pencil $\mathfrak{A}$ and $u^{(0)}, \ldots, u^{(s)}$ is a Jordan chain corresponding to this eigenvalue.

The assertions of Theorem 4.1.1 are also valid for the operator pencil generated by the Neumann problem and their proof is literally the same as that for the mixed problem in Section 4.1. In particular, the spectrum of the pencil $\mathfrak{A}$ consists of isolated eigenvalues with finite algebraic multiplicities, and the equality

$$a(u, v; \lambda) = \overline{a(v, u; -1 - \overline{\lambda})}$$

holds for arbitrary $u, v \in W_2^1(\Omega)^3$. This means that $(\mathfrak{A}(\lambda))^* = \mathfrak{A}(-1 - \overline{\lambda})$.

Let λ_0 be an eigenvalue of the pencil $\mathfrak{A}$ and let u_0 be an eigenvector corresponding to this eigenvalue. The equation

$$\mathfrak{A}(\lambda_0) \, u^{(1)} = -\mathfrak{A}'(\lambda_0) \, u_0$$

for the generalized eigenvector $u^{(1)}$ is solvable if and only if $\mathfrak{A}'(\lambda_0) \, u_0$ is orthogonal to the kernel of the operator $(\mathfrak{A}(\lambda_0))^* = \mathfrak{A}(-1 - \overline{\lambda}_0)$ or, what is the same, if

$$a^{(1)}(u_0, v; \lambda_0) = 0$$

for all eigenvectors of the pencil $\mathfrak{A}$ corresponding to the eigenvalue $-1 - \overline{\lambda}_0$.

4.3.2. The case of the half-space. Let $\mathcal{K}$ be the half-space

$$\mathbb{R}^3_+ = \{(x_1, x_2, x_3) : x_3 > 0\}.$$

In this case we denote the above introduced operator pencil by $\mathfrak{A}_0$. We leave to the reader the proof of the following well-known assertion concerning positively homogeneous solutions of the Neumann problem for the Lamé system in $\mathbb{R}^3_+$ (compare with Theorems 10.3.1–10.3.3).

LEMMA 4.3.1. *The pencil $\mathfrak{A}_0$ has the eigenvalues*

$$\lambda_j = j \quad and \quad \mu_j = -j - 1,$$

$j = 0, 1, 2, \ldots$. The eigenvectors corresponding to the eigenvalues λ_j are the traces on S^2_+ of homogeneous polynomials of degree j which are solutions of the problem (4.3.1), (4.3.2), while the eigenvectors corresponding to the eigenvalues μ_j are traces of derivatives of the Poisson kernels, i.e., of solutions to the problems

$$\Delta P_j + (1 - 2\nu)^{-1} \nabla \nabla \cdot P_j = 0 \qquad in \ \mathbb{R}^3_+,$$
$$\sigma_{i,3}(P_j) = c\,\delta_{i,j}\,\delta(x') \qquad for \ x_3 = 0, \ i = 1, 2, 3.$$

Other eigenvalues of the pencil $\mathfrak{A}_0$ do not exist.

4.3.3. Eigenvalues in the strip $-1 \le \operatorname{Re}\lambda \le 0$. In particular, it follows from Lemma 4.3.1 that 0 and -1 are the only eigenvalues of the pencil $\mathfrak{A}_0$ in the strip $-1 \le \operatorname{Re}\lambda \le 0$. We show that this is also true for the pencil $\mathfrak{A}$ if $\mathcal{K}$ is an arbitrary cone given by the inequality $x_3 > \phi(x_1, x_2)$ with a function ϕ which is smooth and positively homogeneous of degree 1 on $\mathbb{R}^2 \backslash \{0\}$.

THEOREM 4.3.1. *The strip $-1 \le \operatorname{Re}\lambda \le 0$ contains exactly two eigenvalues $\lambda_0 = 0$ and $\lambda_1 = -1$. Both eigenvalues have the geometric and algebraic multiplicity 3. Generalized eigenvectors to these eigenvalues do not exist. The eigenvectors corresponding to the eigenvalue λ_0 are the constant vectors in $\mathbb{C}^3$.*

Proof: We show first that $\lambda_0 = 0$ is the only eigenvalue of the pencil $\mathfrak{A}$ on the line $\operatorname{Re}\lambda = 0$. Let $U(x) = r^\lambda u(\omega)$ be a solution of problem (4.3.1), (4.3.2), where $\operatorname{Re}\lambda = 0$. Using the Green formula (4.1.3) and the equality

$$\sum_{i,j=1}^3 \sigma_{i,j}(U)\,\varepsilon_{i,j}(\overline{V}) = \sum_{i,j=1}^3 \varepsilon_{i,j}(U)\,\sigma_{i,j}(\overline{V}),$$

we obtain

$$(4.3.8) \qquad \int\limits_{\substack{\mathcal{K} \\ 1/2 < |x| < 2}} \frac{\partial}{\partial x_3} \sum_{i,j=1}^3 \sigma_{i,j}(U) \cdot \varepsilon_{i,j}(\overline{U})\, dx$$

$$= 2\operatorname{Re} \sum_{i,j=1}^3 \int\limits_{\substack{\mathcal{K} \\ 1/2 < |x| < 2}} \sigma_{i,j}(U) \cdot \varepsilon_{i,j}\left(\frac{\partial \overline{U}}{\partial x_3}\right) dx$$

$$= 2\operatorname{Re}\left(\int\limits_{S_{1/2}} \sum_{i,j=1}^3 \sigma_{i,j}(U)\, n_j\, \frac{\partial \overline{U}_i}{\partial x_3}\, d\sigma + \int\limits_{S_2} \sum_{i,j=1}^3 \sigma_{i,j}(U)\, n_j\, \frac{\partial \overline{U}_i}{\partial x_3}\, d\sigma \right),$$

where $S_{1/2}$, S_2 are the intersections of the cone $\mathcal{K}$ with the spheres $|x| = 1/2$ and $|x| = 2$, respectively. Since $U(x) = r^\lambda u(\omega)$ with $\operatorname{Re}\lambda = 0$, the right side of (4.3.8) vanishes. Hence,

$$
\begin{aligned}
0 &= \int_{\substack{\mathcal{K} \\ 1/2<|x|<2}} \frac{\partial}{\partial x_3} \sum_{i,j=1}^{3} \sigma_{i,j}(U) \cdot \varepsilon_{i,j}(\overline{U})\, dx = \int_{\substack{\partial\mathcal{K} \\ 1/2<|x|<2}} n_3 \sum_{i,j=1}^{3} \sigma_{i,j}(U)\,\varepsilon_{i,j}(\overline{U})\, d\sigma \\
&\quad + \int_{S_{1/2}} n_3 \sum_{i,j=1}^{3} \sigma_{i,j}(U)\,\varepsilon_{i,j}(\overline{U})\, d\sigma + \int_{S_2} n_3 \sum_{i,j=1}^{3} \sigma_{i,j}(U)\,\varepsilon_{i,j}(\overline{U})\, d\sigma \\
&= \int_{\substack{\partial\mathcal{K} \\ 1/2<|x|<2}} n_3 \sum_{i,j=1}^{3} \sigma_{i,j}(U)\,\varepsilon_{i,j}(\overline{U})\, d\sigma \\
&= 2\mu \int_{\substack{\partial\mathcal{K} \\ 1/2<|x|<2}} n_3 \left(\frac{\nu}{1-2\nu}|\nabla\cdot U|^2 + \sum_{i,j=1}^{3} |\varepsilon_{i,j}(U)|^2 \right) d\sigma.
\end{aligned}
$$

By our assumptions on $\mathcal{K}$, the component n_3 of n is negative on $\partial\mathcal{K}\backslash\{0\}$. Consequently, we get

$$\nabla \cdot U = 0 \quad \text{on } \partial\mathcal{K}\backslash\{0\} \qquad \text{and} \qquad \varepsilon_{i,j}(U) = 0 \quad \text{on } \partial\mathcal{K}\backslash\{0\} \text{ for } i,j = 1,2,3.$$

Applying the operator $\nabla\cdot$ to equation (4.3.1), we obtain $\Delta\nabla\cdot U = 0$ in $\mathcal{K}$. Here $\nabla\cdot U$ is a function of the form $r^{\lambda-1} v(\omega)$. Consequently, $v = 0$ or $\lambda(\lambda - 1)$ is an eigenvalue of the Dirichlet problem for the operator $-\delta$ in Ω (see Section 2.2). Since the eigenvalues of $-\delta$ with Dirichlet boundary conditions on $\partial\Omega$ are real and positive, the last can be excluded. Hence $\nabla\cdot U = 0$. This implies $\Delta\varepsilon_{i,j}(U) = 0$ and, by the same arguments as before, $\varepsilon_{i,j} = 0$ for $i,j = 1,2,3$. Thus, we obtain $U = const \in \mathbb{C}^3$ and $\lambda = 0$.

Now we prove that there are no generalized eigenvectors to the eigenvalue $\lambda_0 = 0$. For this we have to show that there are no solutions of problem (4.3.1), (4.3.2) which have the form $U(x) = c\log r + v(\omega)$ with a nonzero constant vector c. Let U be a solution of this form. Since the derivatives $\partial U/\partial x_i$ have the form $r^{-1} w_i(\omega)$, it can be shown, in the same way as above, that $\varepsilon_{i,j}(U) = 0$. Consequently, $c = 0$, i.e., there are no generalized eigenvectors.

Thus, we have shown that $\lambda_0 = 0$ is the only eigenvalue on the line $\operatorname{Re}\lambda = 0$ and that this eigenvalue has the geometric and algebraic multiplicity 3. From the equality $(\mathfrak{A}(\lambda))^* = \mathfrak{A}(-1 - \overline{\lambda})$ and from Theorem 1.1.7 it follows that $\lambda_1 = -1$ is the only eigenvalue on the line $\operatorname{Re}\lambda = -1$ and that this eigenvalue has also the geometric and algebraic multiplicity 3.

Let $\mathcal{K}_t$ be the cone $\{(x_1, x_2, x_3) \in \mathbb{R}^3 : x_3 > t\,\phi(x_1, x_2)\}$, $0 \le t \le 1$, and let Ω_t be the intersection of $\mathcal{K}_t$ with the unit sphere S^2. We denote by $\mathfrak{A}_t$ the operator pencil generated by the problem

$$\Delta U + (1 - 2\nu)^{-1} \nabla\nabla\cdot U = 0 \qquad \text{in } \mathcal{K}_t,$$

$$\sum_{j=1}^{3} \sigma_{i,j}(U)\,n_j = 0 \qquad \text{on } \partial\mathcal{K}_t\backslash\{0\}, \ i = 1,2,3,$$

i.e., the operator $\mathfrak{A}_t(\lambda) : W_2^1(\Omega_t)^3 \to (W_2^1(\Omega_t)^*)^3$ is determined by the sesquilinear form (4.3.5), where $\mathcal{K}$ has to be replaced by $\mathcal{K}_t$. In particular, $\mathfrak{A}_1 = \mathfrak{A}$ and $\mathfrak{A}_0$ is the operator pencil generated by problem (4.3.1), (4.3.2) in the half-space. The operator pencils $\mathfrak{A}$ and $\mathfrak{A}_t$ have only the eigenvalues $\lambda_0 = 0$ and $\lambda_1 = -1$ on the lines $\operatorname{Re} \lambda = 0$ and $\operatorname{Re} \lambda = -1$. Both eigenvalues have algebraic multiplicity 3. There exists a set of diffeomorphisms $\psi_t : \Omega_t \to \Omega_0$ infinitely differentiable with respect to the parameter $t \in [0, 1]$ which transform the operators $\mathfrak{A}_t(\lambda)$ into the operators

$$\tilde{\mathfrak{A}}_t(\lambda) : W_2^1(\Omega_0)^3 \to (W_2^1(\Omega_0)^*)^3.$$

The pencils $\mathfrak{A}_t$ and $\tilde{\mathfrak{A}}_t$ have the same eigenvalues with the same geometric and algebraic multiplicities. Consequently, on the lines $\operatorname{Re} \lambda = 0$ and $\operatorname{Re} \lambda = -1$ there are only the eigenvalues $\lambda_0 = 0$ and $\lambda_1 = -1$ of the pencils $\tilde{\mathfrak{A}}_t$. Moreover, by Lemma 4.3.1 there are no eigenvalues of the pencil $\tilde{\mathfrak{A}}_0$ in the strip $-1 < \operatorname{Re} \lambda < 0$. Applying Theorem 1.1.4, we obtain that this strip does not contain eigenvalues of the pencil $\tilde{\mathfrak{A}}_1$. This completes the proof. $\blacksquare$

REMARK 4.3.1. Theorem 4.3.1 remains true if the domain Ω which is cut out by $\mathcal{K}$ on the unit sphere has a finite number of angles not equal to 0 and 2π on the boundary. Then the gradients of the eigen- and generalized eigenvectors (with finite energy integrals) have the order $O(\rho^{-1/2+\varepsilon})$, $\varepsilon > 0$, at the corners of the contour $\partial\Omega$ (here ρ denotes the distance to the corner). This allows one to use the same arguments as in the proof of Theorem 4.3.1.

4.3.4. The Neumann problem in a circular cone. Now let $\mathcal{K}$ be the circular cone $\{x : x/|x| \in \Omega_\alpha\}$, where $\Omega_\alpha = \{\omega \in S^2 : 0 \leq \theta < \alpha, \ 0 \leq \varphi < 2\pi\}$. In the same way as for the Dirichlet problem it is possible to establish a transcendental equation for real eigenvalues of the pencil $\mathfrak{A}$ by means of Lemma 3.7.1.

We write the components of the vector function (3.7.6) in the form

$$u_r = \cos(m\varphi)\, \vec{m}_r^+(\theta) \cdot \vec{a}, \quad u_\theta = \cos(m\varphi)\, \vec{m}_\theta^+(\theta) \cdot \vec{a}, \quad u_\varphi = \sin(m\varphi)\, \vec{m}_\varphi^+(\theta) \cdot \vec{a},$$

where $\vec{m}_r^+$, $\vec{m}_\theta^+$, $\vec{m}_\varphi^+$ are the rows of the matrix $M_m^+(\lambda, \theta)$ introduced in Section 3.7 and $\vec{a} = (a, b, c)$. Analogously, the components of the vector function (3.7.7) have the form

$$u_r = \sin(m\varphi)\, \vec{m}_r^-(\theta) \cdot \vec{a}, \quad u_\theta = \sin(m\varphi)\, \vec{m}_\theta^-(\theta) \cdot \vec{a}, \quad u_\varphi = \cos(m\varphi)\, \vec{m}_\varphi^-(\theta) \cdot \vec{a},$$

where $\vec{m}_r^-$, $\vec{m}_\theta^-$, $\vec{m}_\varphi^-$ are the rows of the matrix $M_m^-(\lambda, \theta)$. The components of the corresponding displacement field are

(4.3.9) $$U_r = r^\lambda u_r, \quad U_\theta = r^\lambda u_\theta, \quad U_\varphi = r^\lambda u_\varphi.$$

We write the Neumann boundary condition $\sigma n|_{\theta=\alpha} = 0$ in spherical coordinates:

(4.3.10) $$\frac{E}{\nu+1}\left(\varepsilon_{\theta r}, \ \varepsilon_{\theta\theta} + \frac{\nu}{1-2\nu}\left(\varepsilon_{rr} + \varepsilon_{\theta\theta} + \varepsilon_{\varphi\varphi}\right), \ \varepsilon_{\theta\varphi}\right)\bigg|_{\theta=\alpha} = 0.$$

Here E is the Young modulus, and the components of the strain tensor in spherical coordinates have the following form for selection (3.7.6):

$$\varepsilon_{rr} = \cos(m\varphi)\, r^{\lambda-1} E_{rr}(\theta), \quad \varepsilon_{r\theta} = \cos(m\varphi)\, r^{\lambda-1} E_{r\theta}(\theta),$$
$$\varepsilon_{\theta\theta} = \cos(m\varphi)\, r^{\lambda-1} E_{\theta\theta}(\theta), \quad \varepsilon_{\varphi\varphi} = \cos(m\varphi)\, r^{\lambda-1} E_{\varphi\varphi}(\theta),$$
$$\varepsilon_{\theta\varphi} = \sin(m\varphi)\, r^{\lambda-1} E_{\theta\varphi}(\theta),$$

where

$$E_{rr} = \lambda \vec{m}_r^+ \cdot \vec{a}, \quad E_{r\theta} = \frac{1}{2}\left(\partial_\theta \vec{m}_r^+ + (\lambda - 1)\vec{m}_\theta^+\right) \cdot \vec{a}, \quad E_{\theta\theta} = \left(\partial_\theta \vec{m}_\theta^+ + \vec{m}_r^+\right) \cdot \vec{a},$$

$$E_{\varphi\varphi} = \left(\frac{m}{\sin\theta}\vec{m}_\varphi^+ + \vec{m}_r^+ + \cot\theta\,\vec{m}_\theta^+\right) \cdot \vec{a},$$

$$E_{\theta\varphi} = \frac{1}{2}\left(\partial_\theta \vec{m}_\varphi^+ - \cot\theta\,\vec{m}_\varphi^+ - \frac{m}{\sin\theta}\vec{m}_\theta^+\right) \cdot \vec{a}$$

(cf. (4.1.7)). For selection (3.7.7) we have

$$\varepsilon_{rr} = \sin(m\varphi)\,r^{\lambda-1}\,E_{rr}(\theta), \quad \varepsilon_{r\theta} = \sin(m\varphi)\,r^{\lambda-1}\,E_{r\theta}(\theta),$$

$$\varepsilon_{\theta\theta} = \sin(m\varphi)\,r^{\lambda-1}\,E_{\theta\theta}(\theta), \quad \varepsilon_{\varphi\varphi} = \sin(m\varphi)\,r^{\lambda-1}\,E_{\varphi\varphi}(\theta),$$

$$\varepsilon_{\theta\varphi} = \cos(m\varphi)\,r^{\lambda-1}\,E_{\theta\varphi}(\theta),$$

where

$$E_{rr} = \lambda \vec{m}_r^- \cdot \vec{a}, \quad E_{r\theta} = \frac{1}{2}\left(\partial_\theta \vec{m}_r^- + (\lambda - 1)\vec{m}_\theta^-\right) \cdot \vec{a}, \quad E_{\theta\theta} = \left(\partial_\theta \vec{m}_\theta^- + \vec{m}_r^-\right) \cdot \vec{a},$$

$$E_{\varphi\varphi} = \left(-\frac{m}{\sin\theta}\vec{m}_\varphi^- + \vec{m}_r^- + \cot\theta\,\vec{m}_\theta^-\right) \cdot \vec{a},$$

$$E_{\theta\varphi} = \frac{1}{2}\left(\partial_\theta \vec{m}_\varphi^- - \cot\theta\,\vec{m}_\varphi^- + \frac{m}{\sin\theta}\vec{m}_\theta^-\right) \cdot \vec{a}.$$

Thus, (4.3.10) is satisfied if

(4.3.11) $$\frac{E}{\nu+1}\left(E_{\theta r},\ E_{\theta\theta} + \frac{\nu}{1-2\nu}\left(E_{rr} + E_{\theta\theta} + E_{\varphi\varphi}\right),\ E_{\theta\varphi}\right)\bigg|_{\theta=\alpha} = 0.$$

Let the matrices $N_m^\pm$ be given by

$$N_m^\pm(\lambda,\theta) = \begin{pmatrix} \partial_\theta \vec{m}_r^\pm + (\lambda-1)\,\vec{m}_\theta^\pm \\ (2-2\nu)\,\partial_\theta \vec{m}_\theta^\pm + 2(1+\nu\lambda)\,\vec{m}_r^\pm + 2\nu\left(\cot\theta\,\vec{m}_\theta^\pm \pm m\,(\sin\theta)^{-1}\vec{m}_\varphi^\pm\right) \\ \partial_\theta \vec{m}_\varphi^\pm - \cot\theta\,\vec{m}_\varphi^\pm \mp m\,(\sin\theta)^{-1}\vec{m}_\theta^\pm \end{pmatrix}.$$

Furthermore, let D be the matrix

$$D = \frac{E}{\nu+1}\begin{pmatrix} 1 & 0 & 0 \\ 0 & (1-2\nu)^{-1} & 0 \\ 0 & 0 & 1 \end{pmatrix}.$$

Then (4.3.11) is equivalent to

$$D\,N_m^\pm(\lambda,\alpha)\,\vec{a} = 0.$$

Here the upper sign corresponds to the selection (3.7.6), and the lower sign to (3.7.7). Therefore, nontrivial solutions of (4.3.1), (4.3.2) of the form (4.3.9) exist if λ is a solution of

(4.3.12) $$\frac{1}{1-2\nu}\left(\frac{E}{\nu+1}\right)^3 \det N_m^\pm(\lambda,\alpha) = 0.$$

The matrix N_m^+ is explicitly given by

$$N_m^+(\lambda,\theta) = P_\lambda^{-m}(\cos\theta)\,A_m(\lambda,\theta) + P_{\lambda+1}^{-m}(\cos\theta)\,B_m(\lambda,\theta),$$

where A_m is the matrix with the columns

$$\begin{pmatrix} -\lambda(\lambda+1-m)\,(\sin\theta)^{-1} \\ 2(1-2\nu)(\lambda+1-m)\,(\sin\theta)^{-2}\cos\theta \\ 0 \end{pmatrix},\qquad \begin{pmatrix} m(\lambda+1-m)\,(\sin\theta)^{-1} \\ -2(\lambda+1-m)\,(\sin\theta)^{-2}\cos\theta \\ (\lambda+1-m)(\lambda-2\cot^2\theta) \end{pmatrix},$$

$$\begin{pmatrix} (1-2\nu)\,\lambda\sin\theta-(\lambda+1)(\lambda-2\nu-2)\,(\sin\theta)^{-1}\cos^2\theta \\ 2\cos\theta\Big(\lambda^2+4\nu(\lambda-\nu-2)-\lambda-3-\big((2-2\nu)(\lambda+1)^2-\lambda+1-m^2\big)(\sin\theta)^{-2}\Big) \\ m\big(2-4\nu-(\lambda+2)\cot\theta(1+\cot\theta)\big) \end{pmatrix}$$

and B_m is the matrix with the columns

$$\begin{pmatrix} \lambda(\lambda+1)\cot\theta \\ 2m^2(\sin\theta)^{-2}-2(1-2\nu)(\lambda+1)\,\big(\lambda+1+\cot^2\theta\big) \\ -m\cot\theta \end{pmatrix},$$

$$\begin{pmatrix} -2m\cot\theta \\ 2(\lambda+2\nu)(\sin\theta)^{-2}-2(1-2\nu)\lambda \\ \big(m^2+(1-m)(\lambda+1)\big)\cot\theta \end{pmatrix},\qquad \begin{pmatrix} (\lambda+m+1)(\lambda+2\nu-2)\cot\theta \\ (2-4\nu)\,\big(\lambda+m+1((2-2\nu-\cot^2\theta)\big) \\ 0 \end{pmatrix}.$$

Note that real eigenvalues of the pencil $\mathfrak{A}$ which are solutions of (4.3.12) with $m = 1, 2, \ldots$ have at least multiplicity 2 and two linearly independent eigensolutions of the form (3.7.6) and (3.7.7). For $m = 0$ in condition (4.3.12) there exist two different, simple eigenvalues, one corresponding to torsion and the other to axisymmetric eigenstates, respectively.

The following lemma lists the eigenvectors corresponding to the eigenvalue $\lambda = 1$ for $\alpha = \pi/2$ and $\alpha = \pi$ together with the values of m in (4.3.12).

LEMMA 4.3.2. *The multiplicity of $\lambda = 1$ for $\alpha = \pi$ is 9. A system of eigenvectors is the restriction to S^2 of the following vectors*

$$\begin{pmatrix} x_1 \\ x_2 \\ 0 \end{pmatrix},\ \begin{pmatrix} x_2 \\ -x_1 \\ 0 \end{pmatrix},\ \begin{pmatrix} 0 \\ 0 \\ x_3 \end{pmatrix} \qquad (m=0),$$

$$\begin{pmatrix} x_3 \\ 0 \\ 0 \end{pmatrix},\ \begin{pmatrix} 0 \\ x_3 \\ 0 \end{pmatrix},\ \begin{pmatrix} 0 \\ 0 \\ x_1 \end{pmatrix},\ \begin{pmatrix} 0 \\ 0 \\ x_2 \end{pmatrix} \qquad (m=1),$$

$$\begin{pmatrix} x_2 \\ x_1 \\ 0 \end{pmatrix},\ \begin{pmatrix} -x_1 \\ x_2 \\ 0 \end{pmatrix} \qquad (m=2).$$

In the case $\alpha = \pi/2$ the multiplicity of $\lambda = 1$ equals 6. The eigenvectors are the restriction to S^2 of

$$\begin{pmatrix} x_2 \\ -x_1 \\ 0 \end{pmatrix},\ \begin{pmatrix} x_1 \\ x_2 \\ \frac{2\nu}{1-\nu}x_3 \end{pmatrix} \quad (m=0),\qquad \begin{pmatrix} x_3 \\ 0 \\ -x_1 \end{pmatrix},\ \begin{pmatrix} 0 \\ x_3 \\ -x_2 \end{pmatrix} \quad (m=1),$$

and

$$\begin{pmatrix} x_2 \\ x_1 \\ 0 \end{pmatrix},\ \begin{pmatrix} -x_1 \\ x_2 \\ 0 \end{pmatrix} \quad (m=2)$$

For all opening angles α, the eigenvalue $\lambda = 1$ has multiplicity ≥ 3. Three eigen-vectors are given by the restrictions to S^2 of

$$\begin{pmatrix} x_2 \\ -x_1 \\ 0 \end{pmatrix} \quad (m = 0), \qquad \begin{pmatrix} x_3 \\ 0 \\ -x_1 \end{pmatrix}, \begin{pmatrix} 0 \\ x_3 \\ -x_2 \end{pmatrix} \quad (m = 1).$$

The figures below show the dependence of the eigenvalues less than 1 on the opening $\alpha \in [\frac{\pi}{2}, \pi]$. In Figure 10 the first eigenvalue obtained from equation (4.3.12) with $m = 0$ is depicted. This eigenvalue corresponds to axisymmetric eigenstates and has already been obtained by Bažant and Keer [12]. It is always less than 1 (and, therefore, induces infinite stresses) and behaves monotonically with respect ν, the minimal value being attained for $\nu = \frac{1}{2}$. As ν approaches -1, the eigenvalue curves tend to $\lambda = 1$. The singular nature of this eigenvalue is already indicated by equation (4.3.12).

The second eigenvalue corresponding to $m = 0$ has an elastic torsion eigenstate. It does not depend on ν, branches off 2 at $\alpha = \pi/2$ and joins 1 for $\alpha = \pi$.

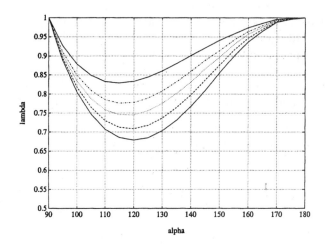

FIGURE 10. Eigenvalues of $\mathfrak{A}$ for $m = 0$, $\nu = 0$, 0.2, 0.3, 0.4, 0.475

Figure 11 shows the real eigenvalues less than 1 corresponding to $m = 1$ for several values of ν between 0.2 and 0.5. All branches decrease monotonically as α increases and enter the interval $(0, 1)$ at some critical angle $\alpha(\nu)$ $(\alpha(\nu) \approx \frac{3}{4}\pi$ for $0.25 \leq \nu \leq 0.4)$, attain an absolute minimum and increase again to $\lambda = 1$ at $\alpha = \pi$. Their minimal values are lower than the axisymmetric eigenvalues, so that, in particular, for slender notches they cause the dominant stress singularity. Note that above 1 the eigenvalues become complex.

For $m = 2$ there is a branch of double eigenvalues. The eigenvalues of this branch are also real and less than 1 for $\frac{\pi}{2} < \alpha < \pi$. Figure 12 shows these eigenvalues for several values of ν. The eigenvalue curves increase monotonically as ν increases and cause, for some α and ν (e.g., $\alpha = 155°$, $\nu = 0.2$), the dominant singularity.

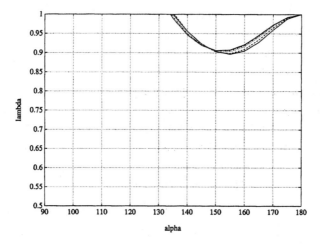

FIGURE 11. Eigenvalues of $\mathfrak{A}$ for $m = 1$, $\nu = 0.2, 0.25, 0.3, 0.4, 0.5$

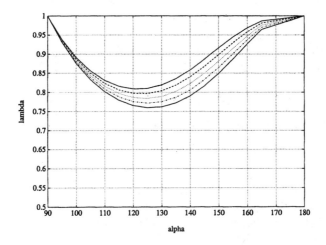

FIGURE 12. Eigenvalues of $\mathfrak{A}$ for $m = 2$, $\nu = 0.2, 0.25, 0.3, 0.4, 0.5$

4.4. Angular crack in an anisotropic elastic space

4.4.1. Statement of the problem. Let H be a plane angle with vertex $x = 0$ and opening α which is placed in the plane $x_2 = 0$. Furthermore, let the arc h be the intersection of H with the unit sphere S^2. We consider the Neumann problem

$$(4.4.1) \qquad \sum_{i=1}^{3} \partial_{x_i} \sigma_{i,j}(U) = 0 \quad \text{in } \mathcal{K} = \mathbb{R}^3 \backslash \overline{H},$$

$$(4.4.2) \qquad \sigma_{2,j}(U) = 0 \quad \text{on } H, \; j = 1, 2, 3.$$

The stress and strain tensors are related by Hooke's law

$$(4.4.3) \qquad \sigma_{i,j} = \sum_{k,h=1}^{3} a_{ij,kh}\, \varepsilon_{k,h}\,,$$

where $a_{ij,kh}$ are real coefficients with the symmetry property $a_{ij,kh} = a_{ji,kh} = a_{kh,ij}$ and with the property of ellipticity

$$(4.4.4) \qquad \sum_{i,j,k,h} a_{ij,kh}\, \varepsilon_{i,j}\, \varepsilon_{k,h} \geq c \sum_{i,j} \varepsilon_{i,j}^2, \quad c > 0,$$

which has to be valid for all real $\varepsilon_{i,j}$ such that $\varepsilon_{i,j} = \varepsilon_{j,i}$. We associate with this problem the sesquilinear form (4.3.5), where $U(x) = r^\lambda u(\omega)$, $V(x) = r^{-1-\overline{\lambda}} v(\omega)$, $u, v \in W_2^1(\Omega)^3$, $\Omega = S^2 \backslash \overline{h}$. Here, in contrast to Section 4.3, the stress tensor is given by (4.4.3). The sesquilinear form $a(\cdot, \cdot; \lambda)$ generates the operator $\mathfrak{A}(\lambda)$ by (4.3.6) and (4.3.7).

4.4.2. Properties of the pencil $\mathfrak{A}$.

THEOREM 4.4.1. *All assertions of Theorem* 4.1.1 *are valid for the pencil* $\mathfrak{A}$.

Proof: Let us prove first the second assertion. For $\lambda = -\frac{1}{2} + it$ the quadratic form $a(u, u; \lambda)$ is given by (4.3.5), where $V = U$. Using (4.4.4), one can show that the right-hand side of (4.3.5) for $\lambda = -\frac{1}{2} + it$ and $V = U$ can be estimated from below by the same expression as in the case of isotropic elasticity and, therefore, the positivity of $\mathfrak{A}(-\frac{1}{2} + it)$ follows from the analogous property for the pencil generated by the Neumann problem to Lamé's system.

To prove the first assertion, it suffices to note that the operator $\mathfrak{A}(-\frac{1}{2} + it)$ is invertible and the operator $\mathfrak{A}(\lambda) - \mathfrak{A}(-\frac{1}{2} + it)$ acting from $W_2^1(\Omega)^3$ into $(W_2^1(\Omega)^*)^3$ is compact for arbitrary λ.

Assertions 3) and 4) follow from Theorems 1.1.1 and 1.2.2. $\blacksquare$

4.4.3. Eigenvalues on the line Re $\lambda = 0$.

LEMMA 4.4.1. *The number* $\lambda_0 = 0$ *is the only eigenvalue of the pencil* $\mathfrak{A}$ *on the line* Re $\lambda = 0$.

Proof: For convenience, we assume that one of the sides of H coincides with the positive x_3-axis. The unit vector directed along the bisectrix of the angle H is denoted by b. Let G_δ be the union of two geodesic δ-neighborhoods of the endpoints of the arc $h = H \cap S^2$ and let

$$\mathcal{K}_\delta = \left\{ x \,:\, 1/2 < |x| < 2, \; x/|x| \notin G_\delta \cup h \right\},$$

where δ is a small positive number.

Suppose $U(x) = |x|^\lambda u(x/|x|)$ is a solution of problem (4.4.1), (4.4.2) with Re $\lambda = 0$. Then

$$0 = \int_{\mathcal{K}_\delta} \sum_{i,j=1}^{3} \partial_{x_j}\, \sigma_{i,j}(U)\, \partial_b \overline{U}_i \, dx$$

$$= \int_{\partial \mathcal{K}_\delta} \sum_{i,j=1}^{3} n_j\, \sigma_{i,j}(U)\, \partial_b \overline{U}_i \, d\sigma - \int_{\mathcal{K}_\delta} \sum_{i,j=1}^{3} \sigma_{i,j}(U)\, \partial_b \varepsilon_{i,j}(\overline{U}) \, dx,$$

where $n = (n_1, n_2, n_3)$ is the exterior normal to $\partial \mathcal{K}_\delta$. Consequently,

$$(4.4.5) \qquad 0 = \mathrm{Re} \int_{\partial \mathcal{K}_\delta} \sum_{i,j=1}^{3} \left(n_j\, \sigma_{i,j}(U)\, \partial_b \overline{U}_i - \frac{1}{2}\, n_b\, \sigma_{i,j}(U)\, \varepsilon_{i,j}(\overline{U}) \right) d\sigma.$$

The integrand vanishes on the surface H and the sum of the integrals over the parts of the spheres $|x| = 1/2$ and $|x| = 2$ equals zero. Thus, the integration in (4.4.5) is carried out only over the two components of the surface $\{x : 1/2 < |x| < 2, \ x/|x| \in \partial G_\delta\}$ and we arrive at the equality

$$(4.4.6) \qquad \mathrm{Re} \int\limits_{\partial G_\delta} \sum_{i,j=1}^{3} \left(n_j \, \sigma_{i,j}(U) \, \partial_b \overline{U}_i - \frac{1}{2} \, n_b \, \sigma_{i,j}(U) \, \varepsilon_{i,j}(\overline{U}) \right) ds = 0.$$

Next we are going to find the asymptotics for the integrand near the north pole of the sphere S^2. Under corresponding choice of the x_1-axis, the principal part of the boundary value problem (4.4.1), (4.4.2) is the operator of generalized plane elasticity in the (x_1, x_2)-plane with a cut along the semi-axis $\mathbf{R}_-^1 = \{(x_1, x_2) \in \mathbb{R}^2 : x_2 = 0, \ x_1 < 0\}$. Therefore,

$$(4.4.7) \qquad u(x/|x|) = L(x_1, x_2) + v(x_1, x_2) + O\big((x_1^2 + x_2^2)^{(1+\sigma)/2}\big),$$

where $\sigma \in (0, 1/2)$, L is a linear vector function and $v = (v_1, v_2, v_3)$ is a positively homogeneous of degree $1/2$ vector function which satisfies the equations

$$\partial_{x_1} \sigma_{1,j}(v) + \partial_{x_2} \sigma_{2,j}(v) = 0 \quad \text{on } \mathbb{R}^2 \backslash \mathbf{R}_-^1 \, ,$$

$$\partial_{i,2}(v) = 0 \quad \text{on } \mathbf{R}_-^1 \, , \ i = 1, 2, 3.$$

The vector function v has the form

$$v(x_1, x_2) = r^{1/2} \left(K_I \, G_I(\varphi) + K_{II} \, G_{II}(\varphi) + K_{III} \, G_{III}(\varphi) \right),$$

where K_I, K_{II}, K_{III} are so-called stress intensity factors, r, φ are polar coordinates, and G_I, G_{II}, G_{III} are smooth vector functions on the interval $[-\pi, \pi]$. From (4.4.7) we immediately obtain that the integrand in (4.4.6) has the form

$$(4.4.8)$$

$$\sin \frac{\alpha}{2} \left(\frac{n_1}{2} \sum_{i,j=1}^{3} \sigma_{i,j}(v) \varepsilon_{i,j}(\overline{v}) - \sum_{i=1}^{3} \big(n_1 \sigma_{i,1}(v) \partial_{x_1}(\overline{v}_i) + n_2 \sigma_{i,2}(v) \partial_{x_1} \overline{v}_i \big) \right) + O(r^{1/2}),$$

where α is the opening of the angle H. It is known that the integral

$$\int\limits_{r=\delta} \left(\frac{n_1}{2} \sum_{i,j=1}^{3} \sigma_{i,j}(w) \varepsilon_{i,j}(w) - \sum_{i=1}^{3} \big(n_1 \sigma_{i,1}(w) \partial_{x_1}(w_i) + n_2 \sigma_{i,2}(w) \partial_{x_1} w_i \big) \right) \delta \, d\varphi,$$

which characterizes the energy flux across the circumference $r = \delta$, does not depend on δ and represents a positive definite quadratic form of the stress intensity factors corresponding to the real-valued displacement field $w = (w_1, w_2, w_3)$ (see Nazarov and Plamenevskiĭ [**207**, Sect.7.4], Nazarov [**204**, Sect.5]). Therefore, the integral of the real part of the expression in braces in (4.4.8) taken over the circumference $r = \delta$ does not depend on δ and is positive, provided the vector function v does not vanish identically. What we said before about the neighborhood of the north pole holds also for the other end of the arc h. Thus, we conclude from (4.4.6) that $u \in W_2^2(S^2 \backslash h)$.

Let X denote the subspace of the eigenvectors corresponding to the eigenvalue λ, $\mathrm{Re}\,\lambda = 0$. Due to assertion 4) of Theorem 4.1.1 (by Theorem 4.4.1 this assertion als also true for the pencil $\mathfrak{A}$), the number $\lambda - 1 = -1 - \overline{\lambda}$ is an eigenvalue of the pencil $\mathfrak{A}$ with the same geometric and algebraic multiplicities as λ. The subspace generated by the eigenvalue $\lambda - 1$ is denoted by Y. Its dimension coincides with $\dim X$. We introduce linear operators acting from X into Y by the equalities

$$Au = r^{1-\lambda} \partial_{x_1}(r^\lambda u), \quad Bu = r^{1-\lambda} \partial_{x_3}(r^\lambda u), \quad u \in X.$$

(It can be easily seen that the vector functions Au and Bu belong to $W_2^1(S^2 \backslash h)$ and that they are eigenvectors of the pencil $\mathfrak{A}$ corresponding to the eigenvalue $\lambda - 1$.)

Suppose first that the kernel of B is nontrivial. Then $\partial_{x_3}(r^\lambda u) = 0$ for some vector function $u \in X$, $u \neq 0$. Hence $r^\lambda u = f(x_1, x_2)$, where f is a vector function which is positively homogeneous of degree λ and which satisfies the two-dimensional elasticity problem in the plane with a cut. Consequently, $f = const$ and $\lambda = 0$.

Now let $\ker B = \{0\}$. By β we denote an eigenvalue of the operator $B^{-1}A$ and by u the corresponding eigenfunction. Then

$$(\partial_{x_1} - \beta \partial_{x_3})(r^\lambda u) = (A - \beta B)u = 0.$$

If β is real, then, using an orthogonal change of variables in the plane $x_2 = 0$, we conclude that the kernel of B is nontrivial and, as we have shown above, that $\lambda = 0$. If $\operatorname{Im} \beta \neq 0$, then $r^\lambda u = f(x_3 + \beta x_1, x_2)$, where f is a vector function which is positively homogeneous of degree λ and analytic in the variable $x_3 + \operatorname{Re} \beta x_1 + i \operatorname{Im} \beta x_1$ for all $x_2 \neq 0$. The boundedness and analyticity of f imply $f = const$. and $\lambda = 0$. The proof is complete. ∎

4.4.4. Eigenvectors corresponding to the eigenvalue $\lambda = 0$. Obviously, the number $\lambda = 0$ is an eigenvalue of the pencil $\mathfrak{A}$ and the constant vectors are eigenvectors corresponding to this eigenvalue.

LEMMA 4.4.2. *The subspace of the eigenvectors corresponding to the eigenvalue $\lambda = 0$ consists of all constant vectors. There are no generalized eigenvectors corresponding to this eigenvalue.*

Proof: Suppose that besides constant vectors c_1, c_2, c_3 there are additional nonconstant eigenvectors $u_1, \ldots, u_n$ corresponding to the eigenvalue $\lambda = 0$ and that the vectors $c_1, c_2, c_3, u_1, \ldots, u_n$ form a basis in the subspace of all eigenvectors. Furthermore, we assume that for some $l = 0, 1, 2, 3$ the vectors $c_1, \ldots, c_l$ have generalized eigenvectors, while the vectors $c_{l+1}, \ldots, c_3$ and their linear combinations do not have generalized eigenvectors. (The values $l = 0$, $l = 3$ and $n = 0$ are included; in these cases we mean that corresponding sets are empty.) We have to show that $l = n = 0$.

By our assumptions, there exist solutions of problem (4.4.1), (4.4.2) of the form

$$U_k(x) = c_k \log |x| + v_k(x/|x|), \qquad k = 1, \ldots, l.$$

As it was shown in the proof of Lemma 4.4.1, the vector functions $u_1, \ldots, u_n$ belong to the space $W_2^2(S^2 \backslash h)^3$. In analogy to this proof we obtain $v_k \in W_2^2(S^2 \backslash h)^3$. Since $\partial U_k / \partial x_1$ and $\partial U_k / \partial x_3$, $k = 1, \ldots, l$ are solutions of problem (4.4.1), (4.4.2), we conclude that the functions $r \, \partial U_k / \partial x_j$, $j = 1, 3$, $k = 1, \ldots, l$, are eigenvectors corresponding to the eigenvalue $\lambda = -1$. The same is true for the functions $r \, \partial u_k / \partial x_j$, $j = 1, 3$, $k = 1, \ldots, n$

Let $\mathcal{Z}_l$ denote the linear span of the vector functions $U_1, \ldots, U_l, u_1, \ldots, u_n$. We introduce the operators $\mathcal{A}$, $\mathcal{B}$ acting from $\mathcal{Z}_l$ into the space Y of the eigenvectors corresponding to $\lambda = -1$ by the formulas

$$\mathcal{A}w = r \, \partial_{x_1} w, \quad \mathcal{B}w = r \, \partial_{x_3} w, \quad w \in \mathcal{Z}_l.$$

If the kernel of the operator $\mathcal{B}$ is nontrivial, then $\partial w/\partial x_3 = 0$ for some vector function $w \in \mathcal{Z}_l$, $w \neq 0$. If $\ker \mathcal{B} = \{0\}$ and $\mathcal{R}(\mathcal{A}) \subset \mathcal{R}(\mathcal{B})$, then (cf. proof of Lemma 4.4.1) we have

$$(\partial_{x_1} - \beta \partial_{x_3})w = 0 \quad \text{for some } \beta \in \mathbb{C}, \ w \in \mathcal{Z}_l \backslash \{0\}.$$

Using the same arguments as in the proof of Lemma 4.4.1, we get $w = const.$ in both cases. Since $\mathcal{Z}_l$ does not contain constant vectors, we conclude that $\ker \mathcal{B} = \{0\}$, $\mathcal{R}(\mathcal{A}) \supset \mathcal{R}(\mathcal{B})$, $\mathcal{R}(\mathcal{A}) \neq \mathcal{R}(\mathcal{B})$.

For $l \neq 0$ or $n \neq 0$ this implies that the vectors

$$r\partial_{x_3}U_1, \dots, r\partial_{x_3}U_l, r\partial_{x_3}u_1, \dots, r\partial_{x_3}u_n$$

are linearly independent and that not all of the vectors

$$r\partial_{x_1}U_1, \dots, r\partial_{x_1}U_l, r\partial_{x_1}u_1, \dots, r\partial_{x_1}u_n$$

belong to $\mathcal{R}(\mathcal{B})$. In particular, we get $\dim Y > l + n$ and, since $\dim Y = \dim X = n + 3$, it follows that $l < 3$.

Let c be a constant vector. If v is an arbitrary element of the subspace Y of the form $v = r\partial w/\partial x_s$, where $s = 1, 3$, and w is one of the vectors $U_1, \dots, U_l$, $u_1, \dots, u_n$, then we obtain

$$2 \log 2 \, a^{(1)}(c, v; 0) = \int\limits_{\substack{\mathbb{R}^3 \backslash H \\ 1/2 < |x| < 2}} \sum_{i,j=1}^{3} \sigma_{i,j}(c \log |x|) \, \varepsilon_{i,j}(\partial_{x_s}\overline{w}) \, dx$$

$$= - \int\limits_{\substack{\mathbb{R}^3 \backslash H \\ 1/2 < |x| < 2}} \sum_{i,j=1}^{3} \sigma_{i,j}(c x_s |x|^{-2}) \, \varepsilon_{i,j}(\overline{w}) \, dx$$

$$= \int\limits_{\substack{\mathbb{R}^3 \backslash H \\ 1/2 < |x| < 2}} \sum_{i,j=1}^{3} \left(c x_s |x|^{-2} \right)_i \overline{\partial_{x_j}\sigma_{i,j}(w)} \, dx$$

and, consequently,

(4.4.9) $a^{(1)}(c, v; 0) = 0.$

If $l \neq 0$ or $n \neq 0$, then the codimension of the linear span of the vectors $r\partial U_k/\partial x_s$ and $r\partial u_j/\partial x_s$, $s = 1, 3$, $k = 1, \dots, l$, $j = 1, \dots, n$, in the space Y does not exceed $(n + 3) - (n + l + 1) = 2 - l$. This implies the existence of a nonzero linear combination c of the vectors $c_{l+1}, \dots, c_3$ for which (4.4.9) holds for all $v \in Y$. The latter means that this vector c has generalized eigenvectors. Since this contradicts our assumptions, it follows that $l = 0$ and $n = 0$, i.e., the space X consists only of constant vectors which posses no generalized eigenvectors. ∎

4.4.5. The spectrum in the strip $-1 \leq \mathrm{Re}\,\lambda \leq 0$. Now we can prove the main result of this section.

THEOREM 4.4.2. *The spectrum of the pencil $\mathfrak{A}$ in the strip $-1 \leq \mathrm{Re}\,\lambda \leq 0$ consists of two eigenvalues $\lambda_0 = 0$ and $\lambda_1 = -1$. The geometric and algebraic multiplicities of both eigenvalues are equal to 3. The subspace of the eigenvectors corresponding to the eigenvalue $\lambda = 0$ consists of the constant vectors.*

Proof: If $\alpha = \pi$, then by using the Fourier transformation with respect to the variable x_3, one can easily derive the absence of the spectrum in the strip $-1 < \operatorname{Re}\lambda < 0$. An arbitrary plane crack H can be continuously deformed into the crack with $\alpha = \pi$. The number of eigenvalues in the strip $-1 < \operatorname{Re}\lambda < 0$ is independent under such deformation (cf. proof of Theorem 4.3.1). Since this strip contains no points of the spectrum for $\alpha = \pi$, it contains no eigenvalues for an arbitrary crack. The assertions of the theorem concerning the eigenvalue $\lambda = 0$ follow from Lemmas 4.4.1 and 4.4.2, and the assertions for $\lambda = -1$ are a consequence of item 4) of Theorem 4.1.1 which is valid by Theorem 4.4.1. ∎

4.5. Notes

Section 4.1. The mixed problem considered in Section 4.1 was investigated by Kozlov, Maz'ya and Roßmann in [137].

Section 4.2. Different boundary value problems for the Lamé system in plane domains with angular points, including the Neumann and mixed problems, were studied by Vorovich and Kopasenko [262], Grisvard [84] [86], Sändig and Richter [233], Nicaise [208, 209, 210], Costabel and Dauge [31], Eck, Nazarov and Wendland [50].

Section 4.3. Theorem 4.3.1 describing the spectrum of the pencil corresponding to the Neumann problem for the Lamé system in the strip $-1 \leq \operatorname{Re}\lambda \leq 0$ is due to Kozlov and Maz'ya [127] (see also their survey article [132]). The results of Section 4.3.4 on the the Neumann problem in a rotational cone are taken from the preprint [139] by Kozlov, Maz'ya and Schwab.

Section 4.4. The material is borrowed from Kozlov and Maz'ya in [128].

Finally, let us give some publications dealing with the numerical calculation of singularities of solutions of problems of elasticity theory at singular boundary points. We mention here the works by Benthem [18, 19], Bažant and Estenssoro [13], Noble, Hussain and Pu [216], Leguillon and Sanchez-Palencia [151], Wendland, Schmitz and Bumb [266], Hsiao, Stephan and Wendland [91], Ghahremani [68], Schmitz, Volk and Wendland [235], Leguillon [150], Andersson, Falk, Babuška and von Petersdorff [5]), Dimitrov, Andrä and Schnack [48].

CHAPTER 5

The Dirichlet problem for the Stokes system

In this chapter we study the first boundary value problem for the Stokes system of linearized hydrodynamics:

(5.0.1) $$-\Delta U + \nabla P = 0, \quad \nabla \cdot U = 0 \quad \text{in } \mathcal{K},$$

(5.0.2) $$U = 0 \quad \text{on } \partial\mathcal{K} \setminus \{0\},$$

where $\mathcal{K}$ is a plane angle or a three-dimensional cone. We are interested in special solutions (U, P) which have the form

(5.0.3) $$U(x) = r^{\lambda_0} \sum_{k=0}^{s} \frac{(\log r)^k}{k!} u^{(s-k)}(\omega), \quad P(x) = r^{\lambda_0-1} \sum_{k=0}^{s} \frac{(\log r)^k}{k!} p^{(s-k)}(\omega),$$

where $r = |x|$, $\omega = x/|x|$ and λ_0 is a complex number.

In Section 5.1 we assume that $\mathcal{K}$ is a plane angle and, similarly to the case of the Lamé system treated in Section 3.1, describe solutions (5.0.3). The more complicated situation when $\mathcal{K}$ is a three-dimensional cone $\{x : 0 < r < \infty, \omega \in \Omega \subset S^2\}$ is investigated in Sections 5.2–5.7, where we study the pencil $\mathcal{L}$ generated by (5.0.1), (5.0.2):

$$\mathcal{L}(\lambda) \begin{pmatrix} u \\ p \end{pmatrix} = \begin{pmatrix} r^{2-\lambda} \big(-\Delta(r^\lambda u) + \nabla(r^{\lambda-1}p) \big) \\ r^{1-\lambda} \nabla \cdot (r^\lambda u) \end{pmatrix}$$

with (u, p) being a vector function on Ω. The vector function (5.0.3) is a solution of (5.0.1), (5.0.2) if and only if λ_0 is an eigenvalue of the pencil $\mathcal{L}$, $(u^{(0)}, p^{(0)})$ is an eigenvector and $(u^{(j)}, p^{(j)})$, $j = 1, \ldots, s$ are generalized eigenvectors of $\mathcal{L}$ corresponding to λ_0. The eigenvalues of $\mathcal{L}$ are placed symmetrically with respect to the line $\operatorname{Re}\lambda = -1/2$.

Section 5.2 deals with basic properties of the pencil $\mathcal{L}$. For example, here the discreteness of the spectrum and the absence of eigenvalues on the line $\operatorname{Re}\lambda = -1/2$ are proven. We also describe connections between generalized eigenvectors corresponding to the eigenvalues λ and $-1-\lambda$. As a byproduct we find the following analog of the Kelvin transform for harmonic functions. If (U, P) solves the Stokes system (5.0.1), then (V, Q) defined by

(5.0.4) $$V(x) = |x|^{-1} \left(I + \Xi(x|x|^{-1}) + \big(I - 2\Xi(x|x|^{-1}) \big) x \cdot \partial_x \right) U\big(x|x|^{-2}\big),$$

where $\Xi(\omega) = \{\omega_i \omega_j\}_{i,j=1}^{3}$, and

(5.0.5) $$Q(x) = |x|^{-3}\left(-2 - 4\,x \cdot \partial_x \right) x \cdot U\big(x|x|^{-2}\big) + |x|^{-1}(1 + x \cdot \partial_x)\, P\big(x|x|^{-2}\big)$$

is a solution of (5.0.1) as well.

In Section 5.3 we show that the region

$$\Big\{ \lambda \in \mathbb{C} : (\operatorname{Re}\lambda + 1/2)^2 - (\operatorname{Im}\lambda)^2 < (F(\Omega))^2 \Big\}$$

contains only real eigenvalues of the pencil $\mathcal{L}$. Here F is a special case of the set function F_ν (see (3.3.2)) corresponding to $\nu = 1/2$. This function was investigated in detail in Chapter 3. In particular, we checked there that $F(\Omega) \geq 3/2$ and that $F(\Omega)$ is nonincreasing with respect to Ω. Another result of Section 5.3 is the nonexistence of generalized eigenfunctions to real eigenvalues in the strip

$$\left\{ \lambda \in \mathbb{C} \,:\, |\operatorname{Re}\lambda + 1/2| < F(\Omega),\ \lambda \neq 1,\ \lambda \neq -2 \right\}.$$

Therefore, the logarithmic terms in the solutions (5.0.3) do not occur for these eigenvalues.

The eigenvalues $\lambda = 1$ and $\lambda = -2$, which have the same geometric and algebraic multiplicities, are considered in Section 5.4. Obviously, the pair $(u, p) = (0, 1)$ is an eigenvector corresponding to $\lambda = 1$. We obtain necessary and sufficient conditions for the existence of additional eigenvectors as well as generalized eigenvectors. To be more precise we show that $\lambda = 1$ has at least two eigenvalues if and only if the scalar problem

$$(\delta + 6)\, w = 0, \quad w \in \overset{\circ}{W}{}^1_2(\Omega), \quad \int_\Omega w(\omega)\, d\omega = 0$$

has a nontrivial solution. Here δ is the Laplace-Beltrami operator on S^2. Moreover, we prove that $\lambda = 1$ has a generalized eigenvector if and only if the problem

$$(\delta + 6)\, w = 1, \quad w \in \overset{\circ}{W}{}^1_2(\Omega), \quad \int_\Omega w(\omega)\, d\omega = 0$$

is solvable and other generalized eigenvectors do not exist. We obtain similar assertions concerning the eigenvalue $\lambda = -2$.

It follows from what we said above that there are only real eigenvalues in the strip $-1/2 \leq \operatorname{Re}\lambda \leq 1$. In Section 5.5 we deal with the eigenvalues in the interval $[-1/2, 1)$. We introduce variational principles for these eigenvalues which show that they are nonincreasing functions of the domain Ω. The monotonicity of the eigenvalues of the Stokes pencil may be used to estimate them by those for right circular cones. A transcendental equation for the eigenvalues of the Stokes pencil for circular cones is written in Section 5.6.

Additional information on the eigenvalues of $\mathcal{L}$ can be derived by the use of the eigenvalues $\mathcal{N}_1(\Omega) \leq \mathcal{N}_2(\Omega) \leq \ldots$ of the quadratic form

$$\int_\Omega |\nabla_\omega v|^2\, d\omega$$

defined on all functions which vanish on $\partial\Omega$ and are orthogonal to 1 in $L_2(\Omega)$. In Section 5.4 some estimates for the eigenvalues of $\mathcal{L}$ formulated in terms of $\mathcal{N}_k(\Omega)$ are obtained. It is proved for example, that the interval $[-1/2, 1)$ contains exactly k positive eigenvalues of $\mathcal{L}$ in the case $\mathcal{N}_k(\Omega) < 6 \leq \mathcal{N}_{k+1}(\Omega)$. If, on the other hand, $\mathcal{N}_1(\Omega) \geq 6$, then the strip $-1/2 \leq \operatorname{Re}\lambda < 1$ is free of the spectrum of the pencil $\mathcal{L}$. If $\mathcal{N}_1(\Omega) > 6$, which is the case of Ω situated in a hemisphere, then the strip $-1/2 \leq \operatorname{Re}\lambda \leq 1$ contains exactly one simple eigenvalue $\lambda = 1$ of $\mathcal{L}$. For the case $\mathcal{N}_1(\Omega) \leq 6$ we prove in Subsection 5.5.4 that the strip

$$\left|\operatorname{Re}\lambda + \frac{1}{2}\right| < \frac{1}{2}\left(13 - 4\big(13 - 2\mathcal{N}_1(\Omega)\big)^{1/2}\right)^{1/2}$$

contains no eigenvalues of $\mathcal{L}$. The bound in the right-hand side is the best possible one for all Ω with either $\mathcal{N}_1(\Omega) = 6$ or $\mathcal{N}_1(\Omega) = 2$.

According to Theorem 5.5.6, the strip

(5.0.6)
$$\left| \operatorname{Re} \lambda + \frac{1}{2} \right| \le \frac{1}{2} + \frac{M}{M+4}$$

does not intersect the spectrum of $\mathcal{L}$. Remind that M is a positive number such that $M(M+1)$ is the first eigenvalue of the Dirichlet problem for the operator $-\delta$ on Ω.

We denote the spherical domain Ω corresponding to the right circular cone with solid opening angle 2α by Ω_α, i.e.

$$\Omega_\alpha = \{ \omega \in \Omega \ : \ 0 \le \varphi < 2\pi, \ 0 \le \theta < \alpha \},$$

and by S_+^2 the upper hemisphere in $\mathbb{R}^3$. The results of Section 5.5 imply the following assertions on the spectrum of the pencil $\mathcal{L}$:

(i) If $S_+^2 \subset \Omega \subset \Omega_{2\pi/3}$, then the strip $-1/2 < \operatorname{Re} \lambda \le 1$ contains exactly 3 eigenvalues. One of them is $\lambda = 1$ with multiplicity one.

(ii) If $\Omega_{2\pi/3} \subset \Omega$, then the strip $-\frac{1}{2} < \operatorname{Re} \lambda \le 1$ contains exactly 4 eigenvalues. One of them is $\lambda = 1$ with multiplicity one.

In the case that $S_+^2 \subset \Omega$ and the boundary $\partial\Omega$ is smooth this along with the general Theorem 1.4.4 leads to the asymptotics near the vertex for solutions U, P of the Dirichlet problem for the Stokes system with smooth data. For $\Omega \subset S^2$, we denote by $\lambda_j(\Omega)$, $j = 1, 2, \ldots$ the eigenvalues of $\mathcal{L}$ with $\operatorname{Re} \lambda_j(\Omega) \ge 0$ ordered with respect to increasing real part and with multiplicity counted.

(i) If $S_+^2 \subset \Omega \subset \Omega_{2\pi/3}$, then

$$U = r^{\lambda_1(\Omega)} u_1^{(0)}(\omega) + r^{\lambda_2(\Omega)} u_2^{(0)}(\omega) + O(r),$$
$$P = r^{\lambda_1(\Omega)-1} p_1^{(0)}(\omega) + r^{\lambda_2(\Omega)-1} p_2^{(0)}(\omega) + O(1)$$

with $0.4996 < \lambda_1(\Omega) \le \lambda_2(\Omega) < 1$. Here the lower bound for λ_1 results from numerical values for $\lambda_1(\Omega_{2\pi/3})$.

(ii) If $\Omega_{2\pi/3} \subset \Omega$, then

$$U = \sum_{j=1}^{3} r^{\lambda_j(\Omega)} u_j^{(0)}(\omega) + O(r),$$

$$P = \sum_{j=1}^{3} r^{\lambda_j(\Omega)-1} p_j^{(0)}(\omega) + O(1),$$

where $0 \le \lambda_1(\Omega) \le \lambda_2(\Omega) \le \lambda_3(\Omega) < 1$. This is in contrast to the Dirichlet problem for linear elasticity in piecewise smooth domains. For that problem, according to Theorem 3.7.1, in the case $S_+^2 \subset \Omega$ there are exactly three terms in the asymptotics which give rise to unbounded stresses at the vertex.

Another important application concerns the L_2-summability of the second derivatives of the velocity vector subject to the Dirichlet problem for the inhomogeneous Stokes system

(5.0.7) $-\Delta U + \nabla P = F, \quad \nabla \cdot U = 0 \text{ in } \mathcal{G}, \quad U = 0 \text{ on } \partial\mathcal{G},$

where $\mathcal{G}$ is a bounded three-dimensional domain with a conic vertex O on $\partial\mathcal{G}$. By Theorem 1.4.3, one needs to verify that there are no eigenvalues of $\mathcal{L}$ in the strip $-3/2 \le \operatorname{Re} \lambda \le 1/2$. A geometrical assumption for this is the inclusion $\Omega \subset \Omega_{\alpha_*}$, where $2\alpha^*$ is the solid opening angle of a right circular cone for which $\lambda_1(\Omega_{\alpha^*}) = 1/2$ ($\alpha^* \approx 0.6665\pi$, see Section 5.6). Therefore, the variational solution

$(U,P) \in W_2^1(\mathcal{G})^3 \times L_2(\mathcal{G})$ of (5.0.7) belongs to the space $W_2^2(\mathcal{G})^3 \times W_2^1(\mathcal{G})$ if $F \in L_2(\mathcal{G})^3$.

One might also observe that the nonlinear term in the Navier–Stokes system

$$-\Delta U + \nabla P + U \cdot \nabla U = F, \quad \nabla \cdot U = 0 \quad \text{in } \mathcal{G}$$

is not strong enough to violate the validity of the last result for this system with zero Dirichlet data. However we leave aside the proof of this fact (compare with Maz'ya's and Plamenevskiĭ's paper [**189**]).

We say a few words about the remaining two sections of this chapters. The goal of Section 5.7 is the complete description of the eigenvalues of $\mathcal{L}$ and the construction of the corresponding eigenvectors and generalized eigenvectors in the case when the cone $\mathcal{K}$ is a dihedral angle. Here we rely heavily on the previous results concerning the plane case and on the above mentioned Kelvin transform.

In the last Section 5.8 we state more regularity results for the Dirichlet problem posed for the Stokes and the Navier–Stokes systems in domains with piecewise smooth boundaries. In order to obtain these results one should use spectral properties of the pencil $\mathcal{L}$ studied in this chapter.

5.1. The Dirichlet problem for the Stokes system in an angle

This section deals with the operator pencil generated by the Dirichlet problem for the Stokes system in a plane angle. We will see that, with the exception of $\lambda = 0$, the spectrum of this pencil coincides with the spectrum of the pencil generated by the Neumann problem for the Lamé system.

5.1.1. The operator pencil generated by the Dirichlet problem for the Stokes system. Let $\mathcal{K} = \{(x_1, x_2) \in \mathbb{R}^2 : 0 < r < \infty, |\varphi| < \alpha/2\}$ be an angle with aperture $\alpha \in (0, 2\pi]$. We consider the Stokes system

$$(5.1.1) \qquad \begin{cases} -(\partial_{x_1}^2 + \partial_{x_2}^2)\, U_1 + \partial_{x_1} P &=& 0 \quad \text{in } \mathcal{K}, \\[2mm] -(\partial_{x_1}^2 + \partial_{x_2}^2)\, U_2 + \partial_{x_2} P &=& 0 \quad \text{in } \mathcal{K}, \\[2mm] \partial_{x_1} U_1 + \partial_{x_2} U_2 &=& 0 \quad \text{in } \mathcal{K} \end{cases}$$

for the vector function (U_1, U_2, P) with the Dirichlet boundary condition

$$(5.1.2) \qquad\qquad U_1 = U_2 = 0 \qquad \text{on } \partial\mathcal{K}\backslash\{0\}.$$

Let U_r, U_φ be the polar components of the vector function $U = (U_1, U_2)$ defined by formula (3.1.3). Then the system (5.1.1) can be written as

$$-\frac{1}{r^2}\Big((r\partial_r)^2\, U_r + \partial_\varphi^2\, U_r - 2\partial_\varphi\, U_\varphi - U_r\Big) + \partial_r\, P = 0,$$

$$-\frac{1}{r^2}\Big((r\partial_r)^2\, U_\varphi + \partial_\varphi^2\, U_\varphi + 2\partial_\varphi\, U_r - U_\varphi\Big) + \frac{1}{r}\partial_\varphi\, P = 0,$$

$$\partial_r\, U_r + \frac{1}{r}(U_r + \partial_\varphi\, U_\varphi) = 0,$$

where $0 < r < \infty$, $-\alpha/2 < \varphi < \alpha/2$. We are interested in solutions (U_r, U_φ, P) which have the form

$$(5.1.3) \qquad \begin{pmatrix} U_r(r, \varphi) \\ U_\varphi(r, \varphi) \end{pmatrix} = r^\lambda \begin{pmatrix} u_r(\varphi) \\ u_\varphi(\varphi) \end{pmatrix}, \qquad P(r, \varphi) = r^{\lambda-1} p(\varphi).$$

Obviously, the functions u_r, u_φ, and p have to satisfy the system of ordinary differential equations

(5.1.4)
$$
\begin{cases}
-u_r'' + (1 - \lambda^2)\, u_r + 2\, u_\varphi' + (\lambda - 1)\, p &= 0 \quad \text{for } \varphi \in (-\alpha/2, +\alpha/2), \\[2mm]
-u_\varphi'' + (1 - \lambda^2)\, u_\varphi - 2\, u_r' + p' &= 0 \quad \text{for } \varphi \in (-\alpha/2, +\alpha/2), \\[2mm]
u_\varphi' + (\lambda + 1)\, u_r &= 0 \quad \text{for } \varphi \in (-\alpha/2, +\alpha/2)
\end{cases}
$$

and the boundary conditions

(5.1.5)
$$
u_r(\pm\alpha/2) = u_\varphi(\pm\alpha/2) = 0.
$$

Let $\mathcal{L}(\lambda)$ be the operator on the left-hand side of (5.1.4) and let $\mathcal{B}$ be the mapping $(u_r, u_\varphi) \to \big(u_r(\pm\alpha/2), u_\varphi(\pm\alpha/2)\big)$. Then by $\mathfrak{A}(\lambda)$ we denote the operator

$$
W_2^2(-\alpha/2, +\alpha/2)^2 \times W_2^1(-\alpha/2, +\alpha/2) \ni \begin{pmatrix} u_r \\ u_\varphi \\ p \end{pmatrix}
$$

$$
\to \left(\mathcal{L}(\lambda) \begin{pmatrix} u_r \\ u_\varphi \\ p \end{pmatrix}, \mathcal{B}\begin{pmatrix} u_r \\ u_\varphi \end{pmatrix} \right) \in L_2(-\alpha/2, +\alpha/2)^2 \times W_2^1(-\alpha/2, +\alpha/2) \times \mathbb{C}^4.
$$

This operator is Fredholm for all complex λ and invertible for sufficiently large purely imaginary λ. Consequently, by Theorem 1.1.1, the spectrum of the pencil $\mathfrak{A}$ is discrete and consists of eigenvalues with finite algebraic multiplicities.

5.1.2. Equations for the eigenvalues. The general solution of the system (5.1.4) for $\lambda \neq 0$ is a linear combination of the following four independent solutions:

$$
\begin{pmatrix} \cos(1+\lambda)\varphi \\ -\sin(1+\lambda)\varphi \\ 0 \end{pmatrix}, \qquad
\begin{pmatrix} \sin(1+\lambda)\varphi \\ \cos(1+\lambda)\varphi \\ 0 \end{pmatrix},
$$

$$
\begin{pmatrix} (1-\lambda)\cos(1-\lambda)\varphi \\ -(1+\lambda)\sin(1-\lambda)\varphi \\ -4\lambda\cos(1-\lambda)\varphi \end{pmatrix}, \qquad
\begin{pmatrix} (1-\lambda)\sin(1-\lambda)\varphi \\ (1+\lambda)\cos(1-\lambda)\varphi \\ -4\lambda\sin(1-\lambda)\varphi \end{pmatrix}.
$$

In order to include the case $\lambda = 0$ we introduce another basis:

$$
\begin{pmatrix} u_r^{(1)} \\ u_\varphi^{(1)} \\ p^{(1)} \end{pmatrix} = \begin{pmatrix} \cos(1+\lambda)\varphi \\ -\sin(1+\lambda)\varphi \\ 0 \end{pmatrix}, \qquad
\begin{pmatrix} u_r^{(2)} \\ u_\varphi^{(2)} \\ p^{(2)} \end{pmatrix} = \begin{pmatrix} \sin(1+\lambda)\varphi \\ \cos(1+\lambda)\varphi \\ 0 \end{pmatrix},
$$

$$
\begin{pmatrix} u_r^{(3)} \\ u_\varphi^{(3)} \\ p^{(3)} \end{pmatrix} = \frac{1}{\lambda} \begin{pmatrix} (1-\lambda)\cos(1-\lambda)\varphi - \cos(1+\lambda)\varphi \\ -(1+\lambda)\sin(1-\lambda)\varphi + \sin(1+\lambda)\varphi \\ -4\lambda\cos(1-\lambda)\varphi \end{pmatrix},
$$

$$
\begin{pmatrix} u_r^{(4)} \\ u_\varphi^{(4)} \\ p^{(4)} \end{pmatrix} = \frac{1}{\lambda} \begin{pmatrix} (1-\lambda)\sin(1-\lambda)\varphi - \sin(1+\lambda)\varphi \\ (1+\lambda)\cos(1-\lambda)\varphi - \cos(1+\lambda)\varphi \\ -4\lambda\sin(1-\lambda)\varphi \end{pmatrix}.
$$

For $\lambda = 0$ we have

$$\begin{pmatrix} u_r^{(3)} \\ u_\varphi^{(3)} \\ p^{(3)} \end{pmatrix} = \begin{pmatrix} -\cos\varphi + 2\varphi\sin\varphi \\ -\sin\varphi + 2\varphi\cos\varphi \\ -4\cos\varphi \end{pmatrix}, \qquad \begin{pmatrix} u_r^{(4)} \\ u_\varphi^{(4)} \\ p^{(4)} \end{pmatrix} = \begin{pmatrix} -\sin\varphi - 2\varphi\cos\varphi \\ \cos\varphi + 2\varphi\sin\varphi \\ -4\sin\varphi \end{pmatrix}.$$

We represent the general solution of the system (5.1.4) in the form

$$(5.1.6) \qquad \begin{pmatrix} u_r \\ u_\varphi \\ p \end{pmatrix} = \sum_{k=1}^{4} c_k \begin{pmatrix} u_r^{(k)} \\ u_\varphi^{(k)} \\ p^{(k)} \end{pmatrix}.$$

Then the boundary condition (5.1.5) yields

$$\sum_{k=1}^{4} c_k \begin{pmatrix} u_r^{(k)}(\lambda, \pm\alpha/2) \\ u_\varphi^{(k)}(\lambda, \pm\alpha/2) \end{pmatrix} = 0.$$

Since the components $u_r^{(1)}$, $u_\varphi^{(2)}$, $u_r^{(3)}$, $u_\varphi^{(4)}$ are even functions of the variable φ and the components $u_\varphi^{(1)}$, $u_r^{(2)}$, $u_\varphi^{(3)}$, $u_r^{(4)}$ are odd functions, the last algebraic system is equivalent to the following two algebraic systems:

$$(5.1.7) \qquad c_1 \begin{pmatrix} u_r^{(1)}(\lambda, \alpha/2) \\ u_\varphi^{(1)}(\lambda, \alpha/2) \end{pmatrix} + c_3 \begin{pmatrix} u_r^{(3)}(\lambda, \alpha/2) \\ u_\varphi^{(3)}(\lambda, \alpha/2) \end{pmatrix} = 0,$$

$$(5.1.8) \qquad c_2 \begin{pmatrix} u_r^{(2)}(\lambda, \alpha/2) \\ u_\varphi^{(2)}(\lambda, \alpha/2) \end{pmatrix} + c_4 \begin{pmatrix} u_r^{(4)}(\lambda, \alpha/2) \\ u_\varphi^{(4)}(\lambda, \alpha/2) \end{pmatrix} = 0.$$

The coefficients determinants of these algebraic systems are

$$d_-(\lambda) = \lambda^{-1}(\sin\lambda\alpha - \lambda\sin\alpha)$$

and

$$d_+(\lambda) = \lambda^{-1}(\sin\lambda\alpha + \lambda\sin\alpha),$$

respectively. Obviously, $d_\mp(0) = \alpha \mp \sin\alpha \neq 0$. Therefore, $\lambda = 0$ does not belong to the spectrum of the operator pencil $\mathfrak{A}$. Every eigenvalue is a solutions of one of the equations

$$(5.1.9) \qquad \sin\lambda\alpha \mp \lambda\sin\alpha = 0.$$

These equations are the same as for the eigenvalues of the operator pencil generated by the Neumann problem for the Lamé system (cf. (4.2.4), (4.2.5)).

5.1.3. Eigenvectors and generalized eigenvectors. Let λ be a solution of the equation $d_-(\lambda) = 0$. Then every solution of the system (5.1.7) has the form

$$\begin{aligned} c_1 &= A_1 \lambda^{-1}\left(\cos\frac{(1+\lambda)\alpha}{2} - (1-\lambda)\cos\frac{(1-\lambda)\alpha}{2}\right) \\ &\quad + A_2 \lambda^{-1}\left(\sin\frac{(1+\lambda)\alpha}{2} - (1+\lambda)\sin\frac{(1-\lambda)\alpha}{2}\right), \\ c_3 &= A_1 \cos\frac{(1+\lambda)\alpha}{2} + A_2 \sin\frac{(1+\lambda)\alpha}{2}. \end{aligned}$$

To avoid the simultaneous vanishing of the coefficients c_1 and c_3, we choose the coefficients A_1, A_2 as follows:

$$A_1 = -\lambda \cos \frac{(1+\lambda)\alpha}{2}, \quad A_2 = -\lambda \sin \frac{(1+\lambda)\alpha}{2}.$$

Then

(5.1.10) $$c_1 = \cos \lambda\alpha - \lambda \cos \alpha - 1, \quad c_3 = -\lambda$$

and we obtain the eigenvector

(5.1.11) $$\begin{pmatrix} u_r^- \\ u_\varphi^- \\ p^- \end{pmatrix} = c_1 \begin{pmatrix} u_r^{(1)} \\ u_\varphi^{(1)} \\ p^{(1)} \end{pmatrix} + c_3 \begin{pmatrix} u_r^{(3)} \\ u_\varphi^{(3)} \\ p^{(3)} \end{pmatrix}$$

$$= -\begin{pmatrix} (1-\lambda)\cos(1-\lambda)\varphi \\ -(1+\lambda)\sin(1-\lambda)\varphi \\ 4\lambda \cos(1-\lambda)\varphi \end{pmatrix} + (\cos\lambda\alpha - \lambda\cos\alpha) \begin{pmatrix} \cos(1+\lambda)\varphi \\ -\sin(1+\lambda)\varphi \\ 0 \end{pmatrix}.$$

which corresponds to an eigenvalue λ satisfying the equality $d_-(\lambda) = 0$. If λ is a solution of the equation $d_+(\lambda) = 0$, then, analogously, the eigenvector

(5.1.12) $$\begin{pmatrix} u_r^+ \\ u_\varphi^+ \\ p^+ \end{pmatrix} = -\begin{pmatrix} (1-\lambda)\sin(1-\lambda)\varphi \\ (1+\lambda)\cos(1-\lambda)\varphi \\ 4\lambda \sin(1-\lambda)\varphi \end{pmatrix}$$

$$+ (\cos\lambda\alpha + \lambda\cos\alpha) \begin{pmatrix} \sin(1+\lambda)\varphi \\ \cos(1+\lambda)\varphi \\ 0 \end{pmatrix}$$

holds. Since every of the systems (5.1.7), (5.1.8) has not more than one linearly independent solution, we have obtained all eigenvectors of the pencil $\mathfrak{A}$ in this way.

For example, the eigenspace corresponding to the eigenvalue $\lambda = 1$ is spanned by the eigenvector

(5.1.13) $$\begin{pmatrix} u_r^- \\ u_\varphi^- \\ p^- \end{pmatrix} = \begin{pmatrix} 0 \\ 0 \\ 4 \end{pmatrix} \quad \text{if } \alpha \neq \pi,\ \alpha \neq 2\pi.$$

If $\alpha = \pi$ or $\alpha = 2\pi$, then additionally the eigenvector

$$\begin{pmatrix} u_r^+ \\ u_\varphi^+ \\ p^+ \end{pmatrix} = \begin{pmatrix} 2\cos\alpha \sin 2\varphi \\ 2(\cos\alpha \cos 2\varphi - 1) \\ 0 \end{pmatrix}$$

occurs. (The Cartesian components of this vector are $u_1^+ = 0$, $u_2^+ = -4\cos\varphi$ for $\alpha = \pi$ and $u_1^+ = 4\sin\varphi$, $u_2^+ = 0$ for $\alpha = 2\pi$.)

Now we deal with the existence of generalized eigenvectors. As in the case of the Lamé system, the algebraic multiplicity of any eigenvalue λ_0 of the pencil $\mathfrak{A}$ coincides with the multiplicity of the root $\lambda = \lambda_0$ of the equation $d_+(\lambda)\,d_-(\lambda) = 0$. Common roots of the equations $d_+(\lambda) = 0$ and $d_-(\lambda) = 0$ exist only in the cases $\alpha = \pi$ and $\alpha = 2\pi$. For $\alpha = \pi$ the spectrum of the pencil $\mathfrak{A}$ consists of the eigenvalues $\pm 1, \pm 2, \ldots$, while in the case $\alpha = 2\pi$ the numbers $\pm k/2$ ($k = 1, 2, \ldots$) are eigenvalues. In both cases there are two eigenvectors (5.1.11), (5.1.12) to every eigenvalue, and generalized eigenvectors do not exist.

Analogously to Lemmas 3.1.3 and 3.1.4, the following results hold.

LEMMA 5.1.1. 1) *There are no generalized eigenvectors to nonreal eigenvalues.*
2) *The algebraic multiplicities of all eigenvalues are not greater than two.*

Proof: 1) Let one of the equalities $d_-(\lambda_0) = d'_-(\lambda_0) = 0$, $d_+(\lambda_0) = d'_+(\lambda_0) = 0$ be satisfied. Then we get

$$(5.1.14) \qquad\qquad \alpha \cos \lambda_0 \alpha = \pm \sin \alpha.$$

Hence $\cos \lambda_0 \alpha$ is real, i.e., either $\sin(\operatorname{Re} \lambda_0 \, \alpha) = 0$ or $\sinh(\operatorname{Im} \lambda_0 \, \alpha) = 0$. The equality $\sin(\operatorname{Re} \lambda_0 \, \alpha) = 0$ cannot be valid. Otherwise, we get $\cos \lambda_0 \alpha = \pm \cosh(\operatorname{Im} \lambda_0 \, \alpha)$ and, therefore, $|\cos \lambda_0 \alpha| > 1$ for $\operatorname{Im} \lambda_0 \neq 0$. This contradicts (5.1.14). Consequently, $\sinh(\operatorname{Im} \lambda_0 \, \alpha) = 0$, i.e., $\operatorname{Im} \lambda_0 = 0$. Thus, only real eigenvalues may have algebraic multiplicity greater than one.

2) Suppose that λ_0 is an eigenvalue satisfying the equalities $d_-(\lambda_0) = d'_-(\lambda_0) = d''_-(\lambda_0) = 0$. As we have shown above, λ_0 is real. Furthermore, by the equality $d''_-(\lambda_0) = -\lambda_0^{-1} \alpha^2 \sin \lambda_0 \alpha$, we have $\sin \lambda_0 \alpha = 0$. Since $\sin \lambda_0 \alpha = \lambda_0 \sin \alpha$, this yields $\sin \alpha = 0$. This contradicts the equality $d'_-(\lambda_0) = 0$. Analogously, the equalities $d_+(\lambda_0) = d'_+(\lambda_0) = d''_+(\lambda_0) = 0$ lead to a contradiction. ∎

LEMMA 5.1.2. *If λ_0 is a multiple root of one of the equations $d_-(\lambda) = 0$ or $d_+(\lambda_0) = 0$, then a generalized eigenvector to (5.1.11) and (5.1.12), respectively, can be obtained by the formula*

$$(5.1.15) \qquad \begin{pmatrix} v_r^{\mp}(\lambda_0, \varphi) \\ v_\varphi^{\mp}(\lambda_0, \varphi) \\ q^{\mp}(\lambda_0, \varphi) \end{pmatrix} = \frac{\partial}{\partial \lambda} \begin{pmatrix} u_r^{\mp}(\lambda_0, \varphi) \\ u_\varphi^{\mp}(\lambda_0, \varphi) \\ p^{\mp}(\lambda_0, \varphi) \end{pmatrix} \Bigg|_{\lambda = \lambda_0}.$$

Proof: Let $c_1 = c_1(\lambda)$, $c_3 = c_3(\lambda)$ be the functions defined in (5.1.10). It can be easily verified that

$$c_1(\lambda) \begin{pmatrix} u_r^{(1)}(\lambda, \alpha/2) \\ u_\varphi^{(1)}(\lambda, \alpha/2) \end{pmatrix} + c_3(\lambda) \begin{pmatrix} u_r^{(3)}(\lambda, \alpha/2) \\ u_\varphi^{(3)}(\lambda, \alpha/2) \end{pmatrix} = \lambda \, d_-(\lambda) \begin{pmatrix} \cos(1 + \lambda)\alpha/2 \\ -\sin(1 + \lambda)\alpha/2 \end{pmatrix}$$

for all $\lambda \in \mathbb{C}$. Using the fact that $u_r^{(1)}$, $u_r^{(3)}$ are even functions of the variable φ and $u_\varphi^{(1)}$, $u_\varphi^{(3)}$ are odd functions, we get

$$c_1(\lambda) \begin{pmatrix} u_r^{(1)}(\lambda, -\alpha/2) \\ u_\varphi^{(1)}(\lambda, -\alpha/2) \end{pmatrix} + c_3(\lambda) \begin{pmatrix} u_r^{(3)}(\lambda, -\alpha/2) \\ u_\varphi^{(3)}(\lambda, -\alpha/2) \end{pmatrix} = \lambda \, d_-(\lambda) \begin{pmatrix} \cos(1 + \lambda)\alpha/2 \\ \sin(1 + \lambda)\alpha/2 \end{pmatrix}.$$

Since $(u_r^{(1)}, u_\varphi^{(1)}, p^{(1)})$ and $(u_r^{(3)}, u_\varphi^{(3)}, p^{(3)})$ are solutions of the system (5.1.4), the last two equalities yield

$$(5.1.16) \quad \mathfrak{A}(\lambda) \begin{pmatrix} u_r^- \\ u_\varphi^- \\ p^- \end{pmatrix} = \lambda \, d_-(\lambda) \left(0, \begin{pmatrix} \cos(1 + \lambda)\alpha/2 \\ -\sin(1 + \lambda)\alpha/2 \end{pmatrix}, \begin{pmatrix} \cos(1 + \lambda)\alpha/2 \\ \sin(1 + \lambda)\alpha/2 \end{pmatrix} \right).$$

Let $\lambda = \lambda_0$ be a multiple root of the equation $d_-(\lambda) = 0$. Differentiating (5.1.16) with respect to λ and setting $\lambda = \lambda_0$, we get

$$\mathfrak{A}'(\lambda_0) \begin{pmatrix} u_r^-(\lambda_0, \varphi) \\ u_\varphi^-(\lambda_0, \varphi) \\ p^-(\lambda_0, \varphi) \end{pmatrix} + \mathfrak{A}(\lambda_0) \begin{pmatrix} v_r^-(\lambda_0, \varphi) \\ v_\varphi^-(\lambda_0, \varphi) \\ q^-(\lambda_0, \varphi) \end{pmatrix} = 0.$$

This proves the lemma for the case "−". The proof for the case "+" proceeds analogously. ∎

We describe the angles α for which generalized eigenvectors exist. Obviously, every multiple root of the equation $d_{\mp}(\lambda) = 0$ satisfies

$$\sin \lambda \alpha \mp \lambda \sin \alpha = \alpha \cos \lambda \alpha \mp \sin \alpha = 0.$$

Consequently, $\tan \lambda \alpha = \lambda \alpha$ and $\alpha^{-1} \sin \alpha = \pm(\lambda^2 \alpha^2 + 1)^{-1/2}$. Hence the sequence of solutions x_n of the equation $\tan x = x$ generates a denumerable set $\{\alpha\}$ of angles satisfying the equation $\alpha^{-1} \sin \alpha = \pm(x_n^2 + 1)^{-1/2}$. The corresponding eigenvalues with index 2 are

$$\lambda = \pm \left(\frac{1}{\sin^2 \alpha} - \frac{1}{\alpha^2} \right)^{1/2}.$$

Note that the points π and 2π are accumulation points of the mentioned set $\{\alpha\}$.

Example. The eigenvector (5.1.13) corresponding to the eigenvalue $\lambda_0 = 1$ has a generalized eigenvector if and only if $\alpha = \alpha_*$, where α_* is the unique solution of the equation $\tan \alpha = \alpha$ in the interval $(0, 2\pi]$, $\alpha_* \approx 1.4303\pi$. By Lemma 5.1.2, this generalized eigenvector is equal to

(5.1.17)
$$\begin{pmatrix} v_r^-(\varphi) \\ v_\varphi^-(\varphi) \\ q^-(\varphi) \end{pmatrix} = \begin{pmatrix} 1 \\ -2\varphi \\ 4 \end{pmatrix} - \frac{1}{\cos \alpha_*} \begin{pmatrix} \cos 2\varphi \\ -\sin 2\varphi \\ 0 \end{pmatrix}.$$

5.1.4. Distribution of eigenvalues. In the following two theorems some properties of the spectrum of $\mathfrak{A}$ are listed. Since the numbers λ, $-\lambda$, $\overline{\lambda}$, and $-\overline{\lambda}$ are simultaneously eigenvalues, we can restrict ourselves to the eigenvalues in the half-plane $\operatorname{Re} \lambda \geq 0$.

Repeating the proof of Theorem 3.1.1 (where ν has to be replaced by $1/2$), we obtain the following results.

THEOREM 5.1.1. *Let $\alpha \neq \pi$ and $\alpha \neq 2\pi$. Then the following assertions are true.*

1) There are no eigenvalues of the pencil $\mathfrak{A}$ on the lines $\operatorname{Re} \lambda = k\pi/\alpha$, $k = 0, 1, 2, \ldots$.

2) For every odd k, $k \geq 1$, the strip

(5.1.18)
$$k \frac{\pi}{\alpha} < \operatorname{Re} \lambda < (k+1) \frac{\pi}{\alpha}$$

contains two eigenvalues of $\mathfrak{A}$ which are solutions of the equation $d_+(\lambda) = 0$ for $\alpha < \pi$ and of the equation $d_-(\lambda) = 0$ for $\alpha > \pi$.

3) If k is an even number, $k \geq 2$, then there are two eigenvalues of $\mathfrak{A}$ in the strip (5.1.18) which satisfy the equality $d_-(\lambda) = 0$ for $\alpha < \pi$ and the equality $d_+(\lambda) = 0$ for $\alpha > \pi$.

4) The strip

(5.1.19)
$$0 < \operatorname{Re} \lambda < \frac{\pi}{\alpha}$$

contains exactly one eigenvalue. This eigenvalue is a solution of the equation $d_-(\lambda) = 0$ if $\alpha < \pi$ and of the equation $d_+(\lambda) = 0$ if $\alpha > \pi$.

For the formulation of the next theorem we need the auxiliary functions $\xi_\pm = \xi_\pm(\alpha)$ introduced in Section 4.2. Let $\xi_+(\alpha)$ be the smallest positive solution of the equation

$$\frac{\sin \xi}{\xi} + \frac{\sin \alpha}{\alpha} = 0$$

and $\xi_-(\alpha)$ the smallest positive solution of the equation

$$\frac{1}{\xi - \alpha}\left(\frac{\sin\xi}{\xi} - \frac{\sin\alpha}{\alpha}\right) = 0.$$

Note that the function ξ_+ is defined for $\alpha \in [\alpha_0, 2\pi]$, where α_0 is the smallest positive root of the equation $\alpha^{-1}\sin\alpha = -\cos\alpha_*$, $\alpha_0 \approx 0.8128\pi$, and the function ξ_- is defined on the interval $[\alpha_1, 2\pi]$, where α_1 is the smallest positive root of the equation $\alpha^{-1}\sin\alpha = \cos\alpha_{**}$ and α_{**} is the second positive root of the equation $\tan\alpha = \alpha$, $\alpha_{**} \approx 2.4590\pi$, $\alpha_1 \approx 0.88397\pi$. Other elementary properties of the functions ξ_+ and ξ_- are given in Section 4.2 after Theorem 4.2.1.

According to Theorem 4.2.2, the following results hold.

THEOREM 5.1.2. 1) *If* $\alpha \in (0, \pi)$, *then the spectrum of the operator pencil* $\mathfrak{A}$ *in the strip* $0 < \operatorname{Re}\lambda < \pi/\alpha$ *consists of the unique and simple real eigenvalue* $\lambda = 1$.

2) *If* $\alpha \in (\pi, 2\pi)$, *then the spectrum of* $\mathfrak{A}$ *in the strip* $0 < \operatorname{Re}\lambda < 2\pi/\alpha$ *consists of the real eigenvalues*

$$\lambda_+(\alpha) = \frac{\xi_+(\alpha)}{\alpha}, \quad 1, \quad \lambda_-(\alpha) = \frac{\xi_-(\alpha)}{\alpha}.$$

3) *The functions* λ_+, λ_- *defined on the intervals* $[\alpha_0, 2\pi]$ *and* $[\alpha_1, 2\pi]$, *respectively, are strictly decreasing and satisfy the following inequalities:*

$$\lambda_+(\alpha) < 1 < \lambda_-(\alpha) < \frac{2\pi}{\alpha} \quad \text{for } \alpha \in (\pi, \alpha_*),$$

$$\lambda_+(\alpha) < \lambda_-(\alpha) < 1 \quad \text{for } \alpha \in (\alpha_*, 2\pi).$$

Moreover, $\lambda_+(\pi) = 1$, $\lambda_-(\pi) = 2$, $\lambda_-(\alpha_*) = 1$, *and* $\lambda_+(2\pi) = \lambda_-(2\pi) = 1/2$.

5.2. The operator pencil generated by the Dirichlet problem in a cone

Let $\mathcal{K}$ be the cone $\{x \in \mathbb{R}^3 : 0 < |x| < \infty, \ x/|x| \in \Omega\}$. We consider the Dirichlet problem (5.0.1), (5.0.2) for the Stokes system. Throughout this chapter, we assume that Ω is a Lipschitz domain on the sphere S^2. We seek special solutions (U, P) of the form (5.0.3), where $u^{(k)} \in \overset{\circ}{W}{}^1_2(\Omega)^3$, and $p^{(k)} \in L_2(\Omega)^3$ for $k = 0, \ldots, s$. For this we have to study the spectrum of a certain pencil of differential operators on a subdomain of the sphere. First we introduce this pencil and prove some basic properties.

5.2.1. Definition of the pencil $\mathcal{L}$. Let $\mathcal{L}(\lambda)$ be the operator

$$(5.2.1) \qquad \overset{\circ}{W}{}^1_2(\Omega)^3 \times L_2(\Omega) \ni \begin{pmatrix} u \\ p \end{pmatrix}$$

$$\to \begin{pmatrix} r^{2-\lambda}\big(-\Delta(r^\lambda u) + \nabla(r^{\lambda-1}p)\big) \\ r^{1-\lambda}\nabla \cdot (r^\lambda u) \end{pmatrix} \in W_2^{-1}(\Omega)^3 \times L_2(\Omega),$$

By Theorem 1.1.5, the functions (5.0.3) form a solution of problem (5.0.1), (5.0.2) if and only if λ_0 is an eigenvalue of the pencil $\mathcal{L}$ and

$$\begin{pmatrix} u^{(0)} \\ p^{(0)} \end{pmatrix}, \begin{pmatrix} u^{(1)} \\ p^{(1)} \end{pmatrix}, \ldots, \begin{pmatrix} u^{(s)} \\ p^{(s)} \end{pmatrix}$$

is a Jordan chain corresponding to this eigenvalue.

As in in the preceding chapters, we will systematically use the spherical components u_r, u_θ u_φ of the vector function (u_1, u_2, u_3) and use the same notations as in Section 3.2.

Then the system $\mathcal{L}(\lambda) \begin{pmatrix} u \\ p \end{pmatrix} = 0$ has the form

(5.2.2) $(\lambda - 1)p - (\delta + \lambda^2 + \lambda - 2)u_r + \dfrac{2}{\sin\theta}\left(\partial_\theta(\sin\theta\, u_\theta) + \partial_\varphi u_\varphi\right) = 0,$

(5.2.3) $\partial_\theta p - \left(\delta + \lambda^2 + \lambda - \dfrac{1}{\sin^2\theta}\right)u_\theta + 2\dfrac{\cot\theta}{\sin\theta}\partial_\varphi u_\varphi - 2\partial_\theta u_r = 0,$

(5.2.4) $\dfrac{1}{\sin\theta}\partial_\varphi p - \left(\delta + \lambda^2 + \lambda - \dfrac{1}{\sin^2\theta}\right)u_\varphi - 2\dfrac{\cot\theta}{\sin\theta}\partial_\varphi u_\theta - \dfrac{2}{\sin\theta}\partial_\varphi u_r = 0,$

(5.2.5) $-\dfrac{1}{\sin\theta}\left(\partial_\theta(\sin\theta\, u_\theta) + \partial_\varphi u_\varphi\right) - (\lambda + 2)u_r = 0,$

where δ denotes the Beltrami operator on the sphere S^2. We denote the matrix operator on the left in (5.2.2)–(5.2.5) also by $\mathcal{L}(\lambda)$.

Let $Q(\cdot,\cdot)$ be the form (3.2.4). We introduce the parameter-depending sesquilinear form

(5.2.6) $a\left(\begin{pmatrix} u \\ p \end{pmatrix}, \begin{pmatrix} v \\ q \end{pmatrix}; \lambda\right) = Q(u_\omega, v_\omega) + \displaystyle\int_\Omega (\nabla_\omega u_r) \cdot \nabla_\omega \overline{v}_r \, d\omega$

$\qquad - \displaystyle\int_\Omega \left((\lambda + 2)(\lambda - 1)\, u_r \overline{v}_r + (\lambda^2 + \lambda - 1)\, u_\omega \cdot \overline{v}_\omega\right) d\omega$

$\qquad -2 \displaystyle\int_\Omega \left((\nabla_\omega u_r) \cdot \overline{v}_\omega - (\nabla_\omega \cdot u_\omega)\, \overline{v}_r\right) d\omega$

$\qquad - \displaystyle\int_\Omega \left(p\left(\nabla_\omega \cdot \overline{v}_\omega + (1 - \lambda)\, \overline{v}_r\right) + \left(\nabla_\omega \cdot u_\omega + (\lambda + 2)\, u_r\right)\overline{q}\right) d\omega$

on $\overset{\circ}{W}{}^1_2(\Omega) \times \overset{\circ}{h}{}^1_2(\Omega) \times L_2(\Omega)$ which is connected with the operator

$$\mathcal{L}(\lambda) : \overset{\circ}{W}{}^1_2(\Omega) \times \overset{\circ}{h}{}^1_2(\Omega) \times L_2(\Omega) \to \left(\overset{\circ}{W}{}^1_2(\Omega) \times \overset{\circ}{h}{}^1_2(\Omega)\right)^* \times L_2(\Omega)$$

by the equality

$$a\left(\begin{pmatrix} u \\ p \end{pmatrix}, \begin{pmatrix} v \\ q \end{pmatrix}; \lambda\right) = \left(\mathcal{L}(\lambda)\begin{pmatrix} u \\ p \end{pmatrix}, \begin{pmatrix} v \\ q \end{pmatrix}\right)_{L_2(\Omega)^4}.$$

By (5.2.6) and by the equality

(5.2.7) $\displaystyle\int_\Omega (\nabla_\omega u_r) \cdot \overline{v}_\omega \, d\omega = -\int_\Omega u_r \, \nabla_\omega \cdot \overline{v}_\omega \, d\omega,$

we have

(5.2.8) $a\left(\begin{pmatrix} u \\ p \end{pmatrix}, \begin{pmatrix} v \\ q \end{pmatrix}; \lambda\right) = \overline{a\left(\begin{pmatrix} v \\ q \end{pmatrix}, \begin{pmatrix} u \\ p \end{pmatrix}; -1 - \overline{\lambda}\right)}$

for all (u,p), $(v,q) \in \overset{\circ}{W}{}^1_2(\Omega) \times \overset{\circ}{h}{}^1_2(\Omega) \times L_2(\Omega)$ and, consequently, $\mathcal{L}(\lambda)^* = \mathcal{L}(-1 - \overline{\lambda})$.

5.2.2. Basic properties of the pencil $\mathcal{L}$. Without proof we state the following well-known lemma which is used below. For the proof we refer to the book of Temam [253, Prop.I.1.2,Le.I.2.4].

LEMMA 5.2.1. *Assume that $\Omega \subset S^2$ is open and Lipschitz. Then*

1) *if a distribution p has a spherical gradient $\nabla_\omega p$ in $(\overset{\circ}{h}{}^1_2(\Omega))^*$, then $p \in L_2(\Omega)$,*

2) *the spherical divergence operator*

$$u_\omega \to \nabla_\omega \cdot u_\omega$$

maps the space $\overset{\circ}{h}{}^1_2(\Omega)$ *onto* $L_2(\Omega)/\mathbb{R}$.

As a consequence of the last lemma, we obtain the following result.

LEMMA 5.2.2. *Let g be an arbitrary L_2 function on Ω. Then the equation*

$$(\lambda + 2)\, u_r + \nabla_\omega \cdot u_\omega = g$$

has a solution $(u_r, u_\omega) \in \overset{\circ}{W}{}^1_2(\Omega) \times \overset{\circ}{h}{}^1_2(\Omega)$ *satisfying the inequality*

$$|\lambda + 2|\, \|u_r\|_{W^1_2(\Omega)} + \|u_\omega\|_{h^1_2(\Omega)} \le c\, \|g\|_{L_2(\Omega)}$$

with a constant c independent of g and λ.

Proof: First let $\int_\Omega g\, d\omega = 0$. Then, according to Lemma 5.2.1, the equation $\nabla_\omega \cdot u_\omega = g$ has a solution $u_\omega \in \overset{\circ}{h}{}^1_2(\Omega)$. Thus, the vector function $(0, u_\omega)$ satisfies the desired equation.

In the case $\int_\Omega g\, d\omega = c \ne 0$ we set $u_r = c\,(\lambda + 2)^{-1}\,\psi$, where ψ is an arbitrary function in $\overset{\circ}{W}{}^1_2(\Omega)$ such that $\int_\Omega \psi\, d\omega = 1$. Then the integral of the function $g - (\lambda + 2)\,u_r$ over Ω vanishes, and we can solve the equation $\nabla_\omega \cdot u_\omega = g - (\lambda + 2)u_r$. The so obtained vector function (u_r, u_ω) has the desired properties. ∎

LEMMA 5.2.3. *Let* $\operatorname{Re}\lambda = -1/2$, *and let* $|\lambda|$ *be sufficiently large. Then for every functional* $f \in \big(\overset{\circ}{W}{}^1_2(\Omega) \times \overset{\circ}{h}{}^1_2(\Omega) \big)^*$ *and every function* $g \in L_2(\Omega)$ *there exists a solution* $(u, p) \in \overset{\circ}{W}{}^1_2(\Omega) \times \overset{\circ}{h}{}^1_2(\Omega) \times L_2(\Omega)$ *of the equation*

(5.2.9)
$$\mathcal{L}(\lambda) \begin{pmatrix} u \\ p \end{pmatrix} = \begin{pmatrix} f \\ g \end{pmatrix}.$$

Proof: By Lemma 5.2.2, there exists a vector function $u^{(0)} = (u_r^{(0)}, u_\omega^{(0)}) \in \overset{\circ}{W}{}^1_2(\Omega) \times \overset{\circ}{h}{}^1_2(\Omega)$ such that $(\lambda + 2)\, u_r^{(0)} + \nabla_\omega \cdot u_\omega^{(0)} = g$. Then we have

$$\mathcal{L}(\lambda) \begin{pmatrix} u^{(0)} \\ 0 \end{pmatrix} = \begin{pmatrix} F \\ g \end{pmatrix},$$

where $F \in \big(\overset{\circ}{W}{}^1_2(\Omega) \times \overset{\circ}{h}{}^1_2(\Omega) \big)^*$ is given by the left-hand sides of (5.2.2)–(5.2.4) with $p = 0$ and $u = u^{(0)}$.

We show that there exists a vector function $(u^{(1)}, p) \in \overset{\circ}{W}{}^1_2(\Omega) \times \overset{\circ}{h}{}^1_2(\Omega) \times L_2(\Omega)$ such that

(5.2.10)
$$\mathcal{L}(\lambda) \begin{pmatrix} u^{(1)} \\ p \end{pmatrix} = \begin{pmatrix} f - F \\ 0 \end{pmatrix}.$$

Using (5.2.6) and (5.2.7), we get

$$a\left(\begin{pmatrix} u \\ 0 \end{pmatrix}, \begin{pmatrix} u \\ 0 \end{pmatrix}; -\frac{1}{2} + it \right) = Q(u_\omega, u_\omega) + \int_\Omega \left(|\nabla_\omega u_r|^2 + (\frac{9}{4} + t^2)\,|u_r|^2 \right) d\omega$$

$$+ \int_\Omega \left((\frac{5}{4} + t^2)\,|u_\omega|^2 - 4\operatorname{Re}(\nabla_\omega u_r) \cdot \overline{u}_\omega \right) d\omega$$

for arbitrary $u_r \in \overset{\circ}{W}{}^1_2(\Omega)$, $u_\omega \in \overset{\circ}{h}{}^1_2(\Omega)$. Hence for $\operatorname{Re}\lambda = -1/2$, $|\lambda|$ sufficiently large, it holds

(5.2.11) $$\left| a\left(\begin{pmatrix} u \\ 0 \end{pmatrix}, \begin{pmatrix} u \\ 0 \end{pmatrix}; \lambda \right) \right| \geq c \left(\|u_r\|^2_{W^1_2(\Omega)} + \|u_\omega\|^2_{h^1_2(\Omega)} \right)$$

with a constant c independent of $u_r \in \overset{\circ}{W}{}^1_2(\Omega)$, $u_\omega \in \overset{\circ}{h}{}^1_2(\Omega)$. In particular, (5.2.11) is satisfied for all (u_r, u_ω) from the subspace

$$\mathcal{H}_\lambda \overset{def}{=} \{ (u_r, u_\omega) \in \overset{\circ}{W}{}^1_2(\Omega) \times \overset{\circ}{h}{}^1_2(\Omega) \; : \; (\lambda + 2)u_r + \nabla_\omega \cdot u_\omega = 0 \}.$$

Consequently, by Lax-Milgram's theorem, there exists a solution $u^{(1)} \in \mathcal{H}_\lambda$ of the equation

(5.2.12) $$a\left(\begin{pmatrix} u^{(1)} \\ 0 \end{pmatrix}, \begin{pmatrix} v \\ 0 \end{pmatrix}; \lambda \right) = (f - F, v)_{L_2(\Omega)^3}, \qquad v \in \mathcal{H}_\lambda.$$

We consider the functional

(5.2.13) $$v \to G(v) \overset{def}{=} a\left(\begin{pmatrix} u^{(1)} \\ 0 \end{pmatrix}, \begin{pmatrix} v \\ 0 \end{pmatrix}; \lambda \right) - (f - F, v)_{L_2(\Omega)^3}$$

on $\overset{\circ}{W}{}^1_2(\Omega) \times \overset{\circ}{h}{}^1_2(\Omega)$. Let $v = (v_r, v_\omega)$ be an arbitrary function in $\overset{\circ}{W}{}^1_2(\Omega) \times \overset{\circ}{h}{}^1_2(\Omega)$. Since the functional G vanishes on the subspace $\mathcal{H}_\lambda$, we have

$$G(v) = G(w)$$

for all $w \in \overset{\circ}{W}{}^1_2(\Omega) \times \overset{\circ}{h}{}^1_2(\Omega)$ such that $(\lambda + 2)\, w_r + \nabla_\omega \cdot w_\omega = (\lambda + 2)\, v_r + \nabla_\omega \cdot v_\omega$. Due to Lemma 5.2.2, we can choose $w = (w_r, w_\omega)$ such that

$$\|w_r\|_{W^1_2(\Omega)} + \|w_\omega\|_{h^1_2(\Omega)} \leq c \big\| (\lambda + 2)\, v_r + \nabla_\omega \cdot v_\omega \big\|_{L_2(\Omega)}.$$

Therefore, $G(v)$ can be understood as a linear and continuous functional applied to $(\lambda + 2)\, v_r + \nabla_\omega \cdot v_\omega$. By the Riesz theorem, there exists a function $p \in L_2(\Omega)$ such that

$$G(v) = \int_\Omega p \cdot \overline{\left((\lambda + 2)v_r + \nabla_\omega \cdot v_\omega \right)}\, d\omega$$

for all $v \in \mathcal{H}_s$. Using (5.2.13) and the last equality, we obtain

$$\begin{aligned} a\left(\begin{pmatrix} u^{(1)} \\ p \end{pmatrix}, \begin{pmatrix} v \\ q \end{pmatrix}; \lambda \right) &= a\left(\begin{pmatrix} u^{(1)} \\ 0 \end{pmatrix}, \begin{pmatrix} v \\ 0 \end{pmatrix}; \lambda \right) - \int_\Omega p \cdot \overline{\left((\lambda + 2)v_r + \nabla_\omega \cdot v_\omega \right)}\, d\omega \\ &= (f - F, v)_{L_2(\Omega)^3}. \end{aligned}$$

Thus, the vector function $(u^{(1)}, p)$ satisfies (5.2.10). This proves the lemma. ∎

THEOREM 5.2.1. *1) The operator $\mathcal{L}(\lambda)$ is Fredholm for all λ.*

2) The spectrum of the pencil $\mathcal{L}$ consists of isolated eigenvalues with finite algebraic multiplicities.

3) The line $\operatorname{Re}\lambda = -1/2$ does not contain eigenvalues of the pencil $\mathcal{L}$.

4) The number λ_0 is an eigenvalue of the pencil $\mathcal{L}$ if and only if $-1 - \overline{\lambda}_0$ is an eigenvalue of this pencil. The geometric, algebraic and partial multiplicities of the eigenvalues λ_0 and $-1 - \overline{\lambda}_0$ coincide.

Proof: We show first that the kernel of the operator $\mathcal{L}(\lambda)$ is trivial for $\operatorname{Re}\lambda = -1/2$. Let (u, p) be an element of the kernel of $\mathcal{L}(\lambda)$, where λ is a number on the line $\operatorname{Re}\lambda = -1/2$. Then, by (5.2.5), we have

$$(\lambda + 2)\, u_r + \nabla_\omega \cdot u_\omega = 0 \quad \text{in } \Omega.$$

Consequently, by (5.2.6), (5.2.7) and by the equality $\overline{\lambda} + 2 = 1 - \lambda$, we obtain

$$
a\left(\begin{pmatrix} u \\ p \end{pmatrix}, \begin{pmatrix} u \\ p \end{pmatrix}; \lambda\right) = Q(u_\omega, u_\omega) + \int_\Omega \left(\left|\nabla_\omega u_r - \frac{1}{2} u_\omega\right|^2 + |\lambda + 2|^2 \, |u_r|^2 \right) d\omega
$$

$$
+ \int_\Omega \left(\left(|\lambda + 2|^2 - \frac{5}{4}\right) |u_\omega|^2 + 3 \operatorname{Re}\left(\nabla_\omega \cdot u_\omega\right) \overline{u}_r \right) d\omega.
$$

Using (3.2.10) and replacing $\nabla_\omega \cdot u_\omega$ by $-(\lambda + 2)u_r$, we arrive at the inequality

$$
0 = a\left(\begin{pmatrix} u \\ p \end{pmatrix}, \begin{pmatrix} u \\ p \end{pmatrix}; \lambda\right) \geq \int_\Omega \left(\left|\nabla_\omega u_r - \frac{1}{2} u_\omega\right|^2 + \left(|\lambda + 2|^2 - \frac{9}{4}\right) |u_\omega|^2 \right) d\omega
$$

$$
+ \int_\Omega \left(2\,|\lambda + 2|^2 - 3\operatorname{Re}\left(\lambda + 2\right) \right) |u_r|^2 \, d\omega.
$$

This implies $u_r = 0$ and $u_\omega = 0$ if $\lambda \neq -1/2$. In the case $\lambda = -1/2$ we obtain

$$\nabla_\omega u_r - \frac{1}{2} u_\omega = 0, \qquad \frac{3}{2} u_r + \nabla_\omega \cdot u_\omega = 0.$$

From this we also conclude, in the same way as in the proof of Lemma 3.2.6, that $u_r = 0$ and $u_\omega = 0$.

Furthermore, for $u = 0$ and for arbitrary $v_r \in \overset{\circ}{W}{}^1_2(\Omega)$, $v_\omega \in \overset{\circ}{h}{}^1_2(\Omega)$ we get

$$
a\left(\begin{pmatrix} 0 \\ p \end{pmatrix}, \begin{pmatrix} v \\ 0 \end{pmatrix}; \lambda\right) = - \int_\Omega p\left((1 - \lambda)\overline{v}_r + \nabla_\omega \cdot \overline{v}_\omega \right) d\omega = 0.
$$

This together with Lemma 5.2.2 implies $p = 0$. Thus, the kernel of the operator $\mathcal{L}(\lambda)$ is trivial for $\operatorname{Re}\lambda = -1/2$.

If moreover $|\lambda|$ is sufficiently large, then, by Lemma 5.2.3, the operator $\mathcal{L}(\lambda)$ is an isomorphism. Since $\mathcal{L}(\lambda) - \mathcal{L}(\mu)$ is a compact operator for arbitrary λ, μ, it follows that the operator $\mathcal{L}(\lambda)$ is Fredholm for every complex λ. This proves assertion 1).

The second assertion follows from 1), from the invertibility of the operator $\mathcal{L}(\lambda)$ for $\operatorname{Re}\lambda = -1/2$, $|\lambda|$ sufficiently large, and from Theorem 1.1.1.

The proof of the third assertion is contained in the proof of assertion 1).

Finally, assertion 4) follows from the equality $\mathcal{L}(\lambda)^* = \mathcal{L}(-1 - \overline{\lambda})$. ∎

5.2.3. Relations between eigenvectors corresponding to the eigenvalues λ and $-1 - \lambda$. Since the coefficients of the Stokes system are real, it follows from assertion 4) of Theorem 5.2.1 that the numbers λ and $-1 - \lambda$ are simultaneously eigenvalues of the pencil $\mathcal{L}$ or not. Our goal is to derive a relation between the eigenvectors and generalized eigenvectors corresponding to these eigenvalues.

We introduce the matrix

(5.2.14) $$\mathcal{S}(\lambda) = \begin{pmatrix} \lambda+2 & 0 & 0 & 0 \\ 0 & 1-\lambda & 0 & 0 \\ 0 & 0 & 1-\lambda & 0 \\ 2+4\lambda & 0 & 0 & 1-\lambda \end{pmatrix}.$$

From (5.2.6) and (5.2.7) it follows that

$$a\left(\begin{pmatrix} u \\ p \end{pmatrix}, \mathcal{S}(\bar\lambda)\begin{pmatrix} v \\ q \end{pmatrix}; \lambda\right) = a\left(\mathcal{S}(\lambda)\begin{pmatrix} u \\ p \end{pmatrix}, \begin{pmatrix} v \\ q \end{pmatrix}; -1-\lambda\right).$$

for all $u, v \in \overset{\circ}{W}{}^1_2(\Omega) \times \overset{\circ}{h}{}^1_2(\Omega)$, $p, q \in L_2(\Omega)$. Consequently, there is the equality

(5.2.15) $$\mathcal{S}(\lambda)^t \mathcal{L}(\lambda) = \mathcal{L}(-1-\lambda)\mathcal{S}(\lambda)$$

for arbitrary $\lambda \in \mathbb{C}$, where $\mathcal{S}(\lambda)^t$ is the transposed matrix to $\mathcal{S}(\lambda)$. This leads to the following assertions.

THEOREM 5.2.2. *Let λ_0 be an eigenvalue of the pencil $\mathcal{L}$ and let the vector functions $(u^{(0)}, p^{(0)}), \ldots, (u^{(s)}, p^{(s)})$ form a Jordan chain of this pencil corresponding to λ_0.*

1) *If* $\lambda_0 \notin \{1, -2\}$ *or* $\lambda_0 \in \{1, -2\}$ *and* $\mathcal{S}(\lambda_0)\begin{pmatrix} u^{(0)} \\ p^{(0)} \end{pmatrix} \neq 0,$

then $-1-\lambda_0$ is also an eigenvalue of the pencil $\mathcal{L}$ and the vector functions $\begin{pmatrix} v^{(0)} \\ q^{(0)} \end{pmatrix} = S(\lambda_0)\begin{pmatrix} u^{(0)} \\ p^{(0)} \end{pmatrix}$,

$$\begin{pmatrix} v^{(k)} \\ q^{(k)} \end{pmatrix} = (-1)^k \left(\mathcal{S}(\lambda_0)\begin{pmatrix} u^{(k)} \\ p^{(k)} \end{pmatrix} + \begin{pmatrix} 1 & 0 & 0 & 0 \\ 0 & -1 & 0 & 0 \\ 0 & 0 & -1 & 0 \\ 4 & 0 & 0 & -1 \end{pmatrix}\begin{pmatrix} u^{(k-1)} \\ p^{(k-1)} \end{pmatrix} \right),$$

$k = 1, \ldots, s$, form a Jordan chain corresponding to $-1-\lambda_0$.

2) *If $\lambda_0 \in \{1, -2\}$, $\mathcal{S}(\lambda_0)\begin{pmatrix} u^{(0)} \\ p^{(0)} \end{pmatrix} = 0$ and $s \geq 1$, then $-1-\lambda_0$ is also an eigenvalue and the vector functions*

$$\begin{pmatrix} v^{(k)} \\ q^{(k)} \end{pmatrix} = (-1)^k \left(\mathcal{S}(\lambda_0)\begin{pmatrix} u^{(k)} \\ p^{(k)} \end{pmatrix} + \begin{pmatrix} 1 & 0 & 0 & 0 \\ 0 & -1 & 0 & 0 \\ 0 & 0 & -1 & 0 \\ 4 & 0 & 0 & -1 \end{pmatrix}\begin{pmatrix} u^{(k-1)} \\ p^{(k-1)} \end{pmatrix} \right),$$

$k = 1, \ldots, s$, form a Jordan chain corresponding to this eigenvalue.

Proof: First note that

$$\mathcal{S}(\lambda_0)\begin{pmatrix} u^{(1)} \\ p^{(1)} \end{pmatrix} + \begin{pmatrix} 1 & 0 & 0 & 0 \\ 0 & -1 & 0 & 0 \\ 0 & 0 & -1 & 0 \\ 4 & 0 & 0 & -1 \end{pmatrix}\begin{pmatrix} u^{(0)} \\ p^{(0)} \end{pmatrix} \neq 0$$

if $\lambda_0 \in \{1, -2\}$ and $\mathcal{S}(\lambda_0)\begin{pmatrix} u^{(0)} \\ p^{(0)} \end{pmatrix} = 0$.

By the assumptions of the lemma, the vector functions $(u^{(j)}, p^{(j)})$ satisfy the equations

$$\sum_{j=0}^{k} \frac{1}{(k-j)!} \mathcal{L}^{(k-j)}(\lambda_0)\begin{pmatrix} u^{(j)} \\ p^{(j)} \end{pmatrix} = 0 \qquad \text{for } k = 0, 1, \ldots, s,$$

where $\mathcal{L}^{(j)}(\lambda) = d^j \mathcal{L}(\lambda)/d\lambda^j$. We have to show that

$$(5.2.16) \qquad \sum_{j=0}^{k} \frac{1}{(k-j)!}\, \mathcal{L}^{(k-j)}(-1-\lambda_0) \binom{v^{(j)}}{q^{(j)}}$$

$$= (-1)^k \, \mathcal{L}(-1-\lambda_0)\, \mathcal{S}(\lambda_0) \binom{u^{(k)}}{p^{(k)}}$$

$$+ \sum_{j=0}^{k-1} \frac{(-1)^j}{(k-j)!} \mathcal{L}(-1-\lambda_0)\, \mathcal{S}(\lambda_0) \binom{u^{(j)}}{p^{(j)}}$$

$$- \sum_{j=0}^{k-1} \frac{(-1)^j}{(k-j-1)!}\, \mathcal{L}^{(k-j-1)}(-1-\lambda_0)\, \mathcal{S}'(\lambda_0) \binom{u^{(j)}}{p^{(j)}} = 0.$$

Differentiating (5.2.15) with respect to λ, we obtain

$$\mathcal{S}(\lambda)^t\, \mathcal{L}^{(k-j)}(\lambda) + (k-j)\, \mathcal{S}'(\lambda)^t\, \mathcal{L}^{(k-j-1)}(\lambda)$$

$$= (-1)^{k-j}\left(\mathcal{L}^{(k-j)}(-1-\lambda)\, \mathcal{S}(\lambda) - (k-j)\, \mathcal{L}^{(k-j-1)}(-1-\lambda)\, \mathcal{S}'(\lambda)\right)$$

for $j \leq k-1$. Hence the left side of (5.2.16) is equal to

$$(-1)^k \mathcal{S}(\lambda_0)^t \sum_{j=0}^{k} \frac{1}{(k-j)!}\, \mathcal{L}^{(k-j)}(\lambda_0) \binom{u^{(j)}}{p^{(j)}}$$

$$+ (-1)^k\, \mathcal{S}'(\lambda_0)^t \sum_{j=0}^{k-1} \frac{1}{(k-1-j)!}\, \mathcal{L}^{(k-1-j)}(\lambda_0) \binom{u^{(j)}}{p^{(j)}} = 0.$$

The theorem is proved. ∎

REMARK 5.2.1. In the case $\lambda_0 \neq 1$, $\lambda_0 \neq -2$ the formulas given in the first part of Theorem 5.2.2 yield a one-to-one correspondence between eigenvectors and generalized eigenvectors of the pencil $\mathcal{L}$ corresponding to the eigenvalues λ_0 and $-1 - \lambda_0$.

5.2.4. The Kelvin transform for the Stokes system. The identity (5.2.15) leads to the following analogue of the Kelvin transform. Let (U, P) be a solution of (5.0.1) in the domain $\mathcal{G} \subset \mathbb{R}^3$, then the vector function (5.0.4), (5.0.5) solves the Stokes system in the domain $\mathcal{G}_{in}$ obtained from $\mathcal{G}$ by the inversion $x \to x\,|x|^{-2}$.

Indeed, we have

$$\mathcal{L}(r\partial_r) \binom{JU}{P} = 0,$$

where J is the matrix (3.2.3). By (5.2.15), we have

$$\mathcal{L}(-1 - r\partial_r)\, \mathcal{S}(r\partial_r) \binom{JU}{P} = 0.$$

Replacing r by r^{-1}, we obtain

$$\mathcal{L}(r\partial_r)\, r^{-1} \mathcal{S}(-r\partial_r) \begin{pmatrix} J & 0 \\ 0 & 1 \end{pmatrix} \binom{U(x/|x|^2)}{P(x/|x|^2)} = 0.$$

Hence the vector function

$$\begin{pmatrix} J^t & 0 \\ 0 & 1 \end{pmatrix} r^{-1} \mathcal{S}(-r\partial_r) \begin{pmatrix} J & 0 \\ 0 & 1 \end{pmatrix} \begin{pmatrix} U(x/|x|^2) \\ P(x/|x|^2) \end{pmatrix}$$

solves the homogeneous Stokes system in $\mathcal{G}_{in}$. Using the definition of J and $\mathcal{S}$, one can directly verify that the above vector function coincides with (5.0.4), (5.0.5).

5.3. Properties of real eigenvalues

We prove in this section that in a certain strip of the complex plane centered about the line $\operatorname{Re}\lambda = -1/2$ there are only real eigenvalues of the pencil $\mathcal{L}$. Another result is the nonexistence of generalized eigenvectors to eigenvalues $\lambda \neq 1$, $\lambda \neq -2$ in the interior of this strip.

5.3.1. A strip which contains only real eigenvalues. Note that the form $a(\cdot, \cdot; \lambda)$ is not symmetric. However, it is possible to obtain a symmetric form corresponding to the same operator $\mathcal{L}(\lambda)$ as follows.

LEMMA 5.3.1. *If $\lambda \neq 1$, then*

$$a\left(\begin{pmatrix} u \\ p \end{pmatrix}, J_\lambda \begin{pmatrix} v \\ q \end{pmatrix}; \lambda\right) = \overline{a\left(\begin{pmatrix} v \\ q \end{pmatrix}, J_{\bar\lambda} \begin{pmatrix} u \\ p \end{pmatrix}; \bar\lambda\right)},$$

where

(5.3.1) $$J_\lambda = \begin{pmatrix} c & 0 & 0 & 0 \\ 0 & 1 & 0 & 0 \\ 0 & 0 & 1 & 0 \\ 2c-2 & 0 & 0 & 1 \end{pmatrix} \quad and \quad c = \frac{\lambda+2}{1-\bar\lambda}.$$

Proof: Using (5.2.6) and (5.2.7), we obtain

(5.3.2) $$a\left(\begin{pmatrix} u \\ p \end{pmatrix}, J_\lambda \begin{pmatrix} v \\ q \end{pmatrix}; \lambda\right) = Q(u_\omega, u_\omega) + \int_\Omega \bar c \left(\nabla_\omega u_r\right) \cdot \nabla_\omega \bar v_r \, d\omega$$

$$+ \int_\Omega \left(|\lambda+2|^2 + 2\left(\lambda+2\right)\left(1-\bar c\right)\right) u_r \, \bar v_r \, d\omega$$

$$+ \int_\Omega \left((1-\lambda-\lambda^2)\, u_\omega \cdot \bar v_\omega + 2\left(\nabla_\omega \cdot u_\omega\right)\bar v_r + 2u_r \nabla_\omega \cdot \bar v_\omega\right) d\omega$$

$$- \int_\Omega \left(p\left(\nabla_\omega \cdot \bar v_\omega + (\bar\lambda+2)\,\bar v_r\right) + \left(\nabla_\omega \cdot u_\omega + (\lambda+2)\,u_r\right)\bar q\right) d\omega.$$

This implies the assertion of the lemma. ∎

The following theorem characterizes a strip in the complex plane, where the spectrum of the pencil $\mathcal{L}$ is real.

THEOREM 5.3.1. *The strip $|\operatorname{Re}\lambda + \tfrac{1}{2}| \leq F(\Omega)$, where*

(5.3.3) $$F(\Omega) = \sqrt{\frac{\Gamma(\Omega)}{2} + \frac{7}{4}},$$

and

$$\Gamma(\Omega) = \inf_{0 \neq u_\omega \in \overset{\circ}{h}{}_2^1(\Omega)} \frac{Q(u_\omega, u_\omega) + \int_\Omega |\nabla_\omega \cdot u_\omega|^2 \, d\omega}{\int_\Omega |u_\omega|^2 \, d\omega},$$

may contain only real eigenvalues of the pencil $\mathcal{L}$.

Proof: We assume that $\operatorname{Im} \lambda \neq 0$ and consider the equation

$$(5.3.4) \qquad a\left(\begin{pmatrix} u \\ p \end{pmatrix}, J_\lambda \begin{pmatrix} u \\ p \end{pmatrix}; \lambda \right) = 0.$$

Taking the imaginary part of (5.3.4) we obtain

$$\operatorname{Im} \bar{c} \int_\Omega |\nabla_\omega u_r|^2 \, d\omega - \operatorname{Im} \lambda (2\operatorname{Re} \lambda + 1) \int_\Omega |u_\omega|^2$$

$$+ 2 \operatorname{Im}\left((\lambda + 2)(1 - \bar{c}) \right) \int_\Omega |u_r|^2 \, d\omega = 0.$$

A simple calculation yields

$$(5.3.5) \qquad -\frac{1}{|1-\lambda|^2} \int_\Omega |\nabla_\omega u_r|^2 \, d\omega + \int_\Omega |u_\omega|^2 \, d\omega + \frac{6}{|1-\lambda|^2} \int_\Omega |u_r|^2 \, d\omega = 0.$$

Now taking the real part of (5.3.4) results in

$$Q(u_\omega, u_\omega) + \frac{(\operatorname{Im} \lambda)^2 - (\operatorname{Re} \lambda)^2 - \operatorname{Re} \lambda + 2}{|1 - \lambda|^2} \int_\Omega |\nabla_\omega u_r|^2 \, d\omega$$

$$+ |\lambda + 2|^2 \int_\Omega |u_r|^2 \, d\omega - \left((\operatorname{Re} \lambda)^2 - (\operatorname{Im} \lambda)^2 + \operatorname{Re} \lambda - 1 \right) \int_\Omega |u_\omega|^2 \, d\omega$$

$$+ 4\operatorname{Re} \int_\Omega \nabla_\omega \cdot u_\omega \, \bar{u}_r \, d\omega + 2(2\operatorname{Re} \lambda + 1)\left(1 + \frac{3(\operatorname{Re} \lambda - 1)}{|1 - \lambda|^2} \right) \int_\Omega |u_r|^2 \, d\omega = 0.$$

We eliminate $\int_\Omega |\nabla_\omega u_r|^2 \, d\omega$ by means of (5.3.5) and obtain after simplification:

$$Q(u_\omega, u_\omega) + \left(2(\operatorname{Im}\lambda)^2 - 2(\operatorname{Re}\lambda)^2 - 2\operatorname{Re}\lambda + 3 \right) \int_\Omega |u_\omega|^2 \, d\omega$$

$$+ |\lambda + 2|^2 \int_\Omega |u_r|^2 \, d\omega = 0.$$

Using the equality

$$\int_\Omega |\nabla_\omega \cdot u_\omega|^2 \, d\omega = |\lambda + 2|^2 \int_\Omega |u_r|^2 \, d\omega$$

and the inequality

$$Q(u_\omega, u_\omega) + \int_\Omega |\nabla_\omega \cdot u_\omega|^2 \, d\omega \geq \Gamma(\Omega) \int_\Omega |u_\omega|^2 \, d\omega,$$

we get

$$\left(2\operatorname{Re}\lambda\,(\operatorname{Re}\lambda+1)-3-\Gamma(\Omega)-2(\operatorname{Im}\lambda)^2\right)\int_\Omega|u_\omega|^2 d\omega \geq 0$$

which yields a contradiction to $\operatorname{Im}\lambda\neq 0$ if

$$\left(\operatorname{Re}\lambda+\frac{1}{2}\right)^2-(\operatorname{Im}\lambda)^2 < (F(\Omega))^2.$$

This proves the theorem. ∎

REMARK 5.3.1. We actually proved the following stronger assertion: If there are any eigenvalues λ in the domain

$$\left\{\lambda:\ \left(\operatorname{Re}\lambda+\frac{1}{2}\right)^2-(\operatorname{Im}\lambda)^2 < (F(\Omega))^2\right\},$$

then they must be real.

REMARK 5.3.2. The quantity $F(\Omega)$ defined by (5.3.3) coincides with $F_\nu(\Omega)$ in Chapter 3 (see formula (3.3.2)) if $\nu = 1/2$. Therefore, the set function F has the following properties (cf. Corollary 3.4.1):

1) For all $\Omega\subset S^2$ it holds $F(\Omega)\geq 3/2$.

2) The function F decreases as the set Ω increases, i.e., if $\Omega_0\subset\Omega_1\subset S^2$, then $F(\Omega_0)\geq F(\Omega_1)$. If, moreover, Ω_1 is connected and $\Omega_0\Subset\Omega_1$, then even the strict inequality is valid. (Here the notation $\Omega_0\Subset\Omega_1$ means that $\Omega_0\subset\Omega_1$ and $\Omega_1\backslash\Omega_0$ contains a nonzero open set.)

3) If $\Omega\subset S^2$ and $\operatorname{cap}(S^2\backslash\Omega) > 0$, where

$$\operatorname{cap}(K) = \inf\left\{\|u\|^2_{W^1_2(S^2)} : u\in C^\infty(S^2), u|_K\geq 1\right\}$$

is the Wiener capacity on S^2, then $F(\Omega) > 3/2$.

All these assertions are valid without our general assumption that Ω is a Lipschitz domain.

5.3.2. Absence of generalized eigenvectors. Now we study the existence of generalized eigenfunctions of the pencil $\mathcal{L}$ for real eigenvalues in the interval $-F(\Omega)-1/2 < \lambda < F(\Omega)-1/2$. Due to assertion 4) of Theorem 5.2.1, it is sufficient to investigate the spectrum of the pencil $\mathcal{L}$ in the interval $(-1/2, F(\Omega)-1/2)$. Throughout this subsection, we will assume that $\lambda\neq 1$ and postpone the investigation of the case $\lambda = 1$ to the next section.

Let $(u^{(0)}, p^{(0)})$ be an eigenvector to the real eigenvalue λ_0 and let $(u^{(1)}, p^{(1)})$ be a corresponding generalized eigenvector. Then $(u^{(1)}, p^{(1)})$ satisfies the equation

$$a\left(\begin{pmatrix} u^{(1)}\\ p^{(1)} \end{pmatrix}, J_\lambda\begin{pmatrix} v\\ q \end{pmatrix};\lambda\right)\Big|_{\lambda=\lambda_0} = -\frac{d}{d\lambda}a\left(\begin{pmatrix} u^{(0)}\\ p^{(0)} \end{pmatrix}, J_\lambda\begin{pmatrix} v\\ q \end{pmatrix};\lambda\right)\Big|_{\lambda=\lambda_0}$$

for all $(v,q)\in\overset{\circ}{W}{}^1_2(\Omega)\times\overset{\circ}{h}{}^1_2(\Omega)\times L_2(\Omega)$. Since the sesquilinear form a has a nontrivial kernel for $\lambda=\lambda_0$ and since $a(\cdot,J_\lambda\cdot;\,\lambda)$ is symmetric, we have the following sufficient condition for the existence of generalized eigenfunctions:

(5.3.6) $$\frac{d}{d\lambda}a\left(\begin{pmatrix} u^{(0)}\\ p^{(0)} \end{pmatrix}, J_\lambda\begin{pmatrix} u^{(0)}\\ p^{(0)} \end{pmatrix};\lambda\right)\Big|_{\lambda=\lambda_0} = 0.$$

The following lemma shows that this is not true for real eigenvalues λ in the interval $(-\frac{1}{2}, F(\Omega)-\frac{1}{2})$, $\lambda\neq 1$.

LEMMA 5.3.2. *Assume that* $(u,p) \in \overset{\circ}{W}{}^1_2(\Omega) \times \overset{\circ}{h}{}^1_2(\Omega) \times L_2(\Omega)$, $(u,p) \neq 0$, *satisfies the equality*

$$(5.3.7) \qquad a\left(\begin{pmatrix} u \\ p \end{pmatrix}, J_\lambda\begin{pmatrix} v \\ q \end{pmatrix}; \lambda\right) = 0$$

for all $(v,q) \in \overset{\circ}{W}{}^1_2(\Omega) \times \overset{\circ}{h}{}^1_2(\Omega) \times L_2(\Omega)$. *Then*

$$(5.3.8) \qquad \frac{d}{d\lambda} a\left(\begin{pmatrix} u \\ p \end{pmatrix}, J_\lambda\begin{pmatrix} u \\ p \end{pmatrix}; \lambda\right) \begin{cases} < 0 & if & \lambda \in (-\tfrac{1}{2}, 1), \\ > 0 & if & \lambda \in (1, F(\Omega) - \tfrac{1}{2}), \end{cases}$$

where the set function F *is defined by* (5.3.3).

Proof: We have

$$(5.3.9) \qquad \frac{d}{d\lambda} a\left(\begin{pmatrix} u \\ p \end{pmatrix}, J_\lambda\begin{pmatrix} u \\ p \end{pmatrix}; \lambda\right) = -(2\lambda+1) \int_\Omega \left(c\,|u_r|^2 + |u_\omega|^2\right) d\omega$$

$$+ \int_\Omega \left(c\,p\,\bar{u}_r - u_r\,\bar{p} - (2c-2)\,|u_r|^2\right) d\omega.$$

We remove u_ω in equation (5.2.2) by means of (5.2.5). Then we get

$$(5.3.10) \qquad (1-\lambda) \int_\Omega p\,\bar{u}_r\,d\omega = -\int_\Omega \left(\delta + (\lambda+1)(\lambda+2)\right) u_r \cdot \bar{u}_r\,d\omega.$$

This equation can be used to eliminate p in (5.3.9). After simplification we arrive at

$$\frac{d}{d\lambda} a\left(\begin{pmatrix} u \\ p \end{pmatrix}, J_\lambda\begin{pmatrix} u \\ p \end{pmatrix}; \lambda\right)$$

$$= \frac{c-1}{1-\lambda}\left(\int_\Omega |\nabla_\omega u_r|^2\,d\omega - 6\int_\Omega |u_r|^2\,d\omega - (1-\lambda^2)\int_\Omega |u_\omega|^2\,d\omega\right).$$

Evaluating the left hand side of (5.3.7) by (5.2.6) and using (5.2.5), we obtain

$$(5.3.11) \qquad c\frac{1-\lambda}{c-\lambda}\frac{d}{d\lambda} a\left(\begin{pmatrix} u \\ p \end{pmatrix}, J_\lambda\begin{pmatrix} u \\ p \end{pmatrix}; \lambda\right) - a\left(\begin{pmatrix} u \\ p \end{pmatrix}, J_\lambda\begin{pmatrix} u \\ p \end{pmatrix}; \lambda\right)$$

$$= -(\lambda+2)^2 \int_\Omega |u_r|^2\,d\omega + (2\lambda^2 + 2\lambda - 3)\int_\Omega |u_\omega|^2\,d\omega - Q(u_\omega, u_\omega).$$

From (5.2.5) we get the equality

$$\int_\Omega |\nabla_\omega \cdot u_\omega|^2\,d\omega = (\lambda+2)^2 \int_\Omega |u_r|^2\,d\omega$$

which implies in particular that $u_\omega \neq 0$. Therefore, the left-hand side of (5.3.11) is bounded from above by the quantity

$$\left(2\lambda^2 + 2\lambda - 3 - \Gamma(\Omega)\right) \int_\Omega |u_\omega|^2\,d\omega.$$

which is less than zero if $(\lambda + 1/2)^2 < F(\Omega)^2$. Together with (5.3.7) this implies (5.3.8). ∎

Using Lemma 5.3.2, we are able to prove the following assertion.

THEOREM 5.3.2. *The pencil $\mathcal{L}$ has no generalized eigenfunctions for eigenvalues $\lambda \in (-\frac{1}{2}, 1) \cup (1, F(\Omega) - \frac{1}{2})$.*

Proof: If λ_0 is an eigenvalue and (u_0, p_0), (u_1, p_1) are eigen- and generalized eigenfunctions, respectively, then

$$a\left(\left(\begin{array}{c} u^{(0)} \\ p^{(0)} \end{array}\right), \left(\begin{array}{c} v \\ q \end{array}\right); \lambda\right)\Big|_{\lambda=\lambda_0} = 0$$

and

$$a\left(\left(\begin{array}{c} u^{(1)} \\ p^{(1)} \end{array}\right), \left(\begin{array}{c} v \\ q \end{array}\right); \lambda\right)\Big|_{\lambda=\lambda_0} = -\frac{d}{d\lambda} a\left(\left(\begin{array}{c} u^{(0)} \\ p^{(0)} \end{array}\right), \left(\begin{array}{c} v \\ q \end{array}\right); \lambda\right)\Big|_{\lambda=\lambda_0}$$

for all $(v, q) \in \overset{\circ}{W}{}_2^1(\Omega) \times \overset{\circ}{h}{}_2^1(\Omega) \times L_2(\Omega)$. As $\lambda_0 \neq 1$ we may substitute $J_\lambda \left(\begin{array}{c} u^{(0)} \\ p^{(0)} \end{array}\right)$

for $\left(\begin{array}{c} v \\ q \end{array}\right)$ and the theorem follows from Lemma 5.3.2. ∎

5.4. The eigenvalues $\lambda=1$ and $\lambda=-2$

Obviously, the vector function $(u_r, u_\omega, p) = (0, 0, 1)$ is a solution of problem (5.2.2)–(5.2.5) with $\lambda = 0$. Consequently, the spectrum of the pencil $\mathcal{L}$ contains the point $\lambda = 1$, and $(0, 0, 1)$ is an eigenvector corresponding to this eigenvalue. In Sections 5.4.1 and 5.4.2 we derive necessary and sufficient conditions for the existence of additional eigenvectors and generalized eigenvectors corresponding to $\lambda = 1$. In Section 5.4.3 we obtain all eigenvectors and generalized eigenvectors for the case of a circular cone. In the last subsection we describe the eigenvectors and generalized eigenvectors corresponding to the eigenvalue $\lambda = -2$.

5.4.1. An auxiliary problem. We consider the auxiliary problem

$$(5.4.1) \qquad \partial_\theta p - (\delta + 2) u_\theta + \frac{u_\theta}{\sin^2 \theta} + 2\frac{\cot \theta}{\sin \theta} \partial_\varphi u_\varphi = f \quad \text{in } \Omega,$$

$$(5.4.2) \qquad \frac{1}{\sin \theta} \partial_\varphi p - (\delta + 2) u_\varphi + \frac{u_\varphi}{\sin^2 \theta} - 2\frac{\cot \theta}{\sin \theta} \partial_\varphi u_\varphi = g \quad \text{in } \Omega,$$

$$(5.4.3) \qquad -\frac{1}{\sin \theta} \partial_\theta(\sin \theta \, u_\theta) - \frac{1}{\sin \theta} \partial_\varphi u_\varphi = \psi \quad \text{in } \Omega$$

with the boundary conditions

$$u_\theta = u_\varphi = 0 \quad \text{on } \partial\Omega.$$

Note that (5.4.1)–(5.4.3) coincide with the equations (5.2.3)–(5.2.5) for $\lambda = 1$, $f = 2\partial_\theta u_r$, $g = 2(\sin \theta)^{-1} \partial_\varphi u_r$, $\psi = 3u_r$.

Taking the inner product in $L_2(\Omega)$ of the equation (5.4.1) with v_θ, of the equation (5.4.2) with v_φ and of (5.4.3) with q, where (v_θ, v_φ, q) is an arbitrary vector function from $\overset{\circ}{h}{}_2^1(\Omega) \times L_2(\Omega)$, and combining the resulting expressions, we arrive at the identity

$$(5.4.4) \qquad Q(u_\omega, v_\omega) - \int_\Omega u_\omega \bar{v}_\omega \, d\omega - \int_\Omega \left(p \nabla_\omega \cdot \bar{v}_\omega + (\nabla_\omega \cdot u_\omega) \bar{q}\right) d\omega$$

$$= \int_\Omega (f\bar{v}_\theta + g\bar{v}_\varphi + \psi\bar{q}) \, d\omega.$$

The following lemma on the solvability of the auxiliary problem is a straightforward analogue of the well-known result for the plane Stokes problem (see Temam [253]).

LEMMA 5.4.1. *Let* $\Omega \neq S^2$, $(f,g) \in \overset{\circ}{h}{}^1_2(\Omega)^*$, $\psi \in L_2(\Omega)$. *For the solvability of problem* (5.4.1)–(5.4.3) *in the space* $\overset{\circ}{h}{}^1_2(\Omega) \times L_2(\Omega)$ *it is necessary and sufficient that*

$$(5.4.5) \qquad\qquad \int_\Omega \psi \, d\omega = 0.$$

Under this condition, the solution (u_θ, u_φ, p) *is unique up to a vector* $(0, 0, const.)$.

Proof: 1) Necessity of the condition (5.4.5): For $v_\omega = 0$, $q = 1$ the integral identity (5.4.4) takes the form $\int_\Omega \psi \, d\omega = - \int_\Omega \nabla_\omega \cdot u_\omega \, d\omega = 0$.

2) Uniqueness of the solution: If Ω is a Lipschitz domain and $\Omega \neq S^2$, then we get, by Remark 5.3.2, that $\Gamma(\Omega) > 1$ and, consequently, that

$$Q(u_\omega, u_\omega) - \int_\Omega |u_\omega|^2 \, d\omega \geq c_0 \, Q(u_\omega, u_\omega)$$

for some $c_0 > 0$. Therefore, if $f = g = \psi = 0$, it follows that $u_\omega = 0$ and $p = const.$

3) Existence of a solution: First we remark that, by the second part of Lemma 5.2.1, there exists a vector function $\tilde{u}_\omega = (\tilde{u}_\theta, \tilde{u}_\varphi) \in \overset{\circ}{h}{}^1_2(\Omega)$ such that $\nabla_\omega \cdot \tilde{u}_\omega = \psi$ on Ω. Consequently, it is sufficient to show the existence of a solution of (5.4.1)–(5.4.3) in the case $\psi = 0$.

Assume that the vector function v_ω satisfies the condition $\nabla_\omega \cdot v_\omega = 0$. Since also $\nabla_\omega \cdot u_\omega = 0$, we obtain from (5.4.4)

$$(5.4.6) \qquad Q(u_\omega, v_\omega) - \int_\Omega u_\omega \bar{v}_\omega \, d\omega = \int_\Omega (f \, \bar{v}_\theta + g \, \bar{v}_\varphi) \, d\omega.$$

By Riesz's theorem on the representation of linear functionals in a Hilbert space, there exists a vector function $u_\omega \in \overset{\circ}{h}{}^1_2(\Omega)$ which satisfies the condition $\nabla_\omega \cdot u_\omega = 0$ and (5.4.6) for all $v_\omega \in \overset{\circ}{h}{}^1_2(\Omega)$ such that $\nabla_\omega \cdot v_\omega = 0$. Now let v_ω be an arbitrary vector function in $\overset{\circ}{h}{}^1_2(\Omega)$. Then

$$Q(u_\omega, v_\omega) - \int_\Omega u_\omega \bar{v}_\omega \, d\omega - \int_\Omega (f \, \bar{v}_\theta + g \, \bar{v}_\varphi) \, d\omega$$

is a continuous linear functional on $\nabla_\omega \cdot v_\omega \in L_2(\Omega)$. Therefore, there exists a function $p \in L_2(\Omega)$ such that

$$Q(u_\omega, v_\omega) - \int_\Omega u_\omega \bar{v}_\omega \, d\omega - \int_\Omega p \nabla_\omega \cdot \bar{v}_\omega \, d\omega = \int_\Omega (f \, \bar{v}_\theta + g \, \bar{v}_\varphi) \, d\omega$$

for all $v_\omega \in \overset{\circ}{h}{}^1_2(\Omega)$. Thus, the existence of a solution is shown. ∎

5.4.2. The multiplicity of the eigenvalue $\lambda = 1$. Now we can characterize the algebraic multiplicity of the eigenvalue $\lambda = 1$ and construct the corresponding system of eigenvectors and generalized eigenvectors.

THEOREM 5.4.1. *Let* $\Omega \neq S^2$. *Then the number* $\lambda = 1$ *is always an eigenvalue with the corresponding eigenvector* $(0,0,0,1)$. *Furthermore, the following assertions are valid.*

1) *If the problem*

(5.4.7) $$(\delta + 6)w = 0, \qquad w \in \overset{\circ}{W}^1_2(\Omega),$$

$$\int_\Omega w \, d\omega = 0$$

has only the trivial solution, then there are no other eigenvectors. If, however, the problem (5.4.7) *has a nontrivial solution* w, *then there exists another eigenvector* (w, u_ω, p), *where* (u_ω, p) *is a solution of* (5.4.1)–(5.4.3) *with the right-hand sides*

$$f = 2\partial_\theta w, \qquad g = 2(\sin\theta)^{-1}\partial_\varphi w, \qquad \psi = 3w.$$

There are no further eigenvectors.

2) *Let* $(u^{(0)}, p^{(0)})$ *denote an eigenvector corresponding to the eigenvalue* $\lambda = 1$. *Then for the existence of a generalized eigenvector it is necessary and sufficient that* $u^{(0)} = 0$, $p^{(0)} = const.$ *and that the problem*

(5.4.8) $$(\delta + 6)w = p^{(0)}, \qquad w \in \overset{\circ}{W}^1_2(\Omega),$$

with the constraint

$$\int_\Omega w \, d\omega = 0$$

is solvable. The generalized eigenvector has the form (w, u_ω, p), *where* (u_ω, p) *is a solution of the problem* (5.4.1)–(5.4.3) *with the right-hand side*

$$f = 2\partial_\theta w, \qquad g = 2(\sin\theta)^{-1}\partial_\varphi w, \qquad \psi = 3w.$$

Further generalized eigenvectors do not exist.

Proof: 1) If the vector function (u, p) is an eigenvector, then it follows from (5.2.2) and (5.2.5) that the function $w = u_r$ is a solution of the problem (5.4.7) and the vector function (u_ω, p) can be obtained by solving the boundary value problem (5.4.1)–(5.4.3) with the right-hand sides given above. The remaining part of the statement 1) can be verified easily.

2) Suppose the eigenvector $(u^{(0)}, p^{(0)})$ has a generalized eigenvector $(u^{(1)}, p^{(1)})$. If $u_r^{(0)} = 0$, then it follows from (5.2.3)–(5.2.5) and Lemma 5.2.1 that $u_\omega^{(0)} = 0$ and $p^{(0)} = const.$ We assume that $u_r^{(0)} \neq 0$. Then the function $u_r^{(1)} \in \overset{\circ}{W}^1_2(\Omega)$ satisfies the equation

$$(\delta + 6)u_r^{(1)} = p^{(0)} - \frac{d}{d\lambda}\left(\delta + (\lambda+1)(\lambda+2)\right)u_r^{(0)}\Big|_{\lambda=1}.$$

Now, as $(\delta + 6)u_r^{(0)} = 0$, it holds

(5.4.9) $$0 = \int_\Omega (\delta+6)\, u_r^{(1)}\, \overline{u_r^{(0)}}\, d\omega = \int_\Omega p^{(0)}\, \overline{u_r^{(0)}}\, d\omega - 5\int_\Omega |u_r^{(0)}|^2\, d\omega.$$

Further, inserting

$$\begin{pmatrix} u \\ p \end{pmatrix} = \begin{pmatrix} u^{(0)} \\ p^{(0)} \end{pmatrix}, \quad \begin{pmatrix} v \\ q \end{pmatrix} = \begin{pmatrix} 0 \\ u_\omega^{(0)} \\ 0 \end{pmatrix}$$

and $\lambda = 1$ into (5.3.6), we obtain

(5.4.10)

$$Q(u_\omega^{(0)}, u_\omega^{(0)}) - \int_\Omega |u_\omega^{(0)}|^2 \, d\omega + 2 \int_\Omega u_r^{(0)} \, \nabla_\omega \cdot \overline{u_\omega^{(0)}} \, d\omega - \int_\Omega p^{(0)} \, \nabla_\omega \cdot \overline{u_\omega^{(0)}} \, d\omega = 0.$$

Inserting (5.4.9) into (5.4.10) and utilizing $\nabla_\omega \cdot u_\omega^{(0)} = -3u_r^{(0)}$, we arrive at

$$Q(u_\omega^{(0)}, u_\omega^{(0)}) - \int_\Omega |u_\omega^{(0)}|^2 \, d\omega + 9 \int_\Omega |u_r^{(0)}|^2 \, d\omega = 0.$$

From this from Lemma 3.2.3 we conclude that $u_r^{(0)} = 0$. Consequently, by the first part of the proof, $(u_\omega^{(0)}, p)$ is a solution of the homogeneous problem (5.4.1)–(5.4.3). Lemma 5.4.1 implies $u_\omega^{(0)} = 0$, $p^{(0)} = const$. It can be easily seen that the eigenvector $(u^{(0)}, p^{(0)}) = (0, const)$ admits a generalized eigenvector $(u^{(1)}, p^{(1)})$ if and only if the problem (5.4.8) is solvable. The generalized eigenvector is then obtained as described in the theorem.

Now we show that there are no generalized eigenvectors of order ≥ 2 which correspond to the eigenvector $(u^{(0)}, p^{(0)})$. We assume the contrary. Then the component $u_r^{(2)} \neq 0$ satisfies the relations

(5.4.11) $$(\delta + 6)u_r^{(2)} = p^{(1)} - \frac{d}{d\lambda}\Big(\delta + (\lambda+1)(\lambda+2)\Big)u_r^{(1)}\Big|_{\lambda=1},$$

$u_r^{(2)} \in \mathring{W}_2^1(\Omega)$, and

(5.4.12) $$\int_\Omega u_r^{(2)} \, d\omega = 0.$$

Multiplying (5.4.11) by $\overline{u_r^{(1)}}$, integrating over the domain Ω and using the equalities $(\delta + 6)u_r^{(1)} = p^{(0)} = const$ and (5.4.12), we obtain the equality

$$\int_\Omega p^{(1)} \, \overline{u_r^{(1)}} \, d\omega = 5 \int_\Omega |u_r^{(1)}|^2 \, d\omega.$$

The vector function $(u^{(1)}, p^{(1)})$ satisfies the equations (5.2.3)–(5.2.5) for $\lambda = 1$. Therefore, also (5.4.10) remains true for $(u^{(1)}, p^{(1)})$ and $\nabla_\omega \cdot u_\omega^{(1)} = -3u_r^{(1)}$. Reasoning as above, we arrive at

$$Q(u_\omega^{(1)}, u_\omega^{(1)}) - \int_\Omega |u_\omega^{(1)}|^2 \, d\omega + 9 \int_\Omega |u_r^{(1)}|^2 \, d\omega = 0.$$

From this and from Lemma 3.2.3 we conclude that $u_r^{(1)} = 0$ and, since $(u_\omega^{(1)}, p^{(1)})$ solves the homogeneous problem (5.4.1)–(5.4.3), $u_\omega^{(1)} = 0$. This leads to a contradiction. ∎

The preceding theorem shows that the investigation of the eigenvalue $\lambda = 1$ of the pencil $\mathcal{L}$ is closely related to the following spectral problem:

Find a pair $(w,c) \in \overset{\circ}{W^1_2}(\Omega) \times \mathbb{C}$ such that

(5.4.13) $$(\delta + \mathcal{N})w = c \quad \text{on } \Omega, \quad \int_\Omega w\, d\omega = 0$$

where $\mathcal{N} \in \mathbb{C}$. More precisely, we are interested in the spectral problem for the operator induced by the quadratic form

$$\mathcal{X}(\Omega) \ni w \rightarrow \int_\Omega |\nabla_\omega w|^2\, d\omega,$$

on the space $L_2(\Omega) \ominus \{1\}$, where

(5.4.14) $$\mathcal{X}(\Omega) = \Big\{ w \in \overset{\circ}{W^1_2}(\Omega) : \int_\Omega w\, d\omega = 0 \Big\}.$$

We denote the nondecreasing sequence of nonnegative eigenvalues of this operator by $\{\mathcal{N}_j\}_{j \geq 1}$ and the corresponding eigenfunctions by w_j. Then it is obvious that the pair (w_j, c_j), where c_j is defined by

$$c_j = \int_\Omega \delta w_j\, d\omega,$$

solves (5.4.13) for $\mathcal{N} = \mathcal{N}_j$.

Now Theorem 5.4.1 above implies the following characterization of the algebraic multiplicity of the eigenvalue $\lambda = 1$ of the pencil $\mathcal{L}$.

THEOREM 5.4.2. *Suppose $\mathcal{N} = 6$ is an eigenvalue of problem (5.4.13) with multiplicity $s \geq 0$. Then the algebraic multiplicity of the eigenvalue $\lambda = 1$ of the pencil $\mathcal{L}$ is equal to $s + 1$.*

Furthermore, the eigenvalue $\mathcal{N}_j$ can be characterized by the variational principle

(5.4.15) $$\mathcal{N}_j = \max_{\mathcal{X}_j} \inf_{w \in \mathcal{X}_j \backslash \{0\}} \frac{\displaystyle\int_\Omega |\nabla w|^2\, d\omega}{\displaystyle\int_\Omega |w|^2\, d\omega},$$

where the maximum is taken over all subspaces $\mathcal{X}_j \subset \mathcal{X}(\Omega)$ of codimension $j - 1$. This implies, in particular, that the numbers $\mathcal{N}_j = \mathcal{N}_j(\Omega)$ are nondecreasing as the domain decreases. Moreover, we get $\mathcal{N}_j(\Omega_1) > \mathcal{N}_j(\Omega_2)$ if $\Omega_1 \subset \Omega_2$.

In the case $\Omega = S^2$ the eigenvalue $\mathcal{N}_j$ is equal to the $(j+1)$st eigenvalue of the operator $-\delta$ on S^2. Hence,

(5.4.16) $$\mathcal{N}_1(S^2) = \mathcal{N}_2(S^2) = \mathcal{N}_3(S^2) = 2, \ \mathcal{N}_4(S^2) = 6.$$

Furthermore, in this case the pairs (w_j, c_j) are given by

(5.4.17) $$w_1 = \cos\theta, \ w_2 = \sin\theta \sin\varphi, \ w_3 = \sin\theta \cos\varphi, \ c_1 = c_2 = c_3 = 0.$$

By the monotonicity of $\mathcal{N}_k$ with respect to Ω, we have $\mathcal{N}_k(\Omega) > 6$ for $k \geq 4$ and $\Omega \subset S^2$.

5.4.3. Eigenvectors and generalized eigenvectors corresponding to the eigenvalue $\lambda = 1$ for circular cones. In the following lemma the values $\mathcal{N}_k$ are estimated for domains corresponding to right circular cones, i.e., for the case

$$\Omega = \Omega_\alpha \overset{def}{=} \{\omega \in S^2 \,:\, 0 \leq \theta < \alpha, \, 0 \leq \varphi < 2\pi\}, \qquad \alpha \in (0, \pi].$$

LEMMA 5.4.2. 1) *The eigenvalues* $\mathcal{N}_j$ *satisfy*

(5.4.18) $\mathcal{N}_1(\Omega_{\frac{2\pi}{3}}) = \mathcal{N}_2(\Omega_{\frac{2\pi}{3}}) < 6, \ \mathcal{N}_3(\Omega_{\frac{2\pi}{3}}) = 6,$

(5.4.19) $\mathcal{N}_1(S_+^2) = \mathcal{N}_2(S_+^2) = 6, \ \mathcal{N}_3(S_+^2) > 6.$

2) *Let* $\Omega \subset S^2$. *If* $\Omega_{\frac{2\pi}{3}} \subset \Omega, \ S_+^2 \subset \Omega \subset \Omega_{\frac{2\pi}{3}}$ *or* $\Omega \subset S_+^2$, *then* $\mathcal{N}_k(\Omega) \neq 6$ *for* $k = 1, 2, \ldots$.

Proof: We decompose the space $\overset{\circ}{W}{}^1_2(\Omega_\alpha)$ into an orthogonal sum $\overset{\circ}{W}{}^1_2(\Omega_\alpha) = Y_\alpha \oplus Z_\alpha$, where the space Z_α consists of all axisymmetric functions which are independent of φ. Then all functions in Y_α are orthogonal to 1. We set

$$y(\alpha) = \inf \left\{ \int_{\Omega_\alpha} |\nabla_\omega w|^2 \, d\omega \,:\, w \in Y_\alpha, \, \|w\|_{L_2(\Omega_\alpha)} = 1 \right\}.$$

It is clear that $y(\alpha)$ is an eigenvalue for the Dirichlet problem of the operator $-\delta$ on Ω_α and its multiplicity equals two. Now, since $y(\frac{\pi}{2}) = 6$, we have $y(\frac{2\pi}{3}) < 6$.

We denote by $\{z_j(\alpha)\}_{j \geq 1}$ a nondecreasing sequence of eigenvalues of the operator on $L_2(\Omega)$ which is induced by the quadratic form

$$\left\{ v \in Z_\alpha \,:\, \int_{\Omega_\alpha} v \, d\omega = 0 \right\} \ni w \to \int_{\Omega_\alpha} |\nabla_\omega w|^2 \, d\omega.$$

Relations (5.4.18), (5.4.19) will follow from the monotonicity of $\mathcal{N}_j(\Omega)$ with respect to Ω if we can show that $z_1(\frac{2\pi}{3}) = 6$. By (5.4.16) and (5.4.17), we have $z_1(\pi) = 2$ and $z_2(\pi) \geq 6$. From this and from the strict monotonicity of the functions $z_k(\alpha)$ we infer that $z_1(\alpha) > 2$ and that the interval $(2, 6]$ contains at least $z_1(\alpha)$ for $\alpha \in (0, \pi)$.

Obviously, the pair $(w, c) = (2\cos 2\theta + 1, 2)$ satisfies the equations

$$\frac{1}{\sin\theta}(\sin\theta w')' + 6w = c, \qquad w\left(\frac{2\pi}{3}\right) = 0, \qquad \int_0^{2\pi/3} w \sin\theta \, d\theta = 0.$$

Hence, 6 is among the numbers $z_k(\frac{2\pi}{3})$ and, therefore, $z_1(\frac{2\pi}{3}) = 6$. This shows part 1) of the lemma.

The statement 2) follows from the monotonicity of $\mathcal{N}_1, \mathcal{N}_2, \mathcal{N}_3$ and from (5.4.16), (5.4.18) and (5.4.19). ∎

In the following theorem we give explicit expressions for the eigen- and generalized eigenvectors corresponding to the eigenvalue $\lambda = 1$ in the particular case of right circular cones.

THEOREM 5.4.3. *Let* $\alpha \in (0, \pi)$.

1) *If* $\alpha \neq \frac{\pi}{2}, \frac{2\pi}{3}$, *then the eigenvalue* $\lambda = 1$ *of the pencil* $\mathcal{L}$ *is simple and the corresponding eigenvector is* $(0, 0, 0, 1)$.

2) *If $\alpha = \frac{\pi}{2}$, then exactly three linearly independent eigenvectors correspond to the eigenvalue $\lambda = 1$:*

$$\begin{pmatrix} u_r \\ u_\theta \\ u_\varphi \\ p \end{pmatrix} = \begin{pmatrix} 0 \\ 0 \\ 0 \\ 1 \end{pmatrix}, \quad \begin{pmatrix} \cos\theta\sin\theta\cos\varphi \\ \cos 2\theta\cos\varphi \\ -\sin\theta\sin\varphi \\ 0 \end{pmatrix}, \quad \begin{pmatrix} \cos\theta\sin\theta\sin\varphi \\ \cos 2\theta\sin\varphi \\ \cos\theta\cos\varphi \\ 0 \end{pmatrix}.$$

There are no generalized eigenvectors.

3) *If $\alpha = \frac{2\pi}{3}$, then exactly one eigenvector $(0,0,0,1)$ and the generalized eigenvector*

(5.4.20)
$$\begin{pmatrix} u_r^{(1)} \\ u_\theta^{(1)} \\ u_\varphi^{(1)} \\ p^{(1)} \end{pmatrix} = \begin{pmatrix} \cos 2\theta + \frac{1}{2} \\ (\cos 3\theta - 1)/2\sin\theta \\ 0 \\ \log(\cos\theta + 1) \end{pmatrix}$$

correspond to the eigenvalue $\lambda = 1$

Proof: 1) If $\alpha \neq \frac{\pi}{2}, \frac{2\pi}{3}$ then, by statement 2) of Lemma 5.4.2, the problem (5.4.13) has nontrivial solutions for $\mathcal{N} = 6$. Referring to Theorem 5.4.1 completes the proof of assertion 1).

2) If the cone coincides with the half-space $\mathbb{R}_+^3 = \{(x_1, x_2, x_3) \in \mathbb{R}^3 : x_3 > 0\}$, then the solutions $(U, P) \in W_2^1(\Omega)^3 \times L_2(\Omega)$ of problem (5.0.1), (5.0.2) with zero boundary conditions ($U = 0$ for $x_3 = 0$) of the form (5.0.3) for $\lambda = 1$ are the vector functions $(0,0,0,1)$, $(x_3,0,0,0)$, $(0, x_3, 0, 0)$. This proves 2).

3) By (5.4.18), problem (5.4.13) has a unique solution $(w, c) = (c\cos 2\theta + \frac{c}{2}, c)$. Applying Theorem 5.4.1, we obtain the unique eigenvector $(0,0,0,1)$. Furthermore, we get the existence of a generalized eigenvector. One immediately verifies that it is the one given in (5.4.20). The theorem is proved. ∎

5.4.4. The eigenvalue $\lambda = -2$. From part 4) of Theorem 5.2.1 it follows that $\lambda = -2$ is an eigenvalue of the pencil $\mathcal{L}$ with the same geometric and algebraic multiplicities as the eigenvalue $\lambda = 1$. The following theorem contains information on the eigenvectors and generalized eigenvectors corresponding to this eigenvalue.

THEOREM 5.4.4. *Let $\Omega \neq S^2$. Then the following assertions are valid:*

1) *The number $\lambda = -2$ is always an eigenvalue. The eigenvectors corresponding to this eigenvalue are exhausted by vector functions of the form $(u_r, 0, p)$, where $u_r \in \overset{\circ}{W}_2^1(\Omega)$, $p \in L_2(\Omega)$,*

$$2u_r - p = c \quad and \quad \delta u_r + 6u_r = 3c \quad on \ \Omega$$

with a constant c.

2) *If the problem*

(5.4.21)
$$(\delta + 6)v = -3, \qquad \int_\Omega v \, d\omega = 0$$

has no solutions in $\overset{\circ}{W}_2^1(\Omega)$, then generalized eigenvectors corresponding to the eigenvalue $\lambda = -2$ do not exist.

3) *Let problem (5.4.21) be solvable. Then generalized eigenvectors exist only for the eigenvectors $(u_r^{(0)}, 0, 2u_r^{(0)} + 1)$ (and nonzero multiples of these) such that $\int_\Omega u_r^{(0)} \, d\omega = 0$ and $\int_\Omega u_r^{(0)} \, \overline{w} \, d\omega = 0$ for all $w \in \overset{\circ}{W}_2^1(\Omega)$, $\delta w + 6w = 0$.*

Suppose that the vector function $(u_\omega^{(1)}, p)$ is a solution of the auxiliary problem (5.4.1)–(5.4.3) *with* $f = g = 0$, $\psi = u_r^{(0)}$, *and* $u_r^{(1)} \in \overset{\circ}{W}{}_2^1(\Omega)$ *is a solution of the problem*

$$(5.4.22) \qquad \delta u_r^{(1)} + 6\, u_r^{(1)} = 3\, u_r^{(0)} - 3p + 1,$$

then the vector $(u_r^{(1)}, u_\omega^{(1)}, p + 2u_r^{(1)})$ *is a generalized eigenvector to the eigenvector* $(u_r^{(0)}, 0, 2u_r^{(0)} + 1)$. *In this way all generalized eigenvectors can be represented. Second generalized eigenvectors do not exist.*

Proof: 1) Let (u_r, u_ω, p) be an eigenvector corresponding to the eigenvalue $\lambda = -2$. Then, by Theorem 5.2.2, the vector

$$\mathcal{S}(-2) \begin{pmatrix} u_r \\ u_\omega \\ p \end{pmatrix} = 3 \begin{pmatrix} 0 \\ u_\omega \\ p - 2u_r \end{pmatrix}$$

is either zero or an eigenvector corresponding to the eigenvalue $\lambda = 1$. In both cases, according to Theorem 5.4.1 and Lemma 5.4.1, we have $u_\omega = 0$ and $2u_r - p = c = const$. This implies

$$0 = a \left(\begin{pmatrix} u_r \\ 0 \\ p \end{pmatrix}, \begin{pmatrix} v_r \\ 0 \\ q \end{pmatrix}; -2 \right) = \int_\Omega \left((\nabla_\omega u_r) \cdot \nabla_\omega \overline{v}_r - 3p\,\overline{v}_r \right) d\omega$$

for all $v_r \in \overset{\circ}{W}{}_2^1(\Omega)$ and, consequently, $-\delta u_r = 3p = 3\,(2u_r - c)$.

2) Suppose that the vector functions $(u^{(0)}, p^{(0)})$, $(u^{(1)}, p^{(1)})$ form a Jordan chain corresponding to the eigenvalue $\lambda = -2$. Then

$$(5.4.23) \qquad a \left(\begin{pmatrix} u^{(1)} \\ p^{(1)} \end{pmatrix}, \begin{pmatrix} v \\ q \end{pmatrix}; -2 \right) + a^{(1)} \left(\begin{pmatrix} u^{(0)} \\ p^{(0)} \end{pmatrix}, \begin{pmatrix} v \\ q \end{pmatrix}; -2 \right) = 0$$

for all $v = (v_r, v_\omega) \in \overset{\circ}{W}{}_2^1(\Omega) \times \overset{\circ}{h}{}_2^1(\Omega)$, $q \in L_2(\Omega)$. Here

$$a \left(\begin{pmatrix} u^{(1)} \\ p^{(1)} \end{pmatrix}, \begin{pmatrix} v \\ q \end{pmatrix}; -2 \right) = Q(u_\omega^{(1)}, v_\omega) + \int_\Omega \left((\nabla_\omega u_r^{(1)}) \cdot \nabla_\omega \overline{v}_r - u_\omega^{(1)} \cdot \overline{v}_\omega \right) d\omega$$

$$-2 \int_\Omega \left((\nabla_\omega u_r^{(1)}) \cdot \overline{v}_\omega - (\nabla_\omega \cdot u_\omega^{(1)}) \overline{v}_r \right) d\omega$$

$$- \int_\Omega \left(p^{(1)} \left(\nabla_\omega \cdot \overline{v}_\omega + 3\,\overline{v}_r \right) + (\nabla_\omega \cdot u_\omega^{(1)}) \overline{q} \right) d\omega$$

and, since $u_\omega^{(0)} = 0$,

$$a^{(1)} \left(\begin{pmatrix} u^{(0)} \\ p^{(0)} \end{pmatrix}, \begin{pmatrix} v \\ q \end{pmatrix}; -2 \right) = \int_\Omega \left(3u_r^{(0)} \overline{v}_r + p^{(0)} \overline{v}_r - u_r^{(0)} \overline{q} \right) d\omega.$$

Inserting $v = 0$ into (5.4.23), we obtain

$$(5.4.24) \qquad \nabla_\omega \cdot u_\omega^{(1)} + u_r^{(0)} = 0.$$

If we set $v_r = 0$, $q = 0$ in (5.4.23), we get

$$(5.4.25) \qquad Q(u_\omega^{(1)}, v_\omega) - \int_\Omega \left(u_\omega^{(1)} \cdot \overline{v}_\omega + 2\,(\nabla_\omega u_r^{(1)}) \cdot \overline{v}_\omega + p^{(1)} \nabla_\omega \cdot \overline{v}_\omega \right) d\omega = 0.$$

Furthermore, the equality

$$\int_{\Omega} \left((\nabla_\omega u_r^{(1)}) \cdot \nabla_\omega \overline{v}_r + 2 \left(\nabla_\omega \cdot u_\omega^{(1)} \right) \overline{v}_r - 3 p^{(1)} \, \overline{v}_r + (3 u_r^{(0)} + p^{(0)}) \, \overline{v}_r \right) d\omega = 0$$

holds, if one sets $v_\omega = 0$, $q = 0$ in (5.4.23). From the last equality together with the equalities $p^{(0)} = 2 u_r^{(0)} - c$ and (5.4.24) it follows that

$$(5.4.26) \qquad \int_{\Omega} \left((\nabla_\omega u_r^{(1)}) \cdot \nabla_\omega \overline{v}_r + (3 u_r^{(0)} - 3 p^{(1)} - c) \, \overline{v}_r \right) d\omega = 0$$

Obviously, the equations (5.4.24)–(5.4.25) are equivalent to (5.4.23).

By (5.4.24), (5.4.25), we have

$$(5.4.27) \qquad Q(u_\omega^{(1)}, v_\omega) - \int_{\Omega} \left(u_\omega^{(1)} \cdot \overline{v}_\omega + p^{(1)} \, \nabla_\omega \cdot \overline{v}_\omega + (\nabla_\omega \cdot u_\omega^{(1)}) \, \overline{q} \right) d\omega$$

$$= \int_{\Omega} \left(2 \left(\nabla_\omega u_r^{(1)} \right) \cdot \overline{v}_\omega + u_r^{(0)} \, \overline{q} \right) d\omega$$

for all $v_\omega \in \overset{\circ}{W}{}^1_2(\Omega)$, $q \in L_2(\Omega)$, i.e., $(u_\omega^{(1)}, p^{(1)})$ is a solution of the auxiliary problem (5.4.4) with $f = 2 \partial_\theta u_r^{(1)}$, $g = 2 (\sin\theta)^{-1} \partial_\varphi u_r^{(1)}$, $\psi = u_r^{(0)}$. According to Lemma 5.4.1, this problem is solvable if and only if $\int_\Omega u_r^{(0)} \, d\omega = 0$.

From (5.4.26) it follows that $\delta u_r^{(1)} = 3 u_r^{(0)} - 3 p^{(1)} - c$ or, what is the same,

$$(5.4.28) \qquad \delta u_r^{(1)} + 6 u_r^{(1)} = 3 u_r^{(0)} - 3 p - c$$

with $p = p^{(1)} - 2 u_r^{(1)}$. Furthermore, (5.4.27) is equivalent to

$$Q(u_\omega^{(1)}, v_\omega) - \int_{\Omega} \left(u_\omega^{(1)} \, \overline{v}_\omega + p \, \nabla_\omega \cdot \overline{v}_\omega + (\nabla_\omega \cdot u_\omega^{(1)}) \, \overline{q} \right) d\omega = \int_{\Omega} u_r^{(0)} \, \overline{q} \, d\omega.$$

This means that $(u_\omega^{(1)}, p)$ is a solution of the auxiliary problem (5.4.4) with $f = g = 0$, $\psi = u_r^{(0)}$. Under our assumption on $u_r^{(0)}$, this solution exists.

We show that problem (5.4.28) is solvable if and only if $c \neq 0$ and $u_r^{(0)}$ is orthogonal to all $w \in \overset{\circ}{W}{}^1_2(\Omega)$ such that $\delta w + 6 w = 0$. Let $c = 0$. Then (5.4.28) implies

$$3 \int_{\Omega} (u_r^{(0)} - p) \, \overline{u_r^{(0)}} \, d\omega = \int_{\Omega} (\delta u_r^{(1)} + 6 u_r^{(1)}) \, \overline{u_r^{(0)}} \, d\omega = \int_{\Omega} u_r^{(1)} \, \overline{(\delta u_r^{(0)} + 6 u_r^{(0)})} \, d\omega = 0$$

i.e., $\int_\Omega p \, \overline{u_r^{(0)}} \, d\omega = \int_\Omega |u_r^{(0)}|^2 \, d\omega$. From this and from (5.4.24), (5.4.25) it follows that

$$Q(u_\omega^{(1)}, u_\omega^{(1)}) - \int_{\Omega} |u_\omega^{(1)}|^2 \, d\omega = \int_{\Omega} \left(2 (\nabla_\omega u_r^{(1)}) \cdot \overline{u_\omega^{(1)}} + p^{(1)} \, \nabla_\omega \cdot \overline{u_\omega^{(1)}} \right) d\omega$$

$$= \int_{\Omega} \left(p^{(1)} - 2 u_r^{(1)} \right) \nabla_\omega \cdot \overline{u_\omega^{(1)}} \, d\omega = - \int_{\Omega} p \, \overline{u_r^{(0)}} \, d\omega = - \int_{\Omega} |u_r^{(0)}|^2 \, d\omega.$$

Due to (3.2.11), the left-hand side of the last equality is nonnegative. Consequently, $u_r^{(0)} = 0$ and, according to assertion 1), the vector $(u^{(0)}, p^{(0)})$ is zero. Therefore, for the solvability of (5.4.28) it is necessary that $c \neq 0$. Then $-u_r^{(0)}/c$ is a solution of problem (5.4.21). This proves, in particular, the second assertion of the theorem.

Let w be a function from $\overset{\circ}{W}{}_2^1(\Omega)$ satisfying the equality $\delta w + 6w = 0$. Then, by the solvability of problem (5.4.21), we have $\int_\Omega w \, d\omega = 0$. Hence, the equation (5.4.28) is solvable if and only if

$$(5.4.29) \qquad \int_\Omega (u_r^{(0)} - p) \, \overline{w} \, d\omega = 0$$

for all $w \in \overset{\circ}{W}{}_2^1(\Omega)$, $\delta w + 6w = 0$. From (5.4.25) it follows that

$$Q(u_\omega^{(1)}, v_\omega) - \int_\Omega \left(u_\omega^{(1)} \cdot \overline{v}_\omega + p \, \nabla_\omega \cdot v_\omega \right) d\omega = 0.$$

Setting $v_\omega = \nabla_\omega f$, where $f \in C_0^\infty(\Omega)$, and using the equality

$$Q(u_\omega, \nabla_\omega f) = \int_\Omega (\nabla_\omega \cdot u_\omega) \, (\delta \overline{f} + \overline{f}) \, d\omega,$$

we get

$$\int_\Omega \left((\nabla_\omega \cdot u_\omega^{(1)}) \, (\delta \overline{f} + 2\overline{f}) - p \, \delta \overline{f} \right) d\omega = 0.$$

It can be easily verified (by passing to the limit) that the same is true for an arbitrary eigenfunction f of the operator $-\delta$. Therefore, putting $f = w$ and using (5.4.24), we obtain

$$\int_\Omega (4u_r^{(0)} + 6p) \, \overline{w} \, d\omega = 0$$

Consequently, condition (5.4.29) is equivalent to the equality $\int_\Omega u_r^{(0)} \, \overline{w} \, d\omega = 0$. The proof is complete. ∎

5.5. A variational principle for real eigenvalues

Now we derive a variational principle for real eigenvalues of the operator pencil $\mathcal{L}$. This variational principle implies a new characterization of real eigenvalues which allows one, in particular, to deduce that real eigenvalues in the interval $[-\frac{1}{2}, 1)$ are monotonous with respect to the set Ω.

5.5.1. Description of real eigenvalues of the Stokes pencil by the eigenvalues of a selfadjoint operator. Suppose that $\lambda \in [-\frac{1}{2}, 1)$ is an eigenvalue of the pencil $\mathcal{L}$ and (u_r, u_ω, p) is a corresponding eigenvector. Then from (5.2.2) and (5.2.5) we obtain

$$(5.5.1) \qquad u_r = -\frac{1}{\lambda + 2} \, \nabla_\omega \cdot u_\omega,$$

and

$$(5.5.2) \qquad p = \frac{1}{(\lambda + 2)(1 - \lambda)} \, \delta \nabla_\omega \cdot u_\omega + \frac{\lambda + 1}{1 - \lambda} \, \nabla_\omega \cdot u_\omega.$$

Imposing the same relations for (v_r, v_ω, q) and eliminating the functions u_r, v_r, p, q in the sesquilinear form (5.3.2), we arrive at the sesquilinear form

$$b_\omega(u_\omega, v_\omega; \lambda) = Q(u_\omega, v_\omega) + \frac{1}{(1-\lambda)(\lambda+2)} \int_\Omega \nabla_\omega(\nabla_\omega \cdot u_\omega) \cdot \nabla_\omega(\nabla_\omega \cdot \overline{v}_\omega) \, d\omega$$

$$- \frac{\lambda^2 + \lambda + 4}{(1-\lambda)(\lambda+2)} \int_\Omega (\nabla_\omega \cdot u_\omega) \nabla_\omega \cdot \overline{v}_\omega \, d\omega + (1 - \lambda - \lambda^2) \int_\Omega u_\omega \cdot \overline{v}_\omega \, d\omega$$

which is defined for u_ω, v_ω in the space

$$\mathcal{Y}(\Omega) = \{u_\omega \in \overset{\circ}{h}{}_2^1(\Omega) \, : \, \nabla_\omega \cdot u_\omega \in \overset{\circ}{W}{}_2^1(\Omega)\}.$$

Clearly, u_ω satisfies

(5.5.3) $b_\omega(u_\omega, v_\omega, \lambda) = 0$ for arbitrary $v_\omega \in \mathcal{Y}(\Omega)$.

Conversely, if $\lambda \in [-\frac{1}{2}, 1)$ and $u_\omega \in \mathcal{Y}(\Omega)$ solves (5.5.3), then λ is an eigenvalue of the pencil $\mathcal{L}$ and the vector (u_r, u_ω, p), where u_r, p are given by (5.5.1) and (5.5.2), is a corresponding eigenvector. Indeed, the vector u_ω satisfies (5.2.3), (5.2.4), where u_r and p are replaced by the right-hand sides of (5.5.1) and (5.5.2). This together with (5.5.1), (5.5.2) leads to the required assertion.

Therefore, it suffices to consider the eigenvalues and eigenvectors of problem (5.5.3).

In order to apply the results of Section 1.3, we verify that the form

(5.5.4) $b(\cdot, \cdot, \lambda) = (1 - \lambda)(\lambda + 2) \, b_\omega(\cdot, \cdot, \lambda)$

satisfies the conditions (I)–(III) in Section 1.3. In our case $\alpha = -1/2$, $\beta = 1$, $\mathcal{H}_+ = \mathcal{Y}(\Omega)$ and $\mathcal{H} = L_2(\Omega)^2$. Condition (I) is evident. The validity of conditions (II) and (III) is proved in the following two lemmas.

LEMMA 5.5.1. *The form $b(u_\omega, u_\omega, -1/2)$ is positive for $u_\omega \neq 0$.*

Proof: Obviously,

$$b_\omega(u_\omega, u_\omega, -1/2) = Q(u_\omega, u_\omega) + \int_\Omega (|u_\omega|^2 - |\nabla_\omega \cdot u_\omega|^2) \, d\omega$$

$$+ \frac{4}{9} \int_\Omega \left| \nabla_\omega(\nabla_\omega \cdot u_\omega) + \frac{3}{4} u_\omega \right|^2 d\omega.$$

Using (3.2.7), we get

$$b_\omega(u_\omega, u_\omega, -1/2) = \int_\Omega \left| \frac{1}{\sin\theta} \partial_\varphi u_\theta - \cot\theta \, u_\varphi - \partial_\theta u_\varphi \right|^2 d\omega$$

$$+ \frac{4}{9} \int_\Omega \left| \nabla_\omega(\nabla_\omega \cdot u_\omega) + \frac{3}{4} u_\omega \right|^2 d\omega.$$

We show that the last integral does not vanish for $u_\omega \neq 0$. Indeed, suppose that $\nabla_\omega(\nabla_\omega \cdot u_\omega) + \frac{3}{4} u_\omega = 0$. This implies

$$\delta v + \frac{3}{4} v = 0 \quad \text{and} \quad \int_\Omega v \, d\omega = 0,$$

where $v = \nabla_\omega \cdot u_\omega$. This is impossible since

$$\int_{S^2} |\nabla_\omega v|^2 \, d\omega \geq 2 \int_{S^2} |v|^2 \, d\omega \quad \text{for all } v \in W_2^1(S^2), \ \int_{S^2} v \, d\omega = 0.$$

The last inequality follows from the fact that the second eigenvalue of the operator $-\delta$ on the sphere S^2 is equal to two (see Section 2.2.2). ∎

LEMMA 5.5.2. *If* (5.5.3) *is satisfied for certain* $\lambda = \lambda_0 \in (-\frac{1}{2}, 1)$, $u_\omega \in \mathcal{Y}(\Omega)$, *then*

$$\frac{d}{d\lambda} b_\omega(u_\omega, u_\omega, \lambda)\Big|_{\lambda = \lambda_0} < 0.$$

Proof: We have

$$b_\omega(u_\omega, u_\omega, \lambda) = a\left(\binom{u}{p}, J_\lambda \binom{u}{p}; \lambda\right),$$

where u_r, p are defined by (5.5.1) and (5.5.2). Differentiating this equality with respect to λ, we get

$$\frac{d}{d\lambda} b_\omega(u_\omega, u_\omega, \lambda) = \frac{\partial}{\partial \lambda} a\left(\binom{u}{p}, J_\lambda \binom{u}{p}; \lambda\right) - 2\operatorname{Re} \int_\Omega \frac{\partial p}{\partial \lambda} \left(\nabla_\omega \cdot \overline{u}_\omega + (\lambda + 2)\,\overline{u}_r\right) d\omega$$

$$-2\operatorname{Re} \int_\Omega \frac{\partial u_r}{\partial \lambda} \left(c\,\delta\overline{u}_r + c\,\lambda\,(\lambda + 5)\,\overline{u}_r - 2\,\nabla_\omega \cdot \overline{u}_\omega + (\lambda + 2)\,p\right) d\omega.$$

By (5.2.2) and (5.2.5), the last two integrals equal zero. Now the result follows from (5.3.6). ∎

Since the sesquilinear forms b and b_ω are connected by the relation (5.5.4), we obtain the following assertion.

COROLLARY 5.5.1. *The assertions of the preceding lemma are also valid for the form* b.

Now we introduce the functions

(5.5.5) $$\mu_j(\lambda) = \max_{\{V\}} \min_{V \setminus \{0\}} \frac{b(u_\omega, u_\omega; \lambda)}{\int_\Omega |u_\omega|^2 \, d\omega},$$

where the maximum is taken over all subspaces $V \subset \mathcal{Y}$ of codimension $\geq j - 1$. Since conditions (I)–(III) of Section 1.3 are satisfied, the results of Theorems 1.3.2–1.3.4 hold. In particular, we get the following result.

THEOREM 5.5.1. 1) *The number* $\lambda^* \in [-\frac{1}{2}, 1)$ *is an eigenvalue of the pencil* $\mathcal{L}$ *with multiplicity* ν *if and only if there exists an integer* $k \geq 1$ *such that*

(5.5.6) $$\mu_j(\lambda^*) = 0 \quad \text{for } j = k, k+1, ..., k+\nu-1,$$

where μ_j *is as in* (5.5.5). *The corresponding eigenspace is spanned by the functions*

(5.5.7) $$\begin{pmatrix} u_r^{(j)} \\ u_\omega^{(j)} \\ p^{(j)} \end{pmatrix} = \begin{pmatrix} -\dfrac{1}{\lambda^* + 2} \nabla_\omega \cdot u_\omega^{(j)} \\ u_\omega^{(j)} \\ \dfrac{1}{(\lambda^* + 2)(1 - \lambda^*)} \delta\nabla_\omega \cdot u_\omega^{(j)} + \dfrac{\lambda^* + 1}{1 - \lambda^*} \nabla_\omega \cdot u_\omega^{(j)} \end{pmatrix}$$

for $j = k, k+1, \ldots, k + \nu - 1$, *where* $u_\omega^{(j)} = u_\omega^{(j)}(\lambda)$ *denotes the j-th eigenfunction of the variational problem* (5.5.5).

2) *For all* $j, j \geq 1$, *the equation* $\mu_j(\lambda) = 0$ *has not more than one zero in the interval* $[-\frac{1}{2}, 1)$. *If* $\mu_j(\lambda_0) = 0$ *for* $\lambda_0 \in (-\frac{1}{2}, 1)$, *then* $\mu_j(\lambda) > 0$ *for* $\lambda \in [-\frac{1}{2}, \lambda_0)$ *and* $\mu_j(\lambda) < 0$ *for* $\lambda \in (\lambda_0, 1)$.

5.5.2. Variational principles for the smallest positive eigenvalue of the pencil $\mathcal{L}$. Now we consider the smallest real eigenvalue of the pencil $\mathcal{L}$ on the interval $[-1/2, 1)$. Let λ_0 be defined as

(5.5.8)
$$\lambda_0 = \sup\left\{ \lambda \in [-\frac{1}{2}, 1) : \ b_\omega(u_\omega, u_\omega; \mu) > 0 \ \text{ for all } \mu \in [-\frac{1}{2}, \lambda), 0 \neq u_\omega \in \mathcal{Y} \right\},$$

and let $R(u_\omega)$, $u_\omega \in \mathcal{Y}$, be the smallest root of the function

$$\lambda \to b(u_\omega, u_\omega; \lambda)$$

which is located in the interval $[-\frac{1}{2}, 1)$. If such root does not exist, then we set $R(u_\omega) = 1$.

The following assertions are direct consequences of Lemmas 1.3.1 and 1.3.2.

LEMMA 5.5.3. *If* $\lambda_0 < 1$, *then* λ_0 *is the smallest eigenvalue of the pencil* $\mathcal{L}$ *in the interval* $[-1/2, 1)$.

LEMMA 5.5.4. *The smallest real eigenvalue of* $\mathcal{L}$ *in the interval* $[-\frac{1}{2}, 1)$ *is given by*

(5.5.9)
$$\lambda_0 = \inf_{0 \neq u_\omega \in \mathcal{Y}} \{R(u_\omega)\}.$$

Now we define

$$T(u_\omega) = R(u_\omega)\left(R(u_\omega) + 1\right).$$

Then, by the definition of the form b_ω, the quantity $T = T(u_\omega)$ is the smallest nonnegative root of the equation

(5.5.10)
$$T^2 - (\alpha + 3)T + \beta + 2 = 0,$$

where

(5.5.11)
$$\alpha = \frac{Q(u_\omega, u_\omega) + \int_\Omega |\nabla_\omega \cdot u_\omega|^2 \, d\omega}{\int_\Omega |u_\omega|^2 \, d\omega}$$

and

(5.5.12)
$$\beta = \frac{2Q(u_\omega, u_\omega) + \int_\Omega |\nabla_\omega (\nabla_\omega \cdot u_\omega)|^2 \, d\omega - 4\int_\Omega |\nabla_\omega \cdot u_\omega|^2 \, d\omega}{\int_\Omega |u_\omega|^2 \, d\omega}.$$

As a consequence of Lemma 5.5.4, we get the following result.

THEOREM 5.5.2. *The smallest non-negative eigenvalue* λ_0 *of the pencil* $\mathcal{L}$ *in the interval* $[-1/2, 1)$ *can be characterized by the equality*

(5.5.13)
$$\lambda_0(\lambda_0 + 1) = \inf_{0 \neq u_\omega \in \mathcal{Y}} \{T(u_\omega)\}.$$

5.5.3. Consequences of the variational principles. We collect information on eigenvalues of the pencil $\mathcal{L}$ which can be deduced from the variational principles established above. From part 2) of Theorem 1.3.3 we obtain the following assertion.

LEMMA 5.5.5. *Suppose that for some $\mu \in (-\frac{1}{2}, 1)$ there exists a subspace $Y_0 \subset \mathcal{Y}(\Omega)$ of dimension k such that*

$$(5.5.14) \qquad b(u_\omega, u_\omega; \mu) < 0 \quad \text{for all } u_\omega \in Y_0 \backslash \{0\}.$$

Then the interval $[-\frac{1}{2}, \mu)$ contains at least k eigenvalues of the pencil $\mathcal{L}$. Moreover, if k denotes the maximal dimension of the subspaces Y_0 for which the inequality (5.5.14) holds, then the interval $[-\frac{1}{2}, \mu)$ contains exactly k eigenvalues of the pencil $\mathcal{L}$.

In the following lemma we give the analogue of Lemma 5.5.5 for the case $\mu = 1$. We note that in this case

$$b(u_\omega, u_\omega; 1) = \int_\Omega \left(|\nabla_\omega(\nabla_\omega \cdot u_\omega)|^2 - 6|\nabla_\omega \cdot u_\omega|^2 \right) d\omega$$

THEOREM 5.5.3. *Let k be the maximal dimension of the subspace X of $\mathcal{X}(\Omega)$, defined by (5.4.14), for which*

$$(5.5.15) \qquad \int_\Omega (|\nabla_\omega v|^2 - 6|v|^2) \, d\omega < 0 \quad \text{for all } v \in X \backslash \{0\}$$

holds. Then the interval $[-\frac{1}{2}, 1)$ contains exactly k eigenvalues of the pencil $\mathcal{L}$.

Proof: Due to the definition of the number k, there is an orthogonal in $L_2(\Omega)$ decomposition

$$\mathcal{X}(\Omega) = X_k \oplus H_k,$$

where the inequality (5.5.15) holds on X_k and

$$\int_\Omega (|\nabla_\omega v|^2 - 6|v|^2) \, d\omega \geq 0 \quad \text{for all } v \in H_k.$$

Let $v_1, ..., v_k$ be a basis of X_k. According to the second part of Lemma 5.2.1, the equations

$$\nabla_\omega \cdot u_\omega = v_j \quad \text{on } \Omega, \quad 1 \leq j \leq k,$$

are solvable in the space $\overset{\circ}{h}{}^1_2(\Omega)$. We denote the solutions by $u_\omega^{(j)}$. Clearly, $u_\omega^{(j)} \in \mathcal{Y}(\Omega)$. Let Y_k be the linear hull of the vectors $u_\omega^{(j)}$ and let Z_k be the set of elements $u_\omega \in \mathcal{Y}(\Omega)$ for which $\nabla_\omega \cdot u_\omega \in H_k$ holds.

Obviously, both Y_k and Z_k are closed and $Y_k \cap Z_k = \{0\}$. Furthermore, if u_ω is an arbitrary element of $\mathcal{Y}(\Omega)$, v is the projection of the function $\nabla_\omega \cdot u_\omega$ onto X_k, and $u'_\omega \in Y_k$ is the solution of the equation $\nabla_\omega \cdot u'_\omega = v$, then $u_\omega - u'_\omega \in Z_k$. Therefore, the space $\mathcal{Y}(\Omega)$ coincides with the direct sum $Y_k \oplus Z_k$.

We write the form b_ω as follows:

$$b_\omega(u_\omega, u_\omega; \mu) = Q(u_\omega, u_\omega) + \frac{1}{(1-\mu)(\mu+2)} \int_\Omega \left(|\nabla_\omega(\nabla_\omega \cdot u_\omega)|^2 - 6|\nabla_\omega \cdot u_\omega|^2 \right) d\omega$$

$$+ \int_\Omega |\nabla_\omega \cdot u_\omega|^2 \, d\omega - \int_\Omega |u_\omega|^2 \, d\omega + (1-\mu)(\mu+2) \int_\Omega |u_\omega|^2 \, d\omega.$$

On the finite dimensional subspace Y_k we have the estimate

$$Q(u_\omega, u_\omega) + \int_\Omega |u_\omega|^2 \, d\omega \le c \int_\Omega |\nabla_\omega \cdot u_\omega|^2 \, d\omega.$$

From this and from (3.2.11) it follows that for all $\mu \in (0,1)$ which are sufficiently close to 1, the inequalities

$$b_\omega(u_\omega, u_\omega; \mu) \quad < \quad 0 \quad \text{for } u_\omega \in Y_k \backslash \{0\},$$
$$b_\omega(u_\omega, u_\omega; \mu) \quad \ge \quad 0 \quad \text{for } u_\omega \in Z_k$$

are valid. Now the assertion of the theorem follows from Lemma 5.5.5. ∎

Now we prove the monotonicity of the eigenvalues of the pencil $\mathcal{L}$ with respect to the domain Ω

THEOREM 5.5.4. *Let* $\Omega_1 \subset \Omega_2$ *and* $\lambda_j(\Omega_1) < 1$. *Then* $\lambda_j(\Omega_2) \le \lambda_j(\Omega_1)$. *If, moreover,* $\Omega_1 \Subset \Omega_2$, *then* $\lambda_j(\Omega_2) < \lambda_j(\Omega_1)$.

Proof: We apply Theorem 1.3.5. In our case, the role of the forms a and $\hat{a}$ is played by the form b with the domains $\mathcal{Y}(\Omega_2)$ and $\mathcal{Y}(\Omega_1)$, respectively. Since $\mathcal{Y}(\Omega_1) \subset \mathcal{Y}(\Omega_2)$, we obtain that the functions $\lambda_j(\Omega)$ are nonincreasing with respect to Ω (see the first part of Theorem 1.3.5).

In order to apply the second part of Theorem 1.3.5, we have to verify that the equality

$$b(u_\omega, v_\omega; \lambda) = 0 \quad \text{for all } v \in \mathcal{Y}(\Omega_2),$$

where $u_\omega \in \mathcal{Y}(\Omega_1)$ and $\lambda \in (-\frac{1}{2}, 1)$, implies $u_\omega = 0$. From the above equality it follows that u_ω satisfies (5.2.3) and (5.2.4) in the domain Ω_2, where u_r and p are defined by (5.5.1), (5.5.2). This system is elliptic in the sense of Douglis-Nirenberg and has analytic coefficients. Therefore, u_ω is analytic in Ω_2. Since $u_\omega = 0$ in $\Omega_2 \backslash \Omega_1$, we conclude that $u_\omega = 0$. Reference to part 2) of Theorem 1.3.5 completes the proof. ∎

Furthermore, from Theorems 5.5.3 and 5.4.2 we conclude the following result.

COROLLARY 5.5.2. *If* $\mathcal{N}_1(\Omega) \ge 6$, *then the strip* $-\frac{1}{2} < \text{Re}\lambda \le 1$ *contains the single eigenvalue* $\lambda = 1$ *of the pencil* $\mathcal{L}$. *This eigenvalue is simple if* $\mathcal{N}_1(\Omega) > 6$.

Now we give precise results for the number of eigenvalues in the strip $-1/2 < \text{Re }\lambda < 1$ and their multiplicities in dependence on the spherical domain Ω.

THEOREM 5.5.5. 1) *If* $\Omega \subset S_+^2$, *then the strip* $-\frac{1}{2} < \text{Re}\lambda \le 1$ *contains only the simple eigenvalue* $\lambda = 1$.

2) *If* $S_+^2 \subset \Omega \subset \Omega_{2\pi/3}$, *then the strip* $-\frac{1}{2} < \text{Re}\lambda \le 1$ *contains exactly 3 eigenvalues. One of them is* $\lambda = 1$ *with multiplicity one.*

3) *If* $\Omega_{2\pi/3} \subset \Omega$, *then the strip* $-1/2 < \operatorname{Re} \lambda \leq 1$ *contains exactly 4 eigenvalues.* *One of them is* $\lambda = 1$ *with multiplicity one.*

Proof: Item 1) follows from the first assertion of Lemma 5.4.2, from the monotonicity of $\mathcal{N}_k(\Omega)$ with respect to Ω and from Corollary 5.5.2 above. Items 2) and 3) are consequences of the monotonicity of $\mathcal{N}_k(\Omega)$ and of Lemma 5.4.2. ∎

5.5.4. The energy strip. A hydrodynamical analog of Corollary 3.6.1 is the following statement.

THEOREM 5.5.6. *There are no eigenvalues of* $\mathcal{L}$ *in the strip*

$$(5.5.16) \qquad \left| \operatorname{Re} \lambda + \frac{1}{2} \right| \leq \frac{1}{2} + \frac{M}{M+4},$$

where M *is the same as in Corollary* 3.6.1.

Proof: We note that the system (3.6.2), (3.6.3) with $\nu = 1/2$ becomes the Stokes system. Similarly to Section 3.6, we denote by $t(M)$ the smallest positive root of the equation $\phi(t) = 2(2t + 1)$ or, what is the same,

$$t(t+3)(2t+1) - (1-t)(M-t)(M+t+1) = 0.$$

Clearly, $t(M) < \min(1, M)$. Repeating the proof of Theorem 3.6.1 for $\nu = 1/2$, we show the absence of eigenvalues of $\mathcal{L}$ in the strip $\left| \operatorname{Re} \lambda + 1/2 \right| \leq t(M) + 1/2$. It remains to use the estimate $t(M) > M(M+4)^{-1}$ obtained in the proof of Corollary 3.6.1, where $\nu = 1/2$. ∎

We close the present section with a result of the same nature stated in terms of $\mathcal{N}_1$. The case $\mathcal{N}_1 \geq 6$ is treated in Corollary 5.5.2. The following assertion for the case $\mathcal{N}_1 \leq 6$ is a consequence of the variational principle (5.5.13).

THEOREM 5.5.7. *If* $\mathcal{N}_1 \leq 6$, *then the strip*

$$(5.5.17) \qquad |\operatorname{Re} \lambda + \frac{1}{2}| < \frac{1}{2} \left(13 - 4(13 - 2\mathcal{N}_1)^{1/2} \right)^{1/2}$$

does not contain eigenvalues of the pencil $\mathcal{L}$. *This estimate is sharp for* $\mathcal{N}_1 = 2$ *and* $\mathcal{N}_1 = 6$.

Proof: From the definition of the quantities β and $\mathcal{N}_1$ we obtain that

$$(5.5.18) \qquad \beta \geq \frac{2Q(u_\omega, u_\omega) + (\mathcal{N}_1 - 4) \displaystyle\int_\Omega |\nabla_\omega \cdot u_\omega|^2 \, d\omega}{\displaystyle\int_\Omega |u_\omega|^2 \, d\omega}.$$

Now, as $\mathcal{N}_1 \geq 2$, it follows from (5.5.18) and from the inequality (3.2.10) that $\beta \geq 0$.

If the equation (5.5.10) does not have roots in the interval $[0, 2)$ for all $u_\omega \in \mathcal{Y}, u_\omega \neq 0$, then $T(u_\omega) = 2$ and the strip (5.5.17) does not contain eigenvalues of the pencil $\mathcal{L}$.

Suppose now that there exists an element $u_\omega \in \mathcal{Y}\backslash\{0\}$ such that the equation (5.5.10) has a root on $[0, 2)$. Then

$$T(u_\omega) = \frac{1}{2} \left(\alpha + 3 - \left((\alpha - 1)^2 + 4(2\alpha - \beta) \right)^{1/2} \right).$$

By (5.5.11) and (5.5.12), we have

$$2\alpha - \beta \leq (6 - \mathcal{N}_1)\tau,$$

where

$$\tau = \frac{\displaystyle\int_\Omega |\nabla_\omega \cdot u_\omega|^2 \, d\omega}{\displaystyle\int_\Omega |u_\omega|^2 \, d\omega}.$$

Therefore,

(5.5.19) $$T(u_\omega) \geq \frac{1}{2}\left(\alpha + 3 - \left((\alpha-1)^2 + 4(6 - \mathcal{N}_1)\tau\right)^{1/2}\right).$$

Using the inequality (3.2.11), we obtain $\alpha \geq 1 + \tau$. Moreover, from (3.2.10) it follows that $\alpha \geq 2\tau - 1$. Consequently,

$$\alpha \geq (2 - \varepsilon)\tau + 2\varepsilon - 1$$

for all $\varepsilon \in [0,1]$. Due to the monotonicity of the right-hand side of (5.5.19), we get the estimate

$$T(u_\omega) \geq (1 - \frac{\varepsilon}{2})\tau + \varepsilon + 1 - \left(\left((1 - \frac{\varepsilon}{2})\tau + \varepsilon - 1\right)^2 + (6 - \mathcal{N}_1)\tau\right)^{1/2}.$$

We denote the right hand side of this inequality by $f(\tau)$ and remark that this function is monotone. Furthermore, we have $f(0) = 2\varepsilon$ and $f(\infty) = 2 - (6 - \mathcal{N}_1)/(2 - \varepsilon)$. Hence

$$T(u_\omega) \geq \min\{2\varepsilon,\, 2 - (6 - \mathcal{N}_1)/(2 - \varepsilon)\}$$

for all $\varepsilon \in [0,1]$. Since the maximum with respect to ε of the right hand side is attained for

$$\varepsilon = \frac{1}{2}\left(3 - (13 - 2\mathcal{N}_1)^{1/2}\right),$$

we have $T(u_\omega) \geq 3 - (13 - 2\mathcal{N}_1)^{1/2}$. Now it remains to refer to Theorem 5.5.2 to complete the proof. ∎

5.6. Eigenvalues in the case of right circular cones

Here we derive a transcendental equation for the eigenvalues of the pencil $\mathcal{L}$ in the case of a domain on the sphere which corresponds to a right circular cone of solid opening angle 2α.

5.6.1. A transcendental equation for the eigenvalues. Let $\mathcal{K}$ be a right circular cone, i.e., $\Omega = \mathcal{K} \cap S^2$ has the form

$$\Omega = \Omega_\alpha = \{\omega \in S^2 : 0 \leq \theta < \alpha,\ 0 \leq \varphi < 2\pi\}.$$

In this case we can separate variables in (5.2.2)–(5.2.5) as follows

(5.6.1) $$\begin{pmatrix} u_r \\ u_\theta \\ u_\varphi \\ p \end{pmatrix} = \begin{pmatrix} f_m^{(1)}(\theta)\,\cos(m\varphi) \\ g_m^{(1)}(\theta)\,\cos(m\varphi) \\ h_m^{(1)}(\theta)\,\sin(m\varphi) \\ p_m^{(1)}(\theta)\,\cos(m\varphi) \end{pmatrix},$$

$$(5.6.2) \qquad \begin{pmatrix} u_r \\ u_\theta \\ u_\varphi \\ p \end{pmatrix} = \begin{pmatrix} f_m^{(2)}(\theta)\,\sin(m\varphi) \\ g_m^{(2)}(\theta)\,\sin(m\varphi) \\ h_m^{(2)}(\theta)\,\cos(m\varphi) \\ p_m^{(2)}(\theta)\,\sin(m\varphi) \end{pmatrix}$$

where $m = 0, 1, 2, 3, \ldots$. For $m \geq 1$ the vector function (5.6.1) is a solution of (5.2.2)–(5.2.5) if and only if (5.6.2) is a solution of these equations and

$$f_m^{(2)} = f_m^{(1)}, \quad g_m^{(2)} = g_m^{(1)}, \quad h_m^{(2)} = -h_m^{(1)}, \quad p_m^{(2)} = p_m^{(1)}.$$

We derive a transcendental equation for the eigenvalues of the pencil $\mathcal{L}$ from the following representation of solutions (U, P) of the homogeneous system (5.0.1) in terms of three harmonic functions Ψ, Θ, Λ, which is the analogue of the Boussinesq representation [23] of solutions of the homogeneous system of linear three-dimensional elastostatics (see Section 3.3):

$$(5.6.3) \qquad \begin{cases} U = a\nabla\Psi + 2b\nabla \times (\Theta\vec{e}_3) + c(\nabla(x_3\Lambda) - 2\Lambda\vec{e}_3), \\ P = 2c\partial_{x_3}\Lambda, \end{cases}$$

where ∇ is the gradient in Cartesian coordinates, $\vec{e}_3$ denotes the unit vector in x_3 direction and a, b, c are arbitrary constants. In spherical coordinates this representation becomes

$$\begin{pmatrix} U_r \\ U_\theta \\ U_\varphi \end{pmatrix} = a \begin{pmatrix} \partial_r\Psi \\ r^{-1}\partial_\theta\Psi \\ (r\sin\theta)^{-1}\partial_\varphi\Psi \end{pmatrix} + 2\frac{b}{r} \begin{pmatrix} \partial_\varphi\Theta \\ \cot\theta\,\partial_\varphi\Theta \\ -\sin\theta\,\partial_r(r\Theta) - \partial_\theta(\Theta\cos\theta) \end{pmatrix}$$
$$+ c \begin{pmatrix} \cos\theta\left(\partial_r(r\Lambda) - 2\Lambda\right) \\ \partial_\theta(\Lambda\cos\theta) + 2\Lambda\sin\theta \\ \cot\theta\,\partial_\varphi\Lambda \end{pmatrix}$$

and

$$P = 2c\left(\cos\theta\,\partial_r\Lambda - \frac{\sin\theta}{r}\partial_\theta\Lambda\right),$$

respectively. Based on (5.6.1) and (5.6.3), we select the potentials

$$(5.6.4) \qquad \begin{aligned} \Psi_m &= r^{\lambda+1} P_{\lambda+1}^{-m}(\cos\theta)\,\cos(m\varphi), \\ \Theta_m &= r^{\lambda+1} P_{\lambda+1}^{-m}(\cos\theta)\,\sin(m\varphi), \\ \Lambda_m &= r^{\lambda} P_{\lambda}^{-m}(\cos\theta)\,\cos(m\varphi). \end{aligned}$$

Then the functions $f_m^{(1)}, g_m^{(1)}$ and $h_m^{(1)}$ in (5.6.1) take the form

$$(5.6.5) \qquad \begin{pmatrix} f_m^{(1)}(\theta) \\ g_m^{(1)}(\theta) \\ h_m^{(1)}(\theta) \end{pmatrix} = M_\lambda^{(m)}(\theta) \begin{pmatrix} a \\ b \\ c \end{pmatrix},$$

where $M_\lambda^{(m)}(\theta)$ is the matrix with the columns

$$\begin{pmatrix} (\lambda+1) P_{\lambda+1}^{-m} \\ (\lambda+1)\cot\theta\, P_{\lambda+1}^{-m} - (\lambda+1-m)(\sin\theta)^{-1} P_\lambda^{-m} \\ -m(\sin\theta)^{-1} P_{\lambda+1}^{-m} \end{pmatrix},$$

$$\begin{pmatrix} 2m\, P_{\lambda+1}^{-m} \\ 2m\, \cot\theta\, P_{\lambda+1}^{-m} \\ 2(\lambda+1)\, \cot\theta\, P_{\lambda}^{-m} - 2(\lambda+1)\,(\sin\theta)^{-1}\, P_{\lambda+1}^{-m} \end{pmatrix},$$

$$\begin{pmatrix} (\lambda-1)\,\cos\theta\, P_{\lambda}^{-m} \\ (\lambda+m+1)\,\cot\theta\, P_{\lambda+1}^{-m} + \Big(\sin\theta - (\lambda+1)\,\cos\theta\,\cot\theta\Big)\, P_{\lambda}^{-m} \\ -m\,\cot\theta\, P_{\lambda}^{-m} \end{pmatrix}$$

and P_{μ}^{-m} are the associated Legendre functions of first kind. Furthermore,

$$(5.6.6) \qquad p_m^{(1)}(\theta) = 2c\Big(\lambda\cos\theta\, P_{\lambda}^{-m}(\cos\theta) - \sin\theta\,\partial_\theta(P_{\lambda}^{-m}(\cos\theta))\Big).$$

The homogeneous Dirichlet condition for the velocity components u_r, u_θ, u_φ at $\theta = \alpha$ gives the following condition for the existence of nontrivial solutions of (5.6.1):

$$(5.6.7) \qquad \det(M_\lambda^{(m)}(\alpha)) = 0$$

which is obtained from (5.6.1), (5.6.5), (5.6.6) and from the eigenvector

$$(a,b,c) \in \ker(M_\lambda^{(m)}(\alpha)).$$

Condition (5.6.7) is the desired transcendental equation for the eigenvalues λ of the pencil $\mathcal{L}$ which was solved numerically. Figure 11 below shows the eigenvalues $\lambda_1(\Omega_\alpha)$ and $\lambda_2(\Omega_\alpha)$ obtained for $m = 1$ and 0, respectively, in the case $\pi/2 \le \alpha \le \pi$. As we remarked earlier (see Theorem 5.4.1), $\lambda = 1$ is an eigenvalue of $\mathcal{L}$ for all $\alpha \in (0,\pi]$, corresponding to zero velocity U and constant pressure P. It is obtained in (5.6.3) for $\Psi = \Theta = 0$ and $\Lambda = x_3$.

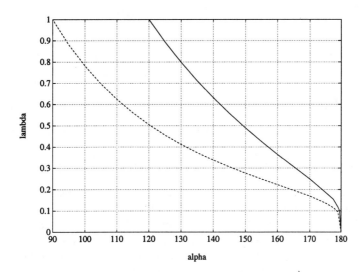

FIGURE 13. Eigenvalues $\lambda_k(\Omega_\alpha) \in [0,1)$ for $\pi/2 \le \alpha \le \pi$. Solid line: simple eigenvalue corresponding to axisymmetric eigenstates, dashed line: twofold eigenvalue corresponding to the first spherical harmonic.

Furthermore, it is interesting to observe that the double eigenvalue $\lambda_1(\Omega_\alpha)$ corresponding to $m = 1$ (with the eigenvectors given by (5.6.1) and (5.6.2)) lies below the eigenvalue $\lambda_2(\Omega_\alpha)$ corresponding to $m = 0$. The latter eigenvalue and $\lambda = 1$ arise from two coalescing simple eigenvalues of the pencil corresponding to the Dirichlet problem of linear elastostatics as the Poisson ratio tends to $1/2$ from below, i.e. in the incompressible limit.

5.6.2. The case $\alpha = 2\pi/3$. By Theorem 5.4.3, there correspond exactly one eigenvector $(0, 0, 0, 1)$ and the generalized eigenvector (5.4.20) to the eigenvalue $\lambda_2(\Omega_{2\pi/3}) = 1$. Figure 11 indicates that $\lambda_1(\Omega_{2\pi/3})$ is either equal or very close to $1/2$. To determine $\lambda_1(\Omega_{2\pi/3})$, we evaluate

$$\det M_{1/2}^{(1)}(\frac{2\pi}{3}) = \frac{1}{12}\Big(17\,(P_{3/2}^{-1}(-\frac{1}{2}))^2\,P_{1/2}^{-1}(-\frac{1}{2})$$
$$+22\,P_{3/2}^{-1}(-\frac{1}{2})\,P_{1/2}^{-1}(-\frac{1}{2}) - 80\,(P_{3/2}^{-1}(-\frac{1}{2}))^3 + (P_{1/2}^{-1}(-\frac{1}{2}))^3\Big).$$

Using

$$P_{3/2}^{-1}(-1/2) = F\Big(\frac{5}{2}, -\frac{3}{2}; 2; \frac{3}{4}\Big), \quad P_{1/2}^{-1}(-1/2) = F\Big(\frac{3}{2}, -\frac{1}{2}; 2; \frac{3}{4}\Big),$$

it was found by Kozlov, Maz'ya and Schwab [**141**], with accurate series expansions for the hypergeometric function, that

$$0 \neq \det M_{1/2}^{(1)}(\frac{2\pi}{3}) = -0.0010545...$$

with all digits shown correct, i.e., $\lambda_1(\Omega_{2\pi/3}) \neq \frac{1}{2}$. A bisection method showed the inclusion

$$0.666496\pi < \alpha^* < 0.666507\pi$$

for the critical angle α^* for which $\lambda_1(\Omega_{\alpha^*}) = 1/2$. Hence $\alpha^* < 2\pi/3$. Furthermore, by means of a bisection method, the inclusion $0.49967 < \lambda_1(\Omega_{2\pi/3}) < 0.49968$ was obtained in [**141**].

5.7. The Dirichlet problem for the Stokes system in a dihedron

Now we consider the special case when the cone $\mathcal{K}$ coincides with the dihedral angle $\mathcal{K} = \mathcal{K}_\alpha \times \mathbb{R}^1$, where $\alpha \in (0, 2\pi]$,

$$\mathcal{K}_\alpha = \{(x_1, x_2) \in \mathbb{R}^2 \,:\, 0 < \rho < \infty, \, \varphi \in (-\alpha/2, \alpha/2)\},$$

and (ρ, φ) are the polar coordinates in the (x_1, x_2)-plane. We construct all solutions to the homogeneous problem (5.0.1), (5.0.2) which have the form (5.0.3). This means that we get a description of all eigenvalues, eigenvectors and generalized eigenvectors of the operator pencil $\mathcal{L} = \mathcal{L}_\alpha$ defined on the subdomain $\Omega_\alpha = \mathcal{K} \cap S^2$ of the unit sphere. The construction of such solutions is done in several steps. First we describe all solutions of the form (5.0.3) with $\operatorname{Re}\lambda > -1/2$ which are independent of x_3. These solutions can be directly expressed by the eigenvectors and generalized eigenvectors of the pencil generated by the plane Stokes problem. Next we give a description of all other solutions of the form (5.0.3) with $\operatorname{Re}\lambda > -1/2$. The solutions of this form with $\operatorname{Re}\lambda < -1/2$ are constructed by means of the relation between the eigenvectors and generalized eigenvectors corresponding to the eigenvalues λ and $-1 - \lambda$ which was given in Theorem 5.2.2. Since this relation is not one-to-one in the case $\lambda = 1$, we consider the eigenvalue $\lambda = -2$ separately in the last subsection.

5.7.1. Special solutions of the homogeneous Stokes system which are independent of x_3.

First we give an explicit description of all solutions of the problem (5.0.1), (5.0.2) which have the form (5.0.3) and are independent of x_3.

Obviously, x_3-independent solutions $U = U(x_1, x_2)$, $P = P(x_1, x_2)$ of the homogeneous system (5.0.1) satisfy the equations

$$(5.7.1) \qquad \begin{cases} -\left(\partial_{x_1}^2 + \partial_{x_2}^2\right)U_1 + \partial_{x_1}P = 0, \\[2mm] -\left(\partial_{x_1}^2 + \partial_{x_2}^2\right)U_2 + \partial_{x_2}P = 0, \\[2mm] \partial_{x_1}U_1 + \partial_{x_2}U_2 = 0 \end{cases}$$

and

$$(5.7.2) \qquad \left(\partial_{x_1}^2 + \partial_{x_2}^2\right)U_3 = 0.$$

in the plane angle $\mathcal{K}_\alpha$. The boundary conditions (5.0.2) on $\partial\mathcal{K}_\alpha\backslash\{0\}$ take the form

$$(5.7.3) \qquad\qquad U_1 = U_2 = 0,$$
$$(5.7.4) \qquad\qquad U_3 = 0.$$

Thus, the system is split into the two-dimensional Stokes system for (U_1, U_2, P) and the Dirichlet problem to the Laplace operator for U_3. We seek all x_3-independent solutions of (5.7.1)–(5.7.2) which have the form

$$U(x_1, x_2) = \rho^\lambda \sum_{k=0}^{s} \frac{1}{(s-k)!}(\log\rho)^{s-k}u^{(k)}(\varphi), \quad u^{(k)} \in \overset{\circ}{W}{}_2^1(\Omega_\alpha)^3,$$

$$P(x_1, x_2) = \rho^{\lambda-1} \sum_{k=0}^{s} \frac{1}{(s-k)!}(\log\rho)^{s-k}p^{(k)}(\varphi), \quad p^{(k)} \in L_2(\Omega_\alpha).$$

As it was shown in Section 2.1, the solutions to the boundary value problem (5.7.2), (5.7.4) of the form

$$U_3 = \rho^\lambda \sum_{k=0}^{s} \frac{1}{(s-k)!}(\log\rho)^{s-k}u_3^{(k)}(\varphi), \quad u_3^{(k)} \in \overset{\circ}{W}{}_2^1(\Omega_\alpha)$$

are exhausted (up to a constant factor) by the functions

$$\rho^{\pm\pi j/\alpha}\sin\frac{\pi j(\varphi + \alpha/2)}{\alpha}, \qquad j = 1, 2, \ldots.$$

Power-logarithmic solutions of the boundary value problem (5.7.1), (5.7.2) have been studied in Section 5.1. Here the polar components

$$U_\rho(\rho, \varphi) = U_1\cos\varphi + U_2\sin\varphi \quad\text{and}\quad U_\varphi(\rho, \varphi) = -U_1\sin\varphi + U_2\cos\varphi$$

were used instead of U_1 and U_2. From (5.7.1), (5.7.3) it follows that the vector function (U_ρ, U_φ, P) is a solution of the system

$$(5.7.5) \qquad \begin{cases} -\frac{1}{\rho^2}\left((\rho\partial_\rho)^2 U_\rho + \partial_\varphi^2 U_\rho - 2\partial_\varphi U_\varphi - U_\rho\right) + \partial_\rho P = 0, \\[2mm] -\frac{1}{\rho^2}\left((\rho\partial_\rho)^2 U_\varphi + \partial_\varphi^2 U_\varphi + 2\partial_\varphi U_\rho - U_\varphi\right) + \frac{1}{\rho}\partial_\varphi P = 0, \\[2mm] \partial_\rho U_\rho + \frac{1}{\rho}\left(U_\rho + \partial_\varphi U_\varphi\right) \qquad\qquad = 0 \end{cases}$$

for $-\alpha/2 < \varphi < \alpha/2$ with the boundary conditions

(5.7.6) $U_\rho(\rho, \varphi) = U_\varphi(\rho, \varphi) = 0$ for $\varphi = \pm\alpha/2$.

By what has been shown in Section 5.1, all solutions of this problem having the form

$$\begin{pmatrix} U_\rho(\rho, \varphi) \\ U_\varphi(\rho, \varphi) \end{pmatrix} = \rho^\lambda \begin{pmatrix} u_\rho(\varphi) \\ u_\varphi(\varphi) \end{pmatrix}, \qquad P(\rho, \varphi) = \rho^{\lambda-1} p(\varphi)$$

are exhausted (up to constant factors) by the functions

(5.7.7) $$\begin{pmatrix} \rho^\lambda u_\rho^\pm(\varphi) \\ \rho^\lambda u_\varphi^\pm(\varphi) \\ \rho^{\lambda-1} p^\pm(\varphi) \end{pmatrix},$$

where λ is a nonzero solution of the equation

$$d_\pm(\lambda) \overset{def}{=} \sin(\lambda\alpha) \pm \lambda \sin\alpha = 0$$

and the vector functions $(u_\rho^\pm, u_\varphi^\pm, p^\pm)$ are given by (5.1.11) and (5.1.12) with ρ instead of r. Furthermore, for all multiple roots of the equations $d_\pm(\lambda) = 0$, the vector functions

(5.7.8) $$\begin{pmatrix} \rho^\lambda \left(u_\rho^\pm \log\rho + v_\rho^\pm \right) \\ \rho^\lambda \left(u_\varphi^\pm \log\rho + v_\varphi^\pm \right) \\ \rho^{\lambda-1} \left(p^\pm \log\rho + q^\pm \right) \end{pmatrix},$$ where $$\begin{pmatrix} v_\rho^\pm \\ v_\varphi^\pm \\ q^\pm \end{pmatrix} = \frac{\partial}{\partial\lambda} \begin{pmatrix} u_\rho^\pm \\ u_\varphi^\pm \\ p^\pm \end{pmatrix},$$

and all linear combinations of the vector functions (5.7.7) and (5.7.8) are solutions of problem (5.7.5), (5.7.6). Other power-logarithmic solutions of this problem do not exist.

Note that the traces of (5.7.7) and (5.7.8) on Ω_α belong to $\overset{\circ}{W}{}_2^1(\Omega_\alpha)$ if $\operatorname{Re}\lambda > 0$. Thus, we obtain the following description of solutions to the homogeneous problem (5.0.1), (5.0.2) which have the form (5.0.3) with $u^{(k)} \in \overset{\circ}{W}{}_2^1(\Omega_\alpha)$, $p^{(k)} \in L_2(\Omega_\alpha)$ and which are independent of x_3.

THEOREM 5.7.1. *All x_3-independent solutions to problem (5.0.1), (5.0.2) of the form (5.0.3) with $\operatorname{Re}\lambda \geq 0$ can be described in the following way:*

1) *Let either $\alpha = \pi$ or $\alpha = 2\pi$. Then such solutions exist only for $\lambda = \pi j/\alpha$, $j = 1, 2, \ldots$. They are linear combinations of the vector-valued functions*

$$\begin{pmatrix} \rho^\lambda u_\rho^- \\ \rho^\lambda u_\varphi^- \\ 0 \\ \rho^{\lambda-1} p^- \end{pmatrix}, \begin{pmatrix} \rho^\lambda u_\rho^+ \\ \rho^\lambda u_\varphi^+ \\ 0 \\ \rho^{\lambda-1} p^+ \end{pmatrix}, \begin{pmatrix} 0 \\ 0 \\ \rho^\lambda \sin\left(\pi j(\varphi + \alpha/2)/\alpha\right) \\ 0 \end{pmatrix}.$$

2) *Let $\alpha \neq \pi$, $\alpha \neq 2\pi$ and let $\lambda = \pi j/\alpha$, $j = 1, 2, \ldots$. Then these solutions are the vector-valued functions*

$$c \begin{pmatrix} 0 \\ 0 \\ \rho^\lambda \sin\left(\pi(\varphi + \alpha/2)/\alpha\right) \\ 0 \end{pmatrix},$$

where c is an arbitrary constant.

3) *Let* $\alpha \neq \pi$, $\alpha \neq 2\pi$ *and let* λ *be a simple root of one of the equations* $d_+(\lambda) = 0$ *or* $d_-(\lambda) = 0$ *(clearly, λ cannot be a root of both equations). Then the solutions are*

$$(5.7.9) \qquad c \begin{pmatrix} \rho^\lambda u_\rho^\pm \\ \rho^\lambda u_\varphi^\pm \\ 0 \\ \rho^{\lambda-1} p^\pm \end{pmatrix}.$$

If λ is a multiple root, then the solutions are linear combinations of the vector functions (5.7.9) *and*

$$\begin{pmatrix} \rho^\lambda \left(u_\rho^\pm \log \rho + v_\rho^\pm \right) \\ \rho^\lambda \left(u_\varphi^\pm \log \rho + v_\varphi^\pm \right) \\ 0 \\ \rho^{\lambda-1} \left(p^\pm \log \rho + q^\pm \right) \end{pmatrix}.$$

There are no other solutions of problem (5.0.1), (5.0.2) *which have the form* (5.0.3) *and which are independent of x_3.*

Given a fixed α, we numerate the above described solutions of the form (5.0.3) which are independent of x_3 and have a finite energy integral in a neighborhood of the origin:

$$(5.7.10) \qquad \begin{pmatrix} U^{(j)}(\rho,\varphi) \\ P^{(j)}(\rho,\varphi) \end{pmatrix} = \begin{pmatrix} \rho^{\lambda_j} \left(u^{(j)}(\varphi) \log \rho + v^{(j)}(\varphi) \right) \\ \rho^{\lambda_j-1} \left(p^{(j)}(\varphi) \log \rho + q^{(j)}(\varphi) \right) \end{pmatrix}, \quad j = 1, 2, \dots,$$

where $\begin{pmatrix} u^{(j)} \\ p^{(j)} \end{pmatrix}$ is either an eigenvector of the pencil $\mathcal{L}$ corresponding to the eigenvalue λ_j or 0. (In the first case $\begin{pmatrix} v^{(j)} \\ q^{(j)} \end{pmatrix}$ is a generalized eigenvector, and in the second case it is an eigenvector). We note that $\operatorname{Re} \lambda_j > 0$.

5.7.2. The case $\operatorname{Re} \lambda > -1/2$. Now we construct solutions to the problem (5.0.1), (5.0.2) which have the form (5.0.3) with $\operatorname{Re} \lambda > -1/2$ and which are not necessarily independent of x_3. First we seek solutions of the form

$$(5.7.11) \qquad \begin{pmatrix} U(x) \\ P(x) \end{pmatrix} = \sum_{k=0}^m x_3^k \begin{pmatrix} \rho^{\mu+m-k} U_k(\varphi, \log \rho) \\ \rho^{\mu+m-1-k} P_k(\varphi, \log \rho) \end{pmatrix}$$

where $m + \mu = \lambda$, $U_k(\varphi, z)$ is a vector-polynomial in z which coefficients are smooth vector functions on $[-\alpha/2, \alpha/2]$, $P_k(\varphi, z)$ is a polynomial in z with smooth coefficients for $\varphi \in [-\alpha/2, \alpha/2]$. Since the traces of the vector-valued function U and of the function P on Ω_α belong to the spaces $W_2^1(\Omega_\alpha)^3$ and $L_2(\Omega_\alpha)$, respectively, it holds $\operatorname{Re} \mu \geq 0$.

We set

$$\Psi^{(k)}(x_1, x_2) = \rho^{\mu+m-k} U_k(\varphi, \log \rho),$$
$$\Phi^{(k)}(x_1, x_2) = \rho^{\mu+m-1-k} P_k(\varphi, \log \rho).$$

Then the equalities (5.7.1) are equivalent to the following system on $\mathcal{K}_\alpha$:

$$
\begin{aligned}
-\left(\partial_{x_1}^2 + \partial_{x_2}^2\right)\Psi_1^{(k)} + \partial_{x_1}\Phi^{(k)} &= (k+2)(k+1)\Psi_1^{(k+2)}, \\
-\left(\partial_{x_1}^2 + \partial_{x_2}^2\right)\Psi_2^{(k)} + \partial_{x_2}\Phi^{(k)} &= (k+2)(k+1)\Psi_2^{(k+2)} \\
\partial_{x_1}\Psi_1^{(k)} + \partial_{x_2}\Psi_2^{(k)} &= -(k+1)\Psi_3^{(k+1)}, \\
-\left(\partial_{x_1}^2 + \partial_{x_2}^2\right)\Psi_3^{(k)} &= -(k+1)\Psi_3^{(k+1)}
\end{aligned}
$$

for $k = 0, \ldots, m$. Here we have set $\Psi^{(k)} = 0$, $\Phi^{(k)} = 0$ if $k > m$. From the boundary conditions (5.0.2) it follows that

$$
\Psi^{(k)} = 0 \quad \text{on } \partial\mathcal{K}_\alpha \backslash \{0\}.
$$

This, in particular, implies that $\left(\Psi_1^{(m)}, \Psi_2^{(m)}, \Phi^{(m)}\right)$ is a solution of the problem (5.7.1), (5.7.3) and $\Psi_3^{(m)}$ is a solution of the problem (5.7.2), (5.7.4). These solutions were described in Theorem 5.7.1. The vector functions $\left(\Psi_1^{(k)}, \Psi_2^{(k)}, \Phi^{(k)}\right)$ and the functions $\Psi_3^{(k)}$, $k = m-1, m-2, \ldots, 1$, can be recurrently calculated from the above systems by means of Theorem 1.1.6.

Let $j = 1, 2, \ldots$ and $m = 0, 1, \ldots$. By $\begin{pmatrix} U^{(j,m)} \\ P^{(j,m)} \end{pmatrix}$ we denote one of the solutions of the form (5.7.11), where the coefficient of x_3^m is equal to the vector-valued function (5.7.10). Using the induction in m we arrive at the following assertion.

LEMMA 5.7.1. *Let $\mu \geq 0$ and let the vector function (5.7.11) be a solution of problem (5.0.1), (5.0.2). Then*

$$
\begin{pmatrix} U(x) \\ P(x) \end{pmatrix} = \sum_{\mu_j + \ell = \mu + m} C_{j,\ell} \begin{pmatrix} U^{(j,\ell)}(x) \\ P^{(j,\ell)}(x) \end{pmatrix}
$$

5.7.3. Two auxiliary assertions.

LEMMA 5.7.2. *Suppose that (U, P) is a vector function of the form (5.0.3) such that $\zeta U \in \overset{\circ}{W}{}_2^1(\mathcal{K})^3$ and $\zeta P \in L_2(\mathcal{K})$ for every function $\zeta \in C_0^\infty(\overline{\mathcal{K}} \backslash \{0\})$. Then $u^{(k)} \in \overset{\circ}{W}{}_2^1(\Omega_\alpha)^3$ and $p^{(k)} \in L_2(\Omega_\alpha)^3$ for $k = 0, 1, \ldots, s$.*

Proof: From (5.0.3) we obtain the algebraic system

$$
\sum_{k=0}^{s} \frac{1}{(s-k)!}\,(\log 2^j r)^{s-k}\, u^{(k)}(\omega) = (2^j r)^{-\lambda_0} U(2^j x), \quad j = 0, 1, \ldots, s,
$$

for the functions $u^{(k)}$. Since the coefficients determinant of this system is nonzero (multiplying this determinant by $1! \cdots s!$, one obtains the Vandermonde determinant), the functions $u^{(k)}$ in (5.0.3) can be represented in terms of the functions $U(2^j x)$, $1 < |x| < 2$, $j = 0, 1, \ldots, s$. Hence, from our assumption on U it follows that $u^{(k)} \in \overset{\circ}{W}{}_2^1(\Omega_\alpha)^3$ for $k = 0, 1, \ldots s$. Analogously, we obtain $p^{(k)} \in L_2(\Omega_\alpha)$. ∎

LEMMA 5.7.3. *If (U, P) is a solution of problem (5.0.1), (5.0.2) of the form (5.0.3), where $\lambda \in \mathbb{C}$, $u^{(k)} \in \overset{\circ}{W}{}_2^1(\Omega_\alpha)^3$, $p^{(k)} \in L_2(\Omega_\alpha)$, then the vector functions $\left(\partial_{x_3}^\ell U, \partial_{x_3}^\ell P\right)$ have the same properties for an arbitrary positive integer ℓ.*

Proof: Let ζ, η be an arbitrary functions from $C_0^\infty(\overline{\mathcal{K}}\backslash\{0\})$ such that $\zeta\eta = \zeta$. We show that $\zeta\partial_{x_3}U \in \overset{\circ}{W}{}_2^1(\mathcal{K})^3$ and $\zeta\partial_{x_3}P \in L_2(\mathcal{K})$. From (5.0.1), (5.0.2) it follows that the functions $V = \zeta U$ and $Q = \zeta P$ satisfy the equations

$$(5.7.12) \qquad -\Delta V + \nabla Q = F, \quad \nabla\cdot V = G \quad \text{in } \mathcal{K},$$

where $F = -2(\nabla\zeta\cdot\nabla)\,U - U\,\Delta\zeta + P\,\nabla\zeta$ and $G = U\cdot\nabla\zeta$. For an arbitrary function Φ in $\mathcal{K}$ let Φ_ε denote the mollification of Φ in x_3 with radius ε:

$$\Phi_\varepsilon(x_1,x_2,\cdot) = \Phi(x_1,x_2,\cdot) * h_\varepsilon\,,$$

where $h_\varepsilon(x_3) = \varepsilon^{-1}\,h(x_3/\varepsilon)$, $h(t) = c\exp\big(-1/(1-t^2)\big)$ for $-1 < t < 1$, $h(t) = 0$ for $|t| \ge 1$ and c is such that $\int_{-1}^{+1} h(t)\,dt = 1$. Then, by (5.7.12), we have

$$-\Delta V_\varepsilon + \nabla Q_\varepsilon = F_\varepsilon\,, \quad \nabla\cdot V_\varepsilon = G_\varepsilon$$

which implies

$$(5.7.13) \qquad -\Delta\,\partial_{x_3}V_\varepsilon + \nabla\,\partial_{x_3}Q_\varepsilon = \partial_{x_3}F_\varepsilon\,, \quad \nabla\cdot\partial_{x_3}V_\varepsilon = \partial_{x_3}G_\varepsilon\,.$$

Multiplying the first equation of (5.7.13) by $\partial_{x_3}\overline{V}_\varepsilon$ and integrating over $\mathcal{K}$, we arrive at

$$\|\nabla\partial_{x_3}V_\varepsilon\|_{L_2(\mathcal{K})^3}^2 + \big(\nabla\partial_{x_3}Q_\varepsilon\,,\partial_{x_3}V_\varepsilon\big)_{\mathcal{K}} = \big(\partial_{x_3}F_\varepsilon\,,\partial_{x_3}V_\varepsilon\big)_{\mathcal{K}}.$$

Since

$$\big(\nabla\partial_{x_3}Q_\varepsilon\,,\partial_{x_3}V_\varepsilon\big)_{\mathcal{K}} = -\big(\partial_{x_3}Q_\varepsilon\,,\nabla\cdot\partial_{x_3}V_\varepsilon\big)_{\mathcal{K}} = -\big(\partial_{x_3}Q_\varepsilon\,,\partial_{x_3}G_\varepsilon\big)_{\mathcal{K}},$$

this implies

$$(5.7.14) \quad \|\partial_{x_3}V_\varepsilon\|_{\overset{\circ}{W}{}_2^1(\mathcal{K})^3}^2 \le \|\partial_{x_3}F_\varepsilon\|_{\overset{\circ}{W}{}_2^{-1}(\mathcal{K})^3}\,\|\partial_{x_3}V_\varepsilon\|_{\overset{\circ}{W}{}_2^1(\mathcal{K})^3} + \big|\big(\partial_{x_3}Q_\varepsilon\,,\partial_{x_3}G_\varepsilon\big)_{\mathcal{K}}\big|.$$

There exists a vector function $W \in \overset{\circ}{W}{}_2^1(\mathcal{K})^3$ with compact support such that $\nabla\cdot W = \partial_{x_3}G_\varepsilon$ and

$$\|W\|_{\overset{\circ}{W}{}_2^1(\mathcal{K})^3} \le c\,\|\partial_{x_3}G_\varepsilon\|_{L_2(\mathcal{K})}\,.$$

Here the constant c is independent of G and ε. Multiplying now the first equation of (5.7.13) by $\overline{W}$ and integrating over $\mathcal{K}$, we obtain

$$\big(\nabla\partial_{x_3}V_\varepsilon\,,\nabla W\big)_{\mathcal{K}} - \big(\partial_{x_3}Q_\varepsilon\,,\partial_{x_3}G_\varepsilon\big)_{\mathcal{K}} = \big(\partial_{x_3}F_\varepsilon\,,W\big)_{\mathcal{K}}$$

and, therefore,

(5.7.15)
$$\big|\big(\partial_{x_3}Q_\varepsilon\,,\partial_{x_3}G_\varepsilon\big)_{\mathcal{K}}\big| \le c\left(\|\partial_{x_3}F_\varepsilon\|_{W_2^{-1}(\mathcal{K})^3} + \|\partial_{x_3}V_\varepsilon\|_{\overset{\circ}{W}{}_2^1(\mathcal{K})^3}\right)\|\partial_{x_3}G_\varepsilon\|_{L_2(\mathcal{K})}\,.$$

From (5.7.14) and (5.7.15) it follows that

$$\|\partial_{x_3}V_\varepsilon\|_{\overset{\circ}{W}{}_2^1(\mathcal{K})^3} \le c\left(\|\partial_{x_3}F_\varepsilon\|_{W_2^{-1}(\mathcal{K})^3} + \|\partial_{x_3}G_\varepsilon\|_{L_2(\mathcal{K})}\right)$$

with a constant c is independent of ε. The right-hand side of the last inequality tends to

$$c\left(\|\partial_{x_3}F\|_{W_2^{-1}(\mathcal{K})^3} + \|\partial_{x_3}G\|_{L_2(\mathcal{K})}\right)$$

as $\varepsilon \to 0$ and can be estimated by the norms of ηU and ηP in $\overset{\circ}{W}{}_2^1(\mathcal{K})^3$ and $L_2(\mathcal{K})$, respectively. Hence $\partial_{x_3}V \in \overset{\circ}{W}{}_2^1(\mathcal{K})^3$. Then, by (5.7.12), we have $\nabla\partial_{x_3}Q = \Delta\partial_{x_3}V + \partial_{x_3}F \in W_2^{-1}(\mathcal{K})^3$ and, therefore, $\partial_{x_3}Q \in L_2(\mathcal{K})$. This together with

Lemma 5.7.2 proves that the vector function $\partial_{x_3}(U,P)$ has the form (5.0.3) with $u^{(k)} \in \overset{\circ}{W}^1_2(\Omega_\alpha)^3$, $p^{(k)} \in L_2(\Omega_\alpha)$. By induction, the same follows for the derivatives $\partial^\ell_{x_3}(U,P)$, $\ell = 2, 3, \ldots$. ∎

5.7.4. The case $\mathrm{Re}\,\lambda > -1/2$. Continuation. Here we describe all solutions to the Dirichlet problem for the homogeneous Stokes system which have the form (5.0.3) with $\mathrm{Re}\,\lambda > -1/2$.

THEOREM 5.7.2. *The vector function (U,P) of the form (5.0.3) with $\mathrm{Re}\,\lambda > -1/2$, $u^{(k)} \in \overset{\circ}{W}^1_2(\Omega_\alpha)^3$, $p^{(k)} \in L_2(\Omega_\alpha)$ is a solution of the problem (5.0.1), (5.0.2) if and only if $\lambda = \mu_\sigma + \ell$ for some $\sigma \geq 1$, $\ell \geq 0$ and (U,P) is a linear combination of the vector functions $(U^{(j,m)}, P^{(j,m)})$ with $\mu_j + m = \lambda$.*

Proof: By Lemma 5.7.3, the following representations are valid for all multi-indices $\beta = (\beta_1, \beta_2, \beta_3)$, $\beta_1 + \beta_2 \leq 1$:

$$(5.7.16) \qquad \partial^\beta_x U(x) = r^{\lambda - |\beta|} \sum_{k=0}^\kappa (\log r)^k \, \phi_k(\omega),$$

$$(5.7.17) \qquad \partial^{\beta_3}_{x_3} P(x) = r^{\lambda - 1 - \beta_3} \sum_{k=0}^\kappa \log r^k \, \psi_k(\omega),$$

where $\phi_k \in W_2^{1 - \beta_1 - \beta_2}(\Omega_\alpha)^3$, $\psi_k \in L_2(\Omega_\alpha)$.

Let $F(x_1, x_2, \zeta)$ and $G(x_1, x_2, \zeta)$ be the Fourier transforms in x_3 of the vector function U and of the function P, respectively. Since

$$(5.7.18) \qquad F(x_1, x_2, \zeta) = |\zeta|^{-2M} \int_\mathbb{R} e^{i\zeta x_3} (-\partial^2_{x_3})^M U(x_1, x_2, x_3) \, dx_3$$

and

$$(5.7.19) \qquad G(x_1, x_2, \zeta) = |\zeta|^{-2M} \int_\mathbb{R} e^{i\zeta x_3} (-\partial^2_{x_3})^M P(x_1, x_2, x_3) \, dx_3,$$

for $\zeta \neq 0$, $M = 0, 1, \ldots$, it follows that the vector-valued function $\zeta \to F(\cdot, \cdot, \zeta)$ belongs to $C^\infty(\mathbb{R}^2 \backslash \{0\}; \overset{\circ}{W}^1_{2,loc}(\mathcal{K}_\alpha, 0)^3)$ and the function $\zeta \to G(\cdot, \cdot, \zeta)$ belongs to $C^\infty(\mathbb{R}^1 \backslash \{0\}; L_{2,loc}(\mathcal{K}_\alpha, 0))$. For all $\zeta \neq 0$ we have

$$(5.7.20) \qquad -\left(\partial^2_{x_1} + \partial^2_{x_2} - |\zeta|^2\right) F + \begin{pmatrix} \partial_{x_1} \\ \partial_{x_2} \\ i\zeta \end{pmatrix} G = 0 \quad \text{on } \mathcal{K}_\alpha,$$

$$(5.7.21) \qquad \partial_{x_1} F_1 + \partial_{x_2} F_2 + i\zeta F_3 = 0 \quad \text{on } \mathcal{K}_\alpha.$$

We show that for each $\zeta \neq 0$ the vector-valued function $F(\cdot, \cdot, \zeta)$ belongs to the space $\overset{\circ}{W}^1_2(\mathcal{K}_\alpha)^3$ and $G(\cdot, \cdot, \zeta)$ belongs to $L_2(\mathcal{K}_\alpha)$. From (5.7.16)–(5.7.19) we get the estimates

$$\int_{-\alpha/2}^{\alpha/2} \left| (\partial^{\beta_1}_{x_1} \partial^{\beta_2}_{x_2} F)(\rho, \varphi, \zeta) \right|^2 d\varphi \leq C_N(\zeta) \, \rho^{-N},$$

$$\int_{-\alpha/2}^{\alpha/2} \left| G(\rho, \varphi, \zeta) \right|^2 d\varphi \leq C_N(\zeta) \, \rho^{-N},$$

where $\beta_1 + \beta_2 \leq 1$, $N = 0, 1, \ldots$. This implies the square summability of the functions $\partial_{x_1}^{\beta_1} \partial_{x_2}^{\beta_2} F$ and G on the set $\{(x_1, x_2) \in \mathcal{K}_\alpha : \rho > 1\}$ for all $\zeta \neq 0$.

Let $\chi \in C_0^\infty(\mathbb{R}^1)$, $\chi(x_3) = 0$ for $|x_3| > 1$ and $\chi(x_3) = 1$ for $|x_3| < 1/2$. By F_0, G_0 and F_∞, G_∞ we denote the Fourier transforms in x_3 of the functions χU, χP and $(1 - \chi)U$, $(1 - \chi)P$, respectively. Then for $\beta_1 + \beta_2 \leq 1$ we have

$$(5.7.22) \qquad \int\limits_{C_\alpha} \int\limits_{\mathbb{R}} \left| \partial_{x_1}^{\beta_1} \partial_{x_2}^{\beta_2} F_0 \right|^2 d\zeta\, dx_1\, dx_2 = c \int\limits_{C_\alpha} \int\limits_{\mathbb{R}} \left| \partial_{x_1}^{\beta_1} \partial_{x_2}^{\beta_2} (\chi U) \right|^2 dx_3\, dx_1\, dx_2,$$

$$(5.7.23) \qquad \int\limits_{C_\alpha} \int\limits_{\mathbb{R}} |G_0|^2 d\zeta\, dx_1\, dx_2 = c \int\limits_{C_\alpha} \int\limits_{\mathbb{R}} |\chi P|^2 dx_3\, dx_1\, dx_2,$$

where $C_\alpha = \{(x_1, x_2) \in K_\alpha : \rho < 1\}$. The condition $\operatorname{Re} \lambda > -1/2$ implies that the right-hand sides of (5.7.22) (5.7.23) are finite.

We represent F_∞, G_∞ in the form

$$F_\infty(x_1, x_2, \zeta) = |\zeta|^{-2M} \int\limits_{\mathbb{R}} e^{i\zeta x_3} \left(-\partial_{x_3}^2 \right)^M \left((1 - \chi)U \right) dx_3,$$

$$G_\infty(x_1, x_2, \zeta) = |\zeta|^{-2M} \int\limits_{\mathbb{R}} e^{i\zeta x_3} \left(-\partial_{x_3}^2 \right)^M \left((1 - \chi)P \right) dx_3,$$

where M is a sufficiently large number. Using Parseval's theorem we get

$$\int\limits_{C_\alpha} \int\limits_{\mathbb{R}} |\zeta|^{4M} \left| \partial_{x_1}^{\beta_1} \partial_{x_2}^{\beta_2} F_\infty \right|^2 d\zeta\, dx_1\, dx_2$$

$$\leq c \sum_{\sigma + \tau = 2M} \int\limits_{C_\alpha} \int\limits_{\mathbb{R}} \left| \partial_{x_3}^\sigma (1 - \chi) \right|^2 \left| \partial_{x_3}^\tau \partial_{x_1}^{\beta_1} \partial_{x_2}^{\beta_2} U \right|^2 dx_3\, dx_1\, dx_2$$

and

$$\int\limits_{C_\alpha} \int\limits_{\mathbb{R}} |\zeta|^{4M} |G_\infty|^2 d\zeta\, dx_1\, dx_2 \leq c \sum_{\sigma + \tau = 2M} \int\limits_{C_\alpha} \int\limits_{\mathbb{R}} \left| \partial_{x_3}^\sigma (1 - \chi) \right|^2 \left| \partial_{x_3}^\tau P \right|^2 dx_3\, dx_1\, dx_2.$$

(The finiteness of the right-hand sides follows from (5.7.16), (5.7.17).) From these estimates we conclude that

$$(5.7.24) \qquad F \in L_{2,loc}\big(\mathbb{R}^1 \backslash \{0\}; \overset{\circ}{W}_2^1(\mathcal{K}_\alpha)^3\big), \quad G \in L_{2,loc}\big(\mathbb{R}^1 \backslash \{0\}; L_2(\mathcal{K}_\alpha)\big).$$

In particular, we have

$$F(\cdot, \cdot, \zeta) \in \overset{\circ}{W}_2^1(\mathcal{K}_\alpha)^3, \quad G(\cdot, \cdot, \zeta) \in L_2(K_\alpha)$$

for almost all ζ.

Multiplying both sides of (5.7.20) by $F(\cdot, \cdot, \zeta)$, integrating by parts and using (5.7.21), we obtain

$$\int\limits_{\mathcal{K}_\alpha} \left(|\partial_{x_1} F|^2 + |\partial_{x_2} F|^2 + |\zeta|^2 |F|^2 \right) dx_1\, dx_2 = 0.$$

Consequently, $F = 0$ and, by (5.7.20) $G = 0$ for almost all ζ. The inclusion (5.7.24) implies $F = 0$, $G = 0$ for $x_3 \neq 0$. Therefore,

$$F(x_1, x_2, \zeta) = \sum f_k(x_1, x_2)\, \delta^{(k)}(\zeta) \quad \text{and} \quad G(x_1, x_2, \zeta) = \sum g_k(x_1, x_2)\, \delta^{(k)}(\zeta).$$

Thus, the functions U and P have the form (5.7.11). Using Lemma 5.7.1, we complete the proof. ∎

REMARK 5.7.1. Using Theorem 5.7.2 one can calculate the algebraic multiplicities of the eigenvalues. In fact, if λ is an eigenvalue of the operator pencil $\mathcal{L}$ with $\operatorname{Re}\lambda > -1/2$, then its algebraic multiplicity is equal to the number of various representations of λ in the form $\lambda = \mu_\sigma + \ell$, where $\sigma \geq 1$, $\ell \geq 0$. Using assertions 3) and 4) of Theorem 5.2.1, we conclude that the eigenvalues $-1 - \lambda$, $\bar{\lambda}$, and $-1 - \bar{\lambda}$ have the same algebraic multiplicity.

5.7.5. Homogeneous solutions of the form (5.0.3) with $\operatorname{Re}\lambda < -1/2$. Next we describe the solutions to problem (5.0.1), (5.0.2) which have the form (5.0.3), where $\operatorname{Re}\lambda < -1/2$. We start with the construction of positively homogeneous solutions of degree λ, $\operatorname{Re}\lambda < -1/2$, $\lambda \neq -2$.

Here we use the fact that the numbers λ and $-1 - \lambda$ are simultaneously eigenvalues and the relation between the corresponding eigenvectors given in Theorem 5.2.2.

Let

$$
\begin{pmatrix} U \\ P \end{pmatrix} = \begin{pmatrix} \rho^\mu\, u_\rho(\varphi) \\ \rho^\mu u_\varphi(\varphi) \\ \rho^\mu\, u_3(\varphi) \\ \rho^{\mu-1}\, p(\varphi) \end{pmatrix}, \qquad \operatorname{Re}\mu > -1/2,\ \mu \neq 1,
$$

be a solution of problem (5.0.1), (5.0.2). We rewrite this solution in the spherical coordinate system (r, θ, φ):

$$
\begin{pmatrix} U_r \\ U_\theta \\ U_\varphi \\ P \end{pmatrix} = \begin{pmatrix} r^\mu\,(\sin\theta)^\mu\,\big(\cos\theta\, u_3(\varphi) + \sin\theta\, u_\rho(\varphi)\big) \\ r^\mu\,(\sin\theta)^\mu\,\big(-\sin\theta\, u_3(\varphi) + \cos\theta\, u_\rho(\varphi)\big) \\ r^\mu\,(\sin\theta)^\mu\, u_\varphi(\varphi) \\ r^{\mu-1}\,(\sin\theta)^{\mu-1}\, p(\varphi) \end{pmatrix}.
$$

According to Theorem 5.2.2, the vector function

$$
\begin{pmatrix} V_r \\ V_\theta \\ V_\varphi \\ Q \end{pmatrix} = \begin{pmatrix} r^{-1-\mu} v_r \\ r^{-1-\mu} v_\theta \\ r^{-1-\mu} v_\varphi \\ r^{-2-\mu} q \end{pmatrix},
$$

where

(5.7.25)
$$
\begin{pmatrix} v_r \\ v_\theta \\ v_\varphi \end{pmatrix} = \begin{pmatrix} (\mu + 2)\,(\sin\theta)^\mu\,\big(\cos\theta\, u_3(\varphi) + \sin\theta\, u_\rho(\varphi)\big) \\ (1 - \mu)\,(\sin\theta)^\mu\,\big(-\sin\theta\, u_3(\varphi) + \cos\theta\, u_\rho(\varphi)\big) \\ (1 - \mu)\,(\sin\theta)^\mu\, u_\varphi(\varphi) \end{pmatrix}
$$

and

(5.7.26)
$$
q = (\sin\theta)^{\mu-1}\Big((1 - \mu)\, p(\varphi) + (2 + 4\mu)\,\sin\theta\,\big(\cos\theta\, u_3(\varphi) + \sin\theta\, u_\rho(\varphi)\big)\Big),
$$

is a solution of the problem (5.0.1), (5.0.2). Consequently, the vector function $(v_r, v_\theta, v_\varphi, q)$ is an eigenvector corresponding to the eigenvalue $\lambda = -1 - \mu$. By Lemma 5.7.3, the derivatives of the vector function $(V_r, V_\theta, V_\varphi, Q)$ with respect to x_3 are also solutions of (5.0.1), (5.0.2). In order to calculate these derivatives, we rewrite the vector $(V_r, V_\theta, V_\varphi, Q)^t$ in the form

$$
\begin{pmatrix}
r^{-2-2\mu}\,(\mu+2)\,\rho^\mu\big(x_3\,u_3(\varphi)+\rho\,u_\rho(\varphi)\big) \\
r^{-2-2\mu}\,(1-\mu)\,\rho^\mu\big(-\rho\,u_3(\varphi)+x_3\,u_\rho(\varphi)\big) \\
r^{-1-2\mu}\,(1-\mu)\,\rho^\mu\,u_\varphi(\varphi) \\
r^{-3-2\mu}\,\rho^{\mu-1}\Big((1-\mu)\,r^2\,P(\varphi)+(2+4\mu)\,\rho\,\big(x_3\,u_3(\varphi)+\rho\,u_\rho(\varphi)\big)\Big)
\end{pmatrix}.
$$

Hence

$$
\partial_{x_3}^k
\begin{pmatrix}
V_r \\
V_\theta \\
V_\varphi
\end{pmatrix}
=
\begin{pmatrix}
(\mu+2)\,\rho^\mu\big(-\tfrac{1}{2\mu}\,\pi_{k+1}^{(\mu)}\,u_3(\varphi)+\pi_k^{(\mu+1)}\,\rho\,u_\rho(\varphi)\big) \\
(1-\mu)\,\rho^\mu\big(-\pi_k^{(\mu+1)}\,\rho\,u_3(\varphi)-\tfrac{1}{2\mu}\,\pi_{k+1}^{(\mu)}\,u_\rho(\varphi)\big) \\
(1-\mu)\,\rho^\mu\,\pi_k^{(\mu)}\,u_\varphi(\varphi)
\end{pmatrix}
$$

and $\partial_{x_3}^k Q$ coincides with the function

$$
\rho^{\mu-1}\Big((1-\mu)\,\pi_k^{(\mu+1/2)}\,p(\varphi)-2\rho\,\pi_{k+1}^{(\mu+1/2)}\,u_3(\varphi)+(2+4\mu)\,\rho^2\,\pi_k^{(\mu+3/2)}\,u_\rho(\varphi)\Big),
$$

where we used the notation

(5.7.27)
$$
\pi_k^{(\kappa)}(x)=\partial_{x_3}^k(r^{-2\kappa}).
$$

As a consequence of the Rodrigues formula for the Gegenbauer polynomials (see Whittaker and Watson [267, Sect.10.9]) the following representation for $\pi_k^{(\kappa)}$ holds:

(5.7.28)
$$
\pi_k^{(\kappa)}(x)=r^{-2\kappa-k}\,\Pi_k^{(\kappa)}(\cos\theta)
$$

where

$$
\Pi_k^{(\kappa)}(\cos\theta)=\frac{2^k k!\,(1-k-\kappa)_k}{(1-2k-2\kappa)_k}\sum_{m=0}^{k/2}\frac{(1/2-k-\kappa)_{k-m}}{m!\,(k-2m)!}\,(1-\cos^2\theta)^m\,(2\cos\theta)^{k-2m}
$$

(Here $(\alpha)_n=\alpha\,(\alpha+1)\cdots(\alpha+n-1)$ for $n=1,2,\dots$ and $(\alpha)_0=1$). Consequently,

$$
\partial_{x_3}
\begin{pmatrix}
V_r \\
V_\theta \\
V_\varphi \\
Q
\end{pmatrix}
=
\begin{pmatrix}
r^{-1-\mu-k}\,v_{kr} \\
r^{-1-\mu-k}\,v_{k\theta} \\
r^{-1-\mu-k}\,v_{k\varphi} \\
r^{-2-\mu-k}\,q_k
\end{pmatrix},
$$

where
(5.7.29)

$$
\begin{cases}
v_{kr}=(\mu+2)\,(\sin\theta)^\mu\Big(-\dfrac{1}{2\mu}\,\Pi_{k+1}^{(\mu)}(\cos\theta)\,u_3(\varphi)+\Pi_k^{(\mu+1)}(\cos\theta)\,\sin\theta\,u_\rho(\varphi)\Big), \\[2ex]
v_{k\theta}=(1-\mu)\,(\sin\theta)^\mu\Big(-\Pi_k^{(\mu+1)}(\cos\theta)\,\sin\theta\,u_3(\varphi)-\dfrac{1}{2\mu}\,\Pi_{k+1}^{(\mu)}(\cos\theta)\,u_\rho(\varphi)\Big), \\[2ex]
v_{k\varphi}=(1-\mu)\,(\sin\theta)^\mu\,\Pi_k^{(\mu)}(\cos\theta)\,u_\rho(\varphi), \\[2ex]
q_k=(\sin\theta)^{\mu-1}\Big((1-\mu)\,\Pi_k^{(\mu+1/2)}(\cos\theta)\,p(\varphi)-2\sin\theta\,\Pi_{k+1}^{(\nu+1/2)}(\cos\theta)\,u_3(\varphi) \\[1ex]
\qquad\qquad +(2+4\mu)\,\sin^2\theta\,\Pi_k^{(\mu+3/2)}(\cos\theta)\,u_\rho(\varphi)\Big).
\end{cases}
$$

By Theorem 5.2.2, the vector function

$$
(5.7.30) \qquad
\begin{pmatrix} w_{kr} \\ w_{k\theta} \\ w_{k\varphi} \\ s_k \end{pmatrix}
=
\begin{pmatrix}
(1 - \mu - k)\, v_{kr}(\varphi) \\
(2 + \mu + k)\, v_{k\theta}(\varphi) \\
(2 + \mu + k)\, v_{k\varphi}(\varphi) \\
(2 + \mu + k)\, q_k(\varphi) - (2 + 4\mu + 4k)\, v_{kr}(\varphi)
\end{pmatrix}
$$

is the eigenfunction corresponding to $\lambda = \mu + k$.

5.7.6. Generalized eigenvectors of the operator pencil $\mathcal{L}$. Let $\operatorname{Re}\mu > -1/2$, $\mu \neq 1$, and let

$$
(5.7.31) \qquad
\begin{pmatrix} U \\ P \end{pmatrix}
=
\begin{pmatrix}
\rho^{\mu}\, \big(u(\varphi)\, \log\rho + v(\varphi)\big) \\
\rho^{\mu - 1}\, \big(p(\varphi)\, \log\rho + q(\varphi)\big)
\end{pmatrix}
$$

be a non-trivial solution to the Dirichlet problem for the system (5.0.1). This means that $\mu = \lambda_j$ for some $j \geq 0$, (u, p) is the corresponding eigenvector of the operator pencil $\mathcal{L}$, and (v, q) is its generalized eigenvector.

The solution (5.7.31) takes the following form in the spherical coordinates

$$
\begin{pmatrix} U_r \\ U_\theta \\ U_\varphi \\ P \end{pmatrix}
=
\begin{pmatrix}
r^{\mu}\, \big(\mathring{u}_r\, \log\rho + \mathring{v}_r \big) \\
r^{\mu}\, \big(\mathring{u}_\theta\, \log\rho + \mathring{v}_\theta \big) \\
r^{\mu}\, \big(\mathring{u}_\varphi\, \log\rho + \mathring{v}_\varphi \big) \\
r^{\mu - 1}\, \big(\mathring{p}\, \log\rho + \mathring{q} \big)
\end{pmatrix},
$$

where

$$
(5.7.32) \qquad
\begin{pmatrix} \mathring{u}_r \\ \mathring{u}_\theta \\ \mathring{u}_\varphi \\ \mathring{p} \end{pmatrix}
=
\begin{pmatrix}
(\sin\theta)^{\mu}\, (\cos\theta\, u_3 + \sin\theta\, u_\rho) \\
(\sin\theta)^{\mu}\, (-\sin\theta\, u_3 + \cos\theta\, u_\rho) \\
(\sin\theta)^{\mu} u_\varphi \\
(\sin\theta)^{\mu - 1} p
\end{pmatrix},
$$

$$
(5.7.33) \qquad
\begin{pmatrix} \mathring{v}_r \\ \mathring{v}_\theta \\ \mathring{v}_\varphi \\ \mathring{q} \end{pmatrix}
=
\begin{pmatrix}
(\sin\theta)^{\mu}\, (\cos\theta\, v_3 + \sin\theta\, v_\rho) \\
(\sin\theta)^{\mu}\, (-\sin\theta\, v_3 + \cos\theta\, v_\rho) \\
(\sin\theta)^{\mu} v_\varphi \\
(\sin\theta)^{\mu - 1} q
\end{pmatrix}.
$$

According to Theorem 5.2.2, the vector $(V_r, V_\theta, V_\varphi, Q)$, where

$$
\begin{aligned}
V_r &= r^{-1-\mu} \Big((\mu + 2)\, (2\, \mathring{u}_r\, \log r - \mathring{u}_r\, \log\rho - \mathring{v}_r\,) - \mathring{u}_r\, \Big), \\
V_\theta &= r^{-1-\mu} \Big((1 - \mu)\, (2\, \mathring{u}_\theta\, \log r - \mathring{u}_\theta\, \log\rho - \mathring{v}_\theta\,) + \mathring{u}_\theta\, \Big), \\
V_\varphi &= r^{-1-\mu} \Big((1 - \mu)\, (2\, \mathring{u}_\varphi\, \log r - \mathring{u}_\varphi\, \log\rho - \mathring{v}_\varphi\,) + \mathring{u}_\varphi\, \Big), \\
Q &= r^{-2-\mu} \Big((1 - \mu)\, (2\, \mathring{p}\, \log r - \mathring{p}\, \log\rho - \mathring{q}\,) \\
&\qquad + (2 + 4\mu)\, (2\, \mathring{u}_r\, \log r - \mathring{u}_r\, \log\rho - \mathring{v}_r\,) + \mathring{p} - 4\, \mathring{u}_r\, \Big)
\end{aligned}
$$

is a solution of the problem (5.0.1), (5.0.2). This implies that the vector function $\left(v_r^{(1)}, v_\theta^{(1)}, v_\varphi^{(1)}, q^{(1)}\right)$, where

$$
\begin{aligned}
v_r^{(1)} &= -(\sin\theta)^\mu \Big((\cos\theta\, u_z + \sin\theta\, u_\rho)\,(1 + (\mu+2)\log\sin\theta) \\
&\qquad + (\mu+2)\,(\cos\theta\, v_3 + \sin\theta\, v_\rho) \Big),
\end{aligned}
$$

$$
\begin{aligned}
v_\theta^{(1)} &= (\sin\theta)^\mu \Big((-\sin\theta\, u_3 + \cos\theta\, u_\rho)\,(1 - (1-\mu)\log\sin\theta) \\
&\qquad - (1-\mu)\,(-\sin\theta\, u_3 + \cos\theta\, v_\rho) \Big),
\end{aligned}
$$

$$
v_\phi^{(1)} = (\sin\theta)^\mu \Big(u_\varphi\,(1 - (1-\mu)\log\sin\theta) - (1-\mu)\,v_\varphi \Big),
$$

$$
\begin{aligned}
q^{(1)} &= (\sin\theta)^{\mu-1} \Big(p\,(1 - (1-\mu)\log\sin\theta) \\
&\qquad - \sin\theta\,(\cos\theta\, u_3 + \sin\theta\, u_\rho)\,(4 + (2+4\mu)\log\sin\theta) \\
&\qquad - (1-\mu)\,q - (2+4\mu)\,\sin\theta\,(\cos\theta\, v_3 + \sin\theta\, v_\rho) \Big),
\end{aligned}
$$

is the generalized eigenvector for the eigenvector $(v_r, v_\theta, v_\varphi, q)$ defined by (5.7.25), (5.7.26) corresponding to the eigenvalue $\lambda = -1 - \mu$.

In order to construct generalized eigenvectors corresponding to the eigenvalue $-1 - \mu - k$, $k = 1, 2, \ldots$, one can apply the operator $\partial_{x_3}^k$ to the vector function $(V_r, V_\theta, V_\varphi, Q)$. We note that the vector $(V_r, V_\theta, V_\varphi)$ can be written in the form

$$
r^{-2-2\mu} \big((\alpha x_3 + \beta)\log r + \gamma x_3 + \delta \big),
$$

where $\alpha, \beta, \gamma, \delta$ are vectors depending only on the variable (x_1, x_2). Similarly, we can write

$$
Q = r^{-3-2\mu}\big((\alpha_1 x_3 + \beta_1)\log r + \gamma_1 x_3 + \delta_1 \big) + r^{-1-2\mu}\big(\alpha_2 \log r + \beta_2 \big),
$$

where $\alpha_1, \beta_1, \gamma_1, \delta_1, \alpha_2, \beta_2$ depend only on (x_1, x_2). Now we can differentiate with respect to x_3 using (5.7.27). For example,

$$
\partial_{x_3}^k V = \alpha\,\partial_\mu\Big(\frac{\pi_{k+1}^{(\mu)}}{4\mu}\Big) - \beta\,\partial_\mu\Big(\frac{\pi_k^{(\mu+1)}}{2}\Big) - \gamma\,\frac{\pi_{k+1}^{(\mu)}}{2\mu} + \delta\,\pi_k^{(\mu+1)}.
$$

In a similar way, we can express $\partial_{x_3}^k Q$. Then we use the formula (5.7.28) for $\pi_k^{(\kappa)}$. After certain algebraic transformations, we get the solution in the form

$$
\partial_{x_3}^k \begin{pmatrix} V \\ Q \end{pmatrix} = \begin{pmatrix} r^{-1-\mu-k}\,\big(v_k(\omega)\log r + v_k^{(1)}(\omega) \big) \\ r^{-2-\mu-k}\,\big(q_k(\omega)\log r + q_k^{(1)}(\omega) \big) \end{pmatrix}.
$$

The eigenvector (v_k, q_k) is defined by (5.7.29); an explicit representation for the generalized eigenvector $(v_k^{(1)}, q_k^{(1)})$ is rather complicated and we omit it here.

If $\lambda = \mu + k$, then, by Theorem 5.2.2, the eigenvector (5.7.30) has the generalized eigenvector

$$
\begin{pmatrix}
(\mu + k - 1)\,v_{kr}^{(1)} - v_{kr} \\
-(2 + \mu + k)\,v_{k\theta}^{(1)} + v_{k\theta} \\
-(2 + \mu + k)\,v_{k\varphi}^{(1)} + v_{k\varphi} \\
-(2 + \mu + k)\,q_k^{(1)} + q_k + (4\mu + 4k + 2)\,v_{kr}^{(1)} - 4v_{kr}
\end{pmatrix}.
$$

In this way we have obtained a complete description of the set of the eigenvectors and generalized eigenvectors corresponding to eigenvalues $\lambda \neq -2$ (see Remark 5.7.1).

5.7.7. The eigenvalue $\lambda = -2$. For $\mu = 1$ the formulas (5.7.25), (5.7.26) give only the eigenfunctions of the operator pencil $\mathcal{L}$ corresponding to the eigenvalue $\lambda = -2$ which have the form $(u_r, 0, 0, 2u_r)$. In order to find other eigenfunctions we solve the equation $\mathcal{L}(-2)\begin{pmatrix} u \\ p \end{pmatrix} = 0$ directly. For this purpose, we use Theorem 5.4.4.

First we find those $\alpha \in (0, 2\pi]$ for which 6 is an eigenvalue of the Dirichlet problem for the operator $-\delta$ in Ω_α and determine the corresponding eigenfunctions. The number 6 is an eigenvalue of $-\delta$ if 2 is an eigenvalue of the pencil $-\delta - \lambda(\lambda+1)$ which is generated by the Dirichlet problem for the Laplace equation in the dihedron $\mathcal{K}_\alpha \times \mathbb{R}$ (see Section 2.2). Applying Theorem 10.5.3, we obtain $j + k\pi/\alpha = 2$ for some integer $j \geq 0$, $k > 0$. Therefore, α takes one of the values $\pi/2$, π, $3\pi/2$, 2π and k may be either 0 or 1. Thus, there are only two functions $\rho^2 \sin(2\varphi + k\pi/2)$ and $x_3 \rho \sin(\varphi + k\pi/2)$ which linear combinations may give the eigenfunctions. These eigenfunctions are given in the following table.

α	$\pi/2$	π	$3\pi/2$	2π
v_1	$(\sin\theta)^2 \cos 2\varphi$	$(\sin\theta)^2 \sin 2\varphi$	$(\sin\theta)^2 \cos 2\varphi$	$(\sin\theta)^2 \sin 2\varphi$
v_2		$\cos\theta \sin\theta \cos\varphi$		$\cos\theta \sin\theta \sin\varphi$

Only the eigenfunctions in the table which correspond to the angles $\pi/2$ and $3\pi/2$ are not orthogonal to 1.

In order to use Theorem 5.4.4, it is necessary to find the solution of the Dirichlet problem

$$(\delta + 6)v = -3 \quad \text{on } \Omega, \quad v = 0 \quad \text{on } \partial\Omega$$

for $\alpha \neq \pi/2$ and $\alpha \neq 3\pi/2$. If we write this solution in the form $v = (\sin\theta)^2 w(\varphi)$, we get the problem

$$w''(\varphi) + 4w(\varphi) = -3 \quad \text{for } |\varphi| < \alpha/2, \qquad w(\pm\alpha/2) = 0.$$

Consequently,

$$(5.7.34) \qquad v(\omega) = -\frac{3}{4}(\sin\theta)^2 \left(1 - \frac{\cos 2\varphi}{\cos\alpha}\right).$$

The orthogonality of v to the corresponding eigenfunctions for $\alpha = \pi$ or $\alpha = 2\pi$ can be readily verified. From this and from Theorem 5.4.4 we obtain the following description of the subspace of eigenvectors corresponding to the the eigenvalue $\lambda = -2$.

LEMMA 5.7.4. 1) *If $\alpha \notin \{\pi/2, \pi, 3\pi/2, 2\pi\}$, then the eigenfunctions of the operator pencil $\mathcal{L}$ corresponding to the eigenvalue $\lambda = -2$ are multiples of the vector*

$$(5.7.35) \qquad \begin{pmatrix} u_r^{(0)} \\ u_\omega^{(0)} \\ p^{(0)} \end{pmatrix} = \begin{pmatrix} -\frac{3}{4}(\sin\theta)^2\left(1 - \frac{\cos 2\varphi}{\cos\alpha}\right) \\ 0 \\ -\frac{3}{2}(\sin\theta)^2\left(1 - \frac{\cos 2\varphi}{\cos\alpha}\right) + 1 \end{pmatrix}.$$

2) *If either $\alpha = \pi/2$ or $\alpha = 3\pi/2$, then the eigenvectors are multiples of*

$$\begin{pmatrix} u_r^{(0)} \\ u_\omega^{(0)} \\ p^{(0)} \end{pmatrix} = \begin{pmatrix} (\sin\theta)^2 \cos 2\varphi \\ 0 \\ 2(\sin\theta)^2 \cos 2\varphi \end{pmatrix}.$$

3) *If $\alpha = \pi$, then the subspace of the eigenvectors $(u_r^{(0)}, u_\omega^{(0)}, p^{(0)})$ is generated by the vector (5.7.35) and by the vectors*

$$\begin{pmatrix} (\sin\theta)^2 \sin 2\varphi \\ 0 \\ 2(\sin\theta)^2 \sin 2\varphi \end{pmatrix}, \quad \begin{pmatrix} \cos\theta \sin\theta \cos\varphi \\ 0 \\ 2\cos\theta \sin\theta \cos\varphi \end{pmatrix}.$$

The same is valid for $\alpha = 2\pi$ if $\cos\varphi$ is replaced by $\sin\varphi$ in the formula for the last vector.

The eigenvectors corresponding to $\lambda = -2 - k$ are

$$r^{-2+k} \partial_{x_3}^k \left(r^2 \left(u_r^{(0)}, u_\omega^{(0)}, p^{(0)} \right) \right)$$

and the eigenvectors corresponding to $\lambda = 1 + k$ are

$$r^{-2+k} \partial_{x_3}^k \left(r^2 S(-2-k) \left(u_r^{(0)}, u_\omega^{(0)}, p^{(0)} \right) \right).$$

Now we deal with the generalized eigenvectors corresponding to the eigenvalue $\lambda = -2$. Since the matrix $S(-1-\lambda)$ degenerates for $\lambda = -2$, Remark 5.2.1 gives no way to find generalized eigenvectors. By Theorem 5.4.4, only the eigenvector $(v, 0, 0, 2v + 1)^t$ (and nonzero multiples of this), where v is defined by (5.7.29), may have a generalized eigenvector. Moreover, it exists only under the condition of orthogonality of v to 1 in $L_2(\Omega)$. From (5.7.34) it follows that this condition is valid for $\alpha = \alpha_*$, where α_* is the unique solution of the equation $\tan\alpha = \alpha$ on the interval $(0, 2\pi)$. As it has been mentioned above, the number 6 is not an eigenvalue of the Dirichlet problem for the operator $-\delta$ on Ω_{α_*}.

Now we turn to a description of the generalized eigenvector. Let the vector $\left(w_r^{(1)}, w_\theta^{(1)}, w_\varphi^{(1)}, s^{(1)} \right)$ be a generalized eigenvector associated to $(v, 0, 0, 2v + 1)$. Theorem 5.2.2 implies that

$$(5.7.36) \quad \begin{pmatrix} 0 & 0 & 0 & 0 \\ 0 & -3 & 0 & 0 \\ 0 & 0 & -3 & 0 \\ 6 & 0 & 0 & -3 \end{pmatrix} \begin{pmatrix} w_r^{(1)} \\ w_\theta^{(1)} \\ w_\varphi^{(1)} \\ s^{(1)} \end{pmatrix} + \begin{pmatrix} -1 & 0 & 0 & 0 \\ 0 & 1 & 0 & 0 \\ 0 & 0 & 1 & 0 \\ -4 & 0 & 0 & 1 \end{pmatrix} \begin{pmatrix} v \\ 0 \\ 0 \\ 2v + 1 \end{pmatrix}$$

is a generalized eigenvector corresponding to the eigenvector

$$\begin{pmatrix} 0 & 0 & 0 & 0 \\ 0 & 3 & 0 & 0 \\ 0 & 0 & 3 & 0 \\ -6 & 0 & 0 & 3 \end{pmatrix} \begin{pmatrix} v \\ 0 \\ 0 \\ 2v + 1 \end{pmatrix} = \begin{pmatrix} 0 \\ 0 \\ 0 \\ 3 \end{pmatrix}$$

and to the eigenvalue $\lambda = 1$. By (5.7.32), (5.7.33), and (5.1.17), the generalized eigenvector has the form

$$\begin{pmatrix} \frac{3}{4}\sin^2\theta\left(1 - \frac{\cos 2\varphi}{\cos\alpha_*}\right) \\ \frac{3}{4}\sin\theta\cos\theta\left(1 - \frac{\cos 2\varphi}{\cos\alpha_*}\right) \\ \frac{3}{4}\sin\theta\left(-2\varphi + \frac{\sin 2\varphi}{\cos\alpha_*}\right) \\ 3 \end{pmatrix}.$$

Since this vector can differ from (5.7.36) only by the eigenvector $(0,0,0,c)$, it follows that

$$w_\theta^{(1)} = -\frac{1}{4}\sin\theta\cos\theta\left(1 - \frac{\cos 2\varphi}{\cos\alpha_*}\right),$$

$$w_\varphi^{(1)} = -\frac{1}{4}\sin\theta\left(-2\varphi + \frac{\sin 2\varphi}{\cos\alpha_*}\right),$$

$$3\left(2w_r^{(1)} - s^{(1)}\right) + 1 - 2v = c.$$

Using the last equality and (5.4.22), where $u_r^{(1)}, p^{(1)}, u_r^{(0)}, p^{(0)}$ are replaced by $w_r^{(1)}$, $s^{(1)}, v, 2v+1$, respectively, we obtain the equation

$$(\delta + 6)\, w_r^{(1)} = 5v + c \quad \text{in } \Omega.$$

Consequently, the generalized eigenvector corresponding to $(v, 0, 0, 2v+1)$ is

$$\begin{pmatrix} w_r^{(1)} \\ w_\theta^{(1)} \\ w_\varphi^{(1)} \\ s^{(1)} \end{pmatrix} = \begin{pmatrix} 5h \\ -\frac{1}{4}\sin\theta\cos\theta\left(1 - \frac{\cos 2\varphi}{\cos\alpha_*}\right) \\ -\frac{1}{4}\sin\theta\left(-2\varphi + \frac{\sin 2\varphi}{\cos\alpha_*}\right) \\ 10h + \frac{1}{3} - \frac{2}{3}v \end{pmatrix} - \frac{c}{3}\begin{pmatrix} v \\ 0 \\ 0 \\ 2v+1 \end{pmatrix},$$

where h is a solution of the boundary value problem

$$(\delta + 6)\, h = v \quad \text{in } \Omega, \qquad h = 0 \quad \text{on } \partial\Omega$$

and v is defined by (5.7.34). This problem is uniquely solvable, since 6 is not an eigenvalue of the operator $-\delta$ for $\alpha = \alpha_*$.

5.8. Stokes and Navier–Stokes systems in domains with piecewise smooth boundaries

It is well known (see, e.g., the books by Girault, Raviart [69] and Ladyzhen-skaya [148]) that the problem

(5.8.1) $-\Delta U + \nabla P = F, \ \nabla \cdot U = 0 \text{ in } \mathcal{G}, \quad U = 0 \ \text{ on } \partial\mathcal{G}$

in an arbitrary bounded domain $\mathcal{G}$ of dimension $n = 2$ or $n = 3$ has a solution $(U, P) \in \overset{\circ}{W}{}_2^1(\mathcal{G})^n \times L_2(\mathcal{G})$ for arbitrary $F \in W_2^{-1}(\mathcal{G})^n$. The velocity vector U is uniquely determined and the pressure P is unique up to a constant term. In this section we suppose that $\mathcal{G}$ is a domain of polygonal or polyhedral type. We apply results on operator pencils obtained in the present chapter to derive various regularity properties of the solutions (U, P).

5.8.1. The Stokes system in plane domains with corners. Let $\mathcal{G}$ be a bounded plane domain with d angle vertices $x^{(1)}, \ldots, x^{(d)}$ on the boundary. Outside the union $\mathcal{S} = \{x^{(1)}, \ldots, x^{(d)}\}$, the boundary $\partial\mathcal{G}$ is assumed to be smooth. We suppose that for every $x^{(\tau)}$ there exist a neighborhood $\mathcal{U}_\tau$ where $\mathcal{G}$ coincides with a plane angle $\mathcal{K}_\tau$. The opening of $\mathcal{K}_\tau$ is denoted by α_τ. Furthermore, let $V^l_{p,\vec{\beta}}(\mathcal{G})$ and $N^{l,\sigma}_{\vec{\beta}}(\mathcal{G})$, where $\vec{\beta} = (\beta_1, \ldots, \beta_d) \in \mathbb{R}^d$, be the weighted spaces introduced in Section 1.4.

Using Theorems 1.4.3 and 5.1.2, we obtain

THEOREM 5.8.1. *Any solution* $(U, P) \in \overset{\circ}{W}{}^1_2(\mathcal{G})^2 \times L_2(\mathcal{G})$ *of problem* (5.8.1) *belongs to the space* $V^l_{p,\vec{\beta}}(\mathcal{G})^2 \times V^{l-1}_{p,\vec{\beta}}(\mathcal{G})$ *if* $F \in V^{l-2}_{p,\vec{\beta}}(\mathcal{G})^2$, $l \geq 2$, *and*

$$(5.8.2) \qquad |l - \beta_\tau - 2/p| < \begin{cases} 1 & \text{for } \alpha_\tau < \pi, \\ \xi_+(\alpha_\tau)/\alpha_\tau & \text{for } \alpha_\tau > \pi. \end{cases}$$

Here $\xi_+(\alpha)$ *is the smallest positive solution of the equation* $\xi^{-1}\sin\xi + \alpha^{-1}\sin\alpha = 0$. *Condition* (5.8.2) *with* $l + \sigma - \beta_\tau$ *instead of* $l - \beta_\tau + 2/p$ *guarantees that* $(U, P) \in N^{l,\sigma}_{\vec{\beta}}(\mathcal{G})^2 \times N^{l-1,\sigma}_{\vec{\beta}}(\mathcal{G})$ *if* $F \in N^{l-2,\sigma}_{\vec{\beta}}(\mathcal{G})^2$.

Furthermore, by means of Theorems 1.4.4 and 5.1.2, the asymptotics of the solution $U \in \overset{\circ}{W}{}^1_2(\mathcal{G})^2$ near the corner point $x^{(\tau)}$ can be described. Suppose, for simplicity, $F = 0$ near $x^{(\tau)}$. Then for x in a vicinity of $x^{(\tau)}$ we have

$$U(x) = o\big(r^{\pi/\alpha_\tau}\big), \quad P(x) = \text{const} + o\big(r^{-1+\pi/\alpha_\tau}\big) \quad \text{if } \alpha_\tau < \pi \ .$$

If $\pi < \alpha_\tau < 2\pi$ and $\alpha_\tau \neq \alpha_*$ (here α_* is the solution of the equation $\tan\alpha = \alpha$ in the interval $(\pi, 2\pi)$), then the asymptotics

$$\binom{U_r}{U_\varphi} = c_+ r^{\frac{\xi_+(\alpha_\tau)}{\alpha_\tau}} \binom{u_r^+}{u_\varphi^+} + c_- r^{\frac{\xi_-(\alpha_\tau)}{\alpha_\tau}} \binom{u_r^-}{u_\varphi^-} + o\big(r^{2\pi/\alpha_\tau}\big),$$

$$P = c + c_+ r^{\frac{\xi_+(\alpha_\tau)}{\alpha_\tau}-1} p^+ + c_- r^{\frac{\xi_-(\alpha_\tau)}{\alpha_\tau}-1} p^- + o\big(r^{-1+2\pi/\alpha_\tau}\big)$$

holds. Here we use the polar coordinates (r, φ) in a neighborhood $x^{(\tau)}$. Furthermore, c, c_+ and c_- are constants, $\xi_\pm(\alpha)$ are the smallest roots of the equations $\xi^{-1}\sin\xi \pm \alpha^{-1}\sin\alpha = 0$ not equal to α, and the vectors $(u_r^\pm, u_\varphi^\pm, p^\pm)$ are defined by (5.1.11) and (5.1.12), respectively. In the case $\alpha_\tau = \alpha_*$, when $\xi_-(\alpha_\tau) = \alpha_\tau$, the asymptotics contains an additional logarithmic term. By Theorem 5.1.2, the functions $\alpha \to \xi_\pm(\alpha)/\alpha$ are strictly decreasing, and we have $\xi_\pm(\alpha)/\alpha > 1/2$ for $\alpha < 2\pi$.

5.8.2. The Stokes system in a domain of polyhedral type. Let $\mathcal{G}$ be the same bounded domain in $\mathbb{R}^3$ with piecewise smooth boundary $\partial\mathcal{G}$ as in Subsection 3.6.2. We use the same notation as in in Subsection 3.6.2. In particular, we denote by Γ_j, $j = 1, \ldots, T$, the faces of $\mathcal{G}$, by E_k, $k = 1, \ldots, q$, the edges, and by $x^{(\tau)}$, $\tau = 1, \ldots, d$, the vertices. Outside the set $\mathcal{S}$ of the edges E_k and vertices $x^{(\tau)}$, the boundary is assumed to be smooth.

For every $x \in \bigcup E_k$ let α_x be the angle at x between the tangential half-planes to $\partial\mathcal{G}\backslash\mathcal{S}$. If $\alpha_x < \pi$, then we set $\theta(x) = 1$. In the case $\pi < \alpha_x < 2\pi$ let $\theta(x)$ be the smallest positive solution of the equation

$$\sin(\theta\alpha_x) + \theta\sin\alpha_x = 0.$$

To every vertex $x^{(\tau)}$ we associate the operator pencil $\mathcal{L}_\tau$ generated by the Dirichlet problem for the Stokes system in the cone $\mathcal{K}_\tau$. By μ_τ we denote the smallest real part of eigenvalues of $\mathcal{L}_\tau$, situated in the half-plane $\mathrm{Re}\,\lambda > -1/2$.

1. **Regularity results for the solutions.** We consider the solutions $(U,P) \in \overset{\circ}{W}{}_2^1(\mathcal{G})^3 \times L_2(\mathcal{G})$ of problem (5.8.1). The following analogue to Theorem 3.8.4 was proved by Maz'ya and Plamenevskiĭ [**189**, Th.6.1].

THEOREM 5.8.2. *Suppose that* $F \in V_{\vec{\beta},\vec{\delta}}^{l-2,p}(\mathcal{G})^3$, *where*

$$(5.8.3) \qquad |l - \delta_k - 2/p| < \inf_{x \in E_k} \theta(x) \quad and \quad |l - \beta_\tau - 3/p + 1/2| < \mu_\tau + 1/2$$

for $k = 1,\dots,q$, $\tau = 1,\dots,d$. *Then the solution* (U,P) *of problem* (5.8.1) *belongs to the space* $V_{\vec{\beta},\vec{\delta}}^{l,p}(\mathcal{G})^3 \times V_{\vec{\beta},\vec{\delta}}^{l-1,p}(\mathcal{G})$.

  If $F \in N_{\vec{\beta},\vec{\delta}}^{l-2,\sigma}(\mathcal{G})^3$, *where*

$$(5.8.4) \qquad |l + \sigma - \delta_k| < \inf_{x \in E_k} \theta(x) \quad and \quad |l + \sigma - \beta_\tau + 1/2| < \mu_\tau + 1/2$$

for $k = 1,\dots,q$, $\tau = 1,\dots,d$. *Then the solution* (U,P) *of problem* (5.8.1) *belongs to the space* $N_{\vec{\beta},\vec{\delta}}^{l,\sigma}(\mathcal{G})^3 \times N_{\vec{\beta},\vec{\delta}}^{l-1,\sigma}(\mathcal{G})$.

We note that, by Theorem 5.5.5, $\mu_\tau = 1$ if $\mathcal{K}_\tau$ is contained in a half-space. It follows from Corollary 3.4.2 that

$$(5.8.5) \qquad\qquad\qquad \mu_\tau > \frac{M_\tau}{M_\tau + 4},$$

where M_τ is the same as in Section 3.6. Another estimate can be found in Theorem 5.5.6. More estimates for μ_τ follow from the monotonicity of real eigenvalues with respect to Ω_τ (see Theorem 5.5.4). If, for example, $\mathcal{K}_\tau$ contains a half-space we see that $0 < \mu_\tau \le 1$ and μ_τ is a monotone non-increasing function of $\mathcal{K}_\tau$.

The assertions of Theorem 5.8.2 are also true for solutions $(U,P) \in \overset{\circ}{W}{}_2^1(\mathcal{G})^3 \times L_2(\mathcal{G})$ of the Dirichlet problem for the Navier-Stokes system

$$-\Delta U + \nabla P + U \cdot \nabla U = F, \qquad \nabla \cdot U = 0 \quad \text{in } \mathcal{G}$$

(see [**189**, Th.10.3,10.4]).

2. **Estimates of the Green matrix.** The Green matrix of problem (5.8.1) is defined as a matrix-function $(x,y) \to \big(G_{i,j}(x,y)\big)_{i,j=1}^3$ such that the vectors $(G_{1,j},\dots,G_{4,j})$, $j = 1,2,3$, are solutions of the problem

$$-\Delta_x G_{i,j}(x,y) + \partial_{x_i} G_{4,j}(x,y) = \delta_{i,j}\,\delta(x-y) \quad \text{for } (x,y) \in \mathcal{G} \times \mathcal{G},\ i = 1,2,3,$$

$$\sum_{i=1}^3 \partial_{x_i} G_{i,j}(x,y) = 0 \ \text{ for } (x,y) \in \mathcal{G} \times \mathcal{G},$$

$$G_{i,j}(x,y) = 0 \ \text{ for } x \in \partial\mathcal{G}\backslash\mathcal{S},\ y \in \mathcal{G},\ i = 1,2,3,$$

while the vector $(G_{1,4}, \dots, G_{4,4})$ is a solution of the problem

$$-\Delta_x G_{i,4}(x,y) + \partial_{x_i} G_{4,4}(x,y) = 0 \ \text{ for } (x,y) \in \mathcal{G} \times \mathcal{G}, \ i = 1, 2, 3,$$

$$\sum_{i=1}^{3} \partial_{x_i} G_{i,4}(x,y) = \delta(x-y) \ \text{ for } (x,y) \in \mathcal{G} \times \mathcal{G},$$

$$G_{i,4}(x,y) = 0 \ \text{ for } x \in \partial\mathcal{G} \backslash \mathcal{S}, \ y \in \mathcal{G}, \ i = 1, 2, 3.$$

Here we mean solutions for which the functions $x \to (G_{1,j}(x,y), \dots, G_{4,j}(x,y))$ belong to the space $W_2^1(\mathcal{G} \backslash \overline{\mathcal{U}})^3 \times L_2(\mathcal{G} \backslash \overline{\mathcal{U}})$ for every $y \in \mathcal{G}$ and every neighborhood $\mathcal{U}$ of y. By [189], the components $G_{i,j}$ of the Green matrix satisfy the following estimates in a neighborhood of the vertex $x^{(\tau)}$.

THEOREM 5.8.3. 1) *If* $2\rho_\tau(x) < \rho_\tau(y)$, *then* (3.8.8) *is valid for the functions* $G_{i,j}$, $i, j = 1, \dots, 3$, $\theta_k = \inf_{x \in E_k} \theta(x)$. *Furthermore,*

$$|G_{4,j}(x,y)| \le c \frac{\rho_\tau(x)^{\mu_\tau - 1 - \varepsilon}}{\rho_\tau(y)^{\mu_\tau + 1 - \varepsilon}} \prod_{\{k : x^{(\tau)} \in \overline{E}_k\}} \left(\frac{r_k(x)}{\rho_\tau(x)}\right)^{\theta_k - 1 - \varepsilon} \prod_{\{k : x^{(\tau)} \in \overline{E}_k\}} \left(\frac{r_k(y)}{\rho_\tau(y)}\right)^{\theta_k - \varepsilon},$$

$$|G_{i,4}(x,y)| \le c \frac{\rho_\tau(x)^{\mu_\tau - \varepsilon}}{\rho_\tau(y)^{\mu_\tau + 2 - \varepsilon}} \prod_{\{k : x^{(\tau)} \in \overline{E}_k\}} \left(\frac{r_k(x)}{\rho_\tau(x)}\right)^{\theta_k - \varepsilon} \prod_{\{k : x^{(\tau)} \in \overline{E}_k\}} \left(\frac{r_k(y)}{\rho_\tau(y)}\right)^{\theta_k - 1 - \varepsilon},$$

for $i, j = 1, 2, 3$ *and*

$$|G_{4,4}(x,y)| \le c \frac{\rho_\tau(x)^{\mu_\tau - 1 - \varepsilon}}{\rho_\tau(y)^{\mu_\tau + 2 - \varepsilon}} \prod_{\{k : x^{(\tau)} \in \overline{E}_k\}} \left(\frac{r_k(x)}{\rho_\tau(x)}\right)^{\theta_k - 1 - \varepsilon} \prod_{\{k : x^{(\tau)} \in \overline{E}_k\}} \left(\frac{r_k(y)}{\rho_\tau(y)}\right)^{\theta_k - 1 - \varepsilon},$$

The expression, which arises if one changes the places of x *and* y *on the right-hand side of these estimates, majorizes the functions* $|G_{i,j}|$, $i, j = 1, \dots, 4$, *in the case* $2\rho_\tau(y) < \rho_\tau(x)$.

2) *If* $\rho_\tau(x)/2 < \rho_\tau(y) < 2\rho_\tau(x)$ *and* $|x - y| > \min(r(x), r(y))$, *then the estimate* (3.8.9) *is valid for the functions* $G_{i,j}$, $i, j = 1, 2, 3$, *while*

$$|G_{4,j}(x,y)| \le \frac{c}{|x-y|^2} \prod_{\{k : x^{(\tau)} \in \overline{E}_k\}} \left(\frac{r_k(x)}{|x-y|}\right)^{\theta_k - 1 - \varepsilon} \prod_{\{k : x^{(\tau)} \in \overline{E}_k\}} \left(\frac{r_k(y)}{|x-y|}\right)^{\theta_k - \varepsilon},$$

$$|G_{i,4}(x,y)| \le \frac{c}{|x-y|^2} \prod_{\{k : x^{(\tau)} \in \overline{E}_k\}} \left(\frac{r_k(x)}{|x-y|}\right)^{\theta_k - \varepsilon} \prod_{\{k : x^{(\tau)} \in \overline{E}_k\}} \left(\frac{r_k(y)}{|x-y|}\right)^{\theta_k - 1 - \varepsilon},$$

$$|G_{4,4}(x,y)| \le \frac{c}{|x-y|^3} \prod_{\{k : x^{(\tau)} \in \overline{E}_k\}} \left(\frac{r_k(x)}{|x-y|}\right)^{\theta_k - 1 - \varepsilon} \prod_{\{k : x^{(\tau)} \in \overline{E}_k\}} \left(\frac{r_k(y)}{|x-y|}\right)^{\theta_k - 1 - \varepsilon}.$$

3) *If* $|x - y| < \min(r(x), r(y))$, *then* $|G_{i,j}(x,y)| \le c|x-y|^{-1}$ *for* $i, j = 1, 2, 3$.

Analogous estimates are valid for the derivatives of the functions $G_{i,j}$, $i, j = 1, \dots, 4$.

Clearly, we can replace μ_τ and θ_k in the above estimates by their lower bounds in order to get more visible dependence on the geometry of singularities of $\partial\mathcal{G}$. In particular, estimate (5.8.5) can be of use, as well as other information on μ_τ obtained in this chapter.

3. L_∞-estimates. Now we consider the problem

(5.8.6) $\qquad\qquad -\Delta U + \nabla P = 0, \ \nabla \cdot U = 0 \text{ in } \mathcal{G}, \quad U = \Phi \text{ on } \partial\mathcal{G}.$

We suppose that Φ belongs to the space $L^\infty_{\vec{\beta},\vec{\delta}}(\partial\mathcal{G})^3$ which was introduced in Section 3.6. The solution of this problem is defined as follows. Let $\{\Phi^{(k)}\} \subset C^\infty_0(\partial\mathcal{G}\backslash\mathcal{S})^3$ be a sequence such that

$$\Phi^{(k)} \to \Phi \text{ a.e. on } \partial\mathcal{G} \quad \text{and} \quad \|\Phi^{(k)}\|_{L^\infty_{\vec{\beta},\vec{\delta}}(\mathcal{G})^3} \le \text{const}.$$

The vector functions $(U^{(k)}, P^{(k)}) \in W_2^1(\mathcal{G})^3 \times L_2(\mathcal{G})$ with the components

$$U_i^{(k)}(x) = \int\limits_{\partial\mathcal{G}\backslash\mathcal{S}} \sum_{j=1}^{3} \frac{\partial G_{i,j}}{\partial\nu_\xi}(x,y)\, \Phi_j^{(k)}(\xi)\, d\sigma_\xi\ ,$$

$$P^{(k)}(x) = \int\limits_{\partial\mathcal{G}\backslash\mathcal{S}} \sum_{j=1}^{3} \frac{\partial G_{4,j}}{\partial\nu_\xi}(x,y)\, \Phi_j^{(k)}(\xi)\, d\sigma_\xi$$

are solutions of problem (5.8.6) with the Dirichlet data $\Phi^{(k)}$. By means of Theorem 5.8.3, it can be shown that the sequence $(U^{(k)}, P^{(k)})$ converges in every point of $\mathcal{G}$ if $\delta_k < \inf_{x\in E_k} \theta(x)$ for $k = 1,\ldots,q$ and $\beta_\tau < 1 + \mu_\tau$ for $\tau = 1,\ldots,d$. The limit (U, P) of the sequence $\{(U^{(k)}, P^{(k)})\}$ is called solution of problem (5.8.6).

THEOREM 5.8.4. [**189**, Th.9.1] *(see also* [**188**, Th.2] *in the case of a domain with conic points) Let (U,P) be a solution of problem (5.8.6), where $\Phi \in L^\infty_{\vec{\beta},\vec{\delta}}(\partial\mathcal{G})$,*

$$|\delta_k| < \inf_{x\in E_k} \theta(x) \quad \text{and} \quad |\beta_\tau - 1/2| < \mu_\tau + 1/2$$

for $k = 1,\ldots,q$ and $\tau = 1,\ldots,d$. Then

(5.8.7) $\qquad\qquad\qquad \|U\|_{L^\infty_{\vec{\beta},\vec{\delta}}(\mathcal{G})^3} \le c\,\|\Phi\|_{L^\infty_{\vec{\beta},\vec{\delta}}(\partial\mathcal{G})^3}.$

Here the constant c is independent of Φ.

Since $\inf_{x\in E_k} \theta(x)$ and μ_τ are positive one may set $\vec{\beta} = \vec{0}$ and $\vec{\delta} = \vec{0}$ in the last theorem. Then one obtains the estimate

$$\|U\|_{L_\infty(\mathcal{G})^3} \le c\,\|\Phi\|_{L_\infty(\partial\mathcal{G})^3}.$$

Furthermore, $\Phi \in C(\partial\mathcal{G})^3$ implies $U \in C(\overline{\mathcal{G}})^3$.

The last sentence is also true for the solutions of the Navier-Stokes system, that is if $(U,P) \in W_2^1(\mathcal{G})^3 \times L_2(\mathcal{G})$ is a solution of the problem

$$-\Delta U + \nabla P + U \cdot \nabla U = 0, \quad \nabla \cdot U = 0 \quad \text{in } \mathcal{G},$$
$$U = \Phi \quad \text{on } \partial\mathcal{G},$$

where $\Phi \in C(\partial\mathcal{G})^3$, then $U \in C(\overline{\mathcal{G}})^3$ (see [**189**, Th.10.8]).

5.9. Notes

Section 5.1. We mention the papers by Kellogg and Osborn [**101**], Osborn [**222**], Lozi [**157**] and Dauge [**38, 42**] dedicated to the Stokes system in domains with corners.

The equation (5.1.9) for the eigenvalues of the pencil generated by the Dirichlet problems for the Stokes system is the same as for the Lamé system with $\nu = 1/2$. Therefore, information on the spectrum of this pencil can be also found in works on linear elasticity (see the references at the end of Chapter 3).

Sections 5.2–5.6. The estimate (5.0.6) for the width of the energy strip of the pencil generated by the Dirichlet problem for the Stokes system in a three-dimensional cone was proved in the paper [**188**] by Maz'ya and Plamenevskiĭ. For convex cones it was improved by Dauge [**42**]. The results of Sections 5.2–5.6 were obtained in Kozlov, Maz'ya and Schwab [**141**]. For the case of a rotational cone we refer additionally to [**139**].

Section 5.7. For dihedral angles the eigenvalues, eigenvectors and generalized eigenvectors of the corresponding pencil were described by Kozlov and Maz'ya in [**134**].

Section 5.8. Regularity results for the Dirichlet problem to the Stokes system in polygonal domains as well as asymptotic formulas for the solutions near the corners can be found in the above cited papers [**38, 42, 101, 157, 222**] and in the book of Grisvard [**86**]. For example, Kellogg and Osborn [**101**] proved that the solution of problem (5.8.1) in a convex polygon $\mathcal{G}$ belongs to $W_2^2(\mathcal{G})^2 \times W_2^1(\mathcal{G})$ if $F \in L_2(\mathcal{G})^2$.

Eigenvalues of the pencil corresponding to the two-dimensional problem are also of importance for the description of the behavior of solutions to the Stokes and Navier-Stokes systems near edges of three-dimensional domains. Maz'ya and Plamenevskiĭ's paper [**176**] contains existence theorems in different function spaces for the solutions of the Dirichlet problem to the Navier-Stokes system in domains with edges and asymptotic formulas for the solution.

The Dirichlet problem for the Stokes and Navier-Stokes systems in three-dimensional domains with conic points was studied by the same authors in [**188**]. This paper includes, in particular, weighted L_p, L_∞ and Hölder estimates for the solutions. The results of this paper were extended in [**189**] to the Stokes and Navier-Stokes systems in domains of polyhedral type. (We discuss the main results of [**189**] in Section 5.8.) Dauge [**42**] proved that the solution of problem (5.8.1) belongs to the space $W_2^2(\mathcal{G})^3 \times W_2^1(\mathcal{G})$ if $\mathcal{G}$ is a convex polyhedron and $F \in W_2^2(\mathcal{G})^3$. A L_p theory for the Dirichlet problem to the Stokes system in nonsmooth domains was also developed by Fabes, Kenig and Verchota [**57**] and Deuring [**47**]. In [**57**] the existence of solutions is proved for arbitrary Lipschitz domains and boundary data in L_p.

CHAPTER 6

Other boundary value problems for the Stokes system in a cone

In this chapter we deal with the spectral properties of operator pencils which are generated by some boundary value problems for the Stokes system

$$-\Delta U + \nabla P = F, \qquad \nabla \cdot U = 0$$

in a three-dimensional cone $\mathcal{K}$. In Sections 6.1 and 6.2 we consider a problem, where the following conditions are given on parts of the boundary:

(i) $U = 0$

(ii) $U_n = 0$, $\quad \varepsilon_{n,\tau}(U) = 0$

(iii) $U_\tau = 0$, $\quad -P + 2\,\varepsilon_{n,n}(U) = 0$.

Here $n = (n_1, n_2, n_3)$ is the exterior normal to $\partial\mathcal{K}$, $U_n = U \cdot n$ is the normal component of the velocity $U = (U_1, U_2, U_3)$, $U_\tau = U - U_n n$ is the tangential component of the vector U, $\varepsilon(U) = \{\varepsilon_{ij}(U)\}$ is the strain tensor (see Introduction to Chapter 4),

$$\varepsilon_{nn} = \sum_{i,j=1}^{3} \varepsilon_{ij}(U)\, n_i\, n_j$$

is the normal component of the vector $\varepsilon_n(U) = \varepsilon(U)n$, and $\varepsilon_{n\tau}(U)$ is the tangential component of the vector $\varepsilon_n(U)$.

The boundary conditions (i)-(iii) appear in the analysis of the steady-state motion of a viscous fluid in a vessel with a free surface having contact with the walls of the vessel (see, for example, the papers by Maz'ya, Plamenevskiĭ, Stupyalis [190] and Solonnikov [247]).

We investigate spectral properties of the pencil associated to a boundary value problem by modifying the method used in the foregoing chapter. In the beginning we obtain some basic results as, for example, the discreteness of the spectrum and the absence of eigenvalues on the line $\operatorname{Re}\lambda = -1/2$. Analogously to the Dirichlet problem, we show that the eigenvalues in the strip $-2 \leq \operatorname{Re}\lambda \leq 1$ are real and that the eigenvectors corresponding to eigenvalues in the interior of this strip do not have generalized eigenvectors.

From this result one can deduce that the solution of the boundary value problem has the following asymptotics near a conic vertex if the right-hand sides in the Stokes system and in the boundary conditions vanish near this vertex:

$$U = \sum_{j=1}^{J} c_j\, r^{\lambda_j} u^{(j)}(\omega) + O(r), \qquad P = \sum_{j=1}^{J} c_j\, r^{\lambda_j - 1} p^{(j)}(\omega) + O(\log r)$$

Here λ_j are real eigenvalues in the interval $(-1/2, 1)$, and $(u^{(j)}, p^{(j)})$ are corresponding eigenvectors. In order to determine the regularity of weak solutions, it is useful to estimate the smallest eigenvalue in the interval $(-1/2, 1]$. At the end of Section 6.1 we show that there are no eigenvalues in the interval $(-1/2, 0)$ and that $\lambda = 0$ is an eigenvalue if and only if there exists a constant vector $(U, P) = (c, 0)$ satisfying the zero boundary conditions. This result implies the Hölder continuity of the velocity U in a neighborhood of the vertex.

In contrast to the Dirichlet problem, the number $\lambda = 1$ is not always an eigenvalue. In Section 6.2 we give conditions ensuring that $\lambda = 1$ belongs to the spectrum and determine all eigenvectors and generalized eigenvectors associated to this eigenvalue. A variational principle derived at the end of Section 6.2 enables one to obtain sharper estimates for the smallest eigenvalue in the interval $(-1/2, 1)$. A consequence of this principle is the monotonicity (with respect to the cone $\mathcal{K}$) of the eigenvalues in $(-1/2, 1)$ if only the boundary conditions (i) and (iii) appear.

Section 6.3 is dedicated to the Neumann problem for the Stokes system in a three-dimensional cone $\mathcal{K}$ which is given by the inequality $x_3 > \phi(x_1, x_2)$, where ϕ is a smooth function on $\mathbb{R}^2 \backslash \{0\}$ positively homogeneous of degree 1. We prove that the operator pencil generated by this problem has only the eigenvalues 0 and -1 in the strip $-1 \leq \operatorname{Re} \lambda \leq 0$ and that both eigenvalues have geometric and algebraic multiplicity 3. The eigenvectors corresponding to $\lambda = 0$ have the form $(u, p) = (c, 0)$, where $c \in \mathbb{C}^3$. From this result, it follows, for example, that the velocity vector U is Hölder continuous in a neighborhood of a conic vertex.

6.1. A mixed boundary value problem for the Stokes system

6.1.1. The operator pencil generated by the mixed boundary value problem.
Let $\mathcal{K}$ be the cone $\{x = (x_1, x_2, x_3) \in \mathbb{R}^3 : x/|x| \in \Omega\}$, where Ω is a domain on the unit sphere with Lipschitz boundary $\partial\Omega = \overline{\gamma}_1 \cup \ldots \cup \overline{\gamma}_N$. Here $\gamma_1, \ldots, \gamma_N$ are pairwise disjoint open connected arcs on the sphere S^2. Then the boundary of $\mathcal{K}$ is the union of faces, $\partial\mathcal{K} = \overline{\Gamma}_1 \cup \cdots \cup \overline{\Gamma}_N$, where $\Gamma_k = \{x : x/|x| \in \gamma_k\}$.

We divide the set of the indices $1, \ldots, N$ into three subsets I_0, I_n, I_τ. Our goal is to find solutions of the system

$$(6.1.1) \qquad -\Delta U + \nabla P = 0, \qquad \nabla \cdot U = 0 \qquad \text{in } \mathcal{K}$$

satisfying the boundary conditions

$$(6.1.2) \qquad \begin{cases} U = 0 \ \text{ on } \Gamma_k \ \text{ for } k \in I_0, \\ U_n = 0, \ \varepsilon_{n,\tau}(U) = 0 \text{ on } \Gamma_k \ \text{ for } k \in I_n, \\ U_\tau = 0, \ -P + 2 \displaystyle\sum_{i,j=1}^{3} \varepsilon_{ij}(U)\, n_i\, n_j = 0 \ \text{ on } \Gamma_k \ \text{ for } k \in I_\tau \end{cases}$$

which will be understood in a generalized sense.

In order to introduce the notion of generalized solutions, we need the following Green formula

$$(6.1.3) \qquad \int_{\mathcal{K}} \left(2 \sum_{i,j=1}^{3} \varepsilon_{ij}(U) \cdot \varepsilon_{ij}(\overline{V}) - P \nabla \cdot \overline{V} - (\nabla \cdot U) \overline{Q} \right) dx$$

$$= \int_{\mathcal{K}} \left(\left(-\Delta U - \nabla \nabla \cdot U + \nabla P \right) \overline{V} - \nabla \cdot U \cdot \overline{Q} \right) dx$$

$$+ \sum_{j=1}^{3} \int_{\partial \mathcal{K} \backslash \{0\}} \left(-P n_j + 2 \sum_{i=1}^{3} \varepsilon_{ij}(U) n_i \right) \overline{V}_j \, d\sigma.$$

If (U, P) is a formal solution of problem (6.1.1), (6.1.2), (V, Q) vanishes for large and small $|x|$, and V satisfies the conditions

$$(6.1.4) \qquad \begin{cases} V = 0 \text{ on } \Gamma_k & \text{for } k \in I_0, \\ V_n = 0 \text{ on } \Gamma_k & \text{for } k \in I_n, \\ V_\tau = 0 \text{ on } \Gamma_k & \text{for } k \in I_\tau, \end{cases}$$

then (6.1.3) implies

$$(6.1.5) \qquad \int_{\mathcal{K}} \left(2 \sum_{i,j=1}^{3} \varepsilon_{ij}(U) \cdot \varepsilon_{ij}(\overline{V}) - P \nabla \cdot \overline{V} - (\nabla \cdot U) \overline{Q} \right) dx = 0.$$

Hence it is natural to define generalized solutions by means of this integral identity.

DEFINITION 6.1.1. Let $\mathcal{H}$ be the space of all vector functions $u \in W_2^1(\Omega)^3$ satisfying the conditions

    (i) $u = 0$ on γ_k for $k \in I_0$,
    (ii) $u_n = 0$ on γ_k for $k \in I_n$,
    (iii) $u_\tau = 0$ on γ_k for $k \in I_\tau$,

where $u_n = u \cdot n$, n denotes the exterior normal to $\partial \mathcal{K}$, and u_τ is the projection of the vector u onto the tangent plane to $\partial \mathcal{K}$. We say that the pair (U, P) given by

$$(6.1.6) \qquad U(x) = r^{\lambda_0} \sum_{k=0}^{s} \frac{1}{k!} (\log r)^k u^{(s-k)}(\omega), \qquad u^{(s-k)} \in \mathcal{H},$$

$$(6.1.7) \qquad P(x) = r^{\lambda_0 - 1} \sum_{k=0}^{s} \frac{1}{k!} (\log r)^k p^{(s-k)}(\omega), \qquad p^{(s-k)} \in L_2(\Omega)$$

is a *generalized solution* of (6.1.1), (6.1.2) if (6.1.5) is satisfied for all $(V, Q) \in W_2^1(\mathcal{K})^3 \times L_2(\mathcal{K})$ with compact support in $\overline{\mathcal{K}} \backslash \{0\}$ such that V satisfies the boundary conditions (6.1.4).

Note that the space $\mathcal{H}$ in the above definition coincides with the space $\mathcal{H}$ which was introduced in Section 4.1.

In what follows, we use the notations introduced in Section 3.2. In terms of the spherical components of the vector function U, the integral identity (6.1.5) takes the form

$$\int\limits_{0}^{\infty} \int\limits_{\Omega} \left(2\varepsilon_{rr}(U)\,\varepsilon_{rr}(\overline{V}) + 2\varepsilon_{\theta\theta}(U)\,\varepsilon_{\theta\theta}(\overline{V}) + 2\varepsilon_{\varphi\varphi}(U)\,\varepsilon_{\varphi\varphi}(\overline{V}) + 4\,\varepsilon_{r\theta}(U)\,\varepsilon_{r\theta}(\overline{V}) \right.$$

$$+ 4\,\varepsilon_{r\varphi}(U)\,\varepsilon_{r\varphi}(\overline{V}) + 4\,\varepsilon_{\theta\varphi}(U)\,\varepsilon_{\theta\varphi}(\overline{V}) - P\left(\varepsilon_{rr}(\overline{V}) + \varepsilon_{\theta\theta}(\overline{V}) + \varepsilon_{\varphi\varphi}(\overline{V}) \right)$$

$$\left. - \left(\varepsilon_{rr}(U) + \varepsilon_{\theta\theta}(U) + \varepsilon_{\varphi\varphi}(U) \right) \overline{Q} \right) r^2 \, d\omega \, dr = 0.$$

Here ε_{rr}, $\varepsilon_{\varphi,\varphi}$, $\varepsilon_{\theta,\theta}$, $\varepsilon_{r\varphi}$, $\varepsilon_{r,\theta}$, and $\varepsilon_{\theta,\varphi}$ are the spherical components of the strain tensor which are defined by (4.1.7). We introduce the sesquilinear form

$$a\!\left(\begin{pmatrix} u \\ p \end{pmatrix}, \begin{pmatrix} v \\ q \end{pmatrix}; \lambda \right) = \frac{1}{2\log 2} \int\limits_{1/2}^{2} \int\limits_{\Omega} \left(2\,\varepsilon_{rr}(U)\,\varepsilon_{rr}(\overline{V}) + 2\,\varepsilon_{\theta\theta}(U)\,\varepsilon_{\theta\theta}(\overline{V}) \right.$$

$$+ 2\varepsilon_{\varphi\varphi}(U)\,\varepsilon_{\varphi\varphi}(\overline{V}) + 4\varepsilon_{r\theta}(U)\,\varepsilon_{r\theta}(\overline{V}) + 4\varepsilon_{r\varphi}(U)\,\varepsilon_{r\varphi}(\overline{V})$$

$$+ 4\,\varepsilon_{\theta\varphi}(U)\,\varepsilon_{\theta\varphi}(\overline{V}) - P\left(\varepsilon_{rr}(\overline{V}) + \varepsilon_{\theta\theta}(\overline{V}) + \varepsilon_{\varphi\varphi}(\overline{V}) \right)$$

$$\left. - \left(\varepsilon_{rr}(U) + \varepsilon_{\theta\theta}(U) + \varepsilon_{\varphi\varphi}(U) \right) \overline{Q} \right) r^2 \, d\omega \, dr,$$

where

$$U = r^\lambda\, u(\omega), \quad P = r^{\lambda-1}\, p(\omega), \quad V = r^{-1-\overline{\lambda}}\, v(\omega), \quad Q = r^{-2-\overline{\lambda}}\, q(\omega),$$

$d\omega = \sin\theta \, d\theta \, d\varphi$. Using (4.1.7), we find

$$(6.1.8) \quad a\!\left(\begin{pmatrix} u \\ p \end{pmatrix}, \begin{pmatrix} v \\ q \end{pmatrix}; \lambda \right) = [u_\omega, v_\omega] + \int\limits_{\Omega} \left((\nabla_\omega u_r) \cdot \nabla_\omega \overline{v}_r \right.$$

$$+ (\lambda + 2)(1 - \lambda)\left(2\,u_r\,\overline{v}_r + u_\omega \cdot \overline{v}_\omega \right) + 2u_r\,\nabla_\omega \cdot \overline{v}_\omega$$

$$+ 2(\nabla_\omega \cdot u_\omega)\,\overline{v}_r - (1 - \lambda)\,u_\omega \cdot \nabla_\omega \overline{v}_r - (\lambda + 2)\,(\nabla_\omega u_r) \cdot \overline{v}_\omega$$

$$\left. - p\left((1 - \lambda)\,\overline{v}_r + \nabla_\omega \cdot \overline{v}_\omega \right) - \left((\lambda + 2)\,u_r + \nabla_\omega \cdot u_\omega \right) \overline{q} \right) d\omega,$$

where the sesquilinear form $[u_\omega, v_\omega]$ is defined by (4.1.10).

We define the space $\mathcal{H}_s$ as the set of all vector functions $(u_r, u_\omega) \in W_2^1(\Omega) \times h_2^1(\Omega)$ such that the boundary conditions (i)–(iii) in Definition 6.1.1 are satisfied, i.e., we have $(u_1, u_2, u_3) \in \mathcal{H}$ if and only if $(u_r, u_\omega) \in \mathcal{H}_s$.

The form $a(\cdot, \cdot; \lambda)$ is continuous on $\left(\mathcal{H}_s \times L_2(\Omega) \right) \times \left(\mathcal{H}_s \times L_2(\Omega) \right)$ and generates the continuous operator

$$\mathfrak{A}(\lambda) \,:\, \mathcal{H}_s \times L_2(\Omega) \to \mathcal{H}_s^* \times L_2(\Omega)$$

by the equality

$$\left(\mathfrak{A}(\lambda) \begin{pmatrix} u \\ p \end{pmatrix}, \begin{pmatrix} v \\ q \end{pmatrix} \right)_{L_2(\Omega)^4} = a\!\left(\begin{pmatrix} u \\ p \end{pmatrix}, \begin{pmatrix} v \\ q \end{pmatrix}; \lambda \right), \qquad u, v \in \mathcal{H}_s, \ p, q \in L_2(\Omega).$$

Obviously, $\mathfrak{A}(\lambda)$ depends polynomially on the complex parameter λ. As a consequence of Theorem 1.2.3, the following statement holds.

LEMMA 6.1.1. *The functions (6.1.6) and (6.1.7) form a solution of the integral identity (6.1.5) if and only if λ_0 is an eigenvalue of the pencil $\mathfrak{A}$ and the vector functions $(u^{(0)}, p^{(0)}), \ldots, (u^{(s)}, p^{(s)})$ form a Jordan chain corresponding to this eigenvalue.*

6.1.2. Basic properties of the pencil $\mathfrak{A}$. We prove that the operator $\mathfrak{A}(\lambda)$ is Fredholm for every fixed λ and invertible for at least one λ. We start with the following analogue of Lemma 5.2.3.

LEMMA 6.1.2. *Let $\mathrm{Re}\,\lambda = -1/2$. Then for every $f \in \mathcal{H}_s^*$, $g \in L_2(\Omega)$ there exists a vector function $(u, p) \in \mathcal{H}_s \times L_2(\Omega)$ satisfying the equation*

$$(6.1.9) \qquad a\left(\begin{pmatrix} u \\ p \end{pmatrix}, \begin{pmatrix} v \\ q \end{pmatrix}; \lambda\right) = (f, v)_\Omega + \int_\Omega g\, \overline{q}\, d\omega$$

for all $v \in \mathcal{H}_s$, $q \in L_2(\Omega)$. Here $(\cdot, \cdot)_\Omega$ denotes the continuation of the scalar product in $L_2(\Omega)^3$ to $\mathcal{H}_s^* \times \mathcal{H}_s$.

Proof: By Lemma 5.2.2, there exists a vector function $u^{(0)} = (u_r^{(0)}, u_\omega^{(0)}) \in \mathcal{H}_s$ satisfying the equation $(\lambda + 2)\, u_r^{(0)} + \nabla_\omega \cdot u_\omega^{(0)} = -g$. Then, according to (6.1.8), we have

$$a\left(\begin{pmatrix} u^{(0)} \\ 0 \end{pmatrix}, \begin{pmatrix} v \\ q \end{pmatrix}; \lambda\right) = (F, v)_\Omega + \int_\Omega g\, \overline{q}\, d\omega,$$

where

$$(F, v)_\Omega = [u_\omega^{(0)}, v_\omega] + \int_\Omega \Bigg((\nabla_\omega u_r^{(0)}) \cdot \nabla_\omega \overline{v}_r + (\lambda + 2)(1 - \lambda)\left(2 u_r^{(0)}\, \overline{v}_r + u_\omega^{(0)} \cdot \overline{v}_\omega\right)$$

$$+ 2 u_r^{(0)}\, \nabla_\omega \cdot \overline{v}_\omega + 2(\nabla_\omega \cdot u_\omega^{(0)})\, \overline{v}_r$$

$$- (1 - \lambda)\, u_\omega^{(0)} \cdot \nabla_\omega \overline{v}_r - (\lambda + 2)(\nabla_\omega u_r^{(0)}) \cdot \overline{v}_\omega \Bigg)\, d\omega.$$

Obviously, F is a continuous functional on $\mathcal{H}_s$.

We show now that there exists a vector function $(u^{(1)}, p) \in \mathcal{H}_s \times L_2(\Omega)$ such that

$$(\lambda + 2)\, u_r^{(1)} + \nabla_\omega \cdot u_\omega^{(1)} = 0$$

and

$$a\left(\begin{pmatrix} u^{(1)} \\ p \end{pmatrix}, \begin{pmatrix} v \\ q \end{pmatrix}; \lambda\right) = (f - F, v)_\Omega$$

for $v \in \mathcal{H}_s$, $q \in L_2(\Omega)$. Then the vector function (u, p) with $u = u^{(0)} + u^{(1)}$ satisfies (6.1.9).

From (4.1.13) it follows that

$$a\left(\begin{pmatrix} u \\ 0 \end{pmatrix}, \begin{pmatrix} u \\ 0 \end{pmatrix}; \lambda\right) \geq c_0 \left(\|u_r\|_{W_2^1(\Omega)}^2 + \|u_\omega\|_{h_2^1(\Omega)}^2\right) - c_1(\lambda)\left(\|u_r\|_{L_2(\Omega)}^2 + \|u_\omega\|_{L_2(\Omega)}^2\right),$$

where c_0 and $c_1(\lambda)$ are positive constants independent of u_r and u_ω. Furthermore, for $\mathrm{Re}\,\lambda = -1/2$ we have

$$a\left(\binom{u}{0}, \binom{u}{0}; \lambda\right) = \int\limits_\Omega \left(2\,|\partial_\theta u_\theta + u_r|^2 + 2\left|\frac{1}{\sin\theta}\partial_\varphi u_\varphi + u_\theta\cot\theta + u_r\right|^2 \right.$$

$$+ 2\,|\lambda|^2\,|u_r|^2 + \left|\partial_\theta u_\varphi + \frac{1}{\sin\theta}\partial_\varphi u_\theta - \cot\theta\,u_\varphi\right|^2$$

$$\left. + \left|\partial_\theta u_r - (1-\lambda)u_\theta\right|^2 + \left|\frac{1}{\sin\theta}\partial_\varphi u_r - (1-\lambda)\,u_\varphi\right|^2 \right) d\omega.$$

Since this form vanishes only for $u = 0$, we get

$$(6.1.10) \qquad\qquad \left|a\left(\binom{u}{0}, \binom{u}{0}; \lambda\right)\right| \geq c\,\|u\|_{\mathcal{H}_s}^2$$

for $\mathrm{Re}\,\lambda = -1/2$, where the constant c depends on λ but not on u. In particular, (6.1.10) is satisfied for all $u \in \mathcal{H}_s^{(\lambda)} \overset{def}{=} \{u \in \mathcal{H}_s \;:\; (\lambda+2)u_r + \nabla_\omega \cdot u_\omega = 0\}$. Consequently, by Lax-Milgram's theorem, there exists a solution $u^{(1)} \in \mathcal{H}_s^{(\lambda)}$ of the equation

$$(6.1.11) \qquad a\left(\binom{u^{(1)}}{0}, \binom{v}{0}; \lambda\right) = (f - F, v)_\Omega, \qquad v \in \mathcal{H}_s^{(\lambda)}.$$

Analogously to the proof of Lemma 5.2.3, we can show that there exists a function $p \in L_2(\Omega)$ such that

$$a\left(\binom{u^{(1)}}{0}, \binom{v}{0}; \lambda\right) - (f - F, v)_\Omega = \int\limits_\Omega p\left((1-\lambda)\,\overline{v}_r + \nabla_\omega \cdot \overline{v}_\omega\right) d\omega$$

for all $v \in \mathcal{H}_s$. Then we have

$$a\left(\binom{u^{(1)}}{p}, \binom{v}{q}; \lambda\right) \;=\; a\left(\binom{u^{(1)}}{0}, \binom{v}{0}; \lambda\right) - \int\limits_\Omega p \cdot \left((1-\lambda)\overline{v}_r + \nabla_\omega \cdot \overline{v}_\omega\right) d\omega$$

$$\;=\; (f - F, v)_\Omega$$

for all $v \in \mathcal{H}_s$. This proves the lemma. ∎

Now we are able to prove the following analogue of Theorem 5.2.1.

THEOREM 6.1.1. 1) *The operator $\mathfrak{A}(\lambda)$ is Fredholm for all λ.*

2) *The spectrum of the pencil $\mathfrak{A}$ consists of isolated eigenvalues with finite algebraic multiplicities.*

3) *There are no eigenvalues of the pencil $\mathfrak{A}$ on the line $\mathrm{Re}\,\lambda = -1/2$.*

4) *The number λ_0 is an eigenvalue of the pencil $\mathfrak{A}$ if and only if $-1 - \overline{\lambda}_0$ is an eigenvalue of this pencil. The geometric, algebraic and partial multiplicities of the eigenvalues λ_0 and $-1 - \overline{\lambda}_0$ coincide.*

Proof: First we show that the kernel of the operator $\mathfrak{A}(\lambda)$ is trivial for $\mathrm{Re}\,\lambda = -1/2$. Let λ be a number on the line $\mathrm{Re}\,\lambda = -1/2$, and let (u, p) be an element of the kernel of $\mathfrak{A}(\lambda)$. Then

$$0 = a\left(\binom{u}{p}, \binom{0}{q}; \lambda\right) = -\int\limits_\Omega \left((\lambda+2)\,u_r + \nabla_\omega \cdot u_\omega\right) \overline{q}\,d\omega$$

for all $q \in L_2(\Omega)$. This implies

(6.1.12) $(\lambda + 2)\, u_r + \nabla_\omega \cdot u_\omega = 0$ in Ω.

Using (6.1.12) and the equality $\overline{\lambda} + 2 = 1 - \lambda$, we obtain

$$0 = a\left(\binom{u}{p}, \binom{u}{p}; \lambda\right) = [u_\omega, u_\omega] - \int_\Omega |\nabla_\omega \cdot u_\omega|^2\, d\omega + \int_\Omega \left|\nabla_\omega u_r - (1-\lambda) u_\omega\right|^2\, d\omega$$

$$+ \int_\Omega \left(\left|\nabla_\omega \cdot u_\omega + 2 u_r\right|^2 + \left(|1-\lambda|^2 - 2\right)|u_r|^2\right)\, d\omega.$$

Due to the inequalities (4.1.12) and $|1-\lambda|^2 \geq 9/4$, the last equality is satisfied only if $u_r = 0$ and $u_\omega = 0$. Furthermore, for $u = 0$ we get

$$a\left(\binom{u}{p}, \binom{v}{0}; \lambda\right) = -\int_\Omega p\left((1-\lambda)\overline{v}_r + \nabla_\omega \cdot \overline{v}_\omega\right)\, d\omega = 0, \qquad (v_r, v_\omega) \in \mathcal{H}_s.$$

In particular, the last equality is valid for all $v_r \in \overset{\circ}{W}{}_2^1(\Omega)$, $v_\omega = 0$. From this we conclude that $p = 0$. Thus, the kernel of the operator $\mathfrak{A}(\lambda)$ is trivial for $\mathrm{Re}\,\lambda = -1/2$. This result together with Lemma 6.1.2 implies that $\mathfrak{A}(\lambda)$ is an isomorphism for $\mathrm{Re}\,\lambda = -1/2$. Since $\mathfrak{A}(\lambda) - \mathfrak{A}(\mu)$ is a compact operator for arbitrary λ, μ, it follows that the operator $\mathfrak{A}(\lambda)$ is Fredholm for every complex λ. This proves assertions 1) and 3).

The second assertion follows from 1), from the invertibility of $\mathfrak{A}(-1/2)$ and from Theorem 1.1.1.

Finally, it can be easily seen that

(6.1.13) $a\left(\binom{u}{p}, \binom{v}{q}; \lambda\right) = \overline{a\left(\binom{v}{q}, \binom{u}{p}; -1-\overline{\lambda}\right)},$

i.e., $\mathfrak{A}(\lambda)^* = \mathfrak{A}(-1-\overline{\lambda})$. As a consequence of this equality, assertion 4) holds. ■

6.1.3. Properties of the space $\mathcal{H}_s$. In the sequel, the following properties of the spaces $\mathcal{H}$ and $\mathcal{H}_s$ will play an important role (cf. Lemma 4.1.3).

LEMMA 6.1.3. 1) *The subspace $\mathcal{H}_s$ admits the representation*

(6.1.14) $\mathcal{H}_s = \mathcal{H}_s^r \times \mathcal{H}_s^\omega,$

where $\mathcal{H}_s^r$ is a subspace of $W_2^1(\Omega)$, $\overset{\circ}{W}{}_2^1(\Omega) \subset \mathcal{H}_s^r$, and $\mathcal{H}_s^\omega$ is a subspace of $h_2^1(\Omega)$, $\overset{\circ}{h}{}_2^1(\Omega) \subset \mathcal{H}_s^\omega$.

2) *For all $u, v \in \mathcal{H}$ the equality*

(6.1.15) $\displaystyle\int_{\partial\Omega} u_n\, \overline{v}_r\, d\omega' = 0$

or, equivalently,

(6.1.16) $\displaystyle\int_\Omega \left((\nabla_\omega \cdot u_\omega)\overline{v}_r + u_\omega \cdot \nabla_\omega \overline{v}_r\right)\, d\omega = 0$

is valid.

REMARK 6.1.1. In the following, we will only use the properties (6.1.14) and (6.1.15) of the space $\mathcal{H}$. All results of Sections 6.1 and 6.2 are also true for other subspaces $\mathcal{H} \subset W_2^1(\Omega) \times h_2^1(\Omega)$ satisfying these conditions.

Due to (6.1.15), the sesquilinear form $a(\cdot, \cdot; \lambda)$ can be written on $\left(\mathcal{H}_s \times L_2(\Omega)\right)^2$ in the form

$$(6.1.17) \quad a\left(\begin{pmatrix} u \\ p \end{pmatrix}, \begin{pmatrix} v \\ q \end{pmatrix}; \lambda\right) = [u_\omega, v_\omega] + \int_\Omega \nabla_\omega u_r \cdot \nabla_\omega \bar{v}_r \, d\omega$$

$$+ \int_\Omega (\lambda + 2)(1 - \lambda)\left(2 u_r \bar{v}_r + u_\omega \cdot \bar{v}_\omega\right)$$

$$+ \int_\Omega \left((\lambda + 4) u_r \nabla_\omega \cdot \bar{v}_\omega + (3 - \lambda)(\nabla_\omega \cdot u_\omega) \bar{v}_r\right) d\omega$$

$$- \int_\Omega \left(p\left((1 - \lambda)\bar{v}_r + \nabla_\omega \cdot \bar{v}_\omega\right) + \left((\lambda + 2) u_r + \nabla_\omega \cdot u_\omega\right)\bar{q}\right) d\omega.$$

6.1.4. Connection between eigenvectors corresponding to the eigenvalues λ and $-1 - \lambda$. Let $\mathcal{S}(\lambda)$ be the matrix defined by (5.2.14). According to (6.1.14), the vector function $\mathcal{S}(\lambda)\begin{pmatrix} u \\ p \end{pmatrix}$ belongs to $\mathcal{H}_s \times L_2(\Omega)$ if $(u_r, u_\omega) \in \mathcal{H}_s$, $p \in L_2(\Omega)$. Moreover, we have

$$a\left(\begin{pmatrix} u \\ p \end{pmatrix}, \mathcal{S}(\bar{\lambda})\begin{pmatrix} v \\ q \end{pmatrix}; \lambda\right) = a\left(\mathcal{S}(\lambda)\begin{pmatrix} u \\ p \end{pmatrix}, \begin{pmatrix} v \\ q \end{pmatrix}; -1 - \lambda\right).$$

for all $u, v \in \mathcal{H}_s$, $p, q \in L_2(\Omega)$. Consequently, there is the equality

$$(6.1.18) \qquad \mathcal{S}(\lambda)^t \, \mathfrak{A}(\lambda) = \mathfrak{A}(-1 - \lambda) \, \mathcal{S}(\lambda)$$

for arbitrary $\lambda \in \mathbb{C}$, where $\mathcal{S}(\lambda)^t$ is the transposed matrix to $\mathcal{S}(\lambda)$. Thus, analogously to Theorem 5.2.2, the following statement holds.

THEOREM 6.1.2. *Suppose that λ_0 is an eigenvalue of the pencil $\mathfrak{A}$ and that the vector functions $(u^{(0)}, p^{(0)}), \ldots, (u^{(s)}, p^{(s)})$ form a Jordan chain corresponding to this eigenvalue.*

1) If $\lambda_0 \notin \{1, -2\}$ or $\lambda_0 \in \{1, -2\}$ and $\mathcal{S}(\lambda_0)\begin{pmatrix} u^{(0)} \\ p^{(0)} \end{pmatrix} \neq 0$, then $-1 - \lambda_0$ is also an eigenvalue of the pencil $\mathfrak{A}$ and the vector functions $\begin{pmatrix} v^{(0)} \\ q^{(0)} \end{pmatrix} = S(\lambda_0)\begin{pmatrix} u^{(0)} \\ p^{(0)} \end{pmatrix}$,

$$\begin{pmatrix} v^{(k)} \\ q^{(k)} \end{pmatrix} = (-1)^k \left(\mathcal{S}(\lambda_0)\begin{pmatrix} u^{(k)} \\ p^{(k)} \end{pmatrix} + \begin{pmatrix} 1 & 0 & 0 & 0 \\ 0 & -1 & 0 & 0 \\ 0 & 0 & -1 & 0 \\ 4 & 0 & 0 & -1 \end{pmatrix}\begin{pmatrix} u^{(k-1)} \\ p^{(k-1)} \end{pmatrix}\right),$$

$k = 1, \ldots, s$, form a Jordan chain corresponding to $-1 - \lambda_0$.

2) If $\lambda_0 \in \{1, -2\}$, $\mathcal{S}(\lambda_0)\begin{pmatrix} u^{(0)} \\ p^{(0)} \end{pmatrix} = 0$ and $s \geq 1$, then $-1 - \lambda_0$ is also an eigenvalue and the vector functions

$$\begin{pmatrix} v^{(k)} \\ q^{(k)} \end{pmatrix} = (-1)^k \left(\mathcal{S}(\lambda_0)\begin{pmatrix} u^{(k)} \\ p^{(k)} \end{pmatrix} + \begin{pmatrix} 1 & 0 & 0 & 0 \\ 0 & -1 & 0 & 0 \\ 0 & 0 & -1 & 0 \\ 4 & 0 & 0 & -1 \end{pmatrix}\begin{pmatrix} u^{(k-1)} \\ p^{(k-1)} \end{pmatrix}\right),$$

$k = 1, \ldots, s$, form a Jordan chain corresponding to this eigenvalue.

6.1.5. Eigenvalues in the strip $-2 \leq \operatorname{Re}\lambda \leq 1$. The sesquilinear form
(6.1.17) is not symmetric in its arguments. In order to obtain a symmetric form,
we introduce the matrix

$$J_\lambda = \begin{pmatrix} c & 0 & 0 & 0 \\ 0 & 1 & 0 & 0 \\ 0 & 0 & 1 & 0 \\ 2c-2 & 0 & 0 & 1 \end{pmatrix}, \qquad c = \frac{\lambda+2}{1-\overline{\lambda}},$$

which was already used in Chapter 5. Obviously, the sesquilinear form

$$(6.1.19) \qquad s\left(\begin{pmatrix} u \\ p \end{pmatrix}, \begin{pmatrix} v \\ q \end{pmatrix}; \lambda\right) \overset{def}{=} a\left(\begin{pmatrix} u \\ p \end{pmatrix}, J_\lambda \begin{pmatrix} v \\ q \end{pmatrix}; \lambda\right)$$

$$= [u_\omega, v_\omega] + \int_\Omega \left(\overline{c}(\nabla_\omega u_r)\cdot\nabla_\omega \overline{v}_r + 2(\lambda+2)(1-\overline{c}\lambda)u_r\,\overline{v}_r\right)d\omega$$

$$+ \int_\Omega (\lambda+2)(1-\lambda)u_\omega\cdot\overline{v}_\omega\,d\omega$$

$$+ \int_\Omega \left((\lambda+4)u_r\nabla_\omega\cdot\overline{v}_\omega + (\overline{\lambda}+4)(\nabla_\omega\cdot u_\omega)\overline{v}_r\right)d\omega$$

$$- \int_\Omega \left(p\left((\overline{\lambda}+2)\overline{v}_r + \nabla_\omega\cdot\overline{v}_\omega\right) + ((\lambda+2)u_r + \nabla_\omega\cdot u_\omega)\overline{q}\right)d\omega$$

is symmetric.

THEOREM 6.1.3. *The strip*

$$(6.1.20) \qquad \left|\operatorname{Re}\lambda + \frac{1}{2}\right| \leq \frac{3}{2}$$

contains only real eigenvalues of the pencil $\mathfrak{A}$.

Proof: Let λ be a complex eigenvalue in the strip (6.1.20) such that $\operatorname{Re}\lambda \neq -1/2$, $\operatorname{Im}\lambda \neq 0$ and let (u,p) be an eigenvector corresponding to this eigenvalue. Using (6.1.12), we obtain

$$0 = s\left(\begin{pmatrix} u \\ p \end{pmatrix}, \begin{pmatrix} u \\ p \end{pmatrix}; \lambda\right) = [u_\omega, u_\omega] + \int_\Omega \left(\frac{\overline{\lambda}+2}{1-\lambda}|\nabla_\omega u_r|^2 - 6\frac{\overline{\lambda}+2}{1-\lambda}|u_r|^2\right.$$

$$\left. + (\lambda+2)(1-\lambda)|u_\omega|^2\right)d\omega.$$

Since

$$\operatorname{Im} s\left(\begin{pmatrix} u \\ p \end{pmatrix}, \begin{pmatrix} u \\ p \end{pmatrix}; \lambda\right) = \operatorname{Im}\lambda\,(2\operatorname{Re}\lambda+1)\int_\Omega \left(\frac{1}{|1-\lambda|^2}|\nabla_\omega u_r|^2\right.$$

$$\left. - \frac{6}{|1-\lambda|^2}|u_r|^2 - |u_\omega|^2\right)d\omega,$$

and

$$\operatorname{Re} s\left(\begin{pmatrix} u \\ p \end{pmatrix}, \begin{pmatrix} u \\ p \end{pmatrix}; \lambda\right) = [u_\omega, u_\omega] + \frac{(1 - \operatorname{Re}\lambda)(2 + \operatorname{Re}\lambda) + (\operatorname{Im}\lambda)^2}{|1 - \lambda|^2} \int_\Omega |\nabla_\omega u_r|^2 \, d\omega$$

$$- 6\frac{(1 - \operatorname{Re}\lambda)(2 + \operatorname{Re}\lambda) + (\operatorname{Im}\lambda)^2}{|1 - \lambda|^2} \int_\Omega |u_r|^2 \, d\omega$$

$$+ \left((1 - \operatorname{Re}\lambda)(2 + \operatorname{Re}\lambda) + (\operatorname{Im}\lambda)^2\right) \int_\Omega |u_\omega|^2 \, d\omega,$$

we get

$$\operatorname{Re} s\left(\begin{pmatrix} u \\ p \end{pmatrix}, \begin{pmatrix} u \\ p \end{pmatrix}; \lambda\right) - \frac{(1 - \operatorname{Re}\lambda)(2 + \operatorname{Re}\lambda) + (\operatorname{Im}\lambda)^2}{\operatorname{Im}\lambda\,(2\operatorname{Re}\lambda + 1)} \operatorname{Im} s\left(\begin{pmatrix} u \\ p \end{pmatrix}, \begin{pmatrix} u \\ p \end{pmatrix}; \lambda\right)$$

$$= [u_\omega, u_\omega] + 2\left((1 - \operatorname{Re}\lambda)(2 + \operatorname{Re}\lambda) + (\operatorname{Im}\lambda)^2\right) \int_\Omega |u_\omega|^2 \, d\omega.$$

If λ satisfies inequality (6.1.20), then the right-hand side of the last equation may be zero only in the case $u_\omega = 0$. Then from (6.1.12) it follows that $u_r = 0$. Hence the form (6.1.17) is only zero if

$$\int_\Omega p\left((1 - \lambda)\bar{v}_r + \nabla_\omega \cdot \bar{v}_\omega\right) d\omega = 0 \qquad \text{for all } v \in \mathcal{H}_s.$$

Since the last equality is satisfied, in particular, for arbitrary $v_r \in \overset{\circ}{W}{}^1_2(\Omega)$ and $v_\omega = 0$, we obtain $p = 0$. Thus, the theorem is proved for the case $\operatorname{Re}\lambda \neq -1/2$. For $\operatorname{Re}\lambda = -1/2$ the assertion of the theorem follows from the third part of Theorem 6.1.1. ∎

6.1.6. Absence of generalized eigenvectors. Our goal is to show that there are no generalized eigenvectors to eigenvalues in the strip $-2 < \operatorname{Re}\lambda < 1$. Since the eigenvalues in this strip are real (see Theorem 6.1.3), we consider the sesquilinear form (6.1.19) in this subsection as a form depending on the real parameter λ.

LEMMA 6.1.4. *If λ_0 is a real eigenvalue of the operator pencil $\mathfrak{A}$ in the interval $(-1/2, 1)$ and $(u^{(0)}, p^{(0)})$ is an eigenvector corresponding to this eigenvalue, then*

$$(6.1.21) \qquad \frac{d}{d\lambda} a\left(\begin{pmatrix} u^{(0)} \\ p^{(0)} \end{pmatrix}, J_\lambda \begin{pmatrix} u^{(0)} \\ p^{(0)} \end{pmatrix}; \lambda\right)\Big|_{\lambda=\lambda_0} < 0$$

Proof: The left side of (6.1.21) is equal to

$$(6.1.22) \quad \frac{d}{d\lambda} s\left(\begin{pmatrix} u^{(0)} \\ p^{(0)} \end{pmatrix}, \begin{pmatrix} u^{(0)} \\ p^{(0)} \end{pmatrix}; \lambda\right)\Big|_{\lambda=\lambda_0}$$

$$= \int_\Omega \left(\frac{3}{(1-\lambda_0)^2} |\nabla_\omega u_r^{(0)}|^2 + 2\left(2\lambda_0 + 5 - \frac{\lambda_0 + 2}{1 - \lambda_0} - 3\frac{\lambda_0 + 2}{(1 - \lambda_0)^2}\right)|u_r^{(0)}|^2\right.$$

$$- (2\lambda_0 + 1)|u_\omega^{(0)}|^2 + u_r^{(0)} \nabla_\omega \cdot \overline{u_\omega^{(0)}} + (\nabla_\omega \cdot u_\omega^{(0)})\overline{u_r^{(0)}}$$

$$\left. - p^{(0)}\overline{u_r^{(0)}} - u_r^{(0)}\overline{p^{(0)}}\right) d\omega.$$

Moreover, since $(u^{(0)}, p^{(0)})$ is an eigenvector, we have

$$(6.1.23) \qquad a\left(\begin{pmatrix} u^{(0)} \\ p^{(0)} \end{pmatrix}, \begin{pmatrix} v \\ q \end{pmatrix}; \lambda_0\right) = 0 \qquad \text{for all } v \in \mathcal{H}_s,\ q \in L_2(\Omega).$$

Inserting $v = 0$ into (6.1.23), we get

$$(6.1.24) \qquad (\lambda_0 + 2)\, u_r^{(0)} + \nabla_\omega \cdot u_\omega^{(0)} = 0 \quad \text{in } \Omega.$$

Furthermore, substituting $v = (u_r^{(0)}, 0)$, $q = 0$ into (6.1.23), we arrive at the equality

$$(6.1.25) \qquad \int_\Omega \Big(|\nabla_\omega u_r^{(0)}|^2 + 2\,(\lambda_0 + 2)\,(1 - \lambda_0)\,|u_r^{(0)}|^2$$

$$+ (3 - \lambda_0)\,(\nabla_\omega \cdot u_\omega^{(0)})\,\overline{u_r^{(0)}} - (1 - \lambda_0)\, p^{(0)}\,\overline{u_r^{(0)}} \Big)\, d\omega = 0.$$

Multiplying the real part of the left-hand side of (6.1.25) by $2\,(\lambda_0 - 1)^{-1}$ and adding this to the right-hand side of (6.1.22), we find that

$$\frac{d}{d\lambda}\, s\left(\begin{pmatrix} u^{(0)} \\ p^{(0)} \end{pmatrix}, \begin{pmatrix} u^{(0)} \\ p^{(0)} \end{pmatrix}; \lambda\right)\Big|_{\lambda=\lambda_0}$$

$$= \int_\Omega \left(\frac{1 + 2\lambda_0}{(1 - \lambda_0)^2}\,|\nabla_\omega u_r^{(0)}|^2 - 2\left(\frac{1 + 2\lambda_0}{1 - \lambda_0} + 3\,\frac{\lambda_0 + 2}{(1 - \lambda_0)^2} \right) |u_r^{(0)}|^2 \right.$$

$$\left. - (2\lambda_0 + 1)\,|u_\omega^{(0)}|^2 - \frac{2}{1 - \lambda_0}\left(u_r^{(0)}\,\overline{\nabla_\omega \cdot u_\omega^{(0)}} + (\nabla_\omega \cdot u_\omega^{(0)})\,u_r^{(0)} \right) \right) d\omega.$$

We eliminate $\nabla_\omega \cdot u_\omega^{(0)}$ in the last equation by means of (6.1.24). Then we obtain

$$\frac{d}{d\lambda}\, s\left(\begin{pmatrix} u^{(0)} \\ p^{(0)} \end{pmatrix}, \begin{pmatrix} u^{(0)} \\ p^{(0)} \end{pmatrix}; \lambda\right)\Big|_{\lambda=\lambda_0}$$

$$= (1 + 2\lambda_0) \int_\Omega \left(\frac{1}{(1 - \lambda_0)^2}\,|\nabla_\omega u_r^{(0)}|^2 - \frac{6}{(1 - \lambda_0)^2}\,|u_r^{(0)}|^2 - |u_\omega^{(0)}|^2 \right) d\omega.$$

Since $s\left(\begin{pmatrix} u^{(0)} \\ p^{(0)} \end{pmatrix}, \begin{pmatrix} u^{(0)} \\ p^{(0)} \end{pmatrix}; \lambda_0\right) = 0$, it follows that

$$-\frac{(\lambda_0 + 2)\,(1 - \lambda_0)}{1 + 2\lambda_0}\,\frac{d}{d\lambda}\, s\left(\begin{pmatrix} u^{(0)} \\ p^{(0)} \end{pmatrix}, \begin{pmatrix} u^{(0)} \\ p^{(0)} \end{pmatrix}; \lambda\right)\Big|_{\lambda=\lambda_0}$$

$$= s\left(\begin{pmatrix} u^{(0)} \\ p^{(0)} \end{pmatrix}, \begin{pmatrix} u^{(0)} \\ p^{(0)} \end{pmatrix}; \lambda_0\right) - \frac{(\lambda_0 + 2)\,(1 - \lambda_0)}{1 + 2\lambda_0}\,\frac{d}{d\lambda}\, s\left(\begin{pmatrix} u^{(0)} \\ p^{(0)} \end{pmatrix}, \begin{pmatrix} u^{(0)} \\ p^{(0)} \end{pmatrix}; \lambda\right)\Big|_{\lambda=\lambda_0}$$

$$= [u_\omega, u_\omega] + 2\,(\lambda_0 + 2)\,(1 - \lambda_0) \int_\Omega |u_\omega^{(0)}|^2\, d\omega.$$

The right-hand side of the last equality is nonnegative and is equal to zero only if $u_\omega^{(0)} = 0$. In this case (6.1.24) implies $u_r^{(0)} = 0$ and, analogously to the proof of Theorem 6.1.1, we get $p^{(0)} = 0$. Hence λ_0 is not an eigenvalue. This proves the lemma. ∎

THEOREM 6.1.4. *The eigenfunctions of the operator pencil $\mathfrak{A}$ which correspond to eigenvalues in the strip $|\operatorname{Re}\lambda + 1/2| < 3/2$ do not have generalized eigenfunctions.*

Proof: By the last assertion of Theorem 6.1.1 and by Theorem 6.1.3, it suffices to prove the theorem for real eigenvalues in the interval $(-1/2, 1)$.

Let $\lambda_0 \in (-1/2, 1)$ be an eigenvalue of the pencil $\mathfrak{A}$, $(u^{(0)}, p^{(0)})$ an eigenvector to this eigenvalue, and $(u^{(1)}, p^{(1)})$ a corresponding generalized eigenvector. Then the equations

$$(6.1.26) \quad a\left(\begin{pmatrix} u^{(0)} \\ p^{(0)} \end{pmatrix}, \begin{pmatrix} v \\ q \end{pmatrix}; \lambda_0\right) = 0,$$

$$(6.1.27) \quad a\left(\begin{pmatrix} u^{(1)} \\ p^{(1)} \end{pmatrix}, \begin{pmatrix} v \\ q \end{pmatrix}; \lambda_0\right) = -\frac{d}{d\lambda} a\left(\begin{pmatrix} u^{(0)} \\ p^{(0)} \end{pmatrix}, \begin{pmatrix} v \\ q \end{pmatrix}; \lambda\right)\Big|_{\lambda=\lambda_0}$$

are satisfied for all $(v, q) \in \mathcal{H}_s \times L_2(\Omega)$. Inserting $\begin{pmatrix} v \\ q \end{pmatrix} = J_{\lambda_0} \begin{pmatrix} u^{(1)} \\ p^{(1)} \end{pmatrix}$ into (6.1.26), we get

$$s\left(\begin{pmatrix} u^{(0)} \\ p^{(0)} \end{pmatrix}, \begin{pmatrix} u^{(1)} \\ p^{(1)} \end{pmatrix}; \lambda_0\right) = 0.$$

By the symmetry of the form s, we conclude from this that

$$(6.1.28) \quad 0 = s\left(\begin{pmatrix} u^{(1)} \\ p^{(1)} \end{pmatrix}, \begin{pmatrix} u^{(0)} \\ p^{(0)} \end{pmatrix}; \lambda_0\right) = a\left(\begin{pmatrix} u^{(1)} \\ p^{(1)} \end{pmatrix}, J_{\lambda_0} \begin{pmatrix} u^{(0)} \\ p^{(0)} \end{pmatrix}; \lambda_0\right).$$

Using (6.1.28) and inserting $\begin{pmatrix} v \\ q \end{pmatrix} = J_{\lambda_0} \begin{pmatrix} u^{(0)} \\ p^{(0)} \end{pmatrix}$ into (6.1.27), we obtain

$$\frac{d}{d\lambda} a\left(\begin{pmatrix} u^{(0)} \\ p^{(0)} \end{pmatrix}, J_{\lambda_0} \begin{pmatrix} u^{(0)} \\ p^{(0)} \end{pmatrix}; \lambda\right)\Big|_{\lambda=\lambda_0} = 0.$$

From this and (6.1.26) we get

$$\frac{d}{d\lambda} a\left(\begin{pmatrix} u^{(0)} \\ p^{(0)} \end{pmatrix}, J_\lambda \begin{pmatrix} u^{(0)} \\ p^{(0)} \end{pmatrix}; \lambda\right)\Big|_{\lambda=\lambda_0} = 0.$$

Since this contradicts Lemma 6.1.4, the theorem is proved. ∎

6.1.7. The strip $-1 \leq \operatorname{Re}\lambda \leq 0$. The following theorem describes the spectrum of the pencil $\mathfrak{A}$ in the strip $-1 \leq \operatorname{Re}\lambda \leq 0$.

THEOREM 6.1.5. 1) *The set*

$$\{\lambda \in \mathbb{C} : -1 \leq \operatorname{Re}\lambda \leq 0\}\backslash\{0, -1\}$$

does not contain eigenvalues of the pencil $\mathfrak{A}$.

2) *Let $\mathcal{H}_c$ be the set of constant vectors $u = (u_1, u_2, u_3) \in \mathbb{C}^3$ which belong to $\mathcal{H}$. The numbers 0 and -1 are eigenvalues if and only if $\mathcal{H}_c \neq \{0\}$. Both values have the same geometric and algebraic multiplicities. The set of the eigenvectors to $\lambda_0 = 0$ coincides with the set $\{(u, 0) : u \in \mathcal{H}_c\}$, while generalized eigenvectors do not exist.*

Proof: Due to Theorems 6.1.1 and 6.1.3, we can restrict ourselves in the proof to real eigenvalues in the interval $[-1/2, 0]$.

Suppose that (u, p) is an eigenvector to the eigenvalue $\lambda \in [-1/2, 0]$. Then, according to (6.1.12), we get

$$(6.1.29) \quad 0 = s\left(\binom{u}{p}, \binom{u}{p}; \lambda\right)$$

$$= [u_\omega, u_\omega] + \int_\Omega c\left(|\nabla_\omega u_r|^2 + (\lambda + 2)(1 - \lambda)\,|u_\omega|^2 - (\lambda + 2)(\nabla_\omega u_r)\,\overline{u}_\omega\right.$$

$$- (\lambda + 2)\,u_\omega \cdot \nabla_\omega \overline{u}_r + 2\,(\lambda + 2)(1 - c\lambda)\,|u_r|^2$$

$$\left. + 2\,u_r\,\nabla_\omega \cdot \overline{u}_\omega + 2\,(\nabla_\omega \cdot u_\omega)\,\overline{u}_r\right) d\omega$$

$$= [u_\omega, u_\omega] - \int_\Omega |\nabla_\omega \cdot u_\omega|^2\,d\omega + \int_\Omega c\,|\nabla_\omega u_r - (1 - \lambda)u_\omega|^2\,d\omega$$

$$+ \int_\Omega |\nabla_\omega \cdot u_\omega + 2u_r|^2\,d\omega + 2\left((\lambda + 2)\,(1 - c\lambda) - 2\right) \int_\Omega |u_r|^2\,d\omega,$$

where $c = (\lambda + 2)/(1 - \lambda) > 0$ and

$$(\lambda + 2)\,(1 - c\lambda) - 2 = -\frac{\lambda}{1 - \lambda}\,(\lambda^2 + 5\lambda + 3) \begin{cases} = 0 & \text{for } \lambda = 0, \\ > 0 & \text{for } \lambda \in [-1/2, 0). \end{cases}$$

In the case $\lambda \in [-1/2, 0)$ we conclude from (4.1.15) and (6.1.29) that $u_r = 0$ and $u_\omega = 0$. Consequently, for arbitrary $v_r \in \overset{\circ}{W}{}^1_2(\Omega)$ and $v_\omega = 0$ we have

$$-(\lambda + 2) \int_\Omega p\,\overline{v}_r\,d\omega = s\left(\binom{u}{p}, \binom{v}{0}; 0\right) = 0$$

and, therefore, $p = 0$. This proves the first part of the theorem.

If $\lambda = 0$, then (6.1.29) together with (4.1.15) yields

$$(6.1.30) \qquad\qquad \nabla_\omega \cdot u_\omega + 2u_r = 0,$$

$$(6.1.31) \qquad\qquad \nabla_\omega u_r - u_\omega = 0,$$

$$(6.1.32) \qquad\qquad \partial_\theta u_\theta - \frac{1}{\sin\theta}\partial_\varphi u_\varphi - \cot\theta\,u_\theta = 0,$$

$$(6.1.33) \qquad\qquad \partial_\theta u_\varphi + \frac{1}{\sin\theta}\partial_\varphi u_\theta - \cot\theta\,u_\varphi = 0.$$

Summing up (6.1.30) and (6.1.32), we get $u_r + \partial_\theta u_\theta = 0$. Furthermore, by (6.1.31), we have $u_\theta = \partial_\theta u_r$ and, consequently, $u_r + \partial_\theta^2 u_r = 0$. Hence the functions u_r, u_θ have the form

$$\begin{aligned} u_r &= c_1(\varphi)\cos\theta + c_2(\varphi)\sin\theta, \\ u_\theta &= c_2(\varphi)\cos\theta - c_1(\varphi)\sin\theta. \end{aligned}$$

Inserting $u_\omega = \nabla_\omega u_r$ (i.e., $u_\theta = \partial_\theta u_r$, $u_\varphi = (\sin\theta)^{-1}\,\partial_\varphi u_r$) into (6.1.33), we arrive at

$$\partial_\theta\left((\sin\theta)^{-1}\,\partial_\varphi u_r\right) = 0.$$

Consequently, the function c_1 in the representation of u_r is constant. Moreover, it holds $u_\varphi = (\sin\theta)^{-1}\,\partial_\varphi u_r = c_2'(\varphi)$. We insert this representation for u_φ and the representations given above for u_r and u_θ into (6.1.30). Then we obtain

$$c_2(\varphi) + c_2''(\varphi) = 0.$$

This implies $c_2 = a \cos\varphi + b \sin\varphi$ and

$$
\begin{pmatrix} u_r \\ u_\theta \\ u_\varphi \end{pmatrix} = \begin{pmatrix} \sin\theta\cos\varphi & \sin\theta\sin\varphi & \cos\theta \\ \cos\theta\cos\varphi & \cos\theta\sin\varphi & -\sin\theta \\ -\sin\varphi & \cos\varphi & 0 \end{pmatrix} \begin{pmatrix} a \\ b \\ c_1 \end{pmatrix} = J \cdot \begin{pmatrix} a \\ b \\ c_1 \end{pmatrix},
$$

where a, b, c_1 are constants. Furthermore, for $v_\omega = 0$, $q = 0$ we have

$$
s\left(\begin{pmatrix} u \\ p \end{pmatrix}, \begin{pmatrix} v \\ q \end{pmatrix}; 0\right) = 2\int_\Omega \left(\nabla_\omega u_r \cdot \nabla_\omega \bar{v}_r + (2u_r + 2\nabla_\omega \cdot u_\omega - p)\bar{v}_r\right) d\omega = 0
$$

Using (6.1.30), (6.1.31) and (6.1.16), we get $\int_\Omega p\,\bar{v}_r\,d\omega = 0$ for all $v_r \in \overset{\circ}{W}{}^1_2(\Omega)$, i.e., $p = 0$.

Thus, $\lambda_0 = 0$ is the only possible eigenvalue in the strip $-1/2 \le \operatorname{Re}\lambda \le 0$ and the corresponding eigenvectors (in the Cartesian coordinate system) have the form $(u, 0)$, where $u \in \mathcal{H}_c$.

The assertions of the theorem concerning the number $\lambda_1 = -1$ follow from Theorem 6.1.1. The proof is complete. ∎

REMARK 6.1.2. The condition $\mathcal{H}_c \ne \{0\}$ in the second part of Theorem 6.1.5 is satisfied only in the following three cases:

1. $\mathcal{K}$ is a dihedron bounded by two half-planes Γ_+, Γ_- and $\mathcal{H} = \{u \in W^1_2(\Omega)^3 : u_n = 0 \text{ on } \gamma_\pm\}$, where $\gamma_\pm = \Gamma_\pm \cap S^2$
2. $\mathcal{K}$ is a dihedron bounded by two half-planes Γ_+, Γ_- which are orthogonal to each other and $\mathcal{H} = \{u \in W^1_2(\Omega)^3 : u_n = 0 \text{ on } \gamma_+, u_\tau = 0 \text{ on } \gamma_-\}$
3. $\mathcal{K}$ is a half-space bounded by a plane Γ and $\mathcal{H} = \{u \in W^1_2(\Omega)^3, u_\tau = 0 \text{ on } \gamma = \Gamma \cap S^2\}$

In all other cases the numbers 0 and -1 do not belong to the spectrum of the pencil $\mathfrak{A}$.

6.2. Real eigenvalues of the pencil to the mixed problem

We continue to study the operator pencil $\mathfrak{A}$ introduced in the preceding section. As we have shown there, the eigenvalues of $\mathfrak{A}$ in the strip $-2 \le \operatorname{Re}\lambda \le 1$ are real and the interval $(-1, 0)$ is free of eigenvalues. In this section we study the eigenvalues in the interval $[0, 1]$. We describe all eigenvectors and generalized eigenvectors corresponding to $\lambda = 1$ when this number is an eigenvalue of the pencil $\mathfrak{A}$. Furthermore, we derive a variational principle for the eigenvalues of $\mathfrak{A}$ in the interval $[0, 1)$ and find the total multiplicity of the spectrum in $[0, 1)$. We show that, in the case of the boundary conditions (i) and (iii) on different parts of $\partial\mathcal{K}$, this multiplicity is an increasing function of $\mathcal{K}$.

6.2.1. Classification of subspaces of $W^1_2(\Omega)^3$.
Some of the results of this section depend on the validity of an additional condition on the subspace $\mathcal{H}$.

DEFINITION 6.2.1. The space $\mathcal{H} \subset W^1_2(\Omega)^3$ is said to be a *subspace of the first kind* if

$$
\int_{\partial\Omega} u_n \, d\omega' = 0 \qquad \text{for all } u \in \mathcal{H}.
$$

Otherwise, $\mathcal{H}$ will be called a *subspace of the second kind*.

REMARK 6.2.1. Obviously, $\mathcal{H}$ is a subspace of the first kind if $I_\tau = \emptyset$ and a subspace of the second kind if $I_\tau \neq \emptyset$.

Furthermore, we note that

$$\int\limits_{\partial\Omega} u_n \, d\omega' = \int\limits_{\Omega} \nabla_\omega \cdot u_\omega \, d\omega$$

for arbitrary $u \in W_2^1(\Omega)^3$. This equality holds if we set $v_r = 1$ in the second part of the proof of Lemma 4.1.3.

LEMMA 6.2.1. *Let f be an arbitrary function in $L_2(\Omega)$.*
1) *If $\mathcal{H}$ is a subspace of the first kind, then for the solvability of the equation*

(6.2.1) $$\nabla_\omega \cdot u_\omega = f$$

in space $\mathcal{H}_s^\omega$ it is necessary and sufficient that

(6.2.2) $$\int\limits_{\Omega} f \, d\omega = 0.$$

2) *If $\mathcal{H}$ is a subspace of the second kind, then the equation (6.2.1) is always solvable in $\mathcal{H}_s^\omega$.*
3) *There exists a constant c such that in both cases a solution $u_\omega \in \mathcal{H}_s^\omega$ of the equation (6.2.1) can be chosen which satisfies the estimate*

(6.2.3) $$\|u_\omega\|_{h_2^1(\Omega)} \leq c \|f\|_{L_2(\Omega)}.$$

Proof: 1) It is evident that the condition (6.2.2) is necessary for the solvability of the equation (6.2.1). By Lemma 5.2.1, this condition is also sufficient for the solvability of (6.2.1) in the space $\overset{\circ}{h}_2^1(\Omega)$. This proves the first assertion.

2) Let $u_\omega^{(0)} \in \mathcal{H}_s^\omega$ and $\int_\Omega \nabla_\omega \cdot u_\omega^{(0)} \, d\omega \neq 0$. We write the solution $u_\omega \in \mathcal{H}_s^\omega$ of (6.2.1) in the form

$$u_\omega = u_\omega^{(1)} + c\, u_\omega^{(0)}, \qquad \text{where} \quad c = \left(\int\limits_\Omega \nabla_\omega \cdot u_\omega^{(0)} \, d\omega\right)^{-1} \cdot \int\limits_\Omega f \, d\omega.$$

Then an equation of the form $\nabla_\omega \cdot u_\omega^{(1)} = g$ holds for $u_\omega^{(1)}$, where g satisfies the condition (6.2.2). This equation is solvable in $\overset{\circ}{h}_2^1(\Omega)$. Thus, the second assertion is proved.

3) We denote by $\mathcal{X}_0$ the space $\{u_\omega \in \mathcal{H}_s^\omega : \nabla_\omega \cdot u_\omega = 0\}$ and by $\mathcal{X}_1$ any direct complement in $\mathcal{H}_s^\omega$. Then, by assertions 1) and 2), the operator $u_\omega \to \nabla_\omega \cdot u_\omega$ is an injective mapping from $\mathcal{X}_1$ onto the space $\{f \in L_2(\Omega) : \int_\Omega f \, d\omega = 0\}$ if $\mathcal{H}$ is a subspace of the first kind and onto $L_2(\Omega)$ if $\mathcal{H}$ is a subspace of the second kind. Consequently, if we choose the solution of (6.2.1) from the space $\mathcal{X}_1$, the estimate (6.2.3) with a constant c independent of f is satisfied. The proof is complete. ∎

6.2.2. An auxiliary problem. Now we deal with the number $\lambda = 1$ which, by Theorem 6.1.3, is the only possible eigenvalue on the line $\operatorname{Re}\lambda = 1$.

For $\lambda = 1$ the sesquilinear form (6.1.17) takes the form

$$(6.2.4) \qquad a\left(\begin{pmatrix} u \\ p \end{pmatrix}, \begin{pmatrix} v \\ q \end{pmatrix}; 1\right) = [u_\omega, v_\omega] + \int_\Omega \left(\nabla_\omega u_r \cdot \nabla_\omega \overline{v}_r + 5\, u_r\, \nabla_\omega \cdot \overline{v}_\omega\right) d\omega$$

$$+ \int_\Omega \left(2\,(\nabla_\omega \cdot u_\omega)\,\overline{v}_r - p\,\nabla_\omega \cdot \overline{v}_\omega - (3\,u_r + \nabla_\omega \cdot u_\omega)\,\overline{q}\right) d\omega.$$

Consequently, every eigenvector (u, p) corresponding to the eigenvalue $\lambda = 1$ satisfies

$$(6.2.5) \qquad\qquad 3\,u_r + \nabla_\omega \cdot u_\omega = 0 \quad \text{in } \Omega.$$

Furthermore, inserting $v_\omega = 0$, $q = 0$ into (6.2.4) and using (6.2.5), we obtain

$$(6.2.6) \qquad \int_\Omega \left((\nabla_\omega u_r) \cdot \nabla_\omega \overline{v}_r - 6\,u_r\,\overline{v}_r\right) d\omega = 0 \qquad \text{for all } v_r \in \mathcal{H}_s^\tau.$$

From (6.2.4)–(6.2.6) it follows that

$$(6.2.7) \qquad [u_\omega, v_\omega] + \int_\Omega \left(5\,u_r\,\nabla_\omega \cdot \overline{v}_\omega - p\,\nabla_\omega \cdot \overline{v}_\omega\right) d\omega = 0 \quad \text{for all } v_\omega \in \mathcal{H}_s^\omega.$$

The equation (6.2.7) together with (6.2.5) can be considered as a variational problem for the vector function (u_ω, p). We consider the following generalization of this problem.

PROBLEM P. *Let the functions* $f, g \in L_2(\Omega)$ *and the vector function* $F \in L_2(\Omega)^2$ *be given. We seek a vector function* $(u_\omega, p) \in \mathcal{H}_s^\omega \times L_2(\Omega)$ *which satisfies the integral identity*

$$(6.2.8) \qquad [u_\omega, v_\omega] + \int_\Omega \left(-p\,\nabla_\omega \cdot \overline{v}_\omega + 2\,(\nabla_\omega \cdot u_\omega)\,\overline{q}\right) d\omega$$

$$= \int_\Omega \left(f\,\nabla_\omega \cdot \overline{v}_\omega + F \cdot \overline{v}_\omega + g\,\overline{q}\right) d\omega$$

for all $v_\omega \in \mathcal{H}_s^\omega$, $q \in L_2(\Omega)$.

REMARK 6.2.2. From the integral identity (6.2.8) it follows that

$$(6.2.9) \qquad\qquad 2\,\nabla_\omega \cdot u_\omega = g \qquad \text{in } \Omega.$$

Therefore, (6.2.8) can be written in the form

$$(6.2.10) \qquad [u_\omega, v_\omega] - \int_\Omega p\,\nabla_\omega \cdot \overline{v}_\omega \, d\omega = \int_\Omega \left(f\,\nabla_\omega \cdot \overline{v}_\omega + F \cdot \overline{v}_\omega\right) d\omega,$$

where v_ω is an arbitrary element of the space $\mathcal{H}_s^\omega$. Clearly, the integral identity (6.2.8) is equivalent to the validity of the equality (6.2.9) and the relation (6.2.10).

6.2.3. Solvability of the auxiliary problem. Before turning to the solvability of Problem P, we prove the following lemma.

LEMMA 6.2.2. *The set of the vector functions* $u_\omega \in h_2^1(\Omega)$ *satisfying the equality*

$$(6.2.11) \qquad\qquad [u_\omega, u_\omega] = 0$$

is exhausted by linear combinations of the vector functions

$$(6.2.12) \quad u_\omega^{(1)} = \begin{pmatrix} \sin\varphi \\ \cos\theta\,\cos\varphi \end{pmatrix}, \quad u_\omega^{(2)} = \begin{pmatrix} \cos\varphi \\ -\cos\theta\,\sin\varphi \end{pmatrix}, \quad u_\omega^{(3)} = \begin{pmatrix} 0 \\ \sin\theta \end{pmatrix}.$$

Proof: Equation (6.2.11) is equivalent to

$$(6.2.13) \qquad\qquad \partial_\theta u_\theta = 0 \quad \text{in } \Omega,$$

$$(6.2.14) \qquad\qquad \partial_\varphi u_\varphi + \cos\theta\, u_\theta = 0 \quad \text{in } \Omega,$$

$$(6.2.15) \qquad\qquad \partial_\theta\left(\frac{u_\varphi}{\sin\theta}\right) + \frac{1}{\sin^2\theta}\partial_\varphi u_\theta = 0 \quad \text{in } \Omega.$$

From (6.2.13) it follows that $u_\theta = u_\theta(\varphi)$. Hence from (6.2.14) we conclude that

$$u_\varphi = -\cos\theta\, F(\varphi) + h(\theta), \qquad \text{where } F' = u_\theta.$$

Using now (6.2.15), we find

$$u_\varphi = \cos\theta\,\partial_\varphi u_\theta + \sin\theta\, g(\varphi).$$

Consequently,

$$\cos\theta\,\big(\partial_\varphi u_\theta + F(\varphi)\big) + \sin\theta\, g(\varphi) = h(\theta).$$

This implies

$$F = a\,\cos\varphi + b\,\sin\varphi + c_1, \quad g = c_2, \quad h(\theta) = c_1\,\cos\theta + c_2\,\sin\theta,$$

where a, b, c_1, c_2 are constants. Thus, we get

$$u_\theta(\varphi) = b\,\cos\varphi - a\,\sin\varphi, \qquad u_\varphi = -\cos\theta\,(a\,\cos\varphi + b\,\sin\varphi) + c_2\,\sin\theta.$$

This proves the lemma. ∎

REMARK 6.2.3. 1) By (4.1.12), we have $\nabla_\omega \cdot u_\omega = 0$ for every vector function $u_\omega \in h_2^1(\Omega)$ satisfying (6.2.11).

2) Let $u_\omega^{(j)}$, $j = 1, 2, 3$, be the vectors (6.2.12). Then the vectors $r\begin{pmatrix} 0 \\ u_\omega^{(j)} \end{pmatrix}$, $j = 1, 2, 3$, have the form of rigid body rotations

$$\begin{pmatrix} 0 \\ x_3 \\ -x_2 \end{pmatrix}, \quad \begin{pmatrix} x_3 \\ 0 \\ -x_1 \end{pmatrix}, \quad \begin{pmatrix} -x_2 \\ x_1 \\ 0 \end{pmatrix}$$

in the Cartesian system of coordinates.

We denote by $\mathcal{H}_0$ the set of all linear combinations of the vector functions (6.2.12) which belong to the space $\mathcal{H}_s^\omega$.

LEMMA 6.2.3. *Let* $f = g = 0$ *and* $F = 0$. *Then the following assertions are valid for the solutions of* Problem P.

1) *If* $\mathcal{H}$ *is a subspace of the first kind, then the solutions of* Problem P *are exhausted by the vector functions* (u_ω, p), *where* $u_\omega \in \mathcal{H}_0$, $p = const \in \mathbb{C}$.

2) *If* $\mathcal{H}$ *is a subspace of the second kind, then the solutions of* Problem P *are exhausted by the vector functions* $(u_\omega, 0)$, *where* $u_\omega \in \mathcal{H}_0$.

Proof: For $f = g = 0$, $F = 0$ it follows from (6.2.8) that $\nabla_\omega \cdot u_\omega = 0$ and

$$[u_\omega, v_\omega] - \int_\Omega p \, \nabla_\omega \cdot \overline{v}_\omega \, d\omega = 0 \qquad \text{for all } v \in \mathcal{H}_s^\omega$$

(see Remark 6.2.2). Consequently, $[u_\omega, u_\omega] = 0$. This implies $[u_\omega, v_\omega] = 0$ for all $v_\omega \in \mathcal{H}_s^\omega$ and

$$\int_\Omega p \, \nabla_\omega \cdot \overline{v}_\omega \, d\omega = 0 \qquad \text{for all } v_\omega \in \mathcal{H}_s^\omega.$$

Since $\overset{\circ}{h}{}^1_2(\Omega) \subset \mathcal{H}_s^\omega$, the last equality can be satisfied only for constant p. Now the assertions of the lemma follow from Definition 6.2.1 and Lemma 6.2.2. $\blacksquare$

Next we consider the inhomogeneous Problem P.

LEMMA 6.2.4. *Let* $f, g \in L_2(\Omega)$ *and* $F \in L_2(\Omega)^2$.
1) *If* $\mathcal{H}$ *is a subspace of the first kind, then for the solvability of* Problem P *it is necessary and sufficient that*

$$(6.2.16) \qquad \int_\Omega g \, d\omega = 0 \qquad \text{and} \qquad \int_\Omega F \cdot \overline{v}_\omega \, d\omega = 0 \quad \text{for all } v_\omega \in \mathcal{H}_0.$$

2) *If* $\mathcal{H}$ *is a subspace of the second kind, then the condition*

$$(6.2.17) \qquad \int_\Omega F \cdot \overline{v}_\omega \, d\omega = 0 \quad \text{for all } v_\omega \in \mathcal{H}_0.$$

is necessary and sufficient for the solvability of Problem P.

Proof: The proof of the necessity of conditions (6.2.16), (6.2.17) is obvious. We prove the sufficiency of these conditions.

Let $\mathcal{H}$ be a subspace of the first kind. As we have mentioned in Remark 6.2.2, the solution (u_ω, p) of Problem P satisfies the equality $2\nabla_\omega \cdot u_\omega = g$. Using the first condition in (6.2.16) and the first part of Lemma 6.2.1, we obtain the existence of a solution $u_\omega^{(0)}$ of the equation $\nabla_\omega \cdot u_\omega = g$ in the space $\mathcal{H}_s^\omega$. We write the solution of Problem P in the form $u_\omega = u_\omega^{(1)} + u_\omega^{(0)}$. Then for the determination of $u_\omega^{(1)}$ and p we have the equation

$$(6.2.18) \qquad [u_\omega^{(1)}, v_\omega] - \int_\Omega p \, \nabla_\omega \cdot \overline{v}_\omega \, d\omega$$

$$= -[u_\omega^{(0)}, v_\omega] + \int_\Omega \left(f \, \nabla_\omega \cdot \overline{v}_\omega + F \cdot \overline{v}_\omega \right) d\omega$$

which has to be satisfied for all $v_\omega \in \mathcal{H}_s^\omega$. Furthermore, $u_\omega^{(1)}$ has to satisfy the equality $\nabla_\omega \cdot u_\omega^{(1)} = 0$ in Ω.

We introduce the space $\mathcal{X} = \{u_\omega \in \mathcal{H}_s^\omega : \nabla_\omega \cdot u_\omega = 0\}$. It is evident that $\mathcal{H}_0 \subset \mathcal{X}$. Let $\mathcal{X}_0$ be the direct complement of $\mathcal{H}_0$ in $\mathcal{X}$. Then $[\cdot, \cdot]$ is a scalar product in $\mathcal{X}_0$. We seek the element $u_\omega^{(1)}$ in the space $\mathcal{X}_0$. Inserting $v_\omega \in \mathcal{X}_0$ into (6.2.18), we obtain

$$(6.2.19) \qquad [u_\omega^{(1)}, v_\omega] = -[u_\omega^{(0)}, v_\omega] + \int_\Omega F \cdot \overline{v}_\omega \, d\omega \qquad \text{for all } v_\omega \in \mathcal{X}_0.$$

The right-hand side of (6.2.19) is a bounded antilinear functional in $\mathcal{X}_0$. Hence the Riesz theorem implies the existence of an element $u_\omega^{(1)} \in \mathcal{X}_0$ satisfying (6.2.19). Furthermore, due to the second condition of (6.2.16), the equality (6.2.19) holds for arbitrary $v_\omega \in \mathcal{H}_0$. Consequently, (6.2.19) is satisfied for all $v_\omega \in \mathcal{X}$.

Let $\mathcal{X}_1$ be the direct complement of the subspace $\mathcal{X}$ in $\mathcal{H}_s^\omega$ and let $\mathcal{Y} = \{g \in L_2(\Omega) : \int_\Omega g \, d\omega = 0\}$. By the first part of Lemma 6.2.1, the operator $u_\omega \to \nabla_\omega \cdot u_\omega$ is an isomorphism from $\mathcal{X}_1$ onto $\mathcal{Y}$. We represent the element $v_\omega \in \mathcal{H}_s^\omega$ as a sum $v_\omega = v_\omega^{(0)} + v_\omega^{(1)}$, where $v_\omega^{(0)} \in \mathcal{X}$, $v_\omega^{(1)} \in \mathcal{X}_1$. Then the integral identity (6.2.18) is satisfied if

$$(6.2.20) \qquad -\int_\Omega p \, \nabla_\omega \cdot \overline{v_\omega^{(1)}} \, d\omega = -[u_\omega^{(0)}, v_\omega^{(1)}] + \int_\Omega F \cdot \overline{v_\omega^{(1)}} \, d\omega$$
$$- [u_\omega^{(1)}, v_\omega^{(1)}] + \int_\Omega f \, \nabla_\omega \cdot \overline{v_\omega^{(1)}} \, d\omega.$$

The right-hand side of (6.2.20) is a functional on $\mathcal{X}_1$ and, consequently, also on $\mathcal{Y}$. Therefore, there exists an element $p \in \mathcal{Y}$ satisfying (6.2.20) for all $v_\omega^{(1)} \in \mathcal{X}_1$. Thus, we have proved part 1) of the lemma. The proof of the second part proceeds analogously. ∎

6.2.4. The eigenvalue $\lambda = 1$ when $\mathcal{H}$ is a subspace of the first kind. Now we are able to give a description of the eigenfunctions and generalized eigenfunctions corresponding to the eigenvalue $\lambda = 1$. We start with the case when $\mathcal{H}$ is a subspace of the first kind.

THEOREM 6.2.1. *Suppose that $\Omega \neq S^2$ and $\mathcal{H}$ is a subspace of the first kind. Then the number $\lambda = 1$ is always an eigenvalue of the pencil $\mathfrak{A}$ and the vector functions $(0, u_\omega, c)$, where $u_\omega \in \mathcal{H}_0$, $c \in \mathbb{C}$, are eigenvectors corresponding to this eigenvalue. Moreover, the following assertions are true.*
1) If the problem

$$(6.2.21) \qquad \begin{cases} \displaystyle\int_\Omega \left(\nabla_\omega w \cdot \nabla_\omega \overline{v} - 6 \, w \, \overline{v} \right) d\omega = 0 \quad \text{for all } v \in \mathcal{H}_s^r, \\[2mm] \displaystyle\int_\Omega w \, d\omega = 0 \end{cases}$$

has only the trivial solution in the space $\mathcal{H}_s^r$, then there are no other eigenvectors. If the problem (6.2.21) has a nontrivial solution $w \in \mathcal{H}_s^r$, then there are the additional eigenvectors (w, u_ω, p), where (u_ω, p) is a solution of Problem P with $f = -5w$, $g = -6w$, $F = 0$. Other eigenvectors to the eigenvalue $\lambda = 1$ do not occur.
2) For the existence of a generalized eigenvector to the eigenvalue $\lambda = 1$ and the eigenvector $(u_r^{(0)}, u_\omega^{(0)}, p^{(0)})$ it is necessary and sufficient that $u_r^{(0)} = 0$, $u_\omega^{(0)} = 0$,

$p^{(0)} = const$ *and that the problem*

(6.2.22)
$$
\begin{cases}
\displaystyle\int_\Omega \left(\nabla_\omega w \cdot \nabla_\omega \overline{v} - 6\,w\,\overline{v}\right) d\omega = -\int_\Omega p^{(0)}\,\overline{v}\,d\omega \quad for\ all\ v \in \mathcal{H}_s^r, \\[2ex]
\displaystyle\int_\Omega w\,d\omega = 0
\end{cases}
$$

has a solution $w \in \mathcal{H}_s^r$. *In this case the corresponding generalized eigenvector has the form* (w, u_ω, p), *where* (u_ω, p) *is a solution of* Problem P *with* $f = -5w$, $g = -6w$, $F = 0$. *Other generalized eigenvectors do not exist.*

Proof: Let (u_r, u_ω, p) be an eigenvector to the eigenvalue $\lambda = 1$. If $u_r = 0$, then, by (6.2.5), we get $\nabla_\omega \cdot u_\omega = 0$ and from (6.2.7) it follows that

$$
[u_\omega, u_\omega] - \int_\Omega p\,\nabla_\omega \cdot v_\omega\,d\omega = 0 \quad \text{for all } v_\omega \in \mathcal{H}_s^\omega.
$$

Using the first part of Lemma 6.2.3, we conclude that $u_\omega \in \mathcal{H}_0$ and $p = const$.

Now let u_r be nonzero. Then, by (6.2.5) and (6.2.6), the function u_r is a solution of problem (6.2.21), and from (6.2.5), (6.2.7) it follows that the vector function (u_ω, p) is a solution of the problem

(6.2.23)
$$
\begin{cases}
\nabla_\omega \cdot u_\omega = -3u_r \\[1ex]
[u_\omega, v_\omega] - \displaystyle\int_\Omega p\,\nabla_\omega \cdot \overline{v}_\omega\,d\omega = -5\int_\Omega u_r\,\nabla_\omega \cdot \overline{v}_\omega\,d\omega \quad \text{for all } v_\omega \in \mathcal{H}_s^\omega.
\end{cases}
$$

This gives the required properties of the eigenvectors.

We investigate the existence of generalized eigenvectors. Let $(u_r^{(0)}, u_\omega^{(0)}, p^{(0)})$ be an eigenvector to the eigenvalue $\lambda = 1$ and let $(u_r^{(1)}, u_\omega^{(1)}, p^{(1)})$ be a corresponding generalized eigenvector. Differentiating (6.1.17) with respect to λ, we obtain

$$
\frac{d}{d\lambda}\,a\!\left(\begin{pmatrix} u \\ p \end{pmatrix}, \begin{pmatrix} v \\ q \end{pmatrix}; \lambda\right)\bigg|_{\lambda=1} = \int_\Omega \Big(-6\,u_r\,\overline{v}_r - 3\,u_\omega \cdot \overline{v}_\omega + u_r\,\nabla_\omega \cdot \overline{v}_\omega
$$

$$
-(\nabla_\omega \cdot u_\omega)\,\overline{v}_r + p\,\overline{v}_r - u_r\,\overline{q}\Big)\,d\omega.
$$

Consequently, the generalized eigenfunction satisfies the equation

(6.2.24)
$$
0 = a\!\left(\begin{pmatrix} u^{(1)} \\ p^{(1)} \end{pmatrix}, \begin{pmatrix} v \\ q \end{pmatrix}; 1\right) + \frac{d}{d\lambda}\,a\!\left(\begin{pmatrix} u^{(0)} \\ p^{(0)} \end{pmatrix}, \begin{pmatrix} v \\ q \end{pmatrix}; \lambda\right)\bigg|_{\lambda=1}
$$

$$
= [u_\omega^{(1)}, v_\omega] + \int_\Omega \left((\nabla_\omega u_r^{(1)}) \cdot \nabla_\omega \overline{v}_r + 2\,(\nabla_\omega \cdot u_\omega^{(1)})\,\overline{v}_r\right) d\omega
$$

$$
+ \int_\Omega \left(5\,u_r^{(1)}\,\nabla_\omega \cdot \overline{v}_\omega - p^{(1)}\,\nabla_\omega \cdot \overline{v}_\omega - (3u_r^{(1)} + \nabla_\omega \cdot u_\omega^{(1)})\,\overline{q}\right) d\omega
$$

$$
+ \int_\Omega \left(-6\,u_r^{(0)}\,\overline{v}_r - (\nabla_\omega \cdot u_\omega^{(0)})\,\overline{v}_r + p^{(0)}\,\overline{v}_r\right) d\omega
$$

$$
+ \int_\Omega \left(-3\,u_\omega^{(0)} \cdot \overline{v}_\omega + u_r^{(0)}\,\nabla_\omega \cdot \overline{v}_\omega - u_r^{(0)}\,\overline{q}\right) d\omega
$$

for all $v_r \in \mathcal{H}_s^r$, $v_\omega \in \mathcal{H}_s^\omega$, $q \in L_2(\Omega)$. From this we obtain

(6.2.25) $$3\,u_r^{(1)} + \nabla_\omega \cdot u_\omega^{(1)} + u_r^{(0)} = 0 \qquad \text{in } \Omega$$

and

(6.2.26) $$\int_\Omega \left((\nabla_\omega u_r^{(1)}) \cdot \nabla_\omega \overline{v}_r + 2\,(\nabla_\omega \cdot u_\omega^{(1)})\,\overline{v}_r - 6\,u_r^{(0)}\,\overline{v}_r \right.$$
$$\left. - (\nabla_\omega \cdot u_\omega^{(0)})\,\overline{v}_r + p^{(0)}\,\overline{v}_r \right) d\omega = 0$$

for all $v_r \in \mathcal{H}_s^r$. Using the equalities $\nabla_\omega \cdot u_\omega^{(0)} = -3u_r^{(0)}$ and $\nabla_\omega \cdot u_\omega^{(1)} = -3u_r^{(1)} - u_r^{(0)}$, we find that (6.2.26) is equivalent to

(6.2.27) $$\int_\Omega \left((\nabla_\omega u_r^{(1)}) \cdot \nabla_\omega \overline{v}_r - 6\,u_r^{(1)}\,\overline{v}_r \right) d\omega = \int_\Omega \left(5\,u_r^{(0)}\,\overline{v}_r - p^{(0)}\,\overline{v}_r \right) d\omega$$

for $v_r \in \mathcal{H}_s^r$. Substituting $v_r = u_r^{(0)}$ into (6.2.27) and using (6.2.6) for $u_r = u_r^{(0)}$, we obtain

(6.2.28) $$\int_\Omega \left(5\,|u_r^{(0)}|^2 - p^{(0)}\,\overline{u_r^{(0)}} \right) d\omega = 0.$$

Furthermore, by (6.2.7), we have

$$[u_\omega^{(0)}, u_\omega^{(0)}] + \int_\Omega \left(5\,u_r^{(0)}\,\nabla_\omega \cdot \overline{u_\omega^{(0)}} - p^{(0)}\,\nabla_\omega \cdot \overline{u_\omega^{(0)}} \right) d\omega = 0$$

The last equality together with $\nabla_\omega \cdot u_\omega^{(0)} = -3u_r^{(0)}$ and (6.2.28) implies $[u_\omega^{(0)}, u_\omega^{(0)}] = 0$. Therefore, $u_\omega^{(0)} \in \mathcal{H}_0$. Moreover, we get $u_r^{(0)} = -\frac{1}{3}\nabla_\omega \cdot u_\omega^{(0)} = 0$ (see Remark 6.2.3). As it has been shown in the beginning of the proof, this implies $p^{(0)} = const$.

Inserting $v_r = 0$, $q = 0$ into (6.2.24), we find that

(6.2.29) $$[u_\omega^{(1)}, v_\omega] + \int_\Omega \left(5\,u_r^{(1)}\,\nabla_\omega \cdot \overline{v}_\omega - p^{(1)}\,\nabla_\omega \cdot \overline{v}_\omega - 3\,u_\omega^{(0)} \cdot \overline{v}_\omega \right) d\omega = 0$$

for all $v_\omega \in \mathcal{H}_s^\omega$. In particular, if $v_\omega \in \mathcal{H}_0$, then

$$\int_\Omega u_\omega^{(0)} \cdot \overline{v}_\omega\, d\omega = 0$$

and, therefore, $u_\omega^{(0)} = 0$. Thus, we have shown that only eigenvectors of the form $(0, 0, p^{(0)})$ may have generalized eigenvectors. In this case the following relations for the generalized eigenvectors hold (see (6.2.25), (6.2.27), (6.2.29)):

(6.2.30) $$3\,u_r^{(1)} + \nabla_\omega \cdot u_\omega^{(1)} = 0 \quad \text{in } \Omega,$$

(6.2.31) $$\int_\Omega \left((\nabla_\omega u_r^{(1)}) \cdot \nabla_\omega \overline{v}_r - 6\,u_r^{(1)}\,\overline{v}_r + p^{(0)}\,\overline{v}_r \right) d\omega = 0 \quad \text{for all } v_r \in \mathcal{H}_s^r,$$

(6.2.32) $$[u_\omega^{(1)}, v_\omega] + \int_\Omega \left(5\,u_r^{(1)} - p^{(1)} \right) \nabla_\omega \cdot \overline{v}_\omega\, d\omega = 0 \quad \text{for all } v_\omega \in \mathcal{H}_s^\omega.$$

From (6.2.30), (6.2.31) it follows that $u_r^{(1)}$ is a solution of the problem (6.2.22), while (6.2.32) is equivalent to Problem P for $f = -5u_r^{(1)}$, $g = -6u_r^{(1)}$, $F = 0$.

We show that there are no second generalized eigenvectors. Let $(u_r^{(1)}, u_\omega^{(1)}, p^{(1)})$ be a generalized eigenvector to the eigenvector $(0, 0, p^{(0)})$, $p^{(0)} = const$. Suppose there exists a second generalized eigenvector $(u_r^{(2)}, u_\omega^{(2)}, p^{(2)})$. Since

$$\frac{d^2}{d\lambda^2} a\left(\begin{pmatrix} u^{(0)} \\ p^{(0)} \end{pmatrix}, \begin{pmatrix} v \\ q \end{pmatrix}; \lambda\right)\Big|_{\lambda=1} = -2 \int\limits_\Omega \left(2\,u_r^{(0)}\,\overline{v}_r + u_\omega^{(0)} \cdot \overline{v}_\omega\right) d\omega = 0,$$

the equality (6.2.24) with $(u_r^{(2)}, u_\omega^{(2)})$ instead of $(u_r^{(1)}, u_\omega^{(1)})$ and $(u_r^{(1)}, u_\omega^{(1)})$ instead of $(u_r^{(0)}, u_\omega^{(0)})$ holds. This implies (cf. (6.2.25), (6.2.27))

(6.2.33) $$3\,u_r^{(2)} + \nabla_\omega \cdot u_\omega^{(2)} + u_r^{(1)} = 0 \quad \text{in } \Omega$$

and

(6.2.34) $$\int\limits_\Omega \left((\nabla_\omega u_r^{(2)}) \cdot \nabla_\omega \overline{v}_r - 6\,u_r^{(2)}\,\overline{v}_r\right) d\omega = \int\limits_\Omega \left(5\,u_r^{(1)}\,\overline{v}_r - p^{(1)}\,\overline{v}_r\right) d\omega$$

for all $v_r \in \mathcal{H}_s^r$. From (6.2.25), (6.2.33) and from the properties of the space $\mathcal{H}_s^\omega$ it follows that $\int_\Omega u_r^{(2)}\, d\omega = 0$. Hence, by (6.2.27), we have

$$\int\limits_\Omega \left(\nabla_\omega u_r^{(2)} \cdot \nabla_\omega \overline{u_r^{(1)}} - 6\,u_r^{(2)}\,\overline{u_r^{(1)}}\right) d\omega = -\overline{p^{(0)}} \int\limits_\Omega u_r^{(2)}\, d\omega = 0.$$

Consequently, inserting $v_r = u_r^{(1)}$ into (6.2.34), we obtain

(6.2.35) $$\int\limits_\Omega \left(5\,|u_r^{(1)}|^2 - p^{(1)}\,\overline{u_r^{(1)}}\right) d\omega = 0.$$

Furthermore, using (6.2.30) and (6.2.32), we get

$$[u_\omega^{(1)}, u_\omega^{(1)}] - 3 \int\limits_\Omega \left(5\,|u_r^{(1)}|^2 - p^{(1)}\,\overline{u_r^{(1)}}\right) d\omega = 0.$$

From this and from (6.2.35) we conclude that $[u_\omega^{(1)}, u_\omega^{(1)}] = 0$, i.e., $u_\omega^{(1)} \in \mathcal{H}_0$. According to (6.2.30) and Remark 6.2.3, this implies $u_r^{(1)} = 0$. Thus, second generalized eigenvectors do not exist. The proof is complete. ∎

6.2.5. The eigenvalue $\lambda = 1$ when $\mathcal{H}$ is a subspace of the second kind.

THEOREM 6.2.2. *Let* $\Omega \neq S^2$ *and let* $\mathcal{H}$ *be a subspace of the second kind.*

1) *If* $\mathcal{H}_0 \neq \{0\}$*, then the number* $\lambda = 1$ *is always an eigenvalue of the pencil* $\mathfrak{A}$*. The vectors* $(0, u_\omega, 0)$*, where* $u_\omega \in \mathcal{H}_0$*, are eigenvectors to this eigenvalue. If the problem*

(6.2.36) $$\int\limits_\Omega \left(\nabla_\omega w \cdot \nabla_\omega \overline{v} - 6\,w\,\overline{v}\right) d\omega = 0 \quad \text{for all } v \in \mathcal{H}_s^r$$

has only the trivial solution $w = 0$ *in* $\mathcal{H}_s^r$*, then there are no other eigenvectors. If the problem* (6.2.36) *has a nontrivial solution* $w \in \mathcal{H}_s^r$*, then there is also the eigenvector* (w, u_ω, p)*, where* (u_ω, p) *is a solution of* Problem P *with* $f = -5w$*,* $g = -6w$*,* $F = 0$*. Generalized eigenvectors do not occur.*

2) *If* $\mathcal{H}_0 = \{0\}$*, then* $\lambda = 1$ *is an eigenvalue if and only if the problem* (6.2.36) *has a nontrivial solution* $w \in \mathcal{H}_s^r$*. Then there is the eigenvector* (w, u_ω, p)*, where* (u_ω, p) *is a solution of* Problem P *with* $f = -5w$*,* $g = -6w$*,* $F = 0$*. Generalized eigenvectors do not occur.*

Proof: First we investigate the existence and the structure of the eigenvectors. Let (u_r, u_ω, p) be an eigenvector of the pencil $\mathfrak{A}$ corresponding to the eigenvalue $\lambda = 1$. Then, analogously to the proof of Theorem 6.2.1, the functions u_r, u_ω, p satisfy (6.2.23). If $u_r = 0$, then, by the second part of Lemma 6.2.3, we get $u_\omega \in \mathcal{H}_0$ and $p = 0$. Consequently, eigenvectors of the form $(0, u_\omega, p)$ exist if and only if $\mathcal{H}_0 \neq \{0\}$. Suppose that $u_r \neq 0$. Then, by (6.2.6), the function u_r is a nontrivial solution of the problem (6.2.36), while (u_ω, p) is a solution of the problem (6.2.23). This proves the assertions of the theorem concerning the eigenvectors to the eigenvalue $\lambda = 1$.

Now we show that generalized eigenvectors do not exist. To this end, we assume that $(u_r^{(0)}, u_\omega^{(0)}, p^{(0)})$ is an eigenvector to the eigenvalue $\lambda = 1$ and that $(u_r^{(1)}, u_\omega^{(1)}, p^{(1)})$ is a corresponding generalized eigenvector. Analogously to the proof of Theorem 6.2.1, we find that $u_\omega^{(0)} = 0$ and, consequently, $\nabla_\omega \cdot u_\omega^{(0)} = 0$. Hence, by (6.1.24), we have $u_r^{(0)} = 0$. As we have shown in the beginning of the proof, this implies $p^{(0)} = 0$. Furthermore, analogously to the proof of Theorem 6.2.1, equality (6.2.29) is satisfied. In particular, we have

$$\int_\Omega u_\omega^{(0)} \cdot \overline{v}_\omega \, d\omega = 0 \qquad \text{for all } v_\omega \in \mathcal{H}_0$$

and, therefore, $u_\omega^{(0)} = 0$. Thus, the vector $(u_r^{(0)}, u_\omega^{(0)}, p^{(0)})$ is not an eigenvector. This proves the theorem. ∎

6.2.6. A variational principle for real eigenvalues. In this subsection we get a variational principle for real eigenvalues of the operator pencil $\mathfrak{A}$ in the interval $-1/2 \leq \lambda < 1$. Since the derivation of this variational principle is analogous to Chapter 5, we give the formulation of the following lemmas and theorems without proofs.

Eliminating the function u_r and p in the quadratic form $s(\cdot, \cdot; \lambda)$ by the equations (5.5.1) and (5.5.2), we obtain the form

$$b_\omega(u_\omega, u_\omega; \lambda) = [u_\omega, u_\omega] + \int_\Omega \frac{1}{(1-\lambda)(\lambda+2)} |\nabla_\omega(\nabla_\omega \cdot u_\omega)|^2 \, d\omega$$

$$+ \int_\Omega \left(2 \frac{5 - 6\lambda - 2\lambda^2}{(\lambda+2)(1-\lambda)} |\nabla_\omega \cdot u_\omega|^2 + (\lambda+2)(1-\lambda) |u_\omega|^2 \right) d\omega$$

which is defined on the space

$$\mathcal{Y} = \{u_\omega \in \mathcal{H}_s^\omega : \nabla_\omega \cdot u_\omega \in \mathcal{H}_s^r\}.$$

Furthermore, let

$$b(u_\omega, u_\omega; \lambda) = (1-\lambda)(\lambda+2) \, b_\omega(u_\omega, u_\omega; \lambda).$$

This form satisfies the conditions (I)–(III) in Section 1.3, where $\alpha = -1/2$, $\beta = 1$, $\mathcal{H} = L_2(\Omega)^2$, $\mathcal{H}_+ = \mathcal{Y}$. We introduce the functions

(6.2.37)
$$\mu_j(\lambda) = \max_{\{V\}} \min_{u_\omega \in V \setminus \{0\}} \frac{b(u_\omega, u_\omega; \lambda)}{\|u_\omega\|^2_{L_2(\Omega)^2}},$$

where the maximum is taken over all subspaces $V \subset \mathcal{Y}$ of codimension $\geq j - 1$. The functions μ_j are continuous in the interval $[-\frac{1}{2}, 1)$ and $\mu_j(-1/2) > 0$. By Theorems 1.3.2 and 1.3.3, the following statement holds.

THEOREM 6.2.3. *All assertions of Theorem 5.5.1 are also valid for the eigenvalues of the pencil* $\mathfrak{A}$ *and for the above introduced function* μ_j.

Due to Theorem 6.2.3, the assertions on the eigenvalue $\lambda = 1$ can be used to obtain information on the eigenvalues in the interval $(-1/2, 1)$.

We consider first the case when $\mathcal{H}_s$ is a subspace of the first kind (i.e., $I_\tau = \emptyset$, compare with Theorem 5.5.3).

THEOREM 6.2.4. *Let* $\mathcal{H}_s$ *be a subspace of the first kind. If* k *is the maximal dimension of the subspace* $\mathcal{X} \subset \{v \in \mathcal{H}_s^r : \int_\Omega v \, d\omega = 0\}$ *for which*

$$\int_\Omega \left(|\nabla_\omega v|^2 - 6\,|v|^2 \right) d\omega < 0 \quad \text{for all } v \in \mathcal{X} \backslash \{0\},$$

then the interval $[-1/2, 1)$ *contains exactly* k *eigenvalues of the pencil* $\mathfrak{A}$.

Analogously, the following assertion holds in the case when $\mathcal{H}_s$ is a subspace of the second kind (i.e., $I_\tau \neq \emptyset$.)

THEOREM 6.2.5. *Let* $\mathcal{H}_s$ *be a subspace of the second kind. If* k *is the maximal dimension of the subspace* $\mathcal{X} \subset \mathcal{H}_s^r$ *for which*

$$\int_\Omega \left(|\nabla_\omega v|^2 - 6\,|v|^2 \right) d\omega < 0 \quad \text{for all } v \in \mathcal{X} \backslash \{0\},$$

then the interval $[-1/2, 1)$ *contains exactly* k *eigenvalues of the pencil* $\mathfrak{A}$.

As a consequence of Theorem 6.2.5, the following assertion holds.

COROLLARY 6.2.1. *In the case* $\mathcal{H}_s^r = \overset{\circ}{W}{}_2^1(\Omega)$ *(this takes place, for example, if only the boundary conditions (i), (iii) appear in the definition of the space* $\mathcal{H}$*), the number of the eigenvalues in* $[-1/2, 1)$ *monotonically depends on the domain* $\Omega \subset S^2$.

6.2.7. Examples.

1) Let $I_\tau \neq \emptyset$, $I_n = \emptyset$. Then $\mathcal{H}_s^r = \overset{\circ}{W}{}_2^1(\Omega)$, and $\mathcal{H}_s$ is a subspace of the second kind. Furthermore, let $\mathcal{D}_\alpha$ be a dihedral angle with opening α.

THEOREM 6.2.6. *If* $\mathcal{K} \subset \mathcal{D}_\alpha$ *with* $\alpha \leq \pi/2$, *and* $\mathcal{D}_\alpha \backslash \overline{\mathcal{K}} \neq \emptyset$, *then the strip* $-1/2 \leq \operatorname{Re}\lambda \leq 1$ *does not contain eigenvalues of the pencil* $\mathfrak{A}$.

Proof: Let $\Omega_\alpha = \mathcal{D}_\alpha \cap S^2$ and let δ be the Beltrami operator on S^2. Then the smallest positive eigenvalue of the operator pencil

$$-\delta - \lambda(\lambda + 1) : \overset{\circ}{W}{}_2^1(\Omega_\alpha) \to W_2^{-1}(\Omega_\alpha)$$

is π/α (see Theorem 10.5.3). Hence

$$(6.2.38) \qquad \int_{\Omega_\alpha} |\nabla_\omega v|^2 \, d\omega \geq 6 \int_{\Omega_\alpha} |v|^2 \, d\omega, \quad v \in \overset{\circ}{W}{}_2^1(\Omega_\alpha),$$

provided $\alpha \leq \pi/2$. Since $\Omega \subset \Omega_\alpha$ and $\Omega_\alpha \backslash \overline{\Omega} \neq \emptyset$, it follows that the inequality (6.2.38) is valid for $v \in \overset{\circ}{W}{}_2^1(\Omega)$ and the equality in (6.2.38) may be valid only for $v = 0$. Using Theorems 6.2.2 and 6.2.5, we obtain the assertion of the theorem. ∎

2) We consider a convex polyhedral cone $\mathcal{K}$ and suppose that $\partial\mathcal{K}\backslash\{0\}$ consists of a finite number of flat open faces $\Gamma_1,\ldots,\Gamma_N$ and rays. The convexity of $\mathcal{K}$ implies that the angles between two adjacent faces are less than π. We assume that the angles between Γ_1 and adjoining faces are less than $\pi/2$.

Furthermore, let the boundary condition (ii) be prescribed on Γ_1, while the Dirichlet condition is given on the remaining faces.

THEOREM 6.2.7. *Under the above assumptions, the operator pencil* $\mathfrak{A}$ *has only one eigenvalue* $\lambda_1 = 1$ *in the strip* $-1/2 \leq \operatorname{Re}\lambda \leq 1$. *This eigenvalue has only the eigenvectors* $(0,0,0,c)$, *where* c *is a constant. There are no generalized eigenvectors to this eigenvalue.*

Proof: We can suppose that the face Γ_1 is located in the plane $x_3 = 0$ and the cone $\mathcal{K}$ is placed in the half-space $x_3 < 0$. Let $\mathcal{K}_+$ and $\mathcal{K}_s$ be the following cones:

$$\mathcal{K}_+ = \{x = (x_1, x_2, x_3) : (x_1, x_2, -x_3) \in \mathcal{K}\}, \qquad \mathcal{K}_s = \mathcal{K} \cup \mathcal{K}_+ \cup \Gamma_1.$$

Obviously, the cone $\mathcal{K}_s$ is also convex. By $\mathfrak{A}_s$ we denote the operator pencil associated with the Dirichlet problem for the Stokes system.

We consider the auxiliary spectral problem generated by the ratio of quadratic forms:

$$(6.2.39) \qquad \frac{\int_\Omega |\nabla_\omega v|^2\, d\omega}{\int_\Omega |v|^2\, d\omega}, \qquad v \in \mathcal{X}\backslash\{0\},$$

where $\Omega = \mathcal{K} \cap S^2$ and

$$\mathcal{X} = \left\{ w \in W_2^1(\Omega) : w = 0 \text{ on } \gamma_2 \cap \gamma_3 \cap \cdots \cap \gamma_N \text{ and } \int_\Omega w\, d\omega = 0 \right\}$$

(here $\gamma_j = \Gamma_j \cap S^2$). Setting

$$(6.2.40) \qquad v_s(\omega) = \begin{cases} v(\omega) & \text{if } \omega_3 < 0, \\ v(\omega_1, \omega_2, -\omega_3) & \text{if } \omega_3 > 0, \end{cases}$$

we obtain an extension of $v \in \mathcal{X}$ onto $\Omega_s = \mathcal{K}_s \cap S^2$. It is evident that

$$v_s \in \left\{ w \in \overset{\circ}{W}_2^1(\Omega_s) : \int_{\Omega_s} w\, d\omega = 0 \right\}.$$

Using this extension operator, one can show that the least eigenvalue of the operator induced by (6.2.39) is not greater than the infimum of the functional (6.2.39), where Ω, v are replaced by Ω_s, v_s, respectively, and v_s is defined by (6.2.40).

Since Ω_s is placed in a half-sphere, it follows from Lemma 5.4.2 that the first eigenvalue of the spectral problem in Ω_s is greater than 6. Hence the same is true for the spectral problem in Ω. Now the assertion of the theorem follows from Theorems 6.2.1 and 6.2.4. ∎

6.3. The Neumann problem for the Stokes system

Let $\mathcal{K}$ be the three-dimensional cone $\{x = (x_1, x_2, x_3) : x_3 > \phi(x_1, x_2)\}$ with a function ϕ which is positively homogeneous of degree 1 and smooth on $\mathbb{R}^2\backslash\{0\}$. We consider the boundary value problem

$$(6.3.1) \qquad -\Delta U + \nabla P = 0, \qquad \nabla \cdot U = 0 \quad \text{in } \mathcal{K},$$

$$(6.3.2) \qquad -P\,n + 2\,\varepsilon_n(U) = 0 \quad \text{on } \partial\mathcal{K}\backslash\{0\},$$

where $\varepsilon_n(U)$ denotes the vector

$$\Big(\sum_{j=1}^{3} \varepsilon_{ij}(U)\, n_j\Big)_{i=1}^{3}$$

and n is the exterior normal to $\partial\mathcal{K}\backslash\{0\}$. We show that the spectrum of the corresponding operator pencil in the strip $-1 \le \operatorname{Re}\lambda \le 0$ consists only of the eigenvalues -1 and 0, that both eigenvalues have geometric multiplicity 3, and that they have no generalized eigenvectors.

6.3.1. The operator pencil generated by the Neumann problem. We are interested in solutions of the Neumann problem (6.3.1), (6.3.2) which have the form (6.1.6), (6.1.7). This means we seek vector functions (6.1.6), (6.1.7) satisfying the integral identity

$$(6.3.3) \qquad \int_{\mathcal{K}} \Big(2 \sum_{i,j=1}^{3} \varepsilon_{ij}(U)\cdot\varepsilon_{ij}(\overline{V}) - P\,\nabla\cdot\overline{V} - (\nabla\cdot U)\,\overline{Q}\Big)\, dx = 0$$

for all $(V,Q) \in W_2^1(\mathcal{K})^3 \times L_2(\mathcal{K})$ with compact support in $\overline{\mathcal{K}}\backslash\{0\}$.

To this end, we introduce the parameter-depending sesquilinear form

$$a\Big(\binom{u}{p}, \binom{v}{q}; \lambda\Big)$$

$$= \frac{1}{2\log 2} \int_{\substack{\mathcal{K}\\ 1/2<|x|<2}} \Big(2 \sum_{i,j=1}^{3} \varepsilon_{ij}(U)\cdot\varepsilon_{ij}(\overline{V}) - P\,\nabla\cdot\overline{V} - (\nabla\cdot U)\,\overline{Q}\Big)\, dx,$$

where

$$U = r^{\lambda}\,u(\omega), \quad P = r^{\lambda-1}\,p(\omega), \quad V = r^{-1-\overline{\lambda}}\,v(\omega), \quad Q = r^{-2-\overline{\lambda}}\,q(\omega),$$

$u, v \in W_2^1(\Omega)^3$, and $p, q \in L_2(\Omega)$.

The form $a(\cdot,\cdot;\lambda)$ generates the linear and continuous operator

$$\mathfrak{A}(\lambda)\ :\ W_2^1(\Omega)^3 \times L_2(\Omega) \to (W_2^1(\Omega)^3)^* \times L_2(\Omega)$$

by the equality

$$\Big(\mathfrak{A}(\lambda)\binom{u}{p}, \binom{v}{q}\Big)_{L_2(\Omega)^4} = a\Big(\binom{u}{p}, \binom{v}{q}; \lambda\Big), \qquad u, v \in W_2^1(\Omega)^3,\ p, q \in L_2(\Omega).$$

Then the vector function (6.1.6), (6.1.7) is a solution of the equation (6.3.3) if and only if λ_0 is an eigenvalue of the operator pencil $\mathfrak{A}$ and the vector functions $(u^{(0)}, p^{(0)}), \dots, (u^{(s)}, p^{(s)})$ form a Jordan chain corresponding to this eigenvalue.

Furthermore, analogously to Theorem 6.1.1, the following assertions can be proved.

THEOREM 6.3.1. 1) *The operator $\mathfrak{A}(\lambda)$ is Fredholm for all $\lambda \in \mathbb{C}$.*

2) *The spectrum of the pencil $\mathfrak{A}$ consists of isolated eigenvalues with finite algebraic multiplicities.*

3) *There are no eigenvalues of the pencil $\mathfrak{A}$ on the line $\operatorname{Re}\lambda = -1/2$.*

4) *The number λ_0 is an eigenvalue of the pencil $\mathfrak{A}$ if and only if $-1 - \overline{\lambda}_0$ is an eigenvalue of this pencil. The geometric, algebraic and partial multiplicities of the eigenvalues λ_0 and $-1 - \overline{\lambda}_0$ coincide.*

6.3.2. The spectrum of the pencil $\mathfrak{A}$ in the strip $-1 \leq \operatorname{Re} \lambda \leq 0$.

THEOREM 6.3.2. *The strip* $-1 \leq \operatorname{Re} \lambda \leq 0$ *contains exactly two eigenvalues of the pencil* $\mathfrak{A}$: $\lambda_0 = 0$ *and* $\lambda_1 = -1$. *Both eigenvalues have algebraic multiplicity 3. The eigenfunctions corresponding to the eigenvalue* $\lambda_0 = 0$ *are* $(c, 0)$, *where* $c \in \mathbb{C}^3$ *is an arbitrary constant vector. Generalized eigenfunctions corresponding to the eigenvalues* λ_0 *and* λ_1 *do not exist.*

Proof: First we show that the line $\operatorname{Re} \lambda = 0$ contains only the eigenvalue $\lambda_0 = 0$ with multiplicity 3. Let $U(x) = r^{\lambda_0} u(\omega)$, $P(x) = r^{\lambda_0 - 1} p(\omega)$, $\operatorname{Re} \lambda_0 = 0$. Then

$$0 = \int\limits_{\substack{\mathcal{K} \\ 1/2 < |x| < 2}} \left(-2 \frac{\partial \varepsilon_{ij}}{\partial x_j} + \frac{\partial p}{\partial x_i} \right) \frac{\partial u_i}{\partial x_3} \, dx = \int\limits_{\substack{\partial \mathcal{K} \\ 1/2 < |x| < 2}} n_3 \sum_{i,j=1}^{3} \varepsilon_{ij}^2 \, d\sigma$$

and, consequently, $\varepsilon_{ij} = 0$ on $\partial \mathcal{K} \backslash \{0\}$. From this and from the boundary condition (6.3.2) it follows that $P = 0$ on $\partial \mathcal{K} \backslash \{0\}$. Applying the operator $\nabla \cdot$ to the first equation of (6.3.1), we conclude that P is a harmonic function. Consequently, $p = 0$ or $\lambda_0(\lambda_0 - 1)$ is an eigenvalue of the Dirichlet problem for the Beltrami operator. Due to the results of Section 2.2, the last can be excluded, i.e., we have $P = 0$. This implies $\Delta \varepsilon_{ij} = 0$. Since $\varepsilon_{ij} = 0$ on $\partial \mathcal{K} \backslash \{0\}$, we obtain, by the same arguments as before, that $\varepsilon_{ij} = 0$ in $\mathcal{K}$ and, therefore, $U = const$. Thus, $\lambda_0 = 0$ is the only eigenvalue of the pencil $\mathfrak{A}$ on the line $\operatorname{Re} \lambda = 0$ and the corresponding eigenfunctions have the form $(c, 0)$, where $c \in \mathbb{C}^3$.

We suppose that there exists a generalized eigenfunction $(u^{(1)}, p^{(1)})$ to the eigenvalue $\lambda_0 = 0$. Then there is a solution (U, P) of problem (6.3.1), (6.3.2) which has the form

$$U = c \log r + u^{(1)}(\omega), \qquad P = r^{-1} p^{(1)}(\omega).$$

Since the derivatives $\partial U / \partial x_j$, $\partial P / \partial x_j$ have the form $r^{-1} v_j(\omega)$ and $r^{-2} w_j(\omega)$, respectively, it can be shown, in the same way as above, that $P = 0$ and $\varepsilon_{ij}(U) = 0$. Consequently, the constant c in the representation of the vector function U vanishes. This means that there are no generalized eigenvectors.

Therefore, the assertions of the theorem concerning the eigenvalue $\lambda_0 = 0$ are proved. The assertions on the eigenvalue $\lambda_1 = -1$ follow from the last item of Theorem 6.3.1. The proof of the absence of eigenvalues in the strip $-1 < \operatorname{Re} \lambda < 0$ proceeds analogously to Theorem 4.3.1. ∎

6.4. Notes

Sections 6.1, 6.2. The results are taken from the paper [138] by Kozlov, Maz'ya and Roßmann. Let us note that all assertions are valid for other boundary value problems to the Stokes system if the subspace $\mathcal{H}$ determining the *stable boundary conditions* (see Lions and Magenes [155, Ch.2,Remark 9.5]) has the properties 1) and 2) of Lemma 6.1.3.

For plane angles, singularities of solutions to the Stokes system with boundary conditions of the form (i), (ii) (see the introduction to this chapter) were considered in papers by Maz'ya, Plamenevskiĭ and Stupyalis [190] and Solonnikov [247]. Equations for the eigenvalues of the operator pencil corresponding to the Stokes system with boundary conditions (i), (ii) or (iii) on the sides of an angle are given in the paper [95] by Kalex. Orlt and Sändig [221] studied a more general class of boundary value problems for the Stokes system in plane polygonal domains and the corresponding operator pencils.

Section 6.3. Theorem 6.3.2 describing the spectrum of the pencil corresponding to the Neumann problem in the strip $-1 \leq \operatorname{Re} \lambda \leq 0$ was proved by Kozlov and Maz'ya [**127**] (see also [**132**]).

CHAPTER 7

The Dirichlet problem for the biharmonic and polyharmonic equations

This chapter is devoted to the study of singularities of solutions to the Dirichlet problem for the biharmonic and polyharmonic equations in angles and cones. We start with the biharmonic equation in the plane case when these singularities can be represented by zeros of a simple entire function. The treatment here is analogous to that in Sections 3.1 and 5.1.

In Section 7.2 we turn to the higher-dimensional case. Let $\mathcal{K}$ be a cone in $\mathbb{R}^n$ and let $\Omega = \mathcal{K} \cap S^{n-1}$. We consider the Dirichlet problem for the biharmonic operator Δ^2 in $\mathcal{K}$. The corresponding operator pencil is given by

$$(7.0.1) \qquad \mathcal{L}(\lambda) = \big(\delta + (\lambda - 2)(\lambda + n - 4)\big)\big(\delta + \lambda(\lambda + n - 2)\big),$$

where δ is the Laplace-Beltrami operator on S^{n-1}.

Let M be a positive number such that $M(M + n - 2)$ is the first eigenvalue of the operator $-\delta$ in Ω with zero Dirichlet data. We introduce the quantity

$$\sigma_0 = 2 - n/2 + \sqrt{(M + n/2 - 1)^2 + 1}.$$

In Sections 7.2.2 and 7.2.3 we show that all eigenvalues of $\mathcal{L}$ in the strip $4 - n - \sigma_0 \leq \mathrm{Re}\,\lambda \leq \sigma_0$ are real and have no generalized eigenfunctions.

We describe the variational principle obtained in Subsection 7.2.4 for the least eigenvalue of $\mathcal{L}$ situated on the half-line $[2 - n/2, \infty)$. Let $u \in \mathring{W}_2^2(\Omega)$ and $u \neq 0$. By $R(u)$ we denote the least root of the equation $(\mathcal{L}(\sigma)u, u)_{L_2(\Omega)} = 0$ in the interval $[2 - n/2, \infty)$. The equation just mentioned is

$$\|\delta u\|_{L_2(\Omega)}^2 + 2\left(\xi - n + 2\right)\left(\delta u, u\right)_{L_2(\Omega)} + \xi\left(\xi - 2(n - 2)\right)\|u\|_{L_2(\Omega)}^2 = 0,$$

where $\xi = \sigma(\sigma + n - 4)$, and hence $R(u)$ can be written explicitly. If there are no roots in this interval, we set $R(u) = \infty$. We use the notation

$$\sigma_1 = \inf_{u \in \mathring{W}_2^2(\Omega),\, u \neq 0} R(u).$$

The variational principle in question states that if the value σ_1 is finite, then it coincides with the least eigenvalue of the pencil $\mathcal{L}$ on the half-line $[2 - n/2, \infty)$. A simple consequence of this principle is the decrease of the least eigenvalue under the increase of Ω.

Estimates for the width of the strip $|\mathrm{Re}\,\lambda + (n - 4)/2| \leq T$, free of eigenvalues, are given in Sections 7.2.4 and 7.2.5. The best result obtained here is the estimate

$$T > \min\{\sigma_0, \sigma_1\} + (n - 4)/2$$

(see Theorem 7.2.4). As a corollary, we deduce that under the condition $\overline{\Omega} \neq S^{n-1}$, the spectrum of $\mathcal{L}$ is placed outside one of the strips:

$$(7.0.2) \qquad |\operatorname{Re}\lambda - 1/2| \leq \max(1/2, M - 1/2) \quad \text{if } n = 3,$$

$$(7.0.3) \qquad |\operatorname{Re}\lambda| \leq \max(1/2, M) \quad \text{if } n = 4,$$

and

$$(7.0.4) \qquad 4 - n - M \leq \operatorname{Re}\lambda \leq M \quad \text{if } n > 4.$$

We give a simple example showing the usefulness of this information for the study of solutions to the Dirichlet problem

$$(7.0.5) \qquad \Delta^2 U = F \text{ in } \mathcal{G}, \quad U = \partial_\nu U = 0 \text{ on } \partial\mathcal{G},$$

where ν is the normal, $F \in C^\infty(\overline{\mathcal{G}})$, and $\mathcal{G}$ is a bounded domain in $\mathbb{R}^n$ which coincides with a cone $\mathcal{K}$ near a boundary point O. The following three facts are direct corollaries of (7.0.2)–(7.0.4) and Theorem 1.4.5.

If $U \in \overset{\circ}{W}_2^2(\mathcal{G})$, then

$$U \in C^{1,\alpha}(\overline{\mathcal{G}}) \quad \text{for } n = 3,$$
$$U \in C^{0,1/2+\alpha}(\overline{\mathcal{G}}) \quad \text{for } n = 4,$$
$$U \in C^{0,\alpha}(\overline{\mathcal{G}}) \quad \text{for } n > 4,$$

where α is a positive number depending on the cone $\mathcal{K}$.

From the variational principle we deduce that, if $\Omega \subset S_+^{n-1}$, then the strip $2 - n \leq \operatorname{Re}\lambda \leq 2$ contains no eigenvalues of $\mathcal{L}$. We also show that if, on the contrary, $S_+^{n-1} \subset \Omega$, then the interior of the last strip contains at least two eigenvalues of $\mathcal{L}$, which are symmetric with respect to the line $\operatorname{Re}\lambda = 2 - n/2$.

Hence from Theorem 1.4.5 we conclude that if $\overline{\mathcal{K}}$ is a proper subset of a closed half-space, then the solution $u \in W_2^2(\mathcal{G})$ of problem (7.0.5) belongs to the space $C^{2,\alpha}(\overline{\mathcal{G}})$ with some positive α. Moreover, assuming that a closed half-space is a proper subset of $\overline{\mathcal{K}}$, we see that there is a solution $U \in W_2^2(\mathcal{G})$ which is not in the space $C^{2,0}(\overline{\mathcal{G}})$.

Other applications of the spectral properties of $\mathcal{L}$ studied in this chapter to the Dirichlet problem for the biharmonic equation are collected in Section 7.4.

In Section 7.3 we study the operator pencil corresponding to the Dirichlet problem for the polyharmonic operator in the cone $\mathcal{K}$. Let the differential operator

$$(7.0.6) \qquad \mathcal{L}(\lambda) : \overset{\circ}{W}_2^m(\Omega) \to W_2^{-m}(\Omega),$$

where $\lambda \in \mathbb{C}$, be defined by the equality

$$\mathcal{L}(\lambda)\, u(\omega) = (-1)^m r^{2m-\lambda} \Delta^m \left(r^\lambda u(\omega) \right).$$

Direct calculations show that

$$(7.0.7) \qquad \mathcal{L}(\lambda) = (-1)^m \prod_{j=0}^{m-1} \left(\delta + (\lambda - 2j)(\lambda - 2j + n - 2) \right).$$

We formulate the main results of this section.

(i) Let $2m \leq n - 4$. Then the strip $m - 2 - n/2 \leq \operatorname{Re}\lambda \leq m + 2 - n/2$ contains no eigenvalues of the pencil $\mathcal{L}$.

(ii) Let $2m = n - 3$. Then the strip $-3 \leq \operatorname{Re} \lambda \leq 0$ is free of the spectrum of $\mathcal{L}$. The wider strip $-3 - 1/2 \leq \operatorname{Re} \lambda \leq 1/2$ contains only real eigenvalues of $\mathcal{L}$, and corresponding eigenfunctions have no generalized eigenfunctions. There is at most one eigenvalue on the interval $(0, 1/2]$ which increases when the domain Ω decreases.

(iii) Let $2m = n - 2$. Then the strip $-2 \leq \operatorname{Re} \lambda \leq 0$ is free of the spectrum of the pencil $\mathcal{L}$. The wider strip $-3 \leq \operatorname{Re} \lambda \leq 1$ contains only real eigenvalues of $\mathcal{L}$ and corresponding eigenfunctions have no generalized eigenfunctions. There is at most one eigenvalue on the interval $(0, 1]$ which increases when the domain Ω decreases.

(iv) Let $2m > n - 2$. Then the strip $m - (n+1)/2 \leq \operatorname{Re} \lambda \leq m - (n-1)/2$ is free of the spectrum of $\mathcal{L}$, the strip $m - 1 - n/2 \leq \operatorname{Re} \lambda \leq m + 1 - n/2$ contains only real eigenvalues of $\mathcal{L}$, and corresponding eigenfunctions have no generalized eigenfunctions. The eigenvalues on the interval $(m - (n-1)/2, m + 1 - n/2]$ increase when the domain Ω decreases.

We give an application of the above information on the spectrum of $\mathcal{L}$. Let $\mathcal{G}$ be the same domain in $\mathbb{R}^n$ as in (7.0.5). Consider the Dirichlet problem

$$(7.0.8) \qquad\qquad (-\Delta)^m U = F \ \text{ in } \mathcal{G},$$

$$(7.0.9) \qquad\qquad \partial_\nu^k U = 0 \ \text{ on } \partial\mathcal{G}, \ k = 0, \dots, m - 1,$$

where $m > 2$. This problem has a unique solution in the Sobolev space $\overset{\circ}{W}_2^m(\mathcal{G})$ if, for example, $F \in C^\infty(\overline{\mathcal{G}})$. The following facts are corollaries of assertions (i)–(iv) and Theorem 1.4.5.

If $U \in \overset{\circ}{W}_2^m(\mathcal{G})$, then

$$U \in C^{0,\alpha} \ \text{ for } n - 4 \leq 2m \leq n - 1,$$

$$U \in C^{0,1/2+\alpha} \ \text{ for } 2m = n.$$

If $2m > n$, then we obtain $U \in C^{m-n/2,1/2+\alpha}$ or $U \in C^{m-n/2+1/2,\alpha}$ for even n and odd n, respectively. Here α is a sufficiently small positive number.

7.1. The Dirichlet problem for the biharmonic equation in an angle

In this section we consider the operator pencil generated by the Dirichlet problem for the biharmonic equation. The equations which determine the eigenvalues of this pencil are almost the same as for the Lamé system. Therefore, the results obtained in Section 3.1 lead to analogous results for the biharmonic equation.

7.1.1. The operator pencil generated by the Dirichlet problem for the biharmonic equation. Let $\mathcal{K} = \{(x_1, x_2) \in \mathbb{R}^2 : 0 < r < \infty, \ |\varphi| < \alpha/2\}$, $\alpha \in (0, 2\pi)$, be a plane angle. We seek solutions of the Dirichlet problem

$$(7.1.1) \qquad\qquad \Delta^2 U = 0 \ \text{ in } \mathcal{K},$$

$$(7.1.2) \qquad\qquad U = 0, \quad \partial_\nu U = 0 \quad \text{for } \varphi = \pm\alpha/2.$$

The biharmonic operator has the form

$$\Delta^2 = r^{-4} \mathcal{L}(\partial_\varphi, r\partial_r)$$

in the polar coordinates r, φ, where

$$\mathcal{L}(\partial_\varphi, \lambda) = \left(\partial_\varphi^2 + (\lambda - 2)^2\right)\left(\partial_\varphi^2 + \lambda^2\right).$$

Therefore, inserting $U(r, \varphi) = r^\lambda u(\varphi)$ into (7.1.1), (7.1.2), we get the spectral problem

(7.1.3) $$\mathcal{L}(\partial_\varphi, \lambda) u = 0 \qquad \text{for } |\varphi| < \alpha/2,$$

(7.1.4) $$\mathcal{B} u = 0,$$

where $\mathcal{B} u = \big(u(\pm\alpha/2), u'(\pm\alpha/2)\big)$. The operator $\mathfrak{A}(\lambda) = \big(\mathcal{L}(\lambda), \mathcal{B}\big)$ realizes a continuous mapping

$$W_2^4((-\alpha/2, +\alpha/2)) \to L_2((-\alpha/2, +\alpha/2)) \times \mathbb{C}^4$$

which is Fredholm for arbitrary complex λ and invertible for large purely imaginary λ. Consequently, the spectrum of the operator pencil $\mathfrak{A}$ is discrete and consists of eigenvalues with finite algebraic multiplicities.

7.1.2. Distribution of eigenvalues in the complex plane. Obviously, $\mathcal{L}(\partial_\varphi, \lambda) = \mathcal{L}(\partial_\varphi, 2 - \lambda)$. For this reason, it suffices to study the spectrum of $\mathfrak{A}$ in the half-plane $\operatorname{Re}\lambda \geq 1$.

If $\lambda \neq 0, 1, 2$, then every solution of the equation $\mathcal{L}(\partial_\varphi, \lambda) u = 0$ is a linear combination of the functions

$$\cos(\lambda\varphi), \quad \sin(\lambda\varphi), \quad \cos((\lambda - 2)\varphi), \quad \sin((\lambda - 2)\varphi),$$

It is more convenient to represent the general solution in the following form which is suitable for arbitrary λ :

(7.1.5) $$u = \sum_{k=1}^4 c_k u_k ,$$

where

$$u_1(\lambda, \varphi) = \cos\lambda\varphi, \qquad\qquad u_2(\lambda, \varphi) = \lambda^{-1} \sin\lambda\varphi,$$

$$u_3(\lambda, \varphi) = \frac{\cos\lambda\varphi - \cos(\lambda - 2)\varphi}{\lambda - 2}, \qquad u_4(\lambda, \varphi) = \frac{(\lambda - 2)\sin\lambda\varphi - \lambda\sin(\lambda - 2)\varphi}{\lambda(\lambda - 1)(\lambda - 2)}.$$

Inserting the function (7.1.5) into the boundary condition (7.1.4), we get the algebraic system

$$\sum_{k=1}^4 c_k u_k(\lambda, \pm\alpha/2) = 0, \qquad \sum_{k=1}^4 c_k u_k'(\lambda, \pm\alpha/2) = 0$$

with the unknowns $c_1, \dots, c_4$ which is equivalent to the following two systems

(7.1.6) $$\left\{ \begin{array}{rcl} c_1 u_1(\lambda, \alpha/2) + c_3 u_3(\lambda, \alpha/2) &=& 0, \\ c_1 u_1'(\lambda, \alpha/2) + c_3 u_3'(\lambda, \alpha/2) &=& 0, \end{array} \right.$$

(7.1.7) $$\left\{ \begin{array}{rcl} c_2 u_2(\lambda, \alpha/2) + c_4 u_4(\lambda, \alpha/2) &=& 0, \\ c_2 u_2'(\lambda, \alpha/2) + c_4 u_4'(\lambda, \alpha/2) &=& 0. \end{array} \right.$$

The coefficients determinants of the systems (7.1.6) and (7.1.7) are

$$d_+(\lambda) = \frac{1}{1 - \lambda}(\sin(\lambda - 1)\alpha + (\lambda - 1)\sin\alpha),$$

$$d_-(\lambda) = \frac{1}{\lambda(1 - \lambda)(\lambda - 2)}(\sin(\lambda - 1)\alpha - (\lambda - 1)\sin\alpha).$$

Every eigenvalue of the pencil $\mathfrak{A}$ is a zero of one of the functions $d_+(\lambda)$ and $d_-(\lambda)$. Since $d_+(1) = -(\alpha + \sin\alpha)$ and $d_-(1) = \sin\alpha - \alpha$, the number $\lambda = 1$ is not an eigenvalue of $\mathfrak{A}$. Moreover,

$$(7.1.8) \qquad d_+(2) = -2\sin\alpha, \quad d'_+(2) = \sin\alpha - \alpha\cos\alpha$$

$$(7.1.9) \qquad d_-(2) = \frac{1}{2}(\alpha\cos\alpha - \sin\alpha), \quad d'_-(2) = -\frac{3}{2}d_-(2) - \frac{\alpha^2}{4}\sin\alpha.$$

Therefore, $\lambda = 2$ is a root of the equation $d_+(\lambda) = 0$ only for $\alpha = \pi$, $\alpha = 2\pi$ and a root of the equation $d_-(\lambda) = 0$ only for $\alpha = \alpha_*$. Here, as in Section 3.1, α_* denotes the unique solution of the equation $\tan\alpha = \alpha$ in the interval $(0, 2\pi]$. In both cases the root $\lambda = 2$ is simple.

If $\lambda \neq 1$, $\lambda \neq 2$, then the equations $d_\pm(\lambda) = 0$ are equivalent to

$$(7.1.10) \qquad \sin(\lambda - 1)\alpha \pm (\lambda - 1)\sin\alpha = 0$$

which we already met (with λ instead of $\lambda - 1$) in Section 4.2 (cf. formulas (4.2.4), (4.2.5)). Hence Theorems 4.2.1–4.2.3 can be reformulated as follows for the operator pencil generated by problem (7.1.1), (7.1.2).

THEOREM 7.1.1. *Let $\alpha \neq \pi$, $\alpha \neq 2\pi$. Then the following assertions hold.*

1) *The lines $\operatorname{Re}\lambda - 1 = k\pi/\alpha$, $k = 0, 1, \dots$, do not contain points of the spectrum of $\mathfrak{A}$.*

2) *For $k = 3, 5, 7, \dots$ the strip*

$$(7.1.11) \qquad k\frac{\pi}{\alpha} < \operatorname{Re}\lambda - 1 < (k+1)\frac{\pi}{\alpha}$$

contains two eigenvalues of $\mathfrak{A}$ which are solutions of the equation $d_+(\lambda) = 0$ for $\alpha < \pi$ and of the equation $d_-(\lambda) = 0$ for $\alpha > \pi$.

3) *For $k = 2, 4, 6, \dots$ the strip (7.1.11) contains two eigenvalues of $\mathfrak{A}$ which satisfy the equality $d_-(\lambda) = 0$ for $\alpha < \pi$ and the equality $d_+(\lambda) = 0$ for $\alpha > \pi$.*

4) *The strip*

$$\frac{\pi}{\alpha} < \operatorname{Re}\lambda - 1 < \frac{2\pi}{\alpha}$$

contains two eigenvalues of $\mathfrak{A}$ satisfying the equality $d_+(\lambda) = 0$ if $\alpha < \pi$ and one eigenvalue satisfying the equality $d_-(\lambda) = 0$ if $\alpha > \pi$.

5) *If $\alpha > \pi$, then the strip*

$$(7.1.12) \qquad 0 < \operatorname{Re}\lambda - 1 < \frac{\pi}{\alpha}$$

contains one eigenvalue satisfying the equation $d_+(\lambda) = 0$. In the case $\alpha < \pi$ the strip (7.1.12) is free of eigenvalues.

THEOREM 7.1.2. *Let α_0, α_*, $z(\alpha)$, $\xi_+(\alpha)$, $\xi_+^{(1)}(\alpha)$, and $\xi_-(\alpha)$ be the same quantities as in Theorems 4.2.2 and 4.2.3.*

1) *If $\alpha \in (0, \alpha_0)$, then the spectrum of the operator pencil $\mathfrak{A}$ in the strip*

$$(7.1.13) \qquad 0 < \operatorname{Re}\lambda - 1 < 2\pi/\alpha$$

consists of two nonreal simple eigenvalues

$$1 + \frac{z(\alpha)}{\alpha} \quad and \quad 1 + \frac{\overline{z(\alpha)}}{\alpha}.$$

2) *For $\alpha = \alpha_0$ the spectrum of $\mathfrak{A}$ in the strip (7.1.13) consists of the unique eigenvalue $1 + \alpha_*/\alpha_0$ which has the algebraic multiplicity 2.*

3) *If $\alpha \in (\alpha_0, \pi)$, then the spectrum of $\mathfrak{A}$ in the strip (7.1.13) consists of the real eigenvalues*

$$1 + \frac{\xi_+(\alpha)}{\alpha} \quad and \quad 1 + \frac{\xi_+^{(1)}(\alpha)}{\alpha}.$$

4) *If $\alpha \in (\pi, 2\pi)$, then the spectrum of $\mathfrak{A}$ in the strip (7.1.13) consists of the eigenvalues $1 + \xi_+(\alpha)/\alpha$ and $1 + \xi_-(\alpha)/\alpha$.*

7.1.3. Eigenfunctions and generalized eigenfunctions of the operator pencil. Now we establish formulas for the eigenfunctions and generalized eigenfunctions of the operator pencil $\mathfrak{A}$. If λ is a root of the equation $d_+(\lambda) = 0$, then the solutions of system (7.1.6) have the form

$$c_1 = \alpha_1 \, u_3(\lambda, \frac{\alpha}{2}) + \alpha_2 \, u_3'(\lambda, \frac{\alpha}{2}),$$

$$c_3 = -\alpha_1 \, u_1(\lambda, \frac{\alpha}{2}) - \alpha_2 \, u_1'(\lambda, \frac{\alpha}{2}).$$

Setting $\alpha_1 = \cos(\lambda\alpha/2)$, $\alpha_2 = -\lambda^{-1}\sin(\lambda\alpha/2)$, we obtain

$$c_1 = \frac{\lambda - (\lambda-1)\cos\alpha - \cos(\lambda-1)\alpha}{\lambda(\lambda-1)},$$

$$c_3 = -1.$$

Therefore, we have found the eigenfunction

$$(7.1.14) \quad u^+(\lambda, \varphi) = c_1 \, u_1(\lambda, \varphi) + c_3 \, u_3(\lambda, \varphi)$$
$$= \frac{\lambda\cos(\lambda-2)\varphi - ((\lambda-1)\cos\alpha + \cos(\lambda-1)\alpha)\cos\lambda\varphi}{\lambda(\lambda-1)}.$$

The function $\mathbb{C} \ni \lambda \to u^+$ is analytic and satisfies the equation $\mathcal{L}(\partial_\varphi, \lambda)u^+ = 0$ and the boundary conditions

$$u^+(\lambda, \pm\alpha/2) = \alpha_2(\lambda) \, d_+(\lambda),$$
$$(\partial_\varphi u^+)(\lambda, \alpha/2) = \mp\alpha_1(\lambda) \, d_+(\lambda).$$

By analogous considerations, we get the following expression for an eigenfunction corresponding to the roots of the equation $d_-(\lambda) = 0$:

$$(7.1.15) \, u^-(\lambda, \varphi) = \frac{\lambda\sin(\lambda-2)\varphi + (\cos(\lambda-1)\alpha - (\lambda-1)\cos\alpha)\sin\lambda\varphi}{\lambda(\lambda-1)(\lambda-2)}.$$

Defining $u^-(\lambda, \cdot)$ for all $\lambda \in \mathbb{C}$ by this equality, we get a function which satisfies the equation $\mathcal{L}(\partial_\varphi, \lambda)u^- = 0$ and the boundary conditions

$$u^-(\lambda, \pm\alpha/2) = \pm\beta_2(\lambda) \, d_-(\lambda),$$
$$(\partial_\varphi u^-)(\lambda, \pm\alpha/2) = -\beta_1(\lambda) \, d_-(\lambda),$$

where $\beta_1(\lambda) = \lambda\sin(\lambda\alpha/2)$, $\beta_2(\lambda) = \cos(\lambda\alpha/2)$.

As it was mentioned before Theorem 7.1.1, there are no generalized eigenvectors to nonreal eigenvalues and no second generalized eigenvectors to all eigenvalues. The eigenvalues for which generalized eigenvectors exist have been described in Section 3.1 after Lemma 3.1.1. In our case we have to set $\gamma = 0$ and to replace λ by $\lambda - 1$ in formula (3.1.35).

The following theorem which can proved analogously to Lemma 3.1.4 gives a formula for the generalized eigenfunctions.

THEOREM 7.1.3. *If λ_0 is a multiple root of the equation $d_\pm(\lambda) = 0$, then the generalized eigenfunction is given by*

$$v^\pm(\lambda_0, \varphi) = (\partial_\lambda u^\pm)(\lambda_0, \varphi).$$

Finally, we consider the cases $\alpha = \pi$ and $\alpha = 2\pi$. In these cases common roots of the equations $d_+(\lambda) = 0$ and $d_-(\lambda) = 0$ exist.

If $\alpha = \pi$, then $3, 4, 5, \ldots$ are simple zeros both of $d_+(\lambda)$ and $d_-(\lambda)$. These numbers are eigenvalues of the pencil $\mathfrak{A}$ with geometric and algebraic multiplicities 2. The corresponding eigenfunctions are given by (7.1.14) and (7.1.15). By (7.1.8), (7.1.9), the number $\lambda = 2$ is a simple eigenvalue which satisfies the equality $d_+(\lambda) = 0$. The eigenfunction corresponding to this eigenvalue is given by (7.1.14). Other eigenvalues in the half-plane $\operatorname{Re}\lambda \geq 1$ do not exist.

Let $\alpha = 2\pi$. Then the eigenvalues in the half-plane $\operatorname{Re}\lambda \geq 1$ form the sequence $\lambda_k = 1 + k/2$ $(k = 1, 2, \ldots)$. All of them, except λ_2, have geometric and algebraic multiplicities 2 and are simple zeros both of $d_+(\lambda)$ and $d_-(\lambda)$. The eigenvalue $\lambda_2 = 2$ is simple. Again the eigenfunctions are given by (7.1.14) and (7.1.15).

7.2. The Dirichlet problem for the biharmonic equation in a cone

Here we study spectral properties of the operator pencil generated by the Dirichlet problem for the biharmonic equation in a cone. We find estimates for the energy strip. They imply, in particular, that the strip $2 - n \leq \operatorname{Re}\lambda \leq 2$ does not contain eigenvalues of this pencil if the cone is situated in a half-space. We show also that in a certain wider strip there are only real eigenvalues whose eigenfunctions have no generalized eigenfunctions. Moreover, these eigenvalues depend monotonically on the cone.

7.2.1. The pencil generated by the biharmonic equation in a cone. Let $\mathcal{K}$ be the cone $\{x \in \mathbb{R}^n : x/|x| \in \Omega\}$, $n \geq 3$, where Ω is a domain on S^{n-1} such that $S^{n-1}\backslash\Omega$ has nonempty interior. We define the space $\overset{\circ}{W}_2^2(\Omega)$ as the closure of $C_0^\infty(\Omega)$ with respect to the norm

$$\|u\|_{\overset{\circ}{W}_2^2(\Omega)} = \|\delta u\|_{L_2(\Omega)}^2,$$

where δ is the Beltrami operator. This space is closely and compactly imbedded into $L_2(\Omega)$. Furthermore, let $W_2^{-2}(\Omega)$ be the dual space of $\overset{\circ}{W}_2^2(\Omega)$ with respect to the scalar product in $L_2(\Omega)$.

We are interested in solutions of the equation

(7.2.1) $$\Delta^2 U = 0 \quad \text{in } \mathcal{K}$$

which have the form

(7.2.2) $$U(x) = r^{\lambda_0} \sum_{k=0}^{s} \frac{1}{k!} (\log r)^s u_{s-k}(\omega), \qquad u_{s-k} \in \overset{\circ}{W}_2^2(\Omega).$$

We introduce the pencil

(7.2.3) $$\mathcal{L}(\lambda) : \overset{\circ}{W}_2^2(\Omega) \to W_2^{-2}(\Omega)$$

defined by

$$\mathcal{L}(\lambda)\, u = r^{4-\lambda} \Delta^2 \left(r^\lambda u(\omega) \right).$$

Using the formula $r^{2-\lambda}\Delta(r^\lambda u(\omega)) = \big(\delta + \lambda(\lambda+n-2)\big)u(\omega)$ (see Section 2.2), we get the representation (7.0.1) for the operator $\mathcal{L}(\lambda)$. By Theorem 1.1.5, the function (7.2.2) satisfies (7.2.1) if and only if λ_0 is an eigenvalue of $\mathcal{L}$, u_0 is an eigenfunction, and $u_1, \ldots, u_s$ are generalized eigenfunctions corresponding to λ_0.

We introduce the sesquilinear form

$$(7.2.4) \quad a(u,v;\lambda) = (\delta u, \delta v)_{L_2(\Omega)} + \lambda(\lambda-2)(\lambda+n-2)(\lambda+n-4)\,(u,v)_{L_2(\Omega)}$$
$$+ \Big(\lambda(\lambda+n-2) + (\lambda-2)(\lambda+n-4)\Big)(\delta u, v)_{L_2(\Omega)}\,.$$

Clearly, $(\mathcal{L}(\lambda)u, v)_{L_2(\Omega)} = a(u,v;\lambda)$ for all $u, v \in \overset{\circ}{W}{}^2_2(\Omega)$. The last equality can be taken as another definition of the operator (7.2.3). It can be directly verified that

$$(7.2.5) \qquad\qquad (\mathcal{L}(\lambda))^* = \mathcal{L}(4 - n - \bar{\lambda})$$

THEOREM 7.2.1. 1) *There exists a constant c depending only on n such that the inequality*

$$a(u,u;\lambda) \geq \frac{1}{2}\left\|\delta u\right\|^2_{L_2(\Omega)} - c\left(|\lambda|+1\right)^4\left\|u\right\|^2_{L_2(\Omega)}$$

is valid for all $u \in \overset{\circ}{W}{}^2_2(\Omega)$.

2) *The operator* $\mathcal{L}(2 - n/2 + it)$ *is positive definite on* $\overset{\circ}{W}{}^2_2(\Omega)$ *for all real t.*

Proof: 1) Estimating $(\delta u, u)$ in (7.2.4) by the Schwarz inequality, we obtain the first assertion.

2) Direct calculation gives

$$a(u,u;2-n/2+it) = \left\|\delta u + (n/2+it)\,(2-n/2+it)\,u\right\|^2_{L_2(\Omega)}.$$

The right hand side vanishes if and only if

$$(7.2.6) \qquad \delta u + (n/2+it)\,(2-n/2+it)\,u = 0, \quad u \in \overset{\circ}{W}{}^2_2(\Omega).$$

Clearly, this impossible for $t \neq 0$. We consider the case $t = 0$. Let first $n > 3$. In this case the equality (7.2.6) is also impossible, since $\frac{n}{2}(2 - \frac{n}{2}) \leq 0$ and $(\delta u, u)_{L_2(\Omega)} < 0$ for $u \neq 0$. Let now $n = 3$. Integrating (7.2.6) over Ω, we obtain $\int_\Omega u\,d\omega = 0$. In order to show that (7.2.6) is not valid in this case, it suffices to prove that

$$-(\delta u, u)_{L_2(S^2)} \geq 2\left\|u\right\|^2_{L_2(S^2)}$$

for $u \in W^1_2(S^2)$, $\int_\Omega u\,d\omega = 0$. The last inequality follows from the fact that the second eigenvalue of the operator $-\delta$ on the sphere S^2 is equal to 2 (see Section 2.2). ∎

From the last theorem and from Theorem 1.2.1 it follows that the spectrum of the pencil $\mathcal{L}$ consists of eigenvalues with finite geometric and algebraic multiplicities. The line $\operatorname{Re}\lambda = 2 - n/2$ is free of eigenvalues and every strip $a \leq \operatorname{Re}\lambda \leq b$ on the complex plane contains only a finite number of eigenvalues of $\mathcal{L}$.

7.2.2. A strip containing only real eigenvalues. We denote by $\mu_0 = \mu_0(\Omega)$ the first eigenvalue of the operator $-\delta$ in the domain Ω with the Dirichlet conditions on $\partial\Omega$. By our assumptions on Ω, we have $\mu_0 > 0$. Let

$$\sigma_0 = \sigma_0(\Omega) = 2 - n/2 + \sqrt{(n/2-1)^2 + 1 + \mu_0(\Omega)}.$$

From the monotonicity of μ_0 (see Theorem 2.2.4) it follows that the set function σ_0 is also monotonous, i.e., $\sigma_0(\Omega_1) \geq \sigma_0(\Omega_2)$ if $\Omega_1 \subset \Omega_2$.

Representing $1 + \mu_0$ as $\tau_0(\tau_0 + n - 2)$ with $\tau_0 > 0$, we obtain that $\sigma_0 = 1 + \tau_0$.

THEOREM 7.2.2. *The strip*

$$4 - n - \sigma_0 \leq \operatorname{Re} \lambda \leq \sigma_0$$

may only contain real eigenvalues of the pencil $\mathcal{L}$.

Proof: The quadratic form

$$(7.2.7) \qquad b(u, u; \lambda) \stackrel{def}{=} -(\delta u, u)_{L_2(\Omega)} - \left(\lambda^2 + (n-4)\lambda - n + 2\right) \|u\|_{L_2(\Omega)}^2$$

is positive definite on $\overset{\circ}{W}{}^1_2(\Omega)$ for real λ satisfying the inequality $\lambda^2 + (n-4)\lambda - n + 2 < \mu_0$ or, what is the same, $4 - n - \sigma_0 < \lambda < \sigma_0$. By Theorem 7.2.1, the line $\operatorname{Re} \lambda = 2 - n/2$ does not contain eigenvalues of $\mathcal{L}$. Let $\lambda_0 = \sigma + it$ be an eigenvalue of the pencil $\mathcal{L}$, where $\sigma \in [4 - n - \sigma_0, \sigma_0]$, $\sigma \neq 2 - n/2$, $t \in \mathbb{R}$. Furthermore, let u be an eigenfunction correspnding to this eigenvalue. Then

$$0 = \operatorname{Im} a(u, u; \lambda_0) = -2t (2\sigma + n - 4) \left(b(u, u; \sigma) + t^2 \|u\|_{L_2(\Omega)}^2\right).$$

The right side of the last equality is nonzero for $t \neq 0$, $\sigma \in [4 - n - \sigma_0, \sigma_0]$, $\sigma \neq 2 - n/2$. Consequently, the eigenvalue λ_0 must be real. This proves the theorem. ∎

7.2.3. On generalized eigenfunctions.

THEOREM 7.2.3. *The eigenfunctions corresponding to eigenvalues in the strip*

$$4 - n - \sigma_0 < \operatorname{Re} \lambda < \sigma_0$$

do not have generalized eigenfunctions.

Proof: By Theorem 7.2.2 and (7.2.5), we may restrict ourselves in the proof to real eigenvalues in the interval $(2 - n/2, \sigma_0)$. For these eigenvalues the quadratic form (7.2.7) is positive definite. Hence

$$(7.2.8) \qquad \frac{d}{d\lambda} \left(\mathcal{L}(\lambda)u, u\right)_{L_2(\Omega)} \Big|_{\lambda=\lambda_0} = -2 (2\lambda_0 + n - 4) b(u, u; \lambda_0) < 0$$

for every eigenvalue $\lambda_0 \in (2 - n/2, \sigma_0)$. Suppose that u_0 is an eigenfunctions and u_1 a generalized eigenfunction corresponding to λ_0. Then it follows from the equality $\mathcal{L}'(\lambda_0) u_0 + \mathcal{L}(\lambda_0) u_1 = 0$ that

$$\left(\mathcal{L}'(\lambda_0)u_0, u_0\right)_{L_2(\Omega)} = -\left(\mathcal{L}(\lambda_0)u_1, u_0\right)_{L_2(\Omega)} = -\left(u_1, \mathcal{L}(\lambda_0)u_0\right)_{L_2(\Omega)} = 0.$$

This contradicts (7.2.8). The theorem is proved. ∎

Note that, by Theorems 7.2.1 and (7.2.8), conditions (I)–(III) of Section 1.3 are satisfied for the form a, where $\mathcal{H} = L_2(\Omega)$, $\mathcal{H}_+ = \overset{\circ}{W}{}^2_2(\Omega)$, $\alpha = 2 - n/2$, $\beta = \sigma_0$. Therefore, all results of Section 1.3 are valid for this case. Using Theorem 1.3.5 and the monotonicity of the space $\overset{\circ}{W}{}^2_2(\Omega)$ with respect to Ω, we obtain, in particular, that the eigenvalues of $\mathcal{L}$ on the interval $(2 - n/2, \sigma_0)$ increase when the domain Ω becomes smaller.

7.2.4. Estimates for the width of the energy strip. We denote by $\sigma_1 = \sigma_1(\Omega)$ the greatest number (including $+\infty$) such that $\sigma_1 > 2 - n/2$ and the operator $\mathcal{L}(\sigma)$ is positive definite for $\sigma \in [2 - n/2, \sigma_1)$ (cf. definition of λ_* in Section 1.3). By the second part of Theorem 7.2.1, this number exists. It can be easily seen that the set function σ_1 is nonincreasing with respect to Ω. Moreover, the following lemma holds.

LEMMA 7.2.1. *Let Ω_1, Ω_2 be domains on the sphere such that $\Omega_1 \subset\subset \Omega_2$. Then $\sigma_1(\Omega_1) > \sigma_1(\Omega_2)$. (The notation $\Omega_1 \subset\subset \Omega_2$ is introduced in Section 3.3.3.)*

Proof: Suppose that $\sigma_1(\Omega_1) = \sigma_1(\Omega_2) = \lambda_0$ and $u \in \overset{\circ}{W}{}_2^2(\Omega_1) \backslash \{0\}$ is such that $\big(\mathcal{L}(\lambda_0)u, u\big)_{L_2(\Omega_1)} = 0$. Extending u by zero to $\Omega_2 \backslash \Omega_1$, we get the same relation for Ω_2. Since the operator $\mathcal{L}(\lambda_0)$ is nonnegative on $L_2(\Omega_2)$, we obtain that $\mathcal{L}(\lambda_0)u = 0$ in Ω_2. The last equation implies that u is analytic in Ω_2 and, hence, $u = 0$. This contradicts our assumption on u. ∎

THEOREM 7.2.4. 1) *The strip*

$$4 - n - \min\{\sigma_0, \sigma_1\} < \operatorname{Re}\lambda < \min\{\sigma_0, \sigma_1\}$$

does not contain eigenvalues of the pencil $\mathcal{L}$.

2) *If $\sigma_1 < \infty$, then σ_1 is an eigenvalue of the pencil $\mathcal{L}$.*

Proof: 1) Due to (7.2.5) and Theorem 7.2.1, it is sufficient to prove that there are no eigenvalues of the pencil $\mathcal{L}$ in the strip $2 - n/2 < \operatorname{Re}\lambda < \min\{\sigma_0, \sigma_1\}$. This follows from Theorem 7.2.2 and from the definition of the quantity σ_1.

2) According to the definition of σ_1, the operator $\mathcal{L}(\sigma_1)$ has the lower bound zero, whence statement 2) follows. ∎

7.2.5. Properties of σ_1. We give another definition of σ_1. For $u \in \overset{\circ}{W}{}_2^2(\Omega)$, $u \neq 0$, let $R(u)$ denote the least root of the equation

$$a(u, u; \sigma) = 0$$

in the interval $[2 - n/2, \infty]$. Lemma 1.3.2 immediately implies the following assertion.

LEMMA 7.2.2. *The quantity σ_1 satisfies*

$$(7.2.9) \qquad \sigma_1(\Omega) = \inf_{u \in \overset{\circ}{W}{}_2^2(\Omega) \backslash \{0\}} R(u).$$

If $\sigma_1 < \infty$, then the infimum is attained on the eigenfunction u_1 of the pencil $\mathfrak{A}$ corresponding to the eigenvalue σ_1.

The equation $a(u, u; \sigma) = 0$ can be written in the variable $\xi = \sigma(\sigma + n - 4)$ as

$$\|\delta u\|_{L_2(\Omega)}^2 + 2\left(\xi - n + 4\right)\left(\delta u, u\right)_{L_2(\Omega)} + \xi\left(\xi - 2n + 4\right)\|u\|_{L_2(\Omega)}^2 = 0.$$

Solving this quadratic equation and expressing σ_1 by ξ, we get the formula

$$(7.2.10) \qquad \sigma_1 = 2 - n/2 + \sqrt{(2 - n/2)^2 + \Xi},$$

where

$$(7.2.11) \qquad \Xi = \inf_{\substack{u \in \overset{\circ}{W}{}_2^2(\Omega) \\ \|u\|_{L_2(\Omega)} = 1,\ D(u) \geq 0}} \left(-(\delta u, u)_{L_2(\Omega)} + n - 2 - \sqrt{D(u)}\right),$$

and
$$D(u) = \left((\delta u, u)_{L_2(\Omega)}\right)^2 - \|\delta u\|^2_{L_2(\Omega)} + (n-2)^2 - 4\,(\delta u, u)_{L_2(\Omega)}\,.$$

LEMMA 7.2.3. *If $n \geq 4$ or $n = 3$ and $\mu_0 \geq 3/4$, then*

(7.2.12)
$$\sigma_1(\Omega) > M,$$

where M is the positive number such that $M(M + n - 2) = \mu_0$.

Proof: By the Schwarz inequality,

(7.2.13)
$$\left|(\delta u, u)_{L_2(\Omega)}\right|^2 \leq \|\delta u\|^2_{L_2(\Omega)}$$

if $\|u\|_{L_2(\Omega)} = 1$. Moreover, equality takes place there only if $\delta u = cu$ with some constant c. But the last relation is impossible for $u \in \overset{\circ}{W}{}^2_2(\Omega)$. Hence we have a strict inequality in (7.2.13) and, therefore,

(7.2.14)
$$D(u) < (n-2)^2 - 4(\delta u, u)_{L_2(\Omega)}\,.$$

Since the right-hand side in (7.2.11) attains its infimum at a certain $u_0 \in \overset{\circ}{W}{}^2_2(\Omega)$, we have, owing to (7.2.14),

(7.2.15)
$$\Xi > -(\delta u_0, u_0)_{L_2(\Omega)} + n - 2 - \sqrt{-4\,(\delta u_0, u_0)_{L_2(\Omega)} + (n-2)^2}\,.$$

If $n \geq 4$, then the right-hand side in (7.2.15) is an increasing function with respect to $-(\delta u_0, u_0)_{L_2(\Omega)}$. Since the last quantity is not less than μ_0, we have

$$\Xi > \mu_0 + n - 2 - \sqrt{4\mu_0 + (n-2)^2} = M(M + n - 4).$$

Therefore, $\sigma_1 > M$.

If $n = 3$, then the function

$$\Lambda \to \Lambda + 1 - \sqrt{4\Lambda + 1}$$

decreases for $\Lambda \in (0, \tfrac{3}{4})$ and increases for $\Lambda > \tfrac{3}{4}$. Hence,

$$\Xi > -\frac{1}{4} \quad \text{if } \mu_0 < \frac{3}{4},$$

$$\Xi > \mu_0 + 1 - \sqrt{4\mu_0 + 1} = M(M - 1) \quad \text{if } \mu_0 \geq \frac{3}{4}\,.$$

The above obtained inequalities for Ξ and (7.2.10) imply (7.2.12). ∎

REMARK 7.2.1. The function σ_1 satisfies also the estimate $\sigma_1(\Omega) > (5-n)/2$. This follows from the monotonicity of the set function σ_1 and from the inequality $|\operatorname{Re}\lambda_j - m + n/2| > 1/2$ for the eigenvalues of the operator pencil generated by the Dirichlet for a $2m$ order strongly elliptic differential operator in a Lipschitz cone which will be proved in Section 11.1 (see Theorem 11.1.1).

In the next theorem we present estimates for the energy strip of the pencil $\mathcal{L}$ in terms of M.

THEOREM 7.2.5. *The strips*

(7.2.16)
$$1 - \max(M, 1) \leq \operatorname{Re}\lambda \leq \max(M, 1) \quad \text{for } n = 3,$$

(7.2.17)
$$-\max(M, 1/2) \leq \operatorname{Re}\lambda \leq \max(M, 1/2) \quad \text{for } n = 4,$$

and

(7.2.18)
$$4 - n - M \leq \operatorname{Re}\lambda \leq M \quad \text{for } n > 4$$

contain no eigenvalues of the pencil $\mathcal{L}$.

Proof: Since $\sigma_0 > M + 1$, the above assertion follows from Theorem 7.2.4, Lemma 7.2.3 and Remark 7.2.1.

7.2.6. On the eigenvalues of $\mathcal{L}$ in the interval $[2 - n/2, \sigma_0)$. The operator $\mathcal{L}(\lambda)$ is selfadjoint and bounded from below for real λ. For $\lambda \geq 2 - n/2$ we denote by $\{\mu_j(\lambda)\}_{j \geq 1}$ a nondecreasing sequence of eigenvalues of $\mathcal{L}(\lambda)$ counted with their multiplicities and by $v_j(\lambda)$ an eigenfunction corresponding to $\mu_j(\lambda)$. We can choose v_j in such a way that the system $\{v_j(\lambda)\}_{j \geq 1}$ forms a orthonormal basis in $L_2(\Omega)$. The functions μ_j are continuous (moreover, piecewise analytic) with respect to λ (see, for example, Kato's book [**96**, Ch.VII,Th.1.8]). From the variational principle it follows that the functions $\mu_j(\lambda)$ are decreasing with respect to Ω. Let N be the integer satisfying $\mu_N(\sigma_0) < 0$ and $\mu_{N+1}(\sigma_0) \geq 0$.

LEMMA 7.2.4. *The interval $2 - n/2 \leq \lambda < \sigma_0$ contains exactly N eigenvalues (counted with their geometric multiplicities) of the operator pencil $\mathcal{L}$. These eigenvalues are zeros of the function μ_j, $j \leq N$, with eigenfunctions v_j. Moreover, every function μ_j has at most one zero at the interval $[2 - n/2, \sigma_0)$.*

Proof: First, we prove that the function μ_j has only one zero on $[2 - n/2, \sigma_0)$. Since $\mu_j(2 - n/2) > 0$ (because of positivity of $(\mathcal{L}(2 - n/2)$, see Theorem 7.2.1) and $\mu_j(\sigma_0) < 0$, the function μ_j has at least one root at the interval $[2 - n/2, \sigma_0)$. Owing to Theorem 1.3.2, the function μ_j has at most one root on this interval.

Second, the number $\lambda^* \in [2 - n/2, \sigma_0)$ is an eigenvalue of the pencil $\mathcal{L}$ if and only if this number is a zero of one of the functions $\mu_1, \ldots, \mu_N$. Clearly, the functions v_j are eigenfunctions of the pencil $\mathcal{L}$ corresponding to the eigenvalue λ^*, and these functions exhaust all eigenfunctions corresponding to λ^*. ∎

COROLLARY 7.2.1. *The eigenvalues of the pencil $\mathcal{L}$ in $[2 - n/2, \sigma_0)$ are decreasing functions with respect to the domain Ω and located in the interval $(0, \sigma_0)$.*

Proof: The assertion follows from Lemma 7.2.4, from the absence of eigenvalues on $[2 - n/2, 0]$ (see Section 7.2.3) and from the monotonicity property of the function μ_j. ∎

7.2.7. The case of a cone contained in a half-space. Using Theorem 7.2.4 and the monotonicity of the set functions σ_0, σ_1, we can prove the following estimate for the width of the energy strip in the case when Ω is contained in the half-sphere S_+^{n-1}.

THEOREM 7.2.6. *If $\Omega \subset S_+^{n-1}$, where S_+^{n-1} denotes the hemisphere $S^{n-1} \cap \mathbb{R}_+^n$, then the strip*

$$2 - n \leq \mathrm{Re}\,\lambda \leq 2$$

is free of eigenvalues of the pencil $\mathcal{L}$.

Proof: By Theorem 2.2.4 and Lemma 7.2.1, the set functions σ_0 and σ_1 are strictly monotonous. Hence, by Theorem 7.2.4, it is sufficient to establish the relations $\sigma_0(S_+^{n-1}) > 2$ and $\sigma_1(S_+^{n-1}) = 2$. The first relation follows immediately from the equality $\mu_0(S_+^{n-1}) = n - 1$ (see Section 2.2) and from the definition of σ_0. Analogously to Theorem 2.2.2, it can be shown that all solutions of the equation $\Delta^2 U = 0$ in the half-space $\mathbb{R}_+^n$ which have the representation $U(x) = r^\lambda u(\omega)$ with

$\operatorname{Re}\lambda > 2 - n/2$, $u \in \overset{\circ}{W}_2^2(S_+^{n-1})$ can be written in the form $U(x) = x_n^2 p(x)$, where p is homogeneous polynomial. From this it follows that the eigenvalues of the pencil $\mathcal{L}$ in the half-plane $\operatorname{Re}\lambda > 2 - n/2$ for $\Omega = S_+^{n-1}$ are exhausted by the values $2, 3, \ldots$. Using the second part of Theorem 7.2.4, we conclude that $\sigma_1(S_+^{n-1}) = 2$. This proves the theorem. ∎

7.3. The polyharmonic operator

Now we deal with the pencil $\mathcal{L}$ corresponding to the Dirichlet problem for the polyharmonic operator $(-\Delta)^m$ in the cone $\mathcal{K} = \{x \in \mathbb{R}^n : x/|x| \in \Omega\}$ which is defined by (7.0.7) for the function $u \in \overset{\circ}{W}_2^m(\Omega)$. Here $\overset{\circ}{W}_2^m(\Omega)$ is the closure of $C_0^\infty(\Omega)$ with respect to the norm

$$\|u\|_{W_2^m(\Omega)} = \left(\int\limits_{\substack{\mathcal{K} \\ 1/2 < |x| < 2}} \sum_{|\alpha| \leq m} |D_x^\alpha u(x)|^2 \, dx \right)^{1/2},$$

where in the integral on the right the function u is extended to $\mathcal{K}$ by $u(x) = u(x/|x|)$. As in Section 7.2, we will assume that $S^{n-1} \backslash \Omega$ has a nonempty interior. We give estimates for the energy strip and for strips containing only real eigenvalues. For these eigenvalues we prove the monotone dependence on the domain Ω.

7.3.1. Some properties of $\mathcal{L}$. We begin with a positivity property of $\mathcal{L}$.

LEMMA 7.3.1. *Let $\lambda = m - n/2 + i\tau$, $\tau \in \mathbb{R}$. Then for all $u \in \overset{\circ}{W}_2^m(\Omega)$ there is the inequality*

$$(7.3.1) \qquad \left(\mathcal{L}(\lambda)u, u\right)_{L_2(\Omega)} \geq c \left(\|u\|_{W_2^m(\Omega)}^2 + |\lambda|^{2m} \|u\|_{L_2(\Omega)}^2 \right),$$

where c is a positive constant.

Proof. 1) Consider first the case $m = 2k$. Then for $\lambda = m - n/2 + i\tau$

$$\mathcal{L}(\lambda) = L_1(\lambda) L_1(\overline{\lambda}),$$

where

$$L_1(\lambda) = \prod_{j=0}^{k-1} \left(\delta + (\lambda - 2j)(\lambda - 2j + n - 2) \right).$$

Hence

$$(7.3.2) \qquad \left(\mathcal{L}(\lambda)u, u\right)_{L_2(\Omega)} = \|L_1(\lambda)u\|_{L_2(\Omega)}^2.$$

Here the kernel of $L_1(\lambda)$ is trivial on $\overset{\circ}{W}_2^m(\Omega)$, since the order of L_1 is equal to m. From (7.3.2) we get (7.3.1) for large $|\tau|$. If $|\tau| \leq C$, where C is a constant, then we have $\|L_1(\lambda)u\|_{L_2(\Omega)} \geq c \|u\|_{W_2^m(\Omega)}$ with some positive c depending on C.

2) Let $m = 2k + 1$. Then

$$\begin{aligned} -\mathcal{L}(\lambda) &= (\delta + (\lambda - 2k)(\lambda - 2k + n - 2)) L_1(\lambda) L_1(\overline{\lambda}) \\ &= (\delta - (1 - n/2)^2 - \tau^2) L_1(\lambda) L_1(\overline{\lambda}), \end{aligned}$$

which implies

$$(7.3.3) \qquad \left(\mathcal{L}(\lambda)u, u\right)_{L_2(\Omega)} = \left((-\delta + (1 - n/2)^2 + \tau^2)L_1(\lambda)u, L_1(\lambda)u\right)_{L_2(\Omega)}.$$

This together with $\ker L_1(\lambda)u = \{0\}$ on $\overset{\circ}{W}{}_2^m(\Omega)$ implies (7.3.1). ∎

The above lemma shows, in particular, that the operator (7.0.6) is isomorphic for $\lambda = m - n/2$. Clearly, the operator

$$\mathcal{L}(\lambda) - \mathcal{L}(m - n/2) : \overset{\circ}{W}{}_2^m(\Omega) \to W_2^{-m}(\Omega)$$

is compact for all $\lambda \in \mathbb{C}$. Therefore, the operator (7.0.6) is Fredholm for every λ. From this we conclude that the spectrum of $\mathcal{L}$ consists of eigenvalues of finite algebraic multiplicity and that the only accumulation point of the spectrum is infinity (see Theorem 1.1.1).

It can be easily verified that

$$\mathcal{L}(\lambda) = \mathcal{L}(2m - n - \lambda) \, .$$

Thus, it suffices to study spectral properties of the pencil $\mathcal{L}$ in the half-plane $\operatorname{Re} \lambda > m - n/2$.

REMARK 7.3.1. In the cases $2m < n$ and $2m \geq n$, n odd, the assertion of Lemma 7.3.1 is true for arbitrary $\Omega \subset S^{n-1}$. This follows from the fact that $L_1(\lambda)u = 0$ for $u \in W_2^m(S^{n-1})$ implies $\Delta^k(r^\lambda u(\omega)) = 0$ on S^{n-1} and, consequently, $u = 0$ (see Theorem 10.2.1).

7.3.2. Eigenvalues on the line $\operatorname{Re} \lambda = m + 1 - n/2$.

LEMMA 7.3.2. *The number $\lambda = m + 1 - n/2 + i\tau$ with nonzero real τ is not an eigenvalue of the pencil $\mathcal{L}$.*

Proof: 1) Let $m = 2k + 1$ and $\lambda = m + 1 - n/2 + i\tau$. Then

$$-\mathcal{L}(\lambda) = \big(\delta + \lambda(\lambda + n - 2)\big) L_2(\lambda) L_2(\bar{\lambda}) \, ,$$

where

$$L_2(\lambda) = \prod_{j=1}^{k} \big(\delta + (\lambda - 2j)(\lambda - 2j + n - 2)\big).$$

Therefore,

(7.3.4) $\operatorname{Im} \big(\mathcal{L}(\lambda)u, u\big)_{L_2(\Omega)} = 2m\tau \, \|L_2(\lambda)u\|_{L_2(\Omega)}^2 \, .$

The right-hand side is only zero if $\tau = 0$ or $L_2(\lambda)u = 0$. Owing to the Dirichlet boundary conditions, this implies $u = 0$. Hence $(\mathcal{L}(\lambda)u, u) \neq 0$ for $\tau \neq 0$ and $u \neq 0$. The result follows for odd m.

2) Now let $m = 2k + 2$. Then

$$\mathcal{L}(\lambda) = \big(\delta + \lambda(\lambda + n - 2)\big) \big(\delta + (\lambda - 2k - 2)(\lambda - 2k + n - 4)\big) L_2(\lambda) L_2(\bar{\lambda}).$$

Since

$$\big(\delta + \lambda(\lambda + n - 2)\big) \big(\delta + (\lambda - 2k - 2)(\lambda - 2k + n - 4)\big)$$
$$= \big(\delta + (i\tau + m)^2 - (1 - n/2)^2\big) \big(\delta - \tau^2 - (1 - n/2)^2\big),$$

we obtain

(7.3.5) $\operatorname{Im}\big(\mathcal{L}(\lambda)u, u\big)_{L_2(\Omega)} = 2m\tau \left(\big(-\delta + (1 - n/2)^2 + \tau^2\big) L_2(\lambda)u, L_2(\lambda)u \right)_{L_2(\Omega)} \, .$

As in the case of odd m, the right hand side vanishes only for $\tau = 0$ or $u = 0$. The proof is complete. ∎

LEMMA 7.3.3. *Let $\lambda_0 = m + 1 - n/2$ be an eigenvalue of the pencil $\mathcal{L}$ and u_0 a corresponding eigenfunction. Then*

(7.3.6)
$$\frac{d}{d\lambda}\big(\mathcal{L}(\lambda)u_0, u_0\big)_{L_2(\Omega)}\Big|_{\lambda=\lambda_0} > 0 \,.$$

Proof: The quadratic form $(\mathcal{L}(\lambda)u_0, u_0)_{L_2(\Omega)}$ is a polynomial in $\lambda = \sigma + i\tau$. We represent it as

$$(\mathcal{L}(\lambda)u_0, u_0)_{L_2(\Omega)} = p(\sigma, \tau) + i\tau\, q(\sigma, \tau),$$

where and p and q are real-valued polynomials with respect to σ and τ. Since

$$\frac{d}{d\lambda}\big(\mathcal{L}(\lambda)u_0, u_0\big)_{L_2(\Omega)} = \frac{d}{i\,d\tau}\big(\mathcal{L}(\lambda)u_0, u_0\big)_{L_2(\Omega)} \,,$$

we have

$$\frac{d}{d\lambda}\big(\mathcal{L}(\lambda)u_0, u_0\big)_{L_2(\Omega)} = \frac{d}{i\,d\tau}p(\sigma, \tau) + q(\sigma, \tau) + \tau\frac{d}{d\tau}q(\sigma, \tau).$$

Using the fact that the left-hand side in this identity is real, we obtain

$$\frac{d}{d\lambda}\big(\mathcal{L}(\lambda)u_0, u_0\big)_{L_2(\Omega)}\Big|_{\lambda=\lambda_0} = q(\lambda_0, 0) \,.$$

Thus,

$$\frac{d}{d\lambda}\big(\mathcal{L}(\lambda)u_0, u_0\big)_{L_2(\Omega)}\Big|_{\lambda=\lambda_0} = \frac{1}{\tau}\,\mathrm{Im}\big(\mathcal{L}(\lambda_0)u_0, u_0\big),$$

and inequality (7.3.6) follows from the positivity (after division by τ) of the right hand sides in (7.3.4) and (7.3.5). ∎

COROLLARY 7.3.1. *If $\lambda_0 = m + 1 - n/2$ is an eigenvalue of the pencil $\mathcal{L}$, then it has no generalized eigenfunctions.*

Proof: Let λ_0 be an eigenvalue and u_0 a corresponding eigenfunction. Then the equation for a generalized eigenfunction u_1 is

$$\mathcal{L}(\lambda_0)u_1 = -\frac{d}{d\lambda}\mathcal{L}(\lambda)u_0|_{\lambda=\lambda_0} \,.$$

In order to show that it is unsolvable, we multiply (in $L_2(\Omega)$) both sides of the equation by u_0. Then the left-hand side vanishes, since u_0 is eigenfunction and the operator $\mathcal{L}(\lambda_0)$ is selfadjoint. But the right-hand side differs from zero by Lemma 7.3.3. This proves the corollary. ∎

7.3.3. Eigenvalues in the strip $m - n/2 \leq \mathrm{Re}\,\lambda \leq m + 1 - n/2$. We consider the operator pencil $\mathcal{L}$ for real λ. Clearly, the operator $\mathcal{L}(\lambda)$ is selfadjoint and semibounded from below. For every $\lambda \geq m - n/2$ we denote by $\{\mu_j(\lambda)\}_{j\geq 1}$ a nondecreasing sequence of eigenvalues of the operator $\mathcal{L}(\lambda)$, counted with their multiplicities. Furthermore, we denote by $v_j(\lambda)$ an eigenfunction of $\mathcal{L}(\lambda)$ corresponding to $\mu_j(\lambda)$. One can suppose that the system $\{v_j(\lambda)\}_{j\geq 1}$ is an orthonormal basis in $L_2(\Omega)$. It is known (see, for example Kato's book [**96**, Ch.VII,Th.1.8]) that the functions μ_j are continuous (moreover piecewise analytic) on $[m - n/2, \infty)$. From the variational principle for $\mu_j(\lambda)$ it follows that the functions μ_j increase when the domain Ω becomes smaller.

Let N be the number for which $\mu_N(m+1-n/2) \leq 0$ and $\mu_{N+1}(m+1-n/2) > 0$.

LEMMA 7.3.4. *The strip*

$$(7.3.7) \qquad\qquad m - n/2 \leq \operatorname{Re}\lambda \leq m + 1 - n/2$$

contains exactly N eigenvalues of the operator pencil $\mathcal{L}$ counted with their geometric multiplicities. All these eigenvalues are real and have no generalized eigenfunctions. Every function μ_j has at most one zero on the interval $[m - n/2, m + 1 - n/2]$.

Proof: 1) We consider the functions $\mu_1, \dots, \mu_N$. By Lemma 7.3.1, the operator $\mathcal{L}(m - n/2)$ is positive. Consequently, $\mu_j(m - n/2) > 0$ for all j. If $j \leq N$, then $\mu_j(m + 1 - n/2) \leq 0$ and, therefore, the function μ_j has at least one zero, say λ_j, on the interval $(m - n/2, m + 1 - n/2]$. Clearly, λ_j is an eigenvalue of the pencil $\mathcal{L}$ and v_j is an eigenfunction corresponding to λ_j. Thus, we have proved that the total geometric multiplicity of the eigenvalues situated on the interval $(m - n/2, m + 1 - n/2]$ is not less than N. Moreover, if one of the functions μ_j, $j \leq N$, has more than one zero on $(m - n/2, m + 1 - n/2]$, then the total geometric multiplicity of eigenvalues of $\mathcal{L}$ in the strip (7.3.7) is greater than N.

2) We consider the family of operator pencils $\mathcal{L}_t(\lambda) = \mathcal{L}(\lambda) + tI$, where $t \geq 0$ and I is the identity operator. Clearly, the set $\{\mu_j(\lambda) + t\}_{j \geq 1}$ represents a nondecreasing sequence of eigenvalues of $\mathcal{L}_t(\lambda)$, counted with their multiplicities. If t is sufficiently large, then there are no eigenvalues of $\mathcal{L}_t$ in the strip (7.3.7).

Now we keep t decreasing. For $t > -\mu_1(m + 1 - n/2)$ this strip is still free of eigenvalues of $\mathcal{L}_t$. Indeed, from Lemmas 7.3.1 and 7.3.2 it follows that there are no eigenvalues on the lines $\operatorname{Re}\lambda = m - n/2$ and $\operatorname{Re}\lambda = m + 1 - n/2$ and there are no eigenvalues in the strip (7.3.7) for large $|\operatorname{Im}\lambda|$ hence, by Corollary 1.1.2, there are no eigenvalues in the strip (7.3.7) for all $t > -\mu_1(m + 1 - n/2)$.

Let $t = -\mu_1(m + 1 - n/2)$. Then the strip $m - n/2 < \operatorname{Re}\lambda < m + 1 - n/2$ is free of eigenvalues of $\mathcal{L}_t$. Otherwise, by Corollary 1.1.2, the same strip should contain eigenvalues of $\mathcal{L}_t$ for some t greater than $-\mu_1(m + 1 - n/2)$. Let κ be such index that $\mu_1(m + 1 - n/2) = \cdots = \mu_\kappa(m + 1 - n/2) = -t$ and $\mu_{\kappa+1}(m + 1 - n/2) > -t$. Then the strip (7.3.7) contains exactly one eigenvalue $\lambda = m + 1 - n/2$ of the pencil $\mathcal{L}_t$ with geometric multiplicity κ. By Lemma 7.3.2, this eigenvalue has no generalized eigenfunctions. Thus, the total algebraic multiplicity of eigenvalues in the strip (7.3.7) is equal to their total geometric multiplicity and equals κ. If we take t a little bit smaller than $-\mu_1(m + 1 - n/2)$, then again both lines $\operatorname{Re}\lambda = m - n/2$ and $\operatorname{Re}\lambda = m + 1 - n/2$ are free of eigenvalues of the pencil $\mathcal{L}_t$ and, by continuous dependence of eigenvalues on a parameter, the total algebraic multiplicity of eigenvalues of $\mathcal{L}_t$ in the strip (7.3.7) is not greater than κ.

The next value of t for which this total algebraic multiplicity changes is $t = -\mu_{\kappa+1}(m + 1 - n/2)$. Let κ_1 be the index such that

$$\mu_{\kappa+1}(m + 1 - n/2) = \cdots = \mu_{\kappa+\kappa_1}(m + 1 - n/2) = -t$$

and $t > -\mu_{\kappa+\kappa_1}(m + 1 - n/2)$. Then, reasoning as above, we conclude that the total algebraic multiplicity of $\mathcal{L}_t$ in the strip (7.3.7) does not exceed $\kappa + \kappa_1$ for this t and for t which is a little bit smaller than $-\mu_{\kappa_1}$. Continuing this procedure, we obtain finally that the total algebraic multiplicity of $\mathcal{L}_0 = \mathcal{L}$ in the strip (7.3.7) is not greater than N. This together with part 1) proves the lemma. ∎

7.3.4. Eigenvalues in the strip $m - n/2 \leq \operatorname{Re}\lambda \leq m + 2 - n/2$.

LEMMA 7.3.5. *Let $m \leq n/2 - 1$. Then the set $\{\lambda = i\tau + m + 2 - n/2\}$, where τ is a nonzero real number, does not contain eigenvalues of the pencil $\mathcal{L}$.*

Proof: 1) Let $m = 2k + 2$ and $\lambda = i\tau + m + 2 - n/2$, $\tau \neq 0$. Then

$$\mathcal{L}(\lambda) = \big(\delta + \lambda(\lambda + n - 2)\big) \big(\delta + (\lambda - 2)(\lambda + n - 4)\big) L_3(\lambda) \, L_3(\overline{\lambda}),$$

where

$$L_3(\lambda) = \prod_{j=2}^{k+1} \big(\delta + (\lambda - 2j)(\lambda - 2j + n - 2)\big).$$

Since

$$\mathrm{Im}\,\big((\delta + \lambda(\lambda + n - 2))(\delta + (\lambda - 2)(\lambda + n - 4))u, u\big)_{L_2(\Omega)}$$

$$= 4m\tau \left((\delta - \tau^2 + m^2 - \frac{n^2}{4} + n - 2)u, u\right)_{L_2(\Omega)},$$

we conclude that

$$\mathrm{Im}\,\big(\mathcal{L}(\lambda)u, u\big)_{L_2(\Omega)} = 4m\tau \left((\delta - \tau^2 + m^2 - \frac{n^2}{4} + n - 2)\, L_3(\lambda)u \,,\, L_3(\lambda)u\right)_{L_2(\Omega)}.$$

The operator $-\delta + \tau^2 - m^2 + \frac{n^2}{4} - n + 2$ is positive if $m \leq n/2 - 1$. Therefore, the left hand side can be zero only if $L_3(\lambda)u = 0$. But, since $u \in \overset{\circ}{W}_2^m(\Omega)$, this implies $u = 0$.

2) We consider the case $m = 2k + 1$. Then

$$-\mathcal{L}(\lambda) = P(\lambda) \, L_4(\lambda) \, L_4(\overline{\lambda}),$$

where

$$
\begin{aligned}
P(\lambda) &= \big(\delta + \lambda(\lambda + n - 2)\big) \big(\delta + (\lambda - 2)(\lambda + n - 4)\big) \\
&\quad \times \big(\delta + (\lambda - 2k - 2)(\lambda - 2k + n - 4)\big), \\
L_4(\lambda) &= \prod_{j=2}^{k} \big(\delta + (\lambda - 2j)(\lambda - 2j + n - 2)\big).
\end{aligned}
$$

After simple calculations, we get

(7.3.8) $\mathrm{Im}\,\big(P(\lambda)v, v\big)_{L_2(\Omega)}$

$$= 4m\tau \left(\big(\delta - \tau^2 - (1 - n/2)^2 + m^2 - 1\big)v \,,\, \big(\delta - \tau^2 - (1 - n/2)^2\big)v\right)_{L_2(\Omega)}.$$

Let $w = (\delta - \tau^2 - (1 - n/2)^2)v$. Then $\|w\|_{L_2(\Omega)} \leq (1 - n/2)^2 \, \|v\|_{L_2(\Omega)}$. Using this estimate together with the Schwarz inequality, we obtain

$$\big|(w, w + (m^2 - 1)v)_{L_2(\Omega)}\big| \leq \left(1 - \frac{m^2 - 1}{(1 - n/2)^2}\right) \|w\|_{L_2(\Omega)}^2.$$

Since $m \leq n/2 - 1$, the left-hand side of the last equality is positive, and (7.3.8) implies

$$-\frac{1}{\tau} \mathrm{Im}\,\big(\mathcal{L}(\lambda)u, u\big)_{L_2(\Omega)} > 0 \quad \text{for } u \in \overset{\circ}{W}_2^m(\Omega), \ u \neq 0.$$

The proof is complete. ∎

The following assertions can be proved analogously to Lemmas 7.3.3, 7.3.4 and Corollary 7.3.1.

LEMMA 7.3.6. *Let $m \leq n/2 - 1$ and let $\lambda_0 = m + 2 - n/2$ be an eigenvalue of the pencil $\mathcal{L}$ and u_0 a corresponding eigenfunction. Then*

$$\frac{d}{d\lambda} \left(\mathcal{L}(\lambda)u_0, u_0 \right)_{L_2(\Omega)} \Big|_{\lambda=\lambda_0} < 0.$$

COROLLARY 7.3.2. *If $\lambda_0 = m + 2 - n/2$ is an eigenvalue of the pencil $\mathcal{L}$, then it has no generalized eigenfunctions.*

Denote by N' the maximal index j such that $\mu_j(m + 2 - n/2) \leq 0$.

LEMMA 7.3.7. *Let $m \leq n/2 - 1$. Then the strip $m - n/2 \leq \operatorname{Re} \lambda \leq m + 2 - n/2$ contains exactly N' eigenvalues of the operator pencil $\mathcal{L}$ counted with their algebraic multiplicities. All these eigenvalues are real and have no generalized eigenfunctions. Every function μ_j has at most one zero in the interval $[m - n/2, m + 2 - n/2]$.*

7.3.5. On real eigenvalues of the pencil $\mathcal{L}$. We denote by $\Lambda = \Lambda(\Omega)$ the smallest real eigenvalue of the pencil $\mathcal{L}$ greater than $m - n/2$. If there are no eigenvalues of $\mathcal{L}$ on $(m - n/2, \infty)$, then we put $\Lambda = \infty$. Clearly, Λ is the smallest zero of the function μ_1 on $(m - n/2, \infty)$. In the following lemma we give some properties of Λ whose proof is quite straightforward.

LEMMA 7.3.8. *Suppose that $\Lambda < \infty$.*

1) The number Λ is the smallest number λ on $(m - n/2, \infty)$ such that the operator $\mathcal{L}(\lambda)$ has non-trivial kernel.

2) Let $R(u)$, where $u \in \overset{\circ}{W}_2^m(\Omega) \backslash \{0\}$, denote the smallest root of the equation $\left(\mathcal{L}(\lambda)u, u \right)_{L_2(\Omega)} = 0$ on the interval $(m - n/2, \infty)$. Then

$$\Lambda = \inf R(u),$$

where the infimum is taken over all nonzero $u \in \overset{\circ}{W}_2^m(\Omega)$.

COROLLARY 7.3.3. *1) The number Λ increases when the domain Ω decreases. Moreover, $\Lambda(\Omega_2) > \Lambda(\Omega_1)$ if $\Omega_2 \subset \Omega_1$ and $\Omega_1 \backslash \overline{\Omega_2}$ is not empty.*

2) If $m - n/2 < 0$, then $\Lambda > 0$.

3) If $m - n/2 \geq 0$, then $\Lambda > m - n/2 + 1/2$.

Proof: 1) The monotonicity of Λ with respect to Ω follows from the second part of Lemma 7.3.8. The strict monotonicity can be proved in the same way as in Theorem 3.5.2.

2) By Remark 7.3.1, the definition of Λ can be applied without changes to all domains in S^{n-1} if $m < n/2$. Moreover, Lemma 7.3.8 is also valid in this case. If $\Omega = S^{n-1}$, then the spectrum of $\mathcal{L}$ consists of the integers $0, 1, \dots$ and $2m - n, 2m - n - 1, \dots$. Therefore, $\Lambda(S^{n-1}) = 0$ and, by part 2) of Lemma 7.3.8, we get $\Lambda(\Omega) > 0$.

3) Let $\mathcal{K}_0$ be the complement of a small closed circular cone, and let Ω_0 be the intersection of $\mathcal{K}_0$ and S^{n-1}. We can suppose that $\Omega \subset \Omega_0$. Clearly, the boundary of $\mathcal{K}_0$ is Lipschitz. Since the strip $|\operatorname{Re} \lambda - m + n/2| \leq 1/2$ is free of eigenvalues of the pencil corresponding to Ω_0 (this will be proved in Section 11.1 for pencils generated by the Dirichlet problem for general $2m$ order elliptic operators in a Lipschitz cone, see Theorem 11.1.1), we have $\Lambda(\Omega_0) > m - n/2 + 1/2$. The same inequality for Ω follows from the second part of Lemma 7.3.8. ∎

7.3.6. The main theorem.

THEOREM 7.3.1. 1) *The strip*

$$(7.3.9) \qquad m - 2 - n/2 \leq \operatorname{Re}\lambda \leq m + 2 - n/2 \quad \text{if } 2m \leq n - 4,$$

$$(7.3.10) \qquad -3 \leq \operatorname{Re}\lambda \leq 0 \quad \text{if } 2m = n - 3$$

$$(7.3.11) \qquad -2 \leq \operatorname{Re}\lambda \leq 0 \quad \text{if } 2m = n - 2$$

and

$$(7.3.12) \qquad m - (n+1)/2 \leq \operatorname{Re}\lambda \leq m - (n-1)/2 \quad \text{if } 2m > n - 2$$

contains no eigenvalues of the pencil $\mathcal{L}$.

2) *Let* $2m = n - 3$. *Then the strip* $-7/2 \leq \operatorname{Re}\lambda \leq 1/2$ *contains only real eigenvalues of* $\mathcal{L}$, *the corresponding eigenfunctions have no generalized eigenfunctions. There is at most one eigenvalue on the interval* $(0, 1/2]$ *which increases when the domain* Ω *decreases.*

3) *If* $2m = n - 2$, *then the strip* $-3 \leq \operatorname{Re}\lambda \leq 1$ *contains only real eigenvalues of* $\mathcal{L}$, *the corresponding eigenfunctions have no generalized eigenfunctions. There is at most one eigenvalue on the interval* $(0, 1]$ *which increases when the domain* Ω *decreases.*

4) *Let* $m \geq n/2$. *Then the strip* $m - 1 - n/2 \leq \operatorname{Re}\lambda \leq m + 1 - n/2$ *contains only real eigenvalues of* $\mathcal{L}$, *the corresponding eigenfunctions have no generalized eigenfunctions. The eigenvalues on the interval* $(m - (n-1)/2, m+1-n/2]$ *increase when the domain* Ω *decreases.*

Proof: 1) Let $2m \leq n - 4$. We introduce the operator pencil $\mathcal{L}_t(\lambda) = \mathcal{L}(\lambda) + tI$, $t \geq 0$. Using Lemma 7.3.1, we obtain

$$\left(\mathcal{L}_t(\lambda)u, u\right)_{L_2(\Omega)} \geq c\, \|u\|^2_{W_2^m(\Omega)} + \left(c\,|\lambda|^{2m} + t\right)\|u\|^2_{L_2(\Omega)}$$

for all λ on the line $\operatorname{Im}\lambda = m - n/2$. This inequality implies that there are no eigenvalues of $\mathcal{L}_t$ for $|\operatorname{Im}\lambda| \geq C$, $\operatorname{Re}\lambda \in [m - n/2, m + 2 - n/2]$, where C is a sufficiently large number. From Lemma 7.3.5 and from the second part of Corollary 7.3.3 it follows that the pencils $\mathcal{L}_t$ have no eigenvalues on the line $\operatorname{Re}\lambda = m+2-n/2$. Hence, by Corollary 1.1.2, all pencils $\mathcal{L}_t$ have the same number of eigenvalues in the strip (7.3.9). Since for large t the pencil $\mathcal{L}_t$ has no eigenvalues in this strip, we conclude that the same is true for $\mathcal{L}$.

The cases $2m = n - 3$ and $2m = n - 2$ are considered analogously. Let $2m > n - 2$. Then, reasoning as in the proof of inequalities (7.3.9) and using assertion 3) of Corollary 7.3.3 instead of 2), we obtain the absence of eigenvalues in the strip (7.3.12).

2) By Lemma 7.3.7, the strip $-7/2 \leq \operatorname{Re}\lambda \leq 1/2$ contains only real eigenvalues of the pencil $\mathcal{L}$ which have no generalized eigenfunctions. By assertion 1), there are no eigenvalues on the interval $[-3, 0]$. If $\Omega = S^{n-1}$, then the interval $[0, 1/2]$ contains exactly one eigenvalue of $\mathcal{L}$, namely $\lambda_0 = 0$ which has multiplicity 1. This eigenvalue is the zero of the functions μ_1. Clearly, this function increases when the domain Ω decreases. Therefore, if $S^{n-1} \setminus \overline{\Omega}$ is nonempty, then the interval $(0, 1/2]$ contains at most one simple eigenvalue which increases when Ω decreases.

The proofs of assertions 3) and 4) are essentially the same as those of 2). ∎

REMARK 7.3.2. Note that the estimate (7.3.12) for the strip free of spectrum is sharp (for $2m = n - 1$ see Section 11.7 and for $2m \geq n$ see Section 11.1). Clearly, the estimates (7.3.12), (7.3.10) and (7.3.9) for $2m = n - 4$ can not be improved for arbitrary cones. It remains an important question. Is it possible to improve the estimate (7.3.9) for $2m < n - 4$, $m > 2$? The conjecture here is that the strip $m - n/2 \leq \lambda \leq 0$ should be free of the eigenvalues of $\mathcal{L}$. This would lead to Hölder continuity of solutions to (7.0.8), (7.0.9) for all m and n.

7.4. The Dirichlet problem for Δ^2 in domains with piecewise smooth boundaries

We consider the problem

$$(7.4.1) \qquad \Delta^2 U = F \ \text{in} \ \mathcal{G}, \quad U = \frac{\partial U}{\partial \nu} = 0 \ \text{on} \ \partial \mathcal{G} \backslash \mathcal{S}$$

in a bounded domain $\mathcal{G}$ with angular or conic points $x^{(1)}, \ldots, x^{(d)}$ on the boundary. We assume that near every point $x^{(\tau)}$ the domain $\mathcal{G}$ coincides with a cone (or angle in the case $n = 2$) $\mathcal{K}_\tau = \{x : (x - x^{(\tau)})/|x - x^{(\tau)}| \in \Omega_\tau\}$.

7.4.1. The biharmonic equation in plane domains with corners.
We start with the case of a plane domain $\mathcal{G}$. By α_τ we denote the interior angle at the point $x^{(\tau)}$.

Using Theorems 1.4.1 and 7.1.2, we obtain the following assertion on the unique solvability in the weighted spaces $V^l_{p,\vec{\beta}}(\mathcal{G})$ and $N^{l,\sigma}_{\vec{\beta}}(\mathcal{G})$ introduced in Section 1.4.

THEOREM 7.4.1. *The boundary value problem* (7.4.1) *is uniquely solvable in* $V^l_{p,\vec{\beta}}(\mathcal{G})$ *for arbitrary* $F \in V^{l-4}_{p,\vec{\beta}}(\mathcal{G})$, $l \geq 4$, *if and only if* l, p *and* $\vec{\beta} = (\beta_1, \ldots, \beta_d)$ *satisfy the inequality*

$$(7.4.2) \qquad |l - \beta_\tau - 1 - 2/p| < \alpha_\tau^{-1} \operatorname{Re} z(\alpha_\tau) \ \text{for} \ \tau = 1, \ldots, d,$$

where $z(\alpha)$ *is the solution of the equation* $z^{-1} \sin z + \alpha^{-1} \sin \alpha$ *with smallest real part. Analogously, the condition*

$$(7.4.3) \qquad |l + \sigma - \beta_\tau - 1| < \alpha_\tau^{-1} \operatorname{Re} z(\alpha_\tau) \ \text{for} \ \tau = 1, \ldots, d,$$

is equivalent to the unique solvability in $N^{l,\sigma}_{\vec{\beta}}(\mathcal{G})$ *for* $F \in N^{l-4,\sigma}_{\vec{\beta}}(\mathcal{G})$, $l \geq 4$.

The properties of the function $z = z(\alpha)$ were studied in Section 4.2. We recall that $z(\alpha)$ is real if $\alpha \geq \alpha_0$, where α_0 is the smallest positive root of the equation $\alpha^{-1} \sin \alpha = -\cos \alpha_*$ and α_* is the smallest positive root of the equation $\tan \alpha = \alpha$. For $\alpha < \alpha_0$ the roots $z(\alpha)$ and $\bar{z}(\alpha)$ are nonreal.

Furthermore, the following statement holds.

THEOREM 7.4.2. *Let* $F \in W_2^{-2}(\mathcal{G})$ *and let* $U \in \overset{\circ}{W}{}_2^2(\mathcal{G})$ *be the unique solution of problem* (7.4.1). *If* $F \in V^{l-4}_{p,\vec{\beta}}(\mathcal{G})$ $(F \in N^{l-4,\sigma}_{\vec{\beta}}(\mathcal{G}))$, $l \geq 4$, *and condition* (7.4.2) *(condition* (7.4.3)*) is satisfied, then* $U \in V^l_{p,\vec{\beta}}(\mathcal{G})$ $(U \in N^{l,\sigma}_{\vec{\beta}}(\mathcal{G}))$.

Analogous results are valid for solutions of problem (7.4.1) in a n-dimensional domain with smooth non-intersecting $(n - 2)$-dimensional edges. Then the role of α_τ is played by the angles α_x at the edge points x (see Subsection 8.7.2).

7.4.2. The biharmonic equation in domains with conic vertices. Now let $\mathcal{G}$ be a bounded domain in $\mathbb{R}^n$, $n \geq 3$, with d conic vertices on $\partial\mathcal{G}$. We denote by $\mathcal{L}_\tau$ the operator pencil generated by the Dirichlet problem for the biharmonic equation in the cone $\mathcal{K}_\tau$.

Applying Theorem 1.4.1, we obtain the following statement, where μ_τ is the minimum of real part of the eigenvalues of $\mathcal{L}_\tau$ situated in the half-plane $\mathrm{Re}\,\lambda > 2 - n/2$.

THEOREM 7.4.3. 1) *The boundary value problem* (7.4.1) *is uniquely solvable in* $V^l_{p,\vec{\beta}}(\mathcal{G})$ *for arbitrary* $F \in V^{l-4}_{p,\vec{\beta}}(\mathcal{G})$, $l \geq 4$, *if and only if and only if*

$$(7.4.4) \qquad 4 - n - \mu_\tau < l - \beta_\tau - n/p < \mu_\tau \ \ for \ \tau = 1, \dots, d.$$

Analogously, if

$$(7.4.5) \qquad 4 - n - \mu_\tau < l + \sigma - \beta_\tau < \mu_\tau \ \ for \ \tau = 1, \dots, d,$$

then problem (7.4.1) *has a unique solution in* $N^{l,\sigma}_{\vec{\beta}}(\mathcal{G})$ *for every* $F \in N^{l-4,\sigma}_{\vec{\beta}}(\mathcal{G})$, $l \geq 4$.

2) *The solution* $U \in \overset{\circ}{W}{}^2_2(\mathcal{G})$ *of problem* (7.4.1) *belongs to* $V^l_{p,\vec{\beta}}(\mathcal{G})$ $(N^{l,\sigma}_{\vec{\beta}}(\mathcal{G}))$, $l \geq 4$, *if* $F \in V^{l-4}_{p,\vec{\beta}}(\mathcal{G})$ *and* $2 - n/2 \leq l - \beta_\tau - n/p < \mu_\tau$ *for* $\tau = 1, \dots, d$ $(F \in N^{l-4,\sigma}_{\vec{\beta}}(\mathcal{G})$ *and* $2 - n/2 \leq l + \sigma - \beta_\tau < \mu_\tau$ *for* $\tau = 1, \dots, d)$.

Lower estimates for μ_τ are given in Sections 7.2.4 and 7.2.5. According to Theorem 7.2.5,

$$\mu_\tau > \max(1, M_\tau) \ \ \text{if } n = 3,$$

$$\mu_\tau > \max(1/2, M_\tau) \ \ \text{if } n = 4$$

and

$$\mu_\tau > M_\tau \ \ \text{if } n > 4,$$

where $M_\tau\,(M_\tau + n - 2)$ is the first eigenvalue of the Dirichlet problem for the operator $-\delta$ on Ω_τ, $M_\tau > 0$. Other lower estimates for μ_τ can be obtained from the monotonicity of real eigenvalues with respect to Ω_τ (see Section 7.2.6). In particular, $\mu_\tau > 2$ if the cones $\mathcal{K}_\tau$ are contained in half-spaces. If, on the contrary, a cone $\mathcal{K}_\tau$ contains a half-space then $0 < \mu_\tau \leq 2$ and μ_τ is a monotone decreasing function of $\mathcal{K}_\tau$.

7.4.3. Maximum principle. Let $\mathcal{G}$ be the same as in the preceding subsection. We consider the problem

$$(7.4.6) \qquad \Delta^2 U = 0 \ \ \text{in } \mathcal{G}, \quad U = \Phi, \ \frac{\partial U}{\partial \nu} = \Psi \ \ \text{on } \partial\mathcal{G} \backslash \mathcal{S}.$$

Maz'ya and Plamenevskiĭ [187] showed that the Miranda-Agmon maximum principle

$$(7.4.7) \qquad \|U\|_{W^1_\infty(\mathcal{G})} \leq c \left(\|\Phi\|_{W^1_\infty(\partial\mathcal{G})} + \|\Psi\|_{L_\infty(\partial\mathcal{G})} \right)$$

holds for all solutions of (7.4.6) if and only if

$$4 - n - \mu_\tau < 1 < \mu_\tau \quad \tau = 1, \dots, d.$$

Since $\mu_\tau > 1$ for $n = 3$, the maximum principle is always valid in the three-dimensional case. Moreover, according to [187], it holds if $n = 2$ and $\mathcal{G}$ is an arbitrary polygon. If $n \geq 4$ and all cones $\mathcal{K}_\tau$ are contained in half-spaces, then $\mu_\tau > 2$, as we said above, and hence (7.4.7) is also true.

The following example shows that the Miranda-Agmon maximum principle (7.4.7) fails, in general, for $n \geq 4$. Let $\mathcal{K}_\tau$ be the cone

$$(7.4.8) \qquad \{x \in \mathbb{R}^4 \ : \ x_4 > |x| \cos{(\alpha/2)}\},$$

where $\alpha \in [\pi, 2\pi)$. Then $\mathcal{L}_\tau$ is the operator of the boundary value problem

$$(7.4.9) \qquad \big(\delta + \lambda(\lambda - 2)\big)\big(\delta + \lambda(\lambda + 2)\big)u = f \ \text{ for } \theta < \alpha/2,$$

$$(7.4.10) \qquad u = \frac{\partial u}{\partial \theta} = 0 \ \text{ for } \theta = \alpha/2,$$

where $\cos\theta = x_4/|x|$. For $\lambda = 1$ equation (7.4.9) yields

$$(7.4.11) \qquad (\delta - 1)(\delta + 3)u = 0.$$

If u depends only on the variable θ, then, according to the representation (2.2.4) of the Beltrami operator, we have

$$\delta u(\theta) = \big(\sin^2\theta\big)^{-1} \frac{d}{d\theta}\left(\sin^2\theta \frac{du}{d\theta}\right).$$

It is easy to verify that $u(\theta) = c_1 \cos\theta + c_2\,\theta/\sin\theta$ is a smooth solution of (7.4.11). The boundary conditions (7.4.10) imply

$$c_1 \cos\frac{\alpha}{2} + c_2 \frac{\alpha}{2\sin{(\alpha/2)}} = 0, \quad -c_1 \sin\frac{\alpha}{2} + c_2 \frac{2\sin{(\alpha/2)} - \alpha\cos{(\alpha/2)}}{2\sin^2{(\alpha/2)}} = 0.$$

The determinant of this system is $(2\sin^2{(\alpha/2)})^{-1}(\alpha\cos\alpha - \sin\alpha)$. Consequently, there exists an eigenfunction of the form $u = u(\theta)$ corresponding to the eigenvalue $\lambda = 1$ if and only if $\alpha = \alpha_*$, where α_* is the root of the equation $\tan\alpha = \alpha$ in the interval $(\pi, 2\pi)$, $\alpha_* \approx 1.4303\,\pi$. Hence, if $\mathcal{G}$ coincides with the cone (7.4.8) in a neighborhood of the vertex $x^{(\tau)} = 0$, then the Miranda-Agmon maximum principle is not valid. Maz'ya and Roßmann [194] showed that for such a domain even smooth solutions of the biharmonic equation do not satisfy (7.4.7) with a constant c independent of U.

7.5. Notes

Section 7.1. Information on singularities of solutions of boundary value problems for the biharmonic equation near angular points can be found in Williams [268], Kondrat'ev [109], Seif [243], Melzer and Rannacher [199], Blum and Rannacher [21], Maz'ya, Morozov and Plamenevskiĭ [168], Maz'ya and Plamenevskiĭ [187], Maslovskaya [162], Grisvard [78, 80], Szabo and Babuška [251].

Section 7.2. The operator pencil generated by the Dirichlet problem for the biharmonic equation in a n-dimensional cone, $n \geq 3$, was first investigated in the paper [187] of Maz'ya and Plamenevskiĭ, where the estimates for the width of the energy strip in Theorem 7.2.5 were essentially proved. Other results of this section are taken from the paper [122] of Kozlov.

Section 7.3. The results were obtained by Kozlov [125].

Section 7.4. Theorems 7.4.1–7.4.3 on the solvability of the Dirichlet problem for the biharmonic equation in domains with angular or conic points and on the smoothness of the solutions follow directly from analogous results of the general theory (see Section 1.4). Let us further note that Dahlberg, Kenig and Verchota [36] obtained existence and uniqueness results for solutions in Lipschitz domains when the boundary values have first derivatives in L_2 and the normal derivative is in L_2. An analogous assertion for L_p spaces was proved by Pipher and Verchota [225].

The validity of the Miranda-Agmon maximum principle (7.4.7) for $n \leq 3$ (and for the case $n \geq 4$, when the cones $\mathcal{K}_\tau$ are contained in half-spaces) was proved by Maz'ya and Plamenevskiĭ [187]. The proof in [187] is based on point estimates for Green's function in a neighborhood of the conic points. The counterexample at the end of Section 7.4 is borrowed from the paper [194] of Maz'ya and Roßmann. Pipher and Verchota [226] proved that the Miranda-Agmon maximum principle is valid for solutions of the biharmonic equation in C^1 domains of arbitrary dimension and in Lipschitz domains of dimension $n \leq 3$. In [227] the same authors showed that the Miranda-Agmon maximum principle is valid for solutions of the polyharmonic equation in three-dimensional Lipschitz domains, while it fails, in general, for dimension $n \geq 4$. Maz'ya and Roßmann [194] obtained necessary and sufficient conditions guaranteeing the validity of this principle for solutions of general $2m$ order strongly elliptic equations in domains with conic points. In [193] they proved this principle for solutions of strongly elliptic equations in three-dimensional domains of polyhedral type.

In papers by Oleĭnik, Yosifyan and Tavkhelidze [219, 220], Kondrat'ev, Kopachek and Oleĭnik [114] pointwise estimates of solutions to the Dirichlet problem for the biharmonic equation near nonregular boundary points were obtained. A survey of these results can be found in Kondrat'ev and Oleĭnik's paper [112]. The regularity of boundary points in the sense of Wiener for the biharmonic and polyharmonic operators was investigated in papers of Maz'ya [163, 165], Maz'ya and Donchev [166].

Part 2

Singularities of solutions to general elliptic equations and systems

CHAPTER 8

The Dirichlet problem for elliptic equations and systems in an angle

In this chapter we are concerned with the singularities of solutions to the Dirichlet problem for elliptic differential equations and systems in a plane angle. In order to describe these singularities, one has to study the eigenvalues, eigenvectors and generalized eigenvectors of certain pencils of ordinary differential operators on an interval. Sections 8.1–8.4 are dedicated to the operator pencil generated by the Dirichlet problem for a $2m$ order differential equation. In Section 8.1 we derive a transcendental equation for the eigenvalues generalizing equation (7.1.10) that we dealt with when studying the biharmonic equation. Although in the general case the corresponding transcendental equation is more complicated, one can derive rather explicit knowledge on their roots. Here is a brief outline of the results obtained in the present chapter.

We consider the Dirichlet problem for a scalar elliptic homogeneous operator $L(\partial_{x_1}, \partial_{x_2})$ with real constant coefficients in a plane angle

$$\mathcal{K}_\alpha = \{(x_1, x_2) \in \mathbb{R}^2 \ : \ 0 < r < \infty, \ \varphi \in (0, \alpha)\},$$

where (r, φ) are polar coordinates and $\alpha \in (0, 2\pi]$. The operator pencil associated to this problem will be denoted by $\mathcal{L}_\alpha$.

If λ is an eigenvalue of $\mathcal{L}_\alpha$, then $\overline{\lambda}$ and $m - 1 - \overline{\lambda}$ are also eigenvalues of $\mathcal{L}_\alpha$. The geometric, partial and algebraic multiplicities of these three eigenvalues coincide. Therefore, it is sufficient to describe the location of eigenvalues and their multiplicities in the half-plane $\operatorname{Re} \lambda \geq m - 1$.

For $\alpha = \pi$ and $\alpha = 2\pi$ all eigenvalues of $\mathcal{L}_\alpha$ can be evaluated explicitly. In fact, we show in Subsection 8.1.5 that

(i) the spectrum of $\mathcal{L}_\pi$ consists of the eigenvalues $m - 1 \pm k$, $k = 1, 2, \ldots$, with the multiplicities k for $k \leq m$ and m for $k > m$,

(ii) the spectrum of $\mathcal{L}_{2\pi}$ consists of the eigenvalues $m - 1 \pm k$, $k = 1, 2, \ldots$, of the same multiplicities as in the case (i) and of the eigenvalues $m - 1/2 \pm k$, $k = 0, 1, \ldots$, of multiplicity m.

In both cases there are no generalized eigenfunctions. If the angle α is close to π or 2π, then it is possible to give explicit asymptotic formulas for eigenvalues of $\mathcal{L}_\alpha$ close to m and to $m - 1/2$ (see Sections 8.2 and 8.3). Let us denote the roots of the equation $L(1, z) = 0$ with positive imaginary parts by $z_1, \ldots, z_m$. If there is a multiple root, then we take into account its multiplicity. The following assertions are contained in Theorems 8.2.1 and 8.3.1.

(i) When α is close to $l\pi$, $l = 1, 2$, there is only one eigenvalue $\lambda_l(\alpha)$ of $\mathcal{L}_\alpha$ located near m and

$$\lambda_l(\alpha) = m + \operatorname{Im}(z_1 + \cdots + z_m)(1 - \alpha/l\pi) + O(|1 - \alpha/l\pi|^2).$$

(ii) When α is close to 2π, there are exactly m eigenvalues $\mu_k(\alpha)$, $k = 1, \ldots, m$, located near $m - 1/2$, and subject to

$$\mu_k(\alpha) = m - 1/2 + \alpha_k(2\pi - \alpha)^{2k-1} + O(|2\pi - \alpha|^{2k}),$$

where α_k are constants written explicitly.

The case $\alpha < \pi$ is treated in Section 8.4, where it is proved that the eigenvalues of $\mathcal{L}_\alpha$ do not coincide with $m + k$, $k = 0, 1, \ldots, m - 1$, and that the strip $m - 1 \leq \operatorname{Re}\lambda \leq m$ contains no eigenvalues of this pencil.

The usefulness of the last statement is illustrated by the Dirichlet problem

$$L(\partial_{x_1}, \partial_{x_2})U = F \ \text{ in } \mathcal{G}, \quad \partial_\nu^j U = 0 \ \text{ on } \partial\mathcal{G}, \ j = 0, \ldots, m-1,$$

where $\mathcal{G}$ is a convex polygon and ν is the normal to $\partial\mathcal{G}$. If $F \in L_2(\mathcal{G})$, then this problem has a solution U in $\overset{\circ}{W}{}_2^m(\mathcal{G})$. Since there are no eigenvalues of the pencil $\mathcal{L}_\alpha$ in the strip $m - 1 \leq \operatorname{Re}\lambda \leq m$, Theorem 1.4.3 implies that $U \in W_2^{m+1}(\mathcal{G})$.

In Section 8.5 we turn to the case $\alpha \in (\pi, 2\pi)$. We prove that

(i) all eigenvalues of $\mathcal{L}_\alpha$ in the strip $m - 1 \leq \operatorname{Re}\lambda \leq m$ are real and situated in $(m - 1/2, m]$,

(ii) all eigenvalues of $\mathcal{L}_\alpha$ in $(m - 1/2, m)$ are simple and strictly decreasing with respect to $\alpha \in (\pi, 2\pi)$,

(iii) the total number of eigenvalues in the interval $(m - 1/2, m)$ changes from m to 1 when α changes from 2π to π.

This result together with Theorem 1.4.4 implies that the variational solution U of the Dirichlet problem for the operator $L(\partial_{x_1}, \partial_{x_2})$ in the angle $\mathcal{K}_\alpha$, $\alpha \in (\pi, 2\pi)$, with the right-hand side in $L_2(\mathcal{K}_\alpha)$, can be represented near the vertex in the asymptotic form

$$U(x_1, x_2) = \sum_{j=1}^{k} c_j r^{\lambda_j} u_j(\varphi) + O(r^{m-\varepsilon}),$$

where ε is an arbitrary positive number, $m - 1/2 < \lambda_1 < \cdots < \lambda_k < m$, $1 \leq k \leq m$, u_j are smooth functions independent of F, and c_j are constants.

It follows from the above results on the eigenvalues of $\mathcal{L}_\alpha$ that the strip $|\operatorname{Re}\lambda - m + 1| < 1/2$ is free of eigenvalues for any α and that, moreover, the same is valid for the strip $|\operatorname{Re}\lambda - m + 1| \leq 1$ if $\alpha < \pi$. As a matter of fact, these properties are not preserved if some coefficients of $L(\partial_{x_1}, \partial_{x_2})$ are nonreal. In Subsection 8.4.3 we give a family of rather simple differential operators with complex coefficients such that the eigenvalues of the associated pencils $\mathcal{L}_\alpha$ approach the line $\operatorname{Re}\lambda = m - 1$.

In Section 8.6 we extend some of the previous results to a class of strongly elliptic second order matrix differential operators by using a quite different method. As usual, we conclude the chapter with some applications to the theory of elliptic equations in domains with piecewise smooth boundary.

8.1. The operator pencil generated by the Dirichlet problem

8.1.1. Formulation of the problem. Let $\mathcal{K}$ be the plane angle

$$\{(x_1, x_2) \in \mathbb{R}^2 : 0 < r < \infty, \ 0 < \varphi < \alpha\},$$

where r, φ are the polar coordinates of the point $x = (x_1, x_2)$. We consider the Dirichlet problem

$$(8.1.1) \quad L(\partial_{x_1}, \partial_{x_2}) U = \sum_{j=0}^{2m} a_j \, \partial_{x_1}^{2m-j} \partial_{x_2}^j U = 0 \qquad \text{on } \mathcal{K},$$

$$(8.1.2) \quad \partial_{x_2}^k U \Big|_{\varphi=0} = (-\sin \alpha \, \partial_{x_1} + \cos \alpha \, \partial_{x_2})^k U \Big|_{\varphi=\alpha} = 0, \quad k = 0, 1, \dots, m-1.$$

Here L is an elliptic differential operator with real constant coefficients a_j. In order to obtain the corresponding operator pencil, we insert the function $U = r^\lambda u(\varphi)$, where $\lambda \in \mathbb{C}$, into (8.1.1), (8.1.2). Then we arrive at

$$(8.1.3) \qquad \mathcal{L}(\lambda) \, u(\varphi) = 0 \qquad \text{for } 0 < \varphi < \alpha,$$

$$(8.1.4) \qquad u^{(k)}(\varphi)|_{\varphi=0} = u^{(k)}(\varphi)|_{\varphi=\alpha} = 0 \quad \text{for } k = 0, 1, \dots, m-1,$$

where $\mathcal{L}(\lambda) = \mathcal{L}(\varphi, \partial_\varphi, \lambda)$ is a parameter-depending ordinary differential operator defined by

$$\mathcal{L}(\varphi, \partial_\varphi, \lambda) \, u(\varphi) = r^{2m-\lambda} \, L(\partial_{x_1}, \partial_{x_2}) \, r^\lambda \, u(\varphi),$$

i.e., $r^{-2m} \, \mathcal{L}(\varphi, \partial_\varphi, r\partial_r) = L(\partial_{x_1}, \partial_{x_2})$.

Example. If $L(\partial_{x_1}, \partial_{x_2}) = (\partial_{x_1}^2 + \partial_{x_2}^2)^m$, then

$$\mathcal{L}(\varphi, \partial_\varphi, \lambda) = \prod_{j=0}^{m-1} \left(\partial_\varphi^2 + (\lambda - j)^2 \right).$$

In general case, $\mathcal{L}(\varphi, \partial_\varphi, \lambda)$ is an ordinary differential operator with variable coefficients which realizes a continuous mapping

$$(8.1.5) \qquad W_2^{2m}((0, \alpha)) \cap \overset{\circ}{W}_2^m((0, \alpha)) \to L_2((0, \alpha))$$

for arbitrary $\lambda \in \mathbb{C}$

We recall some well-known properties of the pencil $\mathcal{L}$ which, according to a result of Agmon, Nirenberg [**3**, Th.5.2] and Agranovich, Vishik [**4**] (see also our book [**136**, Th.3.6.1,Le.6.1.5]), follow from the ellipticity of problem (8.1.1), (8.1.2).

THEOREM 8.1.1. *The operator $\mathcal{L}(\lambda)$ realizes an isomorphism (8.1.5) for all $\lambda \in \mathbb{C}$ except a countable set of isolated points, the eigenvalues of the pencil $\mathcal{L}$. The eigenvalues have finite algebraic multiplicities and are contained, with the possible exception of finitely many, in a double angle with opening less than π containing the real axis.*

THEOREM 8.1.2. *If λ_0 is an eigenvalue of the pencil $\mathcal{L}$, then $\overline{\lambda}_0$, $2m - 2 - \lambda_0$, and $2m - 2 - \overline{\lambda}_0$ are eigenvalues with the same geometric, partial and algebraic multiplicities.*

Proof: Let $U = r^\lambda u(\varphi)$, $V = r^{2m-2-\overline{\lambda}} v(\varphi)$, where u and v are arbitrary functions from $C_0^\infty((0, \alpha))$. Integrating by parts, we get

$$\int_{\substack{\mathcal{K} \\ 1 < |x| < 2}} LU \cdot \overline{V} \, dx = \int_{\substack{\mathcal{K} \\ 1 < |x| < 2}} U \cdot \overline{LV} \, dx$$

(the sum of the integrals on the arcs $|x| = 1$ and $|x| = 2$ vanishes). This implies

$$\int\limits_0^\alpha \mathcal{L}(\varphi, \partial_\varphi, \lambda)\, u(\varphi) \cdot \overline{v(\varphi)}\, d\varphi = \int\limits_0^\alpha u(\varphi) \cdot \overline{\mathcal{L}(\varphi, \partial_\varphi, 2m - 2 - \bar\lambda)\, v(\varphi)}\, d\varphi,$$

i.e., the operator $\mathcal{L}(\varphi, \partial_\varphi, \lambda)$ is adjoint to $\mathcal{L}(\varphi, \partial_\varphi, 2m - 2 - \bar\lambda)$. Consequently, the numbers λ and $2m - 2 - \bar\lambda$ are simultaneously eigenvalues (see Theorem 1.1.7).

Since the coefficients of the operator $\mathcal{L}(\varphi, \partial_\varphi, \lambda)$ are real, the numbers λ and $\bar\lambda$ are simultaneously eigenvalues. This proves the theorem. ∎

Due to Theorem 8.1.2, it suffices to study the eigenvalues in the half-plane $\operatorname{Re}\lambda \geq m - 1$.

8.1.2. A transcendental equation for the eigenvalues. Let the ordinary differential operators $\mathcal{B}_\alpha^{(k)}(\varphi, \partial_\varphi, \lambda)$ be defined by

$$\mathcal{B}_\alpha^{(k)}(\varphi, \partial_\varphi, \lambda)\, u(\varphi) = r^{k-\lambda}(-\sin\alpha\, \partial_{x_1} + \cos\alpha\, \partial_{x_2})^k\, (r^\lambda u(\varphi)).$$

It can be easily seen that the coefficient of the derivative ∂_φ^k of $\mathcal{B}_\alpha^{(k)}$ is equal to one at $\varphi = \alpha$. Hence the boundary conditions (8.1.4) are equivalent to the conditions

(8.1.6)
$$\mathcal{B}_0^{(k)}(\varphi, \partial_\varphi, \lambda) u(\varphi)|_{\varphi=0} = \mathcal{B}_\alpha^{(k)}(\varphi, \partial_\varphi, \lambda) u(\varphi)|_{\varphi=\alpha} = 0, \quad k = 0, 1, \ldots, m-1.$$

For technical reasons, we will use the boundary conditions (8.1.6) instead of the conditions (8.1.4). We denote the operator

$$u \to \left\{ \left(\mathcal{B}_0^{(k)}(\varphi, \partial_\varphi, \lambda) u(\varphi)|_{\varphi=0}\,,\, \mathcal{B}_\alpha^{(k)}(\varphi, \partial_\varphi, \lambda) u(\varphi)|_{\varphi=\alpha}\right) \right\}_{0 \leq k \leq m-1}$$

by $\mathcal{B}(\lambda)$ and the operator

$$W_2^{2m}((0, \alpha)) \ni u \to \left(\mathcal{L}(\varphi, \partial_\varphi, \lambda) u, \mathcal{B}(\lambda) u \right) \in L_2((0, \alpha)) \times \mathbb{C}^{2m}$$

by $\mathfrak{A}(\lambda)$. Obviously, the spectra of the pencils $\mathcal{L}$ and $\mathfrak{A}$ coincide.

For $l = 0, 1, \ldots, 2m - 1$ let $v_l = v_l(\lambda, \varphi)$ be the solution of the Cauchy problem

(8.1.7) $\mathcal{L}(\varphi, \partial_\varphi, \lambda) v_l = 0 \qquad \text{for } \varphi > 0,$

(8.1.8) $\mathcal{B}_0^{(k)}(\varphi, \partial_\varphi, \lambda) v_l \Big|_{\varphi=0} = \delta_{k,l} \qquad \text{for } k = 0, 1, \ldots, 2m - 1.$

The coefficient of ∂_φ^{2m} in the operator $\mathcal{L}(\varphi, \partial_\varphi, \lambda)$ is equal to $L(-\sin\varphi, \cos\varphi) \neq 0$. Hence the functions v_l are analytic in λ, $\lambda \in \mathbb{C}$, and depend on the variable φ real-analytically.

LEMMA 8.1.1. *Let $\Psi(\lambda)$ be the quadratical matrix with the rows*

(8.1.9) $$B_0^{(k)}(\varphi, \partial_\varphi, \lambda)\, (v_0, v_1, \ldots, v_{2m-1}) \Big|_{\varphi=0}$$

and

(8.1.10) $$B_\alpha^{(k)}(\varphi, \partial_\varphi, \lambda)\, (v_0, v_1, \ldots, v_{2m-1}) \Big|_{\varphi=\alpha},$$

$k = 0, 1, \ldots, m - 1$. *Then λ_0 is an eigenvalue of $\mathcal{L}$ if and only if*

(8.1.11) $\det \Psi(\lambda_0) = 0.$

The order of the zero λ_0 of the function $\det \Psi(\lambda)$ coincides with the algebraic multiplicity of the eigenvalue λ_0, while the dimension of the kernel of the matrix $\Psi(\lambda_0)$ coincides with the geometric multiplicity of the eigenvalue λ_0.

Proof: Let $\mathcal{P}(\lambda)$ be the operator

$$\mathbb{C}^{2m} \ni (c_1, \dots, c_{2m}) \to \sum_{l=1}^{2m} c_l \, v_l(\lambda, \varphi) \in \ker \mathcal{L}(\varphi, \partial_\varphi, \lambda).$$

Then the operator function $\lambda \to \mathcal{B}(\lambda)\mathcal{P}(\lambda)$ coincides with the matrix-function Ψ. Consequently, by Lemma 3.1.1, every eigenvalue of the pencil $\mathfrak{A}$ is also an eigenvalue of the matrix-function Ψ. The geometric and algebraic multiplicities of these eigenvalues coincide. Using Lemma 3.1.2, we obtain the assertions of the lemma. ∎

Equation (8.1.11) is the transcendental equation we have sought for the eigenvalues of the pencil $\mathcal{L}$. Since

$$\mathcal{L}(\varphi, \partial_\varphi, \overline{\lambda}) = \overline{\mathcal{L}(\varphi, \partial_\varphi, \lambda)} \quad \text{and} \quad \mathcal{B}_\alpha^{(k)}(\varphi, \partial_\varphi, \overline{\lambda}) = \overline{\mathcal{B}_\alpha^{(k)}(\varphi, \partial_\varphi, \lambda)},$$

it follows that $v_l(\overline{\lambda}, \varphi) = \overline{v_l(\lambda, \varphi)}$ and, consequently,

$$\det \Psi(\overline{\lambda}) = \overline{\det \Psi(\lambda)}.$$

8.1.3. A more explicit form of the transcendental equation. Now our goal is to obtain more explicit expressions for the function $\det \Psi(\lambda)$. Let $z_1, \dots, z_d$ be the distinct roots of the equation $L(1, z) = 0$ with positive imaginary parts, and let $\nu_1, \dots, \nu_d$ be their multiplicities. Since the coefficients of the operator L are real, we have $\nu_1 + \cdots + \nu_d = m$. Let $z_{d+q} = \overline{z}_q$ and $\nu_{d+q} = \nu_q$ for $q = 1, \dots, d$. Then

$$L(\partial_{x_1}, \partial_{x_2}) = a_{2m} \prod_{q=1}^{2d} \left(\partial_{x_2} - z_q \partial_{x_1}\right)^{\nu_q}.$$

We introduce the notation

$$(\mu)_l = \mu \, (\mu + 1) \cdots (\mu + l - 1) \quad \text{for } l = 1, 2, \dots, \qquad (\mu)_0 = 1$$

and set

$$(8.1.12) \qquad w_{q,j}(\lambda, \varphi) = \partial_z^j (\cos \varphi + z \sin \varphi)^\lambda \Big|_{z=z_q}$$

$$= (\lambda - j + 1)_j \, (\sin \varphi)^j \, (\cos \varphi + z_q \sin \varphi)^{\lambda - j}$$

for $q = 1, \dots, d$, $j = 0, \dots, \nu_q - 1$. Here it is assumed that the argument of the function $\varphi \to \cos \varphi + z \sin \varphi$ varies from 0 to 2π for $\operatorname{Im} z > 0$ and from 0 to -2π for $\operatorname{Im} z < 0$.

If $P(\partial_{x_1}, \partial_{x_2})$ is an arbitrary homogeneous differential operator of order μ with constant coefficients and the operator $\mathcal{P}(\varphi, \partial_\varphi, \lambda)$ is defined by the equality

$$\mathcal{P}(\varphi, \partial_\varphi, \lambda) u(\varphi) = r^{\mu - \lambda} \, P(\partial_{x_1}, \partial_{x_2}) \, (r^\lambda u(\varphi)),$$

then

$$(8.1.13) \quad \mathcal{P}(\varphi, \partial_\varphi, \lambda) \, w_{q,j}(\lambda, \varphi) = r^{\mu - \lambda} \, P(\partial_{x_1}, \partial_{x_2}) \left(\partial_z^j (x_1 + z x_2)^\lambda\right)\Big|_{z=z_q}$$

$$= r^{\mu - \lambda} \, (\lambda - \mu + 1)_\mu \, \partial_z^j \Big(P(1, z)(x_1 + z x_2)^{\lambda - \mu}\Big)\Big|_{z=z_q}$$

$$= (\lambda - \mu + 1)_\mu \, \partial_z^j \Big(P(1, z) \, (\cos \varphi + z \sin \varphi)^{\lambda - \mu}\Big)\Big|_{z=z_q}.$$

In particular, it follows from (8.1.13) that the functions $w_{q,j}(\lambda, \varphi)$ are solutions of the equation $\mathcal{L}(\varphi, \partial_\varphi, \lambda)u = 0$. Furthermore, we have

$$(8.1.14) \qquad \mathcal{B}_\alpha^{(k)}(\varphi, \partial_\varphi, \lambda)w_{q,j}(\lambda, \varphi)\big|_{\varphi=\alpha}$$

$$= (\lambda - k + 1)_k \, \partial_z^j \big((-\sin\alpha + z\cos\alpha)^k (\cos\alpha + z\sin\alpha)^{\lambda-k}\big)\big|_{z=z_q} .$$

We write the solutions v_l of the Cauchy problems (8.1.7), (8.1.8) in the form

$$v_l(\lambda, \varphi) = \sum_{q=1}^{2d} \sum_{j=0}^{\nu_q-1} c_{l,q,j}(\lambda) \, w_{q,j}(\lambda, \varphi), \qquad l = 0, 1 \dots, 2m - 1.$$

Using (8.1.14) with $\alpha = 0$, we obtain the following system of equations for the coefficients $c_{l,q,j}(\lambda)$:

$$\sum_{q=1}^{2d} \sum_{j=0}^{\nu_q-1} c_{l,q,j}(\lambda) \, (\lambda - \mu + 1)_\mu \, \partial_z^j z^\mu \big|_{z=z_q} = \delta_{l,\mu} , \qquad l, \mu = 0, 1, \dots, 2m - 1.$$

The matrix T with the rows

$$\partial_z^j \big(1, z, \dots, z^{2m-1}\big)\big|_{z=z_q} , \qquad q = 1, \dots, 2d, \; j = 0, \dots, \nu_q - 1,$$

is invertible. Otherwise, there exist numbers c_μ, not all zero, such that

$$\sum_{\mu=0}^{2m-1} c_\mu \, \partial_z^j z^\mu \big|_{z=z_q} = 0 \qquad \text{for } q = 1, \dots, 2d, \; j = 0, \dots, \nu_q - 1,$$

i.e., the polynomial $\sum c_\mu z^\mu$ of degree $2m - 1$ has $2m$ roots (counting multiplicity). This is only possible if $c_0 = \dots = c_{2m-1} = 0$.

Let $s_{l,q,j}$ be the elements of the inverse matrix to T, $l = 0, 1, \dots, 2m - 1$, $q = 1, \dots, 2d$, $j = 0, \dots, \nu_q - 1$. Then

$$(8.1.15) \qquad c_{l,q,j} = \frac{1}{(\lambda - l + 1)_l} s_{l,q,j} .$$

We set

$$(8.1.16) \qquad \sigma_k(z, \lambda) = (\sin\alpha - z\cos\alpha)^k (\cos\alpha + z\sin\alpha)^{\lambda-k}$$

and denote the matrix with the rows

$$\partial_z^j \big(z^0, \dots, z^{m-1}, \sigma_0(z, \lambda), \dots, \sigma_{m-1}(z, \lambda)\big)\big|_{z=z_q} ,$$

$q = 1, \dots, 2d$, $j = 0, \dots, \nu_q - 1$, by $\mathcal{D}(\lambda)$. According to (8.1.14), we have

$$\mathcal{B}_0^{(k)}(\varphi, \partial_\varphi, \lambda) v_l(\lambda, \varphi)\big|_{\varphi=0} = (\lambda - k + 1)_k \sum_{q,j} c_{l,q,j}(\lambda) \, \partial_z^j z^k \big|_{z=z_q} ,$$

$$\mathcal{B}_\alpha^{(k)}(\varphi, \partial_\varphi, \lambda) v_l(\lambda, \varphi)\big|_{\varphi=\alpha} = (-1)^k (\lambda - k + 1)_k \sum_{q,j} c_{l,q,j}(\lambda) \, \partial_z^j \sigma_k(z, \lambda)\big|_{z=z_q} .$$

From this and (8.1.15) we obtain the formula

$$(8.1.17) \qquad \det \Psi(\lambda) = (-1)^{m(m-1)/2} \det T^{-1} \det \mathcal{D}(\lambda) \prod_{k=0}^{m-1} \frac{1}{(\lambda - k - m + 1)_m} .$$

Thus, the following lemma holds.

LEMMA 8.1.2. *If* $\lambda_0 \neq 0, 1, \ldots, 2m - 2$, *then the number* λ_0 *is an eigenvalue of the pencil* $\mathcal{L}$ *if and only if* $\det \mathcal{D}(\lambda_0) = 0$. *The geometric multiplicity of this eigenvalue is equal to the dimension of the kernel of the matrix* $\mathcal{D}(\lambda_0)$, *while its algebraic multiplicity is equal to the order of the zero* λ_0 *of the function* $\det \mathcal{D}(\lambda)$.

If $\lambda_0 = m - 1 \pm k$, $k \in \{0, 1, \ldots, m - 1\}$, *then* λ_0 *is an eigenvalue of the pencil* $\mathcal{L}$ *if and only if*

$$(8.1.18) \qquad \partial_\lambda^{m-k} \det \mathcal{D}(\lambda)\Big|_{\lambda=\lambda_0} = 0.$$

8.1.4. The transcendental equation for $\alpha \neq \pi$, $\alpha \neq 2\pi$. We give another representation for $\det \mathcal{D}(\lambda)$ in the case $\alpha \neq \pi$, $\alpha \neq 2\pi$. Let $D(\lambda)$ be the matrix with the rows

$$\partial_z^j \left(\sigma(z)^0, \ldots, \sigma(z)^{m-1}, \ \sigma(z)^{\lambda-0}, \ldots, \sigma(z)^{\lambda-m+1} \right)\Big|_{z=z_q},$$

$q = 1, \ldots, 2d$, $j = 0, \ldots, \nu_q - 1$, where

$$\sigma(z) = \cos \alpha + z \sin \alpha.$$

Using the formulas

$$z = -\cot \alpha + (\cos \alpha + z \sin \alpha)/\sin \alpha,$$

$$\sin \alpha - z \cos \alpha = 1/\sin \alpha - (\cos \alpha + z \sin \alpha) \cot \alpha,$$

by means of elementary transformations of the matrix $\mathcal{D}(\lambda)$, we obtain

$$(8.1.19) \qquad \det \mathcal{D}(\lambda) = (\sin \alpha)^{-m(m-1)} \det D(\lambda).$$

Equation (8.1.11) for the eigenvalues, together with the relations (8.1.17) and (8.1.19) will play an important role in the study of the spectrum of the pencil $\mathcal{L}$.

Example. Let $L = \partial^2/\partial_{x_1}^2 + \partial^2/\partial_{x_2}^2$ be the Laplace operator. Then the zeros of $L(1, z)$ are $z_1 = i$ and $z_2 = -i$. In this case the matrices $\mathcal{D}(\lambda)$ and $D(\lambda)$ coincide:

$$\mathcal{D}(\lambda) = D(\lambda) = \begin{pmatrix} 1 & e^{i\lambda\alpha} \\ 1 & e^{-i\lambda\alpha} \end{pmatrix}.$$

The determinant of $\mathcal{D}(\lambda)$ is equal to zero if and only if $\lambda = k\pi/\alpha$, where k is an arbitrary integer. Since $\partial_\lambda \det \mathcal{D}(\lambda)|_{\lambda=0} = -2i\alpha \neq 0$, the number $\lambda = 0$ is not an eigenvalue of the pencil $\mathcal{L}$.

8.1.5. The spectrum of the pencil for the angles π **and** 2π. We give an explicit description of the spectrum of the pencil $\mathcal{L}$ for the cases $\alpha = \pi$ and $\alpha = 2\pi$.

THEOREM 8.1.3. 1) *If* $\alpha = \pi$, *then the spectrum of the pencil* $\mathcal{L}$ *consists of the eigenvalues*

$$\lambda_k = m - 1 + k \qquad (k = \pm 1, \pm 2, \ldots).$$

The geometric and algebraic multiplicities of the eigenvalues λ_k *are equal to* $|k|$ *for* $k = \pm 1, \ldots, \pm(m - 1)$ *and equal to* m *for* $|k| \geq m$.

2) *If* $\alpha = 2\pi$, *then the spectrum of the pencil* $\mathcal{L}$ *consists of the eigenvalues*

$$\lambda_k = m - 1 + k \quad (k = \pm 1, \pm 2, \ldots) \quad and \quad \mu_k = m - \frac{1}{2} + k \quad (k = 0, \pm 1, \pm 2, \ldots).$$

The eigenvalues λ_k *have the same multiplicities as in the case* $\alpha = \pi$, *while the eigenvalues* μ_k *have geometric and algebraic multiplicities* m.

Proof: Let $\alpha = l\pi$, $l = 1, 2$. Then

$$\partial_z^j \sigma_k(z, \lambda)\Big|_{z=z_q} = (-1)^k \, e^{\pm i l \pi \lambda} \, \partial_z^j z^k \Big|_{z=z_q} \qquad \text{for } \pm \operatorname{Im} z_q > 0.$$

We denote by T_0 the square matrix with the rows

$$\partial_z^j \left(1, z, \dots, z^{m-1}\right)\Big|_{z=z_q}, \quad q = 1, \dots, d, \; j = 0, \dots, \nu_k - 1.$$

In the same way as for the matrix T, it can be shown that $\det T_0 \neq 0$. Let $\overline{T}_0$ be the matrix whose elements are the complex conjugates of the elements of the matrix T_0. Then

$$(8.1.20) \qquad \mathcal{D}(\lambda) = (-1)^{m(m-1)/2} \begin{pmatrix} T_0 & e^{il\pi\lambda} T_0 \\ \overline{T}_0 & e^{-il\pi\lambda} \overline{T}_0 \end{pmatrix}$$

and, therefore,

$$(8.1.21) \qquad \det \mathcal{D}(\lambda) = (-1)^{m(m-1)/2} \, |\det T_0|^2 \, (e^{-il\pi\lambda} - e^{il\pi\lambda})^m.$$

Consequently, by Lemma 8.1.2, the numbers $\lambda = m - 1 + k$ for $k = \pm m, \pm(m+1), \dots$ and $\mu_k = m + k - 1/2$ for $k = 0, \pm 1, \dots$ are eigenvalues with algebraic multiplicity m. Furthermore, it follows from (8.1.20) that the dimension of the kernel of the matrix $\mathcal{D}(\lambda)$ is equal to m for these values of λ. Hence there are no generalized eigenfunctions.

We consider the eigenvalues $\lambda_k = m - 1 + k$ for $k = \pm 1, \dots, \pm(m-1)$. From (8.1.17) and (8.1.21) it follows that the multiplicities of the zeros λ_k of the function $\det \Psi(\lambda)$ are equal to $|k|$. Obviously, for $k = 1, 2, \dots, m-1$ the functions

$$x_2^{m-1+k}, \; x_1 \, x_2^{m-2+k}, \dots, x_1^{k-1} \, x_2^m$$

are homogeneous solutions of degree $m - 1 + k$ of the Dirichlet problem (8.1.1), (8.1.2). Hence

$$(\sin \varphi)^{m+k-1}, \; \cos \varphi \, (\sin \varphi)^{m+k-2}, \dots, (\cos \varphi)^{k-1} (\sin \varphi)^m$$

are eigenfunctions of the operator pencil $\mathcal{L}$ corresponding to the eigenvalues $\lambda_k = m - 1 + k$, $k = 1, 2, \dots, m-1$. This means that both the geometric and the algebraic multiplicities of the eigenvalues $\lambda_k = m - 1 + k$ are equal to k for $k = 1, \dots, m-1$. By Theorem 8.1.2, the same is true for the eigenvalues $\lambda_{-k} = m - 1 - k$. The proof is complete. ∎

8.1.6. On the width of the energy strip. We have seen that in the case $\alpha = 2\pi$ the strip $|\operatorname{Re} \lambda - m + 1| < 1/2$ does not contain eigenvalues of the pencil $\mathcal{L}$. Now we show that for $\alpha < 2\pi$ even the closure of the above strip is free of eigenvalues. We start with the following lemma.

LEMMA 8.1.3. *For every positive number ε there exists a number R such that $\det \Psi(\lambda) \neq 0$ if $\alpha \in [\varepsilon, 2\pi]$, $m - 1 \leq \operatorname{Re} \lambda \leq m$, and $|\operatorname{Im} \lambda| \geq R$.*

Proof: 1) We transform the operator $\mathcal{L}(\lambda)$ by means of the coordinate change $\varphi' = \pi\varphi/\alpha$ into an operator $\tilde{\mathcal{L}}(\alpha; \lambda)$ defined on the interval $(0, \pi)$ which is infinitely differentiable with respect to the parameter $\alpha \in [\varepsilon, 2\pi]$. For this operator the condition of ellipticity with parameter (in the sense of Agranovich and Vishik [**4**]) is satisfied uniformly with respect to α. Hence this pencil is invertible for sufficiently large $|\lambda|$ in some sector containing the imaginary axis (see, e.g., our book [**136**, Th.3.6.1]). ∎

THEOREM 8.1.4. *If* $0 < \alpha < 2\pi$, *then the strip* $|\operatorname{Re}\lambda - m + 1| \leq 1/2$ *does not contain eigenvalues of the pencil* $\mathcal{L}$.

Proof: For $\alpha = \pi$ the assertion follows from the first part of Theorem 8.1.3. Therefore, we may assume that $\alpha \neq \pi$. We show that the line $\operatorname{Re}\lambda = m - 1/2$ does not contain eigenvalues of $\mathcal{L}$. For this it suffices to prove that the columns of the matrix $D(\lambda)$ are linearly independent for $\lambda = m - 1/2 + it$, $t \in \mathbb{R}$.

Suppose that $c_0, \dots, c_{m-1}, d_0, \dots, d_{m-1}$ are complex numbers such that

$$(8.1.22) \qquad \sum_{k=0}^{m-1} c_k \, \partial_z^j \sigma(z)^k|_{z=z_q} + \sum_{k=0}^{m-1} d_{m-1-k} \, \partial_z^j \sigma(z)^{m-1/2+it-k}|_{z=z_q} = 0$$

for $q = 1, \dots, 2d$, $j = 0, \dots, \nu_q - 1$, where $\sigma(z) = \cos\alpha + z\sin\alpha$. We have to show that $c_0 = \cdots = c_{m-1} = d_0 = \cdots = d_{m-1} = 0$.

First let $\alpha \in (0, \pi)$. Then we introduce the polynomials

$$P(\zeta) = \sum_{k=0}^{m-1} c_k \, \zeta^k, \qquad Q(\zeta) = \sum_{k=0}^{m-1} d_k \, \zeta^k.$$

Obviously, (8.1.22) is equivalent to

$$\partial_z^j \left(P(\sigma(z)) + \sigma(z)^{it+1/2} Q(\sigma(z)) \right) \Big|_{z=z_q} = 0, \qquad q = 1, \dots, 2d, \ j = 0, \dots, \nu_q - 1.$$

We set $\zeta = \sigma(z)$, $\zeta_q = \sigma(z_q) = \cos\alpha + z_q \sin\alpha$, $\overline{P}(\zeta) = \overline{P(\bar\zeta)}$, and $\overline{Q}(\zeta) = \overline{Q(\bar\zeta)}$. Since $\zeta_{q+d} = \bar\zeta_q$, we get

$$\partial_\zeta^j \left(P(\zeta) + \zeta^{it+1/2} Q(\zeta) \right) \Big|_{\zeta=\zeta_q} = \partial_\zeta^j \left(\overline{P}(\zeta) + \zeta^{-it+1/2} \overline{Q}(\zeta) \right) \Big|_{\zeta=\zeta_q} = 0$$

for $q = 1, \dots, d$, $j = 0, \dots, \nu_q - 1$. This implies

$$\partial_z^j \left(P(\zeta) \overline{P}(\zeta) - \zeta Q(\zeta) \overline{Q}(\zeta) \right) \Big|_{\zeta=\zeta_q} = 0 \qquad \text{for } q = 1, \dots, d, \ j = 0, \dots, \nu_q - 1.$$

Hence the polynomial $P(\zeta)\overline{P}(\zeta) - \zeta Q(\zeta)\overline{Q}(\zeta)$ has m roots in the upper half-plane $\operatorname{Im}\zeta > 0$. However, $P(\zeta)\overline{P}(\zeta) - \zeta Q(\zeta)\overline{Q}(\zeta)$ is a polynomial of degree $2m - 1$ which assumes real values for real ζ. Consequently, $P(\zeta)\overline{P}(\zeta) - \zeta Q(\zeta)\overline{Q}(\zeta) = 0$ for all ζ. From this we conclude that $P = Q = 0$.

Now let $\alpha \in (\pi, 2\pi)$. Then

$$(8.1.23) \qquad \partial_z^j \sigma(z)^{\lambda-k}|_{z=z_q} = e^{\pm i\pi(\lambda-k)} \, \partial_z^j (-\sigma(z))^{\lambda-k}|_{z=z_q} \qquad \text{for } \pm\operatorname{Im}z_q > 0.$$

Here the argument of $-\sigma(z) = -\cos\alpha - z\sin\alpha$ lies in the interval $(-\pi, +\pi)$. We introduce the polynomials

$$P(\zeta) = \sum_{k=0}^{m-1} (-1)^k c_k \, \zeta^k, \qquad Q(\zeta) = \sum_{k=0}^{m-1} (-1)^{m+k-1} d_k \, \zeta^k.$$

Then (8.1.22) and (8.1.23) yield

$$(8.1.24) \qquad \partial_\zeta^j \left(P(\zeta) + i\,e^{-\pi t} \zeta^{it+1/2} Q(\zeta) \right) \Big|_{\zeta=\zeta_q}$$
$$= \partial_\zeta^j \left(\overline{P}(\zeta) + i\,e^{\pi t} \zeta^{-it+1/2} \overline{Q}(\zeta) \right) \Big|_{\zeta=\zeta_q} = 0$$

for $q = 1, \dots, d$, $j = 0, \dots, \nu_q - 1$, where $\zeta_q = -\sigma(z_q) = -\cos\alpha - z_q \sin\alpha$. Analogously to the case $\alpha < \pi$, it follows from (8.1.24) that the polynomial $P(\zeta)\overline{P}(\zeta) + \zeta Q(\zeta)\overline{Q}(\zeta)$ of degree $2m - 1$ has m roots in the upper half-plane. Since this polynomial has real values for real ζ, this is only possible if $P = Q = 0$.

Thus, we have proved that the columns of the matrix $D(\lambda)$ are linearly independent for $\operatorname{Re}\lambda = m - 1/2$ and, consequently, also for $\operatorname{Re}\lambda = m - 3/2$. Using Rouché's theorem and Lemma 8.1.3, we conclude from this that the number of the roots of the equation $\det\Psi(\lambda) = 0$ in the strip $m - 3/2 \leq \operatorname{Re}\lambda \leq m - 1/2$ does not depend on the angle $\alpha \in (0, 2\pi)$. Since this strip does not contain zeros of the function $\det\Psi$ for $\alpha = \pi$ (see Theorem 8.1.3), it follows that there are no zeros of $\det\Psi$ in this strip for arbitrary $\alpha \in (0, 2\pi)$. The theorem is proved. ∎

8.1.7. Eigenvalues on the line $\operatorname{Re}\lambda = m$. We show that m is the only possible eigenvalue of the pencil $\mathcal{L}$ on the line $\operatorname{Re}\lambda = m$.

THEOREM 8.1.5. *Let $\alpha \in (0, 2\pi]$. Then the set $\{\lambda \in \mathbb{C} : \lambda = m + it,\ t \in \mathbb{R}\backslash\{0\}\}$ does not contain eigenvalues of the pencil $\mathcal{L}$.*

Proof: If $\alpha = \pi$ or $\alpha = 2\pi$, then the assertion follows from Theorem 8.1.3. Therefore, we suppose that $\alpha \neq \pi$ and $\alpha \neq 2\pi$. We have to show that the columns

$$U_k(\alpha) = \left(\partial_z^j \sigma(z)^k\big|_{z=z_q} : q = 1, \ldots, 2d,\ j = 0, \ldots, \nu_q - 1\right)$$

and

$$V_k(\alpha, \lambda) = \left(\partial_z^j \sigma(z)^{\lambda-k}\big|_{z=z_q} : q = 1, \ldots, 2d,\ j = 0, \ldots, \nu_q - 1\right),$$

$k = 0, \ldots, m-1$, of the matrix $D(\lambda)$ are linearly independent for $\lambda = m + it$. To this end, we suppose the contrary, i.e., there exist numbers $c_0, \ldots, c_{m-1}$ and $d_1, \ldots, d_m$ such that

$$(8.1.25) \qquad \sum_{k=0}^{m-1}\left(c_k\, U_k(\alpha) + d_{m-k}\, V_k(\alpha, m+it)\right) = 0.$$

First let $\alpha \in (0, \pi)$. We introduce the polynomials

$$P(\zeta) = \sum_{k=0}^{m-1} c_k\, \zeta^k, \qquad Q(\zeta) = \sum_{k=1}^{m} d_k\, \zeta^k.$$

Furthermore, let $\overline{P}(\zeta) = \overline{P(\overline{\zeta})}$, $\overline{Q}(\zeta) = \overline{Q(\overline{\zeta})}$. From (8.1.25) it follows that

$$\partial_z^j\left(P(\sigma(z)) + \sigma(z)^{it}\, Q(\sigma(z))\right)\big|_{z=z_q} = 0$$

for $q = 1, \ldots, 2d$, $j = 0, \ldots, \nu_q - 1$. Setting $\zeta = \sigma(z)$, we get

$$\partial_\zeta^j\left(P(\zeta) + \zeta^{it}\, Q(\zeta)\right)\big|_{\zeta=\zeta_q} = \partial_\zeta^j\left(\overline{P}(\zeta) + \zeta^{-it}\, \overline{Q}(\zeta)\right)\big|_{\zeta=\zeta_q} = 0$$

for $q = 1, \ldots, d$, $j = 0, \ldots, \nu_q - 1$, where $\zeta_q = \sigma(z_q) = \cos\alpha + z_q\sin\alpha$. This implies

$$\partial_\zeta^j\left(P\overline{P}(\zeta) - Q\overline{Q}(\zeta)\right)\big|_{\zeta=\zeta_q} = 0 \qquad \text{for } q = 1, \ldots, d,\ j = 0, \ldots, \nu_q - 1.$$

The polynomial $Q\overline{Q}(\zeta) - P\overline{P}(\zeta)$ of degree $2m$ assumes real values for real ζ and has m roots in the upper half-plane. Hence

$$Q\overline{Q}(\zeta) - P\overline{P}(\zeta) = c\prod_{q=1}^{d}(\zeta - \zeta_q)^{\nu_q}\,(\zeta - \overline{\zeta}_q)^{\nu_q},$$

where c is a real number. The coefficient c is nonnegative, since the leading coefficient of the polynomial $Q\overline{Q} - P\overline{P}$ coincides with the leading coefficient of the polynomial $Q\overline{Q}$. However, by the equality $Q(0) = 0$, we have $Q\overline{Q}(0) - P\overline{P}(0) = -P\overline{P}(0) \leq 0$ and, therefore, $c \leq 0$. Hence we get $Q\overline{Q} - P\overline{P} = 0$. From this it follows, in particular, that $c_0 = d_m = 0$. Suppose that $m \geq 2$ and one of the coefficients $c_1, \ldots, c_{m-1}, d_1, \ldots, d_{m-1}$ is nonzero. Then the vectors

$$U_1(\alpha), \ldots, U_{m-1}(\alpha), V_1(\alpha, m+it), \ldots, V_{m-1}(\alpha, m+it)$$

are linearly dependent. Due to the relation $V_{k+1}(\alpha, \lambda) = V_k(\alpha, \lambda - 1)$, this implies the linear dependence of the vectors

$$U_1(\alpha), \ldots, U_{m-1}(\alpha), V_0(\alpha, m-1+it), \ldots, V_{m-2}(\alpha, m-1+it)$$

which are columns of the matrix $D(m - 1 + it)$. By Lemma 8.1.2 and (8.1.19), we conclude from this that $m - 1 + it$ is an eigenvalue of $\mathcal{L}$. However, the line $\operatorname{Re}\lambda = m - 1$ does not contain eigenvalues (see Theorem 8.1.4).

Now we consider the case $\alpha \in (\pi, 2\pi)$. We introduce the polynomials

$$P(\zeta) = \sum_{k=0}^{m-1} (-1)^k\, c_k\, \zeta^k, \qquad Q(\zeta) = \sum_{k=1}^{m} (-1)^{m+k}\, d_k\, \zeta^k.$$

Using (8.1.23), where the argument of the function $\alpha \to -\cos\alpha - z\sin\alpha$ varies from $-\pi$ to π for $\alpha \in (\pi, 2\pi)$, we rewrite (8.1.25) in the form

$$\partial_\zeta^j\big(P(\zeta) + e^{-\pi t}\, \zeta^{it}\, Q(\zeta)\big)\big|_{\zeta=\zeta_q} = \partial_\zeta^j\big(\overline{P}(\zeta) + e^{\pi t}\, \zeta^{-it}\, \overline{Q}(\zeta)\big)\big|_{\zeta=\zeta_q} = 0$$

for $q = 1, \ldots, d$, $j = 0, \ldots, \nu_q - 1$, where $\zeta_q = -\sigma(z_q) = -\cos\alpha - z_q \sin\alpha$. This relations imply

$$\partial_\zeta^j\big(P\overline{P}(\zeta) - Q\overline{Q}(\zeta)\big)\big|_{\zeta=\zeta_q} = 0 \qquad \text{for } q = 1, \ldots, d,\ j = 0, \ldots, \nu_q - 1.$$

Arguing as in the case $\alpha \in (0, \pi)$, we find that $P\overline{P} = Q\overline{Q}$, whence it follows that $c_0 = d_m = 0$. If one of the coefficients $c_1, \ldots, c_{m-1}, d_1, \ldots, d_{m-1}$ is nonzero then, in analogy to the foregoing, we arrive at the equality $\det D(m - 1 + it) = 0$ which is impossible. Thus, we have proved that the columns of the matrix $D(m + it)$ are linearly independent. This proves the theorem. $\blacksquare$

8.2. An asymptotic formula for the eigenvalue close to m

If $\alpha = \pi$ or $\alpha = 2\pi$, then, by Theorem 8.1.3, there is only one simple eigenvalue $\lambda = m$ of the pencil $\mathcal{L}$ on the line $\operatorname{Re}\lambda = m$. Suppose that the angle α is close to $l\pi$, $l = 1, 2$. We denote the unique eigenvalue of the pencil $\mathcal{L}$ which is situated in a small neighborhood of the point $\lambda = m$ by $\lambda_l^*(\alpha)$. Let again $z_1, \ldots, z_d$ be the distinct roots of the equation $L(1, z) = 0$ with positive imaginary parts and $\nu_1, \ldots, \nu_d$ their multiplicities. Furthermore, we set $z_{d+q} = \overline{z}_q$, and $\nu_{d+q} = \nu_q$ for $q = 1, \ldots, d$.

THEOREM 8.2.1. *If α is close to $l\pi$, $l = 1, 2$, then the formula*

(8.2.1) $$\lambda_l^*(\alpha) = m + \left(1 - \frac{\alpha}{l\pi}\right) \operatorname{Im}\left(\nu_1 z_1 + \cdots + \nu_d z_d\right) + O\big(|1 - \frac{\alpha}{l\pi}|^2\big)$$

is valid.

Proof: Let σ_k be the function (8.1.16). If $\alpha = l\pi - \varepsilon$, $l = 1, 2$, then

$$\partial_z^j \sigma_k(z, \lambda)\Big|_{z=z_q} = (-1)^k\, e^{\pm il\pi\lambda}\, \partial_z^j\Big(\big(\sin\varepsilon + z\cos\varepsilon\big)^k \big(\cos\varepsilon - z\sin\varepsilon\big)^{\lambda-k}\Big)\Big|_{z=z_q}$$

for $\pm\operatorname{Im} z_q > 0$ (here the argument of the function $\varepsilon \to \cos\varepsilon - z\sin\varepsilon$ varies from $-\pi$ to π). We set $\omega = \lambda - m$ and denote by $\mathcal{H}_l(\omega, \varepsilon)$ the matrix with the rows

$$\partial_z^j \Big(z^0, \ldots, z^{m-1}, e^{\pm il\pi\omega} z^0 \tilde{\sigma}(z)^{\omega+m}, \ldots, e^{\pm il\pi\omega} z^{m-1} \tilde{\sigma}(z)^{\omega+1} \Big) \Big|_{z=z_q}$$

for $\pm\operatorname{Im} z_q > 0$, $q = 1, \ldots, 2d$, $j = 0, \ldots, \nu_q - 1$, where

$$\tilde{\sigma}(z) = \tilde{\sigma}(z, \varepsilon) = \cos\varepsilon - z\sin\varepsilon.$$

Since

$$(\sin\varepsilon + z\cos\varepsilon)^k (\cos\varepsilon - z\sin\varepsilon)^{\lambda-k} = \sum_{s=0}^{k} \binom{k}{s} \frac{(\tan\varepsilon)^{k-s}}{(\cos\varepsilon)^s} z^s (\cos\varepsilon - z\sin\varepsilon)^{\lambda-s},$$

we get

(8.2.2) $\det \mathcal{D}(\lambda) = (-1)^{m(m+2l-1)/2} (\cos\varepsilon)^{-m(m-1)/2} \det \mathcal{H}_l(\omega, \varepsilon).$

By (8.1.17) and Theorem 8.1.3, the function

$$f_l(\omega, \varepsilon) \stackrel{def}{=} \omega^{1-m} \det \mathcal{H}_l(\omega, \varepsilon)$$

is smooth and $f_l(0,0) = 0$. Furthermore, using (8.1.21) and (8.2.2), we obtain

$$f_l(\omega, 0) = \omega^{1-m} |\det T_0|^2 \left(e^{-il\pi\omega} - e^{il\pi\omega} \right)^m$$

and, consequently,

(8.2.3) $\partial_\omega f_l(0,0) = |\det T_0|^2 (-2\pi li)^m \neq 0.$

Hence, by the implicit function theorem, there is the representation

(8.2.4) $\lambda_l^*(\alpha) = m - \dfrac{\partial_\varepsilon f_l(0,0)}{\partial_\omega f_l(0,0)} \varepsilon + O(\varepsilon^2).$

We calculate the derivative $\partial_\varepsilon f_l(0,0)$. Since

$$\partial_\varepsilon \partial_z^j \big(z^k \tilde{\sigma}(z, \varepsilon)^{\omega+m-k} \big) \big|_{z=z_q} = -(\omega + m - k) \partial_z^j \big(z^{k+1} \tilde{\sigma}(z, \varepsilon)^{\omega+m-k-1} \big) \big|_{z=z_q},$$

it follows that

$$\partial_\varepsilon \det \mathcal{H}_l(\omega, \varepsilon) \big|_{\varepsilon=0} = -(\omega + 1) \det \Phi_l(\omega, 0),$$

where $\Phi_l(\omega, \varepsilon)$ is the matrix with the rows

$$\partial_z^j \Big(z^0, \ldots, z^{m-1}, e^{\pm il\pi\omega} z^0 \tilde{\sigma}(z)^{\omega+m}, \ldots, e^{\pm il\pi\omega} z^{m-2} \tilde{\sigma}(z)^{\omega+2}, e^{\pm il\pi\omega} z^m \tilde{\sigma}(z)^\omega \Big) \Big|_{z=z_q}$$

for $\pm\operatorname{Im} z_q > 0$, $q = 1, \ldots, 2d$, $j = 0, \ldots, \nu_q - 1$. We denote by $\vec{g} = (g_0, \ldots, g_{m-1})^t$ the solution of the equation

(8.2.5) $T_0 \cdot \vec{g} = \Big(\partial_z^j z^m |_{z=z_q} \; : \; q = 1, \ldots, d, \; j = 0, \ldots, \nu_q - 1 \Big).$

Furthermore, we set

$$M = \begin{pmatrix} & & g_0 \\ I_{m-1} & & g_1 \\ & & \vdots \\ 0 & & g_{m-1} \end{pmatrix},$$

where I_n is the $n \times n$ identity matrix. Then

$$\Phi_l(\omega, 0) = \begin{pmatrix} T_0 & 0 \\ 0 & \overline{T}_0 \end{pmatrix} \cdot \begin{pmatrix} I_m & e^{il\pi\omega} M \\ I_m & e^{-il\pi\omega} \overline{M} \end{pmatrix}.$$

This implies

(8.2.6) $\partial_\varepsilon \det \mathcal{H}_l(\omega, \varepsilon)|_{\varepsilon=0}$

$$= -(\omega + 1) |\det T_0|^2 (e^{-il\pi\omega} - e^{il\pi\omega})^{m-1} (\overline{g}_{m-1} e^{-il\pi\omega} - g_{m-1} e^{il\pi\omega}).$$

Since the vector $(g_0, \ldots, g_{m-1})$ is a solution of the system

$$\partial_z^j \left(z^m - \sum_{k=0}^{m-1} g_k z^k \right)\Big|_{z=z_q} = 0, \qquad q = 1, \ldots, d, \ j = 0, \ldots, \nu_q - 1$$

(see (8.2.5)), the polynomial $z^m - \sum g_k z^k$ has the zeros $z_1, \ldots, z_d$ with multiplicities $\nu_1, \ldots, \nu_d$. Consequently, by Viète's theorem,

$$g_{m-1} = \nu_1 z_1 + \cdots + \nu_d z_d$$

and (8.2.6) yields

(8.2.7) $\partial_\varepsilon f_l(0,0) = |\det T_0|^2 (-2\pi li)^{m-1} 2i \operatorname{Im} (\nu_1 z_1 + \cdots + \nu_d z_d).$

Substituting (8.2.3) and (8.2.7) into (8.2.4), we arrive at formula (8.2.1). The theorem is proved. ∎

8.3. Asymptotic formulas for the eigenvalues close to $m - 1/2$

If $\alpha = 2\pi$, then, by the second part of Theorem 8.1.3, there is the only one eigenvalue $\lambda = m - 1/2$ on the line $\operatorname{Re}\lambda = m - 1/2$ with geometric multiplicity m (there are no generalized eigenvalues in this case). So, if α is close to 2π, there are m eigenvalues of the pencil $\mathcal{L}$ in a neighborhood of $m - 1/2$. Our goal is to obtain an asymptotic formula for them.

8.3.1. A lemma on zeros of polynomials.

LEMMA 8.3.1. Let m and n be integers, $m \geq n \geq 0$, $m \geq 1$. Suppose that P and Q are polynomials with real coefficients, $\deg P \leq m + n - 1$, $\deg Q \leq m - n - 1$, such that the polynomial $P + iQ$ has at least m roots in the upper half-plane $\operatorname{Im} z > 0$. Then $P = Q = 0$.

Proof: It suffices to consider the case when $\deg P = m + n - 1$, $\deg Q = m - n - 1$, and the polynomials P, Q have no common roots.

We represent the polynomial $P + iQ$ in the form

$$P(z) + i Q(z) = \sum_{s=0}^{m+n-1} (p_s + iq_s) z^{m+n-s-1},$$

where p_s and q_s are real numbers, $q_s = 0$ for $s = 0, \ldots, 2n - 1$. Let R_{2k} be the quadratical $(2k) \times (2k)$ matrix with the rows

$$(p_{j-1}, q_{j-1}, p_{j-2}, q_{j-2}, \ldots, p_{j-2k}, q_{j-2k}), \qquad j = 1, \ldots, 2k,$$

where $q_s = p_s = 0$ if $s < 0$. Applying the general Hurwitz Theorem (see Gantmakher [65, Vol.2, p.194]) to the polynomial $P + iQ$, we find that the number of the roots of this polynomial located in the upper half-plane is equal to the number of changes of sign in the sequence $\{1, \det R_2, \ldots, \det R_{2(m+n-1)}\}$. Since $\det R_{2k} = 0$ for $k =$

$1, \ldots, n$, the number of changes of signs in this sequence is not greater than $m - 1$ (the rule for determining the number of changes of signs in the case when some of the determinants R_{2k} are zero is given in [**65**, Vol.2,p.194]). This proves the lemma. ∎

8.3.2. A representation for $\det \mathcal{D}(\lambda)$ if α is close to 2π and λ is close to $m - 1/2$. Let $\alpha = 2\pi - \varepsilon$, where ε is a small positive number. Using the second assertion of Theorem 8.1.3, formula (8.2.1) with $l = 2$, Lemma 8.1.3, and Theorem 8.1.5, we find that the sum of the multiplicities of the zeros of the function $\det \Psi(\lambda)$ located in the strip $m - 1 \leq \operatorname{Re} \lambda \leq m$ is equal to m. Moreover, all these zeros are located near the point $\lambda = m - 1/2$. We set $\tau = \tan \varepsilon$. Then we have the formula

$$\partial_z^j \, \sigma_k(z, \lambda) \Big|_{z = z_q} = e^{\pm 2\pi \lambda i} \, (\cos \varepsilon)^\lambda \, \partial_z^j \Big((-\tau - z)^k \, (1 - \tau z)^{\lambda - k} \Big) \Big|_{z = z_q}$$

for $\pm \operatorname{Im} z_q > 0$. Let $\omega = \lambda - (m - 1/2)$. We denote by $\mathcal{N}(\omega, \tau)$ the matrix with the rows

$$\partial_z^j \Big(z^0, \ldots, z^{m-1}, \psi_0^\pm(\omega, \tau, z), \ldots, \psi_{m-1}^\pm(\omega, \tau, z) \Big) \Big|_{z = z_q}$$

(for $\pm \operatorname{Im} z_q > 0$), $q = 1, \ldots, 2d$, $j = 0, \ldots, \nu_q - 1$, where

$$\psi_k^\pm(\omega, \tau, z) = -e^{\pm 2\pi \omega i} \, z^k \, (1 - \tau z)^{\omega + m - 1/2 - k} \, .$$

Since

$$(-\tau - z)^k \, (1 - \tau z)^{\lambda - k} = (-1)^k \sum_{s=0}^k \binom{k}{s} \Big(z(1 + \tau^2) \Big)^s \, \tau^{k-s} \, (1 - \tau z)^{\omega + m - 1/2 - s},$$

we obtain, by elementary operations on the matrix $\mathcal{D}(\lambda)$, the relation

$$\det \mathcal{D}(\lambda) = (-1)^{m(m-1)/2} \, (\cos \varepsilon)^{m\lambda} \, (1 + \tau^2)^{m(m-1)/2} \, \det \mathcal{N}(\omega, \tau).$$

It can be easily seen that $\det \mathcal{N}(0, 0) = 0$ and

$$(8.3.1) \qquad \det \mathcal{N}(\bar{\omega}, \tau) = (-1)^m \, \det \overline{\mathcal{N}(\omega, \tau)}.$$

Our goal is to find solutions of the equation $\mathcal{N}(\omega, \tau) = 0$ in a neighborhood of the point $(\omega, \tau) = (0, 0)$.

8.3.3. Taylor expansion of $\det \mathcal{N}$.

LEMMA 8.3.2. *The function $\det \mathcal{N}(\omega, \tau)$ can be expanded into a series*

$$(8.3.2) \qquad \det \mathcal{N}(\omega, \tau) = \sum_{l=0}^{m-1} \sum_{k=l^2}^{(l+1)^2} \sum_{s \geq m-l} b_{k,s} \, \tau^k \, \omega^s + \sum_{k \geq m^2, \, s \geq 0} b_{k,s} \, \tau^k \, \omega^s$$

which converges absolutely in a neighborhood of the point $(\omega, \tau) = (0, 0)$, where $i^m \, b_{k,s}$ are real numbers. The coefficients $b_{l^2, m-l}$, $l = 0, 1, \ldots, m$, are nonzero and have the representation

$$(8.3.3) \qquad b_{l^2, m-l} = (-1)^{l(2m+l+1)/2} \, (2\pi i)^{m-l} \Big(\prod_{s=1}^l (\tfrac{3}{2} - s)_{2s-1} \Big) \det J_l \, \det \Lambda_l \, ,$$

where J_l denotes the matrix

$$J_l = \Big(\frac{1}{(k + s - 1)!} \Big)_{k,s=1,\ldots,l}$$

and Λ_l is the matrix with the rows

$$\partial_z^j \left(z^0, \ldots, z^{m+l-1}, -z^0, \ldots, -z^{m-l-1} \right)\Big|_{z=z_q}$$

and

$$\partial_z^j \left(z^0, \ldots, z^{m+l-1}, z^0, \ldots, z^{m-l-1} \right)\Big|_{z=\bar{z}_q},$$

$q = 1, \ldots, d,\ j = 0, 1, \ldots, \nu_q - 1.$

Proof: The function $\det \mathcal{N}(\omega, \tau)$ is analytic in some neighborhood of the point $(\omega, \tau) = (0,0)$. Hence it can be expanded into an absolutely convergent series

(8.3.4) $$\det \mathcal{N}(\omega, \tau) = \sum_{k,s \geq 0} b_{k,s}\, \tau^k\, \omega^s.$$

From (8.3.1) it follows that $i^m\, b_{k,s}$ are real numbers.

Nonzero elements in (8.3.4). We show that the only nonzero terms in the series (8.3.4) are those with $k \geq m^2$ or $l^2 \leq k < (l+1)^2$ and $s \geq m - l$ for some l with $l = 0, \ldots, m-1$.

Using the formula

$$\partial_\tau^h \psi_k^\pm(\omega, \tau, z)\big|_{z=z_q} = (-1)^h\, (\omega + m + 1/2 - k - h)_h\, \psi_{k+h}^\pm(\omega, \tau, z)\big|_{z=z_q}$$

and the formula for differentiating the determinant of a matrix, we get

(8.3.5) $$\partial_\tau^h \det \mathcal{N}(\omega, \tau)$$

$$= (-1)^h \sum_{|\gamma|=h} \frac{h!}{\gamma!} \Big(\prod_{k=0}^{m-1} (\omega + m + 1/2 - k - \gamma_k)_{\gamma_k} \Big) \det \mathcal{N}_\gamma(\omega, \tau),$$

where γ denotes the multi-index $(\gamma_0, \ldots, \gamma_{m-1})$ and $\mathcal{N}_\gamma(\omega, \tau)$ denotes the matrix with the rows

$$\partial_z^j \left(z^0, \ldots, z^{m-1}, \psi_{0+\gamma_0}^\pm(\omega, \tau, z), \ldots, \psi_{m-1+\gamma_{m-1}}^\pm(\omega, \tau, z) \right)\Big|_{z=z_q}$$

(for $\pm \mathrm{Im}\, z_q > 0$), $q = 1, \ldots, 2d,\ j = 0, \ldots, \nu_q - 1.$

We are interested in multi-indices γ for which the determinant of $\mathcal{N}_\gamma$ is nonzero. Let N_k be the vector with the components

$$-e^{\pm 2\pi\omega i}\, \partial_z^j \big(z^k (1 - \tau z)^{\omega+m-1/2-k} \big)\big|_{z=z_q}, \quad q = 1, \ldots, 2d,\ j = 0, \ldots, \nu_q - 1,$$

for $\pm \mathrm{Im}\, z_q > 0$. Then the last m columns of the matrix $\mathcal{N}_\gamma(\omega, \tau)$ are the vectors $N_{\gamma_0}, \ldots, N_{m-1+\gamma_{m-1}}$. We divide these vectors into two groups. The first group contains the vectors $N_{k+\gamma_k}$ with $k + \gamma_k \leq m - 1$, the second gets all the others. Let n be the number of vectors in the second group and let $\{k_1, \ldots, k_n\}$ be the set of the indices determining the vectors $N_{k+\gamma_k}$ of the second group. For the nonvanishing of $\det \mathcal{N}_\gamma$ it is necessary that the vectors in this group are distinct. This leads to the inequality

$$\sum_{\mu=1}^{n} (k_\mu + \gamma_{k_\mu}) \geq \sum_{\mu=0}^{n-1} (m + \mu)$$

for the indices. Since $k_1 + \ldots + k_n \leq (m-1) + \cdots + (m-n)$, we obtain the estimate

(8.3.6) $$\sum_{\mu=1}^{n} \gamma_{k_\mu} \geq \sum_{\mu=0}^{n-1} (m + \mu) - \sum_{\mu=1}^{n} (m - \mu) = n^2.$$

Here equality is only possible if $\{k_1, \ldots, k_n\} = \{m - 1, \ldots, m - n\}$. In particular, we obtain $h = |\gamma| \geq n^2$. If $h = |\gamma| = n^2$, then equality in (8.3.6) holds only in the case $\{0, 1, \ldots, m-1\} \backslash \{k_1, \ldots, k_n\} = \{0, 1, \ldots, m-n-1\}$, $\gamma_0 = \ldots = \gamma_{m-n-1} = 0$.

If we add the $(k + \gamma_k + 1)$-th column of the matrix $\mathcal{N}_\gamma(\omega, 0)$ (i.e., the vector with the elements $\partial_z^j z^{k+\gamma_k}|_{z=z_q}$) to the column $N_{k+\gamma_k}|_{\tau=0}$ of the first group $(k + \gamma_k \leq m-1)$, each component of the new column contains the factor $1 - e^{2\pi\omega i}$ or $1 - e^{-2\pi\omega i}$. Since this operation does not change the determinant of the matrix $\mathcal{N}_\gamma$, we get

$$\det \mathcal{N}_\gamma(\omega, 0) = O(|\omega|^{m-n}).$$

From this and from formula (8.3.5) it follows that

a) $\displaystyle \partial_\tau^h \det \mathcal{N}(\omega, \tau)\Big|_{\tau=0} = O(|\omega|^{m-l+1})$ if $h < l^2$,

b) $\displaystyle \partial_\tau^h \det \mathcal{N}(\omega, \tau)\Big|_{\tau=0} = I_l(\omega) + O(|\omega|^{m-l+1})$ if $h = l^2$

for some $l, l = 0, 1, \ldots, m$, where

(8.3.7)

$$I_l(\omega) = (-1)^{l^2} \sum_{\substack{|\gamma|=l^2 \\ \gamma_1=\ldots=\gamma_{m-l-1}=0}} \frac{(l^2)!}{\gamma!} \left(\prod_{k=m-l}^{m-1} (\omega + m + 1/2 - k - \gamma_k)_{\gamma_k} \right) \det \mathcal{N}_\gamma(\omega, 0).$$

It follows from property a) that the coefficient $b_{h,s}$ in series (8.3.4) is equal to zero if $(l-1)^2 \leq h < l^2$, $s < m - (l-1)$ for some $l = 1, \ldots, m$. This proves formula (8.3.2)

Proof of (8.3.3). The multi-indices γ over which the summation in (8.3.7) is extended admit the representation

$$m - k + \gamma_{m-k} = m - 1 + \beta_k, \qquad k = 1, \ldots, l,$$

where $\beta_1, \ldots, \beta_l$ are pairwise distinct numbers assuming the values $1, 2, \ldots, l$. Arranging the last l columns $N_{m-1+\beta_1}, \ldots, N_{m-1+\beta_l}$ of the matrices $\mathcal{N}_\gamma$ in (8.3.7) in increasing order of their indices (in doing this, we introduce a factor $\delta(\beta)$, where $\delta(\beta) = \pm 1$ depends on the parity of the permutation $(1, \ldots, l) \to (\beta_1, \ldots, \beta_l)$), we obtain

(8.3.8)

$$I_l(\omega) = (-1)^{l(l+1)/2} (l^2)! \sum_\beta \delta(\beta) \left(\prod_{k=1}^l \frac{(\omega + 3/2 - \beta_k)_{k-1+\beta_k}}{(k-1+\beta_k)!} \right) \det \mathcal{N}^{(l)}(\omega, 0),$$

where the summation is extended over all permutations β of the numbers $1, \ldots, l$ and $\mathcal{N}^{(l)}(\omega, \tau)$ is the matrix with the rows

$$\partial_z^j \left(z^0, \ldots, z^{m-1}, \psi_0^\pm, \ldots, \psi_{m-l-1}^\pm, \psi_m^\pm, \ldots, \psi_{m+l-1}^\pm \right)\Big|_{z=z_q}$$

(for $\pm \text{Im}\, z_q > 0$), $q = 1, \ldots, 2d$, $\nu = 0, \ldots, \nu_q - 1$. We introduce the matrix

$$G = \left(\frac{(\omega - s + 3/2)_{k+s-1}}{(k+s-1)!} \right)_{k,s=1,\ldots,l}.$$

Since $(\omega - s + 3/2)_{k+s-1} = (\omega - s + 3/2)_s \, (\omega + 3/2)_{k-1}$, we have

$$
\det G \;=\; \left(\prod_{k=1}^{l} (\omega - k + 3/2)_k \, (\omega + 3/2)_{k-1} \right) \det J_l
$$

$$
\;=\; \left(\prod_{k=1}^{l} (\omega - k + 3/2)_{2k-1} \right) \det J_l
$$

and (8.3.8) can be rewritten in the form

$$
I_l(\omega) \;=\; (-1)^{l(l+1)/2} \, (l^2)! \, \det G \, \det \mathcal{N}^{(l)}
$$

$$
\;=\; (-1)^{l(l+1)/2} \, (l^2)! \left(\prod_{k=1}^{l} (\omega - k + 3/2)_{2k-1} \right) \det \mathcal{N}^{(l)} .
$$

Now we study the function $\det \mathcal{N}^{(l)}(\omega, 0)$. If we add the k-th column of the matrix $\mathcal{N}^{(l)}$ to the $(m+k)$-th column N_{k-1} $(k = 1, \ldots, m-l)$, the elements of the $(m+k)$-th column of the new matrix for $\tau = 0$ are $(1 - e^{\pm 2\pi\omega i})\partial_z^j z^{k-1}|_{z=z_q}$, $q = 1, \ldots, 2d$, $j = 0, \ldots, \nu_q - 1$. Hence we have

$$
\det N^{(l)}(\omega, 0) = (2\pi\omega i)^{m-l} \, (-1)^{lm} \, \det \Lambda_l + O(|\omega|^{m-l+1}).
$$

This proves (8.3.3).

We show that the coefficients $b_{l^2, m-l}$ are nonzero for $l = 0, 1, \ldots, m$. To this end, we prove that

$$
(8.3.9) \qquad \det J_l = (-1)^{l-1} \frac{((l-1)!)^2}{(2l-1)! \, (2l-2)!} \det J_{l-1}.
$$

Let $F(z)$ be the determinant of the $l \times l$-matrix whose first $l-1$ rows are the same as those of the matrix J_l and whose last row is $(z^l/l!, \, z^{l+1}/(l+1)!, \ldots, z^{2l-1}/(2l-1)!)$. It is not difficult to see that the coefficient of z^{2l-1} in the polynomial $F(z)$ is $\det J_{l-1}/(2l-1)!$. Furthermore, the polynomial F satisfies the relations $\partial_z^k F(0) = 0$ for $k = 0, \ldots, l-1$, $F(1) = \det J_l$, $\partial_z^k F(1) = 0$ for $k = 1, \ldots, m-1$. Using these data we can recover the polynomial F using the Hermite interpolation (see Gel'fond [67, p.46]):

$$
F(z) = \det J_l \, z^l \sum_{k=0}^{l-1} \frac{\partial_z^{l-k-1} z^{-l}|_{z=1}}{(l-k-1)!} (z-1)^{l-k-1}.
$$

Therefore, the coefficient of z^{2l-1} in the polynomial F is also equal to

$$
(-1)^{l-1} \frac{(2l-2)!}{((l-1)!)^2} \det J_l.
$$

This implies (8.3.9). In particular, from (8.3.9) it follows that $\det J_l \neq 0$.

Finally we prove that $\det \Lambda_l \neq 0$ for $l = 0, \ldots, m$. Assume that there exist numbers $c_0, \ldots, c_{m+l-1}$ and $p_0, \ldots, p_{m-l-1}$ such that

$$
(8.3.10) \qquad \sum_{k=0}^{m+l-1} c_k \, \partial_z^j z^k|_{z=z_q} + \sum_{k=0}^{m-l-1} p_k \, (-\partial_z^j z^k)|_{z=z_q} \;=\; 0,
$$

$$
(8.3.11) \qquad \sum_{k=0}^{m+l-1} c_k \, \partial_z^j z^k|_{z=\bar{z}_q} + \sum_{k=0}^{m-l-1} p_k \, \partial_z^j z^k|_{z=\bar{z}_q} \;=\; 0
$$

for $q = 1, \ldots, d$, $j = 0, \ldots, \nu_q - 1$. We introduce the polynomials

$$P(z) = \sum_{k=0}^{m+l-1} c_k \, z^k, \qquad Q(z) = \sum_{k=0}^{m-l-1} p_k \, z^k.$$

Equations (8.3.10, (8.3.11) can be written in the form

$$(8.3.12) \qquad \partial_z^j (P(z) - Q(z))\big|_{z=z_q} = \partial_z^j (\overline{P}(z) + \overline{Q}(z))\big|_{z=z_q} = 0$$

for $q = 1, \ldots, d$, $j = 0, \ldots, \nu_q - 1$ (here $\overline{P}(z)$ denotes the polynomial $\overline{P(\overline{z})}$). Let $P_1 = P + \overline{P}$, $iQ_1 = \overline{Q} - Q$. Then it follows from (8.3.12) that the polynomial $P_1 + iQ_1$ has m roots in the upper half-plane. Applying Lemma 8.3.1, we get $P_1 = Q_1 = 0$, i.e., $P = -\overline{P}$, $Q = \overline{Q}$. The equalities $P = \overline{P}$, $Q = -\overline{Q}$ are proved analogously. Thus, $P = Q = 0$ and the columns of Λ_l are linearly independent, i.e., $\det \Lambda_l \neq 0$. The lemma is proved. ∎

8.3.4. Asymptotic formulas for the eigenvalues.

THEOREM 8.3.1. *There exist a positive number ε_0 and real-analytic functions*

$$\mu_k : (2\pi - \varepsilon_0, 2\pi) \to (m - 1/2, m), \qquad k = 1, \ldots, m,$$

such that the spectrum of the pencil $\mathcal{L}$ lying in the strip $m - 1 \leq \operatorname{Re} \lambda \leq m$ is exhausted by the simple eigenvalues $\mu_1(\alpha), \ldots, \mu_m(\alpha)$ for $2\pi - \varepsilon_0 < \alpha < 2\pi$. Moreover, the following asymptotic formula holds as $\alpha \to 2\pi$:

$$(8.3.13) \qquad \mu_k(\alpha) = m - 1/2 + \alpha_k \, (2\pi - \alpha)^{2k-1} + O(|2\pi - \alpha|^{2k}),$$

$k = 1, \ldots, m$. *Here α_k are positive numbers computed as follows. We set*

$$R(z) = \prod_{q=1}^{d} (z - z_q)^{\nu_q} = \sum_{l=0}^{m} (\gamma_l + i\beta_l) \, z^{m-l},$$

where γ_j, β_j are real numbers. Furthermore, we introduce the following matrices: $\nabla_0 = 1$, ∇_{2k} is the $2k \times 2k$ matrix whose j-the row has the form

$$(\gamma_{j-1}, \beta_{j-1}, \gamma_{j-2}, \beta_{j-2}, \ldots, \gamma_{j-2k}, \beta_{j-2k}),$$

where we agree that $\gamma_{-l} = \beta_{-l} = 0$ for $l > 0$. Then

$$(8.3.14) \qquad \alpha_k = -\frac{1}{4\pi} \left(\frac{(k-1)! \, \Gamma(k - 1/2)}{(2k-2)! \, \Gamma(1/2)} \right)^2 \frac{\det \nabla_{2k}}{\det \nabla_{2k-2}}.$$

(Note that the quantity $\det \nabla_{2k}$ is positive for even k and negative for odd k, cf. Gantmakher [65, Vol.2, p.194].)

Proof: We construct the solutions of the equation $\det \mathcal{N}(\omega, \tau) = 0$ in a neighborhood of the point $(\omega, \tau) = (0, 0)$ using a Newton diagram (see Vainberg and Trenogin [254, Ch.1,§2]) which, by Lemma 8.3.2, is generated by the points $(l^2, m - l)$, $l = 0, 1, \ldots, m$, and consists of m segments. The defining equation for each segment of the Newton diagram is linear. Therefore, the equation $\det \mathcal{N}(\omega, \tau) = 0$ has exactly m real solutions $\omega_1(\tau), \ldots, \omega_m(\tau)$ defined for $\tau \in (0, \delta)$ and having the asymptotics

$$(8.3.15) \qquad \omega_l(\tau) = \alpha_l \, \tau^{2l-1} + O(\tau^{2l}), \qquad \text{where } \alpha_l = -b_{l^2, m-l}/b_{(l-1)^2, m-l+1}.$$

According to Theorem 8.1.4, the equation $\det \Psi(\lambda) = 0$ has no solutions in the strip $|\operatorname{Re} \lambda - m + 1| \leq 1/2$ for $\alpha \in (0, 2\pi)$. Consequently, $\omega_l(\tau) > 0$ and $\alpha_l > 0$ for $l = 1, \ldots, m$.

We calculate the constants α_l in (8.3.15). To this end, we establish the connection between the determinants of the matrices Λ_l and Λ_{l-1}. Let the column-vector $\underline{c}^{(l)} = (c_0, \dots, c_{m+l-2}, p_0, \dots, p_{m-l})^t$ be the solution of the equation $\Lambda_{l-1}\underline{c}^{(l)} = \underline{u}^{(m+l-1)}$, where $\underline{u}^{(m+l-1)}$ is the vector with the components $\partial_z^j z^{m+l-1}|_{z=z_q}$, $q = 1, \dots, 2d$, $j = 0, \dots, \nu_q - 1$. Then the matrix

$$\mathcal{M} = \left(\underline{e}^{(1)}, \cdots, \underline{e}^{(m+l-1)}, \underline{c}^{(l)}, \underline{e}^{(m+l)}, \dots, \underline{e}^{(2m-1)}\right),$$

where $\underline{e}^j$ is the j-th unit vector in $\mathbb{R}^{2m}$, satisfies the equality $\Lambda_{l-1}\mathcal{M} = \Lambda_l$. Therefore,

(8.3.16) $$(-1)^{m-l} p_{m-l} \det \Lambda_{l-1} = \det \Lambda_l.$$

Using formulas (8.3.9) and (8.3.16), we find that

$$\alpha_l = -\frac{1}{2\pi i} \frac{(l - 1/2)(1/2)_{l-1}^2}{(l)_l (l)_{l-1}} p_{m-l}.$$

Now we determine the numbers p_{m-l}. Let

$$P(z) = -z^{m+l-1} + \sum_{k=0}^{m+l-2} c_k z^k, \qquad Q(z) = \sum_{k=0}^{m-l} p_k z^k.$$

Then the relation $\Lambda_{l-1}\underline{c}^{(l)} = \underline{u}^{(m+l-1)}$ can be written in the form

(8.3.17) $$\partial_z^j (P(z) - Q(z))|_{z=z_q} = \partial_z^j (\overline{P}(z) + \overline{Q}(z))|_{z=z_q} = 0$$

for $q = 1, \dots, d$, $j = 0, \dots, \nu_q - 1$. The polynomials $P_1 = -i(P - \overline{P})$ and $Q_1 = Q + \overline{Q}$ have real coefficients and $\deg P_1 \leq m + l - 2$, $\deg Q_1 \leq m - l$. It follows from (8.3.17) that the polynomial $P_1 + iQ_1$ has m roots in the upper half-plane. Applying Lemma 8.3.1, we find that $P_1 = Q_1 = 0$. In particular, we deduce from this that $\operatorname{Re} p_{m-l} = 0$. Therefore,

(8.3.18) $$\alpha_l = -\frac{1}{2\pi i} \frac{(l - 1/2)(1/2)_{l-1}^2}{(l)_l (l)_{l-1}} \operatorname{Im} p_{m-l}.$$

Now let $P_2 = (P + \overline{P})/2$, $Q_2 = (Q - \overline{Q})/2i$. Then P_2 and Q_2 are polynomials with real coefficients, and from (8.3.17) it follows that

(8.3.19) $$\partial_z^j (-P_2(z) + iQ_2(z))|_{z=z_q} = 0 \quad \text{for } q = 1, \dots, d, \ j = 0, \dots, \nu_q - 1.$$

In order to calculate the leading coefficient $\operatorname{Im} p_{m-l}$ of the polynomial Q_2, we represent the polynomial $-P_2 + iQ_2$ in the form

(8.3.20) $$-P_2(z) + i\, Q_2(z) = R(z)\, S(z), \qquad \text{where } S(z) = \sum_{k=0}^{l-1} (x_k + iy_k)\, z^{l-1-k}.$$

The representation (8.3.20) is possible by (8.3.19). It can be easily seen that $\gamma_0 = x_0 = 1$ and $\beta_0 = y_0 = 0$. From (8.3.20) we obtain the equalities

(8.3.21) $$\sum_{k+s=n} (\gamma_k\, y_s + \beta_k\, x_s) = 0 \quad \text{for } n = 0, \dots, 2l - 2,$$

(8.3.22) $$\sum_{k+s=2l-1} (\gamma_k\, y_s + \beta_k\, x_s) = \operatorname{Im} p_{m-l},$$

in which we have taken $\gamma_k = \beta_k = 0$ for $k > m$ and $x_s = y_s = 0$ for $s \geq l$. Let $\underline{y}$ be the column-vector $(y_0, x_0, y_1, x_1, \dots, y_{l-1}, x_{l-1})^t$. Then (8.3.21), (8.3.22) can be written in the form $\nabla_{2l}\underline{y} = \operatorname{Im} p_{m-l} \cdot \underline{e}^{(2l)}$, where $\underline{e}^{(2l)}$ is the $(2l)$-th unit vector in $\mathbb{R}^{2l}$. We denote by $\underline{m} = (m_1, \ldots, m_{2l})^t$ the column-vector consisting of the minors corresponding to the elements of the last row of the matrix ∇_{2l}. Since $\gamma_0 = 1$ and $\beta_0 = 0$, it follows that $m_1 = 0$ and $m_2 = \det \nabla_{2l-2}$. Moreover, there is the equality $\nabla_{2l}\underline{m} = \det \nabla_{2l} \cdot \underline{e}^{(2l)}$. Consequently, $\underline{y} = (\operatorname{Im} p_{m-l}/\det \nabla_{2l}) \underline{m}$. Since $x_0 = 1$ and $m_2 = \det \nabla_{2l-2}$, we have $\operatorname{Im} p_{m-l} = \det \nabla_{2l}/\det \nabla_{2l-2}$. From this and from (8.3.18) we obtain formula (8.3.14). The theorem is proved. ∎

8.4. The case of a convex angle

Here we improve the result of Theorem 8.1.4 for $\alpha < \pi$. We show that the strip $|\operatorname{Re}\lambda - m + 1| \leq 1$ does not contain eigenvalues of the pencil $\mathcal{L}$ in this case.

8.4.1. An auxiliary assertion.

LEMMA 8.4.1. *Suppose that m and n are integers, $m \geq 2$, $n = 0, \ldots, m - 2$, and P, Q are polynomials, $\deg P \leq m + n$, $\deg Q \leq m - n - 2$, such that the polynomial $P\overline{Q}$ has real coefficients. If the function*

$$(8.4.1) \qquad F(z) \overset{def}{=} P(z)\,\overline{Q}(z) + z^{n+1}\,Q\overline{Q}(z)\,\log z$$

has at least m roots with positive imaginary parts, then $F = 0$ and, consequently, also $Q = 0$.

Proof: We consider the function F in the complex plane cut along the negative semi-axis and suppose that F has at least m roots in the upper half-plane. Since $F(\overline{z}) = \overline{F(z)}$, it follows that F has at least $2m$ roots with nonzero imaginary part.

We represent the polynomials P and Q in the form $P = L_1 P_1$ and $Q = L_1 L_2 Q_1$, where L_1 and L_2 are polynomials with real coefficients whose roots are situated on the real axis, the polynomials P_1 and L_2 have no common roots, and $Q_1(z) \neq 0$ for $\operatorname{Im} z = 0$. Then $F = (L_1)^2 L_2 F_1$, where

$$F_1(z) = P_1\overline{Q}_1(z) + z^{n+1}\,L_2(z)\,Q_1\overline{Q}_1(z)\,\log z.$$

It can be easily shown that the function F_1, just as F, has at least $2m$ roots with nonzero imaginary parts. We show that F_1 does not vanish on the negative real semi-axis. Indeed, if $F_1(z_0) = 0$, $\operatorname{Im} z_0 = 0$, and $\operatorname{Re} z_0 < 0$, then $L_2(z_0) = 0$ and $P_1(z_0) = 0$, but the polynomials P_1 and L_1 have no common roots.

We consider the contour $\Gamma_{\varepsilon,R} = (Re^{i\pi}, \varepsilon e^{i\pi}) \cup S_\varepsilon \cup (\varepsilon e^{-i\pi}, Re^{-i\pi}) \cup S_R$, where S_ρ is a circle of radius ρ with center at the point 0.

If ε is sufficiently small and R is a large positive number, then the function F_1 does not vanish on $\Gamma_{\varepsilon,R}$ and has at least $2m$ roots inside $\Gamma_{\varepsilon,R}$. We compute the increment of the argument of F_1 on a single passage around the contour $\Gamma_{\varepsilon,R}$ in a counterclockwise direction. Suppose the degrees of the polynomials L_1 and L_2 are equal to l_1 and l_2, respectively. First we show that the modulus of the increment of F_1 on the segment $(Re^{i\pi}, \varepsilon e^{i\pi})$ does not exceed $\pi(l_2 + 1)$. Indeed, the imaginary part of F_1 can change sign only on passing through a root of the function L_2. Thus, as z varies from $Re^{i\pi}$ to $\varepsilon e^{i\pi}$, the imaginary part of F_1 can change sign not more than $l - 2$ times. From this we obtain the required estimate of the increment of the argument. Similarly, it can be proved that the modulus of the increment of F_1 on the segment $(\varepsilon e^{-i\pi}, Re^{-i\pi})$ also does not exceed $\pi(l_2 + 1)$. Separating out the principal part of the function F_1 at zero and infinity, we find that the increment of the argument of F_1 on the contours S_ε, S_R is contained within the limits from $-2\pi(2m - 2 - 2l_1 - l_2) - \delta_1$ to δ_1 and from $-\delta_2$ to $2\pi(2m - 2 - 2l_1 - l_2) + \delta_2$, respectively, where the numbers δ_1 and δ_2 can be made arbitrarily small by the choice of the numbers ε and R. In summary, we find that the modulus of the increment of the argument of F_1 on the contour $\Gamma_{\varepsilon,R}$ does not exceed $2\pi(m-1)$. Since F_1 has $2m$ roots inside $\Gamma_{\varepsilon,R}$, it follows that $F_1 = 0$. The lemma is proved. ∎

8.4.2. The energy strip in the case $\alpha < \pi$.

THEOREM 8.4.1. *Let $\alpha \in (0, \pi)$. Then*
1) the spectrum of the pencil $\mathcal{L}$ does not contain the points $m, m+1, \ldots, 2m-1$,
2) the strip $m - 2 \leq \operatorname{Re}\lambda \leq m$ does not contain eigenvalues of the pencil $\mathcal{L}$.

Proof: 1) Suppose first that $\lambda = 2m - 1$. We prove the linear independence of the vectors $U_k(\alpha)$ and $V_k(\alpha, \lambda)$, $k = 0, \ldots, m-1$, introduced in the proof of Theorem 8.1.5. Let $c_0, \ldots, c_{2m-1}$ be complex numbers such that

$$(8.4.2) \qquad \sum_{k=0}^{m-1} \left(c_k\, U_k(\alpha) + c_{2m-1-k}\, V_k(\alpha, 2m-1) \right) = 0.$$

We introduce the polynomial

$$P(z) = \sum_{k=0}^{2m-1} c_k\, z^k.$$

Then (8.4.2) can be written as

$$\partial_z^j P(z)\Big|_{z=\cos\alpha + z_q \sin\alpha} = 0, \qquad q = 1, \ldots, 2d,\ \ j = 0, \ldots, \nu_q - 1.$$

From this we conclude that $P = 0$. Hence the vectors $U_k(\alpha)$ and $V_k(\alpha, \lambda)$ are linear independent for $\lambda = 2m - 1$.

Next we prove that the number $\lambda = m + n$ does not belong to the spectrum of $\mathcal{L}$ for $n = 0, \ldots, m-2$. Due to Lemma 8.1.2 and (8.1.19), it suffices to show that

$$(8.4.3) \qquad \partial_\lambda^{m-n-1} \det D(\lambda)\Big|_{\lambda=m+n} \neq 0.$$

Since $V_{m-k}(\alpha, m+n) = U_{n+k}(\alpha)$ for $k = 1, \ldots, m-n-1$, the quantity on the left side of (8.4.3) is equal to the determinant of the matrix

$$\left(U_0, \ldots, U_{m-1}, V_0, \ldots, V_n, \partial_\lambda V_{n+1}, \ldots, \partial_\lambda V_{m-1} \right)\Big|_{\lambda=m+n}.$$

We assume that there exist numbers $c_0, \ldots, c_{m+n}$ and $d_0, \ldots, d_{m-n-2}$ such that

$$(8.4.4) \qquad \sum_{k=0}^{m-1} c_k U_k(\alpha) + \sum_{k=m}^{m+n} c_k V_{m+n-k}(\alpha, m+n)$$

$$+ \sum_{k=0}^{m-n-2} d_k \partial_\lambda V_{m-1-k}(\alpha, m+n) = 0.$$

Then we introduce the polynomials

$$P(z) = \sum_{k=0}^{m+n} c_k\, z^k, \qquad Q(z) = i \sum_{k=0}^{m-n-2} d_k\, z^k.$$

Since $\partial_\lambda V_{m-1-k}(\alpha, m+n)$ is the vector with the components

$$\partial_z^j \left((\cos\alpha + z\sin\alpha)^{n+k+1} \log(\cos\alpha + z\sin\alpha) \right)\Big|_{z=z_q},$$

$q = 1, \ldots, 2d$, $j = 0, \ldots, \nu_q - 1$, we can write (8.4.4) in the form

$$(8.4.5) \qquad \begin{cases} \partial_z^j(P(z) + z^{n+1} Q(z) \log z)|_{z=\zeta_q} &= 0, \\[2mm] \partial_z^j(\overline{P}(z) + z^{n+1} \overline{Q}(z) \log z)|_{z=\zeta_q} &= 0, \end{cases}$$

where $\zeta_q = \cos\alpha + z_q \sin\alpha$, $q = 1, \ldots, d$, $j = 0, \ldots, \nu_q - 1$. This implies

$$\partial_z^j \Big(z^{n+1} (P\overline{Q}(z) - \overline{P}Q(z)) \log z \Big)\Big|_{z=\zeta_q} = 0,$$

and, consequently, $\partial_z^j(P\overline{Q}(z) - \overline{P}Q(z))|_{z=\zeta_q} = 0$ for $q = 1, \ldots, d$, $j = 0, \ldots, \nu_q - 1$. The polynomial $P\overline{Q} - \overline{P}Q$ of degree $2m - 2$ assumes real values for real z and has m roots in the upper half-plane. Hence $P\overline{Q} = \overline{P}Q$, i.e., the coefficients of the polynomial $P\overline{Q}$ are real.

We consider the function

$$F(z) \stackrel{def}{=} P(z)\,\overline{Q}(z) + z^{n+1}\, Q(z)\,\overline{Q}(z) \log z.$$

From (8.4.5) it follows that

$$\partial_z^j F(z)|_{z=\zeta_q} = 0 \qquad \text{for } q = 1, \ldots, d, \; j = 0, \ldots, \nu_q - 1.$$

Consequently, by Lemma 8.4.1, the polynomial Q is equal to zero. Since the degree of the polynomial P is less than $2m - 1$, it follows from (8.4.5) that the polynomial P is equal to zero. Hence the numbers $m + n$ are not eigenvalues of $\mathcal{L}$ for $n = 0, \ldots, m - 2$. The first part of the theorem is proved.

2) Let δ be a small positive number. By Lemma 8.1.3, there exists a positive number N such that the function $\lambda \to \det \Psi(\lambda)$ has no zeros in the set

$$\{\lambda \in \mathbb{C} : m - 1 \le \operatorname{Re}\lambda \le m, \; |\operatorname{Im}\lambda| \ge N\}$$

if $\alpha \in [\delta, 2\pi]$. We denote the boundary of the rectangle

$$G_N = \{\lambda \in \mathbb{C} : m - 1 \le \operatorname{Re}\lambda \le m, \; |\operatorname{Im}\lambda| \le N\}$$

by Γ_N. By Theorem 8.1.5, Lemma 8.1.3 and part 1) of Theorem 8.4.1, the function $\det \Psi(\cdot)$ is nonzero on Γ_N if $\alpha \in [\delta, 2\pi]$. Applying Rouché's theorem, we find that the number of zeros of $\det \Psi(\cdot)$ situated in G_N does not depend on $\alpha \in [\delta, \pi]$. If the angle α is close to π, then, by Theorems 8.1.3 and 8.2.1, the strip $m-1 \le \operatorname{Re}\lambda \le m$ is free of zeros of $\det \Psi(\cdot)$. Therefore, for $\alpha \in [\delta, \pi]$ the function $\det \Psi(\cdot)$ is nonzero in the rectangle G_N and, consequently, also in the strip $m - 1 \le \operatorname{Re}\lambda \le m$. From Theorem 8.1.2 we conclude that the strip $m - 2 \le \operatorname{Re}\lambda \le m$ does not contain eigenvalues of $\mathcal{L}$. The proof is complete. ∎

8.4.3. A counterexample. We show that Theorems 8.1.4 and 8.4.1 are not true, in general, if the coefficients of the operator L are nonreal. Let

$$L(\partial_{x_1}, \partial_{x_2}) = \partial_{x_2}^2 + i\,(t - 1)\,\partial_{x_1}\partial_{x_2} + t\,\partial_{x_1}^2 = (\partial_{x_2} - i\,\partial_{x_1})\,(\partial_{x_2} + i\,t\,\partial_{x_1})$$

with a positive real constant t. If $\lambda \ne 0$, then the functions

$$u_1(\lambda, \varphi) = (\cos\varphi + i\sin\varphi)^\lambda \quad \text{and} \quad u_2(\lambda, \varphi) = (\cos\varphi - it\sin\varphi)^\lambda$$

form a complete system of solutions of the equation $\mathcal{L}(\varphi, \partial_\varphi, \lambda)\,u(\varphi) = 0$. It can be easily verified that the number λ_0 is an eigenvalue of the pencil $\mathcal{L}$ if and only if

$$(8.4.6) \qquad \lambda_0^{-1}\big(u_1(\lambda_0, \alpha) - u_2(\lambda_0, \alpha)\big) = 0.$$

Since $\log(\cos\alpha + i\sin\alpha) \ne \log(\cos\alpha - it\sin\alpha)$ (we recall that the argument of the function $\varphi \to \cos\varphi + z\sin\varphi$ varies from 0 to 2π for $\operatorname{Im} z > 0$ and from 0

to -2π for $\operatorname{Im} z < 0$), it follows that the number 0 is not an eigenvalue. We denote by γ, $\gamma \in [-2\pi, 0)$, the argument of the number $\cos\alpha - it\sin\alpha$ and set $\rho = \log(\cos^2\alpha + t\sin^2\alpha)$. Assuming that $\rho \neq 0$ (i.e., $\alpha \neq \pi$, $\alpha \neq 2\pi$, and $t \neq 1$), we find that the solutions of the equation (8.4.6) are exhausted by the numbers λ_k, $k = \pm 1, \pm 2, \ldots$, where

$$\operatorname{Im}\lambda_k = \frac{2k\pi}{\rho + (\alpha - \gamma)^2/\rho}, \qquad \operatorname{Re}\lambda_k = \frac{2k\pi(\alpha - \gamma)}{\rho^2 + (\alpha - \gamma)^2}.$$

Hence, if t is a sufficiently large number and $\alpha \neq \pi$, $\alpha \neq 2\pi$, then the eigenvalues of the pencil $\mathcal{L}$ can lie arbitrarily close to the line $\operatorname{Re}\lambda = 0$.

8.5. The case of a nonconvex angle

Now we assume that the opening α of the angle $\mathcal{K}$ is greater than π. Then, in contrast to the case $\alpha < \pi$, the strip $m - 1 \leq \operatorname{Re}\lambda \leq m$ contains eigenvalues of the pencil $\mathcal{L}$. We prove that all eigenvalues in this strip are real and simple and that they depend monotonically on α.

8.5.1. Auxiliary assertions on the distribution of roots of complex functions. We start with the proof of the following three lemmas.

LEMMA 8.5.1. *Let c be a complex number with nonzero imaginary part, and let P, Q be polynomials with real coefficients. Suppose that the polynomial*

$$P(z) + c\,z^2\,Q(z)$$

has $m \geq 2$ roots in the upper half-plane. Furthermore, we assume that one of the following assumptions is valid:
   (i) $\deg P \leq m - 1$ *and* $\deg Q \leq m - 2$,
   (ii) $\deg P \leq m$ *and* $\deg Q \leq m - 3$.
Then $P = Q = 0$.

Proof: Let $\deg P \leq m - 1$ and $\deg Q \leq m - 2$. Without loss of generality, we may assume that $P(0) \neq 0$, $\deg Q = m - 2$, and the leading coefficient of the polynomial Q is 1. Suppose that the polynomial $P(z) + c\,z^2\,Q(z)$ has the roots $z_1, \ldots, z_m$, which lie in the upper half-plane $\operatorname{Im} z > 0$. Then, by Viète's theorem we have

$$P(0) = (-1)^m c\,z_1 \cdots z_m, \qquad P'(0) = (-1)^{m-1} c\,z_1 \cdots z_m\,(z_1^{-1} + \cdots + z_m^{-1}).$$

Consequently, $P'(0)/P(0) = -(z_1^{-1} + \cdots + z_m^{-1})$ which is impossible, since $\operatorname{Im} z_k^{-1} < 0$ for $k = 1, \ldots, m$. This proves the assertion for case (i). The proof for case (ii) proceeds analogously. ∎

LEMMA 8.5.2. *Let P, Q be polynomials with real coefficients such that $\deg P \leq m$ and $\deg Q \leq m - 3$, $m \geq 2$. Furthermore, we assume that the function*

$$h(z) = P(z) + (\pi i + \log z)\,z^2\,Q(z)$$

has at least m zeros in the upper half-plane. Then $P = Q = 0$.

Proof: We assume at first that the polynomials P and $z^2 Q$ have no common real roots. Then the function h does not vanish on the real axis. Let

$$P(z) = \sum_{j=0}^{k} p_j\,z^j, \qquad Q(z) = \sum_{j=0}^{l} q_j\,z^j,$$

where $k \leq m$, $l \leq m - 3$, $p_k \neq 0$, $q_l \neq 0$. We consider the family of functions

$$h_t(z) = P(z) + (\pi i + t \log z) z^2 Q(z), \qquad 0 \leq t \leq 1.$$

These functions do not vanish for $\operatorname{Im} z = 0$. Moreover, for large $|z|$ the following formulas hold:

$$\begin{aligned}
h_t(z) &= p_k z^k + O(|z|^{k-1} \log |z|) & \text{if } k \geq l + 3, \\
h_t(z) &= \left(q_l (\pi i + t \log z) + p_k\right) z^k + O(|z|^{k-1} \log |z|) & \text{if } k = l + 2, \\
h_t(z) &= q_l (\pi i + t \log z) z^{l+2} + O(|z|^{l+1} \log |z|) & \text{if } k < l + 2.
\end{aligned}$$

Consequently, in all cases for sufficiently large $|z|$ with $\operatorname{Im} z \geq 0$ the functions h_t, $t \in [0,1]$, are nonzero. Using Rouché's theorem, we find that the functions h_t have the same number of zeros in the upper half-plane. Since $h_1 = h$, the polynomial h_0 has at least m zeros in the upper half-plane. Applying Lemma 8.5.1 to h_0, we get $P = Q = 0$.

We consider the case when the polynomials P and $z^2 Q$ have common real roots. Then there are the decompositions

$$P(z) = P_1(z) P_2(z) \quad \text{and} \quad z^2 Q(z) = P_1(z) Q_2(z),$$

where P_1 is a polynomial with real roots, $\deg P_2 \leq m - 1$, $\deg Q_2 \leq m - 2$, and $P_2(z) + i Q_2(z) \neq 0$ for real z. The function $z \to P_2(z) + (\pi i + \log z) Q_2(z)$ has at least m roots in the upper half-plane. Just as above, it can be proved that the polynomial $P_2 + i \pi Q_2$ has m roots with positive imaginary part. From this we conclude that $P_2 = Q_2 = 0$. The lemma is proved. ∎

Similarly, the following lemma can be proved.

LEMMA 8.5.3. *Let P, Q be polynomials with real coefficients, $\deg P \leq m - 1$, $\deg Q \leq m - 2$, $m \geq 2$. Furthermore, let τ be a real number, $\tau \neq 0$, $|\tau| < 1/2$. If the function $h(z) = P(z) + e^{\pi \tau i} z^{\tau+2} Q(z)$ has at least m zeros in the upper half-plane $\operatorname{Im} z > 0$, then $P = Q = 0$.*

Proof: Suppose that the function h has at least m roots with positive imaginary part. We assume first that the polynomials $P(z)$ and $z^2 Q(z)$ have no common real roots. Then $h(z) \neq 0$ for real z. Let

$$P(z) = \sum_{j=0}^{k} p_j z^j, \qquad Q(z) = \sum_{j=0}^{l} q_j z^j,$$

where $k \leq m - 1$, $l \leq m - 2$, $p_k \neq 0$, $q_l \neq 0$. We consider the family of functions

$$h_t(z) = P(z) + e^{\pi \tau i} z^{t\tau+2} Q(z), \qquad 0 \leq t \leq 1,$$

which are nonzero for real z. For large $|z|$ we have

$$\begin{aligned}
h_t(z) &= q_l e^{\pi \tau i} z^{t\tau+l+2} + O(|z|^{l+3/2}) & \text{if } k \leq l + 1, \\
h_t(z) &= \left(q_l e^{\pi \tau i} z^{t\tau} + p_k\right) z^k + O(|z|^{k-1/2}) & \text{if } k = l + 2, \\
h_t(z) &= p_k z^k + O(|z|^{k-1/2}) & \text{if } k > l + 2.
\end{aligned}$$

Consequently, in all cases the functions h_t are nonzero for large $|z|$, $\operatorname{Im} z \geq 0$. Therefore, the functions h_t, $t \in [0,1]$, have the same number of zeros in the upper half-plane as $h_1 = h$. In particular, it follows that the polynomial h_0 has at least m roots with positive imaginary part, which is impossible by Lemma 8.5.1.

Now we consider the case when the polynomials P and $z^2 Q$ have common real roots. Then $P(z) = P_1(z) P_2(z)$, $z^2 Q(z) = P_1(z) Q_2(z)$, where P_1 is a polynomial with real roots and $P_2(z) + i Q_2(z) \neq 0$ for real z. Furthermore, the function $z \to P_2(z) + e^{\pi \tau i} z^\tau Q_2(z)$ has at least m roots in the upper half-plane $\operatorname{Im} z > 0$. Reasoning as above, we find that the polynomial $P_2 + e^{\pi \tau i} Q_2$ has at least m roots with positive imaginary parts. This implies $P_2 = Q_2 = 0$. The lemma is proved. ∎

8.5.2. A property of the function Ψ. Let $\Psi = \Psi(\alpha, \lambda)$ be the matrix introduced in Lemma 8.1.1.

LEMMA 8.5.4. *If* $\det \Psi(\alpha, \lambda) = 0$ *for some* $\alpha \in [\pi, 2\pi)$, $\lambda \in (m - 1/2, m + 1/2)$, *then* $\partial_\alpha \det \Psi(\alpha, \lambda) \neq 0$.

Proof: We consider three cases.

1. *The case* $\alpha = \pi$, $\lambda \in (m - 1/2, m + 1/2)$. By Theorem 8.1.3, the number $\lambda = m$ is the only solution of the equation $\det \Psi(\pi, \lambda) = 0$ in the interval $(m - 1/2, m + 1/2)$. Furthermore, by (8.1.17) and (8.2.2), we have

$$\det \Psi(\alpha, \lambda) = (-1)^{m(m+1)/2} \det T^{-1} (\cos \alpha)^{-m(m-1)/2} f_1(\lambda - m, \pi - \alpha)$$
$$\times (\lambda - m)^{m-1} \prod_{k=0}^{m-1} \frac{1}{(\lambda - k - m + 1)_m},$$

where f_1 is the function introduced in the proof of Theorem 8.2.1. From (8.2.7) it follows that $\partial_\alpha f_1(\lambda - m, \pi - \alpha) \neq 0$ for $\alpha = \pi$, $\lambda = m$. Consequently, $\partial_\alpha \det \Psi(\alpha, \lambda) \neq 0$ for $\alpha = \pi$, $\lambda = m$.

2. *The case* $\alpha \in (\pi, 2\pi)$, $\lambda = m$. If $\det \Psi(\alpha, m) = 0$ for some $\alpha \in (\pi, 2\pi)$, then

$$(8.5.1) \qquad \partial_\lambda^{m-1} \det D(\alpha, \lambda)\big|_{\lambda = m} = 0$$

(cf. (8.1.18), (8.1.19)). Let $U_k(\alpha)$, $V_k(\alpha, \lambda)$ be the the column-vectors introduced in the proof of Theorem 8.1.5. Since $V_{m-k}(\alpha, m) = U_k(\alpha)$ for $k = 1, \dots, m - 1$, the left side of (8.5.1) is equal to the determinant of the matrix with the columns

$$(8.5.2) \qquad U_0(\alpha), \dots, U_m(\alpha), \partial_\lambda V_1(\alpha, \lambda)\big|_{\lambda = m}, \dots, \partial_\lambda V_{m-1}(\alpha, \lambda)\big|_{\lambda = m}.$$

Thus, (8.5.1) is satisfied if and only if the vectors (8.5.2) are linearly dependent.

We assume that $\partial_\alpha \det \Psi(\alpha, m) = 0$ or, what is the same, that

$$(8.5.3) \qquad \partial_\alpha \partial_\lambda^{m-1} \det D(\alpha, \lambda)\big|_{\lambda = m} = 0.$$

Since the vector $\partial_\alpha U_k(\alpha)$ is a linear combination of $U_k(\alpha)$ and $U_{k-1}(\alpha)$, $k = 1, \dots, m - 1$, and the vector $\partial_\alpha \partial_\lambda V_{m-k}(\alpha, \lambda)\big|_{\lambda = m}$ is a linear combination of $U_k(\alpha)$, $U_{k-1}(\alpha)$, $\partial_\lambda V_{m-k}(\alpha, \lambda)\big|_{\lambda = m}$, and $\partial_\lambda V_{m-k+1}(\alpha, \lambda)\big|_{\lambda = m}$, it follows from (8.5.3) that the vectors
$$(8.5.4)$$
$$U_0(\alpha), \dots, U_m(\alpha), \partial_\lambda V_1(\alpha, \lambda)\big|_{\lambda = m}, \dots, \partial_\lambda V_{m-2}(\alpha, \lambda)\big|_{\lambda = m}, \partial_\lambda V_m(\alpha, \lambda)\big|_{\lambda = m}$$

are linearly dependent.

We prove that the vectors

$$(8.5.5) \qquad U_0(\alpha), \dots, U_{m-1}(\alpha), \partial_\lambda V_1(\alpha, \lambda)\big|_{\lambda = m}, \dots, \partial_\lambda V_m(\alpha, \lambda)\big|_{\lambda = m}$$

are linearly independent. Indeed, since $V_k(\alpha, \lambda) = V_{k-1}(\alpha, \lambda - 1)$, the linear dependence of the vectors (8.5.5) is equivalent to the vanishing of the determinant of the matrix with the columns

$$U_0(\alpha), \dots, U_{m-1}(\alpha), \partial_\lambda V_0(\alpha, \lambda)\big|_{\lambda = m-1}, \dots, \partial_\lambda V_{m-1}(\alpha, \lambda)\big|_{\lambda = m-1}.$$

However, by the relation $V_k(\alpha, m-1) = U_{m-1-k}(\alpha)$, $k = 0, \ldots, m-1$, the determinant just mentioned equals $\partial_\lambda^m D(\alpha, \lambda)|_{\lambda=m-1}$, which is nonzero by (8.1.19), Lemma 8.1.2, and Theorem 8.1.4.

From the linear dependence of the vectors (8.5.2), (8.5.4) and the linear independence of the vectors (8.5.5) it follows that

$$U_0(\alpha), \ldots, U_m(\alpha), \partial_\lambda V_1(\alpha, \lambda)|_{\lambda=m}, \ldots, \partial_\lambda V_{m-2}(\alpha, \lambda)|_{\lambda=m}$$

are linearly dependent, i.e., there exist numbers $c_0, \ldots, c_m$, $d_1, \ldots, d_{m-2}$, not all zero, such that

$$(8.5.6) \qquad \sum_{k=0}^{m} c_k\, U_k(\alpha) + \sum_{k=1}^{m-2} d_k\, \partial_\lambda V_k(\alpha, \lambda)|_{\lambda=m} = 0.$$

We introduce the polynomials

$$P(z) = \sum_{k=0}^{m} c_k\, (-z)^k, \qquad Q(z) = \sum_{k=1}^{m-2} d_k\, (-z)^{m-2-k}.$$

Since

$$\partial_\lambda \partial_z^j \sigma(z)^{\lambda-k}|_{\lambda=m, z=z_q}$$
$$= \partial_z^j \left(i(\pm \pi i + \log(-\cos\alpha - z\sin\alpha))(\cos\alpha + z\sin\alpha)^{m-k} \right)\Big|_{z=z_q}$$

for $\pm \operatorname{Im} z_q > 0$ (the principal value of the logarithm is taken here), the equality (8.5.6) can be written in the form

$$(8.5.7) \qquad \partial_\zeta^j \Big(-iP(\zeta) + \zeta^2(\pi i + \log\zeta)\, Q(\zeta) \Big)\Big|_{\zeta=\zeta_q}$$
$$= \partial_\zeta^j \Big(i\overline{P}(\zeta) + \zeta^2(\pi i + \log\zeta)\, \overline{Q}(\zeta) \Big)\Big|_{\zeta=\zeta_q} = 0,$$

where $\zeta_q = -\sigma(z_q) = -\cos\alpha - z_q\sin\alpha$, $q = 1, \ldots, d$, $j = 0, \ldots, \nu_q - 1$. We consider the function

$$h(\zeta) = P_1(\zeta) + \zeta^2\, (\pi i + \log\zeta)\, Q_1(\zeta),$$

where $P_1 = i(\overline{P} - P)$ and $Q_1 = Q + \overline{Q}$ are polynomials with real coefficients. By (8.5.7), the function h has m zeros with positive imaginary parts. Applying Lemma 8.5.2 to this function, we find that $P_1 = Q_1 = 0$. Similarly, it can be proved that $P + \overline{P} = Q - \overline{Q} = 0$ and, consequently, $P = Q = 0$. This proves the lemma for $\lambda = m$.

3. *The case $\alpha \in (\pi, 2\pi)$, $\lambda \in (m - 1/2, m + 1/2)$, $\lambda \neq m$.* If $\det \Psi(\alpha, \lambda) = 0$, then the vectors

$$(8.5.8) \qquad U_0(\alpha), \ldots, U_{m-1}(\alpha), V_0(\alpha, \lambda), \ldots, V_{m-1}(\alpha, \lambda)$$

are linearly independent. If, moreover, $\partial_\alpha \det \Psi(\alpha, \lambda) = 0$, then

$$(8.5.9) \qquad \partial_\alpha \det D(\alpha, \lambda) = 0.$$

Since the vector $\partial_\alpha U_k(\alpha)$, $k = 1, \ldots, m-1$, is a linear combination of $U_k(\alpha)$ and $U_{k-1}(\alpha)$ and the vector $\partial_\alpha V_k(\alpha, \lambda)$ is a linear combination of $V_k(\alpha, \lambda)$ and $V_{k+1}(\alpha, \lambda)$, the equality (8.5.9) is equivalent to the linear dependence of the vectors

$$(8.5.10) \qquad U_0(\alpha), \ldots, U_{m-1}(\alpha), V_0(\alpha, \lambda), \ldots, V_{m-2}(\alpha, \lambda), V_m(\alpha, \lambda).$$

By Theorem 8.1.4, the number $\lambda - 1$ does not belong to the spectrum of $\mathcal{L}$. Consequently, the determinant of $D(\alpha, \lambda - 1)$ is nonzero. From this and from the relation $V_k(\alpha, \lambda) = V_{k-1}(\alpha, \lambda - 1)$ we conclude that the vectors

$$(8.5.11) \qquad U_0(\alpha), \dots, U_{m-1}(\alpha), V_1(\alpha, \lambda), \dots, V_m(\alpha, \lambda)$$

are linearly independent. By the linear dependence of the vectors (8.5.8) and (8.5.10) and the linear independence of the vectors (8.5.11), there exist numbers $c_0, \dots, c_{m-1}, d_0, \dots, d_{m-2}$, not all zero, such that

$$(8.5.12) \qquad \sum_{k=0}^{m-1} c_k \, U_k(\alpha) + \sum_{k=0}^{m-2} d_k \, V_{m-2-k}(\alpha, \lambda) = 0.$$

We set $\tau = \lambda - m$. Then $\tau \in (-1/2, +1/2)$, $\tau \neq 0$, and the components of the vector V_k are

$$\partial_z^j \sigma(z)^{\lambda-k}\big|_{z=z_q} = (-1)^{m-k} \, e^{\pm \pi \tau i} \partial_z^j (-\cos\alpha - z\sin\alpha)^{\tau+m-k}\big|_{z=z_q} \quad \text{for } \pm \operatorname{Im} z_q > 0,$$

where the argument of the function $-\cos\alpha - z\sin\alpha$ lies in the interval $(-\pi, \pi)$. We introduce the polynomials

$$P(z) = \sum_{k=0}^{m-1} c_k \, (-z)^k, \qquad Q(z) = (-1)^m \sum_{k=0}^{m-2} d_k \, (-z)^k.$$

Then (8.5.12) can be written in the form

$$(8.5.13) \quad \partial_\zeta^j \big(P(\zeta) + e^{\pi\tau i} \, \zeta^{\tau+2} \, Q(\zeta) \big)\big|_{\zeta=\zeta_q} = \partial_\zeta^j \big(\overline{P}(\zeta) + e^{\pi\tau i} \, \zeta^{\tau+2} \, \overline{Q}(\zeta) \big)\big|_{\zeta=\zeta_q} = 0,$$

where $\zeta_q = -\sigma(z_q) = -\cos\alpha - z_q \sin\alpha$, $q = 1, \dots, d$, $j = 0, \dots, \nu_q - 1$. Let $P_1 = P + \overline{P}$, $Q_1 = Q + \overline{Q}$. Then, according to (8.5.13), the function $P_1(\zeta) + e^{\pi\tau i} \zeta^{\tau+2} Q_1(\zeta)$ has m roots in the upper half-plane. Applying Lemma 8.5.3, we find that $P_1 = Q_1 = 0$, i.e., $P = -\overline{P}$, $Q = -\overline{Q}$. Similarly, we can prove that $P = \overline{P}$, $Q = \overline{Q}$. Thus, $P = Q = 0$. This proves the lemma for $\alpha \in (\pi, 2\pi)$, $\lambda \neq m$. ∎

8.5.3. Eigenvalues in the strip $m - 1/2 < \operatorname{Re}\lambda \leq m$. Now we give a description of the eigenvalues of the operator pencil $\mathcal{L}$ in the strip $m - 1/2 < \operatorname{Re}\lambda \leq m$.

THEOREM 8.5.1. *Let* $\alpha \in [\pi, 2\pi)$. *There exist real-analytic functions* $\alpha_k = \alpha_k(\lambda)$, $\lambda \in (m - 1/2, m]$, $k = 1, \dots, m$, *with values in the interval* $[\pi, 2\pi)$ *which have the following properties.*

1) *The functions* α_k, $k = 1, \dots, m$, *satisfy the inequality* $\alpha_k'(\lambda) < 0$ *for* $\lambda \in (m - 1/2, m)$, *i.e., these functions are decreasing. For all* $\lambda \in (m - 1/2, m]$ *we have*

$$\alpha_1(\lambda) < \alpha_2(\lambda) < \cdots < \alpha_m(\lambda).$$

Furthermore,

$$\alpha_1(m) = \pi, \qquad \alpha_1'(m) = -\pi/\operatorname{Im}(\nu_1 z_1 + \cdots + \nu_d z_d), \quad and$$

$$\lim_{\lambda \to m-1/2} \alpha_k(\lambda) = 2\pi \quad for \; k = 1, \dots, m$$

(see Figure 14 below).

2) *The eigenvalues of the pencil $\mathcal{L}$ lying in the strip $m - 1 \le \operatorname{Re}\lambda \le m$ are the elements of the set*

$$\{\lambda \in (m - 1/2, m] : \alpha_k(\lambda) = \alpha \quad for \ some \ k = 1, \dots, m\}.$$

If the number m is an eigenvalue, then its algebraic multiplicity is at most two. All other eigenvalues in the strip $m - 1 \le \operatorname{Re}\lambda \le m$ are simple.

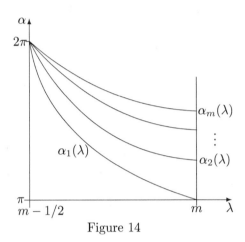

Figure 14

Proof: We set $F(\alpha, \lambda) = \det \Psi(\alpha, \lambda)$. Then $F(\alpha, \overline{\lambda}) = \overline{F(\alpha, \lambda)}$, i.e., the function $F(\alpha, \cdot)$ assumes real values for real λ. By Lemma 8.5.4, the function $F(\cdot, \lambda)$ has only simple zeros for $\lambda \in (m - 1/2, m]$. We show that the number of zeros of the function $F(\cdot, \lambda)$ is independent of $\lambda \in (m - 1/2, m]$. If $\lambda \in (m - 1/2, m)$, then, by Theorem 8.1.3, we have $F(\pi, \lambda) = F(2\pi, \lambda) \ne 0$, i.e., the function $F(\cdot, \lambda)$ is nonzero at the endpoints of the interval $[\pi, 2\pi]$. Since the zeros of the function $F(\cdot, \lambda)$ are simple, it follows from this that the number of zeros is independent of λ for $\lambda \in (m - 1/2, m)$. Now let $\lambda = m$. We denote by $\alpha_1, \dots, \alpha_n$ the roots of the equation $F(\alpha, m) = 0$ on the interval $[\pi, 2\pi)$. By Theorem 8.1.3, one of them (suppose it is α_1 for definiteness) equals π. If we consider the equation $F(\alpha, m) = 0$ on the interval $[\pi, 2\pi]$, then another simple root appears by Theorems 8.1.3 and 8.2.1, namely $\alpha_{n+1} = 2\pi$. Using the implicit function theorem, we conclude that there exist smooth functions $\Phi_k = \Phi_k(\lambda)$, $k = 1, \dots, n + 1$, defined in some neighborhood $(m - \varepsilon, m + \varepsilon)$ of the point $\lambda = m$ with values in a neighborhood of the segment $[\pi, 2\pi]$ such that $\Phi_k(m) = \alpha_k$, $F(\Phi_k(\lambda), \lambda) = 0$ for $k = 1, \dots, n + 1$, and the functions Φ_k exhaust all the roots of the equation $F(\alpha, \lambda) = 0$ for $\lambda \in (m - \varepsilon, m + \varepsilon)$, $\alpha \in [\pi, 2\pi]$. Since $\pi < \alpha_k < 2\pi$ for $k = 2, \dots, n$, it follows that for sufficiently small ε the functions Φ_k, $k = 2, \dots, n$ assume values in the segment $(\pi, 2\pi)$. From the asymptotic formulas (8.2.1) we conclude that the function Φ_1 assumes values on the segment $[\pi, 2\pi)$ for $\lambda \in (m - \varepsilon, m]$, while the values of the function Φ_{n+1} lie outside the interval $[\pi, 2\pi)$ for $\lambda \in (m - \varepsilon, m]$. Therefore, the functions $F(\cdot, \lambda)$ have n zeros on the interval $[\pi, 2\pi)$, if $\lambda \in (m - \varepsilon, m]$. Consequently, for all $\lambda \in (m - 1/2, m]$ the functions $F(\cdot, \lambda)$ have n zeros on the interval $[\pi, 2\pi)$.

Now we show that $n = m$. Indeed, if $\lambda = m - 1/2$, then the equation $F(\alpha, \lambda) = 0$ has the unique root $\alpha_0 = 2\pi$ on the interval $[\pi, 2\pi]$ (cf. Theorems 8.1.3, 8.1.4). As λ increases, this root splits into m distinct roots (cf. Theorem 8.3.1). Hence $n = m$.

We enumerate the zeros of the function $F(\cdot, \lambda)$, $\lambda \in (m - 1/2, m]$, belonging to the interval $[\pi, 2\pi)$ in increasing order:

$$\alpha_1(\lambda) < \alpha_2(\lambda) < \cdots < \alpha_m(\lambda), \quad \lambda \in (m - 1/2, m].$$

It follows from the implicit function theorem that the functions α_k, $k = 1, \dots, m$, are real-analytic in $(m - 1/2, m]$. In addition, $\alpha_1(m) = \pi$ and from formula (8.2.1) we conclude that

$$\alpha_1'(m) = -\pi / \mathrm{Im}\, (\nu_1 z_1 + \cdots \nu_d z_d).$$

Furthermore, by Theorem 8.3.1, we have $\alpha_k(\lambda) \to 2\pi$ as $\lambda \to m - 1/2$, $k = 1, \dots, m$.

It remains to prove part 2) of the theorem and the relations $\alpha_k'(\lambda) < 0$ for $\lambda \in (m - 1/2, m)$, $k = 1, \dots, m$. Assume that for some $\lambda_0 \in (m - 1/2, m)$, $k = 1, \dots, m$, $l \geq 1$ we have $\partial_\lambda^j \alpha_k(\lambda_0) = 0$ for $j = 0, 1, \dots, l - 1$ and $\partial_\lambda^l \alpha_k(\lambda_0) \neq 0$. From this and from the rule for differentiating an implicit function we conclude that the function $\lambda \to F(\alpha_k(\lambda_0), \lambda)$ has a zero of order l at the point $\lambda = \lambda_0$. We call the number l the *multiplicity* of the function α_k at the point λ_0.

Let $\alpha \in [\pi, 2\pi)$ and $k \in \{1, \dots, m\}$. We denote by $\lambda_1, \dots, \lambda_s$ the solutions of the equation $\alpha_k(\lambda) = \alpha$. Furthermore, we denote by l_j the multiplicity of the function α_k at the point λ_j, $j = 1, \dots, s$, and set

$$n_k(\alpha) = l_1 + \cdots + l_s, \quad n(\alpha) = n_1(\alpha) + \cdots + n_m(\alpha).$$

It is not difficult to see that the quantity $n(\alpha)$ equals the number of zeros (counting multiplicity) of the function $F(\alpha, \cdot)$ on the interval $(m - 1/2, m]$. Let $n^*(\alpha)$ be the number of zeros of the function $F(\alpha, \cdot)$ contained in the strip $m - 1/2 \leq \mathrm{Re}\, \lambda \leq m$. Then

(8.5.14) $$n(\alpha) \leq n^*(\alpha) \quad \text{for } \alpha \in [\pi, 2\pi).$$

We set $\omega_k = \alpha_k(m)$, $k = 1, \dots, m$, $\omega_{m+1} = 2\pi$ and assume that the following two properties have been established for all k, $k = 1, \dots, m$:

(i) If $\alpha \in (\omega_k, \omega_{k+1})$, then $n(\alpha) = n^*(\alpha) = k$.

(ii) The function α_k has order 1 or 2 at the point $\lambda = m$. In the first case $n(\omega_k) = n^*(\omega_k) = k$ and in the second case $n(\omega_k) = n^*(\omega_k) = k + 1$.

From this it can be seen that for any $\alpha \in [\pi, 2\pi)$ the function $F(\alpha, \cdot)$ has only real roots in the strip $m - 1/2 \leq \mathrm{Re}\, \lambda \leq m$. Since the range of the function α_k, $k = 1, \dots, m$, contains the interval $[\omega_k, 2\pi)$, it follows from the properties (i) and (ii) that the functions α_k have order 1 at each point of the interval $(m - 1/2, m)$, i.e., $\alpha_k'(\lambda) \neq 0$ for $\lambda \in (m - 1/2, m)$. Using the inequality $\alpha_k(m) < 2\pi = \lim\limits_{\lambda \to m - 1/2} \alpha_k(\lambda)$, we find that $\alpha_k'(\lambda) < 0$ for $\lambda \in (m - 1/2, m)$ from which we arrive at the assertion of the theorem.

We establish properties (i) and (ii). The inequality

(8.5.15) $$n(\alpha) \geq k \quad \text{for} \quad \alpha \in [\omega_k, \omega_{k+1}), \quad k = 1, \dots, m,$$

is obvious. By Theorems 8.1.4 and 8.1.5, the function $F(\alpha, \cdot)$ has no zeros on the line $\mathrm{Re}\, \lambda = m - 1/2$ and in the set $\{\lambda = m + it : t \in \mathbb{R} \backslash \{0\}\}$. Furthermore, by Lemma 8.1.3, the function $F(\alpha, \cdot)$ has no zeros in the strip $m - 1/2 \leq \mathrm{Re}\, \lambda \leq m$ if $|\mathrm{Im}\, \lambda| \geq R$, where R is a sufficiently large number independent of $\alpha \in [\pi, 2\pi)$. Therefore, as α increases from π to 2π, the value of the function $n^*(\cdot)$ may change only if α passes through a zero of the function $F(\cdot, m)$. Consequently, the function $n^*(\cdot)$ is constant on the intervals (ω_k, ω_{k+1}), $k = 1, \dots, m$. By Theorems 8.1.3 and

8.2.1, we have $n^*(\alpha) = 1$ for α close to π and hence for $\alpha \in [\omega_1, \omega_2)$. From (8.5.14), (8.5.15) it follows that $n(\alpha) = n^*(\alpha) = 1$ for $\alpha \in [\omega_1, \omega_2)$. This proves properties (i) and (ii) for $k = 1$.

Suppose that these properties have been proved for $k = l - 1$. We verify them for $k = l$, $l \geq 2$. By Lemma 8.5.4, we have $\partial_\alpha F(\alpha, m)|_{\alpha = \omega_l} \neq 0$. In addition, the function $F(\omega_l, \cdot)$ is not identically zero. Therefore, the Taylor series of the function F has the form

$$F(\alpha, \lambda) = c_0 (\alpha - \omega_l) + c_1 (\lambda - m)^s + \sum_{i \geq 1,\, i+j > 1} c_{i,j}^{(0)} (\alpha - \omega_l)^i (\lambda - m)^j$$

$$+ \sum_{i+j > s,\, j \geq s} c_{i,j}^{(1)} (\alpha - \omega_l)^i (\lambda - m)^j$$

in a neighborhood of the point $(\alpha, \lambda) = (\omega_l, m)$, where c_0, c_1, $c_{i,j}^{(0)}$, and $c_{i,j}^{(1)}$ are real numbers, $c_0 \neq 0$ and $c_1 \neq 0$ for $s \geq 1$. The number s is the multiplicity of the zero $\lambda = m$ for the function $F(\omega_l, \cdot)$ and, consequently, the multiplicity of the function α_l at the point $\lambda = m$. Since $n(\alpha) = n^*(\alpha) = l - 1$ for $\alpha \in (\omega_{l-1}, \omega_l)$, the roots of the function $F(\alpha, \cdot)$, $\alpha \in (\omega_l - \varepsilon, \omega_l)$ are distinct from the point $\lambda = m$ for small $\varepsilon > 0$ and it holds $\alpha_l(\lambda) \geq \omega_l$ for $\lambda \leq m$. From this we conclude that $n^*(\omega_l) = l - 1 + s$ and the quantity c_0/c_1 is positive for odd s and negative for even s. From the obvious inequality $n(\omega_l) \geq l - 1 + s$ and (8.5.14) we obtain

$$(8.5.16) \qquad n(\omega_l) = n^*(\omega_l) = l - 1 + s.$$

We consider the following cases:

1. $s = 1$. Then the point $\lambda = m$ is a simple zero of the function $F(\omega_l, \cdot)$. Since $c_1/c_0 > 0$, it follows that for angles α near ω_l, $\alpha > \omega_l$, this zero lies inside the strip $m - 1/2 \leq \mathrm{Re}\,\lambda \leq m$. Therefore, $n^*(\alpha) = l$ for $\alpha \in (\omega_l, \omega_{l+1})$.

2. $s = 2$. Then $c_0/c_1 < 0$. For $\alpha \in (\omega_l, \omega_l + \varepsilon)$, where ε is a small positive number, the equation $F(\alpha, \lambda) = 0$ has two real solutions

$$\lambda = m \pm |c_0/c_1 (\alpha - \omega_l)|^{1/2} + O(|\alpha - \omega_l|)$$

(see Vainberg and Trenogin [**254**, Ch.1,§2]). Hence, if $\alpha \in (\omega_l, \omega_l + \varepsilon)$, then inside the strip $m - 1/2 \leq \mathrm{Re}\,\lambda \leq m$ there is a unique simple zero of the function $F(\alpha, \cdot)$ close to m. Therefore, $n^*(\alpha) = l$ for $\alpha \in (\omega_l, \omega_{l+1})$.

3. $s \geq 3$. If $\alpha \in (\omega_l - \varepsilon, \omega_l)$, where ε is a small positive number, then the equation $F(\alpha, \lambda) = 0$ has the solution

$$\lambda = m + e^{\pi i (s-1)/s} |c_0/c_1 (\alpha - \omega_l)|^{1/s} + O(|\alpha - \omega_l|^{2/s})$$

for odd s and

$$\lambda = m + e^{\pi i (s+2)/2s} |c_0/c_1 (\alpha - \omega_l)|^{1/s} + O(|\alpha - \omega_l|^{2/s})$$

for even s. In both cases the real part of $\lambda - m$ is negative and the imaginary part is nonzero for $\alpha \in (\omega_l - \varepsilon, \omega_l)$. Consequently, if $\alpha \in (\omega_l - \varepsilon, \omega_l)$, the function $F(\alpha, \cdot)$ has nonreal roots in the strip $m - 1/2 \leq \mathrm{Re}\,\lambda \leq m$. This contradicts the equality $n(\alpha) = n^*(\alpha)$ for $\alpha \in (\omega_{l-1}, \omega_l)$. This means that case 3 is impossible.

Thus, properties (i) and (ii) are proved for $k = l$. The proof of the theorem is complete. ∎

As a consequence of Theorem 8.5.1, we obtain the following results.

COROLLARY 8.5.1. *Let $n(\alpha)$, $\alpha \in (\pi, 2\pi)$, be the number of eigenvalues of the pencil $\mathcal{L}$ located in the strip $m - 1/2 \leq \operatorname{Re}\lambda \leq m$. Then $n(\alpha) = k$ if $\alpha \in (\alpha_k(m), \alpha_{k+1}(m)]$, $k = 1, \ldots, m - 1$, and $n(\alpha) = m$ if $\alpha \in (\alpha_m(m), 2\pi)$.*

COROLLARY 8.5.2. *The number ε_0 in Theorem 8.3.1 can be taken as $2\pi - \alpha_m(m)$. Then the inequalities $m > \mu_1(\alpha) > \cdots > \mu_m(\alpha) > m - 1/2$ hold for $\alpha \in (\alpha_m(m), 2\pi)$.*

8.6. The Dirichlet problem for a second order system

Here we consider the operator pencil corresponding to the Dirichlet problem for a $\ell \times \ell$ system of second order elliptic equations in an angle. We show, in particular, that the strip $|\operatorname{Re}\lambda| \leq 1$ does not contain eigenvalues of this pencil if the opening of the angle is less than π. For opening greater than π (but less than 2π) we prove that both strips $-1 \leq \operatorname{Re}\lambda < 0$ and $0 < \operatorname{Re}\lambda \leq 1$ contain exactly ℓ eigenvalues.

8.6.1. Basic properties of the pencil generated by the Dirichlet problem.
Let $\mathcal{K}$ be the angle $\{(x_1, x_2) \in \mathbb{R}^2 : 0 < r < \infty, 0 < \varphi < \alpha\}$, where r, φ are the polar coordinates in $\mathbb{R}^2$. Furthermore, let L be the differential operator

$$(8.6.1) \qquad L(\partial_{x_1}, \partial_{x_2}) = \sum_{i,j=1}^{2} A_{i,j} \, \partial_{x_i} \, \partial_{x_j} \,,$$

where $A_{i,j}$ are $\ell \times \ell$-matrices and $A_{i,j} = A_{j,i}^*$. We assume that there exists a positive constant c_0 such that

$$(8.6.2) \qquad \sum_{i,j=1}^{2} \left(A_{i,j} f^{(j)}, f^{(i)}\right)_{\mathbb{C}^\ell} \geq c_0 \left(|f^{(1)}|_{\mathbb{C}^\ell}^2 + |f^{(2)}|_{\mathbb{C}^\ell}^2\right)$$

for all $f^{(1)}, f^{(2)} \in \mathbb{C}^\ell$. It is clear that this condition implies the ellipticity of L.

We seek all solutions of the Dirichlet problem

$$(8.6.3) \qquad L(\partial_{x_1}, \partial_{x_2}) \, U = 0 \quad \text{in } \mathcal{K},$$

$$(8.6.4) \qquad U = 0 \quad \text{for } \varphi = 0 \text{ and } \varphi = \alpha,$$

which have the form

$$(8.6.5) \qquad U(x_1, x_2) = r^\lambda \sum_{k=0}^{s} \frac{1}{k!} \, (\log r)^k \, u_{s-k}(\varphi),$$

where $\lambda \in \mathbb{C}$, $u_k \in W_2^2((0, \alpha))^\ell$.

Inserting $U = r^\lambda u(\varphi)$ into (8.6.3), we obtain the differential equation

$$\mathcal{L}(\varphi, \partial_\varphi, \lambda) \, u(\varphi) = 0 \quad \text{for } 0 < \varphi < \alpha,$$

where $\mathcal{L}(\varphi, \partial_\varphi, \lambda) : u \to r^{2-\lambda} L(\partial_{x_1}, \partial_{x_2}) \, (r^\lambda u)$ is a differential operator quadratically depending on the parameter λ. We consider the differential operator $\mathcal{L}(\lambda) = \mathcal{L}(\varphi, \partial_\varphi, \lambda)$ for fixed $\lambda \in \mathbb{C}$ as a continuous mapping

$$(8.6.6) \qquad W_2^2((0, \alpha))^\ell \cap \overset{\circ}{W}{}_2^1((0, \alpha))^\ell \to L_2((0, \alpha))^\ell.$$

THEOREM 8.6.1. 1) *The operator $\mathcal{L}(\lambda)$ realizes an isomorphism (8.6.6) for all $\lambda \in \mathbb{C}$, except a countable set of isolated points, the eigenvalues of the pencil $\mathcal{L}$. The eigenvalues have finite algebraic multiplicities and are contained, with the possible exception of finitely many, in a double angle with opening less than π containing the real axis.*

2) *The number λ is an eigenvalue of the pencil $\mathcal{L}$ if and only if $-\bar{\lambda}$ is an eigenvalue of $\mathcal{L}$. The geometric, partial, and algebraic multiplicities of these eigenvalues coincide.*

3) *There exist no purely imaginary eigenvalues of the pencil $\mathcal{L}$.*

Proof: We prove first that the operator $\mathcal{L}(\lambda)$ is invertible for purely imaginary λ. Let $U = r^{it}u(\varphi)$, where t is an arbitrary real number and $u \in C_0^\infty((0,\alpha))^\ell$. Integrating by parts, we get (because of the cancelling of the integrals on the arcs $|x| = 1$ and $|x| = 2$)

$$-\int_{\substack{\mathcal{K} \\ 1<|x|<2}} \left(L(\partial_x)U, U\right)_{\mathbb{C}^\ell} dx = \int_{\substack{\mathcal{K} \\ 1<|x|<2}} \sum_{i,j=1}^{2} \left(A_{i,j}\, \partial_{x_j} U,\, \partial_{x_i} U\right)_{\mathbb{C}^\ell} dx.$$

By (8.6.2), the right-hand side of the last equality can be estimated from below by

$$\int_{\substack{\mathcal{K} \\ 1<|x|<2}} \sum_{j=1}^{2} \left|\partial_{x_j} U\right|^2 dx.$$

Consequently, we obtain

$$\left|\int_0^\alpha \left(\mathcal{L}(\varphi,\partial_\varphi, it)u,\, u\right)_{\mathbb{C}^\ell} d\varphi\right| \geq c\, \|u\|^2_{W_2^1((0,\alpha))^\ell}.$$

This implies that the operator $\mathcal{L}(it) : \overset{\circ}{W}{}_2^1((0,\alpha))^\ell \to W_2^{-1}((0,\alpha))^\ell$ and, therefore, also the operator (8.6.6) are invertible. Applying Theorem 1.1.1, we get assertions 1) and 3).

Now let $U = r^\lambda u(\varphi)$, $V = r^{-\bar{\lambda}}v(\varphi)$, where λ is an arbitrary complex number and u, v are arbitrary vector functions from $C_0^\infty((0,\alpha))^\ell$. Then, analogously to the first part of the proof, integration by parts gives

$$\int_{\substack{\mathcal{K} \\ 1<|x|<2}} \left(LU, V\right)_{\mathbb{C}^\ell} dx = \int_{\substack{\mathcal{K} \\ 1<|x|<2}} \left(U, LV\right)_{\mathbb{C}^\ell} dx.$$

This implies

$$\int_0^\alpha \left(\mathcal{L}(\varphi,\partial_\varphi,\lambda)\, u(\varphi),\, v(\varphi)\right)_{\mathbb{C}^\ell} d\varphi = \int_0^\alpha \left(u(\varphi),\, \mathcal{L}(\varphi,\partial_\varphi, -\bar{\lambda})\, v(\varphi)\right)_{\mathbb{C}^\ell} d\varphi,$$

Hence the operator $\mathcal{L}(\varphi,\partial_\varphi,\lambda)$ is adjoint to $\mathcal{L}(\varphi,\partial_\varphi, -\bar{\lambda})$. Using Theorem 1.1.7, we obtain assertion 2). ∎

By Theorem 1.1.5, the function (8.6.5) is a solution of problem (8.6.3), (8.6.4) if and only if λ is an eigenvalue and $u_0, u_1, \ldots, u_s$ is a Jordan chain of the pencil $\mathcal{L}$ corresponding to this eigenvalue.

8.6.2. The cases $\alpha = \pi$ and $\alpha = 2\pi$. If $\alpha = \pi$, then the set of all solutions of problem (8.6.3), (8.6.4) having the form (8.6.5) for $\operatorname{Re}\lambda \geq 0$ is exhausted by homogeneous vector-polynomials. Therefore, the spectrum of the pencil $\mathcal{L}$ in the half-plane $\operatorname{Re}\lambda \geq 0$ consists of the eigenvalues $\lambda_k = k$, $k = 1,2,\ldots$, with geometric and algebraic multiplicities equal to ℓ. In this case generalized eigenfunctions do not exist. Due to the second part of Theorem 8.6.1, the spectrum of $\mathcal{L}$ in the half-plane $\operatorname{Re}\lambda \leq 0$ consists of the eigenvalues $\lambda_k = k$, $k = -1, -2, \ldots$, with geometric and algebraic multiplicities equal to ℓ.

In the case $\alpha = 2\pi$ it can be shown in the same way as in the preceding section that the spectrum of the operator pencil $\mathcal{L}$ consists of the eigenvalues $\pm k/2$, $k = 1, 2, \ldots$, with geometric and algebraic multiplicities equal to ℓ. As in the case $\alpha = \pi$, there exist no generalized eigenfunctions.

8.6.3. Eigenvalues of the pencil $\mathcal{L}$ in the strip $-1 \leq \operatorname{Re}\lambda \leq 1$. Now we consider the pencil $\mathcal{L}$ in the case $\alpha \neq \pi$, $\alpha \neq 2\pi$.

LEMMA 8.6.1. *If $\alpha \neq \pi$ and $\alpha \neq 2\pi$, then the line $\operatorname{Re}\lambda = 1$ does not contain eigenvalues of the pencil $\mathcal{L}$.*

Proof: We set

$$B_{i,j} = \frac{1}{2}\left(A_{i,j} + A_{i,j}^*\right).$$

Obviously, $B_{i,j} = B_{j,i} = B_{i,j}^*$. Suppose that the function $U = r^\lambda u(\varphi)$, where $\operatorname{Re}\lambda = 1$ and $u \in W_2^2((0,\alpha))^\ell$, is a solution of problem (8.6.3), (8.6.4). Then

$$(8.6.7) \qquad 0 = \int\limits_{\substack{\mathcal{K}\\1<|x|<2}} \left(L(D_x)U, \Delta U\right)_{\mathbb{C}^\ell} dx = \sum_{i,j=1}^{2} \int\limits_{\substack{\mathcal{K}\\1<|x|<2}} \left(\partial_{x_i} B_{i,j}\partial_{x_j} U, \Delta U\right)_{\mathbb{C}^\ell} dx$$

$$= -\sum_{i,j=1}^{2} \int\limits_{\substack{\mathcal{K}\\1<|x|<2}} \left(B_{i,j}\partial_{x_j} U, \partial_{x_i}\Delta U\right)_{\mathbb{C}^\ell} dx + \sum_{i,j=1}^{2} \int\limits_{1}^{2} \nu_i \left(B_{i,j}\partial_{x_j} U, \Delta U\right)_{\mathbb{C}^\ell} \Big|_{\varphi=0}^{\varphi=\alpha} dr.$$

where (ν_1, ν_2) is the exterior normal to the rays $\varphi = 0$ and $\varphi = \alpha$. (Due to the condition $\operatorname{Re}\lambda = 1$, the integrals over the arcs $r = 1$ and $r = 2$ are cancelled.) Integrating by parts in the first sum on the right of (8.6.7), we get

$$(8.6.8) \qquad -\sum_{i,j,k=1}^{2} \int\limits_{\substack{\mathcal{K}\\1<|x|<2}} \left(B_{i,j}\partial_{x_j}\partial_{x_k} U, \partial_{x_i}\partial_{x_k} U\right)_{\mathbb{C}^\ell} dx$$

$$= \sum_{i,j,k=1}^{2} \int\limits_{1}^{2} \left(\nu_i \left(B_{i,j}\partial_{x_j} U, \partial_{x_k}^2 U\right)_{\mathbb{C}^\ell} - \nu_k \left(B_{i,j}\partial_{x_j} U, \partial_{x_i}\partial_{x_k} U\right)_{\mathbb{C}^\ell}\right)\Big|_{\varphi=0}^{\varphi=\alpha} dr.$$

We show that the right-hand side of the last equality vanishes for $\varphi = 0$. Since $\nu_1 = 0$, $\nu_2 = -1$ for $\varphi = 0$ and $\partial_{x_1} U\big|_{\varphi=0} = \partial_{x_1}^2 U\big|_{\varphi=0} = 0$, the right-hand side of (8.6.8) (for $\varphi = 0$) is equal to

$$\int\limits_{1}^{2} \left(B_{1,2}\partial_{x_2} U, \partial_{x_1}\partial_{x_2} U\right)_{\mathbb{C}^\ell}\big|_{\varphi=0} dx_1$$

and, consequently, also to

$$(8.6.9) \qquad \frac{1}{2}\int\limits_{1}^{2} \partial_{x_1}\left(B_{1,2}\partial_{x_2} U, \partial_{x_2} U\right)_{\mathbb{C}^\ell}\big|_{\varphi=0} dx_1.$$

Here we made use of the fact that the left-hand side of (8.6.8) is real by (8.6.2). Since $\partial_{x_2} U\big|_{\varphi=0} = r^{\lambda-1} u'(0)$ and $\mathrm{Re}\,\lambda = 1$, the scalar product of the vector functions $B_{1,2}\partial_{x_2} U$ and $\partial_{x_2} U$ in $\mathbb{C}^\ell$ is equal to a constant for $\varphi = 0$. Therefore, the expression (8.6.9) vanishes. Thus, we have shown that the right-hand side of (8.6.8) vanishes for $\varphi = 0$. The same can be shown for $\varphi = \alpha$. Consequently,

$$\sum_{i,j,k=1}^{2} \int_{\substack{\mathcal{K} \\ 1<|x|<2}} \big(B_{i,j}\partial_{x_j}\partial_{x_k} U,\, \partial_{x_i}\partial_{x_k} U\big)_{\mathbb{C}^\ell}\, dx = 0.$$

Using (8.6.2), we conclude that $\partial_{x_j}\partial_{x_k} U = 0$ for $j,k = 1,2$. Hence U has the form $U = x_1 f^{(1)} + x_2 f^{(2)}$ with certain $f^{(1)}, f^{(2)} \in \mathbb{C}^\ell$. However, the equalities $U(0) = U(\alpha) = 0$ exclude this representation for $\alpha \neq \pi$, $\alpha \neq 2\pi$. The lemma is proved. ∎

THEOREM 8.6.2. 1) *If $\alpha \in (0,\pi)$, then the strip $|\mathrm{Re}\,\lambda| \leq 1$ does not contain eigenvalues of the operator pencil $\mathcal{L}$.*

2) *If $\alpha \in (\pi, 2\pi)$, then every of the strips $0 < \mathrm{Re}\,\lambda \leq 1$ and $-1 \leq \mathrm{Re}\,\lambda < 0$ contains exactly ℓ eigenvalues (counting multiplicity) of $\mathcal{L}$.*

Proof: By the second part of Theorem 8.6.1, it suffices to prove the statement of the theorem for the strip $0 < \mathrm{Re}\,\lambda \leq 1$. We introduce the differential operator

$$L_t(\partial_{x_1}, \partial_{x_2}) = (1-t)\,\Delta\,I + t\,L(\partial_{x_1}, \partial_{x_2}),$$

where $t \in [0,1]$ and I is the identity matrix. Note that condition (8.6.2) is also true for the coefficients of the operator L_t. Let $\mathcal{L}_t$ denote the operator pencil generated by the Dirichlet problem for the differential operator L_t.

We consider the case $\alpha \in (0,\pi)$. By Theorem 8.6.1 and Lemma 8.6.1, the lines $\mathrm{Re}\,\lambda = 0$ and $\mathrm{Re}\,\lambda = 1$ do not contain eigenvalues of the pencil $\mathcal{L}_t$. Furthermore, it can be easily verified that for large c there are no eigenvalues of the pencil $\mathcal{L}_t$ in the set $\{\lambda : 0 \leq \mathrm{Re}\,\lambda \leq 1,\ |\lambda| > c\}$ (cf. Lemma 8.1.3). Using Corollary 1.1.2, we conclude from this that the operator pencils $\mathcal{L}_t$, $0 \leq t \leq 1$, have the same number of eigenvalues (counting multiplicity) in the strip $0 \leq \mathrm{Re}\,\lambda \leq 1$. Since the operator pencil $\mathcal{L}_0$ has no eigenvalues in this strip (see Section 2.1), the same is true for the pencil $\mathcal{L}_1 = \mathcal{L}$.

Now let $\alpha \in (\pi, 2\pi)$. By the same arguments as above, we conclude that the number of eigenvalues of the operator pencil $\mathcal{L}_t$ in the strip $0 < \mathrm{Re}\,\lambda < 1$ is independent of t. By what has been shown in Section 2.1, the operator pencil $\mathcal{L}_0$ has the only eigenvalue π/α of multiplicity ℓ in this strip. This implies assertion 2) of the theorem. ∎

REMARK 8.6.1. In the case $\ell = 1$ it follows from Theorem 8.1.4 that the strip $|\mathrm{Re}\,\lambda| \leq 1/2$ does not contain eigenvalues of the pencil $\mathcal{L}$ if $\alpha < 2\pi$. Note that this result is also true for arbitrary ℓ. We will prove the corresponding fact for the Dirichlet problem in a n-dimensional cone, $n \geq 2$, in Section 11.1.

8.7. Applications

In this section we give some applications of the results obtained in Sections 8.1–8.5 to the Dirichlet problem for elliptic equations in plane domains with corners and in n-dimensional domains with smooth non-intersecting $(n-2)$-dimensional edges.

8.7.1. Two-dimensional case. Let $\mathcal{G}$ be a bounded open polygon in $\mathbb{R}^2$. By $\mathcal{S}$ we denote the set of its vertices $x^{(\tau)}$, $\tau = 1, \ldots, d$, and by α_τ the opening of the interior angle at the corner $x^{(\tau)}$.

We consider the Dirichlet problem

$$(8.7.1) \qquad L(x, \partial_x)\, U = F \ \text{ in } \mathcal{G}, \quad \partial_\nu^k U = 0 \ \text{ on } \partial\mathcal{G} \backslash \mathcal{S}, \ k = 0, 1, \ldots, m-1,$$

where L is an elliptic differential operator of order $2m$ with infinitely differentiable coefficients on $\overline{\mathcal{G}}$. The coefficients of the principal part of L are assumed to be real.

1. Solvability in weighted Sobolev and Hölder spaces. For every $\tau = 1, \ldots, d$ we introduce local polar coordinates r, φ in a neighborhood of $x^{(\tau)}$ such that φ takes the values 0 and α_τ on the tangents to $\partial\mathcal{G}$ at $x^{(\tau)}$. Let $\mathcal{L}_\tau(\lambda)$ denote the operator pencil

$$W_2^{2m}((0,\alpha_\tau)) \cap \overset{\circ}{W}_2^{m}((0,\alpha_\tau)) \ni u \to r^{2m-\lambda} L^\circ(x^{(\tau)}, \partial_x)\, r^\lambda u(\varphi) \in L_2((0,\alpha_\tau)),$$

where $L^\circ(x^{(\tau)}, \partial_x)$ is the principal part of L with coefficients frozen at $x^{(\tau)}$. The eigenvalues of $\mathcal{L}_\tau$ are roots of the transcendental equation

$$(8.7.2) \qquad\qquad\qquad \det \Psi_\tau(\lambda) = 0,$$

where Ψ_τ is the matrix introduced in Section 8.1 for the operator $L^\circ(x^{(\tau)}, \partial_x)$ and the angle $0 < r < \infty$, $0 < \varphi < \alpha_\tau$.

Let $V_{p,\vec{\beta}}^l(\mathcal{G})$ and $N_{\vec{\beta}}^{l,\sigma}(\mathcal{G})$ be the weighted spaces introduced in Section 1.4. Here l is an integer, $l \geq 0$, $p \in (1,\infty)$, $\sigma \in (0,1)$, and $\vec{\beta} = (\beta_1, \ldots, \beta_d)$ is a real d-tuple. Furthermore, by $V_{p,\vec{\beta}}^{l-1/p}(\partial\mathcal{G})$ and $N_{\vec{\beta}}^{l,\sigma}(\partial\mathcal{G})$ we denote the corresponding trace spaces.

By Theorems 1.4.1 and 1.4.5, the operators

$$(8.7.3) \qquad V_{p,\vec{\beta}}^l(\mathcal{G}) \ni U \to \left(LU, U|_{\partial\mathcal{G}\backslash\mathcal{S}}, \ldots, D_\nu^{m-1} U|_{\partial\mathcal{G}\backslash\mathcal{S}} \right)$$

$$\in V_{p,\vec{\beta}}^{l-2m}(\mathcal{G}) \times \prod_{k=0}^{m-1} V_{p,\vec{\beta}}^{l-k-1/2}(\partial\mathcal{G})$$

and

$$(8.7.4) \qquad N_{\vec{\beta}}^{l,\sigma}(\mathcal{G}) \ni U \to \left(LU, U|_{\partial\mathcal{G}\backslash\mathcal{S}}, \ldots, D_\nu^{m-1} U|_{\partial\mathcal{G}\backslash\mathcal{S}} \right)$$

$$\in N_{\vec{\beta}}^{l-2m,\sigma}(\mathcal{G}) \times \prod_{k=0}^{m-1} N_{\vec{\beta}}^{l-k,\sigma}(\partial\mathcal{G})$$

are Fredholm if for $\tau = 1, \ldots, d$, there are no eigenvalues of $\mathcal{L}_\tau$ on the lines $\operatorname{Re}\lambda = l - \beta_\tau - n/p$ and $\operatorname{Re}\lambda = l + \sigma - \beta_\tau$, respectively.

Let problem (8.7.1) be uniquely solvable in $\overset{\circ}{W}_2^{\,m}(\mathcal{G})$ for all $F \in W_2^{-m}(\mathcal{G})$. Then, as is well known and essentially follows from Theorems 1.4.3 and 1.4.5, the operators (8.7.3) and (8.7.4) are isomorphic if and only if the numbers $l - \beta_\tau - 2/p$ and $l + \sigma - \beta_\tau$ lie in the energy strip of $\mathcal{L}_\tau$:

$$\{\lambda \in \mathbb{C} : |\operatorname{Re}\lambda - m + 1| < \mu_\tau - m + 1\}.$$

(Here μ_τ is the minimum real part of roots of (8.7.2) situated in the half-plane $\operatorname{Re}\lambda = m - 1$.) By Theorem 8.1.4, $\mu_\tau > m - 1/2$ that leads to the bijectivity of operator (8.7.3) if

$$m - 3/2 \leq l - \beta_\tau - 2/p \leq m - 1/2 \quad \text{for } \tau = 1, \ldots, d,$$

as well as the bijectivity of (8.7.4) if

$$m - 3/2 \leq l + \sigma - \beta_\tau \leq m - 1/2 \quad \text{for } \tau = 1, \ldots, d.$$

In the case when all values α_τ are less than π, the above conditions can be replaced by the inequalities $m - 2 \leq l - \beta_\tau - 2/p \leq m$ and $m - 2 \leq l + \sigma - \beta_\tau \leq m$, respectively (see the second part of Theorem 8.4.1). We add that, according to Theorem 8.5.1, μ_τ changes monotonically from $m - 1$ to $m - 1/2$ with respect to the angle $\mathcal{K}_\tau$, $\alpha_\tau \in [\pi, 2\pi]$.

2. Regularity assertions and asymptotics of the solution. The following result stems from Theorems 1.4.3 and 1.4.5.

THEOREM 8.7.1. *Let* $U \in \overset{\circ}{W}{}_2^m(\mathcal{G})$ *be a variational solution of* (8.7.1), *where* $F \in V_{p,\vec{\beta}}^{l-2m}(\mathcal{G})$, $l \geq 2m$. *If*

$$m - 1 \leq l - \beta_\tau - 2/p < \mu_\tau \quad \text{for } \tau = 1, \ldots, d,$$

then $U \in V_{p,\vec{\beta}}^{l}(\mathcal{G})$. *Analogously, if* $F \in N_{\vec{\beta}}^{l-2m,\sigma}(\mathcal{G})$ *and*

$$m - 1 \leq l + \sigma - \beta_\tau < \mu_\tau \quad \text{for } \tau = 1, \ldots, d,$$

then $U \in N_{\vec{\beta}}^{l,\sigma}(\mathcal{G})$.

Replacing μ_τ by its lower bounds mentioned above, we immediately arrive at explicit sufficient conditions for $U \in V_{p,\vec{\beta}}^{l}(\mathcal{G})$ and $U \in N_{\vec{\beta}}^{l,\sigma}(\mathcal{G})$.

As a consequence of Theorems 1.4.4 and 8.5.1 we have

THEOREM 8.7.2. *If* $F \in V_{p,\vec{\beta}}^{l-2m}(\mathcal{G})$, $m - 1/2 < l - \beta_\tau - 2/p < m$, *and* $\alpha_\tau > \pi$, *then in a neighborhood* $\mathcal{U}$ *of* $x^{(\tau)}$ *the variational solution* U *admits the asymptotic representation:*

$$(8.7.5) \qquad U = \sum_{j=1}^{J} c_j \, r^{\lambda_j} \, u_j(\varphi) + W,$$

where $J \leq m$, λ_j *are real roots of* (8.7.2) *in the interval* $(m-1/2, l-\beta_\tau-2/p)$, u_j *are infinitely differentiable functions on* $(0, \alpha_\tau)$, *and* $\eta W \in V_{p,\vec{\beta}}^{l}(\mathcal{G})$ *for all* $\eta \in C_0^\infty(\mathcal{U})$.

In contrast to the general asymptotics given in Theorem 1.4.4, the above decomposition contains only real powers of r and no logarithmic terms.

8.7.2. Higher-dimensional case. Now let $\mathcal{G}$ be a bounded domain in $\mathbb{R}^n$ whose boundary consists of smooth $(n-1)$-dimensional faces $\Gamma_1, \ldots, \Gamma_d$ and smooth non-intersecting $(n-2)$-dimensional edges $M_1, \ldots, M_{d-1}$. We suppose that for each point $\xi \in \mathcal{S} = M_1 \cup \cdots \cup M_{d-1}$ there exist a neighborhood $\mathcal{U}(\xi)$ and a diffeomorphism $\kappa : \mathcal{U}(\xi) \to \mathbb{R}^n$ of class C^∞ such that $\kappa(\mathcal{U} \cap \mathcal{G})$ is the intersection of the unit ball with a dihedron $\mathcal{D}_\xi = K_\xi \times \mathbb{R}^{n-2}$, where K_ξ is a plane angle with vertex at the origin and opening $\alpha_\xi \in (0, 2\pi)$.

We consider the Dirichlet problem (8.7.1), where L is again a $2m$ order elliptic differential operator with infinitely differential coefficients on $\overline{\mathcal{G}}$. The coefficients of the principal part of L are assumed to be real.

For every point $\xi \in M_j$, $j = 1, \ldots, d-1$, we define an operator pencil $\mathcal{L}_\xi$ as follows. Let Γ_{k+}, Γ_{k-} be the faces of $\partial\mathcal{G}$ which are adjoint to ξ. There exist $(n-1)$-dimensional half-spaces $\Gamma^\pm$ tangent to $\Gamma_{k\pm}$ in ξ such that $\overline{\Gamma^+} \cap \overline{\Gamma^-}$ is a $(n-2)$-dimensional subspace tangent to the edge M_j in ξ. We introduce new Cartesian coordinates $(y, z) = (y_1, y_2, z_1, \ldots, z_{n-2})$ such that ξ is the origin of the y, z-coordinate system and Γ^-, Γ^+ coincide with the sets $\{(y, z) : y_1 > 0, y_2 = 0\}$ and $\{(y, z) : \arg(y_1 + iy_2) = \alpha_\xi\}$, respectively. By $L_\xi^\circ(\partial_y, \partial_z)$ we denote the operator which arises from the principal part of L after freezing the coefficients at ξ and passing to the coordinates y, z. Then $\mathcal{L}_\xi(\lambda)$ denotes the operator pencil

$$W_2^{2m}((0, \alpha_\xi)) \cap \overset{\circ}{W}_2^m((0, \alpha_\xi)) \ni u \to r^{2m-\lambda} L_\xi^\circ(\partial_y, 0) \, r^\lambda u(\varphi) \in L_2((0, \alpha_\xi)).$$

Here r, φ are the polar coordinates in the y_1, y_2-plane. It was shown in Section 8.1 that the eigenvalues of the pencil $\mathcal{L}_\xi$ are solutions of the transcendental equation

$$(8.7.6) \qquad\qquad \det \Psi(\xi; \lambda) = 0,$$

where $\Psi(\xi; \lambda)$ is the square matrix introduced in Lemma 8.1.1 (with $\mathcal{L} = \mathcal{L}_\xi$, $\alpha = \alpha_\xi$). By analogy with our previous notation, let us denote by $\mu(\xi)$ the minimum real part of roots of (8.7.6) situated in the half-plane $\operatorname{Re}\lambda = m - 1$.

1. **Solvability in weighted Sobolev spaces.** We introduce the weighted Sobolev space $V_{p,\beta}^l(\mathcal{G})$, where the weight function is a power of the distance to the edges. Let r be the regularized distance to the set $\mathcal{S}$, i.e., $r = r(x)$ is a smooth positive function on $\overline{\mathcal{G}} \backslash \mathcal{S}$ which coincides with the distance to $\mathcal{S}$ if x is situated in a neighborhood of $\mathcal{S}$. Then $V_{p,\beta}^l(\mathcal{G})$ is the closure of the set $C_0^\infty(\overline{\mathcal{G}} \backslash \mathcal{S})$ with respect to the norm

$$\|U\|_{V_{p,\beta}^l(\mathcal{G})} = \left(\int_{\mathcal{G}} \sum_{|\alpha| \le l} r^{p(\beta(\xi) - l + |\alpha|)} |D^\alpha U(x)|^p \, dx \right)^{1/p}.$$

Here l is a nonnegative integer, $1 < p < \infty$, β is a smooth real-valued function on $\mathcal{S}$, and ξ is a nearest point to x on $\mathcal{S}$.

Applying Theorem 10.1 of Maz'ya and Plamenevskiĭ's paper [183] to the Dirichlet problem (8.7.1), we get

THEOREM 8.7.3. *Let $l \ge 2m, 2m+1, \ldots$. The operator*

$$(8.7.7) \qquad V_{p,\beta}^l(\mathcal{G}) \ni U \to \left(LU, U|_{\partial\mathcal{G}\backslash\mathcal{S}}, \ldots, D_\nu^{m-1} U|_{\partial\mathcal{G}\backslash\mathcal{S}}\right)$$

$$\in V_{p,\beta}^{l-2m}(\mathcal{G}) \times \prod_{k=0}^{m-1} V_{p\beta}^{l-k-1/2}(\partial\mathcal{G})$$

is Fredholm if and only if for all $\xi \in \mathcal{S}$

$$(8.7.8) \qquad 2m - 2 - \mu(\xi) < l - \beta(\xi) - 2/p < \mu(\xi) \quad \text{for all } \xi \in \mathcal{S}.$$

If (8.7.8) holds and for an arbitrary $F \in W_2^{-m}(\mathcal{G})$ there exists a unique variational solution $U \in \overset{\circ}{W}_2^m(\mathcal{G})$ of (8.7.1), then the operator in question realizes an isomorphism.

This result complemented by the inequality $\mu(\xi) > 1/2$ guarantees that the operator (8.7.7) is isomorphic if

$$m - 3/2 \leq l - \beta(\xi) - 2/p \leq m - 1/2 \quad \text{for all } \xi \in \mathcal{S}$$

and if for an arbitrary $F \in W_2^{-m}(\mathcal{G})$ there exists a unique variational solution $U \in \overset{\circ}{W}_2^m(\mathcal{G})$ of (8.7.1). If the angle α_ξ is less than π, then $\mu(\xi) > 1$ and therefore the weaker inequalities $m - 2 \leq l - \beta(\xi) - 2/p \leq m$ are sufficient.

2. Regularity assertions and asymptotics of the solution. For the variational solution $U \in \overset{\circ}{W}_2^m(\mathcal{G})$ of problem (8.7.1) we get the same regularity result as in the plane case (see [**183**, Th.10.3]):

THEOREM 8.7.4. *Let* $U \in \overset{\circ}{W}_2^m(\mathcal{G})$ *solve problem* (8.7.1), *where* $F \in V_{p,\beta}^{l-2m}(\mathcal{G})$, $l \geq 2m$. *If* (8.7.8) *holds, then* $U \in V_{p,\beta}^l(\mathcal{G})$.

From this result together with Theorem 8.1.4 we conclude, in particular, that U belongs to the space $V_{p,\beta}^l(\mathcal{G})$ if $F \in V_{p,\beta}^{l-2m}(\mathcal{G})$ and $m-1 \leq l - \beta - 2/p < m + \varepsilon - \frac{1}{2}$. Here ε is a sufficiently small number. If the interior angles at the edges are less than π, then for this statement it is sufficient that $m - 1 \leq l - \beta - 2/p < m + \varepsilon$. If the angle α_ξ is greater than π for a certain edge point ξ, then in a neighborhood of this point an asymptotic decomposition for the solution similar as (8.7.5) holds (see, e.g., the papers of Dauge [**41**], Maz'ya and Roßmann [**192**] or the book of Nazarov and Plamenevskiĭ [**207**, Ch.11]).

3. Solvability in weighted Hölder spaces. Let $N_\beta^{l,\sigma}(\mathcal{G})$ be the weighted Hölder space with the norm

$$\|U\|_{N_\beta^{l,\sigma}(\mathcal{G})} = \sum_{|\alpha| \leq l} \sup_{x \in \mathcal{G}} r^{\beta(\xi) - l - \sigma + |\alpha|} |D^\alpha U(x)|$$

$$+ \sum_{|\alpha| = l} \sup_{x, x' \in \mathcal{G}} |x - x'|^{-\sigma} \left| r(x)^{\beta(\xi)} D^\alpha U(x) - r(x')^{\beta(\xi')} D^\alpha U(x') \right|.$$

Here l is a nonnegative integer, $\sigma \in (0,1)$, β is a smooth real-valued function on $\mathcal{S}$, and ξ, ξ' are nearest points on $\mathcal{S}$ to x and x', respectively.

Essentially the same conditions as in the case of weighted Sobolev spaces guarantee the Fredholm property of the operator of problem (8.7.1) in the class of the just introduced weighted Hölder spaces. More precisely, the following assertion is true (see Maz'ya and Plamenevskiĭ [**185**, Th.3.1]).

THEOREM 8.7.5. *The operator*

$$(8.7.9) \qquad N_\beta^{l,\sigma}(\mathcal{G}) \ni U \to \left(LU, U|_{\partial\mathcal{G}\setminus\mathcal{S}}, \ldots, D_\nu^{m-1} U|_{\partial\mathcal{G}\setminus\mathcal{S}} \right)$$

$$\in N_\beta^{l-2m,\sigma}(\mathcal{G}) \times \prod_{k=0}^{m-1} N_\beta^{l-k,\sigma}(\partial\mathcal{G})$$

is Fredholm if and only if

$$(8.7.10) \qquad 2m - 2 - \mu(\xi) < l + \sigma - \beta(\xi) < \mu(\xi) \quad \text{for all } \xi \in \mathcal{S}.$$

If (8.7.10) *holds and for an arbitrary* $F \in W_2^{-m}(\mathcal{G})$ *there exists a unique variational solution* $U \in \overset{\circ}{W}_2^m(\mathcal{G})$ *of* (8.7.1), *then the operator in question realizes an isomorphism.*

In particular, the operator (8.7.9) is Fredholm if $m-3/2 \le l+\sigma-\beta(\xi) \le m-1/2$ for all $\xi \in \mathcal{S}$. If $\alpha_\xi < \pi$, then even the condition $m - 2 \le l + \sigma - \beta(\xi) \le m$ is sufficient for the Fredholm property.

8.8. Notes

Section 8.1. Explicit equations for the eigenvalues of operator pencils generated by boundary value problems for general elliptic equations in a plane angle were obtained by Maz'ya and Plamenevskiĭ [**184**], Kozlov [**120**], Costabel and Dauge [**29**]. In [**184**] and [**120**] such equations were derived for boundary value problems for $2m$ order elliptic equations, while [**29**] is concerned with general boundary value problems for elliptic (in the sense of Douglis-Nirenberg) systems.

Sections 8.2.–8.5. For the case of the Dirichlet problem, Kozlov [**119, 120, 123, 124**] obtained detailed information about the eigenvalues of the corresponding operator pencil in the strip $m - 2 \le \operatorname{Re}\lambda \le m$. These results are presented in Sections 8.2–8.5.

Section 8.6. The spectrum of the pencil corresponding to the Dirichlet problem for a system of second order equations in the strip $|\operatorname{Re}\lambda| \le 1$ was described by Kozlov and Maz'ya in [**131, 132**].

Section 8.7. In the first subsection we collected some results which follow from the theory of general elliptic boundary value problems in domains with angular and conic points (see Section 1.4 and, for references, Section 1.5).

The theorems on the solvability of the Dirichlet problem in domains with edges and on the smoothness of the solutions stated in the second subsection were proved by Maz'ya and Plamenevskiĭ [**183, 185**]. We give here some additional references. Boundary value problems for second order equations in domains with edges are handled in works of Kondrat'ev [**110, 111**], Maz'ya and Plamenevskiĭ [**175, 178**], Grisvard [**75, 76, 77, 83, 86**], Nikishkin [**213**], Dauge [**40, 43, 44**], Eskin [**53**], Komech and Merzon [**108**], Costabel and Dauge [**30**], Maz'ya and Roßmann [**195**] and others. These works include theorems on the solvability in different function spaces, regularity assertions for the solutions and asymptotic formulas near the edges. A theory for higher order equations was developed by Maz'ya and Plamenevskiĭ [**177, 181, 183, 184, 185**], Maz'ya and Roßmann [**191, 192**], Dauge [**41**], Roßmann [**231**], Nazarov and Plamenevskiĭ [**207**], Nazarov [**206**], Apel and Nicaise [**6**]. Schulze [**237, 242**] studied pseudo-differential operators on manifolds with edges.

CHAPTER 9

Asymptotics of the spectrum of operator pencils generated by general boundary value problems in an angle

The present chapter occupies a somewhat special position in this book, because we are interested here and only here, in the asymptotic distribution of eigenvalues at infinity. In order to give an idea of the subject, we consider the transcendental equation

$$(9.0.1) \qquad \sin^2(\lambda - 1)\alpha = (\lambda - 1)^2 \sin^2 \alpha \,,$$

which, according to Section 7.1, describes singularities of the Dirichlet problem for the biharmonic equation in an angle with opening α. An elementary analysis shows that the roots of (9.0.1) approach either positive or negative semiaxis in the sense of the angular distance so that

$$(9.0.2) \qquad n_{\pm}(t) = \frac{2\alpha}{\pi}t + o(t) \quad \text{as } t \to +\infty,$$

where $n_+(t)$ and $n_-(t)$ are the numbers of the roots with positive and negative real parts in the disk $\{z \in \mathbb{C} : |z| < t\}$.

In this chapter we give a nontrivial generalization of (9.0.2) describing the asymptotic distribution of the eigenvalues of operator pencils generated by a class of elliptic boundary value problems for differential equations in an angle.

We establish the existence of a finite number of rays in the complex plane such that all eigenvalues, except for finitely many, lie near these rays and we obtain an asymptotic formula for the distribution of the eigenvalues as the main result of this chapter.

9.1. The operator pencil generated by a regular boundary value problem

9.1.1. Formulation of the problem. Let $\mathcal{K}$ be the angle $\{(x_1, x_2) \in \mathbb{R}^2 : 0 < r < \infty, 0 < \varphi < \alpha\}$, where r, φ are the polar coordinates of the point $x = (x_1, x_2)$. We consider the boundary value problem

$$(9.1.1) \qquad L(\partial_{x_1}, \partial_{x_2}) U = 0 \quad \text{in } \mathcal{K},$$
$$(9.1.2) \qquad B_{0,k}(\partial_{x_1}, \partial_{x_2}) U = 0 \quad \text{for } \varphi = 0, \ k = 1, \dots, l - m,$$
$$(9.1.3) \qquad B_{1,k}(\partial_{x_1}, \partial_{x_2}) U = 0 \quad \text{for } \varphi = \alpha, \ k = 1, \dots, m.$$

Here L is a homogeneous elliptic differential operator of order l with constant coefficients, $B_{j,k}$ are homogeneous differential operators of order $\mu_{j,k}$ with constant coefficients, and m is an integer number between 0 and l. In the case $m = 0$ the boundary conditions (9.1.3) on the ray $\varphi = \alpha$ do not occur, while the boundary conditions (9.1.2) on the ray $\varphi = 0$ do not occur if $m = l$.

We seek solutions of problem (9.1.1)–(9.1.3) which have the form

$$(9.1.4) \qquad U(x_1, x_2) = r^{\lambda_0} \sum_{k=0}^{s} \frac{1}{k!} (\log r)^k u_{s-k}(\varphi),$$

where λ_0 is a complex number and u_k are infinitely differentiable functions on the interval $[0, \alpha]$.

Let $\mathcal{L}$ and $\mathcal{B}_{j,k}$ be the parameter-depending differential operators defined by the equalities

$$\mathcal{L}(\varphi, \partial_\varphi, \lambda) u(\varphi) = r^{l-\lambda} L(\partial_{x_1}, \partial_{x_2}) (r^\lambda u(\varphi)),$$
$$\mathcal{B}_{j,k}(\varphi, \partial_\varphi, \lambda) u(\varphi) = r^{\mu_{j,k}-\lambda} B_{j,k}(\partial_{x_1}, \partial_{x_2}) (r^\lambda u(\varphi)).$$

By $\mathfrak{A}(\lambda)$ we denote the operator of the boundary value problem

$$(9.1.5) \qquad \begin{cases} \mathcal{L}(\varphi, \partial_\varphi, \lambda) v(\varphi) = f(\varphi) & \text{for } 0 < \varphi < \alpha, \\[2mm] \mathcal{B}_{0,k}(\varphi, \partial_\varphi, \lambda) v(\varphi)|_{\varphi=0} = g_{0,k}, & k = 1, \dots, l - m, \\[2mm] \mathcal{B}_{1,k}(\varphi, \partial_\varphi, \lambda) v(\varphi)|_{\varphi=\alpha} = g_{1,k}, & k = 1, \dots, m. \end{cases}$$

For arbitrary complex λ the operator $\mathfrak{A}(\lambda)$ maps the space $W_2^s(0, \alpha)$ into the space $W_2^{s-l}(0, \alpha) \times \mathbb{C}^m$ if $s \geq l$, $s > \max \mu_{j,k} + 1/2$. By Theorem 1.1.5, the function (9.1.4) is a solution of the homogeneous problem (9.1.1)–(9.1.3) if and only if λ_0 is an eigenvalue of the pencil $\mathfrak{A}$ and $u_0, u_1, \dots, u_s$ is a Jordan chain corresponding to this eigenvalue.

9.1.2. An equation for the eigenvalues. Since the operator L is homogeneous and elliptic, it can be written in the form

$$L(\partial_{x_1}, \partial_{x_2}) = c_0 \prod_{q=1}^{d} (\partial_{x_2} - z_q \partial_{x_1})^{\nu_q},$$

where $z_1, \dots, z_d$ are pairwise distinct complex numbers with nonzero imaginary parts, $\nu_1 + \dots + \nu_d = l$. We set

$$W_{q,j}(x_1, x_2) = \partial_z^j (x_1 + z x_2)^\lambda \big|_{z=z_q} = (\lambda - j + 1)_j \, x_2^j \, (x_1 + z_q x_2)^{\lambda-j},$$
$$w_{q,j}(\lambda, \varphi) = (\lambda - j + 1)_j (\sin \varphi)^j (\cos \varphi + z_q \sin \varphi)^{\lambda-j},$$

where

$$(z)_j = z (z + 1) \cdots (z + j - 1) \text{ for } j = 1, 2, \dots, \qquad (z)_0 = 1.$$

Here it is assumed that the argument of the functions $\cos \varphi + z_q \sin \varphi$ and $x_1 + z_q x_2$ varies from 0 to 2π for $\operatorname{Im} z_q > 0$ and from 0 to -2π for $\operatorname{Im} z_q < 0$. If $\lambda \neq 0, 1, \dots, l - 2$, then the functions $w_{q,j}$, $q = 1, \dots, d$, $j = 0, \dots, \nu_q - 1$, form a basis in the space of solutions of the equation $\mathcal{L}(\varphi, \partial_\varphi, \lambda) u = 0$.

For every homogeneous differential operator B of order μ with constant coefficients we have

$$B(\partial_{x_1}, \partial_{x_2}) W_{q,j}$$
$$= \sum_{\nu=0}^{j} \binom{j}{\nu} (\lambda - \mu - \nu + 1)_{\mu+\nu} \partial_z^{j-\nu} B(1, z)|_{z=z_q} x_2^\nu (x_1 + z_q x_2)^{\lambda-\mu-\nu}$$
$$= (\lambda - \mu + 1)_\mu \, r^{\lambda-\mu} \, \partial_z^j \Big(B(1, z) (\cos \varphi + z \sin \varphi)^{\lambda-\mu} \Big) \Big|_{z=z_q}.$$

This implies

$$\mathcal{B}_{0,k}(\varphi, \partial_\varphi, \lambda)\, w_{q,j}(\lambda, \varphi)|_{\varphi=0} = (\lambda - \mu_{0,k} + 1)_{\mu_{0,k}}\, \partial_z^j \mathcal{B}_{0,k}(1, z)|_{z=z_q}$$

for $k = 1, \ldots, l - m$, $q = 1, \ldots, d$, $j = 0, \ldots, \nu_q - 1$, and

(9.1.6)
$$\mathcal{B}_{1,k}(\varphi, \partial_\varphi, \lambda) w_{q,j}(\lambda, \varphi)|_{\varphi=\alpha}$$
$$= (\lambda - \mu_{1,k} + 1)_{\mu_{1,k}}\, \partial_z^j \Big(B_{1,k}(1, z)\, (\cos\alpha + z\,\sin\alpha)^{\lambda - \mu_{1,k}} \Big) \Big|_{z=z_q}$$

for $k = 1, \ldots, m$, $q = 1, \ldots, d$, $j = 0, \ldots, \nu_q - 1$. We introduce the column vectors

$$U_k = \Big(\partial_z^j B_{0,k}(1, z)|_{z=z_q} \; : \; 1 \le q \le d,\; 0 \le j \le \nu_q - 1 \Big),$$

$k = 1, \ldots, l - m$, and

$$V_k(\lambda) = \Big(\partial_z^j \big(B_{1,k}(1, z)\, (\cos\alpha + z\,\sin\alpha)^{\lambda - \mu_{1,k}} \big) \big|_{z=z_q} \; : \; 1 \le q \le d,\; 0 \le j \le \nu_q - 1 \Big),$$

$k = 1, \ldots, m$. By $\mathcal{D}(\lambda)$ we denote the $l \times l$-matrix with the columns $U_1, \ldots, U_{l-m}$, $V_1, \ldots, V_m$. Then we obtain the following assertion (cf. Lemma 8.1.2).

LEMMA 9.1.1. *Let $\lambda_0 \ne 0, 1, \ldots, l - 2$. Then λ_0 is an eigenvalue of the pencil $\mathfrak{A}$ if and only if*

(9.1.7)
$$\prod_{k=1}^{l-m}(\lambda - \mu_{0,k} + 1)_{\mu_{0,k}} \cdot \prod_{k=1}^{m}(\lambda - \mu_{1,k} + 1)_{\mu_{1,k}} \cdot \det\mathcal{D}(\lambda) = 0.$$

The algebraic multiplicity of the eigenvalue λ_0 is equal to the multiplicity of the root λ_0 of the equation (9.1.7).

REMARK 9.1.1. If $d = 1$ or $m = 0$ or $m = l$, then the pencil $\mathfrak{A}$ has a finite number of eigenvalues and all eigenvalues have finite algebraic multiplicities.

Indeed, for $d = 1$ the determinant of $\mathcal{D}(\lambda)$ has the form $F(\lambda)\, (\cos\alpha + z_1 \sin\alpha)^{m\lambda}$, where $F(\lambda)$ is a polynomial not identically equal to zero, for $m = 0$ the matrix $\mathcal{D}(\lambda)$ does not depend on λ, and for $m = l$ there is the representation

$$\det\mathcal{D}(\lambda) = F(\lambda)\, (\cos\alpha + z_1 \sin\alpha)^{\nu_1\lambda} \cdots (\cos\alpha + z_d \sin\alpha)^{\nu_d\lambda}$$

with a polynomial $F(\lambda)$ not equal to zero. Hence the equation (9.1.7) has only finitely many solutions in these cases.

9.1.3. Regularity of boundary conditions. Let Λ_m be the set of vectors $\kappa = (\kappa_1, \ldots, \kappa_d)$ with integer components κ_q, $0 \le \kappa_q \le \nu_q$, $\kappa_1 + \cdots + \kappa_d = m$. We set

(9.1.8)
$$\tau_q = \log(\cos\alpha + z_q \sin\alpha)$$

and define the polygon $\mathcal{P}_m$ as the convex hull of the points

$$\kappa \cdot \overline{\tau} = \kappa_1 \overline{\tau}_1 + \cdots + \kappa_d \overline{\tau}_d,$$

where $\kappa \in \Lambda_m$. The set of the corners of the polygon $\mathcal{P}_m$ is denoted by Γ_m. If $d = 1$ or $m = 0$ or $m = l$, then the set $\mathcal{P}_m$ consists of one point only. Assume that $m \ne 0$, $m \ne l$, and between the numbers $z_1, \ldots, z_q$ are both such with positive and negative imaginary parts. Then for $\alpha = \pi$ the set $\mathcal{P}_m$ coincides with the line with the endpoints $\pm m\pi i$ on the imaginary axis. If $\alpha = 2\pi$, then $\mathcal{P}_m$ is the line with the endpoints $\pm 2m\pi i$.

LEMMA 9.1.2. 1) *Let κ, κ' be vectors from Λ_m such that $\kappa \cdot \bar{\tau}$, $\kappa' \cdot \bar{\tau} \in \Gamma_m$ for the vector $\tau = (\tau_1, \ldots, \tau_d)$. If $\kappa \cdot \bar{\tau} = \kappa' \cdot \bar{\tau}$, then $\kappa = \kappa'$.*

2) *If $\kappa \in \Lambda_m$, $\kappa \cdot \bar{\tau} \in \Gamma_m$, then for all indices $q = 1, \ldots, d$, with the possible exception of one index, we have $\kappa_q = \nu_q$ or $\kappa_q = 0$.*

3) *Let $\kappa \in \Lambda_m$ and $\nu - \kappa = (\nu_1 - \kappa_1, \ldots, \nu_d - \kappa_d) \in \Lambda_{l-m}$. Furthermore, let $\mathcal{Q}_\kappa$ be the convex hull of all points τ_q with index $q = 1, \ldots, d$ satisfying the condition $\kappa_q \neq 0$. (Analogously, $\mathcal{Q}_{\nu-\kappa}$ is the convex hull of all points τ_q such that $\kappa_q \neq \nu_q$.) Then the number $\kappa \cdot \bar{\tau}$ belongs to the set Γ_m if and only if the intersection $\mathcal{Q}_\kappa \cap \mathcal{Q}_{\nu-\kappa}$ is empty or consists of the unique common corner.*

Proof: 1) Suppose that $\kappa \neq \kappa'$. Then for certain indices j, k the inequalities $\kappa_j < \kappa'_j$ and $\kappa_k > \kappa'_k$ are valid. Without loss of generality, we may assume that $k > j$. We consider the vectors

$$\kappa^{(1)} = (\kappa_1, \ldots, \kappa_{j-1}, \kappa_j + 1, \kappa_{j+1}, \ldots, \kappa_{k-1}, \kappa_k - 1, \kappa_{k+1}, \ldots, \kappa_d)$$

and

$$\kappa^{(2)} = (\kappa'_1, \ldots, \kappa'_{j-1}, \kappa'_j - 1, \kappa'_{j+1}, \ldots, \kappa'_{k-1}, \kappa'_k + 1, \kappa'_{k+1}, \ldots, \kappa'_d)$$

which belong to Λ_m. Furthermore, $\kappa^{(1)} \cdot \bar{\tau} \neq \kappa^{(2)} \cdot \bar{\tau}$. Indeed, if $\kappa^{(1)} \cdot \bar{\tau} = \kappa^{(2)} \cdot \bar{\tau}$, then $(\kappa^{(1)} - \kappa) \cdot \bar{\tau} = (\kappa^{(2)} - \kappa') \cdot \bar{\tau}$ and, consequently, $\tau_j = \tau_k$. However, the numbers $\tau_1, \ldots, \tau_d$ are pairwise distinct. Consequently, $\kappa^{(1)} \cdot \bar{\tau} \neq \kappa^{(2)} \cdot \bar{\tau}$.

From this inequality it follows that the point $\kappa \cdot \bar{\tau} = (\kappa^{(1)} \cdot \bar{\tau} + \kappa^{(2)} \cdot \bar{\tau})/2$ does not belong to the set Γ_m. This proves the first assertion.

2) Let the second assertion be not true. Then, without loss of generality, we may assume that $0 < \kappa_1 < \nu_1$ and $0 < \kappa_2 < \nu_2$. We set

$$\kappa^{(1)} = (\kappa_1 - 1, \kappa_2 + 1, \kappa_3, \ldots, \kappa_d) \quad \text{and} \quad \kappa^{(2)} = (\kappa_1 + 1, \kappa_2 - 1, \kappa_3, \ldots, \kappa_d).$$

These vectors belong to Λ_m and the corresponding complex numbers $\kappa^{(1)} \cdot \bar{\tau}$ and $\kappa^{(2)} \cdot \bar{\tau}$ do not coincide. Therefore, the point $\kappa \cdot \bar{\tau} = (\kappa^{(1)} \cdot \bar{\tau} + \kappa^{(2)} \cdot \bar{\tau})/2$ cannot belong to Γ_m.

3) Let $\kappa \cdot \bar{\tau} \in \Gamma_m$. We suppose that the polygons $\mathcal{Q}_\kappa$, $\mathcal{Q}_{\nu-\kappa}$ have the common point C. Then there exist nonnegative numbers $\beta_1, \ldots, \beta_d$, $\gamma_1, \ldots, \gamma_d$ such that

$$\sum_{q=1}^{d} \gamma_q \kappa_q \bar{\tau}_q = \sum_{q=1}^{d} \beta_q (\nu_q - \kappa_q) \bar{\tau}_q = C,$$

where

(9.1.9)
$$\sum_{q=1}^{d} \gamma_q \kappa_q = \sum_{q=1}^{d} \beta_q (\nu_q - \kappa_q) = 1.$$

This implies

$$\sum_{q=1}^{d} \beta_q (\nu_q - \kappa_q) (\kappa \cdot \bar{\tau} + \bar{\tau}_q - C) = \kappa \cdot \bar{\tau}$$

and

$$\kappa \cdot \bar{\tau} + \bar{\tau}_q - C = \sum_{j=1}^{d} \gamma_j \kappa_j (\kappa \cdot \bar{\tau} + \bar{\tau}_q - \bar{\tau}_j).$$

Consequently,

$$(9.1.10) \qquad \kappa \cdot \overline{\tau} = \sum_{q=1}^{d} \sum_{j=1}^{d} \beta_q \left(\nu_q - \kappa_q \right) \gamma_j \, \kappa_j \left(\kappa \cdot \overline{\tau} + \overline{\tau}_q - \overline{\tau}_j \right).$$

If $\kappa_q \neq \nu_q$ and $\kappa_j \neq 0$, then the point $\kappa \cdot \overline{\tau} + \overline{\tau}_q - \overline{\tau}_j$ belongs to $\mathcal{P}_m$. Since $\kappa \cdot \overline{\tau} \in \Gamma_m$, the equality (9.1.10) can be only valid in the case

$$(9.1.11) \qquad \sum_{q=1}^{d} \beta_q \left(\nu_q - \kappa_q \right) \gamma_q \, \kappa_q = 1.$$

From (9.1.9), (9.1.11) it follows that for a certain index q we have $\beta_q \left(\nu_q - \kappa_q \right) = \gamma_q \, \kappa_q = 1$. Therefore, the intersection of the polygons $\mathcal{Q}_\kappa$, $\mathcal{Q}_{\nu-\kappa}$ contains not more than one point which is a corner of both polygons.

Now let $\kappa \in \Lambda_m$ and let the intersection of $\mathcal{Q}_\kappa$, $\mathcal{Q}_{\nu-\kappa}$ consist of one corner. Then there exist two nonparallel lines g_1 and g_2 such that $\mathcal{Q}_\kappa$ and $\mathcal{Q}_{\nu-\kappa}$ lie on different sides both of g_1 and g_2. We denote by g_1', g_2' lines through the point $\kappa \cdot \overline{\tau}$ which are parallel to g_1 and g_2, respectively. It can be easily shown that the convex hull of the points $\sigma \cdot \overline{\tau}$, where $\sigma \in \Lambda_m$, lies on one side of these lines. Hence $\kappa \cdot \overline{\tau} \in \mathcal{P}_m$. The lemma is proved. ∎

In particular, it follows from Lemma 9.1.2 that for every corner a_0 of the polygon $\mathcal{P}_m$ there exists a unique vector $\kappa \in \Lambda_m$ such that $\kappa \cdot \overline{\tau}$ coincides with a_0. We denote the set of these vectors by Λ_m'. By the third part of the foregoing lemma, the vector κ belongs to the set Λ_m' if and only if $\nu - \kappa = (\nu_1 - \kappa_1, \ldots, \nu_d - \kappa_d) \in \Lambda_{l-m}'$.

We introduce the notion of regularity of boundary conditions.

DEFINITION 9.1.1. Let $\kappa \in \Lambda_m'$ and let $\langle \kappa \rangle$ be the set of all pairs (q, j) of integer numbers such that $1 \leq q \leq d$, $0 \leq j \leq \kappa_q - 1$. The boundary conditions (9.1.2), (9.1.3) are said to be κ-*regular* if the matrices

$$(9.1.12) \qquad \left(\partial_z^j B_{0,k}(1, z)|_{z=z_q} \right)_{1 \leq k \leq l-m, \ (q,j) \in \langle \nu - \kappa \rangle}$$

and

$$(9.1.13) \quad \left(\partial_z^j B_{1,k} \left((\cos \alpha + z \sin \alpha)^{-1}, z(\cos \alpha + z \sin \alpha)^{-1} \right)|_{z=z_q} \right)_{1 \leq k \leq m, \ (q,j) \in \langle \kappa \rangle}$$

are nondegenerate. In this case the boundary conditions (9.1.2) are said to be $(\nu - \kappa)$-regular, while the boundary conditions (9.1.3) are said to be κ-regular.

DEFINITION 9.1.2. The system of the boundary conditions (9.1.2), (9.1.3) is said to be *regular* if it is κ-regular for every $\kappa \in \Lambda_m'$.

It is frequently more convenient to consider the matrix

$$(9.1.14) \qquad \left(\partial_\zeta^j B_{1,k}(-\zeta \sin \alpha + \cos \alpha, \zeta \cos \alpha + \sin \alpha)|_{\zeta=\zeta_q} \right)_{1 \leq k \leq m, \ (q,j) \in \langle \kappa \rangle},$$

where $\zeta_q = (z_q \cos \alpha - \sin \alpha)(z_q \sin \alpha + \cos \alpha)^{-1}$, instead of (9.1.13). The matrix (9.1.14) is invertible simultaneously with the matrix

$$\left(\left((\cos \alpha - \zeta \sin \alpha)^2 \, \partial_\zeta \right)^j B_{1,k} \left(\cos \alpha - \zeta \sin \alpha, \zeta \cos \alpha + \sin \alpha \right)|_{\zeta=\zeta_q} \right)_{1 \leq k \leq m, \ (q,j) \in \langle \kappa \rangle}$$

which arises from the matrix (9.1.13) via the coordinate change

$$\zeta = (z \cos \alpha - \sin \alpha) \cdot (z \sin \alpha + \cos \alpha)^{-1}.$$

Therefore, the regularity of the matrix (9.1.13) is equivalent to the regularity of the matrix (9.1.14).

9.1.4. The relation between κ-regularity and Shapiro-Lopatinskiĭ condition. Suppose that $l = 2m$ and the operator L is properly elliptic. We consider the vectors $\kappa^+ = (\kappa_1^+, \dots, \kappa_d^+)$, $\kappa^- = (\kappa_1^-, \dots, \kappa_d^-)$, where

$$\kappa_q^+ = \nu_q, \ \kappa_q^- = 0 \quad \text{for Im } z_q > 0,$$
$$\kappa_q^+ = 0, \ \kappa_q^- = \nu_q \quad \text{for Im } z_q < 0.$$

Note that $\kappa^+, \kappa^- \in \Lambda_m'$. The validity of the Shapiro-Lopatinskiĭ condition on the sides of the angle $\mathcal{K}$ is equivalent to the κ^+- and κ^--regularity of the boundary conditions (9.1.2), (9.1.3). In particular, this implies that for $\alpha = \pi$, $\alpha = 2\pi$ or for $d = 2$ and arbitrary angle α the regularity of the boundary conditions is equivalent to the validity of the Shapiro-Lopatinskiĭ condition on the sides of the angle.

9.1.5. Examples. Let $\kappa \in \Lambda_m'$. We give some examples for κ-regular boundary conditions on the ray $\varphi = \alpha$, i.e., for operators $B_{1,k}$, $k = 1, \dots, m$, for which the matrix (9.1.14) is regular. To this end, we set

$$p_k(\zeta) = B_{1,k}(-\zeta \sin \alpha + \cos \alpha, \ \zeta \cos \alpha + \sin \alpha)$$

and

$$\zeta_q = (z_q \cos \alpha - \sin \alpha) \cdot (z_q \sin \alpha + \cos \alpha)^{-1}, \qquad q = 1, \dots, d.$$

Example 1: Dirichlet conditions. If the polynomials p_k have degree $k - 1$ for $k = 1, \dots, m$, then the boundary conditions (9.1.3) are κ-regular on the ray $\varphi = \alpha$ for arbitrary κ.

Example 2: Conditions of the Neumann type. Suppose that the polynomials p_k have the form $p_k(\zeta) = p(\zeta) g_k(\zeta)$, where g_k are polynomials of degree $k - 1$, $k = 1, \dots, m$, and p is a polynomial not equal to zero at the points ζ_q with indices q satisfying the condition $\kappa_q \neq 0$. Then the boundary conditions (9.1.3) are κ-regular on the ray $\varphi = \alpha$.

Example 3. Let $p_k(\zeta) = p(\zeta)(g(\zeta))^{k-1}$ for $k = 1, \dots, m$, where p, g are polynomials such that

$$g(\zeta_q) \neq g(\zeta_j) \text{ for } q \neq j, \quad g'(\zeta_q) \neq 0, \quad \text{and} \quad p(\zeta_q) \neq 0$$

(here only the indices q and j are considered for which $\kappa_q \neq 0$, $\kappa_j \neq 0$). Then the boundary conditions (9.1.3) are κ-regular.

Analogously, $(\nu - \kappa)$-regular boundary conditions on the ray $\varphi = 0$ can be constructed. For this in Examples 1–3 the vector κ has to be replaced by $\nu - \kappa$, m by $l - m$, and α has to be replaced by 0.

The κ-regularity of the boundary conditions in Examples 1 and 2 essentially follows from the κ-regularity in Example 3 if we set $g(u) = u$. We prove the κ-regularity of the boundary conditions in Example 3.

Let the matrix (9.1.14) be degenerate. Then there exist numbers $c_1, \dots, c_m$ not all simultaneously equal to zero such that

$$(9.1.15) \qquad \sum_{k=1}^{m} c_k \, \partial_\zeta^j \, p_k(\zeta)|_{\zeta = \zeta_q} = 0 \qquad \text{for } (q, j) \in \langle \kappa \rangle.$$

We introduce the polynomial $h(z) = \sum_{k=1}^{m} c_k \, z^{k-1}$. From our assumptions and from (9.1.15) it follows that

$$\partial_z^j h(z)|_{z = g(\zeta_q)} = 0 \qquad \text{for } (q, j) \in \langle \kappa \rangle.$$

Consequently, the polynomial h has m roots counting multiplicities. Hence $h = 0$ which contradicts our assumption. This proves the κ-regularity of the boundary conditions in Example 3.

9.2. Distribution of the eigenvalues

9.2.1. A property of $\det \mathcal{D}(\lambda)$. If we expand the determinant of $\mathcal{D}(\lambda)$ with respect to its columns $V_1, \dots, V_m$, we arrive at the representation

$$\det \mathcal{D}(\lambda) = \sum_{\kappa \in \Lambda_m} F_\kappa(\lambda) \, e^{\lambda \kappa \cdot \tau},$$

where F_κ are polynomials in λ.

LEMMA 9.2.1. *If $\kappa \in \Lambda'_m$ and the boundary conditions (9.1.2), (9.1.3) are κ-regular, then the polynomial F_κ is not identically equal to zero.*

Proof: By Lemma 9.1.2 we may assume, without loss of generality, that $\kappa_1 = \dots = \kappa_{s-1} = 0$, $\kappa_{s+1} = \nu_{s+1}, \dots, \kappa_d = \nu_d$.

We consider the case $0 < \kappa_s < \nu_s$. Let $\overset{\circ}{U}_k$ be the vectors with length l whose first $\nu_1 + \dots + \nu_s$ components are the numbers $\partial_z^j B_{0,k}(1, z)|_{z = z_q}$, $q = 1, \dots, s$, $j = 0, \dots, \nu_q - 1$, while the others are zero. Analogously, we define $\overset{\circ}{V}_k(\lambda)$ as the l-vector with the first $\nu_1 + \dots + \nu_{s-1}$ components equal to zero and the others equal to

$$(9.2.1) \quad \partial_z^j \big(B_{1,k}(1, z)(\cos \alpha + z \sin \alpha)^{\lambda - \mu_{1,k}} \big)\big|_{z = z_q}, \quad s \leq q \leq d, \ 0 \leq j \leq \nu_q - 1.$$

Then it can be easily verified that the matrix

$$\mathfrak{M}(\lambda) = \left(\overset{\circ}{U}_1, \dots, \overset{\circ}{U}_{l-m}, \overset{\circ}{V}_1, \dots, \overset{\circ}{V}_m \right)$$

satisfies

$$\det \mathfrak{M}(\lambda) = F_\kappa(\lambda) \, e^{\lambda \kappa \cdot \tau}.$$

Dividing the components (9.2.1) of the vector $\overset{\circ}{V}_k(\lambda)$ by $(\cos \alpha + z_q \sin \alpha)^\lambda$, we obtain the vector $V_k^{(1)}$ with the components

$$v_{k;q,j}(\lambda) = \begin{cases} 0 & \text{for } q < s, \\ (\cos \alpha + z_q \sin \alpha)^{-\lambda} \partial_z^j \big(B_{1,k}(1, z)(\cos \alpha + z \sin \alpha)^{\lambda - \mu_{1,k}} \big)\big|_{z = z_q} & \text{else.} \end{cases}$$

Let $M(\lambda)$ be the polynomial matrix consisting of the columns $\overset{\circ}{U}_1, \dots, \overset{\circ}{U}_{l-m}$ and $V_1^{(1)}, \dots, V_1^{(m)}$. Then $F_\kappa(\lambda) = \det M(\lambda)$.

We show that the matrix $M(\lambda)$ is nondegenerate for at least one λ. Suppose that $\det M(\lambda) = 0$ for all $\lambda \in \mathbb{C}$. Since $M(\lambda)$ is a polynomial matrix, there exist polynomials $c_1(\lambda), \dots, c_{l-m}(\lambda), d_1(\lambda), \dots, d_m(\lambda)$ such that

$$(9.2.2) \qquad \sum_{k=1}^{l-m} c_k(\lambda) \overset{\circ}{U}_k + \sum_{k=1}^{m} d_k(\lambda) V_k^{(1)}(\lambda) = 0$$

for all $\lambda \in \mathbb{C}$. We introduce the functions

$$G(z, \lambda) \;=\; \sum_{k=1}^{l-m} c_k(\lambda)\, B_{0,k}(1, z) = \sum_{\sigma=0}^{p} G_\sigma(z)\, \lambda^\sigma,$$

$$H(z, \lambda) \;=\; \sum_{k=1}^{m} d_k(\lambda)\, B_{1,k}(1, z)\, (\cos\alpha + z \sin\alpha)^{-\mu_{1,k}} = \sum_{\sigma=0}^{p'} H_\sigma(z)\, \lambda^\sigma.$$

The coefficients $G_\sigma(z)$ and $H_\sigma(z)$ can be written in the form

$$G_\sigma(z) \;=\; \sum_{k=1}^{l-m} c_{k,\sigma}\, B_{0,k}(1, z),$$

$$H_\sigma(z) \;=\; \sum_{k=1}^{m} d_{k,\sigma}\, B_{1,k}(1, z)\, (\cos\alpha + z \sin\alpha)^{-\mu_{1,k}}.$$

Suppose that the functions $G_p(z)$, $H_{p'}(z)$ are not identically equal to zero. The equality (9.2.2) yields

$$(9.2.3) \quad \partial_z^j G(z, \lambda)|_{z=z_q} = 0 \qquad \text{for } q = 1, \dots, s-1, \; j = 0, \dots, \nu_q - 1,$$

$$(9.2.4) \quad \partial_z^j H(z, \lambda)|_{z=z_q} = 0 \qquad \text{for } q = s+1, \dots, d, \; j = 0, \dots, \nu_q - 1,$$

and

$$(9.2.5) \quad \sum_{\nu=0}^{j} \binom{j}{\nu} \partial_z^\nu H(z, \lambda)\, (\lambda - j + \nu + 1)_{j-\nu} \left(\frac{\sin\alpha}{\cos\alpha + z\sin\alpha} \right)^{j-\nu} \bigg|_{z=z_s}$$

$$+ \,\partial_z^j G(z, \lambda)\big|_{z=z_s} = 0 \quad \text{for } j = 0, \dots, \nu_s - 1.$$

From (9.2.3), (9.2.4) it follows that

$$(9.2.6) \quad \partial_z^j G_p(z)|_{z=z_q} = 0 \qquad \text{for } q = 1, \dots, s-1, \; j = 0, \dots, \nu_q - 1,$$

$$(9.2.7) \quad \partial_z^j H_{p'}(z)|_{z=z_q} = 0 \qquad \text{for } q = s+1, \dots, d, \; j = 0, \dots, \nu_q - 1,$$

If we compare the coefficients of the same powers of λ in (9.2.5), we obtain

$$(9.2.8) \qquad G_p(z_s) = \dots = \partial_z^{p-p'-1} G_p(z)|_{z=z_s} = 0,$$

$$(9.2.9) \qquad H_{p'}(z_s) = \dots = \partial_z^{p'-p+\nu_s-2} H_{p'}(z)|_{z=z_s} = 0.$$

(In the cases $p - p' < 1$ and $p' - p + \nu_s < 2$ the corresponding equations can be omitted.)

Since $\nu_1 + \dots + \nu_d = l$, one of the inequalities $p - p' + \nu_1 + \dots + \nu_s \geq l - m$ and $\nu_{s+1} + \dots + \nu_d + p' - p + \nu_s - 1 \geq m$ must be satisfied. If $p - p' + \nu_1 + \dots + \nu_s \geq l - m$, then (9.2.6), (9.2.8) imply that the matrix (9.1.12) is degenerate. For $\nu_{s+1} + \dots + \nu_d + p' - p + \nu_s - 1 \geq m$ it follows from (9.2.7), (9.2.9) that the matrix (9.1.13) is degenerate. Thus, we have got a contradiction to the assumption of the κ-regularity if $0 < \kappa_s < \nu_s$.

In the same way a contradiction holds if $\kappa_s = 0$ or $\kappa_s = \nu_s$. ∎

9.2.2. Distribution of zeros of quasipolynomials. We present a classical result on the asymptotic behaviour of zeros of functions which have the form

$$f(\lambda) = \sum_{k=1}^{n} p_k(\lambda)\, e^{\alpha_k \lambda},$$

where p_k are polynomials and α_k are complex numbers. If f is a function of the above form, then there exists the limit

$$h_f(\vartheta) = \lim_{r \to \infty} \frac{\ln \left| f(r\, e^{i\vartheta}) \right|}{r}$$

for every real ϑ. The function h_f is called the *indicator* of f. In order to express h_f in terms of α_k, we consider the convex hull $\mathcal{P}$ of the points $\overline{\alpha}_k$, $k = 1, \ldots, n$. Let $a_1, \ldots, a_N$ be the corners of the polygon $\mathcal{P}$ enumerated in clockwise direction. Obviously, there exist indices $k_1, \ldots, k_N$ such that

$$a_j = \overline{\alpha}_{k_j} \quad \text{for } j = 1, \ldots, N.$$

We suppose that the polynomials p_{k_j}, $j = 1, \ldots, N$, are not identically equal to zero. Then

$$h_f(\vartheta) = \max_{1 \le j \le N} \left(\operatorname{Re} a_j\, \cos \vartheta + \operatorname{Im} a_j\, \sin \vartheta \right),$$

i.e., $h_f(\vartheta)$ is the supporting function of the polygon $\mathcal{P}$ (see Levin's book [**153**, Ch.1, Sect.19]). Furthermore, the polygon $\mathcal{P}$ can be characterized by means of the function

$$s(\varsigma, \vartheta) = h'_f(\vartheta) - h'_f(\varsigma) + \int_{\varsigma}^{\vartheta} h_\varsigma(\psi)\, d\psi.$$

It admits the following geometric interpretation. Let

$$\vartheta_j = \arg i(a_{j+1} - a_j) \quad \text{for } j = 1, \ldots, N,$$

where $a_{N+1} = a_1$. If $\varsigma < \vartheta$, $\vartheta - \varsigma < 2\pi$, and ς, ϑ are not equal to $\vartheta_1, \ldots, \vartheta_N$, then (see [**153**, Ch.1, Sect.19])

$$(9.2.10) \qquad s(\varsigma, \vartheta) = \sum \left| a_{j+1} - a_j \right|,$$

where the summation is extended over all j such that $\vartheta_j + 2k\pi \in (\varsigma, \vartheta)$ for a certain integer k.

We denote the number of zeros of f in the sector

$$\varsigma \le \arg \lambda \le \vartheta, \quad |\lambda| \le r,$$

by $n(r, \varsigma, \vartheta)$. The following theorem is a particular case of Theorem 3 in [**153**, Ch.3, Sect.3].

THEOREM 9.2.1. *Let $\varsigma < \vartheta$, $\vartheta - \varsigma < 2\pi$ and let ς, ϑ be not equal to $\vartheta_j + 2k\pi$. We suppose that the polynomials p_{k_j} are not identically equal to zero. Then*

$$\lim_{r \to \infty} \frac{n(r, \varsigma, \vartheta)}{r} = \frac{s(\varsigma, \vartheta)}{2\pi}.$$

9.2.3. An asymptotic formula for the distribution of the eigenvalues.
Now we formulate the main result of this section. Let $a_1, \ldots, a_N$ be the corners
of the polygon $\mathcal{P}_m$ enumerated in clockwise direction. We set $\ell_k = a_{k+1} - a_k$ for
$k = 1, \ldots, N-1$ and $\ell_N = a_1 - a_N$. Furthermore, for every index $k = 1, \ldots, N$ let
$S_k(\varepsilon)$ be the angle

$$S_k(\varepsilon) = \{\lambda \in \mathbb{C} : |\arg \lambda - \arg i\ell_k| \le \varepsilon, \ |\lambda| > 0\},$$

where ε is a positive number. In the sequel, the number ε will be chosen sufficiently
small such that the angles $S_k(\varepsilon)$ and $S_j(\varepsilon)$ do not intersect if $k \ne j$. By $n_k(t,\varepsilon)$ we
denote the number of the eigenvalues (counted with their algebraic multiplicities)
of the pencil $\mathfrak{A}$ which lie in the sector $\{\lambda \in S_k(\varepsilon) : |\lambda| < t\}$, $k = 1, \ldots, N$.

THEOREM 9.2.2. *Let the boundary conditions* (9.1.2), (9.1.3) *be regular and
let ε be a small positive number. Then the eigenvalues of the pencil $\mathfrak{A}$, with the
possible exception of finitely many, lie in the angles $S_k(\varepsilon)$. For the functions n_k of
the distribution of the eigenvalues the formula*

$$(9.2.11) \qquad n_k(t,\varepsilon) = \frac{|\ell_k|}{2\pi} t + o(t), \quad k = 1, \ldots, N,$$

is valid.

Proof: The function $\lambda \to \det \mathcal{D}(\lambda)$ is a quasipolynomial which, by Lemma 9.2.1,
satisfies the condition of Theorem 9.2.1. Applying Theorem 9.2.1 and using formula
(9.2.10), we arrive at the asymptotic formula (9.2.11).

We consider the sector $\omega_1 \le \omega \le \omega_2$, where $\arg(i\ell_k) < \omega_1 < \omega_2 < \arg(i\ell_{k-1})$ for
a certain k. Let $\kappa \in \Lambda'_m$ and $a_k = \kappa \cdot \tau$, where τ is the vector with the components
(9.1.8). Then the relation

$$\det \mathcal{D}(\lambda)\, e^{-\lambda a_k} = F_\kappa(\lambda) + O(e^{-\delta |\lambda|})$$

is valid in the sector $\omega_1 \le \arg \lambda \le \omega_2$, where δ is a positive number. Since the
polynomial F_κ is not identically equal to zero (see Lemma 9.2.1), the last relation
implies the existence of only finitely many zeros of the function $\det \mathcal{D}(\lambda)$ outside
the sector $S_k(\varepsilon)$, $k = 1, \ldots, N$, for arbitrary positive ε. The theorem is proved. ∎

9.2.4. The angles π and 2π. Suppose that the set of the numbers $z_1, \ldots, z_d$
can be divided into two groups: the numbers $z_1, \ldots, z_s$ which have positive imaginary parts and the numbers $z_{s+1}, \ldots, z_d$ which have negative imaginary parts,
where $\nu_1 + \ldots + \nu_s = l - m$. We introduce the matrices

$$A = \left(\partial_z^j B_{0,k}(1,z)|_{z=z_q}\right)_{\substack{k=1,\ldots,l-m, \\ q=1,\ldots,s, \ j=0,\ldots,\nu_q-1}},$$

$$B = \left(\partial_z^j B_{1,k}(1,z)|_{z=z_q}\right)_{\substack{k=1,\ldots,m, \\ q=s+1,\ldots,d, \ j=0,\ldots,\nu_q-1}},$$

$$C = \left(\partial_z^j B_{1,k}(1,z)|_{z=z_q}\right)_{\substack{k=1,\ldots,m, \\ q=1,\ldots,s, \ j=0,\ldots,\nu_q-1}},$$

$$D = \left(\partial_z^j B_{0,k}(1,z)|_{z=z_q}\right)_{\substack{k=1,\ldots,l-m, \\ q=s+1,\ldots,d, \ j=0,\ldots,\nu_q-1}}.$$

From the regularity of the boundary conditions (9.1.2), (9.1.3) it follows that the
matrices A and B are invertible. We calculate the eigenvalues of the pencil $\mathfrak{A}$ which
are not equal to $0, 1 \ldots, l-2$.

First we consider the case $\alpha = \pi$. Then

$$\mathcal{D}(\lambda) = \begin{pmatrix} A & e^{i\pi\lambda}C \\ D & e^{-i\pi\lambda}B \end{pmatrix},$$

and, therefore,

$$\det \mathcal{D}(\lambda) = e^{im\pi\lambda} \det A \det B \det(e^{-2i\pi\lambda}I - B^{-1}CA^{-1}D).$$

We denote by $\mu_1, \dots, \mu_p$, $p \leq m$, the nonzero eigenvalues of the matrix $B^{-1}CA^{-1}D$ and by $\sigma_1, \dots, \sigma_p$ their algebraic multiplicities. Furthermore, let λ_k be a root of the equation $e^{-2i\pi\lambda} = \mu_k$, $k = 1, \dots, p$. Then the pencil $\mathfrak{A}$ has the eigenvalues

$$\lambda_{k,s} = \lambda_k + s, \qquad k = 1, \dots, p, \ s = 0, \pm 1, \dots,$$

where the algebraic multiplicity of the eigenvalue $\lambda_{k,s}$ is equal to σ_k. Furthermore, the roots of the polynomial

(9.2.12) $$\prod_{k=1}^{m}(\lambda - \mu_{1,k} + 1)_{\mu_{1,k}} \cdot \prod_{k=1}^{l-m}(\lambda - \mu_{0,k} + 1)_{\mu_{0,k}}$$

are eigenvalues of the pencil $\mathfrak{A}$. The algebraic multiplicities of these eigenvalues coincide with the multiplicities of the corresponding roots of the polynomial (9.2.12).

Now we consider the case $\alpha = 2\pi$. In this case we have

$$\mathcal{D}(\lambda) = \begin{pmatrix} A & e^{2i\pi\lambda}C \\ D & e^{-2i\pi\lambda}B \end{pmatrix}.$$

Consequently,

$$\det \mathcal{D}(\lambda) = e^{2im\pi\lambda} \det A \det B \det(e^{-4i\pi\lambda}I - B^{-1}CA^{-1}D).$$

From this it follows that the numbers

$$\lambda'_{k,s} = \lambda_k + \frac{s}{2}, \qquad k = 1, \dots, p, \ s = 0, \pm 1, \dots,$$

are eigenvalues of $\mathfrak{A}$ and the algebraic multiplicity of the eigenvalue $\lambda_{k,s}$ is equal to σ_k. Additionally, the roots of the polynomial (9.2.12) are eigenvalues.

9.2.5. Example. We consider the homogeneous equations which describe the plane deformation of an anisotropic body.

(9.2.13) $$\sum_{j=1}^{2} \frac{\partial \sigma_{i,j}}{\partial x_j} = 0 \quad \text{in } \mathcal{K}, \ i = 1, 2,$$

(9.2.14) $$\sigma_{2,2} = \sigma_{1,2} = 0 \quad \text{for } \varphi = 0,$$

(9.2.15) $$\sigma_{1,2}\cos\alpha - \sigma_{1,1}\sin\alpha = \sigma_{2,2}\cos\alpha - \sigma_{2,1}\sin\alpha = 0 \quad \text{for } \varphi = \alpha.$$

The stress tensor $\sigma = (\sigma_{i,j})_{i,j=1,2}$ is connected with the strain tensor $\varepsilon = (\varepsilon_{i,j})_{i,j=1,2}$ by the Hooke law

$$\varepsilon_{i,j} = \sum_{k,l=1}^{2} a_{i,j,k,l}\,\sigma_{k,l},$$

where $a_{i,j,k,l}$ are real numbers such that $a_{i,j,k,l} = a_{k,l,i,j}$, $a_{i,j,k,l} = a_{j,i,k,l}$,

$$\sum_{i,j,k,l=1}^{2} a_{i,j,k,l}\,\sigma_{i,j}\,\sigma_{k,l} \geq c_0 \sum_{i,j=1}^{2} \sigma_{i,j}^2, \qquad c_0 > 0.$$

Recall that σ and ε are symmetric tensors and the tensor ε satisfies the condition

$$(9.2.16) \qquad \frac{\partial^2 \varepsilon_{1,1}}{\partial x_2^2} + \frac{\partial^2 \varepsilon_{2,2}}{\partial x_1^2} = 2\,\frac{\partial^2 \varepsilon_{1,2}}{\partial x_1\,\partial x_2}\,.$$

The general solution of (9.2.13) can be represented by one function U:

$$(9.2.17) \qquad \sigma_{1,1} = \frac{\partial^2 U}{\partial x_2^2}\,, \quad \sigma_{2,2} = \frac{\partial^2 U}{\partial x_1^2}\,, \quad \sigma_{1,2} = -\frac{\partial^2 U}{\partial x_1\,\partial x_2}\,.$$

Inserting these expressions into the Hooke law and the obtained relations into (9.2.16), we arrive at the equation

$$(9.2.18) \qquad Q(\partial_{x_1}, \partial_{x_2})\,U = 0 \quad \text{in } \mathcal{K}.$$

We do not need the explicit form of the differential operator Q. Let us only note that Q is an elliptic operator of fourth order with real coefficients. It is determined by the roots μ_1, μ_2, $\bar{\mu}_1$, $\bar{\mu}_2$ ($\operatorname{Im}\mu_1$, $\operatorname{Im}\mu_2 > 0$) of the equation $Q(1,\mu) = 0$. The boundary conditions (9.2.14), (9.2.15) take the form

$$(9.2.19) \qquad \frac{\partial^2 U}{\partial x_1^2} = \frac{\partial^2 U}{\partial x_1\,\partial x_2} = 0 \quad \text{for } \varphi = 0,$$

$$(9.2.20) \qquad \frac{\partial^2 U}{\partial x_2^2}\sin\alpha + \frac{\partial^2 U}{\partial x_1\,\partial x_2}\cos\alpha = \frac{\partial^2 U}{\partial x_1\,\partial x_2}\sin\alpha + \frac{\partial^2 U}{\partial x_1^2}\cos\alpha = 0$$
$$\text{for } \varphi = \alpha.$$

We check that the boundary operators are regular. The polynomials corresponding to the boundary conditions (9.2.19) are 1 and ζ. The regularity follows from Example 1 in Section 9.1. The polynomials corresponding to conditions (9.2.20) are

$$p_1(\zeta) = \zeta\cos\alpha + \sin\alpha, \quad p_2(\zeta) = -\zeta\sin\alpha + \cos\alpha.$$

Since $p_1(\zeta)\sin\alpha + p_2(\zeta)\cos\alpha = 1$ and $p_1(\zeta)\cos\alpha - p_2(\zeta)\sin\alpha = \zeta$, the regularity of (9.2.20) follows from Example 2 in Section 9.1. Thus, Theorem 9.2.2 is applicable to the operator pencil $\mathfrak{A}$ generated by problem (9.2.18)–(9.2.20).

First let $|\cos\alpha + \mu_2\sin\alpha| \neq |\cos\alpha + \mu_1\sin\alpha|$. Then the polygon $\mathcal{P}_2$ is a rhombus. Consequently, there are four rays (exactly one in every quadrant) such that the spectrum of the pencil is situated near these. We find the ray and the corresponding distribution function for the eigenvalues in the first quadrant. The other rays can be found by means of reflection with respect to the x_1- and x_2-axes. Without loss of generality, let $|\cos\alpha + \mu_2\sin\alpha| > |\cos\alpha + \mu_1\sin\alpha|$. Then the desired ray has the direction

$$\ell_0 = \log\Big(\big(\cos\alpha + \mu_2\sin\alpha\big)\big(\cos\alpha + \bar{\mu}_1\sin\alpha\big)^{-1}\Big).$$

We consider the angle $S(\varepsilon) = \{\lambda \in \mathbb{C} : |\arg\lambda - \arg i\ell_0| \leq \varepsilon,\ |\lambda| > 0\}$. The function $n(t,\varepsilon)$ of the distribution of eigenvalues in the angle $S(\varepsilon)$ has the asymptotics

$$n(t,\varepsilon) = \frac{|\ell_0|}{2\pi}\,t + o(t).$$

If $|\cos\alpha + \mu_2\sin\alpha| = |\cos\alpha + \mu_1\sin\alpha|$, then the polygon $\mathcal{P}_2$ coincides with a line segment parallel to the imaginary axis with the length

$$s = 2\Big(\arg(\cos\alpha + \mu_2\sin\alpha) + \arg(\cos\alpha + \mu_1\sin\alpha)\Big).$$

The spectrum of the pencil $\mathfrak{A}$ is situated near the real axis. If we denote the number of eigenvalues in the sectors $\{\lambda \in \mathbb{C} : |\arg(\pm\lambda)| < \varepsilon,\, 0 < |\lambda| < t\}$ by $n_\pm(t,\varepsilon)$, then we have

$$n_\pm(t,\varepsilon) = \frac{s}{2\pi}\, t + o_\pm(t).$$

9.3. Notes

The notion of κ-regular boundary conditions was introduced by Kozlov [**118**] who also found the asymptotic formula for the eigenvalues in Theorem 9.2.2. This notion proved to be useful in establishing the unique continuation property for solutions of general elliptic boundary value problems at a corner point (Kozlov [**121, 124**]).

CHAPTER 10

The Dirichlet problem for strongly elliptic systems in particular cones

In this chapter we study the singularities of solutions to the Dirichlet problem for strongly elliptic systems of partial differential equations of order $2m$ in a n-dimensional cone $\mathcal{K}$. In other words, we are interested in solutions of the problem

$$(10.0.1) \quad L(D_x)U(x) = 0 \text{ on } \mathcal{K}, \quad \partial_\nu^j U(x) = 0 \text{ on } \partial\mathcal{K} \setminus \{0\}, \ j = 0, \dots, m-1,$$

which have the form

$$(10.0.2) \qquad U(x) = |x|^{\lambda_0} \sum_{k=0}^{s} \frac{1}{k!} (\log|x|)^k \, u_{s-k}(x/|x|)$$

with $u_k \in \overset{\circ}{W}_2^m(\Omega)^\ell$, $\Omega = \mathcal{K} \cap S^{n-1}$. For this we have to find the eigenvalues, eigenfunctions and generalized eigenfunctions of the associated operator pencil $\mathcal{L}$ on Ω. Preliminary information about this pencil is collected in Section 10.1.

There are several cases when all power-logarithmic solutions (10.0.2) can be found. Sections 10.2–10.4 deal with three cases of such a kind.

In Section 10.2 we assume that $\mathcal{K} = \mathbb{R}^n \setminus \{0\}$. We show that the eigenvalues of $\mathcal{L}$ may only be equal to $0, 1, \dots$ or $2m - n$, $2m - n - 1, \dots$ and that the function (10.0.2) is a solution of (10.0.1) if and only if

$$(10.0.3) \qquad U(x) = \sum_{|\alpha|=\lambda_0} p_\alpha \, x^\alpha + \sum_{|\beta|=2m-n-\lambda_0} \partial_x^\beta E(x) \, q_\beta,$$

where E is the fundamental matrix of $L(D_x)$, p_α, q_β are constant vectors and the first sum in the right-hand side of (10.0.3) satisfies (10.0.1). If $\lambda_0 < 0$ or $\lambda_0 > 2m-n$, then the corresponding sum in (10.0.3) is omitted.

Another particular case $\mathcal{K} = \mathbb{R}_+^n = \{x = (x', x_n) : x_n > 0\}$ is considered in Section 10.3. The exponent λ_0 in (10.0.2) may only be equal to $m, m+1, \dots$ or $m - n$, $m - n - 1$, $m - n - 2, \dots$. In order to describe solutions (10.0.2), we introduce the Poisson kernels G_j, $j = 0, \dots, m-1$, as homogeneous solutions of the Dirichlet problem

$$L(D_x) \, G_j(x) = 0 \text{ in } \mathbb{R}_+^n, \quad \partial_{x_n}^k G_j(x) = I_\ell \delta(x') \delta_j^k \text{ for } x_n = 0, \ k = 0, \dots, m-1,$$

where I_ℓ is the $\ell \times \ell$ identity matrix. Then (10.0.2) is a solution of (10.0.1) if and only if

$$(10.0.4) \qquad U(x) = x_n^m \sum_{|\alpha|=\lambda_0-m} p_\alpha \, x^\alpha + \sum_{j=0}^{m-1} \sum_{|\beta|=j+1-n-\lambda_0} \partial_{x'}^\beta G_j(x) \, q_\beta,$$

where p_α, q_β are constant vectors and the first term in the right-hand side of (10.0.4) satisfies (10.0.1). If $\lambda_0 < m$ or $\lambda_0 > j + 1 - n$ for some $j = 0, \dots, m-1$, then the corresponding terms in (10.0.4) are omitted.

We turn to another case considered in Section 10.4 when the spectrum of $\mathcal{L}$ can be given explicitly. Let $\mathfrak{R}$ be the ray $\{x \in \mathbb{R}^n : x' = 0, \ x_n \geq 0\}$. We put $\mathcal{K} = \mathbb{R}^n \backslash \mathfrak{R}$ and $2m \geq n$. We prove the following statements:

(i) Let n be odd. Then the spectrum of $\mathcal{L}$ is exhausted by the eigenvalues $k = 0, \pm 1, \pm 2, \ldots$. If u is an eigenfunction corresponding to k, then

$$(10.0.5) \qquad |x|^k u(x/|x|) = \sum_{|\alpha|=k} p_\alpha \, x^\alpha + \sum_{|\beta|=2m-n-k} \partial_x^\beta E(x) \, q_\beta \, ,$$

where $p_\alpha, q_\beta \in \mathbf{C}^\ell$. If $k < 0$, then the first sum on the right is absent. If $k > 2m-n$, then the second sum is omitted.

(ii) Let n be even. Then the integers k, $k \neq m - n/2$, are eigenvalues of $\mathcal{L}$, and the corresponding eigenfunctions admit the representation (10.0.5). The half-integers

$$m - (n-1)/2 + k, \quad \text{where} \quad k = 0, \pm 1, \pm 2, \ldots ,$$

are also eigenvalues. Their geometric multiplicities are $\ell\binom{m+n/2-1}{n-1}$. Other eigenvalues do not exist.

Both in the cases (i) and (ii) there are no generalized eigenfunctions.

For the operator $L(D_x) = (-\Delta)^m$, $2m \geq n$, all eigenfunctions of the corresponding pencil are presented in Subsection 10.4.5.

In the last Section 10.5 we study solutions (10.0.2) of problem (10.0.1) in the cone $\mathcal{K} = \mathcal{K}_d \times \mathbb{R}^{n-d}$, where $\mathcal{K}_d$ is an arbitrary open cone in $\mathbb{R}^d$, $n > d > 1$. We are able to represent the solutions (10.0.2) by analogous solutions to the Dirichlet problem in $\mathcal{K}_d$.

Needless to say, all power-logarithmic solutions explicitly constructed in this chapter can be used to describe the asymptotic behavior of solutions to the Dirichlet problem near the corresponding boundary singularities.

10.1. Basic properties of the operator pencil generated by the Dirichlet problem

10.1.1. Function spaces. For an arbitrary open set $\mathcal{G} \subset \mathbb{R}^n$ let $W_2^m(\mathcal{G})$ be the Sobolev space with the norm

$$\|u\|_{W_2^m(\mathcal{G})} = \left(\int_{\mathcal{G}} \sum_{|\alpha| \leq m} |D_x^\alpha u(x)|^2 \, dx \right)^{1/2},$$

where $x = (x_1, \ldots, x_n) \in \mathbb{R}^n$, $D_x = -i(\partial_{x_1}, \ldots, \partial_{x_n})$. Furthermore, let $\overset{\circ}{W}{}_2^m(\mathcal{G})$ denote the closure of $C_0^\infty(\mathcal{G})$ in $W_2^m(\mathcal{G})$. Here $C_0^\infty(\mathcal{G})$ is the set of all infinitely differentiable functions on $\mathcal{G}$ whose support is compact and contained in $\mathcal{G}$. The space $W_2^{-m}(\mathcal{G})$ is defined as the dual space of $\overset{\circ}{W}{}_2^m(\mathcal{G})$.

In the sequel, let Ω be an open subset of the $(n-1)$-dimensional unit sphere S^{n-1}, and let $\mathcal{K}$ be the cone $\{x \in \mathbb{R}^n : x/|x| \in \Omega\}$. We set

$$\overset{\circ}{W}{}_{2,loc}^m(\mathcal{K}, 0) = \{U : \eta U \in \overset{\circ}{W}{}_2^m(\mathcal{K}) \text{ for all } \eta \in C_0^\infty(\mathbb{R}^n \backslash \{0\})\},$$
$$W_{2,loc}^{-m}(\mathcal{K}, 0) = \{F : \eta F \in W_2^{-m}(\mathcal{K}) \text{ for all } \eta \in C_0^\infty(\mathbb{R}^n \backslash \{0\})\}.$$

Furthermore, we introduce the following function spaces on the open subset $\Omega \subset S^{n-1}$. By $L_2(\Omega)$ we denote the set of square summable functions on Ω. This space is a Hilbert space with the scalar product

$$(u, v)_{L_2(\Omega)} = \int_\Omega u \cdot \overline{v} \, d\omega,$$

where $d\omega$ is the measure on Ω. We define $W_2^m(\Omega)$ as the space of all functions u on Ω such that $Mu \in L_2(\Omega)$ for every differential operator M of order $\leq m$ on Ω with smooth coefficients. The norm in this space is defined as

$$(10.1.1) \qquad \|u\|_{W_2^m(\Omega)} = \left(\int_{\substack{\mathcal{K} \\ 1/2 < |x| < 2}} \sum_{|\alpha| \leq m} |D_x^\alpha u(x)|^2 \, dx \right)^{1/2},$$

where in the integral the function u is extended to $\mathcal{K}$ by $u(x) = u(x/|x|)$. An equivalent norm holds if the domain of integration $\{x \in \mathcal{K} : 1/2 < |x| < 2\}$ on the right of (10.1.1) is replaced by $\{x \in \mathcal{K} : \varepsilon < |x| < 1/\varepsilon\}$ with arbitrary positive $\varepsilon < 1$. Finally, we denote the closure of the set $C_0^\infty(\Omega)$ with respect to the norm (10.1.1) by $\overset{\circ}{W}_2^m(\Omega)$.

10.1.2. Definition of the operator pencil. Let the differential operator

$$(10.1.2) \qquad L(D_x) = \sum_{|\alpha| = 2m} A_\alpha D_x^\alpha$$

be given, where A_α are constant $\ell \times \ell$ matrices. We suppose that the operator L is *strongly elliptic*, i.e., for arbitrary $\xi \in \mathbb{R}^n$, $f \in \mathbb{C}^\ell$ there is the inequality

$$\mathrm{Re}\,(L(\xi)\,f\,,\,f)_{\mathbb{C}^\ell} \geq c_0\,|\xi|^{2m}\,|f|_{\mathbb{C}^\ell}^2$$

with a positive constant c_0. Obviously, the differential operator L realizes a linear and continuous mapping $\overset{\circ}{W}_{2,loc}^m(\mathcal{K}, 0)^\ell \to W_{2,loc}^{-m}(\mathcal{K}, 0)^\ell$. Our goal is the description of all solutions of the equation

$$(10.1.3) \qquad L(D_x)\,U = 0 \qquad \text{in } \mathcal{K}$$

which have the form

$$(10.1.4) \qquad U(x) = r^{\lambda_0} \sum_{k=0}^{s} \frac{(\log r)^k}{k!}\, u_{s-k}(\omega),$$

where $r = |x|$, $\omega = x/|x|$, and $u_k \in \overset{\circ}{W}_2^m(\Omega)^\ell$ are vector functions on Ω.

LEMMA 10.1.1. *The function* (10.1.4) *belongs to the space* $\overset{\circ}{W}_{2,loc}^m(\mathcal{K}, 0)^\ell$ *if and only if* $u_k \in \overset{\circ}{W}_2^m(\Omega)^\ell$ *for* $k = 0, \dots, s$.

Proof: It is obvious that every function of the form (10.1.4) with $u_k \in \overset{\circ}{W}_2^m(\Omega)^\ell$ belongs to the space $\overset{\circ}{W}_{2,loc}^m(\mathcal{K}, 0)^\ell$. On the other hand, the functions $u_k(\omega)$ in (10.1.4) can be represented in terms of the functions $U(2^j x)$, $1 < |x| < 2$, $j = 0, 1, \dots, s$, since

$$\sum_{k=0}^{s} \frac{(\log 2^j r)^{s-k}}{(s-k)!}\, u_k(\omega) = \frac{U(2^j x)}{(2^j r)^{\lambda_0}}, \qquad j = 0, 1, \dots, s.$$

The coefficients determinant of this algebraic system is nonzero. (Multiplying this determinant by $1! \cdots s!$, we obtain the Vandermonde determinant.) Hence from

$U \in \overset{\circ}{W}_{2,loc}^{m}(\mathcal{K},0)^{\ell}$ it follows that $u_k \in \overset{\circ}{W}_2^m(\Omega)^{\ell}$ for $k = 1, \dots, m$. ∎

As in the two-dimensional case, let the parameter-depending differential operator $\mathcal{L}(\lambda)$ on the unit sphere S^{n-1} be defined by the equality

(10.1.5) $\mathcal{L}(\lambda) u = r^{2m-\lambda} L(D_x) (r^{\lambda} u).$

The operator $\mathcal{L}(\lambda)$ is a polynomial of degree $2m$ in λ. For arbitrary $\lambda \in \mathbb{C}$ the operator $\mathcal{L}(\lambda)$ continuously maps $\overset{\circ}{W}_2^m(\Omega)^{\ell}$ into $W_2^{-m}(\Omega)^{\ell}$. Furthermore,

$$\mathcal{L}(r\partial_r) = r^{2m} L(D_x).$$

From Theorem 1.1.5 it follows that the vector function (10.1.4) is a solution of the equation (10.1.3) if and only if λ_0 is an eigenvalue of the pencil $\mathcal{L}$ and $u_0, \dots, u_s$ is a Jordan chain of $\mathcal{L}$ corresponding to this eigenvalue, i.e.,

(10.1.6) $\displaystyle\sum_{j=0}^{k} \frac{1}{j!} \mathcal{L}^{(j)}(\lambda_0) u_{k-j} = 0 \qquad \text{for } k = 0, 1, \dots, s.$

By Corollary 1.1.3, the dimension of the space of all solutions of the form (10.1.4) is equal to the algebraic multiplicity of the eigenvalue λ_0.

10.1.3. The adjoint operator pencil. Let A_{α}^* denote the adjoint matrix to A_{α}. Then the operator

$$L^+(D_x) = \sum_{|\alpha|=2m} A_{\alpha}^* D_x^{\alpha},$$

is formally adjoint to the differential operator $L(D_x)$. Analogously to $\mathcal{L}(\lambda)$, let the differential operator $\mathcal{L}^+(\lambda)$ on the sphere be defined by the equality

$$\mathcal{L}^+(\lambda) u(\omega) = r^{2m-\lambda} L^+(D_x) (r^{\lambda} u(\omega)).$$

LEMMA 10.1.2. *For all* $\lambda \in \mathbb{C}$ *there is the equality*

(10.1.7) $\mathcal{L}^+(2m - n - \bar{\lambda}) = (\mathcal{L}(\lambda))^*,$

where the operator on the right-hand side is the adjoint operator to $\mathcal{L}(\lambda)$.

Proof: Obviously,

$$\int_{\Omega} \big(\mathcal{L}(\lambda)u, v\big)_{\mathbb{C}^{\ell}} \, d\omega = \frac{1}{2 \, |\log \varepsilon|} \int_{\substack{\mathcal{K} \\ \varepsilon < |x| < 1/\varepsilon}} \big(L(D_x) (r^{\lambda}u), r^{2m-n-\bar{\lambda}}v\big)_{\mathbb{C}^{\ell}} \, dx$$

for all $u, v \in C_0^{\infty}(\Omega)^{\ell}$, where ε is an arbitrary positive real number less than one. We apply Green's formula to the integral on the right-hand side. Then, due to the cancelling of the integrals over the spheres $|x| = \varepsilon$ and $|x| = 1/\varepsilon$, we get

$$\frac{1}{2 \, |\log \varepsilon|} \int_{\substack{\mathcal{K} \\ \varepsilon < |x| < 1/\varepsilon}} \big(L(D_x) (r^{\lambda}u), r^{2m-n-\bar{\lambda}}v\big)_{\mathbb{C}^{\ell}} \, dx$$

$$= \frac{1}{2 \, |\log \varepsilon|} \int_{\substack{\mathcal{K} \\ \varepsilon < |x| < 1/\varepsilon}} \big(r^{\lambda}u, L^+(D_x) (r^{2m-n-\bar{\lambda}}v)\big)_{\mathbb{C}^{\ell}} \, dx$$

$$= \int_{\Omega} \big(u, \mathcal{L}^+(2m - n - \bar{\lambda}) v\big)_{\mathbb{C}^{\ell}} \, d\omega.$$

This proves the lemma. ■

Note that the definition of the operator $\mathcal{L}^+(\lambda)$ does not coincide with the definition of the operator $\mathcal{L}^*(\lambda)$ in Section 1.1. The adjoint (in the sense of Section 1.1) operator pencil $\mathcal{L}^*$ is given by the equality $\mathcal{L}^*(\lambda) = (\mathcal{L}(\overline{\lambda}))^*$. Hence there is the relation $\mathcal{L}^+(\lambda) = \mathcal{L}^*(2m - n - \lambda)$ between the pencils $\mathcal{L}^+$ and $\mathcal{L}^*$.

10.1.4. Discreteness of the spectrum of the pencil $\mathcal{L}$. Let $\mathcal{M}(\lambda)$ be the differential operator on the sphere S^{n-1} which is generated by the polyharmonic operator:

$$(10.1.8) \qquad \mathcal{M}(\lambda)\, v \;=\; r^{2m-\lambda}\,(-\Delta)^m\,(r^\lambda v)$$

$$=\; (-1)^m \prod_{k=0}^{m-1} \big(\delta + (\lambda - 2k)(\lambda - 2k + n - 2) \big)\, v$$

(see Section 7.3). According to Lemma 7.3.1 and Remark 7.3.1, there exists a positive constant c such that

$$(10.1.9) \qquad \int_\Omega \mathcal{M}(\lambda)v \cdot \overline{v}\, d\omega \geq c \left(\|v\|^2_{W_2^m(\Omega)} + |\lambda|^{2m}\, \|v\|^2_{L_2(\Omega)} \right)$$

for all $v \in C_0^\infty(\Omega)$, $\lambda = m - n/2 + it$, $t \in \mathbb{R}$, if one of the following conditions is satisfied:

   (i) $2m < n$,

   (ii) $2m \geq n$ and n odd,

   (iii) $2m \geq n$, n even, and $\Omega \neq S^{n-1}$.

The constant c depends only on m and n.

By means of estimate (10.1.9), the following assertion can be proved.

COROLLARY 10.1.1. *Let one of the conditions* (i), (ii), (iii) *be satisfied. Then for all $U \in C_0^\infty(\mathcal{K})$ there is the inequality*

$$(10.1.10) \qquad \sum_{|\alpha| \leq m} \int_{\mathcal{K}} r^{-2(m-|\alpha|)}\, |D_x^\alpha U|^2\, dx \leq c \sum_{|\alpha|=m} \int_{\mathcal{K}} |D_x^\alpha U|^2\, dx$$

with a constant c independent of U.

Proof: The left-hand side of (10.1.10) can be estimated by the inequality

$$\sum_{|\alpha| \leq m} \int_{\mathcal{K}} r^{-2(m-|\alpha|)}\, |D_x^\alpha U|^2\, dx \leq c \int_0^\infty \sum_{j=0}^m r^{n-1-2m}\, \|(r\partial_r)^j U\|^2_{W_2^{m-j}(\Omega)}\, dr.$$

Here the constant c is independent of U. Furthermore, we have

$$\sum_{|\alpha|=m} \int_{\mathcal{K}} |D_x^\alpha U|^2\, dx \geq c \int_{\mathcal{K}} (-\Delta)^m U \cdot \overline{U}\, dx = c \int_0^\infty \int_\Omega \mathcal{M}(r\partial_r)U \cdot \overline{U}\, d\omega\; r^{n-1-2m}\, dr.$$

Hence, in order to obtain (10.1.10), we have to prove the inequality

$$\int_0^\infty \sum_{j=0}^m r^{n-1-2m}\, \|(r\partial_r)^j U\|^2_{W_2^{m-j}(\Omega)}\, dr \leq c \int_0^\infty \int_\Omega \mathcal{M}(r\partial_r)U \cdot \overline{U}\, d\omega\; r^{n-1-2m}\, dr$$

or, what is the same,

$$\int\limits_{-\infty}^{+\infty} \sum_{j=0}^{m} \|(m - \frac{n}{2} + \partial_t)^j V\|^2_{W_2^{m-j}(\Omega)}\, dt \leq c \int\limits_{-\infty}^{+\infty} \int\limits_{\Omega} \mathcal{M}(m - \frac{n}{2} + \partial_t)V \cdot \overline{V}\, d\omega\, dt,$$

where $t = \log r$, $V = e^{-(m-n/2)t}U$. r by e^t, U by $e^{(m-n/2)t}V$. By the Parseval equality, the last inequality is equivalent to

$$\int\limits_{-\infty}^{+\infty} \sum_{j=0}^{m} \|(m - \frac{n}{2} + i\tau)^j \hat{V}\|^2_{W_2^{m-j}(\Omega)}\, d\tau \leq c \int\limits_{-\infty}^{+\infty} \int\limits_{\Omega} \mathcal{M}(m - \frac{n}{2} + i\tau)\hat{V} \cdot \overline{\hat{V}}\, d\omega\, d\tau,$$

where $\hat{V}$ denotes the Fourier transform with respect to t of V, and follows immediately from

$$\int\limits_{\Omega} \mathcal{M}(\lambda)\hat{V} \cdot \overline{\hat{V}}\, d\omega \geq c \sum_{j=0}^{m} |\lambda|^{2j} \|\hat{V}\|^2_{W_2^{m-j}(\Omega)}$$

with $\lambda = m - n/2 + i\tau$. ∎

We show now that the conditions of Theorem 1.1.1 are satisfied for the operator pencil $\mathcal{L}$. Condition (ii) of Theorem 1.1.1 follows from the estimate given in the following lemma.

LEMMA 10.1.3. *There exist positive numbers c, R, and ε such that the inequality*

$$(10.1.11) \qquad \mathrm{Re} \int\limits_{\Omega} \left(\mathcal{L}(\lambda)u, u\right)_{\mathbb{C}^\ell} d\omega \geq c \sum_{j=0}^{m} |\lambda|^{2j} \|u\|^2_{W_2^{m-j}(\Omega)^\ell}$$

is satisfied for all $u \in C_0^\infty(\Omega)^\ell$, $|\lambda| > R$, $|\mathrm{Re}\,\lambda| < \varepsilon\,|\mathrm{Im}\,\lambda|$.

Proof: Since the operator $L(D_x)$ is strongly elliptic, there is the estimate

$$(-1)^m \int\limits_{\mathcal{K}} (\Delta^m U, U)_{\mathbb{C}^\ell}\, dx \leq c\, \mathrm{Re} \int\limits_{\mathcal{K}} \left(L(D_x)U, U\right)_{\mathbb{C}^\ell} dx$$

for all $U \in C_0^\infty(\mathcal{K})$. We set $U(x) = \varepsilon^{1/2}\, \eta(\varepsilon \log r)\, r^\lambda\, u(\omega)$, where $u \in C_0^\infty(\Omega)^\ell$, $\eta \in C_0^\infty(\mathbb{R})$, $\lambda = m - n/2 + it$, $t \in \mathbb{R}$. Then for $\varepsilon \to 0$ we arrive at the inequality

$$(10.1.12) \quad \int\limits_{\Omega} \left(\mathcal{M}(m - n/2 + it)u, u\right)_{\mathbb{C}^\ell} d\omega \leq c\, \mathrm{Re} \int\limits_{\Omega} \left(\mathcal{L}(m - n/2 + it)u, u\right)_{\mathbb{C}^\ell} d\omega.$$

This and (10.1.9) imply (10.1.11) for $\lambda = m - n/2 + it$ (for large t inequality (10.1.9) is satisfied without conditions (i), (ii) or (iii)).

Now let $\lambda = s + it$, where s and t are real numbers such that $|s - m + n/2| \leq 2\varepsilon\,|t|$. Then for sufficiently large $|t|$ we have

$$\mathrm{Re} \int\limits_{\Omega} \left((\mathcal{L}(\lambda) - \mathcal{L}(m - \frac{n}{2} + it))\, u,\, u\right)_{\mathbb{C}^\ell} \leq c\varepsilon \sum_{j=0}^{m} |\lambda|^{2j} \|u\|^2_{W_2^{m-j}(\Omega)^\ell}.$$

Choosing ε sufficiently small, we obtain (10.1.11) for $|\mathrm{Re}\,\lambda - m + n/2| \leq 2\varepsilon\,|\mathrm{Im}\,\lambda|$ and sufficiently large $\mathrm{Im}\,\lambda$. This proves the lemma. ∎

From (10.1.11) it follows that the operator $\mathcal{L}(\lambda)$ is invertible for $|\lambda| > R$, $|\operatorname{Re}\lambda| < \varepsilon\,|\operatorname{Im}\lambda|$. Furthermore, the operator $\mathcal{L}(\lambda_1) - \mathcal{L}(\lambda_2)$ is compact for arbitrary complex numbers λ_1 and λ_2. Hence the operator $\mathcal{L}(\lambda)$ is Fredholm for arbitrary λ. Using Theorem 1.1.1, we obtain the following result.

THEOREM 10.1.1. *Suppose that the differential operator* (10.1.2) *is strongly elliptic. Then the operator* $\mathcal{L}(\lambda)$ *is an isomorphism for all* $\lambda \in \mathbb{C}$, *with the possible exception of a denumerable set of isolated points. The mentioned denumerable set consists of eigenvalues with finite algebraic multiplicities which are situated, except finitely many, outside a double sector* $|\operatorname{Re}\lambda| < \varepsilon\,|\operatorname{Im}\lambda|$ *containing the imaginary axis.*

Furthermore, as a consequence of Theorem 1.1.7 and Lemma 10.1.2, the following assertion holds.

THEOREM 10.1.2. *The number* λ_0 *is an eigenvalue of the pencil* $\mathcal{L}$ *if and only if* $2m - n - \overline{\lambda}_0$ *is an eigenvalue of the pencil* $\mathcal{L}^+$. *The geometric, partial and algebraic multiplicities of these eigenvalues coincide.*

Finally, from (10.1.9) and (10.1.12) we obtain the following assertion on the nonexistence of eigenvalues on the energy line.

THEOREM 10.1.3. *Let one of the conditions* (i), (ii), (iii) *in the beginning of this subsection be satisfied. Then there are no eigenvalues of the pencil* $\mathcal{L}$ *on the line* $\operatorname{Re}\lambda = m - n/2$.

The only exceptional case for the validity of the assertion in Theorem 10.1.3 is the case $\Omega = S^{n-1}$, n even, and $2m \geq n$. Then the pencil $\mathcal{L}$ has the eigenvalue $\lambda = m - n/2$. The traces of homogeneous polynomials of degree $m - n/2$ on S^{n-1} are eigenfunctions corresponding to these eigenvalues (see Theorems 10.2.2, 10.2.3).

10.2. Elliptic systems in $\mathbb{R}^n$

Let $L(D_x)$ be the same differential operator $L(D_x)$ as in the previous section. We show that all solutions of the system

$$(10.2.1) \qquad L(D_x)\,U = 0 \qquad \text{in } \mathbb{R}^n\backslash\{0\},$$

which have the form (10.1.4), where u_k are smooth vector functions on the sphere S^{n-1}, are linear combinations of homogeneous vector-polynomials and derivatives of the columns of the Green matrix. In this way, we describe the spectrum of the pencil $\mathcal{L}$ introduced in Section 10.1 for the case $\Omega = S^{n-1}$.

10.2.1. Power-logarithmic solutions of elliptic systems. We denote by $G(x)$ the Green matrix of the operator $L(D_x)$, i.e., the solution of the system

$$(10.2.2) \qquad L(D_x)\,G(x) = I_\ell\,\delta(x) \qquad \text{in } \mathbb{R}^n,$$

where I_ℓ is the $\ell \times \ell$ identity matrix and δ is the Dirac function. It is well-known (see John's book [**94**]) that G admits the representation

$$(10.2.3) \quad G(x) = \begin{cases} r^{2m-n}\,Q(\omega) & \text{if } 2m \geq n, \ n \text{ odd, or } 2m < n, \\ R(x)\log r + r^{2m-n}S(\omega) & \text{if } \qquad 2m \geq n, \ n \text{ even,} \end{cases}$$

where Q and S are smooth matrix-functions on S^{n-1}, and R is a homogeneous polynomial matrix of degree $2m - n$.

In the sequel, let $\Pi_k^{(n)}$ be the space of homogeneous polynomials of degree k with n variables, $n \geq 1$. The dimension of this space is equal to $\binom{n+k-1}{n-1}$. For $n = 0$ we set $\Pi_k^{(0)} = \{0\}$.

Obviously, homogeneous polynomials of degree k are functions of the form (10.1.4), where $\lambda_0 = k$ and $s = 0$. We calculate the number of linearly independent homogeneous polynomials of degree k which are solutions of equation (10.2.1). For this we need the following lemma.

LEMMA 10.2.1. *Let $f \in (\Pi_s^{(n)})^\ell$ and $g_k \in (\Pi_{2m+s-k}^{(n-1)})^\ell$ for $k = 0, 1 \ldots, 2m - 1$, where s is an arbitrary nonnegative integer. Then the Cauchy problem*

(10.2.4)
$$\begin{cases} L(D_x)\,p = f \quad in \ \mathbb{R}^n, \\ \\ \partial_{x_n}^k p|_{x_n=0} = g_k \quad on \ \mathbb{R}^{n-1}, \quad k = 0, \ldots, 2m - 1, \end{cases}$$

has a uniquely determined solution $p \in (\Pi_{2m+s}^{(n)})^\ell$.

Proof: Without loss of generality, we may assume that $g_k = 0$. Then every solution $p \in (\Pi_{2m+s}^{(n)})^\ell$ of problem (10.2.4) has the form

$$p(x) = \sum_{j=0}^{s} x_n^{2m+s-j}\, p_j(x'),$$

where $x' = (x_1, \ldots, x_{n-1})$ and $p_j \in (\Pi_j^{(n-1)})^\ell$. Furthermore, the polynomial f can be written in the form

$$f(x) = \sum_{j=0}^{s} x_n^{s-j}\, f_j(x'),$$

where $f_j \in (\Pi_j^{(n-1)})^\ell$. Then we have

$$L(D_x)\,p(x) = \sum_{k=0}^{2m}\sum_{j=k}^{s} \binom{2m+s-j}{2m-k}\, x_n^{k+s-j}\, (-i)^{2m-k}\, L^{(2m-k)}(D_{x'}, 0)\, p_j(x'),$$

where $L^{(k)}(\xi) = \partial_{\xi_n}^k L(\xi)$. Consequently, we get

$$\sum_{k=0}^{\min(2m,s-j)} (-i)^{2m-k} \binom{2m-k+s-j}{2m-k}\, L^{(2m-k)}(D_{x'}, 0)\, p_{k+j}(x') = f_j(x')$$

for $j = 0, 1, \ldots, s$. These equations form a triangular system with the elements $\binom{2m+s-j}{2m} L^{(2m)}$ in the diagonal. Since $L^{(2m)}$ is a constant regular matrix, this system is uniquely solvable. This proves the lemma. ∎

COROLLARY 10.2.1. *The dimension of the space $\{p \in (\Pi_k^{(n)})^\ell \ : \ L(D_x)p = 0\}$, $k \geq 0$, is equal to $\ell\,\kappa(k)$, where*

(10.2.5)
$$\kappa(k) = \begin{cases} \binom{n+k-1}{n-1} & if \ k < 2m, \\ \\ \binom{n+k-1}{n-1} - \binom{n-2m+k-1}{n-1} & if \ k \geq 2m. \end{cases}$$

Proof: For $k < 2m$ the set $\{p \in (\Pi_k^{(n)})^\ell : L(D_x)p = 0\}$ coincides with $(\Pi_k^{(n)})^\ell$. In the case $k \geq 2m$ the mapping $L(D_x) : (\Pi_k^{(n)})^\ell \to (\Pi_{k-2m}^{(n)})^\ell$ is surjective by Lemma 10.2.1. Hence the dimension of its kernel is equal to $\dim(\Pi_k^{(n)})^\ell - \dim(\Pi_{k-2m}^{(n)})^\ell$. This proves (10.2.5) ∎

By Lemma 10.2.1, the numbers $\lambda_k^{(1)} = k$, $k = 0, 1, \ldots$, form a sequence of exponents λ_0 in (10.1.4) which correspond to solutions $U \in (\Pi_k^{(n)})^\ell$ of the system (10.2.1). Moreover, for the exponents $\lambda_k^{(2)} = 2m - n - k$, $k = 0, 1, \ldots$, one can find the solutions

$$(10.2.6) \qquad \sum_{|\alpha|=k} D_x^\alpha G(x) \cdot c_\alpha, \qquad c_\alpha \in \mathbb{C}^\ell.$$

Such linear combinations of the matrix-function $D_x^\alpha G$ are equal to zero in $\mathbb{R}^n \backslash \{0\}$ if and only if there exists a vector-polynomial $p \in (\Pi_{k-2m}^{(n)})^\ell$ such that

$$\sum_{|\alpha|=k} c_\alpha \, \xi^\alpha = L(\xi)\, p(\xi).$$

Hence the number of linearly independent solutions of the form (10.2.6) in $\mathbb{R}^n \backslash \{0\}$ is equal to $\ell \, \kappa(k)$.

The following theorem shows that, additionally to the solutions given above, there are no other solutions of system (10.2.1) which have the form (10.1.4).

THEOREM 10.2.1. *The vector function* (10.1.4) *is a solution of* (10.2.1) *if and only if the following conditions are satisfied:*

(i) $\lambda_0 = k$ *or* $\lambda_0 = 2m - n - k$ *with nonnegative integer* k,
(ii) *the vector function* (10.1.4) *has the form*

$$U(x) = \begin{cases} P_k(x) \;\; if \;\; \lambda_0 = k \geq \max(0, 2m - n + 1), \\[2mm] \displaystyle\sum_{|\alpha|=k} D_x^\alpha G(x)\, c_\alpha \;\; if \;\; \lambda_0 = 2m - n - k < 0, \\[2mm] P_k(x) + \displaystyle\sum_{|\alpha|=2m-n-k} D_x^\alpha G(x)\, c_\alpha \;\; if \;\; \lambda_0 = k \;\; and \;\; 0 \leq k \leq 2m - n. \end{cases}$$

Here $c_\alpha \in \mathbb{C}^\ell$ and P_k is a homogeneous vector-polynomial of degree k satisfying the equation $L(D_x)\, P_k = 0$.

Proof: We assume at first that $\operatorname{Re} \lambda_0 > -n$. Then the function (10.1.4) is locally integrable and belongs to the space $S'(\mathbb{R}^n)^\ell$. Since the distribution $L(D_x)U$ is concentrated in the point $x = 0$, there is the representation

$$L(D_x)\, U(x) = \sum_\alpha c_\alpha D_x^\alpha \, \delta(x) \qquad \text{in } \mathbb{R}^n,$$

where the summation on the right-hand side is extended over a finite set of multi-indices. We set

$$U(x) = \sum_\alpha D_x^\alpha G(x)\, c_\alpha + V(x).$$

Since $L(D_x)\, V = 0$ in $\mathbb{R}^n$ and $V \in S'(\mathbb{R}^n)$, the function V is a vector-polynomial. This proves the theorem for $\operatorname{Re} \lambda_0 > -n$.

Let $\mathcal{L}(\lambda)$ be the differential operator (10.1.5) which maps $\overset{\circ}{W}{}_2^m(S^{n-1})^\ell$ into the space $W_2^{-m}(S^{n-1})^\ell$. We have shown above, in particular, that the spectrum of the operator pencil $\mathcal{L}$ in the half-plane $\operatorname{Re}\lambda \geq 2m$ is exhausted by the eigenvalues $\lambda_k^{(1)} = k$, $k \geq 2m$, with the algebraic multiplicities $\ell\,\kappa(k)$. Applying this assertion to the formally adjoint operator $L^+(D_x)$, we conclude that the same is true for the pencil $\mathcal{L}^+$. Hence from Theorem 10.1.2 it follows that the spectrum of $\mathcal{L}$ in the half-plane $\operatorname{Re}\lambda \leq -n$ is exhausted by the eigenvalues $\lambda_k^{(2)} = 2m - n - k$, $k \geq 2m$, with the algebraic multiplicities $\ell\,\kappa(k)$. Therefore, solutions of the form (10.1.4), where $\operatorname{Re}\lambda_0 \leq -n$, exist only for $\lambda_0 = \lambda_k^{(2)}$ and the dimension of the space of such solutions is equal to $\ell\,\kappa(k)$. As it was shown before this theorem, this number is also equal to the number of linear independent solutions of the form (10.2.6). The theorem is proved. ∎

10.2.2. Spectral properties of the pencil $\mathcal{L}$ in the case $\Omega = S^{n-1}$.

Let the operator $\mathcal{L}(\lambda) : \overset{\circ}{W}{}_2^m(S^{n-1})^\ell \to W_2^{-m}(S^{n-1})^\ell$ be defined by (10.1.5). As an immediate consequence of Theorem 10.2.1, we obtain the following result.

THEOREM 10.2.2. 1) *Let $2m < n$. Then the spectrum of the pencil $\mathcal{L}$ consists of two nonintersecting sequences $\{\lambda_k^{(1)} = k\}_{k=0,1\ldots}$ and $\{\lambda_k^{(2)} = 2m - n - k\}_{k=0,1,\ldots}$. The geometric and algebraic multiplicities of the eigenvalues $\lambda_k^{(1)}$ and $\lambda_k^{(2)}$ are equal to $\ell\,\kappa(k)$. The eigenfunctions corresponding to the eigenvalues $\lambda_k^{(1)}$ are the traces on S^{n-1} of the vector-polynomials from $(\Pi_k^{(n)})^\ell$ which satisfy (10.2.1). The eigenfunctions to the eigenvalues $\lambda_k^{(2)}$ are the traces on S^{n-1} of the solutions of (10.2.1) which have the form (10.2.6).*

2) Let $2m \geq n$. Then the spectrum of $\mathcal{L}$ is formed by the sequences $\{\lambda_k^{(1)}\}_{k>2m-n}$, $\{\lambda_k^{(2)}\}_{k>2m-n}$, and the numbers $0, 1, \ldots, 2m - n$. For the eigenvalues $\lambda_k^{(1)}$, $\lambda_k^{(2)}$ and their eigenfunction the same assertions as in part 1) are true.

The eigenvalues $\lambda = k$, $k = 0, 1, \ldots, 2m - n$, have the algebraic multiplicities

$$\ell\big(\kappa(k) + \kappa(2m - n - k)\big).$$

There are no second generalized eigenvectors to these eigenvalues (i.e., the index of these eigenvalues is not greater than two). If additionally n is an odd number, then there are no generalized eigenvectors. In this case the eigenfunctions corresponding to the eigenvalues $\lambda = k$, $k \in \{0, 1, \ldots, 2m - n\}$, are the traces on S^{n-1} of vector functions of the form

$$P_k(x) + \sum_{|\alpha| = 2m - n - k} D_x^\alpha G(x)\, c_\alpha,$$

where $P_k \in (\Pi_k^{(n)})^\ell$, $c_\alpha \in \mathbb{C}^\ell$.

Theorem 10.2.2 does not answer the question on the number of the generalized eigenvectors in the case $n \leq 2m$, n even, and $\lambda = k$, $k \in \{0, 1, \ldots, 2m - n\}$. An answer is given in the following two theorems.

THEOREM 10.2.3. *Let L be a strongly elliptic operator and let n be an even number, $n \leq 2m$. Then the geometric multiplicity of the eigenvalue $\lambda = m - n/2$ is equal to $\ell\,\kappa(m - n/2)$. The space of the eigenfunctions to this eigenvalue coincides with the restriction of $(\Pi_{m-n/2}^{(n)})^\ell$ to the unit sphere S^{n-1}. Every eigenfunction has exactly one generalized eigenfunction.*

Proof: It is known (see John [**94**, Ch.3]) that the matrix polynomial R in (10.2.3) is determined by the equality

$$(10.2.7) \qquad R(x) = -\frac{1}{(2\pi i)^n \, (2m-n)!} \int\limits_{S^{n-1}} (x \cdot \omega)^{2m-n} \, L^{-1}(\omega) \, d\omega.$$

We show that the equality

$$(10.2.8) \qquad \sum_{|\alpha|=m-n/2} D_x^\alpha R(x) \, c_\alpha = 0 \quad \text{in } \mathbb{R}^n$$

implies $c_\alpha = 0$. Indeed, the left-hand side of (10.2.8) is equal to

$$-\frac{1}{(2\pi i)^{m-n/2} \, (m-n/2)!} \sum_{|\alpha|=m-n/2} \int\limits_{S^{n-1}} (x \cdot \omega)^{m-n/2} \, \omega^\alpha \, L^{-1}(\omega) \, c_\alpha \, d\omega.$$

Differentiating this equality, we obtain

$$\sum_{|\alpha|=m-n/2} \int\limits_{S^{n-1}} \omega^{\alpha+\beta} \, L^{-1}(\omega) \, c_\alpha \, d\omega = 0$$

for all multi-indices β, $|\beta| = m - n/2$. From this it follows that

$$\int\limits_{S^{n-1}} \left(L^{-1}(\omega) \, F(\omega), \, F(\omega) \right)_{\mathbb{C}^\ell} d\omega = 0, \qquad \text{where} \quad F(\omega) = \sum_{|\alpha|=m-n/2} c_\alpha \, \omega^\alpha.$$

Using the strong ellipticity of the operator L, we conclude that $F(\omega) = 0$ on S^{n-1} and, therefore, $c_\alpha = 0$.

Let U be an arbitrary solution of the system (10.2.1) which has the form (10.1.4) with $\lambda_0 = m - n/2$. By the second part of Theorem 10.2.1, there is the representation

$$U(x) = P_{m-n/2}(x) + \sum_{|\alpha|=m-n/2} D_x^\alpha G(x) \, c_\alpha,$$

where $P_{m-n/2} \in (\Pi^{(n)}_{m-n/2})^\ell$, $c_\alpha \in \mathbb{C}^\ell$. Thus,

$$(10.2.9) \qquad U(x) = \sum_{|\alpha|=m-n/2} D_x^\alpha R(x) \, c_\alpha \, \log r + r^{m-n/2} \, \Phi(\omega)$$

with a vector function $\Phi \in C^\infty(S^{n-1})^\ell$. By Theorem 1.1.5, the vector function (10.2.9) is a solution of (10.2.1) if and only if the coefficient of $\log r$ is an eigenvector and Φ is a generalized eigenvector corresponding to the eigenvalue $m - n/2$. We show that every vector-polynomial $p \in (\Pi^{(n)}_{m-n/2})^\ell$ has the representation

$$p(x) = \sum_{|\alpha|=m-n/2} D_x^\alpha R(x) \, c_\alpha,$$

where $c_\alpha \in \mathbb{C}^\ell$. To this end, we consider the mapping

$$\mathbb{C}^{\ell \, \kappa(m-n/2)} \ni \{c_\alpha\}_{|\alpha|=m-n/2} \xrightarrow{T} \sum_{|\alpha|=m-n/2} D_x^\alpha R(x) \, c_\alpha \in (\Pi^{(n)}_{m-n/2})^\ell.$$

As it was shown above, the kernel of T is trivial and $\dim(\Pi^{(n)}_{m-n/2})^\ell = \ell \, \kappa(m-n/2)$. Hence T is an isomorphism.

Thus, every eigenvector from $(\Pi^{(n)}_{m-n/2})^\ell|_{S^{n-1}}$ has a generalized eigenvector. The dimension of the space of these eigenvectors and generalized eigenvectors is equal to $2\ell\kappa(m - n/2)$. According to Theorem 10.2.2, this number coincides with the algebraic multiplicity of the eigenvalue $m - n/2$. Consequently, there are no other generalized eigenvectors. The theorem is proved. ∎

THEOREM 10.2.4. *Let $L(D_x)$ be strongly elliptic, n even, and $2m > n$. Then the geometric multiplicity of the eigenvalue $\lambda = k$, $k \in \{0, 1, \dots, 2m - n\}\backslash\{m - n/2\}$, is equal to*

(10.2.10) $\ell\kappa(2m - n - k)$ *if* $k = 0, \dots, m - n/2 - 1$,

(10.2.11) $\ell\kappa(k)$ *if* $k = m - n/2 + 1, \dots, 2m - n$.

In the first case only eigenvectors from $(\Pi_k^{(n)})^\ell|_{S^{n-1}}$ have generalized eigenvectors. In the second case the set of the eigenvectors coincides with $(\Pi_k^{(n)})^\ell|_{S^{n-1}}$.

Proof: Let $\lambda_0 = k > m - n/2$. By the second part of Theorem 10.2.1, every solution (10.1.4) of the system (10.2.1) can be written in the form

$$U(x) = P_k(x) + \sum_{|\alpha|=2m-n-k} D_x^\alpha G(x)\, c_\alpha,$$

where $P_k \in (\Pi_k^{(n)})^\ell$, $c_\alpha \in \mathbb{C}^\ell$. From (10.2.3) it follows that

(10.2.12) $$U(x) = \sum_{|\alpha|=2m-n-k} D_x^\alpha R(x)\, c_\alpha \log r + r^k\, \Phi(\omega),$$

where $\Phi \in C^\infty(S^{n-1})^\ell$ and R is a $\ell \times \ell$-matrix with elements from $(\Pi_{2m-n}^{(n)})^\ell$ which has the representation (10.2.7). We show that the mapping

$$\mathbb{C}^{\ell\kappa(2m-n-k)} \ni \{c_\alpha\}_{|\alpha|=2m-n-k} \xrightarrow{T_k} \sum_{|\alpha|=2m-n-k} D_x^\alpha R(x)\, c_\alpha \in (\Pi_k^{(n)})^\ell$$

has a trivial kernel. Let β be a multi-index of order $k - m + n/2$. Then

$$D_x^\beta T_k \{c_\alpha\}_{|\alpha|=2m-n-k} = T_{m-n/2} \{c_\gamma'\}_{|\gamma|=m-n/2},$$

where $c_\gamma' = c_\alpha$ if $\gamma = \alpha + \beta$, $c_\gamma' = 0$ else. In the proof of the Theorem 10.2.3 it was shown that the kernel of $T = T_{m-n/2}$ is trivial. Consequently, the same is true for the mapping T_k. From this it follows that the dimension of the range of T_k is equal to $\ell\kappa(2m - n - k)$.

Since the vector function (10.2.12) is a solution of (10.2.1) if and only if the coefficient of $\log r$ is an eigenvector and Φ is a generalized eigenvector, it follows that the dimension of the space of eigenvectors which have generalized eigenvectors is equal to $\ell\kappa(2m-n-k)$. Furthermore, by Theorem 10.2.2, the algebraic multiplicity of the eigenvalue $\lambda = k$ is equal to $\ell\,(\kappa(2m - n - k) + \kappa(k))$. Using $\dim(\Pi_k^{(n)})^\ell = \ell\kappa(k)$, we conclude that the set of eigenvectors coincides with $(\Pi_k^{(n)})^\ell|_{S^{n-1}}$.

Now let $\lambda_0 = k < m - n/2$. We show that the mapping T_k is surjective. Since $T_{m-n/2}$ is an isomorphism, the equation

$$T_{m-n/2} \{c_\gamma\}_{|\gamma|=m-n/2} = \frac{i^{|\beta|}\, \beta!}{(\alpha+\beta)!}\, x^{\alpha+\beta}$$

is solvable for all multi-indices α and β, $|\alpha| = k$, $|\beta| = m - n/2 - k$. The solution $\{c_\gamma\}$ satisfies the equality

$$D_x^\beta T_{m-n/2} \{c_\gamma\}_{|\gamma|=m-n/2} = x^\alpha$$

which can be rewritten in the form

$$T_k \{c'_\sigma\}_{|\sigma|=2m-n-k} = x^\alpha ,$$

where $c'_\sigma = c_\gamma$ if $\sigma = \beta + \gamma$, $c'_\sigma = 0$ else. Since α is an arbitrary multi-index with $|\alpha| = k$, we conclude from this that T_k is surjective.

Hence the coefficient of $\log r$ in (10.2.12) may be an arbitrary polynomial from $(\Pi_k^{(n)})^\ell$. Thus, the dimension of the set of eigenvectors which have generalized eigenvectors is equal to $\ell \kappa(k)$. Using the fact that the algebraic multiplicity of the eigenvalue $\lambda = k$ is equal to $\ell (\kappa(k) + \kappa(2m - n - k))$ (see Theorem 10.2.2), we find that the geometric multiplicity is equal to $\ell \kappa(2m - n - k)$. This proves the theorem. ∎

10.3. The Dirichlet problem in the half-space

Now we consider power-logarithmic solutions of the Dirichlet problem for the operator (10.1.2) in a half-space. We show that they are polynomials or linear combinations of derivatives of the Poisson kernels. This enables us to give a complete description of the spectrum of the pencil $\mathcal{L}$ for the case when Ω is a half-sphere.

10.3.1. Power-logarithmic solutions of the Dirichlet problem. Let L be the operator (10.1.2). We consider the Dirichlet problem

$$(10.3.1) \quad \begin{cases} L(D_x)\, U = 0 & \text{in } \mathbb{R}^n_+, \\[2mm] (\partial^k_{x_n} U)(x',0) = 0 & \text{on } \mathbb{R}^{n-1}, \ k = 0,1,\dots,m-1, \end{cases}$$

in the half-space $\mathbb{R}^n_+ = \{x = (x',x_n) : x' \in \mathbb{R}^{n-1}, x_n > 0\}$. As in the previous section, we are interested in solutions of the form

$$(10.3.2) \quad U(x) = r^{\lambda_0} \sum_{k=0}^{s} \frac{(\log r)^k}{k!} u_{s-k}(\omega),$$

where u_k are smooth vector functions on the closure of the half-sphere S^{n-1}_+. (In the case $\mathcal{K} = \mathbb{R}^n_+$ the generalized formulation of this problem given in the beginning of Section 10.1 is equivalent to the classical.)

THEOREM 10.3.1. *Let* $\operatorname{Re}\lambda_0 > m - n/2$, *and let the vector function* (10.3.2) *be a solution of the Dirichlet problem* (10.3.1). *Then* $\lambda_0 = m + k$ *with a nonnegative integer* k *and*

$$(10.3.3) \quad U(x) = x_n^m\, P_k(x), \qquad \text{where } P_k \in (\Pi_k^{(n)})^\ell.$$

The dimension of the space of these solutions is equal to

$$(10.3.4) \quad \begin{cases} \ell \dbinom{n+k-1}{n-1} & \text{if } k < m, \\[4mm] \ell \displaystyle\sum_{j=0}^{m-1} \dbinom{n-2+k-j}{n-2} & \text{if } k \geq m. \end{cases}$$

Proof: If $\operatorname{Re}\lambda_0 > m - n/2$, then $\zeta U \in \overset{\circ}{W}{}^m_2(\mathbb{R}^n_+)^\ell$ for all $\zeta \in C_0^\infty(\overline{\mathbb{R}^n_+})$. Therefore, from well-known regularity assertions for the generalized solution of the Dirichlet problem it follows that U is infinitely differentiable on $\overline{\mathbb{R}^n_+}$. Consequently, the number λ_0 in (10.3.2) is a nonnegative integer and U is a homogeneous vector-polynomial. Since U satisfies the homogeneous Dirichlet conditions on the hyperplane $x_n = 0$, we get $\lambda_0 = m + k$ with nonnegative integer k and $U(x) = x_n^m P_k(x)$, where $P_k \in (\Pi_k^{(n)})^\ell$.

If $k < m$, then P_k may be an arbitrary polynomial from $(\Pi_k^{(n)})^\ell$. Since the dimension of the space $\Pi_k^{(n)}$ is equal to $\binom{n+k-1}{n-1}$, we conclude that the space of the solutions (10.3.2) has the dimension (10.3.4).

In the case $k \geq m$ the desired formula for the dimension follows from the fact that the space of the solutions (10.3.3) coincides with the space of the solutions to the Cauchy problem (10.2.4), where $f = 0$, $g_0 = \cdots = g_{m-1} = 0$, and $g_{m+j} \in (\Pi_{k-j}^{(n-1)})^\ell$ for $j = 0, \ldots, m-1$. The theorem is proved. ∎

We denote by $G_j(x)$, $j = 0, 1, \ldots, m-1$, the Poisson kernels of the Dirichlet problem in $\mathbb{R}_+^n$, i.e., the solutions of the boundary value problems

$$L(D_x)\, G_j = 0 \quad \text{in } \mathbb{R}_+^n,$$
$$(\partial_{x_n}^j G_j)(x', 0) = I_\ell\, \delta(x') \quad \text{on } \mathbb{R}^{n-1},$$
$$(\partial_{x_n}^k G_j)(x', 0) = 0 \quad \text{on } \mathbb{R}^{n-1}, \quad k \neq j,\ 0 \leq k \leq m-1.$$

As is known (we refer to the works of Agmon, Douglis, Nirenberg [2] and Solonnikov [246]), there is the representation

$$G_j(x) = r^{j+1-n}\, G_j(\omega),$$

where the elements of the matrix $G_j(\omega)$ are smooth functions on $\overline{S_+^{n-1}}$.

THEOREM 10.3.2. *Let* $\operatorname{Re} \lambda_0 < m - n/2$. *Then the vector function (10.3.2) is a solution of the Dirichlet problem (10.3.1) if and only if* $\lambda_0 = m - n - k$ *with a nonnegative integer* k *and*

$$(10.3.5) \qquad U(x) = \sum_{j-|\alpha|=m-k-1} D_{x'}^\alpha G_j(x)\, c_{j,\alpha},$$

where $c_{j,\alpha} \in \mathbb{C}^\ell$. *The dimension of the space of these solutions is equal to (10.3.4).*

Proof: We calculate the dimension of the space of solutions having the form (10.3.5). It follows from the definition of the Poisson kernels G_j that the right-hand side of (10.3.5) is equal to zero only if all coefficients $c_{j,\alpha}$ are equal to zero. Consequently, the dimension of the considered space is equal to the number of all pairs (j, α) satisfying the conditions

$$\max(m - k - 1, 0) \leq j \leq m - 1, \quad |\alpha| = j - m + k + 1.$$

This number equals

$$\ell \sum_{j=\max(m-k-1,0)}^{m-1} \binom{n+j+k-m-1}{n-2},$$

i.e., it coincides with (10.3.4). Using Theorem 10.1.2 and applying Theorem 10.3.1 to the formally adjoint operator $L^+(D_x)$, we conclude that the number of linearly independent solutions (10.3.2) is equal to (10.3.4). Consequently, every solution (10.3.2) coincides with one of the solutions of the form (10.3.5). This proves the theorem. ∎

10.3.2. Spectral properties of the pencil $\mathcal{L}$ in the case when Ω is the half-sphere. As a consequence of Theorems 10.1.3, 10.3.1 and 10.3.2, we obtain the following assertion on the spectrum of the operator pencil $\mathcal{L}$ introduced in Section 10.1.

THEOREM 10.3.3. *The spectrum of the pencil $\mathcal{L}$ generated by the Dirichlet problem* (10.3.1) *consists of two nonintersecting sequences*

$$\{\lambda_k^{(1)} = m + k\}_{k=0,1,\dots} \qquad and \qquad \{\lambda_k^{(2)} = m - n - k\}_{k=0,1,\dots}.$$

The geometric and algebraic multiplicities of these eigenvalues are equal to (10.3.4).

The eigenfunctions corresponding to the eigenvalues $\lambda_k^{(1)}$ are the restrictions to S_+^{n-1} of vector-polynomials of the form (10.3.3) *which are solutions of the system $L(D_x)U = 0$. The eigenfunctions corresponding to the eigenvalues $\lambda_k^{(2)}$ are the restrictions to S_+^{n-1} of vector functions* (10.3.5).

10.4. The Sobolev problem in the exterior of a ray

In this section the exterior of a ray plays the role of the cone $\mathcal{K}$. We assume that $2m \geq n$ and consider the Dirichlet problem with zero Dirichlet data on the ray. For the corresponding operator pencil $\mathcal{L}$ we show that its spectrum coincides with the set

$$\{m - \frac{n-1}{2} + k\}_{k=0,\pm 1,\dots} \qquad \text{if } n \text{ is an odd number}$$

and with the set

$$\{m - \frac{n}{2} + \frac{k}{2}\}_{k=\pm 1,\pm 2\dots} \qquad \text{if } n \text{ is even.}$$

Moreover, we prove that there are no generalized eigenvectors and that the eigenvectors corresponding to integer eigenvalues are smooth on the unit sphere.

In the case of the polyharmonic operator an explicit representation of all eigenvectors will be given.

10.4.1. Formulation of the problem. Let $\mathbf{R}^1$ be the line $\{x = (x', x_n) \in \mathbb{R}^n : x' = 0\}$. Using the Fourier transform with respect to the variable x_n, it can be easily shown that an arbitrary distribution from $S'(\mathbb{R}^n)$ with support on the line $x' = 0$ is a finite sum of distributions $\tau_\alpha(x_n)\,(D_{x'}^\alpha \delta)(x')$, where $\tau_\alpha \in S'(\mathbb{R}^1)$. Consequently, every functional from $W_2^{-m}(\mathbb{R}^n)$ equal to zero on $C_0^\infty(\mathbb{R}^n \backslash \mathbf{R}^1)$ has the form

$$\sum_{|\alpha| \leq m - n/2} \tau_\alpha(x_n)\,(D_{x'}^\alpha \delta)(x'),$$

where $\tau_\alpha \in W_2^{|\alpha| - m + (n-1)/2}(\mathbf{R}^1)$. From this it follows that

(10.4.1) $\overset{\circ}{W}_2^m(\mathbb{R}^n \backslash \mathbf{R}^1) = \left\{ U \in W_2^m(\mathbb{R}^n) : D_{x'}^\alpha U = 0 \text{ on } \mathbf{R}^1 \text{ for } |\alpha| \leq m - n/2 \right\}.$

We denote by $\mathfrak{R}$ the ray $\{x = (x', x_n) \in \mathbb{R}^n : x' = 0, x_n \geq 0\}$ and consider the Sobolev problem

(10.4.2) $L(D_x)\,U = 0 \quad \text{in } \mathbb{R}^n \backslash \mathfrak{R},$

(10.4.3) $D_{x'}^\alpha U = 0 \quad \text{on } \mathfrak{R} \backslash \{0\}, \ |\alpha| \leq m - n/2,$

for the strongly elliptic differential operator (10.1.2).

We say that U is a solution of problem (10.4.2), (10.4.3) if $U \in \overset{\circ}{W}{}_2^m(\mathbb{R}^n \backslash \mathfrak{R}, 0)^\ell$ and (10.4.2) is satisfied in the distributional sense. By (10.4.1), the vector function U belongs to $\overset{\circ}{W}{}_2^m(\mathbb{R}^n \backslash \mathfrak{R}, 0)^\ell$ if and only if $U \in W_{2,loc}^m(\mathbb{R}^n \backslash \{0\})^\ell$ and condition (10.4.3) is satisfied.

Our goal is the description of all solutions which have the form

$$(10.4.4) \qquad U(x) = r^{\lambda_0} \sum_{k=0}^{s} \frac{(\log r)^k}{k!} \, u_{s-k}(\omega),$$

where $u_k \in \overset{\circ}{W}{}_2^m(S^{n-1} \backslash N)$. Here $N = (0, \dots, 0, 1)$ denotes the north pole of the unit sphere.

Let $\mathcal{L}(\lambda)$ and $\mathcal{L}^+(\lambda)$ be the operators introduced in Section 10.1, where $\Omega = S^{n-1} \backslash N$.

10.4.2. Properties of distributions with support on the ray. For the study of the spectrum of the pencil $\mathcal{L}$ we need the following two lemmas concerning distributions which are concentrated on the ray $\mathfrak{R}$. Proofs of the following lemmas are given in the book of Gel'fand and Shilov [**66**].

LEMMA 10.4.1. *Let $F \in S'(\mathbb{R}^n)$ be a positively homogeneous of degree $z \in \mathbb{C}$ distribution with support on the ray $\mathfrak{R}$. Then F has the representation*

$$F = \sum_{|\alpha| \le \mu} a_\alpha \frac{x_{n+}^{z+|\alpha|+n-1}}{\Gamma(z+|\alpha|+n)} D_{x'}^\alpha \, \delta(x'),$$

where a_α are constants.

Here $x_{n+}^{\lambda-1}$ denotes the distribution corresponding to the function $x_{n+}^{\lambda-1}$ which is equal to $x_n^{\lambda-1}$ for $x_n > 0$ and zero for $x_n \le 0$. Recall (see [**66**, Ch.1, Sect.3]) that $x_{n+}^{\lambda-1}$ is an analytic function of $\lambda \in \mathbb{C}$ and that $x_{n+}^{\lambda-1}/\Gamma(\lambda) = \delta^{(k)}(x_n)$ for $\lambda = -k$, $k = 0, 1, \dots$.

LEMMA 10.4.2. *Let $F \in S'(\mathbb{R}^n)$ be a distribution with support on the ray $\mathfrak{R}$ satisfying the equality*

$$F(tx) = t^z \big(F(x) + H(x) \log t \big) \qquad \text{for } t > 0,$$

where H is a positively homogeneous distribution of order $z \in \mathbb{C}$. Then F admits the representation

$$F = \sum_{|\alpha| \le \mu} \Big(a_\alpha + b_\alpha \frac{d}{dz} \Big) \frac{x_{n+}^{z+|\alpha|+n-1}}{\Gamma(z+|\alpha|+n)} \, D_{x'}^\alpha \delta(x'),$$

where a_α and b_α are constants.

10.4.3. Positively homogeneous solutions of the Sobolev problem. Let $2m \ge n$. In order to describe the eigenvalues and eigenfunctions of the operator pencil $\mathcal{L}$, we have to find all functions

$$U(x) = r^\lambda u(\omega), \qquad u \in \overset{\circ}{W}{}_2^m(S^{n-1} \backslash N)^\ell,$$

which are solutions of the equation

$$(10.4.5) \qquad L(D_x) U(x) = 0 \qquad \text{for } x \in \mathbb{R}^n \backslash \mathfrak{R}.$$

As in the previous section, we denote the set of the homogeneous polynomials of degree k with n variables by $\Pi_k^{(n)}$. The following lemma is essentially contained in Lemma 10.2.1.

LEMMA 10.4.3. *Let* $f \in (\Pi_{k-2m}^{(n)})^\ell$, $k \geq 2m$, $n \geq 2$, *and let* $c \in \mathbb{C}^\ell$ *be an arbitrary vector. Then there exists a polynomial* $p \in (\Pi_k^{(n)})^\ell$ *such that*

$$L(D_x)\, p = f \quad in\ \mathbb{R}^n,$$
$$D_{x'}^\alpha p|_{x'=0} = 0 \quad for\ |\alpha| \leq 2m-1,\ \alpha \neq (2m-1,0,\dots,0),$$
$$D_{x_1}^{2m-1} p|_{x'=0} = c\, x_n^{k-2m+1}.$$

Proof: By Lemma 10.2.1, there exists a uniquely determined solution $p \in (\Pi_k^{(n)})^\ell$ of the problem

$$L(D_x)\, p = f \quad in\ \mathbb{R}^n,$$
$$p|_{x_1=0} = \cdots = \partial_{x_1}^{2m-2} p|_{x_1=0} = 0, \quad \partial_{x_1}^{2m-1} p|_{x_1=0} = c\, x_n^{k-2m+1}.$$

Obviously, this polynomial also satisfies the condition $D_{x'}^\alpha p = 0$ on the line $x' = 0$ for $|\alpha| \leq 2m-1$, $\alpha \neq (2m-1,0,\dots,0)$. ∎

In particular, from Lemma 10.4.3 it follows that for every integer $k > m - n/2$ there exists a nontrivial solution $p \in (\Pi_k^{(n)})^\ell$ of the equations

$$L(D_x)\, p = 0 \quad in\ \mathbb{R}^n, \quad D_{x'}^\alpha p|_{x'=0} = 0 \quad for\ |\alpha| \leq m - n/2.$$

(For $m - n/2 < k < 2m-1$ we can choose $p = x_1^k$ and in the case $k \geq 2m$ the existence of p is ensured by the assertion of Lemma 10.4.3 for $f = 0$, $c \neq 0$.) Consequently, every integer number $k > m - n/2$ is an eigenvalue of the operator pencil $\mathcal{L}$.

Now we are interested in solutions $U(x) = r^\lambda u(\omega)$, $\mathrm{Re}\,\lambda \geq m - n/2$, of (10.4.5) which are not polynomials. Since $U \in S'(\mathbb{R}^n)^\ell$, according to Lemma 10.4.1, equation (10.4.5) can be written in the equivalent form

$$(10.4.6) \qquad L(D_x)\, U = \sum_{|\alpha| \leq m-n/2} c_\alpha \frac{x_{n+}^{\lambda+|\alpha|+n-2m-1}}{\Gamma(\lambda+|\alpha|+n-2m)}\, D_{x'}^\alpha \delta(x') \quad in\ \mathbb{R}^n.$$

Applying the Fourier transform $\hat{u}(\xi) = \int e^{ix\xi} u(x)\, dx$ to (10.4.6), we obtain (see [**66**, Ch.2, Sect.2])

$$L(\xi)\, \hat{U}(\xi) = \sum_{|\alpha| \leq m-n/2} e^{i(\lambda+|\alpha|+n-2m)\pi/2}\, (\xi_n + i0)^{-\lambda-|\alpha|-n+2m}\, c_\alpha\, \xi'^\alpha$$

or

$$(10.4.7) \quad \hat{U}(\xi) = \sum_{|\alpha| \leq m-n/2} e^{i(\lambda+|\alpha|+n-2m)\pi/2} (\xi_n + i0)^{-\lambda-|\alpha|-n+2m}\, \xi'^\alpha L^{-1}(\xi)\, c_\alpha.$$

The right-hand side of (10.4.7) is a distribution for $\lambda \neq 0,1,\dots$. The method of regularizing such distributions is given, for example, in [**66**, Ch.4]. Since $\hat{U} \in S'(\mathbb{R}^n)^\ell$, the right-hand side of (10.4.7) belongs to $S'(\mathbb{R}^n)^\ell$ also for the exceptional values of λ. For $\lambda = k$, $k = 0,1,\dots$, this is true if and only if

$$(10.4.8) \qquad \sum_\alpha \int_{S^{n-1}} \omega'^\alpha\, (\omega_n + i0)^{-k-|\alpha|-n+2m}\, \omega^\gamma\, L^{-1}(\omega)\, c_\alpha\, e^{i|\alpha|\pi/2}\, d\omega = 0$$

for $|\gamma| = k$, where

$$\omega' = \xi'/|\xi'|, \quad \omega_n = \xi_n/|\xi|$$

(see [**66**, Ch.3, Sect.3]).

The following two lemmas contain conditions on the coefficients c_α for the validity of (10.4.8).

LEMMA 10.4.4. *Let n be odd and let $k \geq m - n/2$ be an integer. Then (10.4.8) implies*

$$(10.4.9) \qquad c_\alpha = 0 \qquad for\ 2m - n + 1 - k \leq |\alpha| \leq m - (n+1)/2.$$

Proof: If $k = m - (n-1)/2$, then the set of multi-indices α subject to (10.4.9) is empty. Therefore, we may assume that $k \geq m - (n-3)/2$.

Let γ' be the multi-index formed by the first $n-1$ components of the multi-index $\gamma = (\gamma_1, \dots, \gamma_n)$. Then (10.4.8) implies

$$(10.4.10) \qquad \sum_{|\alpha| \leq m-n/2} \int_{S^{n-1}} \omega'^{\alpha+\gamma'} (\omega_n + i0)^{2m-n-|\alpha|-|\gamma'|} L^{-1}(\omega)\, d\omega\, h_\alpha = 0,$$

where $h_\alpha = e^{i|\alpha|\pi/2} c_\alpha \in \mathbb{C}^\ell$, $|\gamma'| \leq k$. We set

$$(10.4.11) \qquad I_{\alpha,\gamma'} = \int_{S^{n-1}} \omega'^{\alpha+\gamma'} (\omega_n + i0)^{2m-n-|\alpha|-|\gamma'|} L^{-1}(\omega)\, d\omega.$$

Then (10.4.10) can be written in the form

$$(10.4.12) \qquad \sum_{|\alpha| \leq m-n/2} I_{\alpha,\gamma'}\, h_\alpha = 0 \qquad for\ |\gamma'| \leq k.$$

After the change of variables $\omega \to -\omega$ in the integral (10.4.11) we get

$$(10.4.13) \qquad I_{\alpha,\gamma'} = -\int_{S^{n-1}} \omega'^{\alpha+\gamma'} (\omega_n - i0)^{2m-n-|\alpha|-|\gamma'|} L^{-1}(\omega)\, d\omega.$$

This and (10.4.11) yields

$$(10.4.14) \qquad I_{\alpha,\gamma'} = 0 \qquad for\ |\alpha| + |\gamma'| \leq 2m - n.$$

Let $2m - n - |\alpha| - |\gamma'| = -s$, $s > 0$. Using (10.4.11) and (10.4.13) together with the relation

$$(t + i0)^{-s} - (t - i0)^{-s} = -2i\frac{\pi(-1)^{s-1}}{(s-1)!} \delta^{(s-1)}(t)$$

(see [**66**, Ch.1, Sect.3]), we find that

$$I_{\alpha,\gamma'} = -i\frac{\pi(-1)^{s-1}}{(s-1)!} \int_{S^{n-1}} \omega'^{\alpha+\gamma'} \delta^{(s-1)}(\omega_n)\, L^{-1}(\omega)\, d\omega.$$

In particular, for $|\alpha| + |\gamma'| = 2m - n + 1$ we have

$$I_{\alpha,\gamma'} = -i\pi \int_{S^{n-1}} \omega'^{\alpha+\gamma'} L^{-1}(\omega', 1)\, d\omega'.$$

We prove (10.4.9) by induction in $|\alpha|$. First let $|\alpha| = m - (n+1)/2$. Then (10.4.12) and (10.4.14) imply

$$\sum_{|\alpha|=m-(n+1)/2} I_{\alpha,\gamma'}\, h_\alpha = 0 \qquad \text{for } |\gamma'| = m - (n-3)/2.$$

We show that the matrix $\left(I_{\alpha,\gamma'}\right)_{|\alpha|=m-(n+1)/2,|\gamma'|=m-(n-3)/2}$ has maximal rank which will obviously imply that $c_\alpha = h_\alpha = 0$ for $|\alpha| = m - (n+1)/2$. For this we consider the square matrix

$$(10.4.15) \qquad \left(\int_{S^{n-2}} \omega'^{\alpha+\beta}\, \omega_1^2\, L^{-1}(\omega',1)\, d\omega' \right)_{|\alpha|=|\beta|=m-(n+1)/2}$$

which is obtained for $\gamma' = (2,0,..,0)+\beta$. Let $\{f_\alpha\}_{|\alpha|=m-(n+1)/2}$ be an arbitrary collection of vectors in $\mathbb{C}^\ell$ not vanishing simultaneously. Since the matrix $\operatorname{Re} L^{-1}(\omega',1)$ is positive definite, we have

$$(10.4.16) \qquad \operatorname{Re} \sum_{|\alpha|=|\beta|=m-(n+1)/2} \left(\int_{S^{n-2}} \omega'^{\alpha+\beta}\, \omega_1^2\, L^{-1}(\omega',1)\, d\omega'\, f_\alpha \,,\, f_\beta \right)_{\mathbb{C}^\ell}$$

$$= \operatorname{Re} \int_{S^{n-2}} \omega_1^2 \left(L^{-1}(\omega',1)P(\omega')\,,\, P(\omega') \right)_{\mathbb{C}^\ell} d\omega > 0,$$

where $P(\omega') = \sum_{|\alpha|=m-(n+1)/2} f_\alpha\, \omega'^\alpha$. From the last equality it follows that the matrix (10.4.15) is regular. Hence, the matrix $(I_{\alpha,\gamma'})$ has maximal rank and we obtain $c_\alpha = 0$ for $|\alpha| = m - (n+1)/2$.

Suppose that $c_\alpha = 0$ for $|\alpha| \geq d > 2m - n + 1 - k$. From (10.4.11), by the induction hypothesis and (10.4.13), we get

$$\sum_{|\alpha|=d-1} I_{\alpha,\gamma'}\, h_\alpha = 0 \qquad \text{for } |\gamma'| = 2m - n - d + 2.$$

(Note that $2m - n - d + 2 \leq k$.) To prove that $h_\alpha = 0$ for $|\alpha| = d - 1$, it suffices to check that the rank of the matrix $(I_{\alpha,\gamma'})_{|\alpha|=d-1,\,|\gamma'|=2m-n-d+2}$ is maximal. To this end, we consider the matrix

$$(10.4.17) \qquad \left(\int_{S^{n-2}} \omega'^{\alpha+\beta}\, \omega_1^{2m-n-2d+3}\, L^{-1}(\omega',1)\, d\omega' \right)_{|\alpha|=|\beta|=d-1}$$

which is obtained for $\gamma' = (2m - n - 2d + 3, 0, .., 0) + \beta$. Here $2m - n - 2d + 3$ is an even positive number. In the same way as for the matrix (10.4.15), it can be verified that the matrix (10.4.17) is regular. Consequently, $c_\alpha = 0$ for $|\alpha| = d - 1$. The proof is complete. ∎

LEMMA 10.4.5. *Let n be even and let $k \geq m-n/2$ be an integer. Then* (10.4.8) *implies that $c_\alpha = 0$ for $|\alpha| \leq m - n/2$.*

Proof: Let $\gamma = (\gamma', \gamma_n)$, $|\gamma'| \leq m - n/2$. Since the integrand in (10.4.8) is an even function, we get

$$(10.4.18) \qquad \sum_{|\alpha|\leq m-n/2} \int_{S^{n-1}_+} \omega'^{\alpha+\beta}\, \omega_n^{2m-n-|\alpha|-|\beta|}\, L^{-1}(\omega)\, d\omega\, e^{i|\alpha|\pi/2}\, c_\alpha = 0$$

for all multi-indices β, $|\beta| \leq m - n/2$. Here S_+^{n-1} denotes the half-sphere $\{\omega \in S^{n-1} : \omega_n > 0\}$. We make the change of variables $\omega = (1 + |\eta|^2)^{-1/2} (\eta, 1)$ or $\eta = \omega'/\omega_n$. Then $d\omega = \omega_n^n d\eta$, and (10.4.18) takes the form

$$(10.4.19) \qquad \sum_{|\alpha| \leq m-n/2} \int_{\mathbb{R}^{n-1}} \eta^{\alpha+\beta} L^{-1}(\eta, 1) \, d\eta \, e^{i|\alpha|\pi/2} \, c_\alpha = 0$$

for $|\beta| \leq m - n/2$. We show that the matrix

$$(10.4.20) \qquad \left(\int_{\mathbb{R}^{n-1}} \eta^{\alpha+\beta} L^{-1}(\eta, 1) \, d\eta \right)_{|\alpha|,|\beta| \leq m-n/2}$$

is nonsingular. Let $\{f_\alpha\}_{|\alpha| \leq m-n/2}$ be an arbitrary collection of vectors in $\mathbb{C}^\ell$ not vanishing simultaneously. Since the matrix $\operatorname{Re} L^{-1}(\eta, 1)$ is positive definite, we have

$$\operatorname{Re} \sum_{|\alpha|,|\beta| \leq m-n/2} \left(\int_{\mathbb{R}^{n-1}} \eta^{\alpha+\beta} L^{-1}(\eta, 1) \, d\eta \, f_\alpha \, , \, f_\beta \right)_{\mathbb{C}^\ell}$$

$$= \operatorname{Re} \int_{\mathbb{R}^{n-1}} \left(L^{-1}(\eta, 1) Q(\eta), Q(\eta) \right) d\eta > 0,$$

where $Q(\eta) = \sum_{|\alpha| \leq m-n/2} f_\alpha \, \eta^\alpha$. From this we conclude that the matrix (10.4.20) is nonsingular and from (10.4.19) it follows that $c_\alpha = 0$ for $|\alpha| = m - n/2$. ∎

10.4.4. Eigenvalues and eigenfunctions of the pencil $\mathcal{L}$. As in Section 10.2, let $G(x)$ be the Green matrix for the operator $L(D_x)$ in $\mathbb{R}^n$ (i.e., the solution of (10.2.2)). Recall that $G(x)$ admits the representation (10.2.3).

Using Lemmas 10.4.1 and 10.4.2, we can prove the following theorem which gives a complete description of the eigenvalues of the operator pencil $\mathcal{L}$ and of the eigenfunctions corresponding to integer eigenvalues.

THEOREM 10.4.1. *Let $2m \geq n$. Then the eigenfunctions of the operator pencil $\mathcal{L}$ have no generalized eigenfunctions. The integer numbers $\mu_k = k$ (if n is even, then $k \neq m - n/2$) are eigenvalues of $\mathcal{L}$. Here the multiplicities of the eigenvalues μ_k and μ_{2m-n-k} coincide. Furthermore, the following assertions are valid:*

1) If n is odd, then the spectrum of the pencil $\mathcal{L}$ is exhausted by the eigenvalues μ_k, $k = 0, \pm 1, \ldots$. For every eigenfunction u corresponding to the eigenvalue μ_k there is the representation

$$(10.4.21) \qquad r^k u(\omega) = \sum_{|\alpha|=k} p_\alpha x^\alpha + \sum_{|\alpha|=2m-n-k} D_x^\alpha G(x) q_\alpha ,$$

where $p_\alpha, q_\alpha \in \mathbb{C}^\ell$. In the case $k < 0$ the first sum is absent, while the second sum is absent in the case $k > 2m - n$. There are no generalized eigenfunctions.

2) Let n be an even number and let u be an eigenfunction corresponding to the eigenvalue μ_k. Then

$$(10.4.22) \quad r^k u(\omega) = \sum_{|\alpha|=k} p_\alpha x^\alpha \quad \text{for } k > m - n/2,$$

$$(10.4.23) \quad r^k u(\omega) = \sum_{|\alpha|=k} p_\alpha x^\alpha + \sum_{|\alpha|=2m-n-k} D_x^\alpha G(x) q_\alpha \quad \text{for } k < m - n/2,$$

where p_α, $q_\alpha \in \mathbb{C}^\ell$. The first sum in (10.4.23) is absent if $k < 0$. Additionally to the eigenvalues $\mu_k = k$, there are also the eigenvalues $\lambda_k = m - (n-1)/2 + k$, $k = 0, \pm 1, \dots$, which have the multiplicities $\ell \binom{m+n/2-1}{n-1}$. Other eigenvalues do not exist. Both the eigenvalues λ_k and μ_k do not have generalized eigenfunctions.

Proof: Let $\lambda = k$, $k = 0, 1, \dots$. Then, according to Lemmas 10.4.4 and 10.4.5, the right-hand side of (10.4.6) contains only derivatives of the Dirac function for odd n and vanishes for even n. Thus, we arrive at (10.4.21) and (10.4.22) for $k \geq m-n/2$.

Let λ be a noninteger. From (10.4.7) we find that

$$(10.4.24) \qquad U(x) = (2\pi)^{-n} \int_{\mathbb{R}^n} e^{-ix\xi} \sum_\alpha e^{i(\lambda+|\alpha|+n-2m)\pi/2} \, \xi'^\alpha$$

$$\times (\xi_n + i0)^{-\lambda-|\alpha|-n+2m} \, L^{-1}(\xi) \, c_\alpha \, d\xi.$$

Using the identity

$$(2\pi)^{-n} \int_{\mathbb{R}^{n-1}} L^{-1}(\xi) \, \xi'^{\alpha+\beta} \, d\xi' = S_{\alpha+\beta} \, |\xi_n|^{-2m+n-1} \, \xi_n^{|\alpha|+|\beta|},$$

where

$$S_{\alpha+\beta} = (2\pi)^{-n} \int_{\mathbb{R}^{n-1}} \xi'^{\alpha+\beta} \, L^{-1}(\xi', 1) \, d\xi',$$

we obtain

$$(10.4.25) \ (-D_{x'})^\beta \, U(x)|_{x'=0} = \sum_\alpha e^{i(|\alpha|-|\beta|)\pi/2} \, \Gamma(-\lambda + |\beta|) \, S_{\alpha+\beta} \, c_\alpha$$

$$\times \left(e^{i(-\lambda-n/2+m)\pi} \, (x_n + i0)^{\lambda-|\beta|} + e^{i(\lambda+n/2-m)\pi} \, (x_n - i0)^{\lambda-|\beta|} \right)$$

for $|\beta| \leq m - n/2$. We seek the vectors c_α from the relations

$$(-D_{x'})^\beta \, U(x)|_{x'=0} \qquad \text{for } x_n > 0, \ |\beta| \leq m - n/2.$$

By (10.4.25), we have the system of algebraic equations

$$(10.4.26) \qquad 2\Gamma(-\lambda + |\beta|) \, \cos(\lambda - m + n/2)\pi \sum_\alpha e^{i(|\alpha|-|\beta|)\pi/2} \, S_{\alpha+\beta} \, c_\alpha = 0,$$

where $|\beta| \leq m - n/2$. As it was shown in the proof of Lemma 10.4.4, the matrix $S_{\alpha+\beta}$ is regular. The diagonal matrix

$$\left(\Gamma(-\lambda + |\beta|) \, \delta_{\alpha,\beta} \, \cos(\lambda - m + n/2)\pi \right)_{|\alpha|,|\beta|\leq m-n/2}$$

is nonsingular for odd dimension n if $\mathrm{Re}\,\lambda \geq m-n/2$ and turns into the zero matrix for even dimension n if λ is half-integer. For other values of λ which differ from integers, this matrix is also nonsingular. Thus, we have proved that the spectrum of the operator pencil $\mathcal{L}$ in the half-plane $\mathrm{Re}\,\lambda \geq m - n/2$ is exhausted by integers for odd n and that the corresponding eigenfunctions have the representation (10.4.21). If n is even, then, additionally to the eigenvalues whose eigenfunctions have the form (10.4.22), there are half-integer eigenvalues of multiplicity $\ell \binom{m+n/2-1}{n-1}$. Other eigenvalues do not exist. The representation (10.4.22) and the equality $D_{x'}^\alpha U = 0$ for $x' = 0$, $|\alpha| \leq m - n/2$ imply that the number $\lambda = m - n/2$ is not an eigenvalue for even n.

Now we show that the eigenfunctions have no generalized eigenfunctions. Let λ be an eigenvalue and let u be an eigenfunction corresponding to this eigenvalue. Suppose there exists a function $v \in \overset{\circ}{W}_2^m(S^{n-1}\backslash N)^\ell$ such that

$$U(x) = r^\lambda \left(u(\omega) \log r + v(\omega) \right)$$

is a solution of the equation (10.4.5). First let $\lambda = k$, where $k \geq m - n/2$ is an integer. Since we proved that $u \in C^\infty(S^{n-1})^\ell$, the distribution $L(D_x)U$ is positively homogeneous of degree $k - 2m$ on $\mathbb{R}^n$ with support on the ray $\mathfrak{R}$. This and Lemma 10.4.1 imply that U satisfies (10.4.6). By Lemmas 10.4.4 and 10.4.5, the coefficients c_α on the right-hand side of (10.4.6) vanish for $|\alpha| \geq 2m - n + 1 - k$ if n is odd and for all α if n is even. Consequently, for even n the function U is a sum of a polynomial and of derivatives of the Green function, and for odd n it is a polynomial. This contradicts the equality $U(x) = r^k \left(u(\omega) \log r + v(\omega) \right)$, $u \neq 0$.

Now let n be even and let $\lambda = k/2$, where k is odd, $k > 2m - n$. Obviously, the distribution $L(D_x)U$ on $\mathbb{R}^n$ satisfies the condition of Lemma 10.4.2 for $z = k/2 - 2m$. Therefore,

$$L(D_x)U = \sum_{|\alpha| \leq m-n/2} \left(c_\alpha + b_\alpha \frac{d}{d\lambda} \right) \frac{x_{n+}^{\lambda+|\alpha|+n/2-2m-1}}{\Gamma(\lambda + |\alpha| + n - 2m)}\bigg|_{\lambda=k/2} D_{x'}^\alpha \, \delta(x'),$$

where c_α and b_α are vectors in $\mathbb{C}^\ell$. Using the same arguments as in the proof of (10.4.26), we get

$$\frac{d}{d\lambda} \sum_{|\alpha| \leq m-n/2} e^{i(|\alpha|-|\beta|)\pi/2} \, \Gamma(-\lambda + |\beta|) \, \cos(\lambda - m + n/2)\pi \, S_{\alpha+\beta} \, b_\alpha$$

$$+ \sum_{|\alpha| \leq m-n/2} e^{i(|\alpha|-|\beta|)\pi/2} \, \Gamma(-\lambda + |\beta|) \, \cos(\lambda - m + n/2)\pi \, S_{\alpha+\beta} \, c_\alpha = 0$$

for $\lambda = k/2$. This implies that $b_\alpha = 0$, i.e., U satisfies (10.4.6). From the representation (10.4.24) for solutions of (10.4.6) we deduce that U is positively homogeneous which contradicts the equality $U = r^{k/2} \left(u(\omega) \log r + v(\omega) \right)$.

Thus, we have proved the theorem for $\operatorname{Re} \lambda \geq m - n/2$. Using Lemma 10.1.2 and the fact that all we established above is also valid for $\mathcal{L}^+(\lambda)$, we get all the assertions of the theorem for $\operatorname{Re} \lambda < m - n/2$, except the representations (10.4.21), (10.4.22) for the eigenfunctions corresponding to the eigenvalues μ_k, $k < m - n/2$.

Let $u \in \overset{\circ}{W}_2^m(S^{n-1}\backslash N)^\ell$ be the eigenfunction corresponding to the eigenvalue μ_k. We show that

$$U(x) = r^k u(\omega) \in S'(\mathbb{R}^n)^\ell.$$

For $k > -n$ this inclusion is obvious. Therefore, we may assume that $k = -n - s$, where s is a nonnegative integer. We have to establish the equalities

$$(10.4.27) \qquad \int_{S^{n-1}} u(\omega) \, \omega^\gamma \, d\omega = 0 \quad \text{for } |\gamma| = s$$

(see Gel'fand and Shilov [**66**, Ch.3, Sec.3]). Applying the assertion of Lemma 10.4.3 to the operator $L^+(D_x)$, we conclude that the problem

$$\mathcal{L}^+(2m + s) \, v = c \omega^\gamma,$$

$$D_{x'}^\alpha \, v = 0 \quad \text{for } \omega = N, \, |\alpha| \leq m - n/2$$

is solvable for all $c \in \mathbb{C}^\ell$, $|\gamma| = s$. Since

$$0 = \int_{S^{n-1}} \big(\mathcal{L}(k)u, v\big)_{\mathbb{C}^\ell} d\omega = \int_{S^{n-1}} \big(u, \mathcal{L}^+(2m + s)v\big)_{\mathbb{C}^\ell} d\omega$$

for all $v \in \overset{\circ}{W}{}_2^m(S^{n-1}\backslash N)^\ell$, we get (10.4.27).

Thus, U is a positively homogeneous distribution of degree k. Consequently, by Lemma 10.4.1, we have the equality

$$L(D_x)\,U = \sum_{|\alpha|=k-2m+n} c_\alpha\, D_x^\alpha\, \delta(x) \qquad \text{on } \mathbb{R}^n$$

which is equivalent to (10.4.21), (10.4.22). The proof of the theorem is complete. ∎

10.4.5. The polyharmonic operator. We consider the Sobolev problem (10.4.2), (10.4.3) for the operator $L(D_x) = (-\Delta)^m$, $2m > n$. The corresponding differential operator on the unit sphere S^{n-1} has the form

$$\mathcal{L}(\lambda) = (-1)^m \prod_{j=0}^{m-1} \Big(\delta + (\lambda - 2j)(\lambda - 2j + n - 2)\Big),$$

where δ is the Beltrami operator on S^{n-1} (see Chapter 7). Obviously,

(10.4.28) $$\mathcal{L}(2m - n - \lambda) = \mathcal{L}(\lambda).$$

Therefore, the eigenfunctions corresponding to the eigenvalues λ and $2m - n - \lambda$ coincide, and we can restrict ourselves to the eigenvalues to the right of the point $m - n/2$.

The case of odd dimension. Let n be an odd number. In this case the spectrum consists of the integers

$$k = m - n/2 + 1/2,\ m - n/2 + 3/2,\ \dots\ .$$

The corresponding eigenfunctions for $k > 2m - n$ are the traces of polynomials of degree k satisfying (10.4.3) on S^{n-1}. For $k \leq 2m - n$ the eigenfunctions have the form

(10.4.29) $$u(\omega) = \sum_{|\alpha|=k} p_\alpha\, \omega^\alpha + \sum_{|\alpha|=2m-n-k} r^{-k}\, D_x^\alpha\, r^{2m-n}\, q_\alpha\,.$$

By (10.4.28), the functions $x^\gamma\, r^{-2m+n+k}$, $|\gamma| = 2m - n - k$, are also eigenfunctions corresponding to k. Therefore, (10.4.21) can be rewritten in the form

$$u(\omega) = \sum_{|\alpha|=k} p_\alpha\, \omega^\alpha + \sum_{|\gamma|=2m-n-k} q_\gamma\, \omega^\gamma.$$

Naturally, the boundary conditions (10.4.3) have to be satisfied.

The case of even dimension. If n is even, then the spectrum consists of integer eigenvalues

$$k = m - n/2 + 1,\ m - n/2 + 2,\ \dots$$

and half-integer eigenvalues

$$\lambda_k = m - n/2 + k + 1/2, \qquad k = 0, 1, \dots\ .$$

The eigenfunctions corresponding to integer eigenvalues are restrictions of polynomials. We consider the eigenvalues λ_k.

LEMMA 10.4.6. *Let n be even, $n \geq 4$. Then*

$$(10.4.30) \qquad \int\limits_{\mathbb{R}^{n-1}} \frac{e^{ix'\xi'}}{(1+|\xi'|^2)^{n/2}}\, d\xi' = \frac{(2\pi)^{n/2}}{\Gamma(n/2-1)}\, e^{-|x'|}.$$

Proof: We denote the left-hand side of (10.4.30) by $I(x')$. Using the equality

$$\int\limits_{S^{n-2}} e^{ix'\xi'}\, d\omega' = 2\pi \left(\frac{|x'|\cdot|\xi'|}{2\pi}\right)^{-(n-3)/2} J_{(n-3)/2}(|x'|\cdot|\xi'|)$$

wich can be found in the book by Stein and Weiss [**249**, Ch.4, §3], where $\omega' = \xi'/|\xi'|$, we get

$$I(x') = 2\pi \left(\frac{|x'|}{2\pi}\right)^{-(n-3)/2} \int\limits_0^\infty \frac{\rho^{(n-1)/2}}{(1+\rho^2)^{n/2}}\, J_{(n-3)/2}(|x'|\rho)\, d\rho.$$

This together with formula (51) in the book by Bateman, Erdélyi et al. [**11**, Vol.2,p.95] implies (10.4.30). ∎

LEMMA 10.4.7. *Let $n \geq 4$ and $k = 0, \pm 1, \ldots$. Then*

$$(10.4.31) \qquad \Delta^{n/2}(x_n + i|x'|)^{k+1/2} = c\,(x_n + i0)^{k-1/2}\,\delta(x'),$$

where $c = -i(k+1/2)\, 2^{n-2}\, \pi^{n/2-1}\, \Gamma(n/2-1)$.

Proof: From (10.4.30) we get

$$(-\Delta_{x'} + 1)^{n/2}\, e^{-|x'|} = c_n\, \delta(x'),$$

where $c_n = 2^{n-2}\, \pi^{n/2-1}\, \Gamma(n/2-1)$. Therefore,

$$(-\Delta_{x'} + \tau^2)^{n/2}\, e^{-|\tau x'|} = c_n\, |\tau|\, \delta(x')$$

for all $\tau \in \mathbb{R}$. Multiplying both sides of this equality by $\tau_+^{-k-3/2}$ and applying the Fourier transform $\mathcal{F}_{\tau \to x_n}$, we obtain (10.4.31). ∎

For $n \geq 4$ let $\theta = \theta(\omega) \in (0, \pi]$ be the angle between the rays $(0, N)$ and $(0, \omega)$, i.e., $\cos\theta = \omega_n$. If $n = 2$, then $\theta \in [0, 2\pi]$ is the polar angle taken counterclockwise from the axis $(0, N)$. We consider the cases $2m = n$ and $2m > n$ separately.

The case $2m = n$. We show that the eigenfunctions corresponding to the eigenvalue $\lambda_k = k + 1/2$, $k \in \mathbb{Z}$, are exhausted by the functions $c \sin(k+1/2)\theta$, $c \in \mathbb{C}$.

If $n \geq 4$, then by Lemma 10.4.7,

$$(10.4.32) \qquad \Delta^{n/2}\, r^{k+1/2}\, \sin(k+1/2)\theta = 0 \qquad \text{on } \mathbb{R}^n\backslash\mathfrak{R}, \quad k \in \mathbb{Z}.$$

For $n = 2$ this equality can be easily verified. Since the functions $\sin(k+1/2)\theta$ are smooth on $S^{n-1}\backslash N$ and belong to $\overset{\circ}{W}{}_2^{n/2}(S^{n-1}\backslash N)$, they are eigenfunctions. By the second part of Theorem 10.4.1, there are no other eigenfunctions.

The case $2m > n$, n even. We check that the functions

$$\sin(m - n/2 + k - 2j + 1/2)\theta, \qquad j = 0, 1, \ldots, m - n/2,$$

satisfy the equation $\mathcal{L}(\lambda_k)\, u = 0$ on $S^{n-1}\backslash N$, where $\lambda_k = m - n/2 + k + 1/2$.

Indeed, (10.4.32) implies

$$\Delta^m \, x_n^j \, r^{m-n/2-j+k+1/2} \, \sin(m - n/2 + k - j + 1/2)\theta = 0$$

for $j = 0, 1, \ldots, m - n/2$. Consequently, the functions

$$(\cos\theta)^j \, \sin(m - n/2 + k - j + 1/2)\theta$$

satisfy the equation $\mathcal{L}(\lambda_k)\, u = 0$ on $S^{n-1}\backslash N$. It can be easily seen that the linear span of the functions $(\cos\theta)^j \, \sin(m - n/2 + k - j + 1/2)\theta$ coincides with the linear span of the functions $\sin(m - n/2 + k - 2j + 1/2)\theta$, $j = 0, \ldots, m - n/2$.

We construct functions

$$(10.4.33) \qquad P_k(\theta) = \sum_{j=0}^{m-n/2} c_j^{(k)} \, \sin(m - n/2 + k - 2j + 1/2)\theta,$$

satisfying the equalities

$$\frac{d^{2s+1}}{d\theta^{2s+1}} P_k(\theta)\Big|_{\theta=0} = 0, \qquad s = 0, 1, \ldots, m - \frac{n}{2} - 1.$$

Inserting (10.4.33) into the last equalities, we get the following algebraic system for the coefficients $c_j^{(k)}$:

$$(10.4.34) \qquad \sum_{j=0}^{m-n/2} c_j^{(k)} \, (m - n/2 + k - 2j + 1/2)^{2s+1} = 0.$$

Putting $c_{m-n/2}^{(k)} = 1$, we obtain a uniquely solvable system.

The functions $P_k(\theta)$ are solutions of the equations $\mathcal{L}(\lambda_k)u = 0$ on $S^{n-1}\backslash N$. Since

$$(10.4.35) \qquad P_k(\theta) = O(\theta^{2m-n+1}) \quad \text{for small } \theta,$$

it follows that

$$(10.4.36) \qquad D_{x'}^\alpha \, P_k(\theta)|_{\theta=0} = 0 \quad \text{for } |\alpha| \leq 2m - n.$$

We set

$$(10.4.37) \qquad u_\alpha(\omega) = r^{-\lambda_k} D_x^\alpha \left(r^{\lambda_k + m - n/2} P_{k+m-n/2}(\theta)\right),$$

where α is an arbitrary multi-index of order $m - n/2$. One can readily verify that $u_\alpha \in \overset{\circ}{W}{}_2^m(S^{n-1}\backslash N)$. Since $V_k(x) = r^{\lambda_k + m - n/2} P_{k+m-n/2}(\theta)$ is a solution of the equation $\Delta^m V = 0$ on $\mathbb{R}^n\backslash\mathfrak{R}$, the functions u_α satisfy the equality $\mathcal{L}(\lambda_k)\, u_\alpha = 0$ on $S^{n-1}\backslash N$. Thus, these functions are eigenfunctions.

We show that the functions u_α are linearly independent on $S^{n-1}\backslash N$. Suppose that

$$(10.4.38) \qquad \sum_{|\alpha|=m-n/2} c_\alpha D_x^\alpha \left(r^{\lambda_k + m - n/2} P_{k+m-n/2}(\theta)\right) = 0$$

outside $\mathfrak{R}$. From (10.4.33), (10.4.34) it follows that

$$(10.4.39) \qquad P_k(\theta) = c\,\theta^{2m-n+1} + O(\theta^{2m-n+3}) \quad \text{as } \theta \to 0,$$

where

$$(10.4.40) \qquad c = (-1)^{m-n/2} \sum_{j=0}^{m-n/2} c_j^{(k)} \, (m - n/2 + k - 2j + 1/2)^{2m-n+1}.$$

The number c does not vanish, since otherwise all the coefficients $c_j^{(k)}$ would vanish by (10.4.34) and (10.4.40). According to (10.4.39), we have

$$(10.4.41) \qquad V_k(x) = c \, x_n^{k-1/2} \, |x'|^{2m-n+1} + F(x', x_n)$$

in a small neighborhood of the point $x' = 0$, $x_n = 1$, where

$$|D_{x'}^{\alpha'} D_{x_n}^{\alpha_n} F(x', x_n)| \le c_{\alpha', \alpha_n} \, |x'|^{2m-n+3-|\alpha'|}.$$

Let

$$\sum_{|\alpha|=m-n/2} c_\alpha D_x^\alpha = \sum_{j=0}^{m-n/2} Q_j(D_{x'}) D_{x_n}^j \, ,$$

where $Q_j(D_{x'})$ are homogeneous differential operators of order $m - n/2 - j$. From (10.4.38) it follows that

$$Q_0(D_{x'}) \, |x'|^{2m-n+1} = 0 \qquad \text{on } \mathbb{R}^{n-1}.$$

Applying the $(n-1)$-dimensional Fourier transform to this equality, we get $Q_0 = 0$. Suppose that the operators $Q_0, Q_1, \ldots, Q_{s-1}$ are zero. Then, by (10.4.38) and (10.4.41), we have $Q_s(D_{x'}) \, |x'|^{2m-n+1} = 0$ or, equivalently, $Q_s = 0$. Thus, the functions u_α are linearly independent.

Since the functions $\{u_\alpha\}_{|\alpha|=m-n/2}$ span a $\binom{m+n/2+1}{n-1}$-dimensional space, it follows from the second part of Theorem 10.4.1 that other eigenfunctions do not exist.

10.5. The Dirichlet problem in a dihedron

In this section the Dirichlet problem for strongly elliptic systems in the cone

$$\mathcal{K} = \mathcal{K}_d \times \mathbb{R}^{n-d}$$

is considered, where $\mathcal{K}_d$ is an arbitrary open cone in $\mathbb{R}^d$ and $n > d > 1$. We describe the eigenvalues, eigenvectors and generalized eigenvectors in terms of those for the Dirichlet problem in the cone $\mathcal{K}_d$.

10.5.1. The operator pencils $\mathcal{L}$ and $\mathcal{L}_d$. We represent $\mathbb{R}^n$, $n \ge 3$, as the Cartesian product $\mathbb{R}^d \times \mathbb{R}^{n-d}$ and use the notation

$$x = (y, z), \quad y = (y_1, \ldots, y_d), \quad z = (z_1, \ldots, z_{n-d}).$$

Let (r, ω), (ρ, ϕ), and (σ, ϑ) be the spherical coordinates in $\mathbb{R}^n$, $\mathbb{R}^d$ and $\mathbb{R}^{n-d}$, respectively, where $r = |x|$, $\rho = |y|$, $\sigma = |z|$, $\omega \in S^{n-1}$, $\phi \in S^{d-1}$, $\vartheta \in S^{n-d-1}$.

We consider the open d-dimensional cone

$$\mathcal{K}_d = \{y \in \mathbb{R}^d \, : \, \rho > 0, \, \phi \in \Omega_d\},$$

where Ω_d is a domain on the sphere S^{d-1}, $\Omega_d \ne S^{d-1}$. Suppose that the cone

$$\mathcal{K} = \{x \in \mathbb{R}^n \, : \, r > 0, \, \omega \in \Omega\}$$

can be represented as the product $\mathcal{K}_d \times \mathbb{R}^{n-d}$. In this case $\omega \in \Omega$ if and only if

$$\omega = \big(\phi \cos \tau, \, \vartheta \sin \tau\big), \quad \text{where } 0 < \tau < \pi/2, \, \phi \in \Omega_d, \, \vartheta \in S^{n-d-1}.$$

We assume again that the differential operator

$$L(D_x) = \sum_{|\alpha|=2m} A_\alpha D_x^\alpha,$$

where A_α are constant $\ell \times \ell$-matrices, is strongly elliptic and denote by $\mathcal{L}$ the operator pencil generated by the Dirichlet problem for L (see Section 10.1).

As we noted in Section 10.1, the vector function (10.1.4) is a solution of the system (10.1.3) if and only if λ is an eigenvalue of the pencil $\mathcal{L}$ and the functions $u_0, \ldots, u_s$ form a Jordan chain corresponding to this eigenvalue. Furthermore, the following assertion holds.

LEMMA 10.5.1. *Let U be a solution of the system (10.1.3) which has the form (10.1.4), where $\lambda \in \mathbb{C}$, $u_k \in \overset{\circ}{W}_2^m(\Omega)^\ell$. Then the function $D_z^\gamma U$ has the same properties for arbitrary multi-index γ.*

Proof: Applying a local energy estimate to the derivative $\partial_{z_j} U_\varepsilon$, where U_ε is a mollification of U in z with radius ε, and passing to the limit as $\varepsilon \to 0$, we obtain that $\partial_{z_j} U$ belongs to the space $\overset{\circ}{W}_2^m(\mathcal{K}, 0)^\ell$ which was introduced in Section 10.1 (cf. proof of Lemma 5.7.3). Obviously, $\partial_{z_j} U$ is a solution of (10.1.3) and has the form (10.1.4). This together with Lemma 10.1.1 yields the desired result for $|\gamma| = 1$. By induction in $|\gamma|$, we get the result for arbitrary multi-indices γ. ∎

Below we need the set of solutions of the equation

(10.5.1) $L(D_y, 0)\, U = 0 \quad$ in $\mathcal{K}_d$

which have the form

(10.5.2) $U(y) = \rho^\lambda \sum_{k=0}^{s} \frac{1}{k!} (\log \rho)^k\, u_{s-k}(\phi),$

where $u_k \in \overset{\circ}{W}_2^m(\Omega_d)^\ell$. Analogously to the definition of the pencil $\mathcal{L}$, we associate the operator pencil

$$\mathcal{L}_d(\lambda) : \overset{\circ}{W}_2^m(\Omega_d)^\ell \to W_2^{-m}(\Omega_d)^\ell$$

with equation (10.5.1). Note that Theorem 10.1.1 is valid for this pencil. Furthermore, by Theorem 10.1.3, there are no eigenvalues of the pencil $\mathcal{L}_d$ on the line $\operatorname{Re}\lambda = m - d/2$.

Our goal is to represent the eigenvalues and eigenfunctions of the pencil $\mathcal{L}$ in terms of the eigenvalues and eigenfunctions of the pencil $\mathcal{L}_d$. We consider the cases $\operatorname{Re}\lambda > m - n/2$ and $\operatorname{Re}\lambda < m - n/2$ separately.

10.5.2. The case $\operatorname{Re}\lambda > m - n/2$. Let $\{\mu_j\}_{j\in\mathbb{Z}}$ be a sequence of eigenvalues of the pencil $\mathcal{L}_d$ listed according to their algebraic multiplicities. We assume that the eigenvalues lying in the half-plane $\operatorname{Re}\mu > m - d/2$ have nonnegative indices, while the indices of the other eigenvalues are negative. Each eigenvalue μ_j generates a solution of problem (10.5.1) which has the form

$$V_j(y) = \rho^{\mu_j}\, Q_j(\phi, \log \rho),$$

where Q_j are polynomials in the second argument with coefficients in $\overset{\circ}{W}_2^m(\Omega_d)^\ell$. Here it can be assumed that to equal eigenvalues μ_j there correspond linearly independent polynomials Q_j.

LEMMA 10.5.2. *Let α be a given $(n-d)$-dimensional multi-index and let j be a given nonnegative integer. Then there exists a solution of the system* (10.1.3) *having the form*

$$(10.5.3) \qquad V_{j,\alpha}(x) = \sum_{\beta \leq \alpha} z^\beta \, \rho^{\mu_j + |\alpha - \beta|} \, Q_{j,\beta}(\phi, \log \rho) \,,$$

where $Q_{j,\beta}$ are polynomials in $\log \rho$ with coefficients in $\overset{\circ}{W}{}_2^m(\Omega_d)^\ell$ and $Q_{j,\alpha}$ coincides with the above introduced polynomial Q_j.

Proof: For the sake of brevity, let the coefficient of z^β on the right-hand side of (10.5.3) be denoted by $\Psi_\beta(y)$. The equality $L(D_x)\, V_{j,\alpha} = 0$ is equivalent to the system of equations

$$(10.5.4) \qquad L(D_y, 0)\, \Psi_\gamma = - \sum_{\gamma < \beta \leq \alpha} \frac{(-i)^{|\beta - \gamma|} \beta!}{(\beta - \gamma)!} \, L^{(\beta - \gamma)}(D_y, 0)\, \Psi_\beta(y) \quad \text{in } \mathcal{K}_d \,,$$

where γ is an arbitrary $(n-d)$-dimensional multi-index satisfying the inequality $\gamma \leq \alpha$ and $L^{(\delta)}(\eta, \zeta) = (\partial_\zeta^\delta L)(\eta, \zeta)$. Suppose all Ψ_β are constructed for $\beta > \gamma$. Then Ψ_γ can be determined by (10.5.4) using Theorem 1.1.6. ∎

REMARK 10.5.1. If among the numbers $\mu_j + 1, \ldots, \mu_j + |\alpha|$ there are no points of the spectrum of the operator pencil $\mathcal{L}_d$, then the vector function $V_{j,\alpha}$ is uniquely defined. Moreover, the degrees of the polynomials $Q_{j,\beta}$, $\beta \leq \alpha$, do not exceed the degree of the polynomial $Q_{j,\alpha}$.

If the collection $\mu_j + 1, \ldots, \mu_j + |\alpha|$ contains s different eigenvalues with the maximal partial multiplicities $\kappa_1, \ldots, \kappa_s$, then the degree of the polynomial $Q_{j,\beta}$ in (10.5.3) does not exceed $\deg Q_j + \kappa_1 + \cdots + \kappa_s$. In this case the vector function $V_{j,\alpha}$ is unique up to a linear combination of solutions $V_{k,\gamma}$ such that $\gamma < \alpha$ and $\mu_k + |\gamma| = \mu_j + |\alpha|$. In the process of subsequent determination of vector functions Ψ_γ satisfying (10.5.4) we will choose any of them if they are not unique. Then to each pair (j, α) there corresponds one and only one solution (10.5.3). Since the coefficients $\rho_j^\mu \, Q_j$ of z^α in (10.5.3) are linearly independent, the same is true for $V_{j,\alpha}$.

LEMMA 10.5.3. *Let the vector function*

$$(10.5.5) \qquad U(x) = \sum_{|\alpha| \leq N} z^\alpha \, \rho^{\lambda - |\alpha|} \, Q_\alpha(\phi, \log \rho),$$

where $\operatorname{Re}\lambda - N > m - d/2$ and Q_α are polynomials in $\log \rho$ with coefficients in $\overset{\circ}{W}{}_2^m(\Omega_d)^\ell$, be a solution of the homogeneous Dirichlet problem for the equation (10.1.3). *Then $\lambda = \mu_q + k$ for some integer $k, q \geq 0$ and*

$$(10.5.6) \qquad U(x) = \sum_{\mu_j + |\alpha| = \mu_q + k} c_{j,\alpha} \, V_{j,\alpha}(x).$$

Proof: It is clear that the coefficients of z^α with $|\alpha| = N$ in (10.5.5) satisfy the equation

$$L(D_y, 0)\, (\rho^{\lambda - N} \, Q_\alpha) = 0 \quad \text{in } \mathcal{K}_d.$$

From this and from the inequality $\operatorname{Re}\lambda - N > m - d/2$ it follows that $\lambda - N = \mu_j$, $j \geq 0$, and

$$\rho^{\lambda-N}\, Q_\alpha(\phi, \log\rho) = \sum_{j:\, \mu_j=\lambda-N} c_{j,\alpha}\, V_j(y),$$

where $c_{j,\alpha}$ are constants. From this and from the equality

$$V_{j,\alpha} = z^\alpha\, V_j + \sum_{\beta<\alpha} z^\beta\, \rho^{\mu_j+|\alpha-\beta|}\, Q_{j,\alpha,\beta}(\phi, \log\rho)$$

(see Lemma 10.5.2) we conclude that

$$U(x) - \sum_{|\alpha|=N}\ \sum_{j:\, \mu_j+|\alpha|=\lambda} c_{j,\alpha}\, V_{j,\alpha}(x)$$

is a function of the form (10.5.5) with $N - 1$ instead of N. Furthermore, this function is a solution of the homogeneous Dirichlet problem for the equation (10.1.3). Subsequently reducing N, we arrive at (10.5.6). ∎

THEOREM 10.5.1. *The vector function* (10.1.4), *where* $\operatorname{Re}\lambda > m - n/2$ *and* $u_k \in \mathring{W}_2^{\,m}(\Omega)^\ell$, *is a solution of system* (10.1.3) *if and only if* $\lambda = \mu_q + k$ *for some nonnegative* k, q *and* U *is a linear combination of vector functions* $V_{j,\alpha}$ *with* $\mu_j + |\alpha| = \mu_q + k$.

Proof: By Lemma 10.5.1, the function $D_z^\gamma U$ is a solution of the homogeneous Dirichlet problem for the system (10.1.3) which can be represented in the form (10.1.4) with coefficients in $\mathring{W}_2^m(\Omega)^\ell$. Here the role of λ is played by $\lambda - |\gamma|$. Hence it follows that

$$(10.5.7) \qquad D_y^\alpha D_z^\gamma U(x) = r^{\lambda-|\alpha|-|\gamma|} \sum_{k=0}^{s} (\log r)^k\, \varphi_k(\omega), \qquad \varphi_k \in W_2^{m-|\alpha|}(\Omega)^\ell,$$

for $|\alpha| \leq m$ and for arbitrary γ.

Let $\hat{U}(y,\zeta) = \int e^{i\zeta\cdot z}\, U(y,z)\, dz$ be the Fourier transform of the function $U(y,z)$ in z and let $\hat{U}(y,\cdot) \in S'(\mathbb{R}^{n-d})^\ell$ for all $y \in \mathcal{K}_d$. Since

$$(10.5.8) \qquad \hat{U}(y,\zeta) = |\zeta|^{-2M} \int_{\mathbb{R}^{n-d}} e^{i\zeta\cdot z}\, (-\Delta_z)^M\, U(y,z)\, dz$$

for $\zeta \neq 0$ and $M = 0,1,\ldots$, the function $\zeta \to \hat{U}(\cdot,\zeta)$ belongs to the class $C^\infty\big(\mathbb{R}^{n-d}\backslash\{0\};\, \mathring{W}_{2,loc}^m(\mathcal{K}_d)^\ell\big)$. Furthermore, $\hat{U}$ is a solution of the equation

$$(10.5.9) \qquad L(D_y, -\zeta)\, \hat{U}(y,\zeta) = 0 \quad \text{in } \mathcal{K}_d$$

for all $\zeta \neq 0$. From (10.5.7) and (10.5.8) it follows that

$$(10.5.10) \qquad \int_{\Omega_d} \big|(D_y^\alpha \hat{U})(|y|\vartheta, \zeta)\big|^2\, d\vartheta \leq c_N(\zeta)\, |y|^{-N},$$

where $|\alpha| \leq m$ and $N = 0,1,\ldots$.

We show that the function $\hat{U}(\cdot,\zeta)$ belongs to the space $\mathring{W}_2^m(\mathcal{K}_d)^\ell$ for any $\zeta \neq 0$. Let $\chi \in C_0^\infty(\mathbb{R}^{n-d})$ be a cut-off function, $\chi = 0$ outside the unit ball and $\chi(z) = 1$ for $|z| \leq 1/2$. By $F_0(y,\zeta)$ and $F_\infty(y,\zeta)$ we denote the Fourier transforms in z of the functions χU and $(1-\chi)U$, respectively. Furthermore, let $\mathcal{K}'_d = \{y \in \mathcal{K}_d : |y| < 1\}$. Then $\hat{U} = F_0 + F_\infty$ and for $|\alpha| \le m$ we have

$$(10.5.11) \qquad \int\limits_{\mathcal{K}'_d} \int\limits_{\mathbb{R}^{n-d}} \left|D_y^\alpha F_0(y,\zeta)\right|^2 d\zeta \, dy = c \int\limits_{\mathcal{K}'_d} \int\limits_{\mathbb{R}^{n-d}} \left|D_y^\alpha(\chi U)\right|^2 dz \, dy.$$

From the condition $\operatorname{Re}\lambda > m - n/2$ and (10.5.7) it follows that the right-hand side of (10.5.11) is finite. We represent F_∞ in the form

$$F_\infty(y,\zeta) = |\zeta|^{-2M} \int\limits_{\mathbb{R}^{n-d}} e^{i\zeta \cdot z} (-\Delta_z)^M \left((1-\chi)U\right) dz,$$

where M is a sufficiently large integer. By Parseval's equality, we have

$$\int\limits_{\mathcal{K}'_d} \int\limits_{\mathbb{R}^{n-d}} |\zeta|^{4M} |D_y^\alpha F_\infty|^2 \, d\zeta \, dy \le c \sum_{|\gamma|+|\delta|=2M} \int\limits_{\mathcal{K}'_d} \int\limits_{\mathbb{R}^{n-d}} \left|D_z^\gamma(1-\chi)\right|^2 \left|D_z^\delta D_y^\alpha U\right|^2 dz \, dy.$$

By (10.5.7), the right-hand side of the last inequality is finite. Thus, $\hat{U}(\cdot,\zeta) \in \overset{\circ}{W}{}_2^m(\mathcal{K}_d)^\ell$ for all $\zeta \ne 0$.

Since $\hat{U}(\cdot,\zeta)$ is a solution of the equation (10.5.9) with a strongly elliptic operator $L(D_y, -\zeta)$, the vector function $\hat{U}(\cdot,\zeta)$ vanishes for $\zeta \ne 0$. Therefore, U has the form

$$U(y,z) = \sum_{|\alpha|\le N} z^\alpha \, \Phi_\alpha(y).$$

From this and (10.1.4) we find that

$$\Phi_\alpha(y) = \rho^{\lambda-|\alpha|} \, Q_\alpha(\phi, \log\rho),$$

where Q_α is a polynomial in the second argument with coefficients in $\overset{\circ}{W}{}_2^m(\Omega_d)^\ell$. Since $U \in \overset{\circ}{W}{}_{2,loc}^m(\mathcal{K})^\ell$, we have $\operatorname{Re}\lambda - N > m - d/2$. Using Lemma 10.5.3, we get the assertion of the theorem. $\blacksquare$

REMARK 10.5.2. From the above theorem and from the linear independence of the functions $V_{j,\alpha}$ (see Remark 10.5.1) it follows that the algebraic multiplicity of the eigenvalue λ of the operator pencil $\mathcal{L}$ with $\operatorname{Re}\lambda > m - n/2$ is equal to the number of pairs (j,α) such that $\lambda = \mu_j + |\alpha|$, $j \ge 0$.

10.5.3. Special solutions of the equation (10.5.9). The theorem just proved gives a description of all solutions of the form (10.1.4) of the problem (10.1.3) for $\operatorname{Re}\lambda > m - n/2$. According to Theorem 10.1.3, there are no solutions with exponent λ on the line $\operatorname{Re}\lambda = m - n/2$. Therefore, it remains to study the case $\operatorname{Re}\lambda < m - n/2$. First we consider equation (10.5.9).

LEMMA 10.5.4. *Let $j < 0$. There exists a solution of the Dirichlet problem for the system (10.5.9) which has the form*

$$(10.5.12) \quad W_j(y,\zeta) = \chi(\rho\zeta) \sum_{|\alpha|\le m-d/2-\operatorname{Re}\mu_j} \zeta^\alpha \rho^{\mu_j+|\alpha|} Q_{j,\alpha}(\phi, \log\rho) + R_j(y,\zeta).$$

Here $\chi \in C_0^\infty(\mathbb{R}^{n-d})$, $\chi = 1$ in a neighborhood of the origin, $Q_{j,\alpha}$ are polynomials in $\log\rho$ with coefficients from $\overset{\circ}{W}{}_2^m(\Omega_d)^\ell$, $Q_{j,0} = Q_j$, and the vector function R_j

belongs to $C^\infty\big(\mathbb{R}^{n-d}\backslash\{0\}; \overset{\circ}{W}{}_2^m(\mathcal{K}_d)^\ell\big)$. *Furthermore, R_j can be represented in the form*

(10.5.13) $$R_j(y,\zeta) = |\zeta|^{-\mu_j} \sum_{k=0}^{N_j} R_{j,k}\big(y|\zeta|, \zeta/|\zeta|\big)\, (\log|\zeta|)^k,$$

where N_j is the largest degree of the polynomials $Q_{j,\alpha}$, and the vector functions $\vartheta \to R_{j,k}(\cdot, \vartheta)$ belong to $C^\infty\big(S^{n-d-1}; \overset{\circ}{W}{}_2^m(\mathcal{K}_d)^\ell\big)$.

Proof: We set

$$\Psi_{j,\alpha}(y) = \rho^{\mu_j + |\alpha|}\, Q_{j,\alpha}(\phi, \log\rho)$$

and seek the functions $\Psi_{j,\alpha}$ from the formal equality

$$L(D_y, -\zeta)\Big(\sum_\alpha \zeta^\alpha\, \Psi_{j,\alpha}(y)\Big) = 0 \quad \text{in } \mathcal{K}_d.$$

Since the left-hand side of the last equality is equal to

$$\sum_\gamma \frac{(-1)^{|\gamma|}}{\gamma!} \sum_\alpha \zeta^{\alpha+\gamma}\, L^{(\gamma)}(D_y, 0)\, \Psi_{j,\alpha} = \sum_\beta \zeta^\beta \sum_{\alpha \le \beta} \frac{(-1)^{|\beta-\alpha|}}{(\beta-\alpha)!}\, L^{(\beta-\alpha)}(D_y, 0)\, \Psi_{j,\alpha},$$

we get

(10.5.14) $$L(D_y, 0)\, \Psi_{j,\beta} = -\sum_{\alpha < \beta} \frac{(-1)^{|\beta-\alpha|}}{(\beta-\alpha)!}\, L^{(\beta-\alpha)}(D_y, 0)\, \Psi_{j,\alpha}(y) \quad \text{on } \mathcal{K}_d$$

for each β. Starting with the vector function $\Psi_{j,0}$ and using Theorem 1.1.6, we can subsequently find all $\Psi_{j,\beta}$ from (10.5.14).

The function R_j in (10.5.12) satisfies the system

(10.5.15) $$L(D_y, -\zeta)\, R_j(y,\zeta) = F_j(y,\zeta) \quad \text{in } \mathcal{K}_d,$$

where F_j admits the representation

(10.5.16) $$F_j(y,\zeta) = \sum_{|\alpha| \le 3m - d/2 - \operatorname{Re}\mu_j} \sum_{|\beta| \le 2m} \chi_{\alpha,\beta}(\rho\zeta)\, \zeta^\alpha\, \rho^{\mu_j + |\alpha| - 2m}\, Q_{j,\alpha}^{(\beta)}(\phi, \log\rho).$$

Here $\chi_{\alpha,\beta} \in C_0^\infty(\mathbb{R}^{n-d}\backslash\{0\})$ and $Q_{j,\alpha}^{(\beta)}$ are polynomials in the second argument with degree not higher than the largest of degrees of the polynomials $Q_{j,\alpha}$ and with coefficients from the class $W_2^{-m}(\Omega_d)^\ell$.

From (10.5.16) we conclude that F_j belongs to $C^\infty\big(\mathbb{R}^{n-d}; W_2^{-m}(\mathcal{K}_d)^\ell\big)$. Moreover, (10.5.16) implies the representation

(10.5.17) $$F_j(y,\zeta) = |\zeta|^{2m - \mu_j} \sum_{k=0}^{N_j} F_{j,k}(y|\zeta|, \zeta/|\zeta|)\, (\log|\zeta|)^k,$$

where the functions $\vartheta \to F_{j,k}(\cdot, \vartheta)$ belong to the class $C^\infty\big(S^{n-d-1}; W_2^{-m}(\mathcal{K}_d)^\ell\big)$.

By the strong ellipticity of the operator $L(D_x)$, the mapping

$$(L(D_y, -\zeta))^{-1} : W_2^{-m}(\mathcal{K}_d)^\ell \to \overset{\circ}{W}{}_2^m(\mathcal{K}_d)^\ell$$

is bounded together with all its derivatives uniformly with respect to ζ, $|\zeta| = 1$. Therefore, the system (10.5.15) has a unique solution in $C^\infty\big(\mathbb{R}^{n-d}\backslash\{0\}; \overset{\circ}{W}{}_2^m(\mathcal{K}_d)^\ell\big)$ which, according to (10.5.17), can be written in the form (10.5.13) with

$$R_{j,k}(\cdot, \vartheta) = (L(D_y, -\vartheta))^{-1}\, F_{j,k}(\cdot, \vartheta).$$

The proof is complete. ∎

REMARK 10.5.3. If among the numbers $\mu_j + 1, \dots, \mu_j + [m - d/2 - \mu_j]$ there are no points of the spectrum of the pencil $\mathcal{L}_d$, then the vector function W_j in Lemma 10.5.4 is uniquely determined and the degrees of the polynomials $Q_{j,\alpha}$ and the number N_j do not exceed the degree of the polynomial Q_j.

If the collection $\mu_j + 1, \dots, \mu_j + [m - d/2 - \mu_j]$ contains eigenvalues of the pencil $\mathcal{L}_d$, then W_j is unique up to a linear combination of solutions $\zeta^\alpha W_k$, $0 \le |\alpha| = \mu_k - \mu_j$. Suppose the collection $\mu_j + 1, \dots, \mu_j + [m - d/2 - \mu_j]$ contains s different eigenvalues with maximal partial multiplicities $\kappa_1, \dots, \kappa_s$. Then the degrees of the polynomials $Q_{j,\alpha}$ and the number N_j do not exceed $\deg Q_j + \kappa_1 + \cdots + \kappa_s$.

Henceforth, we assume that the solutions W_j, $j < 0$, with the properties given in Lemma 10.5.4 are fixed.

REMARK 10.5.4. Since the functions $\rho^{\mu_j} Q_j$ are linearly independent, the same is true for the functions $\zeta^\alpha W_j$, $j < 0$, $|\alpha| \ge 0$.

Now we prove two properties of the function R_j which will be used in the next subsection when we apply the inverse Fourier transform (with respect to the variable ζ) to the functions W_j.

LEMMA 10.5.5. Let R_j be the function introduced in Lemma 10.5.4. Then

$$D_\zeta^\gamma R_j(y, \zeta) = |\zeta|^{-\mu_j - |\gamma|} \sum_{k=0}^{N_j} R_{j,\gamma,k}(y|\zeta|, \zeta/|\zeta|) \, (\log |\zeta|)^k,$$

where the functions $\vartheta \to R_{j,\gamma,k}(\cdot, \vartheta)$ belong to the class $C^\infty \big(S^{n-d-1}; \overset{\circ}{W}{}_2^m (\mathcal{K}_d)^\ell \big)$ and γ is an arbitrary multi-index.

Proof: Formula (10.5.16) implies

(10.5.18) $$D_\zeta^\gamma F_j(y, \zeta) = |\zeta|^{2m - \mu_j - |\gamma|} \sum_{k=0}^{N_j} F_{j,\gamma,k}(y|\zeta|, \zeta/|\zeta|) \, (\log |\zeta|)^k$$

where the functions $\vartheta \to F_{j,\gamma,k}(\cdot, \vartheta)$ belong to $C^\infty \big(S^{n-d-1}; W_2^{-m}(\Omega_d)^\ell \big)$. Using the obvious equality

$$L(D_y, -\zeta) \, (D_\zeta^\gamma R_j) = D_\zeta^\gamma F_j - \sum_{\beta < \gamma} \binom{\gamma}{\beta} i^{|\gamma - \beta|} \, L^{(\gamma - \beta)}(D_y, -\zeta) \, (D_\zeta^\beta R_j)$$

and (10.5.18), we complete the proof by virtue of induction in $|\gamma|$. ∎

LEMMA 10.5.6. Let $R_{j,k}$ be the vector functions in (10.5.13). Then the integrals

$$\int_{\mathcal{K}_d} (1 + \rho^2)^N \left| D_y^\alpha R_{j,k}(y, \vartheta) \right|^2 dy, \quad k = 0, 1, \dots, N_J,$$

are uniformly bounded with respect to ϑ for every positive integer N and every multi-index α, $|\alpha| \le m$.

Proof: Let $F_{j,k} = L(D_y, -\vartheta) \, R_{j,k}$. Then we have

$$L(D_y, -\vartheta) \big((1 + \rho^2)^{1/2} R_{j,k} \big) = (1 + \rho^2)^{1/2} F_{j,k} - \big[L(D_y, -\vartheta), (1 + \rho^2)^{1/2} \big] R_{j,k},$$

where $\big[L(D_y, -\vartheta), (1 + \rho^2)^{1/2}\big]$ is the commutator of $L(D_y, -\vartheta)$ and $(1 + \rho^2)^{1/2}$. From (10.5.16) and (10.5.17) it follows that the function $(1 + \rho^2)^{N/2} F_{j,k}$ belongs to $C^\infty\big(S^{n-d-1}; W_2^m(\mathcal{K}_d)^\ell\big)$ for all N. Using the boundedness of the operator

$$\big[L(D_y, -\vartheta), (1 + \rho^2)^{1/2}\big] : \overset{\circ}{W}_2^m(\mathcal{K}_d)^\ell \to W_2^{-m}(\mathcal{K}_d)^\ell$$

and induction in N, we obtain the boundedness of the norm of $(1 + \rho^2)^{N/2} R_{j,k}$ in $\overset{\circ}{W}_2^m(\mathcal{K}_d)^\ell$. The lemma is proved. ∎

10.5.4. Solutions (10.1.4) with $\operatorname{Re}\lambda < m - n/2$. Let $\check{f}$ denote the inverse Fourier transform of the distribution $f = f(\zeta)$, i.e.,

$$\check{f}(z) = (2\pi)^{d-n} \int\limits_{\mathbb{R}^{n-d}} e^{-iz\cdot\zeta} f(\zeta)\, d\zeta.$$

Then, by (10.5.12), we have

$$(10.5.19) \quad (\check{W}_j - \check{R}_j)(y, z) = \rho^{\mu_j + d - n} \sum_{|\alpha| \leq m - d/2 - \operatorname{Re}\mu_j} (D^\alpha \check{\chi})(z/\rho)\, Q_{j,\alpha}(\phi, \log\rho).$$

Since $\check{\chi} \in S(\mathbb{R}^{n-d})$, all components of derivatives in x of order not higher than m of the vector function (10.5.19) do not exceed

$$(10.5.20) \qquad\qquad r^{-\mu}\, \rho^{M - m + \mu_j + d - n}\, (1 + |\log\rho|)^{N_j}\, q(\phi)$$

for $|x| \leq b < \infty$, where M is an arbitrary positive number and q is a positive function in $L_2(\Omega_d)$. It can be easily verified that (10.5.20) is a square summable function on the set $\{x \in \mathcal{K} : 0 < a \leq |x| \leq b < \infty\}$. Therefore, the function (10.5.19) belongs to $\overset{\circ}{W}_{2,loc}^m(\mathcal{K})^\ell$.

Since $\rho = r\cos\tau$ and $z/\rho = \vartheta\tan\tau$, the right-hand side of (10.5.19) is equal to

$$(r\cos\tau)^{\mu_j + d - n} \sum_{|\alpha| \leq m - d/2 - \operatorname{Re}\mu_j} (D^\alpha \check{\chi})(\vartheta\tan\tau)\, Q_{j,\alpha}\big(\phi, \log(r\cos\tau)\big)$$

$$= (r\cos\tau)^{\mu_j + d - n} \sum_{\nu=0}^{N_j} \frac{(\log r)^\nu}{\nu!} \sum_{|\alpha| \leq m - d/2 - \operatorname{Re}\mu_j} (D^\alpha \check{\chi})(\vartheta\tan\tau)\, Q_{j,\alpha}^{(\nu)}\big(\phi, \log\cos\tau\big),$$

where $Q_{j,\alpha}^{(\nu)}(\phi, t) = \partial_t^\nu Q_{j,\alpha}(\phi, t)$. Hence the function (10.5.19) has the form (10.1.4) with coefficients from $\overset{\circ}{W}_2^m(\Omega)^\ell$.

By Lemma 10.5.6, we have

$$(10.5.21) \sum_{|\alpha| \leq m} \int\limits_{\mathcal{K}_d} |y|^{-2(m - |\alpha|)} \big|D_y^\alpha R_j(y, \zeta)\big|^2 dy$$

$$\leq c\,|\zeta|^{-2\operatorname{Re}\mu_j - d + 2m}\, \big(1 + |\log|\zeta||\big)^{2N_j} \sum_{k=0}^{N_j} \max_{\vartheta \in S^{n-d-1}} \|R_{j,k}(\cdot, \vartheta)\|_{\overset{\circ}{W}_2^m(\mathcal{K}_d)^\ell}^2.$$

Since $-2\operatorname{Re}\mu_j - d + 2m > d - n$, the inverse Fourier transform $\check{R}(y, \cdot) \in S'(\mathbb{R}^{n-d})^\ell$ is defined for almost all $y \in \mathcal{K}_d$. Moreover, by (10.5.21), the function R_j can be considered as an element of the space $S'(\mathbb{R}^{n-d}; \overset{\circ}{L}_2^m(\mathcal{K}_d)^\ell)$, where $\overset{\circ}{L}_2^m(\mathcal{K}_d)$ is the closure of $C_0^\infty(\mathcal{K}_d)$ with respect to the norm

$$\|U\| = \left(\int\limits_{\mathcal{K}_d} \sum_{|\alpha|=m} |D_y^\alpha U(y)|^2 \, dy \right)^{1/2}.$$

The first summand on the right of (10.5.12) belongs to the space

$$(C^\infty \cap S')(\mathbb{R}^{n-d}; \overset{\circ}{W}_{2,loc}^m(\mathcal{K}_d)^\ell).$$

Hence $W_j \in S'(\mathbb{R}^{n-d}; \overset{\circ}{W}_{2,loc}^m(\mathcal{K}_d)^\ell)$ and $\check{W}_j \in S'(\mathbb{R}^{n-d}; \overset{\circ}{W}_{2,loc}^m(\mathcal{K}_d)^\ell)$.

Furthermore, by (10.5.21), W_j belongs to the class $L_{\infty,loc}(\mathbb{R}^{n-d}; \overset{\circ}{W}_{2,loc}^m(\mathcal{K}_d, 0)^\ell)$. Consequently, $L(D_y, -\zeta) W_j \in L_{\infty,loc}(\mathbb{R}^{n-d}; W_{2,loc}^{-m}(\mathcal{K}_d, 0)^\ell)$. Thus, the equation $L(D_y, -\zeta) W_j = 0$ obtained earlier for $\zeta \neq 0$ can be extended to all ζ, i.e., we have $L(D_y, -\zeta) W_j = 0$ in $\mathcal{K}$ in the space $S'(\mathbb{R}^{n-d}; W_{2,loc}^{-m}(\mathcal{K}_d, 0)^\ell)$. From this it follows that $L(D_x) \check{W}_j = 0$ in $\mathcal{K}$ in the sense of the theory of distributions.

We show that $\check{W}_j$ has the form (10.1.4). For this we need the following lemma.

LEMMA 10.5.7. *The vector function $\check{R}_j$ belongs to the space $\overset{\circ}{W}_{2,loc}^m(\mathcal{K})^\ell$ and has the representation (10.1.4) with $s \leq N_j$.*

Proof: Let $R_j = R_j^{(1)} + R_j^{(2)}$, where

$$R_j^{(1)}(y, \zeta) = \chi(\zeta) R_j(y, \zeta), \qquad R_j^{(2)}(y, \zeta) = (1 - \chi(\zeta)) R_j(y, \zeta).$$

Let b be a positive constant and $\mathcal{K}_{d,b} = \{y \in \mathcal{K}_d : |y| < b\}$. Using Parseval's equality, we get

$$(10.5.22) \qquad \int\limits_{\mathcal{K}_{d,b}} |y|^{2(|\alpha|-m)} \int\limits_{\mathbb{R}^{n-d}} |D_y^\alpha D_z^\beta \check{R}_j^{(1)}(y, z)|^2 \, dz \, dy$$

$$= c \int\limits_{\mathbb{R}^{n-d}} |\zeta^\beta \chi(\zeta)|^2 \int\limits_{\mathcal{K}_{d,b}} |y|^{2(|\alpha|-m)} |D_y^\alpha R_j(y, \zeta)|^2 \, dy \, d\zeta$$

for $|\alpha| + |\beta| \leq m$. From (10.5.21) it follows that the right-hand side of (10.5.22) does not exceed

$$c \int\limits_{\mathbb{R}^{n-d}} |\chi(\zeta)|^2 |\zeta|^{-2\mathrm{Re}\,\mu_j - d + 2m + 2|\beta|} (1 + |\log|\zeta||)^{2N_j} \, d\zeta < \infty.$$

We estimate the integral

$$(10.5.23) \qquad \int\limits_{\mathcal{K}_{d,b}} \int\limits_{\mathbb{R}^{n-d}} (|y|^{4N} + |z|^{4N}) |D_y^\alpha D_z^\beta \check{R}_j^{(2)}(y, z)|^2 \, dz \, dy$$

for $|\alpha| + |\beta| \leq m$ and for sufficiently large N. By Parseval's equality, this integral is equal to

$$(10.5.24) \qquad c \int\limits_{\mathcal{K}_{d,b}} \int\limits_{\mathbb{R}^{n-d}} \left| (-\Delta_\zeta)^N \left(\zeta^\beta (1 - \chi(\zeta)) D_y^\alpha R_j(y, \zeta) \right) \right|^2 \, d\zeta \, dy$$

$$+ c \int\limits_{\mathcal{K}_{d,b}} \int\limits_{\mathbb{R}^{n-d}} \left| |y|^{2N} \zeta^\beta (1 - \chi(\zeta)) D_y^\alpha R_j(y, \zeta) \right|^2 \, d\zeta \, dy.$$

Due to Lemma 10.5.6, the first integral does not exceed

$$(10.5.25) \qquad c \int\limits_{|\zeta|>a} |\zeta|^{-2\mu_j-4N+2|\alpha|+2|\beta|-d} \left(1+|\log|\zeta||\right)^{2N_j}$$

$$\times \int\limits_{\mathcal{K}_d} \sum_{k=0}^{N_j} \sum_{|\gamma|\leq 2N} \left|D_y^\alpha R_{j,\gamma,k}(y,\zeta/|\zeta|)\right|^2 dy\, d\zeta,$$

where a and c are positive constants. Since the inner integral is bounded and N is large, the value (10.5.25) is finite.

The second integral in (10.5.24) is not greater than

$$c \int\limits_{|\zeta|>a} |\zeta|^{-2\mu_j-4N+2|\alpha|+2|\beta|-d} \left(1+|\log|\zeta||\right)^{2N_j} \int\limits_{\mathcal{K}_d} |y|^{4N} \sum_{k=0}^{N_j} \left|D_y^\alpha R_{j,k}(y,\zeta/|\zeta|)\right|^2 dy\, d\zeta.$$

This value is also finite because, by Lemma 10.5.6, the inner integral is uniformly bounded with respect to $\zeta/|\zeta|$.

From the boundedness of the integrals (10.5.22) and (10.5.23) it follows that the integral

$$\int\limits_{\substack{\mathcal{K}\\a<|x|<b}} \sum_{|\alpha|\leq m} |D_x^\alpha \check{R}_j(x)|^2 dx$$

is finite.

We verify the zero Dirichlet conditions for $\check{R}_j$ on $\partial\mathcal{K}\backslash\{0\}$. Let

$$R_{j,\varepsilon}(y,\zeta) = \left(1-\chi(\zeta/\varepsilon)\right)\chi(\varepsilon\zeta)\,R_j(y,\zeta),$$

where ε is a sufficiently small positive number. By Lemma 10.5.5, we have $R_{j,\varepsilon} \in S\left(\mathbb{R}^{n-d}; \overset{\circ}{W}{}_2^{\,m}(\mathcal{K}_d)^\ell\right)$ and, therefore, $\check{R}_{j,\varepsilon} \in S\left(\mathbb{R}^{n-d}; \overset{\circ}{W}{}_2^{\,m}(\mathcal{K}_d)^\ell\right)$. Moreover, $\check{R}_{j,\varepsilon}$ belongs to $\overset{\circ}{W}{}_2^{\,m}(\mathcal{K})^\ell$. Replacing R_j by $R_j - R_{j,\varepsilon}$ at the above arguments, we obtain

$$\int\limits_{\mathcal{K}_{d,b}} |y|^{2(|\alpha|-m)} \int\limits_{\mathbb{R}^{n-d}} \left|D_y^\alpha D_z^\beta(\check{R}_j^{(1)}-\check{R}_{j,\varepsilon}^{(1)})\right|^2 dz\, dy \to 0,$$

$$\int\limits_{\mathcal{K}_{d,b}} \int\limits_{\mathbb{R}^{n-d}} \left(|y|^{4N}+|z|^{4N}\right)\left|D_y^\alpha D_z^\beta(\check{R}_j^{(2)}-\check{R}_{j,\varepsilon}^{(2)})\right|^2 dz\, dy \to 0$$

as $\varepsilon \to +0$. Consequently, $\check{R}_j \in \overset{\circ}{W}{}_{2,loc}^{\,m}(\mathcal{K})^\ell$.

Finally, by (10.5.13), we have

$$\check{R}_j(x) = \frac{r^{\mu_j+d-n}}{(2\pi)^{n-d}} \int\limits_{\mathbb{R}^{n-d}} e^{-ir^{-1}z\cdot\zeta} |\zeta|^{-\mu_j} \sum_{k=0}^{N_j} R_{j,k}\left(y|\zeta|/r,\ \zeta/|\zeta|\right)\left(\log\frac{|\zeta|}{r}\right)^k d\zeta$$

which is equivalent to (10.1.4) for $d = N_j$. The lemma is proved. ∎

As a consequence of Lemma 10.5.7, we obtain the following result.

LEMMA 10.5.8. *The function $\check{W}_j$ has the form (10.1.4), where $\lambda = \mu_j + d - n$, $s \leq N_j$, $u_k \in \overset{\circ}{W}{}_2^{\,m}(\Omega)^\ell$, and satisfies the system (10.1.3).*

Proof: We have shown above that $\check{W}_j$ satisfies (10.1.3) and that $\check{W}_j - \check{R}_j$ has the form (10.1.4), where $\lambda = \mu_j + d - n$, $s \leq N_j$, $u_k \in \mathring{W}_2^m(\Omega)^\ell$. This together with Lemma 10.5.7 completes the proof. ∎

The following result is an immediate consequence of Lemmas 10.5.1, 10.5.8 and Remark 10.5.4.

LEMMA 10.5.9. *For any multi-index γ the vector function $D_z^\gamma \check{W}_j$ has the form* (10.1.4), *where $\lambda = \mu_j + d - n - |\gamma|$, $s \leq N_j$, $u_k \in \mathring{W}_2^m(\Omega)^\ell$, and satisfies the system* (10.1.3). *The vector functions $D_z^\gamma \check{W}_j$, where $j < 0$ and γ is an arbitrary multi-index, are linearly independent.*

Now we are able to give a description of all solutions (10.1.4) for $\operatorname{Re} \lambda > m - n/2$.

THEOREM 10.5.2. *Let U be a vector function of the form* (10.1.4), *where $\operatorname{Re} \lambda < m - n/2$, $u_k \in \mathring{W}_2^m(\Omega)^\ell$. Then U is a solution of the system* (10.1.3) *if and only if $\lambda = \mu_q + d - n - k$ for some integers $k \geq 0$, $q < 0$. Moreover, the solution U is linear combination of vector functions $D_z^\gamma \check{W}_j$, where $\mu_j + d - n - |\gamma| = \lambda$.*

Proof: Let $\mathcal{L}^+$ be the adjoint operator pencil to $\mathcal{L}$ (see Section 10.1). Analogously, we define the operator pencil $\mathcal{L}_d^+$ by means of the differential operator $L^+(D_y, 0)$ in $\mathcal{K}_d$. By Theorem 10.1.2, the spectrum of $\mathcal{L}_d^+$ consists of the eigenvalues $2m - d - \overline{\mu}_j$, $j = 0, \pm 1, \ldots$, listed according to their algebraic multiplicities. This and Theorem 10.5.1 imply that the spectrum of $\mathcal{L}^+$ consists of the eigenvalues $\lambda_{j,\gamma}^+ = 2m - d - \overline{\mu}_j + |\gamma|$ in the half-plane $\operatorname{Re} \lambda > m - n/2$, where $j < 0$ and γ is an arbitrary $(n - d)$-dimensional multi-index. Applying again Theorem 10.1.2, we obtain that the spectrum of the pencil $\mathcal{L}$ in the half-plane $\operatorname{Re} \lambda < m - n/2$ consists of the eigenvalues $2m - n - \lambda_{j,\gamma}^+ = \mu_j + d - n - |\gamma|$ numerated with account taken of their multiplicities. Therefore, by Remark 10.5.2, the number of linearly independent solutions of the system (10.1.3) having the form (10.1.4) with $\operatorname{Re} \lambda < m - n/2$ and $u_k \in \mathring{W}_2^m(\Omega)^\ell$ is equal to the number of representations of λ in the form $\mu_j + d - n - |\gamma|$ with $j < 0$, $|\gamma| \geq 0$. As it was shown in Lemma 10.5.9, the solutions $D_z^\gamma \check{W}_j$ are linearly independent. Consequently, all solutions of the system (10.1.3) which have the form (10.1.4) are linear combinations of functions $D_z^\gamma \check{W}_j$. The theorem is proved. ∎

10.5.5. The Dirichlet problem for the Laplace operator. If $L(D_x)$ is the Laplace operator, then it is possible to obtain more explicit interpretations of Theorems 10.5.1 and 10.5.2. Let $L(D_x) = -\Delta_x$. Then $L(D_y, 0) = -\Delta_y$ and

$$\mathcal{L}_d(\mu) = -\delta - \mu(\mu + d - 2),$$

where δ is the Beltrami operator on S^{d-1}. Let $\{\mu_j\}_{j \geq 0}$ be the sequence of positive eigenvalues of the operator pencil $\mathcal{L}_d$ numerated with account taken of their algebraic multiplicities, and let $\{\psi_j\}_{j \geq 0}$ be the corresponding sequence of real eigenfunctions. Note that $2 - d - \mu_j$ is also an eigenvalue for each integer $j \geq 0$ and ψ_j is an eigenfunction corresponding to this eigenvalue.

By $\{\lambda_j\}$ we denote the sequence of eigenvalues of the pencil $\mathcal{L}$ generated by the operator $-\Delta_x$ in the cone $\mathcal{K}$ and by $\{u_j\}_{j \geq 0}$ a sequence of corresponding eigenfunctions. As we have shown in Section 2.2, the eigenvalues λ_j and $2 - n - \lambda_j$ have the same eigenfunctions. In particular, the function $(\cos\tau)^{\mu_j}\,\psi_j(\phi)$ is an eigenfunction of the pencil $\mathcal{L}$ for every $j \geq 0$ which generates two solutions of the form (10.1.4):

$$\rho^{\mu_j}\,\psi_j(\phi) \quad \text{and} \quad U_j(x) = r^{2-n-2\mu_j}\,\rho^{\mu_j}\,\psi_j(\phi).$$

In order to find all eigenfunctions of the operator pencil $\mathcal{L}$ and, therefore, all solutions of (10.1.3) which have the form (10.1.4) it suffices to do this for the eigenvalues $2 - n - \lambda_j$.

First we show that the functions U_j play the role of the functions $\breve{W}_j$ introduced in the preceding subsection. We have

$$(10.5.26) \int\limits_{\mathbb{R}^{n-d}} e^{iz\cdot\zeta}\,U_j(y,z)\,dz = \rho^{\mu_j}\,\psi_j(\phi)\int\limits_{\mathbb{R}^{n-d}} \frac{e^{iz\cdot\zeta}}{(\rho^2 + |z|^2)^{\mu_j-1+n/2}}\,dz$$

$$= \frac{2^{2-d/2\mu_j}\,\pi^{(n-d)/2}}{\Gamma(\mu_j - 1 + n/2)}\,\rho^{2-d-\mu_j}\,\psi_j(\phi)\,(\rho|\zeta|)^{\mu_j-1+d/2}\,K_{1-\mu_j-d/2}(\rho|\zeta|),$$

where K_ν is the modified Bessel function. The function

$$\rho^{1-d/2}\,K_{1-\mu_j-d/2}(\rho)\,\psi_j(\phi)$$

is a solution of the equation $\Delta u - u = 0$ in $\mathcal{K}_d$ and satisfies the zero Dirichlet condition on $\partial\mathcal{K}_d\backslash\{0\}$. Furthermore,

$$K_{1-\mu_j-d/2}(\rho) = \rho^{1-\mu_j-d/2}\,P_k(\rho^2) + O(\rho^\varepsilon), \quad \varepsilon > 0,$$

for $\rho \to 0$, where P_k is a polynomial of degree k, $2k \leq \mu_j - 1 + d/2$. Consequently, the right-hand side of (10.5.26) admits the representation (10.5.12) and can play the role of W_j. This and Theorem 10.1.2 imply that linear combinations of the functions $D_z^\gamma U_j$ exhaust the set of solutions of the form (10.1.4) for $\operatorname{Re}\lambda < 1 - n/2$.

We introduce the polynomials of degree $|\gamma|$

$$Q_{\gamma,\kappa}(z) = \left(|z|^2 + 1\right)^{|\gamma|+\kappa/2}\,D_z^\gamma\left(|z|^2 + 1\right)^{-\kappa/2}$$

and represent them as a sum of homogeneous polynomials:

$$Q_{\gamma,k} = \sum_{q=0}^{[|\gamma|/2]} Q_{\gamma,k}^{(q)}, \quad \text{where } \deg Q_{\gamma,k}^{(q)} = |\gamma| - 2q.$$

Then

$$r^{\kappa+2|\gamma|}\,D_z^\gamma\,r^{-\kappa} = \sum_{q=0}^{[|\gamma|/2]} \rho^{2q}\,Q_{\gamma,k}^{(q)}(z).$$

Consequently, the functions

$$u_{\nu,\gamma}(\omega) = (\cos\tau)^{\mu_\nu}\,\psi_\nu(\phi)\sum_{q=0}^{[|\gamma|/2]}(\cos\tau)^{2q}\,Q_{\gamma,n-2+2\mu_\nu}^{(q)}(\vartheta\sin\tau)$$

where ν and γ satisfy the equality $\mu_\nu + |\gamma| = \mu_j + k$, are eigenfunctions corresponding to the eigenvalues $2 - n - \mu_j - k$, $k = 0, 1, \dots$. By Theorem 10.5.2, there are no other linearly independent eigenfunctions. The same eigenfunctions are generated by the eigenvalue $\mu_j + k$. Thus, we have proved the following theorem.

THEOREM 10.5.3. *Let* $U \in \overset{\circ}{W}{}^1_{2,loc}(\mathcal{K}, 0)$ *be a function of the form* (10.1.4). *Then* U *is a solution of the equation* $-\Delta_x U = 0$ *in* $\mathcal{K}$ *if and only if one of the following two assertions is true:*

(i) $\lambda = \mu_j + k$, $j, k = 0, 1, \dots$, *and U is a linear combination of the functions*

$$\rho^{\mu_\nu}\,\psi_\nu(\phi) \sum_{q=0}^{[|\gamma|/2]} \rho^{2q}\,Q^{(q)}_{\gamma,n-2+2\mu_\nu}(z)\,,$$

where $\mu_\nu + |\gamma| = \mu_j + k$

(ii) $\lambda = 2 - n - \mu_j - k$, $j, k = 0, 1, \dots$, *and U is a linear combination of the functions*

$$\rho^{\mu_\nu}\,\psi_\nu(\phi)\,D_z^\gamma\,r^{2-n-2\mu_\nu} = \rho^{\mu_\nu}\,\psi_\nu(\phi)\,r^{2-n-2\mu_\nu} \sum_{q=0}^{[|\gamma|/2]} \rho^{2q}\,Q^{(q)}_{\gamma,n-2+2\mu_\nu}(z)\,,$$

where $\mu_\nu + |\gamma| = \mu_j + k$.

10.6. Notes

Section 10.1. Modulo presentation, the material is standard.

Sections 10.2 and 10.3. It is well-known that power-logarithmic solutions of an elliptic system in $\mathbb{R}^n \backslash \{0\}$ and of the Dirichlet problem in the half-space are linear combinations of polynomials and derivatives of Green's matrix. We used this fact together with classical results on Green's matrix (see John's book [94] and the papers by Agmon, Douglis, Nirenberg [2] and Solonnikov [246]) to obtain a description of the spectrum of the corresponding operator pencils. We refer also to Pazy's paper [224].

Section 10.4. The eigenvalues, eigenvectors and generalized eigenvectors of the pencil generated by the Sobolev problem for an elliptic system of $2m$ order equations in the exterior of a ray were described in the paper [130] of Kozlov and Maz'ya. For the plane case (the case of a crack) we refer also to the papers by Nazarov [203, 204] and the book by Nazarov and Plamenevskiĭ [207], where more general elliptic problems are considered.

Section 10.5. The material is borrowed from the paper [133] by Kozlov and Maz'ya.

CHAPTER 11

The Dirichlet problem in a cone

In the present chapter, as in the previous one, we deal with singularities of solutions to the Dirichlet problem for strongly elliptic equations and systems of order $2m$ in n-dimensional cones. Some basic properties of the operator pencil $\mathcal{L}$ generated by this problem were studied in Section 10.1, where we proved, in particular, the absence of eigenvalues of $\mathcal{L}$ on the energy line $\operatorname{Re} \lambda = m - n/2$.

We wish to estimate the width of the energy strip $|\operatorname{Re} \lambda - m + n/2| < \delta$, that is the widest strip free of eigenvalues of $\mathcal{L}$ and symmetric with respect to the energy line. It is proved in Section 11.1 that a certain $\delta > 1/2$ suits if the differential operator is selfadjoint and the cone $\mathcal{K}$ has an explicit representation $x_n > \phi(x_1, \dots, x_{n-1})$ in Cartesian coordinates, where ϕ is smooth on $\mathbb{R}^{n-1} \backslash \{0\}$ and positively homogeneous of degree 1. This, along with a general property of elliptic boundary value problems in domains with conic vertices (we refer to the book of Dauge [**41**]) implies square summability of "derivatives" of order $m+1/2+\varepsilon$ for solutions to the Dirichlet problem in these domains. Such an improvement of smoothness by an order greater than $1/2$ is independent both of the cone opening and the differential operator L.

In Section 11.2 we try to weaken the smoothness condition on ϕ. By requiring the mere continuity of ϕ, we are able to prove only that the strip $|\operatorname{Re} \lambda - m + n/2| < 1/2$ contains no eigenvalues. The same assertion concerning the strip $|\operatorname{Re} \lambda - m + n/2| < 1/2 + \varepsilon$, $\varepsilon > 0$, is obtained for elliptic systems of second order differential equations if the restriction of the function ϕ to the sphere S^{n-2} belongs to the Sobolev space $W_2^1(S^{n-2})$.

The condition of selfadjointness of L cannot be removed. In fact, by counterexample 8.4.3, there exists a family of second order strongly elliptic operators with complex coefficients such that a sequence of eigenvalues of the corresponding operator pencils $\mathcal{L}$ tends to the energy line.

In Section 11.4 we consider the Dirichlet problem for second order square elliptic systems in a three-dimensional polyhedral cone. We show that the strip $|\operatorname{Re} \lambda + 1/2| \leq 1$ is free of eigenvalues of $\mathcal{L}$ if all interior angles at the edges, with the possible exception of one of them, are less than π.

In Section 11.5 we pass to the Dirichlet problem for a $2m$ order elliptic equation in the case when $\mathcal{K}$ is the exterior of a thin cone. Here the operator pencil generated by the Dirichlet problem is a small singular perturbation of that generated by the Sobolev problem in the exterior of a ray. Resting on this, we show that an arbitrary neighborhood of an eigenvalue of the pencil associated to the Sobolev problem contains also eigenvalues of the pencil generated by the Dirichlet problem in the exterior of a sufficiently thin cone. From this result it follows, in particular, that the estimate $\delta > 1/2$ obtained in Section 11.1 for the width of the energy strip $|\operatorname{Re} \lambda - m + n/2| < \delta$ is sharp for $2m \geq n$.

Section 11.6 is dedicated to the Dirichlet problem in a cone close to a half-space. Here we derive an asymptotic formula for the eigenvalue situated near the point $\lambda_0 = m$. We wish to make an observation apropos this formula. To begin with, differential operators considered earlier, such as Δ, Δ^2 and the Lamé operator in a cone, as well as an arbitrary elliptic operator with real constant coefficients in an angle, possess the following common property. Their variational solutions with zero Dirichlet data on $\partial\mathcal{K}$ have continuous partial derivatives up to order m on $\overline{\mathcal{K}}$ if $\mathcal{K} \subset \mathbb{R}^n_+$. One might presuppose intuitively that this law is universal for strongly elliptic selfadjoint operators in a cone situated in a half-space. However, we construct a counterexample to this hypothesis based on the asymptotic formula in question.

In Section 11.7 we consider pencils generated by the Dirichlet problem in a cone with small opening. We find an algebraic condition on elliptic differential operators ensuring the nonrealness of all eigenvalues in a sufficiently large strip. Some results concerning the Dirichlet problem in the exterior of a thin cone are given without proofs in Section 11.8.

In the last Section 11.9 we outline known results on the solvability and regularity for solutions to the Dirichlet problem in domains of polyhedral type, taking into account spectral properties of operator pencils investigated in this and previous chapters.

11.1. The case of a "smooth" cone

Additionally to the assumptions of Section 10.1, it is supposed in this section that the operator $L(D_x)$ is formally selfadjoint and

$$(11.1.1) \qquad \mathcal{K} = \{x = (x', x_n) \in \mathbb{R}^n \ : \ x_n > \phi(x')\},$$

where the function ϕ is is infinitely differentiable on $\mathbb{R}^{n-1}\backslash\{0\}$ and positively homogeneous of degree 1. We prove that the strip

$$(11.1.2) \qquad |\operatorname{Re}\lambda - m + n/2| \leq 1/2$$

does not contain eigenvalues of the operator pencil $\mathcal{L}$ introduced in Section 10.1.

11.1.1. Absence of eigenvalues on the line $\operatorname{Re}\lambda = m - (n-1)/2$.
Suppose that the operator (10.1.2) is formally selfadjoint. Then it can be represented in the form

$$(11.1.3) \qquad L(D_x) = \sum_{|\alpha|=|\beta|=m} A_{\alpha,\beta}\, D_x^{\alpha+\beta},$$

where $A_{\alpha,\beta}$ are constant $\ell \times \ell$-matrices satisfying the condition $A_{\alpha,\beta} = A^*_{\beta,\alpha}$. Note that every elliptic formally selfadjoint operator $L(D_x)$ of the given form is strongly elliptic. As in the preceding chapter, let $\mathcal{L}$ denote the operator pencil generated by $L(D_x)$, i.e., the operator $\mathcal{L}(\lambda) : \mathring{W}_2^m(\Omega)^\ell \to W_2^{-m}(\Omega)^\ell$ is defined by (10.1.5). Here Ω is the intersection of the cone $\mathcal{K}$ with the unit sphere S^{n-1}.

Since ϕ is assumed to be smooth on $\mathbb{R}^{n-1}\backslash\{0\}$, there exists the exterior normal $\nu = (\nu_1, \ldots, \nu_n)$ to $\partial\mathcal{K}\backslash\{0\}$. The last component ν_n is given by the equality

$$\nu_n\big(x', \phi(x')\big) = -\Big(1 + \sum_{j=1}^{n-1} |\partial_{x_j}\phi(x')|^2\Big)^{-1/2}.$$

We will prove that the line $\operatorname{Re} \lambda = m - (n-1)/2$ does not contain eigenvalues of the pencil $\mathcal{L}$. For this we need the following lemma.

LEMMA 11.1.1. *Let* $\operatorname{Re} \lambda_0 = m - (n-1)/2$, $u \in C^\infty(\overline{\Omega})^\ell$, *and let the function* $U(x) = r^{\lambda_0} u(\omega)$ *be a solution of the problem*

$$L(D_x)\, U = 0 \quad in \ \mathcal{K}, \qquad \frac{\partial^j U}{\partial \nu^j} = 0 \quad on \ \partial\mathcal{K}\backslash\{0\}, \ j = 0, \dots, m-1.$$

Then $\partial^m U/\partial \nu^m = 0$ *on* $\partial\mathcal{K}\backslash\{0\}$.

Proof: We introduce the sets

$$\mathcal{K}_\varepsilon = \{x \in \mathcal{K} : \varepsilon < |x| < 1\}, \qquad \Gamma_\varepsilon = \{x \in \partial\mathcal{K} : \varepsilon < |x| < 1\},$$

and $\Omega_\varepsilon = \{x \in \mathcal{K} : |x| = \varepsilon\}$. Here ε is an arbitrary positive real number less than one. Integrating by parts and using the equality

$$D_x^\alpha U = (-i)^m \nu^\alpha \frac{\partial^m U}{\partial \nu^m} \qquad on \ \partial\mathcal{K}\backslash\{0\} \quad for \ |\alpha| = m,$$

we get

$$(11.1.4)\operatorname{Re} \int_{\mathcal{K}_\varepsilon} \left(L(D_x)U, \frac{\partial U}{\partial x_n} \right)_{\mathbb{C}^\ell} dx = \operatorname{Re} \int_{\mathcal{K}_\varepsilon} \sum_{|\alpha|=|\beta|=m} \left(A_{\alpha,\beta} D_x^\beta U, \partial_{x_n} D_x^\alpha U \right)_{\mathbb{C}^\ell} dx$$

$$- \int_{\Gamma_\varepsilon} \nu_n \sum_{|\alpha|=|\beta|=m} \left(A_{\alpha,\beta} \nu^{\alpha+\beta} \frac{\partial^m U}{\partial \nu^m}, \frac{\partial^m U}{\partial \nu^m} \right)_{\mathbb{C}^\ell} d\sigma$$

$$+ \int_{\Omega} \Psi_1(x)\, d\omega - \int_{\Omega_\varepsilon} \Psi_1(x)\, d\omega_\varepsilon \,,$$

where Ψ_1 is a positively homogeneous function of degree $1 - n$. Since

$$\frac{1}{2} \sum_{|\alpha|=|\beta|=m} \frac{\partial}{\partial x_n} \left(A_{\alpha,\beta}\, D_x^\beta U, D_x^\alpha U \right)_{\mathbb{C}^\ell} = \operatorname{Re} \sum_{|\alpha|=|\beta|=m} \left(A_{\alpha,\beta}\, D_x^\beta U, D_x^\alpha \frac{\partial U}{\partial x_n} \right)_{\mathbb{C}^\ell},$$

the first term on the right-hand side of (11.1.4) is equal to

$$\frac{1}{2} \int_{\Gamma_\varepsilon} \nu_n \sum_{|\alpha|=|\beta|=m} \left(A_{\alpha,\beta} \nu^{\alpha+\beta} \frac{\partial^m U}{\partial \nu^m}, \frac{\partial^m U}{\partial \nu^m} \right)_{\mathbb{C}^\ell} d\sigma + \int_{\Omega} \Psi_2(x)\, d\omega - \int_{\Omega_\varepsilon} \Psi_2(x)\, d\omega_\varepsilon \,,$$

where Ψ_2 is positively homogeneous of degree $1 - n$. Obviously,

$$\int_{\Omega} \Psi_j(x)\, d\omega = \int_{\Omega_\varepsilon} \Psi_j(x)\, d\omega_\varepsilon \qquad for \ j = 1,\, 2.$$

Hence (11.1.4) implies

$$(11.1.5) \qquad \operatorname{Re} \int_{\mathcal{K}_\varepsilon} \left(L(D_x)U, \frac{\partial U}{\partial x_n} \right)_{\mathbb{C}^\ell} dx$$

$$= -\frac{1}{2} \int_{\Gamma_\varepsilon} \nu_n \sum_{|\alpha|=|\beta|=m} \left(A_{\alpha,\beta} \nu^{\alpha+\beta} \frac{\partial^m U}{\partial \nu^m}, \frac{\partial^m U}{\partial \nu^m} \right)_{\mathbb{C}^\ell} d\sigma.$$

The left-hand side of (11.1.5) vanishes, since U is a solution of the equation (10.1.3). By the ellipticity and selfadjointness of the operator $L(D_x)$, the expression

$$(11.1.6) \qquad \sum_{|\alpha|=|\beta|=m} \left(A_{\alpha,\beta} \, \nu^{\alpha+\beta} \, \frac{\partial^m U}{\partial \nu^m} \, , \, \frac{\partial^m U}{\partial \nu^m} \right)_{\mathbb{C}^\ell}$$

is real and does not change the sign. Furthermore, ν_n is negative on $\partial \mathcal{K} \backslash \{0\}$. Consequently, from (11.1.5) it follows that the expression (11.1.6) vanishes on $\partial \mathcal{K} \backslash \{0\}$. Using the ellipticity of $L(D_x)$, we get $\partial^m U/\partial \nu^m = 0$ on $\partial \mathcal{K} \backslash \{0\}$. ∎

LEMMA 11.1.2. *There are no eigenvalues of $\mathcal{L}$ on the line* $\operatorname{Re} \lambda = m-n/2+1/2$.

Proof: Suppose the assertion of the lemma is not true, i.e., there exists an eigenvalue λ_0 with real part $m-n/2+1/2$. Let $u_1, \dots, u_I$ be a basis in the space of the eigenfunctions of $\mathcal{L}$ corresponding to the eigenvalue λ_0. By the selfadjointness of the operator $L(D_x)$ and Theorem 10.1.2, the number $\lambda_1 = 2m - n - \overline{\lambda}_0 = \lambda_0 - 1$ is also an eigenvalue of $\mathcal{L}$ with the same geometric multiplicity I. We denote by $v_1, \dots, v_I$ a basis in the space of eigenfunctions corresponding to the eigenvalue λ_1 and set

$$U_j(x) = r^{\lambda_0} \, u_j(\omega), \quad V_j(x) = r^{\lambda_1} \, v_j(\omega), \qquad j = 1, \dots, I.$$

The functions $\partial U_j/\partial x_s$ are solutions of the equation $L(D_x)U = 0$ in $\mathcal{K}$. By Lemma 11.1.1, they also satisfy the homogeneous Dirichlet boundary conditions on $\partial \mathcal{K} \backslash \{0\}$. Moreover, these functions are positively homogeneous of degree λ_1. Consequently, there exist constants $c_{j,k}^s$ such that

$$(11.1.7) \qquad \frac{\partial}{\partial x_s} U_j(x) = \sum_{k=1}^{I} c_{j,k}^s \, V_k(x) \qquad \text{for } j = 1, \dots, I, \ s = 1, \dots, n.$$

Let $a = (a_1, \dots, a_n)$ be a nonzero vector such that the matrix

$$(11.1.8) \qquad \left(\sum_{s=1}^{n} c_{j,k}^s \, a_s \right)_{j,k=1,\dots,I}$$

has a nontrivial kernel. Such vector always exists. If the inverse $(d_{j,k})$ of the matrix $(c_{j,k}^1)$ exists, we can set $a = (\mu, 1, 0, \dots, 0)$, where μ is an eigenvalue of the matrix

$$-\left(\sum_{k=1}^{I} d_{j,k} \, c_{k,s}^2 \right)_{j,s=1,\dots,I}.$$

Otherwise, the vector $a = (1, 0, \dots, 0)$ has the desired property.

We choose a nonzero vector $b = (b_1, \dots, b_I)$ from the kernel of the matrix (11.1.8). Then (11.1.7) implies

$$(11.1.9) \qquad \sum_{s=1}^{n} a_s \, \frac{\partial}{\partial x_s} \left(\sum_{j=1}^{I} b_j \, U_j(x) \right) = 0.$$

Now we introduce the function

$$U(x) = b_1 \, U_1(x) + \dots + b_I \, U_I(x) \qquad \text{in } \mathcal{K}$$

and extend this function by zero to $\mathbb{R}^n$. Since $\partial^k U/\partial \nu^k = 0$ on $\partial \mathcal{K} \backslash \{0\}$ for $k = 0, \dots, m$, this extension has continuous derivatives of order $\leq m$ on $\mathbb{R}^n \backslash \{0\}$. Furthermore, for large $|x|$ it grows not faster than a polynomial.

Equation (11.1.9) can be transformed, by a linear change of variables, into one of the following equations

$$(11.1.10) \qquad \frac{\partial}{\partial y_1} U(y) = 0, \qquad \left(\frac{\partial}{\partial y_1} + i \frac{\partial}{\partial y_2} \right) U(y) = 0,$$

Here the cone $\mathcal{K}$ turns into a new cone $\mathcal{K}'$.

Suppose that the first of the equations (11.1.10) holds. If one of the lines $y' = (y_2, \dots, y_n) = const$ crosses the boundary $\partial \mathcal{K}' \backslash \{0\}$ at a nonzero angle, then from the equation $\partial U / \partial y_1 = 0$ and from the fact that $U = 0$ outside $\mathcal{K}'$ it follows that U vanishes in an open subset of $\mathcal{K}'$. Since U is a real-analytic function, we obtain $U = 0$. If there are no lines $y' = const$ crossing $\partial \mathcal{K}' \backslash \{0\}$ at a nonzero angle, then $\mathcal{K}' = \mathbb{R} \times \mathcal{K}_1$, where $\mathcal{K}_1$ is a $(n-1)$-dimensional cone. Since $\mathcal{K}' \backslash \{0\}$ is a smooth manifold, this is only possible if $\mathcal{K}$ is a half-space in $\mathbb{R}^n$. In this case we get a contradiction to Theorem 10.3.1.

Suppose now that the second equation of (11.1.10) is realized. If one of the planes $y'' = (y_3, \dots, y_n) = const$ contains both interior and exterior points of $\mathcal{K}'$, then again integrating the equation $(\partial / \partial y_1 + i \partial / \partial y_2) U = 0$ and using the fact that $U = 0$ outside $\mathcal{K}'$, we get $U = 0$ in an open subset of $\mathcal{K}'$ and, therefore, $U = 0$ everywhere. In the contrary case we have $\mathcal{K}' = \mathbb{R}^2 \times \mathcal{K}_2$, where $\mathcal{K}_2$ is a cone in $\mathbb{R}^{n-2}$. Arguing as before, we conclude that $\mathcal{K}$ is a half-space in $\mathbb{R}^n$. This contradicts Theorem 10.3.1. The proof is complete. ∎

11.1.2. The width of the energy strip.

THEOREM 11.1.1. *Suppose that the cone $\mathcal{K}$ is determined by the inequality $x_n > \phi(x')$ with a positively homogeneous function ϕ of degree 1 which is infinitely differentiable on $\mathbb{R}^{n-1} \backslash \{0\}$. Furthermore, we assume that the operator $L(D_x)$ is elliptic and formally selfadjoint. Then the strip $|\operatorname{Re} \lambda - m + n/2| \leq 1/2$ does not contain eigenvalues of the pencil $\mathcal{L}$.*

Proof: Let $\mathcal{K}_t$ be the cone $\{x = (x', x_n) \in \mathbb{R}^n : x_n > t \phi(x')\}$, $0 \leq t \leq 1$, and let Γ_t be the boundary of $\mathcal{K}_t$. Furthermore, let Ω_t be the intersection of $\mathcal{K}_t$ with the unit sphere S^{n-1}. We denote by $\mathcal{L}_t$ the operator pencil generated by the Dirichlet problem

$$L(D_x) U = 0 \quad \text{in } \mathcal{K}_t, \qquad \frac{\partial^j U}{\partial \nu^j} = 0 \quad \text{on } \partial \mathcal{K}_t \backslash \{0\}, \ j = 0, \dots, m-1.$$

From Lemma 11.1.2, Theorem 10.1.2 and from the selfadjointness of the operator $L(D_x)$ it follows that there are no eigenvalues of $\mathcal{L}_t$ on the lines $\operatorname{Re} \lambda = m - n/2 \pm 1/2$. Using the strong ellipticity of L, it can be proved, analogously to Lemma 10.1.3, that the set

$$\{\lambda \in \mathbb{C} : |\operatorname{Re} \lambda - m + n/2| \leq 1/2, \ |\operatorname{Im} \lambda| \geq c\}$$

does not contain eigenvalues of $\mathcal{L}_t$ if c is a sufficiently large positive number.

If $t = 0$, then the cone $\mathcal{K}_t$ coincides with the half-space. Hence, by Theorems 10.3.1 and 10.3.2, the strip $m - n < \operatorname{Re} \lambda < m$ does not contain eigenvalues of the pencil $\mathcal{L}_0$.

There exists a family of diffeomorphisms $\psi_t : \Omega_t \to \Omega_0$ infinitely differentiable with respect to the parameter $t \in [0, 1]$. These diffeomorphisms transform the operators $\mathcal{L}_t(\lambda)$ into the operators

$$\tilde{\mathcal{L}}_t(\lambda) : \overset{\circ}{W}_2^m(\Omega_0)^\ell \to W_2^{-m}(\Omega_0)^\ell.$$

The pencils $\mathcal{L}_t$ and $\tilde{\mathcal{L}}_t$ have the same eigenvalues with the same algebraic multiplicities. Hence there are no eigenvalues of the pencils $\tilde{\mathcal{L}}_t$, $0 \leq t \leq 1$, on the boundary of the rectangle

$$\Lambda_c = \{\lambda \in \mathbb{C} : |\operatorname{Re}\lambda - m + n/2| < 1/2, \ |\operatorname{Im}\lambda| < c\}.$$

Moreover, no eigenvalues of $\tilde{\mathcal{L}}_0 = \mathcal{L}_0$ are situated in Λ_c. Applying Theorem 1.1.4, we conclude that Λ_c does not contain eigenvalues of the pencils $\tilde{\mathcal{L}}_t$, $0 \leq t \leq 1$. Consequently, the strip $|\operatorname{Re}\lambda - m + n/2| \leq 1/2$ is free of eigenvalues of the pencil $\mathcal{L} = \mathcal{L}_1$. The theorem is proved. ∎

REMARK 11.1.1. We have proved the assertion of Theorem 11.1.1 only under the assumption that the cone $\mathcal{K}$ is determined by the inequality $x_n > \phi(x')$. Not every cone $\mathcal{K} = \{x \in \mathbb{R}^n : x/|x| \in \Omega\}$ has this form (see Figure 1 on p. 4).

11.2. The case of a nonsmooth cone

Now let $\mathcal{K}$ be the cone (11.1.1), where ϕ is an arbitrary continuous and positively homogeneous function of degree 1. We prove that, without the assumption of smoothness of ϕ, the open strip $|\operatorname{Re}\lambda - m + n/2| < 1/2$ does not contain eigenvalues of the pencil $\mathcal{L}$ generated by the Dirichlet problem in the cone.

11.2.1. Behaviour of the spectrum of operator pencils under small variation of the domain. Let $\mathcal{X}$ be a Hilbert space with the scalar product $b_0(\cdot, \cdot)$. Furthermore let $\mathcal{X}_\varepsilon$, $\varepsilon > 0$, be closed subspaces of $\mathcal{X}$ such that the orthogonal projections P_ε from $\mathcal{X}$ onto $\mathcal{X}_\varepsilon$ converge to the identity in the weak sense, i.e.,

$$(11.2.1) \qquad\qquad P_\varepsilon u \to u \qquad \text{for every } u \in \mathcal{X}$$

as ε tends to zero. By $b(\cdot, \cdot; \lambda)$ we denote the following sesquilinear form on $\mathcal{X} \times \mathcal{X}$:

$$b(u, v; \lambda) = \sum_{k=0}^{\mu} b_k(u, v)\,\lambda^k,$$

where $b_k(\cdot, \cdot)$ are sesquilinear forms on $\mathcal{X} \times \mathcal{X}$ and λ is a complex parameter. Suppose that the operators $T_k : \mathcal{X} \to \mathcal{X}$, $k = 1, \dots, \mu$, defined by the equality

$$b_0(T_k u, v) = b_k(u, v) \qquad \text{for all } u, v \in \mathcal{X}$$

are compact. The sesquilinear form $b(\cdot, \cdot; \lambda)$ generates the operators $A(\lambda) : \mathcal{X} \to \mathcal{X}$ and $A_\varepsilon(\lambda) : \mathcal{X}_\varepsilon \to \mathcal{X}_\varepsilon$ by the equalities

$$(11.2.2) \qquad b_0(A(\lambda)u, v) \ = \ b(u, v; \lambda) \qquad \text{for all } u, v \in \mathcal{X},$$

$$(11.2.3) \qquad b_0(A_\varepsilon(\lambda)u, v) \ = \ b(u, v; \lambda) \qquad \text{for all } u, v \in \mathcal{X}_\varepsilon.$$

Obviously,

$$A(\lambda) = I + \sum_{k=1}^{\mu} T_k\,\lambda^k,$$

where I denotes the identity operator in the space $\mathcal{X}$. Our goal is to find a connection between the spectra of the pencils A and A_ε for small ε. To this end, we introduce the operators

$$S_\varepsilon(\lambda) = I - P_\varepsilon + P_\varepsilon A(\lambda) : \mathcal{X} \to \mathcal{X}.$$

LEMMA 11.2.1. *The operator pencils A_ε and S_ε have the same eigenvalues with the same geometric, algebraic, and partial multiplicities.*

Proof: We show that every Jordan chain of the pencil A_ε is a Jordan chain of S_ε and, conversely, every Jordan chain of S_ε is a Jordan chain of A_ε.

1) Let $u_0, \ldots, u_{s-1}$ be a Jordan chain of the pencil A_ε corresponding to the eigenvalue λ_0, i.e.,

$$(11.2.4) \qquad \sum_{k=0}^{j} \frac{1}{k!} A_\varepsilon^{(k)}(\lambda_0) \, u_{j-k} = 0 \quad \text{for } j = 0, 1, \ldots, s-1, \quad u_j \in \mathcal{X}_\varepsilon \, .$$

Since $b_0(A_\varepsilon(\lambda)u, v) = b_0(A(\lambda)u, v)$ for all $u, v \in \mathcal{X}_\varepsilon$, we have

$$(11.2.5) \qquad b_0\big(A_\varepsilon^{(k)}(\lambda)u, v\big) = b_0\big(A^{(k)}(\lambda)u, v\big) \qquad \text{for all } u, v \in \mathcal{X}_\varepsilon.$$

Consequently, (11.2.4) implies

$$b_0\Big(\sum_{k=0}^{j} \frac{1}{k!} A^{(k)}(\lambda_0) \, u_{j-k} \, , \, v \Big) = 0 \qquad \text{for all } v \in \mathcal{X}_\varepsilon,$$

i.e.,

$$(11.2.6) \qquad P_\varepsilon \Big(\sum_{k=0}^{j} \frac{1}{k!} A^{(k)}(\lambda_0) \, u_{j-k} \Big) = 0.$$

Since $u_j \in \mathcal{X}_\varepsilon$, we further have $(I - P_\varepsilon)u_j = 0$ for $j = 0, \ldots, s-1$. Hence (11.2.6) implies

$$(11.2.7) \qquad \sum_{k=0}^{j} \frac{1}{k!} S_\varepsilon^{(k)}(\lambda_0) \, u_{j-k} = 0.$$

Therefore, $u_0, \ldots, u_{s-1}$ is a Jordan chain of the pencil S_ε corresponding to the eigenvalue λ_0.

2) Suppose that $u_0, \ldots, u_{s-1}$ is a Jordan chain of S_ε corresponding to the eigenvalue λ_0. This means that $u_0, \ldots, u_{s-1}$ are elements of the space $\mathcal{X}$ satisfying (11.2.7) for $j = 0, \ldots, s-1$. In particular, for $j = 0$ we get $u_0 = P_\varepsilon\big(I - A(\lambda_0)\big)u_0$ and, therefore, $u_0 \in \mathcal{X}_\varepsilon$. Analogously, it follows that $u_j \in \mathcal{X}_\varepsilon$ for $j = 1, \ldots, s-1$. Thus, $(I - P_\varepsilon)u_j = 0$ for $j = 0, \ldots, s-1$, and (11.2.7) implies (11.2.6). Using (11.2.5), we obtain the equality

$$b_0\Big(\sum_{k=0}^{j} \frac{1}{k!} A_\varepsilon^{(k)}(\lambda_0) \, u_{j-k} \, , \, v \Big) = 0 \qquad \text{for each } v \in \mathcal{X}_\varepsilon$$

which is equivalent to (11.2.4). This proves the lemma. ∎

Now the following assertion holds as a direct consequence of Theorem 1.1.4.

THEOREM 11.2.1. *Let λ_0 be an eigenvalue of the pencil A with the algebraic multiplicity κ and let δ be a sufficiently small real number such that the ball $B_\delta(\lambda_0) = \{\lambda : |\lambda - \lambda_0| < \delta\}$ does not contain other eigenvalues of A. Then there exists a number $\varepsilon_0 > 0$ such that the sum of the algebraic multiplicities of the eigenvalues of A_ε in $B_\delta(\lambda_0)$ is equal to κ if $\varepsilon < \varepsilon_0$.*

Proof: From (11.2.1) and from the compactness of the operator $A(\lambda) - I$ it follows that the operator

$$A(\lambda) - S_\varepsilon(\lambda) = (I - P_\varepsilon)\,(A(\lambda) - I)$$

converges to zero in the space of linear and continuous operators from $\mathcal{X}$ into $\mathcal{X}$. Hence Theorem 1.1.4 implies that the sum of the algebraic multiplicities of eigenvalues of $S_\varepsilon(\lambda)$ in $B_\delta(\lambda_0)$ is equal to κ if $\varepsilon < \varepsilon_0$. Using Lemma 11.2.1, we get the assertion of the theorem. ∎

11.2.2. An estimate for the energy strip. Let $L(D_x)$ be the differential operator (11.1.3) and let $\mathcal{K}$ be the cone $\{x = (x', x_n) : x_n > \phi(x')\}$, where ϕ is positively homogeneous of degree 1. We introduce the sesquilinear form

(11.2.8)

$$a(u, v; \lambda) = \frac{1}{|\log \varepsilon|} \int\limits_{\mathcal{K}_\varepsilon} \sum_{|\alpha|=|\beta|=m} \left(A_{\alpha,\beta}\, D_x^\beta(r^\lambda u)\,,\, D_x^\alpha(r^{2m-n-\bar{\lambda}}v) \right)_{\mathbb{C}^\ell} dx,$$

$$u, v \in \mathcal{X} \overset{def}{=} \overset{\circ}{W}_2^m(\Omega)^\ell,$$

where $\mathcal{K}_\varepsilon = \{x \in \mathcal{K} : \varepsilon < |x| < 1\}$ and ε is an arbitrary positive real number less than one. It can be easily seen that the right-hand side of (11.2.8) is independent of ε. Integrating by parts, we get (by the compensation of the integrals on the spheres $|x| = \varepsilon$ and $|x| = 1$)

$$\int\limits_\Omega \left(\mathcal{L}(\lambda)u, v \right)_{\mathbb{C}^\ell} dx = \frac{1}{|\log \varepsilon|} \int\limits_{\mathcal{K}_\varepsilon} \sum_{|\alpha|=|\beta|=m} \left(L(D_x)\,(r^\lambda u)\,,\, r^{2m-n-\bar{\lambda}}v \right)_{\mathbb{C}^\ell} dx$$

$$= \frac{1}{|\log \varepsilon|} \int\limits_{\mathcal{K}_\varepsilon} \sum_{|\alpha|=|\beta|=m} \left(A_{\alpha,\beta}\, D_x^\beta(r^\lambda u)\,,\, D_x^\alpha(r^{2m-n-\bar{\lambda}}v) \right)_{\mathbb{C}^\ell} dx$$

$$= a(u, v; \lambda) \qquad \text{for all } u, v \in C_0^\infty(\Omega)^\ell.$$

Hence the operator $\mathcal{L}(\lambda) : \mathcal{X} \to \mathcal{X}^*$ introduced in Section 10.1 satisfies the equality

(11.2.9) $$\left(\mathcal{L}(\lambda)u, v \right)_{L_2(\Omega)^\ell} = a(u, v; \lambda) \qquad \text{for all } u, v \in \mathcal{X}.$$

By (10.1.9) and (10.1.12), there exists a positive constant c such that

$$|a(u, u; m - n/2 + it)| \geq c\, \|u\|^2_{W_2^m(\Omega)^\ell}$$

for all $u \in \mathcal{X}$ and for all real t. In particular, it follows from this inequality that

$$b_0(u, v) = a(u, v; m - n/2)$$

is a scalar product in $\mathcal{X} = \overset{\circ}{W}_2^m(\Omega)^\ell$.

Let $\{\phi_\varepsilon\}_{\varepsilon>0}$ be a set of positively homogeneous of degree 1 functions on $\mathbb{R}^{n-1}$ which are smooth in $\mathbb{R}^{n-1}\backslash\{0\}$ and satisfy the inequality $\phi_\varepsilon(x') > \phi(x')$ for $x' \in \mathbb{R}^{n-1}\backslash\{0\}$. Furthermore, let $\mathcal{K}^{(\varepsilon)}$ be the cone $\{x = (x', x_n) : x_n > \phi_\varepsilon(x')\}$ and $\Omega^{(\varepsilon)}$ the intersection of $\mathcal{K}^{(\varepsilon)}$ with the unit sphere S^{n-1}. We suppose that the subdomains $\Omega^{(\varepsilon)}$ of Ω satisfy the condition $0 < \mathrm{dist}\,(\Omega^{(\varepsilon)}, \partial\Omega) < \varepsilon$. If we identify every vector function $u \in \overset{\circ}{W}_2^m(\Omega^{(\varepsilon)})^\ell$ with its zero extension to the domain Ω, we can consider the space $\mathcal{X}_\varepsilon \overset{def}{=} \overset{\circ}{W}_2^m(\Omega^{(\varepsilon)})^\ell$ as a subspace of $\mathcal{X} = \overset{\circ}{W}_2^m(\Omega)^\ell$.

LEMMA 11.2.2. *Let P_ε be the orthogonal projection from $\mathcal{X}$ onto $\mathcal{X}_\varepsilon$. Then $P_\varepsilon u \to u$ for all $u \in \mathcal{X}$ as $\varepsilon \to 0$.*

Proof: Let u be an arbitrary element of the space $\mathcal{X}$ and let $\{u_k\} \subset C_0^\infty(\Omega)^\ell$ be a sequence which converges to u in $\mathcal{X}$. Obviously,

$$
\begin{aligned}
\|(I - P_\varepsilon)u\|_{\mathcal{X}} &\leq \|(I - P_\varepsilon)u_k\|_{\mathcal{X}} + \|(I - P_\varepsilon)(u - u_k)\|_{\mathcal{X}} \\
&\leq \|(I - P_\varepsilon)u_k\|_{\mathcal{X}} + \|u - u_k\|_{\mathcal{X}}.
\end{aligned}
$$

Since $u_k \in C_0^\infty(\Omega)^\ell$, we have $u_k \in \mathcal{X}_\varepsilon$ for sufficiently small ε and fixed k. Hence the norm of $(I - P_\varepsilon)u_k$ in $\mathcal{X}$ tends to zero as $\varepsilon \to 0$. Furthermore, the norm of $u - u_k$ tends to zero as $k \to \infty$. Consequently, we obtain $\lim_{\varepsilon \to 0} \|(I - P_\varepsilon)u\|_{\mathcal{X}} = 0$. This proves the lemma. ∎

Now it is easy to prove the following theorem.

THEOREM 11.2.2. *Suppose that the differential operator $L(D_x)$ is elliptic and formally selfadjoint. Then the strip*

(11.2.10) $$|\operatorname{Re}\lambda - m + n/2| < 1/2$$

does not contain eigenvalues of the pencil $\mathcal{L}$.

Proof: We set

$$b(u, v; \lambda) = a(u, v; \lambda + m - n/2)$$

and define the operators $A(\lambda)$, $A_\varepsilon(\lambda)$ by the relations (11.2.2) and (11.2.3), respectively. Then any number λ_0 is an eigenvalue of the pencil $\mathcal{L}$ if and only if $\lambda_0 - m + n/2$ is an eigenvalue of A. The same is true for the eigenvalues of the pencil $\mathcal{L}_\varepsilon$ generated by the Dirichlet problem in the cone $\mathcal{K}^{(\varepsilon)}$ and the eigenvalues of the pencil A_ε.

We assume that there is an eigenvalue λ_0 of the pencil $\mathcal{L}$ in the strip (11.2.10). Then, by Theorem 11.2.1, this strip contains at least one eigenvalue of the pencil $\mathcal{L}_\varepsilon$ if ε is sufficiently small. This contradicts Theorem 11.1.1. ∎

11.3. Second order systems

11.3.1. Formulation of the main result. Let

(11.3.1) $$L(\partial_x) = -\sum_{i,j=1}^{n} A_{i,j}\, \partial_{x_i}\, \partial_{x_j}\,,$$

where $A_{i,j}$ are constant $\ell \times \ell$-matrices such that $A_{i,j}^* = A_{j,i} = A_{i,j}$. We suppose that the operator (11.3.1) is strongly elliptic, i.e., there exists a positive constant c_0 such that

(11.3.2) $$|(L(\xi)f, f)_{\mathbb{C}^\ell}| = \left(\sum_{i,j=1}^{n} A_{i,j}\xi_i\xi_j\, f, f\right)_{\mathbb{C}^\ell} \geq c_0|\xi|^2|f|^2_{\mathbb{C}^\ell}$$

for all $\xi \in \mathbb{R}$, $f \in \mathbb{C}^\ell$. Furthermore, let

$$\mathcal{K}_\phi = \{x = (x', x_n) \in \mathbb{R}^n : x_n > \phi(x')\}$$

be a cone with vertex at the origin. Here ϕ is a continuous positively homogeneous function of degree 1 on $\mathbb{R}^{n-1}$. By (r, ω) we denote spherical coordinates of the point $x \in \mathbb{R}^n$, i.e., $r = |x|$, $\omega = (\omega_1, \ldots, \omega_n) = x/|x|$. Then the set

$$\Omega_\phi = \{\omega = (\omega', \omega_n) \in S^{n-1} : \omega_n > \phi(\omega')\}$$

is the intersection of the cone $\mathcal{K}_\phi$ with the unit sphere S^{n-1}.

We are interested in solutions of the Dirichlet problem

(11.3.3) $L(\partial_x)U = 0$ in $\mathcal{K}_\phi$, $U = 0$ on $\partial\mathcal{K}_\phi\backslash\{0\}$

which have the form

(11.3.4) $$U(x) = r^{\lambda_0} \sum_{k=0}^{s} \frac{(\log r)^k}{k!}\, u_{s-k}(\omega)\,,$$

where $\lambda_0 \in \mathbb{C}$ and $u_k \in \overset{\circ}{W}^1_2(\Omega_\phi))^\ell$, $k = 0, 1, \dots, s$.

In order to find such solutions, we have to explore the spectrum of the pencil

(11.3.5) $$\mathcal{L}_\phi(\lambda) : \overset{\circ}{W}^1_2(\Omega_\phi)^\ell \to W^{-1}_2(\Omega_\phi)^\ell$$

where

$$\mathcal{L}_\phi(\lambda)u(\omega) = \mathcal{L}(\lambda)\, u(\omega) = r^{2-\lambda} L(\partial_x)\, (r^\lambda u(\omega))   \quad \text{for } u \in \overset{\circ}{W}^1_2(\Omega_\phi)^\ell.$$

Let $a(\cdot, \cdot; \lambda)$ be the following sesquilinear form on $\overset{\circ}{W}^1_2(\Omega_\phi)^\ell \times \overset{\circ}{W}^1_2(\Omega_\phi)^\ell$ quadratically depending on λ:

(11.3.6) $$a(u, v; \lambda) = \frac{1}{|\log \varepsilon|} \int\limits_{\substack{\mathcal{K}_\phi \\ \varepsilon < |x| < 1}} \sum_{i,j=1}^{n} \left(A_{i,j}\partial_{x_j}U, \partial_{x_i}V\right)_{\mathbb{C}^\ell} dx\,,$$

where ε is an arbitrary real number less than one, $U(x) = r^\lambda u(\omega)$, and $V(x) = r^{2-n-\overline{\lambda}}v(\omega)$ (cf. (11.2.8)). Then we have

$$\int\limits_{\Omega_\phi} \left(\mathcal{L}_\phi(\lambda)u, v\right)_{\mathbb{C}^\ell} d\omega = a(u, v; \lambda)   \quad \text{for all } u, v \in \overset{\circ}{W}^1_2(\Omega_\phi)^\ell,$$

where $d\omega$ is the standard measure on S^{n-1}. Note that the right-hand side of (11.3.6) is independent of ε.

We introduce the differential operators

$$\partial_{\omega_j} = r\left(\partial_{x_j} - \omega_j\, \partial_r\right), \qquad j = 1, \dots, n,$$

on the sphere S^{n-1}. It can be easily verified that

$$\mathcal{L}(\lambda) = -\sum_{i,j=1}^{n} A_{i,j}\big((\lambda-1)\omega_i + \partial_{\omega_i}\big)\, (\lambda\omega_j + \partial_{\omega_j}).$$

Since $(n-1)\omega_j - \partial_{\omega_j}$ is the adjoint operator to ∂_{ω_j} in $L_2(\Omega)$, we get

$$a(u, v; \lambda) = \int\limits_{\Omega_\phi} \sum_{i,j=1}^{n} \left(A_{i,j}(\lambda\omega_j + \partial_{\omega_j})u, \big((2-n-\overline{\lambda})\omega_i + \partial_{\omega_i}\big)v\right)_{\mathbb{C}^\ell} d\omega\,.$$

Furthermore,

(11.3.7) $$\begin{cases} \partial_{\omega_j}\partial_{\omega_k} - \partial_{\omega_k}\partial_{\omega_j} = \omega_j\partial_{\omega_k} - \omega_k\partial_{\omega_j}\,, \\[2mm] \displaystyle\sum_{j=1}^{n}\omega_j\partial_{\omega_j} = 0, \qquad \sum_{j=1}^{n}\partial^2_{\omega_j} = \delta\,. \end{cases}$$

Here δ denotes the Beltrami operator on S^{n-1}. The expression

$$\sum_{j=1}^{n} \int_{\Omega_\phi} \left(\partial_{\omega_j} u, \partial_{\omega_j} v \right)_{\mathbb{C}^\ell} d\omega$$

defines a scalar product in $\overset{\circ}{W}_2^1(\Omega_\phi)^\ell$. Besides, we have

$$(11.3.8) \qquad \sum_{j=1}^{n} \int_{\Omega_\phi} \left| \partial_{\omega_j} w \right|^2 d\omega \geq \mu_0(\phi) \int_{\Omega_\phi} |w|^2 d\omega \,,$$

where $\mu_0(\phi)$ is the smallest positive eigenvalue of the operator $-\delta$ on Ω_ϕ with Dirichlet boundary condition.

The main result of this section is the following:

> Let ϕ be a continuous function on $\mathbb{R}^{n-1}$ positively homogeneous of degree 1 such that $\phi|_{S^{n-2}} \in W_2^1(S^{n-2})$. Then the strip $|\operatorname{Re}\lambda - 1 + n/2| \leq 1/2$ is free of eigenvalues of the pencil $\mathcal{L}_\phi$.

The proof of this assertion is given in the following subsections. First we give some general properties of the operator pencil $\mathcal{L}_\phi$ for continuous ϕ. Then we consider the operator pencil $\mathcal{L}_\chi$ for smooth χ satisfying the inequality $\phi(x') \leq \chi(x') \leq \phi(x') + \tau|x'|$. We prove that for sufficiently small τ depending only on the coefficients of L there exists a positive number ε_1 such that all eigenvalues of $\mathcal{L}_\chi$ lie outside the strip $|\operatorname{Re}\lambda - 1 + n/2| \leq (1 + \varepsilon_1)/2$. This number ε_1 depends only on ϕ and on the W_2^1-norm of $\chi|_{S^{n-2}}$. Applying Theorem 11.2.1, we get the above assertion.

11.3.2. Basic properties of the operator pencil $\mathcal{L}_\phi$. In the following theorem we give some properties of the operator pencil $\mathcal{L}_\phi(\lambda)$ which were partially proved in the previous sections.

THEOREM 11.3.1. *Let ϕ be a function continuous and positively homogeneous of degree 1 cn $\mathbb{R}^{n-1}$. Then the following assertions are true.*

1) *The operator (11.3.5) is Fredholm for every $\lambda \in \mathbb{C}$.*

2) *For every λ, $\operatorname{Re}\lambda = 1 - n/2$, the operator (11.3.5) is selfadjoint and positive definite. Moreover, for all λ, $\operatorname{Re}\lambda = 1 - n/2$, and $u \in \overset{\circ}{W}_2^1(\Omega_\phi)$ there is the inequality*

$$(11.3.9) \qquad a(u, u, \lambda) \geq c_0 \int_{\Omega_\phi} \left(\sum_{1 \leq j \leq n} \left| \partial_{\omega_j} u \right|^2 + |\lambda|^2 |u|^2 \right) d\omega \,,$$

where c_0 is the same constant as in (11.3.2).

3) *The spectrum of the operator pencil $\mathcal{L}_\phi$ consists of isolated eigenvalues with finite algebraic multiplicities. The set*

$$\left\{ \lambda \in \mathbb{C} : (\operatorname{Re}\lambda - 1 + n/2)^2 < \frac{c_0}{c_1}\left((1 - n/2)^2 + (\operatorname{Im}\lambda)^2 + \mu_0(\phi) \right) \right\} \,,$$

where

$$(11.3.10) \qquad c_1 = \max_{\omega \in S^{n-1}} \left\| \sum_{i,j=1}^{n} A_{i,j}\, \omega_i\, \omega_j \right\|_{\mathbb{C}^\ell \to \mathbb{C}^\ell} \,,$$

does not contain eigenvalues of the pencil $\mathcal{L}_\phi$.

4) *If λ_0 is an eigenvalue of $\mathcal{L}_\phi$, then $2 - n - \overline{\lambda}_0$ is an eigenvalue, too. The geometric, partial and algebraic multiplicities of λ_0 and $2 - n - \overline{\lambda}_0$ coincide.*

5) *The strip* $|\operatorname{Re}\lambda - 1 + n/2| < 1/2$ *is free of eigenvalues of the pencil* $\mathcal{L}_\phi$.

Proof: For assertions 1), 4) and 5) we refer to Theorems 10.1.1, 10.1.2 and 11.2.2. We prove 2) and 3).

2) Inequality (11.3.2) implies

$$(11.3.11) \qquad \int_{\mathcal{K}_\phi} \sum_{i,j=1}^{n} \left(A_{i,j}\partial_{x_j} U \,,\, \partial_{x_i} U \right)_{\mathbb{C}^\ell} dx \geq c_0 \int_{\mathcal{K}_\phi} \sum_{j=1}^{n} \left| \partial_{x_j} U \right|^2 dx$$

for all $U \in \overset{\circ}{W}{}^1_2(\mathcal{K}_\phi)^\ell$. Let

$$(11.3.12) \qquad \tilde{U}(\lambda,\cdot) = (M_{r\to\lambda}U)(\lambda,\cdot) = (2\pi)^{-1/2} \int_0^\infty r^{-\lambda-1}\, U(r,\cdot)\, dr$$

be the *Mellin transform* of U with respect to the variable t. Using the equalities

$$M_{r\to\lambda}\left(r\partial_r U \right) = \lambda\, M_{r\to\lambda} U\,, \qquad r\partial_{x_j} = \omega_j\, r\partial_r + \partial_{\omega_j}$$

and the Parseval equality

$$\int_0^\infty r^{n-3}\, U(r)\, \overline{V(r)}\, dr = \int_{\mathbb{R}} \tilde{U}(it+1-n/2)\, \overline{\tilde{V}(it+1-n/2)}\, dt,$$

from (11.3.11) we get

$$\int_{\mathbb{R}} a\Big(\tilde{U}(it+1-n/2,\cdot),\, \tilde{U}(it+1-n/2,\cdot); it+1-n/2 \Big)\, dt$$

$$\geq c_0 \int_{\mathbb{R}} \int_{\Omega_\phi} \sum_{j=1}^{n} \left| \Big((it+1-n/2)\omega_j + \partial_{\omega_j} \Big) \tilde{U}(it+1-n/2,\omega) \right|_{\mathbb{C}^\ell}^2 d\omega\, dt\,.$$

Hence we have

$$(11.3.13) \qquad a(u,u;\lambda) \geq c_0 \int_{\Omega_\phi} \sum_{j=1}^{n} \left| (\lambda\omega_j + \partial_{\omega_j})\, u \right|^2 d\omega$$

for all $u \in \overset{\circ}{W}{}^1_2(\Omega_\phi)^\ell$, $\operatorname{Re}\lambda = 1 - n/2$. From this, by means of (11.3.7), we obtain (11.3.9).

3) The first part of assertion 3) follows from 1) and 2). We prove the second part. Let $\lambda = \tau + \lambda_1$, $\lambda_1 = 1 - n/2 + it$, $t \in \mathbb{R}$, $\tau \in \mathbb{R}$. Then

$$a(u,u;\lambda) = a(u,u;\lambda_1) - \tau^2 \sum_{i,j=1}^{n} \int_{\Omega_\phi} \left(A_{i,j}\, \omega_j\, u,\, \omega_i\, u \right)_{\mathbb{C}^\ell} d\omega$$

$$+ \tau \sum_{i,j=1}^{n} \int_{\Omega_\phi} \Big(\big(A_{i,j}\omega_j u, (\lambda_1\omega_i + \partial_{\omega_i})u \big)_{\mathbb{C}^\ell} - \big(A_{i,j}(\lambda_1\omega_j + \partial_{\omega_j})u, \omega_i u \big)_{\mathbb{C}^\ell} \Big)\, d\omega\,.$$

Hence

$$(11.3.14) \qquad \operatorname{Re} a(u,u;\lambda) = a(u,u;\lambda_1) - \tau^2 \int_{\Omega_\phi} (L(\omega)u, u)_{\mathbb{C}^\ell} d\omega$$

and, by virtue of (11.3.8), (11.3.9), we get the second assertion of 3). The proof is complete. ∎

Note that

(11.3.15) $\|A_{i,j}\|_{\mathbb{C}^\ell \to \mathbb{C}^\ell} \leq c_1$ for $i, j = 1, \ldots, n$,

where c_1 is the constant (11.3.10). Indeed, if $i = j$, then (11.3.15) immediately follows from (11.3.10). If $i \neq j$, then, by the positivity of the matrix $L(\omega)$, $\omega \in S^{n-1}$, we get the inequality

$$|(A_{i,j}f, f)|^2 \leq (A_{i,i}f, f)\,(A_{j,j}f, f) \quad \text{for all } f \in \mathbb{C}^\ell$$

which implies (11.3.15).

11.3.3. On eigenfunctions of $\mathcal{L}_\chi$ for smooth χ. In the following three lemmas it is assumed that χ is positively homogeneous of degree 1 and infinitely differentiable on $\mathbb{R}^{n-1} \backslash \{0\}$.

LEMMA 11.3.1. *The following Green formula is satisfied for all $u, v \in W_2^2(\Omega_\chi)^\ell$.*

$$
(11.3.16) \quad \int_{\Omega_\chi} \left(\big(\mathcal{L}(\lambda)u(\omega), v(\omega) \big)_{\mathbb{C}^\ell} - \big(u(\omega), \mathcal{L}(2 - n - \overline{\lambda})v(\omega) \big)_{\mathbb{C}^\ell} \right) d\omega
$$

$$
= - \sum_{i,j=1}^{n} \int_{\partial\Omega_\chi} \Big(A_{i,j}\nu_i(\lambda\omega_j + \partial_{\omega_j})u, v \Big)_{\mathbb{C}^\ell} d\omega
$$

$$
+ \sum_{i,j=1}^{n} \int_{\partial\Omega_\chi} \Big(A_{i,j}\nu_j u, \big((2 - n - \overline{\lambda})\omega_i + \partial_{\omega_i}\big)v \Big)_{\mathbb{C}^\ell} d\omega
$$

Here ν_j denotes the j-th component of the exterior normal ν to $\partial\mathcal{K}_\chi$.

Proof: Let $U = r^\lambda u$ and $V = r^{2-n-\overline{\lambda}}v$ Then, due to the cancellation of the integrals on the spheres $|x| = \varepsilon$ and $|x| = 1$, we get

$$
\int_{\substack{\mathcal{K}_\chi \\ \varepsilon < |x| < 1}} \left(\big(L(\partial_x)U, V \big)_{\mathbb{C}^\ell} - \big(U, L(\partial_x)V \big)_{\mathbb{C}^\ell} \right) dx
$$

$$
= - \sum_{i,j=1}^{n} \int_{\substack{\partial\mathcal{K}_\chi \\ \varepsilon < |x| < 1}} \left(\big(A_{i,j}\nu_i\partial_{x_j}U, V \big)_{\mathbb{C}^\ell} - \big(A_{i,j}\nu_j U, \partial_{x_i}V \big)_{\mathbb{C}^\ell} \right) d\sigma.
$$

This relation yields (11.3.16). ∎

In the sequel, let Ω_t be the subdomain $\{\omega \in S^{n-1} : \omega_n > t|\omega'|\}$ of the unit sphere. Here t is an arbitrary real number.

LEMMA 11.3.2. 1) *There exists a positive constant ε_0 depending only on n, t and on the coefficients $A_{i,j}$ such that the Dirichlet problem*

$$
(11.3.17) \quad
\begin{cases}
\mathcal{L}(\lambda)\,G(\omega, \theta; \lambda) = I_\ell \delta(\omega - \theta), & \omega \in \Omega_t, \\[2mm]
G(\omega, \theta; \lambda) = 0, & \omega \in \partial\Omega_t,
\end{cases}
$$

where I_ℓ denotes the identity matrix in $\mathbb{C}^\ell$, has a unique solution $G(\cdot, \theta; \lambda)$ for arbitrary $\theta \in \Omega_t$ and λ lying in the strip

(11.3.18) $|\operatorname{Re}\lambda - 1 + n/2| < (1 + \varepsilon_0)/2.$

This solution (the Green function) satisfies the estimate

(11.3.19) $\|G(\omega, \theta; \lambda)\|_{\mathbb{C}^\ell \to \mathbb{C}^\ell} \leq \begin{cases} c\,|\omega - \theta|^{3-n} & \text{if } n > 3, \\ c\,(|\log|\omega - \theta||+1) & \text{if } n = 3 \end{cases}$

for all $\omega, \theta \in \Omega_t$ and λ from the strip (11.3.18). The constant c in (11.3.19) does not depend on ω, θ, and λ.

2) Let $\chi(x') \geq t + 1$ for $|x'| = 1$. Furthermore, let $\varepsilon \in [0, \varepsilon_0]$ and let $\lambda_0 = (3 - n + \varepsilon)/2 + it$, $t \in \mathbb{R}$, be an eigenvalue of $\mathcal{L}_\chi$. Then every eigenvector u_0 corresponding to this eigenvalue admits the representation

(11.3.20) $u_0(\theta) = -\int\limits_{\partial\Omega_\chi} \Big(L(\nu)\,\partial_\nu u_0(\omega),\, G(\omega, \theta; 2 - n - \overline{\lambda}_0) \Big)_{\mathbb{C}^\ell} d\omega.$

Proof: By Theorem 11.1.1, there exists a positive constant $\tau > 1/2$ such that the operator $\mathcal{L}(\lambda)$ is an isomorphism $\overset{\circ}{W}_2^1(\Omega_t) \to W_2^{-1}(\Omega_t)$ for $|\operatorname{Re}\lambda - 1 + n/2| < \tau$. Hence the Green function of the problem (11.3.17) exists for each λ lying in this strip (see, e.g., (see, e.g., our book [**136**, Sect.3.5]).

We prove (11.3.19). According to Theorem 2.2 in Maz'ya's and Plamenevskiĭ's paper [**186**] (see also [**136**, Th.8.4.8]), for every $y \in \mathcal{K}$ there exists a solution $G_0(\cdot, y)$ of the problem

$$L(\partial_x)G_0(x, y) = I_\ell\,\delta(x - y), \qquad x \in \mathcal{K},$$
$$G_0(x, y) = 0, \qquad x \in \partial\mathcal{K}$$

in the cone $\mathcal{K} = \{x \in \mathbb{R}^n : \omega = x/|x| \in \Omega_t\}$ satisfying the estimates

(11.3.21) $|G_0(x, y)| \leq c\,(|x - y|^{2-n} + |y|^{2-n})$ if $|x|/2 \leq |y| \leq 2|x|$, $n \geq 3$,

(11.3.22) $|G_0(x, y)| \leq c\,|x|^{1+\tau-\varepsilon-n/2}\,|y|^{1-\tau+\varepsilon-n/2}$ if $|y| > 2|x|$,

(11.3.23) $|G_0(x, y)| \leq c\,|x|^{1-\tau-\varepsilon-n/2}\,|y|^{1+\tau+\varepsilon-n/2}$ if $2|y| < |x|$

with an arbitrarily small positive number ε. Since $r\partial_{x_j} = \omega_j r\partial_r + \partial_{\omega_j}$, we have

$$\begin{aligned} r^2 L(\partial_x) &= -r^2 \sum_{i,j=1}^{n} A_{i,j}\,\partial_{x_i}\,\partial_{x_j} = -\sum_{i,j=1}^{n} A_{i,j}\,(r\partial_{x_i} - \omega_i)\,r\partial_{x_j} \\ &= -\sum_{i,j=1}^{n} A_{i,j}\,\Big(\omega_i(r\partial_r - 1) + \partial_{\omega_i}\Big)\Big(\omega_j r\partial_r + \partial_{\omega_j}\Big). \end{aligned}$$

Applying the Mellin transform (11.3.12) to the equation

$$r^2 L(\partial_x)\,G_0(x, y) = I_\ell\,r^2\,\delta(x - y),$$

we get

$$\mathcal{L}(\lambda)\,\tilde{G}_0(\lambda, \omega, \rho, \theta) = I_\ell\,M_{r \to \lambda}\,r^2\delta(x - y),$$

where $\tilde{G}_0(\lambda, \omega, \rho, \theta) = (2\pi)^{-1/2} \int_0^\infty r^{-\lambda-1} G_0(r\omega, \rho\theta)\, dr$ denotes the Mellin transform of G_0. Furthermore, inserting $U(x) = r^{-\lambda-n+2} u(\omega)$ into the equality

$$U(y) = \int_K \delta(x-y)\, U(x)\, dx = \int_{\Omega_t} \int_0^\infty r^{n-1}\, \delta(x-y)\, U(x)\, dr\, d\omega\,,$$

we obtain

$$\rho^{-\lambda-n+2}\, u(\theta) = \int_{\Omega_t} \Big(\int_0^\infty r^{-\lambda+1}\, \delta(x-y)\, dr \Big)\, u(\omega)\, d\omega\,.$$

This means that $M_{r\to\lambda}\, r^2\delta(x-y) = \rho^{-\lambda-n+2}\, \delta(\omega-\theta)$. Hence the Green function $G(\omega, \theta; \lambda)$ is given by the equality

$$G(\omega, \theta; \lambda) = \rho^{\lambda+n-2}\, \tilde{G}_0(\lambda, \omega, \rho, \theta) = \rho^{\lambda+n-2} \int_0^\infty r^{-\lambda-1}\, G_0(r\omega, \rho\theta)\, dr\,.$$

Here, by (11.3.22), we have

$$\Big| \int_0^{\rho/2} r^{-\lambda-1}\, G_0(r\omega, \rho\theta)\, dr \Big| \;\leq\; c\rho^{1-\tau+\varepsilon-n/2} \int_0^{\rho/2} r^{-\operatorname{Re}\lambda+\tau-\varepsilon-n/2}\, dr = C\,\rho^{-\operatorname{Re}\lambda-n+2}$$

if $\operatorname{Re}\lambda < 1+\tau-\varepsilon$, where $C = c/(-\operatorname{Re}\lambda+1+\tau-\varepsilon-n/2)$. An analogous estimate holds for the integral over the interval $(2\rho, +\infty)$. Furthermore, by (11.3.21), we have

$$\Big| \int_{\rho/2}^{2\rho} r^{-\lambda-1}\, G_0(r\omega, \rho\theta)\, dr \Big| \;\leq\; c\,\rho^{1-n-\operatorname{Re}\lambda} \int_{\rho/2}^{2\rho} \Big(\Big| \frac{r}{\rho}\omega - \theta \Big|^{2-n} + 1 \Big)\, dr$$

$$= c\,\rho^{2-n-\operatorname{Re}\lambda} \int_{1/2}^{2} \big(|r\omega - \theta|^{2-n} + 1 \big)\, dr\,.$$

Using the inequality

$$\int_{1/2}^{2} |r\omega - \theta|^{2-n}\, dr \leq \begin{cases} c\,|\omega-\theta|^{3-n} & \text{if } n > 3, \\ c\,(|\log|\omega-\theta||+1) & \text{if } n = 3, \end{cases}$$

we get the desired inequality for $G(\omega, \theta; \lambda)$ if λ lies in the strip (11.3.18) and ε is less than $2\tau - 1$. ∎

LEMMA 11.3.3. *Let $\lambda_0 = (3-n+\varepsilon)/2 + it$, $t \in \mathbb{R}$, $\varepsilon > 0$, be an eigenvalue of the operator pencil $\mathcal{L}_\chi$. Furthermore, let $u_0 \in (\overset{\circ}{W}{}_2^1(\Omega_\chi))^\ell$ be an eigenvector corresponding to λ_0. Then*

$$(11.3.24) \qquad \int_{\partial\Omega_\chi} |\nu_n|\, |\partial_\nu u_0|_{\mathbb{C}^\ell}^2\, d\omega \leq C\, \frac{c_1^2}{c_0^2}\, \varepsilon\, (1+\varepsilon)^2 \int_{\Omega_\chi} |u_0|_{\mathbb{C}^\ell}^2\, d\omega\,,$$

where the constant C only depends on n.

Proof: Let $U(x) = r^{\lambda_0} u_0(\omega)$, $\mathcal{K}_\chi^{(1)} = \{x \in \mathcal{K}_\chi : 1/2 < |x| < 1\}$ and $\partial \mathcal{K}_\chi^{(1)} = \{x \in \partial \mathcal{K}_\chi : 1/2 < |x| < 1\}$. Then we have

$$
0 = -\mathrm{Re} \int_{\mathcal{K}_\chi^{(1)}} \sum_{i,j=1}^n \left(A_{i,j}\partial_{x_i}\partial_{x_j}U, \partial_{x_n}U\right)_{\mathbb{C}^\ell} dx
$$

$$
= \mathrm{Re} \int_{\partial\mathcal{K}_\chi^{(1)}} \sum_{i,j=1}^n \left(\frac{\nu_n}{2}\left(A_{i,j}\partial_{x_j}U, \partial_{x_i}U\right)_{\mathbb{C}^\ell} - \nu_i\left(A_{i,j}\partial_{x_j}U, \partial_{x_n}U\right)_{\mathbb{C}^\ell}\right) d\sigma
$$

$$
+ \mathrm{Re} \int_{\Omega_\chi} \sum_{i,j=1}^n \left(\frac{1}{2}\omega_n\left(A_{i,j}\partial_{x_j}U, \partial_{x_i}U\right)_{\mathbb{C}^\ell} - \omega_i\left(A_{i,j}\partial_{x_j}U, \partial_{x_n}U\right)_{\mathbb{C}^\ell}\right) d\omega\, r^{n-1}\Big|_{r=1/2}^{r=1}.
$$

Using the equality $\partial_{x_j}U = \nu_j \partial_\nu U$ on $\partial\mathcal{K}_\chi \backslash \{0\}$, we obtain

$$
(11.3.25) \quad \frac{1}{2} \int_{\partial\Omega_\chi} \nu_n \sum_{i,j=1}^n \left(A_{i,j}\nu_i\nu_j\partial_\nu u_0, \partial_\nu u_0\right)_{\mathbb{C}^\ell} d\omega
$$

$$
= (1 - 2^{-\varepsilon})\,\mathrm{Re} \int_{\Omega_\chi} \sum_{i,j=1}^n \left(\frac{\omega_n}{2}\left(A_{i,j}(\lambda_0\omega_j + \partial_{\omega_j})u_0, (\lambda_0\omega_i + \partial_{\omega_i})u_0\right)_{\mathbb{C}^\ell}\right.
$$

$$
\left. - \omega_i\left(A_{i,j}(\lambda_0\omega_j + \partial_{\omega_j})u_0, (\lambda_0\omega_n + \partial_{\omega_n})u_0\right)_{\mathbb{C}^\ell}\right) d\omega.
$$

Applying (11.3.15), we can estimate the right-hand side of (11.3.25) from above by

$$
(11.3.26) \qquad C\, c_1\, \varepsilon\left(\int_{\Omega_\chi} \sum_{j=1}^n \left|(\lambda_1\omega_j + \partial_{\omega_j})u_0\right|^2 d\omega + (1+\varepsilon)^2 \int_{\Omega_\chi} |u_0|^2 d\omega\right),
$$

where $\lambda_1 = 1 - \frac{n}{2} + it = \lambda_0 - \frac{1+\varepsilon}{2}$ and the constant C depends only on n. Using (11.3.13), (11.3.14) and the fact that u_0 is an eigenfunction of $\mathcal{L}_\chi$ corresponding to the eigenvalue λ_0, it can be easily seen that (11.3.26) does not exceed

$$
C\,\frac{c_1}{c_0}\,(c_1 + c_0)\,\varepsilon(1+\varepsilon)^2 \int_{\Omega_\chi} |u_0|^2 d\omega.
$$

Estimating now the left-hand side of (11.3.25) by means of (11.3.2), we arrive at (11.3.24). ∎

11.3.4. The width of the energy strip.

LEMMA 11.3.4. *Let χ, ψ, η be continuous functions on $\mathbb{R}^{n-1}$ positively homogeneous of degree 1 such that*

$$
\chi(x') \le \psi(x') \le \eta(x') \qquad \text{for all } x' \in \mathbb{R}^{n-1} .
$$

We set $\Omega(\chi,\psi) = \{\omega = (\omega', \omega_n) \in S^{n-1} : \chi(\omega') < \omega_n < \psi(\omega')\}$. Analogously, we define the domains $\Omega(\chi,\eta)$, $\Omega(\psi,\eta) \subset S^{n-1}$. Then the following estimate is satisfied for every function $w \in W_2^1(\Omega(\chi,\eta))$:

$$
\int_{\Omega(\chi,\psi)} |w|^2 d\omega \le \frac{16}{3}\rho \log 2 \int_{\Omega(\chi,\eta)} \left|(\lambda_1\omega_n + \partial_{\omega_n})w\right|^2 d\omega + 2\,(1+q^2)\,\kappa \int_{\Omega(\psi,\eta)} |w|^2 d\omega,
$$

where

$$\lambda_1 = 1 - \frac{n}{2} + it, \qquad t \in \mathbb{R},$$

$$\rho = \max_{|x'|=1} (\psi(x') - \chi(x')) (\eta(x') - \chi(x')),$$

$$\kappa = \max_{|x'|=1} (\psi(x') - \chi(x')) (\eta(x') - \psi(x'))^{-1},$$

and

$$q = \max_{|x'|=1} \max \left(|\chi(x')|, |\eta(x')| \right).$$

Proof: If $\chi(x') < x_n < \psi(x') < \xi < \eta(x')$, then

$$|W(x', x_n)| \leq \int_{x_n}^{\xi} |\partial_t W(x', t)| dt + |W(x', \xi)|$$

and, therefore,

$$|W(x', x_n)|^2 \leq 2(\eta(x') - \chi(x')) \int_{\chi(x')}^{\eta(x')} |\partial_t W(x', t)|^2 \, dt + 2 |W(x', \xi)|^2 .$$

Integrating with respect to ξ over the interval $(\psi(x'), \eta(x'))$, we get

$$(11.3.27) \qquad |W(x', x_n)|^2 \leq 2 (\eta(x') - \chi(x')) \int_{\chi(x')}^{\eta(x')} |\partial_t W(x', t)|^2 \, dt$$

$$+ \frac{2}{\eta(x') - \psi(x')} \int_{\psi(x')}^{\eta(x')} |W(x', t)|^2 \, dt .$$

We introduce the set

$$Q(\chi, \psi) = \{ x \in \mathbb{R}^n : 1 < |x'| < 2, \ \chi(x') < x_n < \psi(x') \}.$$

Analogously, we define the sets $Q(\psi, \eta)$ and $Q(\chi, \eta)$. Integrating (11.3.27) at first with respect to x_n over the interval $(\chi(x'), \psi(x'))$ and then with respect to x' over the set $1 < |x'| < 2$, we get

$$\int_{Q(\chi,\psi)} |W(x)|^2 dx \leq 8\rho \int_{Q(\chi,\eta)} |\partial_{x_n} W(x)|^2 \, dx + 2\kappa \int_{Q(\psi,\eta)} |W(x)|^2 dx .$$

If we set $W(x) = r^{\lambda_1} w(\omega)$, we obtain

$$(11.3.28) \quad \frac{3}{2} \int_{\Omega(\chi,\psi)} |w(\omega)|^2 |\omega'|^{-2} d\omega \leq 8 \rho \log 2 \int_{\Omega(\chi,\eta)} |(\lambda_1 \omega_n + \partial_{\omega_n}) w|^2 \, d\omega$$

$$+ 3 \kappa \int_{\Omega(\psi,\eta)} |w(\omega)|^2 |\omega'|^{-2} \, d\omega.$$

Here we have used the fact that $r = |x|$ satisfies the inequality $|\omega'|^{-1} < r < 2|\omega'|^{-1}$ in the domains $Q(\chi, \psi)$, $Q(\chi, \eta)$ and $Q(\psi, \eta)$.

Since $|\omega_n| \leq q|\omega'|$, we have $(1+q^2)^{-1/2} \leq |\omega'| \leq 1$ for all $\omega = (\omega', \omega_n) \in \Omega(\phi, \eta)$. Using these inequalities, we deduce the assertion of the lemma from (11.3.28). ∎

COROLLARY 11.3.1. *Let* ϕ, χ *be continuous functions on* $\mathbb{R}^{n-1}$ *positively homogeneous of degree 1 satisfying the inequalities*

$$(11.3.29) \qquad \phi(x') \leq \chi(x') \leq \phi(x') + \tau|x'| \text{ for all } x' \in \mathbb{R}^{n-1}$$

with a constant $\tau \in (0, 1]$. *Then the following inequality is satisfied for each* $w \in W_2^1(\Omega_\chi)$:

$$(11.3.30) \quad \int_{\Omega_\chi} |w|^2 \, d\omega \leq 15 \left(\tau \int_{\Omega_\chi} |(\lambda_1 \omega_n + \partial_{\omega_n})w|^2 \, d\omega + (1 + \phi_0^2) \int_{\Omega_{\phi+\tau|x'|}} |w|^2 \, d\omega \right),$$

where $\phi_0 = \max_{|x'|=1} |\phi(x')|$.

Proof: We set $\psi(x') = \phi(x') + \tau|x'|$ and $\eta(x') = \phi(x') + (\tau+1)|x'|$ and apply Lemma 11.3.4 to the functions χ, ψ and η. Since $\rho \leq 2\tau$, $\kappa \leq \tau$ and $q \leq \phi_0 + \tau + 1$, we get the inequality

$$\int_{\Omega(\chi,\psi)} |w|^2 d\omega \leq 15\,\tau \int_{\Omega_\chi} |(\lambda_1\omega_n + \partial_{\omega_n})w|^2 \, d\omega + (14 + 15\phi_0^2) \int_{\Omega(\psi,\eta)} |w|^2 \, d\omega$$

which implies (11.3.30). ∎

The last result is used in the proof of the following theorem.

THEOREM 11.3.2. *Let* ϕ *be a fixed continuous function on* $\mathbb{R}^{n-1}$ *which is positively homogeneous of degree 1. Furthermore, let* $\chi \in C^\infty(\mathbb{R}^{n-1}\backslash\{0\})$ *be a positively homogeneous function of degree 1 satisfying the inequalities*

$$\phi(x') \leq \chi(x') \leq \phi(x') + \frac{\tau}{2}|x'|,$$

where $\tau = \frac{1}{30}c_0/c_1$. *Then the operator pencil* $\mathcal{L}_\chi$ *has no eigenvalues in the strip*

$$|\operatorname{Re}\lambda - 1 + n/2| < (1 + \varepsilon_1)/2$$

where $\varepsilon_1 = \left(c \int_{S^{n-1}} (1 + |\nabla_{x'}\chi|^2)dx' \right)^{-1}$. *Here the constant* c *depends only on* n, c_0, c_1 *and on the function* ϕ.

Proof: Let $\lambda_0 = (3 - n + \varepsilon)/2 + it$ be an eigenvalue of the operator pencil $\mathcal{L}_\chi$, where $t \in \mathbb{R}$ and $\varepsilon \in [0, 1]$, and let u_0 be a corresponding eigenvector. By Lemma 11.3.3, we have

$$(11.3.31) \qquad \int_{\partial\Omega_\chi} |\nu_n| \, |\partial_\nu u_0|_{\mathbb{C}^\ell}^2 \, d\omega \leq c\varepsilon \int_{\Omega_\chi} |u_0|_{\mathbb{C}^\ell}^2 \, d\omega.$$

Here and in the sequel, we denote by c constants which depend only on n, c_0, c_1 and ϕ. We set $\lambda_1 = 1 + it - n/2$. Then, by means of (11.3.13) and (11.3.14), we get

$$(11.3.32) \qquad c_0 \sum_{j=1}^{n} \int_{\Omega_\chi} |(\lambda_1 \omega_j + \partial_{\omega_j}) u_0|_{\mathbb{C}^\ell}^2 \, d\omega \leq c_1 \int_{\Omega_\chi} |u_0|_{\mathbb{C}^\ell}^2 \, d\omega.$$

Using Corollary 11.3.1, we can estimate the integral on the right-hand side of (11.3.32). This leads to the inequality

$$c_0 \sum_{j=1}^{n} \int_{\Omega_\chi} |(\lambda_1\omega_j + \partial_{\omega_j})u_0|^2_{\mathbb{C}^\ell} \, d\omega$$

$$\leq 15 \, c_1 \Big(\tau \int_{\Omega_\chi} |(\lambda_1\omega_n + \partial_{\omega_n})u_0|^2_{\mathbb{C}^\ell} \, d\omega + (1 + \phi_0^2) \int_{\Omega_{(\phi + \tau |x'|)}} |u_0|^2_{\mathbb{C}^\ell} \, d\omega \Big).$$

Consequently, we have

$$\sum_{j=1}^{n} \int_{\Omega_\chi} |(\lambda_1\omega_j + \partial_{\omega_j})u_0|^2_{\mathbb{C}^\ell} \, d\omega \leq c \int_{\Omega_{(\phi + \tau |x'|)}} |u_0|^2_{\mathbb{C}^\ell} \, d\omega.$$

This inequality and (11.3.31) together with the estimate

$$\sum_{j=1}^{n} \int_{\Omega_\chi} |(\lambda_1\omega_j + \partial_{\omega_j})u_0|^2_{\mathbb{C}^\ell} \, d\omega \geq (|\lambda_1|^2 + \mu_0(\chi)) \int_{\Omega_\chi} |u_0|^2 \, d\omega$$

yield

(11.3.33) $$\int_{\partial\Omega_\chi} |\nu_n| \, |\partial_\nu u_0|^2_{\mathbb{C}^\ell} \, d\omega \leq c\varepsilon \int_{\Omega_{(\phi + \tau |x'|)}} |u_0|^2_{\mathbb{C}^\ell} \, d\omega.$$

We denote by $G(\omega, \theta; \lambda)$ the Green function of problem (11.3.17) in the domain

$$\Omega = \{\omega \in S^{n-1} : \omega_n > (-\phi_0 - 1)|\omega'|\}.$$

This function exists for each λ lying in the strip $|\operatorname{Re}\lambda - 1 + n/2| \leq (1 + \varepsilon_0)/2$, where ε_0 is a positive number depending only on n, ϕ_0 and $A_{i,j}$. By means of (11.3.19) and (11.3.20), we can estimate the right-hand side of (11.3.33). Then we get

(11.3.34) $$\int_{\partial\Omega_\chi} |\nu_n| \, |\partial_\nu u_0|^2_{\mathbb{C}^\ell} \, d\omega \leq c\varepsilon \Big(\int_{\partial\Omega_\chi} |\partial_\nu u_0|_{\mathbb{C}^\ell} \, d\omega \Big)^2$$

$$\leq c\varepsilon \int_{\partial\Omega_\chi} |\nu_n| \, |\partial_\nu u_0|^2 d\omega \cdot \int_{\partial\Omega_\chi} |\nu_n|^{-1} d\omega.$$

Here

$$\int_{\partial\Omega_\chi} |\nu_n|^{-1} \, d\omega = \frac{n-1}{1 - 2^{1-n}} \int_{\partial\mathcal{K}_\chi^{(1)}} |\nu_n|^{-1} \, d\sigma = \frac{n-1}{1 - 2^{1-n}} \int_{Q_\chi^{(1)}} |\nu_n|^{-2} \, dx',$$

where $\partial\mathcal{K}_\chi^{(1)} = \{x \in \partial\mathcal{K}_\chi : 1/2 < |x| < 1\}$ and

$$Q_\chi^{(1)} = \Big\{ x' \in \mathbb{R}^{n-1} : \frac{1}{2}(1 + \chi(\omega')^2)^{-1/2} < |x'| < (1 + \chi(\omega')^2)^{-1/2} \Big\}.$$

Since $\nu_n(x') = (1 + |\nabla_{x'}\chi(x')|^2)^{-1/2}$ is independent of $|x'|$, we get

$$\int_{\partial\Omega_\chi} |\nu_n|^{-1} d\omega = \frac{n-1}{1 - 2^{1-n}} \int_{S^{n-2}} \left(\int_{\frac{1}{2}(1+\chi(\omega')^2)^{-1/2}}^{(1+\chi(\omega')^2)^{-1/2}} |x'|^{n-2} d|x'| \right) \left(1 + |\nabla\chi(\omega')|^2 \right) d\omega'$$

$$\leq \int_{S^{n-2}} \left(1 + |\nabla\chi(\omega')|^2 \right) d\omega' .$$

Therefore, (11.3.34) is impossible if ε satisfies the inequality

$$\varepsilon < \varepsilon_1 = \left(c \int_{S^{n-2}} \left(1 + |\nabla_{x'}\chi(x')|^2 \right) dx' \right)^{-1} .$$

Thus, there are no eigenvalues of the pencil $\mathcal{L}_\chi$ on the line $\operatorname{Re}\lambda = (3 - n + \varepsilon)/2$ for $0 \leq \varepsilon < \varepsilon_1$. Using parts 4) and 5) of Theorem 11.3.1, we get the assertion of the lemma. ∎

Theorems 11.2.1 and 11.3.2 enable us to prove the following result.

THEOREM 11.3.3. *Let ϕ be a continuous and positively homogeneous function of degree 1 on $\mathbb{R}^{n-1}$ such that the restriction of ϕ to the sphere S^{n-2} belongs to the Sobolev space $W_2^1(S^{n-2})$. Then the strip*

$$(11.3.35) \qquad\qquad \left| \operatorname{Re}\lambda - 1 + \frac{n}{2} \right| \leq \frac{1}{2}$$

does not contain eigenvalues of the pencil $\mathcal{L}_\phi$.

Proof: Let τ be the same positive real number as in Theorem 11.3.2. Furthermore, let $\{\chi_j\} \subset C^\infty(\mathbb{R}^{n-1})$ be a sequence of positively homogeneous functions of degree 1 such that

(i) $\phi(x') \leq \chi_j(x') \leq \phi(x') + \frac{\tau}{2}|x'|$ and $\chi_{j+1}(x') \leq \chi_j(x')$ for all $x' \in \mathbb{R}^{n-1}$.

(ii) $\chi_j\big|_{S^{n-2}} \to \phi\big|_{S^{n-2}}$ in $C(S^{n-2})$ as $j \to \infty$.

(iii) $\big\|\chi_j\big|_{S^{n-2}}\big\|_{W_2^1(S^{n-2})} \leq c$, where c does not depend on j.

Such a sequence $\{\chi_j\}$ can be always found, since ϕ is continuous on $\mathbb{R}^{n-1}$ and the restriction of ϕ to the unit sphere belongs to $W_2^1(S^{n-2})$.

Due to Theorem 11.3.2, there exists a positive number ε_1 which does not depend on j such that the strip

$$|\operatorname{Re}\lambda - 1 + n/2| < (1 + \varepsilon_1)/2$$

is free of eigenvalues of the pencil $\mathcal{L}_{\chi_j}$ for every index j. Let

$$b(u, v; \lambda) = a(u, v; \lambda + 1 - n/2), \quad \mathcal{X} = \overset{\circ}{W}{}_2^1(\Omega_\phi))^\ell, \quad \text{and} \quad \mathcal{X}_j = \overset{\circ}{W}{}_2^1(\Omega_{\chi_j})^\ell .$$

Furthermore, let the operator $A(\lambda)$ be defined by (11.2.2). Then the conditions of Theorem 11.2.1 are satisfied. Consequently, analogously to the proof of Theorem 11.2.2, we obtain that the strip (11.3.35) does not contain eigenvalues of the operator pencil $\mathcal{L}_\phi$. ∎

REMARK 11.3.1. The assumption of Theorem 11.3.3 is satisfied, e.g., if $\phi \in C^{0,1}(\mathbb{R}^{n-1})$. We give another example.

Let Q be an arbitrary point on S^{n-2} and let ϕ be a positively homogeneous function of degree 1 on $\mathbb{R}^{n-1}$ such that $\phi\big|_{S^{n-2}}$ is smooth on $S^{n-2}\backslash Q$ and has the asymptotics

$$\phi(y) = c|y|^\kappa \, , \quad \kappa > 2 - n/2$$

in a neighborhood of the point Q where $y = (y_1, \dots, y_{n-2})$ are local coordinates on the sphere S^{n-2} in a neighborhood of Q with the point Q at the origin. Then the function ϕ satisfies the condition of Theorem 11.3.3.

11.4. Second order systems in a polyhedral cone

This section deals with operator pencils generated by the Dirichlet problem for elliptic systems of second order partial differential equations in a three-dimensional polyhedral cone. The main result of this section is the nonexistence of eigenvalues in the strip $|\operatorname{Re}\lambda + 1/2| \leq 1$ if all, with the possible exception of one, interior angles at the edges are less than π.

11.4.1. Nonexistence of eigenvalues on the line $\operatorname{Re}\lambda = 1/2$. Let $\mathcal{K}$ be a cone in $\mathbb{R}^3$ with vertex at the origin such that

$$\partial\mathcal{K}\backslash\{0\} = \bigcup_{j=1}^{N} \Gamma_j \cup \bigcup_{j=1}^{N} \mathcal{R}_j,$$

where $N > 2$, Γ_j are plane angles, and $\mathcal{R}_j$ are rays. Here it is assumed that the boundary of Γ_j coincides with $\mathcal{R}_j \cup \mathcal{R}_{j+1} \cup \{0\}$ for $j = 1, \dots, N-1$ and with $\mathcal{R}_N \cup \mathcal{R}_0 \cup \{0\}$ if $j = N$. By α_j we denote the interior angle between two plane parts of the boundary $\partial\mathcal{K}$ intersecting along the edge $\mathcal{R}_j$. Again let Ω be the intersection of $\mathcal{K}$ with the unit sphere S^2.

We consider the differential operator

$$L(D_x) = \sum_{i,j=1}^{3} A_{i,j}\, D_{x_i}\, D_{x_j}$$

where $A_{i,j}$ are $\ell \times \ell$ matrices such that $A_{i,j}^* = A_{j,i}$, and assume that there exists a positive constant c_0 such that

$$(11.4.1) \qquad \sum_{i,j=1}^{3} \left(A_{i,j} f^{(j)}, f^{(i)}\right)_{\mathbb{C}^\ell} \geq c_0 \sum_{j=1}^{3} |f^{(j)}|^2_{\mathbb{C}^\ell}$$

for all $f^{(1)}, f^{(2)}, f^{(3)} \in \mathbb{C}^\ell$. It is clear that the operator $L(D_x)$ is strongly elliptic.

We are interested in solutions of the Dirichlet problem

$$(11.4.2) \qquad\qquad\qquad L(D_x)\, U = 0 \quad \text{in } \mathcal{K},$$

$$(11.4.3) \qquad\qquad\qquad U = 0 \quad \text{on } \partial\mathcal{K}\backslash\{0\}.$$

which have the form (10.1.4), where λ_0 is a complex number and $u_k \in \overset{\circ}{W}{}^1_2(\Omega)^\ell$. As it was mentioned in Section 10.1, the vector function (10.1.4) is a solution of problem (11.4.2), (11.4.3) if and only if λ_0 is an eigenvalue of the operator pencil $\mathcal{L}$ defined by (10.1.5) and the vector functions $u_0, \dots, u_s$ form a Jordan chain corresponding to this eigenvalue.

LEMMA 11.4.1. *Let $\alpha_j \in (0, \pi)$ for $j = 2, \dots, N$ and $\alpha_1 \in (0, 2\pi]$. Then the line $\operatorname{Re}\lambda = 1/2$ does not contain eigenvalues of the operator pencil $\mathcal{L}$.*

Proof. Let λ be an eigenvalue of the pencil $\mathcal{L}$ on the line $\operatorname{Re}\lambda = 1/2$ and let $u \in \overset{\circ}{W}{}_2^1(\Omega)^\ell$ be a corresponding eigenfunction. Since the boundary of Ω is smooth outside the points $\omega_k = \mathcal{R}_k \cap \partial\Omega$, $k = 1, \dots N$, the vector function u is smooth outside these points.

It can be easily verified that the principal part of the differential operator $\mathcal{L}(\lambda)$ at ω_k satisfies the conditions in Section 8.5. Using Theorems 8.6.1, 8.6.2 together with well-known results on the asymptotics of solutions of general elliptic boundary value problems near angular points (see Section 1.4), we conclude that

$$(11.4.4) \qquad\qquad u(\omega) = O\big(|\omega - \omega_k|^{1+\delta}\big)$$

in a neighborhood of the point ω_k, $k = 2, \dots, N$, for some positive δ and

$$(11.4.5) \qquad\qquad u(\omega) = O\big(|\omega - \omega_1|^{1/2+\delta}\big)$$

in a neighborhood of ω_1. Moreover, both relations (11.4.4) and (11.4.5) may be differentiated.

We introduce the truncated cone

$$\mathcal{K}_1 = \big\{x \in \mathbb{R}^3 \,:\, 1 < r < 2,\ \omega \in \Omega\big\}$$

and the set

$$(\partial\mathcal{K})_1 = \big\{x \in \partial\mathcal{K} \,:\, 1 < r < 2\big\}.$$

Let $\nu = (\nu_1, \nu_2, \nu_3)$ denote the unit exterior normal to $\partial\mathcal{K}$, and let τ be the vector directed along the ray $\mathcal{R}_1$. Furthermore, let $U(x) = r^\lambda u(\omega)$. Since $L(D_x)U = 0$ in $\mathcal{K}$, we have

$$(11.4.6) \quad 0 = \int_{\mathcal{K}_1} \big(L(D_x)U,\, D_\tau^2 U\big)_{\mathbb{C}^\ell}\, dx = \sum_{i,j=1}^{3} \int_{\mathcal{K}_1} \big(A_{i,j} D_{x_j} D_\tau U,\, D_{x_i} D_\tau U\big)_{\mathbb{C}^\ell}\, dx$$

$$- i \sum_{i,j=1}^{3} \int_{(\partial\mathcal{K})_1} \Big(\nu_i\big(A_{i,j} D_{x_j} U,\, D_\tau^2 U\big)_{\mathbb{C}^\ell} - (\tau,\nu)\big(A_{i,j} D_{x_j} U,\, D_{x_i} D_\tau U\big)_{\mathbb{C}^\ell}\Big)\, d\sigma.$$

Here, we made use of the fact that the integrals over the boundary lying on the spheres $r = 1$ and $r = 2$ mutually cancel. According to (11.4.4), (11.4.5), all the integrals in (11.4.6) make sense.

We show that the last sum in the right-hand side of (11.4.6) vanishes. Since the first integral in (11.4.6) is real, it suffices to prove that

$$(11.4.7) \quad \operatorname{Im} \sum_{i,j=1}^{3} \int_{\Gamma_{n,1}} \Big(\nu_i\big(A_{i,j} D_{x_j} U,\, D_\tau^2 U\big)_{\mathbb{C}^\ell} - (\tau,\nu)\big(A_{i,j} D_{x_j} U,\, D_{x_i} D_\tau U\big)_{\mathbb{C}^\ell}\Big)\, d\sigma$$

$$= 0$$

for $n = 1, \dots, N$, where $\Gamma_{n,1} = \big\{x \in \Gamma_n \,:\, 1 < r < 2\big\}$. Without loss of generality, we can assume that $\Gamma_{n,1}$ lies in the plane $x_3 = 0$ (one can always gain this by virtue of a linear change of coordinates) and that $\nu = (0, 0, 1)$ on $\Gamma_{n,1}$. Then the left-hand side of (11.4.7) is equal to

$$\operatorname{Im} \int_{\Gamma_{n,1}} \left(\left(A_{3,3} D_{x_3} U, D_\tau^2 U \right)_{\mathbb{C}^\ell} - \tau_3 \sum_{i=1}^{3} \left(A_{i,3} D_{x_3} U, D_{x_i} D_\tau U \right)_{\mathbb{C}^\ell} \right) d\sigma$$

$$= \operatorname{Im} \int_{\Gamma_{n,1}} \left(\sum_{j=1}^{2} \tau_j \, D_{x_j} \left(A_{3,3} D_{x_3} U, D_{x_3} U \right)_{\mathbb{C}^\ell} - \sum_{i=1}^{2} \tau_3 D_{x_i} \left(A_{i,3} D_{x_3} U, D_{x_3} U \right)_{\mathbb{C}^\ell} \right) d\sigma$$

$$= 0.$$

Thus, we get

$$\sum_{i,j=1}^{3} \int_{\mathcal{K}_1} \left(A_{i,j} D_{x_j} \partial_\tau U, \partial_{x_i} \partial_\tau U \right)_{\mathbb{C}^\ell} dx = 0.$$

Using estimate (11.4.1), we obtain

$$D_\tau U = c = const. \quad \text{in } \mathcal{K}.$$

Since $D_\tau U$ is a positive homogeneous function of degree $-1/2$, we get $D_\tau U = 0$. If $\mathcal{K}$ is not a dihedral angle, this together with $U = 0$ on $\partial \mathcal{K}$ implies $U = 0$ in $\mathcal{K}$. The lemma is proved. ∎

11.4.2. The width of the energy strip.

THEOREM 11.4.1. *Let $\alpha_j \in (0, \pi)$ for $j = 2, \ldots, N$ and $\alpha_1 \in (0, 2\pi]$. Then the strip*

(11.4.8) $$|\operatorname{Re} \lambda + 1/2| \le 1$$

is free of eigenvalues of the operator pencil $\mathcal{L}$.

Proof: Since the number λ is an eigenvalue of $\mathcal{L}$ simultaneously with $-1 - \bar{\lambda}$, we conclude from Lemma 11.4.1 that the line $\operatorname{Re} \lambda = -3/2$ does not contain eigenvalues of the pencil $\mathcal{L}$.

Let I_ℓ be the $\ell \times \ell$ identity matrix. We set

$$L_t(D_x) = -(1 - t) I_\ell \, \Delta + t \, L(D_x) \quad \text{for } t \in [0, 1].$$

By $\mathcal{L}_t$ we denote the operator pencil generated by the Dirichlet problem for the operator L_t. Note that the coefficients of the operators L_t satisfy condition (11.4.1). Hence, by Lemma 11.4.1, the operator pencils $\mathcal{L}_t$ have no eigenvalues on the lines $\operatorname{Re} \lambda = -3/2$ and $\operatorname{Re} \lambda = 1/2$. Using the strong ellipticity of $L_t(D_x)$, we can prove, analogously to Lemma 10.1.3, that the operator pencils $\mathcal{L}_t$ do not have eigenvalues in the strip $-3/2 \le \operatorname{Re} \lambda \le 1/2$ for large $|\lambda|$. By Theorem 1.1.4, we conclude from this that the operator pencils $\mathcal{L}_t$, $0 \le t \le 1$, have the same number of eigenvalues (counting multiplicities) in the strip (11.4.8). Thus, it suffices to show that the operator pencil $\mathcal{L}_0 = -I_\ell(\delta + \lambda(\lambda + 1))$ has no eigenvalues in the strip (11.4.8).

Let $\mathcal{K}_0$ be the dihedral angle with opening α_1 and with the edge directed along the ray $\mathcal{R}_1$. Furthermore, let $\Omega_0 = \mathcal{K}_0 \cap S^2$. Then, by Theorem 10.5.3, there are no eigenvalues of the operator pencil

$$-\delta - \lambda(\lambda + 1) : \overset{\circ}{W}{}_2^1(\Omega_0) \to W_2^{-1}(\Omega_0)$$

in the strip $|\operatorname{Re} \lambda + 1/2| < 1/2 + \pi/\alpha_1$. Since $\Omega \subset \Omega_0$, we conclude from Theorem 2.2.4 that the strip (11.4.8) is free of eigenvalues of the pencil $\mathcal{L}_0$. The proof is complete. ∎

11.5. Exterior of a thin cone

In this section we consider the Dirichlet problem in a cone $\mathcal{K}$ which is the exterior of a thin cone. We assume here that $2m \geq n$ and prove that the eigenvalues of the corresponding operator pencil are close to those of the Sobolev problem. This shows, in particular, that the estimate for the energy strip given in Theorem 11.1.1 is sharp in the case $2m \geq n$.

11.5.1. Asymptotics of solutions of the Sobolev problem on S^{n-1}.
Let $2m \geq n$ and let

$$(11.5.1) \qquad \mathcal{L}(\lambda) : \overset{\circ}{W}{}_2^m(S^{n-1}\backslash N) \to W_2^{-m}(S^{n-1}\backslash N)$$

be the operator of the problem

$$(11.5.2) \qquad \left\{ \begin{array}{ll} \mathcal{L}(\lambda)\, u(\omega) = f(\omega) & \text{on } S^{n-1}\backslash N, \\ D_{x'}^\alpha u|_{x'=0} = 0 & \text{for } |\alpha| \leq m - n/2 \end{array} \right.$$

which is generated by the Sobolev problem (10.4.2), (10.4.3) in the exterior of the ray $\mathfrak{R} = \{x = (0, x_n) \in \mathbb{R}^n : x_n \geq 0\}$. Here N denotes the north pole on the sphere S^{n-1}. Suppose that λ is a regular point of the pencil (11.5.1). Note that every regular point, except $\lambda = m - n/2$ (if n is even), is noninteger (see Theorem 10.4.1).

We consider the equation

$$(11.5.3) \qquad \mathcal{L}(\lambda)\, u(\omega) = f(\omega) \qquad \text{on } S^{n-1}.$$

By the definition of the differential operator $\mathcal{L}(\lambda)$, this equation is equivalent to

$$L(D_x)\left(r^\lambda u(\omega)\right) = r^{\lambda-2m} f(\omega) \qquad \text{on } \mathbb{R}^n\backslash\{0\}.$$

Properties of the pencil

$$(11.5.4) \qquad \mathcal{L}(\lambda) : W_2^s(S^{n-1})^\ell \to W_2^{s-2m}(S^{n-1})^\ell, \qquad s \in \mathbb{R},$$

were studied in Section 10.2. In particular, the following assertions are valid:

1. The spectrum of the pencil (11.5.4) is exhausted by integers. For noninteger λ the operator (11.5.4) is an isomorphism.
2. If $\lambda = m - n/2$ and n is even, then the condition

$$\int_{S^{n-1}} (f, c\omega^\gamma)_{\mathbb{C}^\ell}\, d\omega = 0 \qquad \text{for all } c \in \mathbb{C}^\ell, \ |\gamma| = m - n/2$$

is necessary and sufficient for the solvability of equation (11.5.3).

For the first assertion we refer to Theorem 10.2.2, while the second assertion follows from the fact that the kernel of the adjoint operator $\mathcal{L}^+(m - n/2)$ is spanned by the functions $c\omega^\gamma$, where $c \in \mathbb{C}^\ell$, $|\gamma| = m - n/2$ (see Theorem 10.2.3).

Let λ be noninteger. Then by $G_\beta(\lambda, \cdot) \in W_2^k(S^{n-1})^\ell$, $k < 2m - |\beta| - (n-1)/2$, we denote the solution of the equation (11.5.3) with $f = I_\ell\, D_{x'}^\beta \delta(x')$, where I_ℓ is the $\ell \times \ell$ identity matrix. Here and below we use the local coordinates x' of the point $x = \left(x', (1 - |x'|^2)^{1/2}\right) \in S^{n-1}$ in a neighborhood of the north pole N.

In the case $\lambda = m - n/2$, n even, $G_\beta(\lambda, \cdot)$ denotes the solution of (11.5.3) with $f = I_\ell\, D_{x'}^\beta \delta(x') + f_\beta$, where f_β are $\ell \times \ell$-matrices with elements from $C^\infty(S^{n-1})$

such that

$$\int_{S^{n-1}} \left(f_\beta, c\omega^\gamma\right) d\omega = -\int_{S^{n-1}} \left(I_\ell D_{x'}^\beta \delta(x'), c\omega^\gamma\right) d\omega$$

for all $c \in \mathbb{C}^{\ell \times \ell}$, $|\gamma| = m - n/2$. Clearly, there exist matrices f_β satisfying this condition.

REMARK 11.5.1. Let $\zeta = \zeta(x')$ be a smooth function with support in a neighborhood of the point $x' = 0$ and $\zeta(x') = 1$ for small $|x'|$. Then the following asymptotic formulas are valid for $G_\beta(\lambda, \cdot)$, $|\beta| \leq 2m - n$:

$$(11.5.5) \quad G_\beta(\lambda, \omega) = \zeta(x') \left(|x'|^{2m-n+1-|\beta|} \Phi_\beta(\lambda, \frac{x'}{|x'|}) + p_{2m-n+1-|\beta|}(\lambda, x')\right)$$
$$+ O(|x'|^{2m-n-|\beta|+3/2})$$

for even n and

$$(11.5.6) \quad G_\beta(\lambda, \omega) = \zeta(x') \left(|x'|^{2m-n+1-|\beta|} \left(\Psi_\beta^{(1)}(\lambda, \frac{x'}{|x'|}) \log|x'| + \Psi_\beta^{(2)}(\lambda, \frac{x'}{|x'|})\right)\right)$$
$$+ \zeta(x') q_{2m-n+1-|\beta|}(\lambda, x') + O(|x'|^{2m-n-|\beta|+3/2})$$

for odd n. Here p_k and q_k are matrix polynomials of degree $\leq k$ and Φ_β, $\Psi_\beta^{(1)}$, $\Psi_\beta^{(2)}$ are smooth functions on S^{n-2}.

LEMMA 11.5.1. 1) Let λ be a regular point of the pencil (11.5.1). Furthermore, let $f \in W_2^{-s}(S^{n-1}\backslash N)^\ell$, $0 \leq s \leq m$. Then problem (11.5.2) has the unique solution

$$u = \sum_{s-(n-1)/2 \leq |\beta| \leq m-n/2} G_\beta(\lambda, \cdot) c_\beta + v,$$

where $v \in W_2^{2m-s}(S^{n-1})^\ell$ and $c_\beta \in \mathbb{C}^\ell$. The function v and the coefficients c_β satisfy the estimate

$$(11.5.7) \quad \sum_{s-(n-1)/2 \leq |\beta| \leq m-n/2} |c_\beta| + \|v\|_{W_2^{2m-s}(S^{n-1})^\ell} \leq c \|f\|_{W_2^{-s}(S^{n-1}\backslash N)^\ell},$$

where c is a constant independent of f.

2) Let λ be a regular point of the pencil (11.5.1) and $f \in W_2^s(S^{n-1})^\ell$, $s \geq 0$. Then problem (11.5.2) has the unique solution

$$u = \sum_{|\beta| \leq m-n/2} G_\beta(\lambda, \cdot) c_\beta + v,$$

where $v \in W_2^{2m+s}(S^{n-1})^\ell$ and $c_\beta \in \mathbb{C}^\ell$. The function v and the coefficients c_β satisfy the estimate

$$\sum_{|\beta| \leq m-n/2} |c_\beta| + \|v\|_{W_2^{2m+s}(S^{n-1})^\ell} \leq c \|f\|_{W_2^s(S^{n-1}\backslash N)^\ell},$$

where c is a constant independent of f.

Proof: We limit considerations to assertion 1), since the proof of the second assertion requires only obvious changes.

We introduce an auxiliary operator which performs a continuous extension of a linear functional f on $\overset{\circ}{W}{}_2^s(S^{n-1}\backslash N)^\ell$ to a linear functional F on $W_2^s(S^{n-1})^\ell$, $s \geq 0$. For arbitrary $f \in W_2^{-s}(S^{n-1}\backslash N)^\ell$ and $\varphi \in W_2^s(S^{n-1})^\ell$, $s \geq 0$, we set

$$(11.5.8) \qquad (F,\varphi)_{L_2(S^{n-1})^\ell} = \left(f\,,\, \varphi - \zeta \sum_{|\alpha|<s-(n-1)/2} \frac{x'^\alpha}{\alpha!}\,(\partial_{x'}^\alpha \varphi)(0) \right)_{L_2(S^{n-1})^\ell}.$$

Here $\zeta = \zeta(x')$ is an arbitrary smooth function with support in a neighborhood of the point $x' = 0$ and $\zeta(x') = 1$ for small $|x'|$. Since

$$\varphi - \zeta \sum_{|\alpha|<s-(n-1)/2} \frac{x'^\alpha}{\alpha!}\,(\partial_{x'}^\alpha \varphi)(0) \in \overset{\circ}{W}{}_2^s(S^{n-1}\backslash N)^\ell,$$

the right-hand side of (11.5.8) is well-defined. Obviously, the functionals f and F coincide on $\overset{\circ}{W}{}_2^s(S^{n-1}\backslash N)^\ell$. Furthermore,

$$\left| (F,\varphi)_{L_2(S^{n-1})^\ell} \right|$$

$$\leq \|f\|_{W_2^{-s}(S^{n-1}\backslash N)^\ell} \left\| \varphi - \zeta \sum_{|\alpha|<s-(n-1)/2} \frac{x'^\alpha}{\alpha!}\,(\partial_{x'}^\alpha \varphi)(0) \right\|_{\overset{\circ}{W}{}_2^s(S^{n-1}\backslash N)^\ell}$$

$$\leq c\,\|f\|_{W_2^{-s}(S^{n-1}\backslash N)^\ell}\,\|\varphi\|_{W_2^s(S^{n-1})^\ell}.$$

Hence the mapping $f \to F$ is continuous from $W_2^{-s}(S^{n-1}\backslash N)^\ell$ into $W_2^{-s}(S^{n-1})^\ell$.

Using this operator, we write (11.5.2) in the form

$$(11.5.9) \qquad \begin{cases} \mathcal{L}(\lambda)\,u(\omega) = F + \displaystyle\sum_{|\beta|\leq m-n/2} c_\beta\, D_{x'}^\beta\, \delta(x') & \text{on } S^{n-1}, \\[2mm] D_{x'}^\alpha u = 0 & \text{for } x' = 0,\ |\alpha| \leq m - n/2. \end{cases}$$

Suppose λ is not an integer. If n is odd, then this assumption is equivalent to the regularity of the point λ, and in the case of even n this assumption excludes only the regular point $\lambda = m - n/2$ (cf. Theorem 10.4.1).

If $|\beta| \leq m - n/2$, then $G_\beta \in W_2^m(S^{n-1})^{\ell\times\ell}$. We introduce the block matrix

$$\left(S_{\alpha,\beta}(\lambda) \right)_{|\alpha|,|\beta|\leq m-n/2} = \left(D_{x'}^\alpha G_\beta(\lambda,x')|_{x'=0} \right)_{|\alpha|,|\beta|\leq m-n/2}$$

and show that it is nonsingular. In fact, if $k_\beta \in \mathbb{C}^\ell$ are vectors such that

$$\sum_{|\beta|\leq m-n/2} S_{\alpha,\beta}(\lambda)\, k_\beta = 0 \qquad \text{for } |\alpha| \leq m - n/2,$$

then the vector-valued function $u = \sum_\beta G_\beta(\lambda,\cdot)\, k_\beta$ is a solution of the equation $\mathcal{L}(\lambda)u = 0$ on $S^{n-1}\backslash N$ which belongs to the space $\overset{\circ}{W}{}_2^m(S^{n-1}\backslash N)^\ell$. Since λ is a regular point of the pencil (11.5.1), it follows that $u = 0$ and, consequently, $k_\beta = 0$ for $|\beta| \leq m - n/2$. This proves the regularity of the matrix $(S_{\alpha,\beta})$.

Let $w \in W_2^{2m-s}(S^{n-1})^\ell$ be a solution of the equation

$$\mathcal{L}(\lambda)\,w = F \qquad \text{on } S^{n-1}.$$

Then the constants c_β in (11.5.9) are given by the equalities

$$D_{x'}^\alpha w|_{x'=0} + S_{\alpha,\beta}\, c_\beta = 0 \qquad \text{for } |\alpha| \leq m - n/2.$$

If $(T_{\beta,\alpha})$ is the inverse matrix of $(S_{\alpha,\beta})$, then we have

$$c_\beta = - \sum_{|\alpha| \leq m-n/2} T_{\beta,\alpha}(\lambda)\, D_{x'}^\alpha w|_{x'=0}\,.$$

This implies the estimate

$$\sum_{|\beta| \leq m-n/2} |c_\beta| \leq c\, \|f\|_{W_2^{-s}(S^{n-1})^\ell}\,.$$

We put

$$\mathcal{F} = F + \sum_{|\beta| < s-(n-1)/2} c_\beta\, D_{x'}^\beta\, \delta(x') \in W_2^{-s}(S^{n-1})^\ell.$$

Then

$$u = v + \sum_{s-(n-1)/2 \leq |\beta| \leq m-n/2} G_\beta(\lambda, \cdot)\, c_\beta\,,$$

where $v \in W_2^{2m-s}(S^{n-1})^\ell$ is a solution of the equation $\mathcal{L}(\lambda)\, v = \mathcal{F}$ on S^{n-1}. This implies the estimate

$$\|v\|_{W_2^{2m-s}(S^{n-1})^\ell} \leq c\, \|\mathcal{F}\|_{W_2^{-s}(S^{n-1})^\ell} \leq c\, \|f\|_{W_2^{-s}(S^{n-1}\backslash N)^\ell}\,.$$

Now we consider the case of even n and $\lambda = m - n/2$. In this case the equation

$$\mathcal{L}(\lambda)\, u(\omega) = F + \sum_{|\beta| \leq m-n/2} c_\beta\, D_{x'}^\beta \delta(x') \qquad \text{on } S^{n-1}$$

is solvable if the orthogonality condition

$$\left(F,\, c\omega^\gamma\right)_{L_2(S^{n-1})^\ell} + \sum_{|\beta| \leq m-n/2} \left(c_\beta\, D_{x'}^\beta \delta(x'),\, c\omega^\gamma\right)_{L_2(S^{n-1})^\ell} = 0$$

is satisfied for $c \in \mathbb{C}^\ell$ and $|\gamma| = m - n/2$. This condition uniquely defines constants c_β for which (11.5.7) is obviously valid.

Using the functions G_β, we can write a special solution of (11.5.9) in the form

$$u = \sum_{|\beta| \leq m-n/2} G_\beta\, c_\beta + v_1\,,$$

where $v_1 \in W_2^{2m-s}(S^{n-1})^\ell$ and the norm of v_1 is estimated by the right-hand side of (11.5.7). Since the nontrivial solutions of the equation $\mathcal{L}(m - n/2)\, u = 0$ on S^{n-1} are exhausted by the functions $c\omega^\gamma$ with $c \in \mathbb{C}^\ell$, $|\gamma| = m - n/2$ (see Theorem 10.2.3), it follows that the general solution of (11.5.9) has the form

$$(11.5.10) \qquad u = \sum_{|\beta| \leq m-n/2} G_\beta\, c_\beta + v_1 + \sum_{|\gamma|=m-n/2} d_\gamma\, \omega^\gamma.$$

The vectors d_γ are uniquely defined from the relations $D_{x'}^\alpha u|_{x'=0} = 0$, $|\alpha| \leq m-n/2$. It is clear that (11.5.10) entails the required representation. The lemma is proved. ∎

11.5.2. Estimates for the eigenvalues of a perturbed problem. It follows from (10.1.9) and (10.1.12) that

$$(11.5.11) \quad \text{Re} \int_{S^{n-1}\backslash N} \left(\mathcal{L}(m - n/2 + it)u, u\right)_{\mathbb{C}^\ell} d\omega \geq c \sum_{j=0}^{m} |t|^{2j} \|u\|^2_{\overset{\circ}{W}_2^{m-j}(S^{n-1}\backslash N)^\ell}$$

for all $u \in C_0^\infty(S^{n-1}\backslash N)^\ell$, $t \in \mathbb{R}$. This, in particular, implies that the operator

$$\mathcal{L}(m - n/2) : \overset{\circ}{W}_2^m(S^{n-1}\backslash N)^\ell \to W_2^{-m}(S^{n-1}\backslash N)^\ell$$

defines an isomorphism. We denote its inverse by $\mathcal{L}^{-1}$.

Let ε be a small positive number and let Ω_ε be a domain on S^{n-1} such that $N \in S^{n-1}\backslash\Omega_\varepsilon$ and the diameter of $S^{n-1}\backslash\Omega_\varepsilon$ is equal to ε. By

$$\mathcal{L}_\varepsilon(\lambda) : \overset{\circ}{W}_2^m(\Omega_\varepsilon)^\ell \to W_2^{-m}(\Omega_\varepsilon)^\ell$$

we denote the operator of the Dirichlet problem

$$\mathcal{L}(\lambda)\, u = f \quad \text{in } \Omega_\varepsilon,$$
$$D^\alpha_{x'} u = 0 \quad \text{on } \partial\Omega_\varepsilon, \; |\alpha| \leq m - 1.$$

Furthermore, let $\mathcal{L}_\varepsilon^{-1}$ denote the inverse operator of $\mathcal{L}_\varepsilon(m - n/2)$. This operator exists by (11.5.11) and maps $W_2^{-m}(\Omega_\varepsilon)^\ell$ onto $\overset{\circ}{W}_2^m(\Omega_\varepsilon)^\ell$. Its norm is bounded uniformly in ε.

Let χ_ε be the operator of restriction to Ω_ε acting from $W_2^{-m}(S^{n-1}\backslash N)^\ell$ into $W_2^{-m}(\Omega_\varepsilon)^\ell$, and let π_ε be the operator of extension by zero acting from $\overset{\circ}{W}_2^m(\Omega_\varepsilon)^\ell$ into $\overset{\circ}{W}_2^m(S^{n-1}\backslash N)^\ell$.

LEMMA 11.5.2. *For all $f \in W_2^{-m+1}(S^{n-1}\backslash N)^\ell$ there is the estimate*

$$\|(\mathcal{L}^{-1} - \pi_\varepsilon\mathcal{L}_\varepsilon^{-1}\chi_\varepsilon)f\|_{\overset{\circ}{W}_2^m(S^{n-1}\backslash N)^\ell} \leq c\varepsilon^{1/2}\|f\|_{W_2^{-m+1}(S^{n-1}\backslash N)^\ell}$$

if n is even and

$$\|(\mathcal{L}^{-1} - \pi_\varepsilon\mathcal{L}_\varepsilon^{-1}\chi_\varepsilon)f\|_{\overset{\circ}{W}_2^m(S^{n-1}\backslash N)^\ell} \leq c|\log\varepsilon|^{-1/2}\|f\|_{W_2^{-m+1}(S^{n-1}\backslash N)^\ell}$$

if n is odd.

Proof: Let f be an arbitrary element of the space $W_2^{-m+1}(S^{n-1}\backslash N)^\ell$. We put

$$u = \mathcal{L}^{-1}f, \qquad u_\varepsilon = \pi_\varepsilon\mathcal{L}_\varepsilon^{-1}\chi_\varepsilon f, \qquad w = u - u_\varepsilon.$$

The function w is the solution of the boundary value problem

$$(11.5.12) \qquad \mathcal{L}(m - n/2)\, w = 0 \quad \text{on } \Omega_\varepsilon, \qquad w - u \in \overset{\circ}{W}_2^m(\Omega_\varepsilon)^\ell.$$

The condition $w - u \in \overset{\circ}{W}_2^m(\Omega_\varepsilon)^\ell$ can be replaced by the inclusion $w - \eta_\varepsilon u \in \overset{\circ}{W}_2^m(\Omega_\varepsilon)^\ell$, where η_ε is a function in $C_0^\infty(S^{n-1})$ equal to unity in a neighborhood of the set $S^{n-1}\backslash\Omega_\varepsilon$. This and (11.5.12) imply that

$$\int_{\Omega_\varepsilon} \left(\mathcal{L}(m - n/2)\,(w - \eta_\varepsilon u),\, w - \eta_\varepsilon u\right)_{\mathbb{C}^\ell} d\omega = \int_{\Omega_\varepsilon} \left(\mathcal{L}(m - n/2)\,\eta_\varepsilon u,\, \eta_\varepsilon u - w\right)_{\mathbb{C}^\ell} d\omega.$$

Applying the Gårding inequality (11.5.11) for $t = 0$, we obtain

$$(11.5.13) \qquad \|w - \eta_\varepsilon u\|_{\overset{\circ}{W}_2^m(\Omega_\varepsilon)^\ell} \leq c\|\eta_\varepsilon u\|_{W_2^m(\Omega_\varepsilon)^\ell}.$$

with a constant c independent of ε. Since $w = \eta_\varepsilon u$ on $S^{n-1}\backslash\Omega_\varepsilon$, by (11.5.13), we have

$$
(11.5.14) \qquad \|w\|_{\overset{\circ}{W}_2^m(S^{n-1}\backslash N)^\ell} \le c\,\|\eta_\varepsilon u\|_{\overset{\circ}{W}_2^m(S^{n-1}\backslash N)^\ell}\,.
$$

Let n be even. Then, by Lemma 11.5.1, there is the representation

$$
u = v + \sum_{|\beta| = m - n/2} G_\beta(\omega)\, c_\beta,
$$

where $v \in W_2^{m+1}(S^{n-1})^\ell$. Moreover,

$$
\|v\|_{W_2^{m+1}(S^{n-1})^\ell} + \sum_{|\beta| = m - n/2} |c_\beta| \le c\,\|f\|_{W_2^{-m+1}(S^{n-1}\backslash N)^\ell}\,.
$$

We introduce the space $V_{2,\sigma}^k(S^{n-1}\backslash N)$, $k = 0, 1, \dots$, $\sigma \in \mathbb{R}$, which consists of all functions $u \in W_{2,loc}^k(S^{n-1}\backslash N)$ for which the integral

$$
\int\limits_{|x'| \le 1/2} \sum_{|\alpha| \le k} |x'|^{2(\sigma - k + |\alpha|)}\,|D_{x'}^\alpha u|^2\, dx'
$$

is finite. By Theorem 2.1 in Maz'ya and Plamenevskiĭ's paper [179] (see also our book [136, Th.7.1.1]), there exists a vector-valued polynomial $p_{m-n/2+1} = p_{m-n/2+1}(x')$ of degree $\le m - n/2 + 1$ with coefficients majorized by the norm of the vector-valued function v in $W_2^{m+1}(S^{n-1})^\ell$ such that

$$
v - \zeta\, p_{m-n/2+1} \in V_{2,\sigma}^{m+1}(S^{n-1}\backslash N)^\ell
$$

for arbitrary $\sigma > 0$. Here $\zeta = \zeta(x')$ is a smooth function with support in a neighborhood of the point $x' = 0$ and $\zeta(x') = 1$ for small $|x'|$. By Remark 11.5.1, the functions G_β have the representation

$$
G_\beta(\omega) = \zeta\left(Q_{m-n/2+1}^{(\beta)}(x') + u_\beta(x')\right) + O(|x'|^{m-n/2+3/2}),
$$

where $Q_{m-n/2+1}^{(\beta)}$ is a polynomial of degree $m - n/2 + 1$ and u_β is positively homogeneous of degree $m - n/2 + 1$. Since $D_{x'}^\alpha u = 0$ at the point N for $|\alpha| \le m - n/2$, it follows that

$$
p_{m-n/2+1} + \sum_{|\beta| = m - n/2} Q_{m-n/2+1}^{(\beta)}\, c_\beta
$$

is a homogeneous polynomial of degree $m - n/2 + 1$. Consequently, we have

$$
u - \zeta\, u^{(1)} \in V_{2,\sigma}^m(S^{n-1}\backslash N)^\ell
$$

for $\sigma > -1$, where $u^{(1)}$ is a function, positively homogeneous of degree $m - n/2 + 1$ satisfying the conditions

$$
|D_{x'}^\alpha u^{(1)}(x')| \le c_\alpha\,\|f\|_{W_2^{-m+1}(S^{n-1}\backslash N)^\ell} \qquad \text{for } |x'| = 1.
$$

Furthermore, (11.5.7) implies the estimate

$$
(11.5.15) \qquad \|u - \zeta u^{(1)}\|_{V_{2,\sigma}^m(S^{n-1}\backslash N)^\ell} \le c\,\|f\|_{W_2^{-m+1}(S^{n-1}\backslash N)^\ell}\,.
$$

Suppose that $\eta_\varepsilon(x') = \eta(x'/\varepsilon)$, where $\eta \in C_0^\infty(\mathbb{R}^{n-1})$, $\eta(x') = 1$ for $|x'| < 1$, and $\eta(x') = 0$ for $|x'| > 2$. Now (11.5.15) entails

$$\|\eta_\varepsilon\,(u - \zeta u^{(1)})\|_{\overset{\circ}{W}_2^m(S^{n-1}\setminus N)^\ell} \le c\,\|\eta_\varepsilon\,(u - \zeta u^{(1)})\|_{V_{2,\sigma}^m(S^{n-1}\setminus N)^\ell}$$

$$\le c\,\varepsilon^{-\sigma}\,\|u - \zeta u^{(1)}\|_{V_{2,\sigma}^m(S^{n-1}\setminus N)^\ell} \le c\,\varepsilon^{-\sigma}\,\|f\|_{W_2^{-m+1}(S^{n-1}\setminus N)^\ell}$$

for $\sigma > -1$. It is readily verified that

$$\|\eta_\varepsilon\zeta u^{(1)}\|_{V_{2,\sigma}^m(S^{n-1}\setminus N)^\ell} \le c\,\varepsilon^{1/2}\,\|f\|_{W_2^{-m+1}(S^{n-1}\setminus N)^\ell}\,.$$

The last two inequalities together with (11.5.14) lead to

$$\|w\|_{\overset{\circ}{W}_2^m(S^{n-1}\setminus N)^\ell} \le c\,\varepsilon^{1/2}\,\|f\|_{W_2^{-m+1}(S^{n-1}\setminus N)^\ell}\,.$$

Thus, we have proved the lemma for even n.

Now let n be an odd number. Then, by Lemma 11.5.1, we have

$$u = v + \sum_{|\beta|=m-(n+1)/2} G_\beta(\omega)\,c_\beta\,,$$

where $v \in W_2^{m+1}(S^{n-1})^\ell$. Moreover,

$$\|v\|_{W_2^{m+1}(S^{n-1})^\ell} + \sum_{|\beta|=m-(n+1)/2} |c_\beta| \le c\,\|f\|_{W_2^{-m+1}(S^{n-1}\setminus N)^\ell}\,.$$

Again, by means of [**179**, Th.2.1], we get

$$v - \zeta\,p_{m-(n-3)/2} \in V_{2,\sigma}^m(S^{n-1}\setminus N)^\ell$$

for $\sigma > -1$, where $p_{m-(n-3)/2}$ is a polynomial of degree $\le m - (n-3)/2$. Hence, by (11.5.6), there exist a polynomial Q of degree $\le m - (n-3)/2$ and smooth functions $\Psi^{(1)}$, $\Psi^{(2)}$ on S^{n-2} such that the function

$$u_1 \overset{def}{=} u - \zeta\left(Q(x') + |x'|^{m-(n-3)/2}\big(\Psi^{(1)}(\omega')\log|x'| + \Psi^{(2)}(\omega')\big)\right)$$

belongs to $V_{2,\sigma}^m(S^{n-1}\setminus N)^\ell$ for $\sigma > -1$. Since $D_{x'}^\alpha u = 0$ at N for $|\alpha| \le m - n/2$, it follows that Q is the sum of homogeneous polynomials of degree $m - (n-1)/2$ and $m - (n-3)/2$. Furthermore, the coefficients of Q, the functions $\Psi^{(1)}$ and $\Psi^{(2)}$ and all their derivatives, and the norm of u_1 in the space $V_{2,\sigma}^m(S^{n-1}\setminus N)^\ell$ are majorized by

$$c\,\|f\|_{W_2^{-m+1}(S^{n-1}\setminus N)^\ell}\,.$$

We put $\eta_\varepsilon(x') = \eta(|\log|x'||/\log\varepsilon)$, where η is a smooth function, $\eta(t) = 1$ for $t \le 2$, and $\eta(t) = 0$ for $t \ge 3$. It is clear that

$$|D_{x'}^\alpha \eta_\varepsilon(x')| \le c_\alpha\,|x'|^{-|\alpha|}\,|\log\varepsilon|^{-1} \qquad \text{for } |\alpha| > 0$$

and that the supports of the derivatives $D_{x'}^\alpha \eta_\varepsilon$, $\alpha \ne 0$, are contained in the domain $\{x' : \varepsilon^3 < |x'| < \varepsilon^2\}$. Therefore,

(11.5.16) $$\|\eta_\varepsilon u\|_{\overset{\circ}{W}_2^m(S^{n-1}\setminus N)^\ell}$$

$$\le c\left(\|\eta_\varepsilon u_1\|_{V_{2,\sigma}^m(S^{n-1}\setminus N)^\ell} + \|\eta_\varepsilon(u - u_1)\|_{\overset{\circ}{W}_2^m(S^{n-1}\setminus N)^\ell}\right).$$

Here for $\sigma > -1$ we have

$$\|\eta_\varepsilon u_1\|_{V_{2,\sigma}^m(S^{n-1}\setminus N)^\ell} \le c\,\varepsilon^{-\sigma}\,\|\eta_\varepsilon u_1\|_{V_{2,\sigma}^m(S^{n-1}\setminus N)^\ell} \le c\,\varepsilon^{-\sigma}\,\|f\|_{W_2^{-m+1}(S^{n-1}\setminus N)^\ell}\,.$$

Furthermore, the above properties of η_ε imply that the second summand on the right in (11.5.16) does not exceed

$$c\,|\log\varepsilon|^{-1/2}\,\|f\|_{W_2^{-m+1}(S^{n-1}\backslash N)^\ell}.$$

Applying (11.5.14) we arrive at the estimate

$$\|w\|_{\overset{\circ}{W}{}_2^m(S^{n-1}\backslash N)^\ell} \le c\,|\log\varepsilon|^{-1/2}\,\|f\|_{W_2^{-m+1}(S^{n-1}\backslash N)^\ell}.$$

This proves the lemma for odd n. ∎

We consider two families of bounded operators in $\overset{\circ}{W}{}_2^m(S^{n-1}\backslash N)^\ell$:

$$\begin{aligned}
\mathfrak{B}(\lambda) &= I + \mathcal{L}^{-1}\big(\mathcal{L}(\lambda) - \mathcal{L}(m - n/2)\big),\\
\mathfrak{B}_\varepsilon(\lambda) &= I + \pi_\varepsilon\,\mathcal{L}_\varepsilon^{-1}\,\chi_\varepsilon\big(\mathcal{L}(\lambda) - \mathcal{L}(m - n/2)\big).
\end{aligned}$$

Obviously, the pencils $\mathcal{L}$ and $\mathfrak{B}$ have the same eigenvalues, eigenfunctions and generalized eigenfunctions. We show that the spectra of the operator pencils $\mathcal{L}_\varepsilon$ and $\mathfrak{B}_\varepsilon$ coincide.

LEMMA 11.5.3. *The functions $u_0, u_1, \dots, u_{s-1}$ form a Jordan chain of the pencil $\mathfrak{B}_\varepsilon$ corresponding to the eigenvalue λ_0 if and only if*

$$u_j = \pi_\varepsilon v_j \quad \text{for } j = 0, \dots, s - 1,$$

where $v_j \in \overset{\circ}{W}{}_2^m(\Omega_\varepsilon)^\ell$ and $v_0, \dots, v_{s-1}$ is a Jordan chain of the pencil $\mathcal{L}_\varepsilon$ corresponding to the eigenvalue λ_0.

Proof: Let $u_0, \dots, u_{s-1}$ be a Jordan chain of $\mathfrak{B}_\varepsilon(\lambda)$ corresponding to the eigenvalue λ_0. Then we have

(11.5.17) $$\sum_{k=0}^{j} \frac{1}{k!}\,\mathfrak{B}_\varepsilon^{(k)}(\lambda_0)\,u_{j-k} = 0$$

for $j = 0, \dots, s - 1$, i.e.,

(11.5.18) $$u_j = -\pi_\varepsilon\,\mathcal{L}_\varepsilon^{-1}\,\chi_\varepsilon\left((\mathcal{L}(\lambda) - \mathcal{L}(m - n/2))\,u_j + \sum_{k=1}^{j}\mathcal{L}^{(k)}(\lambda_0)\,u_{j-k}\right)$$

for $j = 0, 1, \dots, s - 1$. Hence the functions u_j are extensions by zero of functions $v_j \in \overset{\circ}{W}{}_2^m(\Omega_\varepsilon)^\ell$. This implies $\chi_\varepsilon\mathcal{L}(\lambda)u_j = \mathcal{L}_\varepsilon(\lambda)v_j$. Therefore, from (11.5.18) we conclude that

$$\mathcal{L}_\varepsilon^{-1}\sum_{k=0}^{j}\mathcal{L}_\varepsilon^{(k)}(\lambda_0)\,v_{j-k} = 0 \qquad \text{for } j = 0, 1, \dots, s - 1.$$

Consequently, $v_0, v_1, \dots, v_{s-1}$ is a Jordan chain of the pencil $\mathcal{L}_\varepsilon$ corresponding to the eigenvalue λ_0.

Conversely, if $v_0, v_1, \dots, v_{s-1}$ is a Jordan chain of $\mathcal{L}_\varepsilon$ corresponding to the eigenvalue λ_0, we obtain the equalities (11.5.18) for the functions $u_j = \pi_\varepsilon v_j$. These equalities are equivalent to (11.5.17). The proof is complete. ∎

Now it is easy to prove the following theorem.

THEOREM 11.5.1. *Let λ_0 be an eigenvalue of the operator pencil $\mathcal{L}$ generated by the Sobolev problem (11.5.2) and let κ be the algebraic multiplicity of λ_0. Then there exists a constant $c(\lambda_0) > 0$ independent of ε such that the operator pencil $\mathcal{L}_\varepsilon$ has κ eigenvalues (counted with their algebraic multiplicities) in the circle*

$$\{\lambda \in \mathbb{C} : |\lambda - \lambda_0| < c(\lambda_0)\,\varepsilon^{1/2}\} \qquad \text{for even } n,$$

$$\{\lambda \in \mathbb{C} : |\lambda - \lambda_0| < c(\lambda_0)\,|\log\varepsilon|^{-1/2}\} \quad \text{for odd } n.$$

Proof: Let λ_0 be an eigenvalue of the pencil $\mathcal{L}$ and, therefore, also of the pencil $\mathfrak{B}$. Since the operator pencil $\mathcal{L}$ has no generalized eigenfunctions, neither does $\mathfrak{B}$. Consequently, by Theorem 1.1.2, the inverse $\mathfrak{B}(\lambda)^{-1}$ satisfies the estimate

$$\|\mathfrak{B}(\lambda)^{-1}\|_{\overset{\circ}{W}_2^m(S^{n-1}\backslash N)^\ell \to \overset{\circ}{W}_2^m(S^{n-1}\backslash N)^\ell} \le c_0\,|\lambda - \lambda_0|^{-1}$$

for small $|\lambda - \lambda_0|$, $\lambda \neq \lambda_0$. Using Lemma 11.5.2, we find that

$$\|\mathfrak{B}(\lambda) - \mathfrak{B}_\varepsilon(\lambda)\|_{\overset{\circ}{W}_2^m(S^{n-1}\backslash N)^\ell \to \overset{\circ}{W}_2^m(S^{n-1}\backslash N)^\ell} \le \begin{cases} c_1\,\varepsilon^{1/2} & \text{for even } n, \\ c_2\,|\log\varepsilon|^{-1/2} & \text{for odd } n. \end{cases}$$

Since

$$\mathfrak{B}_\varepsilon(\lambda) = \mathfrak{B}(\lambda)\left(I + \mathfrak{B}(\lambda)^{-1}\left(\mathfrak{B}_\varepsilon(\lambda) - \mathfrak{B}(\lambda)\right)\right),$$

it follows that $\mathfrak{B}_\varepsilon(\lambda)$ is invertible for $|\lambda - \lambda_0| = \delta$ if

$$\delta \ge d \overset{def}{=} \begin{cases} 2c_0\,c_1\,\varepsilon^{1/2} & \text{for even } n, \\ 2c_0\,c_2\,|\log\varepsilon|^{-1/2} & \text{for odd } n. \end{cases}$$

Hence from Theorem 1.1.4 we conclude that the sum of the algebraic multiplicities of all eigenvalues of $\mathfrak{B}_\varepsilon$ in the circle $|\lambda - \lambda_0| \le d$ coincides with the algebraic multiplicity of the eigenvalue λ_0 of the pencil $\mathcal{L}$. Applying Lemma 11.5.3, we complete the proof. ∎

Using the fact that $\lambda = m - (n-1)/2$ is an eigenvalue of $\mathcal{L}$ (see Theorem 10.4.1), the following assertion holds as a consequence of Theorem 11.5.1.

COROLLARY 11.5.1. *Let $2m \ge n$. Then for every $\delta > 0$ there exists a number $\varepsilon_0 = \varepsilon_0(\delta) > 0$ such that the strip $|\operatorname{Re}\lambda - m + n/2| < 1/2 + \delta$ contains at least one eigenvalue of the pencil $\mathcal{L}_\varepsilon$ for $\varepsilon < \varepsilon_0$.*

11.6. A cone close to the half-space

Let $\mathcal{K}_\varepsilon$ denote the cone $\{x = (x', x_n) \in \mathbb{R}^n : x_n > \varepsilon\,\phi(x')\}$, where $\varepsilon \in \mathbb{R}$ and ϕ is a given smooth real-valued function in $\mathbb{R}^{n-1}\backslash\{0\}$ which is positively homogeneous of degree one. By Ω_ε we denote the intersection of the cone $\mathcal{K}_\varepsilon$ with the unit sphere S^{n-1}. Furthermore, let

$$L(\partial_x) = \sum_{|\alpha|=2m} a_\alpha\,\partial_x^\alpha$$

be a scalar elliptic differential operator with constant real coefficients. We define the differential operator $\mathcal{L}(\lambda)$ on the unit sphere S^{n-1} by

$$\mathcal{L}(\lambda)\,u(\omega) = r^{2m-\lambda}L(\partial_x)\left(r^\lambda u(\omega)\right),$$

where u is a function on the sphere, and consider the operator pencil

$$\overset{\circ}{W}_2^m(\Omega_\varepsilon) \ni u \to \mathcal{L}(\lambda)\,u \in W_2^{-m}(\Omega_\varepsilon)$$

which is denoted by $\mathcal{L}_\varepsilon(\lambda)$. If $\varepsilon = 0$, then the cone $\mathcal{K}_\varepsilon$ coincides with the half-space $\mathbb{R}^n_+$. The spectrum of the pencil $\mathcal{L}_0$ is explicitly described in Section 10.3. In particular, $\lambda_0 = m$ is an eigenvalue with the eigenfunction $c\,\omega_n^m$. Here we derive an asymptotic formula for the eigenvalue close to m of the pencil $\mathcal{L}_\varepsilon$ when $\varepsilon \to 0$.

11.6.1. A normalization property of eigenfunctions of $\mathcal{L}_0$. There exist homogeneous differential operators T_j of degree j with constant coefficients such that the *Green formula*

$$(11.6.1) \qquad \int_{\mathbb{R}^n_+} L(\partial_x)\,U \cdot \overline{V}\,dx + \sum_{j=0}^{m-1} \int_{\mathbb{R}^{n-1}} \partial_{x_n}^j U \cdot \overline{T_{2m-1-j}(\partial_x)\,V}\,dx'$$

$$= \int_{\mathbb{R}^n_+} U \cdot \overline{L(\partial_x)\,V}\,dx + \sum_{j=0}^{m-1} \int_{\mathbb{R}^{n-1}} T_{2m-1-j}(\partial_x)\,U \cdot \overline{\partial_{x_n}^j V}\,dx'$$

is satisfied for all $U, V \in C_0^\infty(\overline{\mathbb{R}^n_+})$.

Let a denote the coefficient of $\partial_{x_n}^{2m}$ in the differential operator L. Then the coefficient of $\partial_{x_n}^{2m-1-j}$ in the operator T_{2m-1-j} is equal to $(-1)^{j+1}\,a$. We introduce the following parameter-depending differential operators on the sphere S^{n-1}:

$$\mathcal{S}_j(\lambda)\,u(\omega) = r^{j-\lambda}\,\partial_{x_n}^j\left(r^\lambda\,u(\omega)\right),$$
$$\mathcal{T}_{2m-1-j}(\lambda)\,u(\omega) = r^{2m-1-j-\lambda}\,T_{2m-1-j}(\partial_x)\left(r^\lambda\,u(\omega)\right),$$

$j = 0, \ldots, m-1$. From (11.6.1) we obtain the Green formula (see, e.g., our book [**136**, Le.6.1.6])

$$\int_{S^{n-1}_+} \mathcal{L}(\lambda)\,u \cdot \overline{v}\,d\omega + \sum_{j=0}^{m-1} \int_{S^{n-2}} \mathcal{S}_j(\lambda)\,u \cdot \overline{\mathcal{T}_{2m-1-j}(2m-n-\overline{\lambda})v}\,d\omega'$$

$$= \int_{S^{n-1}_+} u \cdot \overline{\mathcal{L}(2m-n-\overline{\lambda})\,v}\,d\omega + \sum_{j=0}^{m-1} \int_{S^{n-2}} \mathcal{T}_{2m-1-j}(\lambda)\,u \cdot \overline{\mathcal{S}_j(2m-n-\overline{\lambda})v}\,d\omega',$$

where $u, v \in C^\infty(\overline{S^{n-1}_+})$. From this formula it follows that

$$(11.6.2) \qquad \left(\mathcal{L}(\lambda)\right)^* = \mathcal{L}(2m-n-\overline{\lambda}).$$

We set $\mathcal{L}^{(k)}(\lambda) = \partial_\lambda^k \mathcal{L}(\lambda)$. Then, according to (11.6.2), we have

$$(11.6.3) \qquad \mathcal{L}^{(k)}(\lambda) = (-1)^k\left(\mathcal{L}^{(k)}(2m-n-\overline{\lambda})\right)^*.$$

By Theorem 10.3.3, the numbers $\lambda = m$ and $\lambda = m-n$ are simple eigenvalues of the pencil $\mathcal{L}_0$. The corresponding eigenvectors are $u_0(\omega) = \left(x_n/|x|\right)^m$ and $v_0(\omega) = G_{m-1}(\omega)$, respectively. Here $G_{m-1}(x) = G_{m-1}(x', x_n)$ is the solution of the boundary value problem

$$L(\partial_x)\,G_{m-1} = 0 \quad \text{in } \mathbb{R}^n_+,$$
$$(\partial_{x_n}^j G_{m-1})(x', 0) = 0 \quad \text{for } x' \in \mathbb{R}^{n-1},\ j = 0, \ldots, m-2,$$
$$(\partial_{x_n}^{m-1} G_{m-1})(x', 0) = \delta(x') \quad \text{for } x' \in \mathbb{R}^{n-1}.$$

LEMMA 11.6.1. *Let u_0, v_0 be the above introduced eigenfunctions. Then*

$$\int\limits_{S_+^{n-1}} \mathcal{L}^{(1)}(m)\, u_0(\omega) \cdot v_0(\omega)\, d\omega = (-1)^m\, m!\, a.$$

Proof: We consider the Dirichlet problem

$$(11.6.4) \qquad\qquad L(\partial_x)\, U = F \quad \text{in } \mathbb{R}_+^n,$$

$$(11.6.5) \qquad\qquad \partial_{x_n}^j\, U = 0 \quad \text{for } x_n = 0, \ j = 0, \ldots, m-1.$$

By $G = G(x,y)$ we denote the Green function, i.e., the solution of problem (11.6.4), (11.6.5) with the right-hand side $F = \delta(x-y)$. Then

$$(11.6.6) \qquad\qquad G_{m-1}(y' - x', y_n) = T_m(\partial_x)\, G(y,x)\big|_{x_n=0}.$$

The solution of problem (11.6.4), (11.6.5) is given by

$$U(y) = \int\limits_{\mathbb{R}_+^n} F(x)\, G(y,x)\, dx.$$

Let $F \in C_0^\infty(\overline{\mathbb{R}_+^n})$, $F = 0$ in a neighborhood of the point $x = 0$. We denote the constant

$$\partial_{y_n}^m U(y)\big|_{y=0} = \int\limits_{\mathbb{R}_+^n} L(\partial_x)\, U(x) \cdot \partial_{y_n}^m G(y,x)\big|_{y=0}\, dx$$

by c. It follows from (11.6.6) that $\partial_{y_n}^m G(y,x)\big|_{y=0} = (-1)^m\, a^{-1}\, G_{m-1}(x)$. Consequently,

$$c = (-1)^m\, a^{-1} \int\limits_{\mathbb{R}_+^n} L(\partial_x)\, U(x) \cdot G_{m-1}(x)\, dx.$$

Let ε be a positive number such that $F(x) = 0$ for $|x| < \varepsilon$, and let the function $\chi_\varepsilon = \chi_\varepsilon(r)$ be equal to zero for $r < \varepsilon$ and equal to one for $r > \varepsilon$. Then

$$c = (-1)^m\, a^{-1} \int\limits_{\mathbb{R}_+^n} U(x) \cdot L(\partial_x)\left(\chi_\varepsilon G_{m-1}\right) dx.$$

It can be easily verified that

$$L(\partial_x)\left(\chi_\varepsilon\, G_{m-1}\right) = r^{-m-n} \sum_{k \geq 1} \frac{1}{k!}\, (r\partial_r)^k\, \chi_\varepsilon(r)\, \mathcal{L}^{(k)}(m-n)\, G_{m-n}(\omega).$$

Since $U(x) = c\, x_n^m / m! + O(r^{m+1})$, we get

$$c = \frac{(-1)^m}{a} \lim_{\varepsilon \to 0} \int\limits_0^\infty \int\limits_{S_+^{n-1}} \frac{c}{m!}\, u_0(\omega) \sum_{k \geq 1} \frac{1}{k!}\, (r\partial_r)^k \chi_\varepsilon(r)\, \mathcal{L}^{(k)}(m-n)\, v_0(\omega)\, d\omega\, \frac{dr}{r}$$

$$= \frac{(-1)^m\, c}{a\, m!} \int\limits_{S_+^{n-1}} u_0(\omega) \cdot \mathcal{L}^{(1)}(m-n)\, v_0(\omega)\, d\omega.$$

Using (11.6.3) for $\lambda = m$ and $k = 1$, we obtain

$$(-1)^m \, m! \, a = - \int\limits_{S_+^{n-1}} \mathcal{L}^{(1)}(m) \, u_0(\omega) \cdot v_0(\omega) \, d\omega.$$

The lemma is proved. ∎

11.6.2. An asymptotic formula for the eigenvalue close to m. Now we consider the pencil $\mathcal{L}_\varepsilon$ generated by the Dirichlet problem in the cone $\mathcal{K}_\varepsilon$. The following theorem contains an asymptotic formula for the eigenvalue λ_ε which is close to the eigenvalue $\lambda_0 = m$ of the pencil $\mathcal{L}_0$.

THEOREM 11.6.1. *The following asymptotic formula is valid:*

$$(11.6.7) \qquad \lambda_\varepsilon = m + \varepsilon \int\limits_{S^{n-2}} \partial_{x_n}^m G_{m-1}(\omega', 0) \, \phi(\omega') \, d\omega' + O(\varepsilon^2).$$

Proof: We seek a number λ_ε and a function $u_\varepsilon \in C^\infty(\Omega_\varepsilon)$ such that

$$(11.6.8) \qquad \mathcal{L}(\lambda_\varepsilon) \, u_\varepsilon = 0 \quad \text{in } \Omega_\varepsilon,$$

$$(11.6.9) \qquad \mathcal{S}_j(\lambda_\varepsilon) \, u_\varepsilon = 0 \quad \text{on } \partial\Omega_\varepsilon, \; j = 0, 1, \dots, m-1,$$

and λ_ε, u_ε have the forms

$$(11.6.10) \qquad \lambda_\varepsilon = m + \varepsilon \lambda_1 + \cdots,$$

$$(11.6.11) \qquad u_\varepsilon(\omega) = u_0(\omega) + \varepsilon \, u_1(\omega) + \cdots,$$

where u_k are smooth functions in $\Omega_0 = S_+^{n-1}$. Since the domain Ω_ε is not necessarily contained in Ω_0, it is assumed that the functions u_k are smoothly extended outside Ω_0. It is easily seen that the following constructions do not depend on the method of extension.

Inserting expressions (11.6.10), (11.6.11) into (11.6.8), (11.6.9) and equating the coefficients of like powers of ε (in this connection, boundary conditions are blown off into $\partial\Omega_0$), we get a recurrence sequence of boundary value problems for the determination of the functions u_k and the numbers λ_k. As usual, the numbers λ_k are found from the condition of the orthogonality of the right-hand sides to the zero of the adjoint problem (i.e., to the function v_0). We restrict ourselves to the determination of the number λ_1 and the function u_1. We find them from the equation

$$(11.6.12) \qquad \mathcal{L}(m) \, u_1 + \lambda_1 \, \mathcal{L}^{(1)}(m) \, u_0 = 0 \quad \text{on } S_+^{n-1}.$$

Since $\mathcal{S}_j(m) \, u_0 \big|_{\partial\Omega_\varepsilon} = O(\varepsilon^2)$ for $j = 0, \dots, m-2$, and

$$\mathcal{S}_{m-1}(m) \, u_0 \big|_{\partial\Omega_\varepsilon} = r^{-1} \, \partial_{x_n}^{m-1} x_n^m \big|_{x_n = \varepsilon \, \phi(x')} = m! \, \varepsilon \, \phi(\omega') + O(\varepsilon^2),$$

the function u_1 must satisfy the boundary conditions

$$(11.6.13) \qquad \mathcal{S}_j(m) \, u_1 = 0 \quad \text{on } S^{n-2}, \; j = 0, \dots, m-2,$$

$$(11.6.14) \qquad \mathcal{S}_{m-1}(m) \, u_1 + m! \, \phi(\omega') = 0 \quad \text{on } S^{n-2}.$$

The space of the solutions of the formally adjoint problem is spanned by the function v_0. Therefore, the condition for the solvability of problem (11.6.12)–(11.6.14) has the following form

$$\lambda_1 \int_{S_+^{n-1}} \mathcal{L}^{(1)}(m)\, u_0 \cdot v_0 \, d\omega + m! \int_{S^{n-2}} \phi(\omega')\, \mathcal{T}_m(m-n)\, v_0 \, d\omega' = 0.$$

Using the equality

$$\begin{aligned}
\mathcal{T}_m(m-n)\, v_0(\omega)\big|_{S^{n-2}} &= r^n\, \mathcal{T}_m(\partial_x)\, G_{m-1}(x)\big|_{x_n=0} \\
&= (-1)^m\, a\, \partial_{x_n}^m\, G_{m-1}(\omega',0)
\end{aligned}$$

and Lemma 11.6.1, we arrive at (11.6.7). The theorem is proved. ∎

11.6.3. Determination of the function $\partial_{x_n}^m G_{m-1}(x',0)$. We seek the function G_{m-1} in the form

$$(11.6.15) \qquad G_{m-1}(x) = (2\pi)^{1-n} \int_{\mathbb{R}^{n-1}} e^{ix'\cdot\xi'}\, P(\xi', x_n)\, d\xi'.$$

Then the function P satisfies the relations

$$(11.6.16) \qquad L(i\xi', \partial_{x_n})\, P(\xi', x_n) = 0 \quad \text{for } x_n > 0,$$

$$(11.6.17) \qquad \partial_{x_n}^j P(\xi', x_n)\big|_{x_n=0} = 0 \quad \text{for } j = 0, 1, \dots, m-2,$$

$$(11.6.18) \qquad \partial_{x_n}^{m-1} P(\xi', x_n)\big|_{x_n=0} = 1.$$

Let $z_1(\xi'), \dots, z_d(\xi')$ denote the roots of the equation $L(\xi', z) = 0$ in the upper half-plane $\operatorname{Im} z > 0$ and let $\nu_1, \dots, \nu_d$ be their multiplicities. Then $\nu_1 + \cdots + \nu_d = m$ and the solution of (11.6.16) has the form

$$P(\xi', x_n) = \sum_{k=1}^d \sum_{l=0}^{\nu_k-1} p_{k,l}(\xi')\, x_n^l\, e^{ix_n z_k(\xi')}.$$

Since $\partial_{x_n}^j \left(x_n^l e^{ix_n z_k(\xi')} \right)\big|_{x_n=0} = \partial_z^l(z^j)\big|_{z=iz_k(\xi')}$, relations (11.6.17) and (11.6.18) are equivalent to

$$(11.6.19) \qquad \sum_{k=1}^d \sum_{l=0}^{\nu_k-1} p_{k,l}(\xi')\, \partial_z^l z^j\big|_{z=iz_k(\xi')} = 0 \quad \text{for } j = 0, \dots, m-2,$$

$$(11.6.20) \qquad \sum_{k=1}^d \sum_{l=0}^{\nu_k-1} p_{k,l}(\xi')\, \partial_z^l z^{m-1}\big|_{z=iz_k(\xi')} = 1.$$

Let $h_j(\xi')$ be the row-vector

$$\left(\partial_z^l z^j\big|_{z=iz_k(\xi')} \right)_{k=1,\dots,d,\ l=0,\dots,\nu_k-1}$$

and let H be the $m \times m$-matrix with the rows $h_0, \dots, h_{m-1}$. Furthermore, let p be the column-vector

$$\left(p_{k,l}(\xi') \right)_{k=1,\dots,d,\ l=0,\dots,\nu_k-1}.$$

Then relations (11.6.19), (11.6.20) can be written in the form

$$H\, p = e_m,$$

where e_m denotes the column of length m with the elements $0, \dots, 0, 1$.

Let the vector $q(\xi') = \big(q_0(\xi'), \dots, q_{m-1}(\xi')\big)$ satisfy the equation $q\,H = h_m$. Then

$$(11.6.21) \qquad \partial_{x_n}^m P(\xi', x_n)\big|_{x_n=0} = h_m(\xi')\,p(\xi') = q(\xi')\,H(\xi')\,p(\xi') = q_{m-1}(\xi').$$

In order to calculate the number $q_{m-1}(\xi')$, we write the equation $q\,H = h_m$ in the coordinate form

$$\partial_z^l \Big(\sum_{s=0}^{m-1} q_s(\xi')\, z^s \Big)\Big|_{z=iz_k(\xi')} = \partial_z^l z^m\big|_{z=iz_k(\xi')}, \quad k=1,\dots,d,\; l=0,\dots,\nu_k-1.$$

From this we conclude that the polynomial

$$z^m - \sum_{s=0}^{m-1} q_s\, z^s$$

has zeros $iz_k(\xi')$ of multiplicities ν_k, $k=1,\dots,d$. Therefore,

$$q_{m-1}(\xi') = i\,\big(\nu_1\, z_1(\xi') + \cdots + \nu_d\, z_d(\xi')\big).$$

Finally, by (11.6.15) and (11.6.21), we have

$$(11.6.22) \quad \partial_{x_n}^m G_{m-1}(x',0) = (2\pi)^{1-n} \int_{\mathbb{R}^{n-1}} e^{ix'\cdot\xi'}\, \partial_{x_n}^m P(\xi', x_n)\big|_{x_n=0}\, d\xi'$$

$$= i\,(2\pi)^{1-n} \int_{\mathbb{R}^{n-1}} e^{ix'\cdot\xi'} \sum_{k=1}^d \nu_k\, z_k(\xi')\, d\xi'.$$

Let us observe that the function q_{m-1} is smooth for $\xi' \neq 0$ and $q_{m-1}(-\xi') = \overline{q_{m-1}(\xi')}$. Therefore, integral (11.6.22) is real.

REMARK 11.6.1. In the case of a strongly elliptic operator $L(\partial_x)$ with complex coefficients, the same argument as in Theorem 11.6.1 leads to the asymptotic formula

$$(11.6.23) \qquad \lambda_\varepsilon = m + \varepsilon \int_{S^{n-2}} \partial_{x_n}^m \overline{G_{m-1}(\omega',0)}\, \phi(\omega')\, d\omega' + O(\varepsilon^2).$$

Moreover, in the proof of formula (11.6.22) the realness of the coefficients of $L(\partial_x)$ has not been used.

11.6.4. Examples.

Example 1. Let $L(\partial_x) = \Delta^m$. Then $q_{m-1}(\xi') = -m\,|\xi'|$. Therefore,

$$\partial_{x_n}^m G_{m-1}(x',0) = m\,\Gamma(n/2)\,\pi^{-n/2}\,|x'|^{-n}.$$

Consequently, the coefficient λ_1 in (11.6.10) is positive in the case $\phi \geq 0$, $\phi \not\equiv 0$.

Example 2. The following example shows that the coefficient λ_1 in (11.6.10) may be negative for positive ϕ. Let

$$L(\partial_x) = \Delta^2 + a\,\partial_{x_1}^4, \quad a > 0.$$

Then $L(\partial_x) = \big(\Delta - ib\partial_{x_1}^2\big)\big(\Delta + ib\partial_{x_1}^2\big)$, where $b^2 = a$, $b > 0$. Therefore,

$$(11.6.24) \qquad z_1(\xi') = i\,\sqrt{|\xi'|^2 - ib\xi_1^2}, \quad z_2(\xi') = i\,\sqrt{|\xi'|^2 + ib\xi_1^2}$$

(here the principal value of the square root is taken). Using the formulas for the Fourier transforms of functions of the form (11.6.24) (see, e.g., Gel'fand and Shilov [**66**]), we get

$$\int_{\mathbb{R}^{n-1}} e^{ix'\cdot\xi'} \sqrt{|\xi'|^2 \mp ib\xi_1^2} \, d\xi'$$

$$= -2^{n-1} \pi^{(n-1)/2} \frac{\Gamma(n/2)}{\Gamma(1/2)\sqrt{1 \mp ib}} \left(\frac{x_1^2}{1 \mp ib} + x_2^2 + \cdots + x_{n-1}^2 \right)^{-n/2}.$$

Consequently,

$$\partial_{x_n}^2 G_1(x', 0) = \frac{2\,\Gamma(n/2)}{\pi^{n/2}} \operatorname{Re}\left(\frac{1}{\sqrt{1+ib}} \left(\frac{x_1^2}{1+ib} + x_2^2 + \cdots + x_{n-1}^2 \right)^{-n/2} \right)$$

and, therefore,

$$\partial_{x_n}^2 G_1(x', 0)\big|_{x_1=0} = \frac{2\,\Gamma(n/2)}{\pi^{n/2}} \operatorname{Re}(1+ib)^{-1/2} \left(x_2^2 + \cdots + x_{n-1}^2 \right)^{-n/2} > 0,$$

(11.6.25) $$\partial_{x_n}^2 G_1(x', 0)\big|_{x_2=\cdots=x_{n-1}=0} = \frac{2\,\Gamma(n/2)}{\pi^{n/2}} \operatorname{Re}(1+ib)^{(n-1)/2}.$$

Let $\alpha = \arg(1 + ib)$. If $\alpha\,(n-1)/2 \in (\pi/2 + 2k\pi,\, 3\pi/2 + 2k\pi)$, $k \in \mathbb{Z}$, then the quantity (11.6.25) is negative. In particular, if $n = 4$ and $\alpha \in (\pi/3,\, \pi/2)$, then

$$\partial_{x_n}^2 G_1(x', 0)\big|_{x_2=\cdots=x_{n-1}=0} < 0.$$

Therefore, the function $\partial_{x_n}^2 G_1(\omega', 0)$ changes sign. Thus, there exists a positive function ϕ such that the coefficient λ_1 in (11.6.10) is negative.

Example 3. The asymptotic formula (11.6.23) takes a particularly simple form if $n = 2$. Let us write $L(\xi)$ as $L(\xi_1, \xi_2)$, and let $\zeta_1^+, \ldots, \zeta_m^+$ and $\zeta_1^-, \ldots, \zeta_m^-$ denote the roots of the polynomial $L(1, \zeta) = 0$ with positive and negative imaginary parts, respectively. We prove that

(11.6.26) $$\lambda_\varepsilon = m + \frac{i\varepsilon}{2\pi} \sum_{k=1}^{m} (\overline{\zeta_k^+} - \overline{\zeta_k^-})\, (\phi(1) + \phi(-1)) + O(\varepsilon^2).$$

By formula (11.6.22) with $n = 2$

$$\partial_{x_2}^m G_{m-1}(x_1, 0) = \frac{i}{2\pi} \left(\sum_{k=1}^{m} \zeta_k^+ \int_0^\infty e^{ix_1\xi_1} \xi_1 \, d\xi_1 + \sum_{k=1}^{m} \zeta_k^- \int_{-\infty}^0 e^{ix_1\xi_1} \xi_1 \, d\xi_1 \right).$$

Hence, for $x_1 \neq 0$

(11.6.27) $$\partial_{x_2}^m G_{m-1}(x_1, 0) = \frac{-i}{2\pi x_1^2} \sum_{k=1}^{m} (\zeta_k^+ - \zeta_k^-).$$

It remains to put (11.6.27) into (11.6.23).

For the operator $L(\partial_x)$ with real coefficients, formula (11.6.26) becomes

(11.6.28) $$\lambda_\varepsilon = m + \frac{\varepsilon}{\pi} \operatorname{Im} \sum_{k=1}^{m} \zeta_k^+ \, (\phi(1) + \phi(-1)) + O(\varepsilon^2),$$

which is in agreement with (8.2.1).

In conclusion, we mention that the real part of the coefficient

$$i \sum_{k=1}^{m} (\overline{\zeta_k^+} - \overline{\zeta_k^-})$$

in (11.6.26) is negative.

11.7. Nonrealness of eigenvalues

As is known (see Chapter 2), all eigenvalues of the operator pencil associated with the Dirichlet Laplacian in a cone are real. For other operators (for example, for the biharmonic operator in a plane angle with Dirichlet boundary conditions), the situation may be different: the spectrum of the corresponding operator pencil may contain nonreal eigenvalues. In the present section we describe a class of operators $L(D_x)$ for which there are no real eigenvalues in a sufficiently large strip if the cone has a small opening. This class includes Δ^{2k} but does not include Δ^{2k-1}, $k = 1, \ldots$.

We assume that $\overline{\mathcal{K}} \backslash \{0\}$ is contained in the half-space $\mathbb{R}_+^n = \{(x', x_n) : x' \in \mathbb{R}^{n-1}, x_n > 0\}$ and denote by $\mathcal{K}_\varepsilon$ the cone $\{x : (\varepsilon^{-1}x', x_n) \in \mathcal{K}\}$, $0 < \varepsilon \leq 1$. The intersection of $\mathcal{K}_\varepsilon$ with the unit sphere is denoted by Ω_ε.

Let $L(D_x)$ be a scalar homogeneous elliptic operator of order $2m$ with constant real coefficients, and let the operator pencil $\mathcal{L}_\varepsilon(\lambda)$ be defined by

$$\mathcal{L}_\varepsilon(\lambda)u = r^{2m-\lambda} L(D_x) r^\lambda u, \quad u \in \overset{\circ}{W}_2^m(\Omega_\varepsilon).$$

We prove the following assertion.

THEOREM 11.7.1. *Suppose that*

(11.7.1) $\operatorname{Re} L(\xi', i\xi_n) > 0 \quad \text{for a.e. } \xi = (\xi', \xi_n) \in \mathbb{R}^n.$

Then for any positive number N there exists $\varepsilon_N > 0$ such that for $\varepsilon \in (0, \varepsilon_N]$ the pencil $\mathcal{L}_\varepsilon$ has no real eigenvalues in the strip $|\operatorname{Re} \lambda| \leq N/\varepsilon$.

Proof: Let $G_\varepsilon = \{x' \in \mathbb{R}^{n-1} : (x', 1) \in \mathcal{K}_\varepsilon\}$. In the region Ω_ε we introduce the local coordinates x' of the point $\omega = (x', 1)(1 + |x'|^2)^{-1/2}$. Furthermore, we set $\lambda_\varepsilon = \lambda/\varepsilon$. Then, after the change of variables $y' = x'/\varepsilon$, the equation $\mathcal{L}_\varepsilon(\lambda_\varepsilon) u = 0$ becomes

$$\big(L(D_{y'}, -i\lambda) + \varepsilon \mathcal{R}_\varepsilon(y', D_{y'}, \lambda)\big) u = 0 \quad \text{on } G_1,$$

where

$$\mathcal{R}_\varepsilon(y', D_{y'}, \lambda) = \sum_{k+|\beta| \leq 2m} q_{k,\beta}(\varepsilon, y') \lambda^k D_{y'}^\beta$$

is a differential operator on $[0, 1] \times \overline{G}_1$ with smooth coefficients $q_{k,\beta}$. The function u satisfies zero Dirichlet conditions on ∂G_1.

We show that the pencil $L(D_{y'}, -i\lambda)$ has no real eigenvalues. In fact, if $L(D_{y'}, -i\lambda) u = 0$ and $\operatorname{Im} \lambda = 0$, then

$$\int_{G_1} L(D_{y'}, -i\lambda) u \cdot \overline{u} \, dy' = \int_{\mathbb{R}^{n-1}} L(\xi', -i\lambda) |\hat{u}|^2 \, d\xi' = 0,$$

where $\hat{u}$ is the Fourier transform of the zero extension of u to $\mathbb{R}^{n-1}$. Hence and by (11.7.1), $u = 0$ as required.

It may be assumed, without loss of generality, that there are no eigenvalues of the pencil $L(D_{y'}, -i\lambda)$ on the lines $\operatorname{Re} \lambda = \pm N$. We denote by $\lambda_1, \ldots, \lambda_m$ the eigenvalues of this pencil in the strip $|\operatorname{Re}\lambda| \leq N$ if there are any. There exists a positive number ε_N such that for all $\varepsilon \in [0, \varepsilon_N]$ the pencil $L + \varepsilon \mathcal{R}_\varepsilon$ has exactly m eigenvalues in this strip. Suppose $D_1, \ldots, D_m$ are disks in the complex plane centered at $\lambda_1 \ldots, \lambda_m$ which do not intersect the real axis. Then, for a sufficiently small ε_N, all eigenvalues of the pencil $L + \varepsilon \mathcal{R}_\varepsilon$ in the strip $|\operatorname{Re}\lambda| \leq N$ are situated in the disks $D_1, \ldots, D_m$ if $\varepsilon \leq \varepsilon_N$. The result follows. ∎

11.8. Further results

In Section 11.5 we studied the singularities of solutions to the Dirichlet problem in the exterior of a thin cone under the assumption that the order $2m$ of the differential operator is not less than the dimension n. The case $2m < n$ was considered in the book [174] by Maz'ya, Nazarov and Plamenevskiĭ.

Let K_ε be the cone

$$K_\varepsilon = \{x = (x', x_n) \in \mathbb{R}^n \; : \; \varepsilon^{-1} x_n^{-1} x' \in G\},$$

where G is a bounded domain in $\mathbb{R}^{n-1}$ containing the origin with smooth boundary ∂G. We consider the operator pencil $\mathcal{L}_\varepsilon$ generated by the Dirichlet problem for the strongly elliptic operator (10.1.2) in the cone $\mathcal{K}_\varepsilon = \mathbb{R}^n \backslash \overline{K}_\varepsilon$ (the definition of this pencil is given in Section 10.1). For small ε there exist eigenvalues $\lambda_j(\varepsilon)$, $j = 1, \ldots, \ell$, of the pencil $\mathcal{L}_\varepsilon$ near the point $\lambda = 0$. In [174, Ch.10] asymptotic formulas for these eigenvalues as $\varepsilon \to 0$ were obtained.

We restrict ourselves first to the scalar case $\ell = 1$ when there is only one eigenvalue $\lambda(\varepsilon)$ near $\lambda = 0$. Let

$$L(\partial_{x'}, 0) = (-1)^m \sum_{|\alpha|=|\beta|=m} A_{\alpha\beta}^{(0)} \partial_{x'}^{\alpha+\beta},$$

where α and β are $(n-1)$-dimensional multi-indices, and

$$k = \int_{\mathbb{R}^{n-1}\backslash G} \sum_{|\alpha|=|\beta|=m} A_{\alpha\beta}^{(0)} \partial_{x'}^\beta w(x') \overline{\partial_{x'}^\alpha w(x')} \, dx'$$

with w being the solution of the Dirichlet problem

$$L(\partial_{x'}, 0)w(x') = 0, \quad \text{for } x' \in \mathbb{R}^{n-1}\backslash G,$$
$$w(x') = 1, \quad \partial_{x'}^\alpha w(x') = 0, \quad \text{for } x' \in \partial G, \; 0 < |\alpha| < m,$$

which vanishes at infinity. Furthermore, let $E(x)$ be the positively homogeneous fundamental solution of the operator $L^+(D_x)$ in $\mathbb{R}^n$.

THEOREM 11.8.1. [174] 1) Let $n - 1 = 2m$. Then

$$\lambda(\varepsilon) = |2\log\varepsilon|^{-1}(1 + o(1)).$$

2) If $n - 1 > 2m$, then

(11.8.1) $$\lambda(\varepsilon) = \varepsilon^{n-1-2m}\left(k\,\overline{E(0, \ldots, 0, 1)} + o(1)\right).$$

The value k is a generalization of the m-harmonic capacity $\operatorname{cap}_m(\Omega)$ of the domain $\Omega \subset \mathbb{R}^{n-1}$. In the special case $L(D_x) = (-\triangle)^m$ the asymptotic formula (11.8.1) takes the form

$$\lambda(\varepsilon) = \varepsilon^{n-1-m}\left(2^{m-2}\pi^{n/2}\Gamma((n-2m)/2)\Gamma(m)^{-1}\operatorname{cap}_m(\omega) + o(1)\right).$$

For the second order elliptic operator

$$L(\partial_x) = -\sum_{j,k=1}^{n} a_{j,k} \frac{\partial^2}{\partial x_j\,\partial x_k}, \quad n > 3,$$

the eigenvalue $\lambda(\varepsilon)$ satisfies

$$\begin{aligned}
\lambda(\varepsilon) &= \varepsilon^{n-3}\Big(\big((n-2)|S^{n-1}|\big)^{-1}\operatorname{cap}(\omega; L(\partial_{x'},0))\,\big(\det(a_{jk})_{j,k=1}^{n-1}\big)^{(2-n)/2} \\
&\quad \times \big(\det(a_{jk})_{j,k=1}^{n}\big)^{(n-3)/2} + o(1)\Big),
\end{aligned}$$

where $|S^{n-1}|$ denotes the area of the $(n-1)$-dimensional unit sphere and

$$\operatorname{cap}(\omega; L(\partial_{x'},0)) = \int_{\mathbb{R}^{n-1}\setminus\omega} \sum_{j,k=1}^{n-1} a_{jk}\,\partial_{x_k} w\,\partial_{x_j}\overline{w}\,dx'$$

with w being the solution of

$$L(\partial_{x'},0)\,w(x') = 0, \quad x' \in \mathbb{R}^{n-1}\setminus G,$$

which equals 1 on ∂G and vanishes at infinity.

Similar asymptotic formulas were obtained in [174, Ch.10] for matrix operators. In particular, for the Lamé operator it was shown that there are exactly three small eigenvalues $\lambda_j(\varepsilon)$, $j = 1,2,3$, with the asymptotics

$$\lambda_j(\varepsilon) = |2\log\varepsilon|^{-1}\big(1 + o(1)\big).$$

The same is true for the Stokes operator.

According to [174, Ch.10], there are no eigenvalues of the pencil $\mathcal{L}_\varepsilon$ in the strips

$$\big\{\lambda \in \mathbb{C} : \operatorname{Re}\lambda \in [2m - n + \varepsilon^{n-1-2m}\mu, \varepsilon^{n-1-2m}\mu]\big\} \quad \text{for } n - 1 > 2m,$$

$$\big\{\lambda \in \mathbb{C} : \operatorname{Re}\lambda \in [-1 - |\log\varepsilon|^{-1}\mu, |\log\varepsilon|^{-1}\mu]\big\} \quad \text{for } n - 1 = 2m,$$

where $\mu < \operatorname{Re}\big(k\,\overline{E(0,\dots,0,1)}\big)$ if $n - 1 > 2m$ and $\mu < 1/2$ if $n - 1 = 2m$.

One can deduce from this fact that in the case $n - 1 = 2m$ all W_2^m-solutions of the Dirichlet problem for the scalar differential operator $L(\partial_x)$ are continuous at the vertex of $\mathcal{K}_\varepsilon$ provided ε is sufficiently small. If the coefficients of L are real, the same is true for an arbitrary Lipschitz cone.

Let us turn to the case $n - 1 > 2m$. If

$$\operatorname{Re}\big(k\,\overline{E(0,\dots,0,1)}\big) > 0,$$

then the above statement about the strip, which is free of the spectrum, guarantees the continuity of solutions to the Dirichlet problem in $\mathcal{K}_\varepsilon$ with a sufficiently small ε. There are examples, which show that the continuity may fail. This may happen in the case $n > 4$ for the second order strongly elliptic operator with complex coefficients

$$L(\partial_x) = (1 + i\beta)\partial_{x_1}^2 + \partial_{x_2}^2 + \cdots + \partial_{x_{n-1}}^2 + \alpha\partial_{x_n}^2,$$

where α and β are some numbers such that $\operatorname{Im}\beta = 0$, $\operatorname{Re}\alpha > 0$ (see [174, Sec.10.6.1]). Another example of the loss of continuity is provided by the operator

$$L(\partial_x) = \Delta^2 + a^2\,\partial_{x_n}^4, \quad n \geq 8,$$

where $(n - 3)\arctan a \in (2\pi, 4\pi)$ (see [169] and [174, Sec.10.6.2]).

11.9. The Dirichlet problem in domains with conic vertices

The contents of the preceding sections together with results of the theory of general elliptic problems in domains with piecewise smooth boundaries can be used to obtain various assertions concerning the Dirichlet problem for $2m$ order elliptic equations (or systems). We illustrate this by a domain with conic vertices.

Let $\mathcal{G}$ in $\mathbb{R}^n$ be the same bounded domain as in Section 1.4. We use the following notation introduced there: $x^{(\tau)}$, $r_\tau(x)$, $r(x)$, $\mathcal{S} = \{x^{(1)}, \dots, x^{(d)}\}$, and $\mathcal{K}_\tau = \{x : (x - x^{(\tau)})/|x - x^{(\tau)}| \in \Omega_\tau\}$. The function spaces $V^l_{p,\vec{\beta}}(\mathcal{G})$, $N^{l,\sigma}_{\vec{\beta}}(\mathcal{G})$, where $\vec{\beta} = (\beta_1, \dots, \beta_d)$, and the corresponding trace spaces were also introduced in Section 1.4.

We consider the Dirichlet problem

(11.9.1) $L(x, D_x)\, U = F$ in $\mathcal{G}$, $\partial^k_\nu U = G_k$ on $\partial\mathcal{G}\backslash\mathcal{S}$, $k = 0, \dots, m - 1$,

where

(11.9.2) $$L(x, D_x) = \sum_{|\alpha| \leq 2m} A_\alpha(x)\, D^\alpha_x$$

is a strongly elliptic differential operator whose coefficients A_α are $\ell \times \ell$ matrices infinitely differentiable in $\overline{\mathcal{G}}$. We assume that the operators $L^\circ(x^{(\tau)}, D_x)$ are formally selfadjoint and the cones $\mathcal{K}_\tau$ are Lipschitz (i.e., $\mathcal{K}_\tau$ has the representation $x_n > \phi(x')$ in a certain Cartesian coordinate system).

Let $\mathfrak{A}_\tau$ be the operator pencil generated by the Dirichlet problem (11.9.1) for $x^{(\tau)}$. (These pencils were introduced in Section 1.4 for general elliptic problems. In the case of problem (11.9.1) we have $t_j = 0$ and $s_i = 2m$.) By Theorem 11.1.1, the eigenvalues of $\mathfrak{A}_\tau$ lie outside the strip

$$|\operatorname{Re}\lambda - m + n/2| \leq 1/2\,.$$

Hence and by using essentially Theorems 1.4.1–1.4.5, we arrive at the following explicit solvability and regularity results.

THEOREM 11.9.1. 1) *Let $l \geq 2m$. Suppose that for arbitrary $F \in W^{-m}_2(\mathcal{G})^\ell$ and for $G_1 = \dots = G_m = 0$ there exists a unique solution $U \in \overset{\circ}{W}{}^m_2(\mathcal{G})^\ell$ of problem (11.9.1). Then the operator of this problem:*

$$V^l_{p,\vec{\beta}}(\mathcal{G})^\ell \to V^{l-2m}_{p,\vec{\beta}}(\mathcal{G})^\ell \times \prod_{k=0}^{m-1} V^{l-k-1/p}_{p,\vec{\beta}}(\partial\mathcal{G})^\ell.$$

is isomorphic if

$$m - \frac{n+1}{2} - \varepsilon \leq l - \beta_\tau - n/p \leq m - \frac{n-1}{2} + \varepsilon$$

for $\tau = 1, \dots, d$. Here and elsewhere in this theorem ε is a sufficient small positive number. Analogously, the operator of problem (11.9.1):

$$N^{l,\sigma}_{\vec{\beta}}(\mathcal{G})^\ell \to N^{l-2m,\sigma}_{\vec{\beta}}(\mathcal{G})^\ell \times \prod_{k=0}^{m-1} N^{l-k,\sigma}_{\vec{\beta}}(\partial\mathcal{G})^\ell.$$

is isomorphic if

$$m - \frac{n+1}{2} - \varepsilon \leq l + \sigma - \beta_\tau \leq m - \frac{n-1}{2} + \varepsilon\,.$$

2) *Let $l \geq 2m$ and let $U \in W_2^m(\mathcal{G})^\ell$ be a variational solution of problem* (11.9.1), *where $F \in V_{p,\vec{\beta}}^{l-2m}(\mathcal{G})^\ell$, $G_k \in V_{p,\vec{\beta}}^{l-k-1/p}(\partial\mathcal{G})^\ell$. If*

$$m - n/2 \leq l - \beta_\tau - n/p \leq m - \frac{n-1}{2} + \varepsilon \quad for \ \tau = 1, \dots, d,$$

then $U \in V_{p,\vec{\beta}}^l(\mathcal{G})$.

If $F \in N_{\vec{\beta}}^{l-2m,\sigma}(\mathcal{G})^\ell$, $G_k \in N_{\vec{\beta}}^{l-k,\sigma}(\partial\mathcal{G})^\ell$, where

$$m - n/2 \leq l + \sigma - \beta_\tau \leq m - \frac{n-1}{2} + \varepsilon \quad for \ \tau = 1, \dots, d,$$

then $U \in N_{\vec{\beta}}^{l,\sigma}(\mathcal{G})$.

11.10. Notes

Sections 11.1.–11.4. The estimate for the width of the energy strip of the pencil corresponding to the Dirichlet problem in a "smooth" cone (Theorem 11.1.1) was proved by Kozlov and Maz'ya [**127**]. The weaker estimate in Theorem 11.2.2 without restrictions on the smoothness of the cone was obtained by Kozlov and Roßmann in [**142**]. In [**143**] the same authors obtained the estimate of Theorem 11.1.1 for the case of a second order system under lower assumptions on the cone (see Theorem 11.3.3). Kozlov and Maz'ya [**131**] improved this estimate for second order systems in a polyhedral cone with edge angles less than π (see Sect. 8.6). Theorem 11.4.1 can be found in their paper [**132**].

Sections 11.5, 11.8. The relation between the spectra of the pencils generated by the Sobolev problem in the exterior of a ray and the Dirichlet problem in the exterior of a thin cone given in Theorem 11.5.1 was obtained by Kozlov and Maz'ya [**130**]. They assumed that the order $2m$ of the differential equation (or system) is not less than the dimension of the domain. Singularities of solutions of the Dirichlet problem in the exterior of a thin cone with dimension $n > 2m$ were investigated by Maz'ya, Nazarov and Plamenevskiĭ [**172, 173, 174**]. They obtained, in particular, the asymptotic formulas for small eigenvalues in Theorem 11.8.1.

Section 11.6. The asymptotic formula for the eigenvalue near m of the operator pencil generated by the Dirichlet problem in a cone which is close to the half-space was obtained by Kozlov [**122**].

Section 11.7. Theorem 11.7.1 is due to Kozlov, Kondrat'ev and Maz'ya [**126**].

CHAPTER 12

The Neumann problem in a cone

In the present chapter we are interested in eigenvalues, eigenvectors and generalized eigenvectors associated to the operator pencil $\mathfrak{A}$ that is generated by the Neumann problem for the operator

$$(12.0.1) \qquad L(\partial_x) = (-1)^m \sum_{|\alpha|=|\beta|=2m} A_{\alpha,\beta}\, \partial_x^{\alpha+\beta}$$

in the cone $\mathcal{K} = \{x \in \mathbb{R}^n : x/|x| \in \Omega\}$. Here $A_{\alpha,\beta}$ are constant $\ell \times \ell$-matrices such that $A_{\alpha,\beta}^* = A_{\beta,\alpha}$, and Ω is a domain on the unit sphere S^{n-1}. We introduce the sesquilinear form

$$(12.0.2) \qquad b(U,V) = \int_{\mathcal{K}} \sum_{|\alpha|=|\beta|=m} \left(A_{\alpha,\beta}\, \partial_x^\beta U\,,\, \partial_x^\alpha V\right)_{\mathbb{C}^\ell} dx\,,$$

where U, V are vector functions in $\mathcal{K}$. Throughout this chapter, the following condition is satisfied.

CONDITION 1 *(local coerciveness).* *There exists a number $\delta > 0$ such that the inequality*

$$(12.0.3) \qquad b(U,U) \geq c \int_{\mathcal{K}} \sum_{|\alpha|=m} |\partial_x^\alpha U|_{\mathbb{C}^\ell}^2\, dx$$

is valid for all vector functions $U \in W_2^m(\mathcal{K})^\ell$ whose supports intersect the unit sphere and have diameters not exceeding δ. Here c is a positive constant.

If $\partial\Omega$ is smooth and

$$(12.0.4) \qquad \sum_{|\alpha|=|\beta|=m} \left(A_{\alpha,\beta}f_\beta\,,\, f_\alpha\right)_{\mathbb{C}^\ell} \geq 0 \quad \text{for all } f_\alpha \in \mathbb{C}^\ell,\ |\alpha| = m,$$

then Condition 1 is satisfied, by Aronszajn's theorem [7], if and only if the matrix

$$\sum_{|\alpha|=|\beta|=0} A_{\alpha,\beta} \left(\xi + i\tau\nu(\omega')\right)^\beta \left(\xi - i\tau\nu(\omega')\right)^\alpha$$

is nonsingular for every $\xi \in \mathbb{R}^n$, $\tau \in \mathbb{R}$, $|\xi| + |\tau| \neq 0$, $\omega' \in \partial\Omega$ (ν is the exterior normal to $\partial\mathcal{K}$).

We always assume in this chapter that the imbedding $W_2^1(\Omega) \subset L_2(\Omega)$ is compact. For this it suffices, for example, that the domain Ω belongs to the class C, i.e., the boundary $\partial\Omega$ admits a local explicit representation by a continuous function. A necessary and sufficient condition as well as more comprehensive sufficient conditions are given in Maz'ya's book [164]. Clearly, the compactness of the imbedding $W_2^1(\Omega) \subset L_2(\Omega)$ implies the compactness of the imbedding $W_2^l(\Omega) \subset L_2(\Omega)$ for arbitrary $l \geq 1$.

In Section 12.1 we show that Condition 1 guarantees discreteness of the spectrum of $\mathfrak{A}$.

It is proved in Section 12.2 that the energy line $\operatorname{Re}\lambda = m - n/2$ contains no eigenvalues of $\mathfrak{A}$ if either $2m < n$ or $2m \geq n$ and n is odd. Moreover, we find that $m - n/2$ is the only eigenvalue on the energy line if $2m \geq n$ and n is even.

These results are obtained under the following condition on the operator L.

CONDITION 2 *(generalized Korn's inequality). For all vector functions $U \in W_2^m(\mathcal{K})^\ell$ vanishing in a neighborhood of the vertex, inequality (12.0.3) is valid with a positive constant c independent of U.*

Obviously, Condition 2 is stronger than Condition 1. In Corollary 12.3.1 it is shown that Condition 1 complemented by (12.0.4) implies the validity of Condition 2 for cones of class $C^{0,1}$ (Lipschitz graph). At the end of Section 12.2 we construct an example showing that under just one condition 1 of local coerciveness, without (12.0.4), the line $\operatorname{Re}\lambda = m - n/2$ may contain nonreal eigenvalues. Thus, Condition 2 does not follow from Condition 1.

The most complete information on the spectrum of $\mathfrak{A}$ is obtained in Section 12.3 under the following condition, more restrictive than Condition 1.

CONDITION 3. *For all vectors $f_\alpha \in \mathbb{C}^\ell$, $|\alpha| = m$, there holds the inequality*

$$\sum_{|\alpha|=|\beta|=m} \left(A_{\alpha,\beta}\, f_\beta\,,\, f_\alpha\right)_{\mathbb{C}^\ell} \geq c \sum_{|\alpha|=m} |f_\alpha|^2_{\mathbb{C}^\ell}\,,$$

where c is a positive constant.

We present this information as

THEOREM. *Let the cone $\mathcal{K}$ be given by the inequality $x_n > \phi(x')$, where ϕ is a smooth function on $\mathbb{R}^{n-1}$. Then*

(i) *For $2m < n-1$ the strip $|\operatorname{Re}\lambda - m + n/2| \leq 1/2$ contains no points of the spectrum of $\mathfrak{A}$.*

(ii) *For $2m \geq n-1$ and even n the strip $|\operatorname{Re}\lambda - m + n/2| \leq 1/2$ contains the single eigenvalue $\lambda_0 = m - n/2$. A vector-valued function u is an eigenfunction if and only if $u = p|_\Omega$, where p is a homogeneous vector polynomial in $\mathbb{R}^n$ of degree $m - n/2$. To each eigenfunction there corresponds exactly one generalized eigenfunction.*

(iii) *For $2m \geq n-1$ and odd n the strip $|\operatorname{Re}\lambda - m + n/2| \leq 1/2$ contains exactly two eigenvalues $\lambda_\pm = m - n/2 \pm 1/2$, with multiplicity equal to*

$$\ell \binom{m+(n-1)/2}{n-1} \quad and \quad \ell\left(\binom{m+(n-1)/2}{n-1} + \binom{m+(n-3)/2}{n-1}\right)$$

respectively. The eigenfunctions corresponding to λ_+ are traces on Ω of homogeneous vector polynomials of degree $(2m - n + 1)/2$. To each eigenvector there corresponds at most one generalized vector.

The statements of this theorem regarding the open strip $|\operatorname{Re}\lambda - m + n/2| < 1/2$ are established in Section 12.3 under somewhat weaker conditions. Namely, it is assumed only that ϕ satisfies the Lipschitz condition and that the operator L is subject to (12.0.4) and Condition 1.

An example at the end of Section 12.3 illustrates the fact that Condition 3 in the above theorem cannot be replaced by Condition 2.

We note that the theorem is sharp for $n = 2$ (see Remark 12.3.1).

In Section 12.4, as an application of the results mentioned above, we derive asymptotic formulae for solutions of the Neumann problem in a domain $\mathcal{G}$ with a conic vertex. We consider two well posed problems. The first is the usual variational formulation of the Neumann problem. For the second one, meaningful only for $2m > n$, we prescribe (in addition to the Neumann conditions) the data for the solution and its derivatives up to order $|\alpha| < m - n/2$ at the vertex. It is shown that for the odd dimension n the solution of the problem in the first formulation has greater smoothness than in the second one.

Finally, in Section 12.5 we give a rather complete information about the spectrum in the strip $|\operatorname{Re}\lambda| \leq 1$ for the pencil generated by the Lamé system of the plane anisotropic elasticity in an angle with Neumann boundary conditions.

12.1. The operator pencil generated by the Neumann problem

12.1.1. Definition of the pencil $\mathfrak{A}$. We are interested in solutions of the Neumann problem for the differential operator (12.0.1) in the cone $\mathcal{K}$ which have the form

$$(12.1.1) \qquad U(x) = r^{\lambda_0} \sum_{k=0}^{s} \frac{1}{k!} (\log r)^k u_{s-k}(\omega),$$

where r, ω are the spherical coordinates of the point x and $u_k \in W_2^m(\Omega)^\ell$ for $k = 0, \dots, s$. This means, we seek solutions (12.1.1) of the equation

$$(12.1.2) \qquad b(U, V) = 0$$

which has to be satisfied for all $V \in W_2^m(\mathcal{K})^\ell$ equal to zero in a neighborhood of the point $x = 0$ and of infinity.

This leads to a spectral problem on the domain Ω which can be described as follows. Let

$$(12.1.3) \qquad Q_\alpha(\omega, \partial_\omega, \lambda)\, u(\omega) = r^{-\lambda+|\alpha|}\, \partial_x^\alpha\left(r^\lambda\, u(\omega)\right).$$

Then we set

$(12.1.4)$

$$a(u, v; \lambda) = \sum_{|\alpha|=|\beta|=m} \int_\Omega \left(A_{\alpha,\beta}\, Q_\beta(\omega, \partial_\omega, \lambda)u,\, Q_\alpha(\omega, \partial_\omega, 2m-n-\bar\lambda)v\right)_{\mathbb{C}^\ell} d\omega.$$

It can be easily verified that the form $a(\cdot, \cdot; \lambda)$ admits the representation

$$(12.1.5) \qquad a(u, v; \lambda) = \frac{1}{2\,|\log\varepsilon|} \int_{\substack{\mathcal{K} \\ \varepsilon<|x|<1/\varepsilon}} \sum_{|\alpha|=|\beta|=m} \left(A_{\alpha,\beta}\partial_x^\beta U,\, \partial_x^\alpha V\right)_{\mathbb{C}^\ell} dx,$$

where $U(x) = r^\lambda\, u(\omega)$, $V(x) = r^{2m-n-\bar\lambda}\, v(\omega)$, and ε is an arbitrary positive number less than one.

The sesquilinear form (12.1.4) generates the linear and continuous operator

$$(12.1.6) \qquad \mathfrak{A}(\lambda)\,:\, W_2^m(\Omega)^\ell \to (W_2^m(\Omega)^\ell)^*$$

polynomially depending on the parameter $\lambda \in \mathbb{C}$.

We show that there is the following relation between solutions of (12.1.2) having the form (12.1.1) and the spectrum of the pencil $\mathfrak{A}$ (cf. Theorem 1.1.5).

LEMMA 12.1.1. *The vector function* (12.1.1) *is a solution of* (12.1.2) *if and only if* λ_0 *is an eigenvalue of the operator pencil* $\mathfrak{A}$ *and* $u_0, \ldots, u_s$ *is a Jordan chain corresponding to this eigenvalue.*

Proof: Let U be a solution of (12.1.2) which has the form (12.1.1). We write the differential operators $Q_\alpha(\omega, \partial_\omega, r\partial_r)$ in the form

$$Q_\alpha(\omega, \partial_\omega, r\partial_r) = \sum_{j=0}^{|\alpha|} \frac{1}{j!} Q_\alpha^{(j)}(0) (r\partial_r)^j,$$

where $Q_\alpha^{(j)}(\lambda) = \partial_\lambda^j Q_\alpha(\omega, \partial_\omega, \lambda)$. Integrating by parts in (12.0.2), we get

$$b(U, V) = \sum_{|\alpha|=|\beta|=m} \int_{\mathcal{K}} \left(A_{\alpha,\beta} r^{-m} Q_\beta(\omega, \partial_\omega, r\partial_r) U, \, r^{-m} Q_\alpha(\omega, \partial_\omega, r\partial_r) V \right)_{\mathbb{C}^\ell} dx$$

$$= \sum_{|\alpha|=|\beta|=m} \sum_{j=0}^{m} \frac{1}{j!} \int_0^\infty \int_\Omega r^{n-2m+\lambda} \left(A_{\alpha,\beta} (2m - n - \lambda_0 - r\partial_r)^j \right.$$

$$\left. \times Q_\beta(\omega, \partial_\omega, \lambda_0 + r\partial_r) \sum_{k=0}^{s} \frac{(\log r)^k}{k!} u_{s-k}(\omega), \, Q_\alpha^{(j)}(0) V \right)_{\mathbb{C}^\ell} \frac{dr}{r} d\omega = 0$$

for $V \in W_2^m(\mathcal{K})^\ell$, $V = 0$ near the origin and infinity. The last equation is satisfied if and only if

$$(12.1.7) \quad \sum_{|\alpha|=|\beta|=m} \sum_{j=0}^{m} \frac{1}{j!} \int_\Omega \left(A_{\alpha,\beta} (2m - n - \lambda_0 - r\partial_r)^j Q_\beta(\omega, \partial_\omega, \lambda_0 + r\partial_r) \right.$$

$$\left. \times \sum_{k=0}^{s} \frac{(\log r)^k}{k!} u_{s-k}(\omega), \, Q_\alpha^{(j)}(0) v(\omega) \right)_{\mathbb{C}^\ell} d\omega = 0$$

for all $v \in W_2^m(\Omega)^\ell$ and $r > 0$. Using the formulas

$$Q_\beta(\omega, \partial_\omega, \lambda_0 + r\partial_r) = \sum_{q=0}^{|\beta|} \frac{1}{q!} Q_\beta^{(q)}(\lambda_0) (r\partial_r)^q,$$

$$(2m - n - \lambda_0 - r\partial_r)^j = \sum_{\mu=0}^{j} \binom{j}{\mu} (2m - n - \lambda_0)^{j-\mu} (-r\partial_r)^\mu,$$

we can write the expression on the left of (12.1.7) in the form

$$\sum_{\alpha,\beta} \sum_{j} \sum_{q} \sum_{\mu} \frac{1}{j! \, q!} \binom{j}{\mu} (2m - n - \lambda_0)^{j-\mu} (-1)^\mu$$

$$\times \int_\Omega \left(A_{\alpha,\beta} Q_\beta^{(q)}(\lambda_0) \sum_{k=0}^{s} \frac{(\log r)^{s-k-\mu-q}}{(s-k-\mu-q)!} u_k(\omega), \, Q_\alpha^{(j)}(0) v(\omega) \right)_{\mathbb{C}^\ell} d\omega$$

$$= \sum_{\alpha,\beta} \sum_{\mu} \sum_{q} \frac{(-1)^\mu}{\mu! \, q!} \int_\Omega \left(A_{\alpha,\beta} Q_\beta^{(q)}(\lambda_0) \right.$$

$$\left. \times \sum_{k=0}^{s} \frac{(\log r)^{s-k-\mu-q}}{(s-k-\mu-q)!} u_k(\omega), \, Q_\alpha^{(\mu)}(2m - n - \overline{\lambda}_0) v(\omega) \right)_{\mathbb{C}^\ell} d\omega.$$

Hence (12.1.7) is satisfied if and only if

$$\sum_{\alpha,\beta} \sum_{\mu+q\le s-\sigma} \frac{(-1)^\mu}{\mu!\,q!} \int_\Omega \left(A_{\alpha,\beta} Q_\beta^{(q)}(\lambda_0)\, u_{s-\mu-q-\sigma}(\omega)\,,\, Q_\alpha^{(\mu)}(2m-n-\overline{\lambda}_0)v(\omega) \right)_{\mathbb{C}^\ell} d\omega$$

vanishes for $\sigma = 0,1,\dots,s$ or, what is the same, if

$$\sum_{j=0}^{s-\sigma} \frac{1}{j!} a^{(j)}(u_{s-\sigma-j}, v; \lambda_0) = 0$$

for $\sigma = 0,1,\dots,s$. This proves the lemma. ∎

12.1.2. Basic properties of the pencil $\mathfrak{A}$. Let

$$(12.1.8) \qquad b_0(U,V) \overset{def}{=} \int_\mathcal{K} \sum_{|\alpha|=m} \left(\partial_x^\alpha U, \partial_x^\alpha V \right)_{\mathbb{C}^\ell} dx$$

and let a_0 be the corresponding parameter-dependent sesquilinear form on $W_2^m(\Omega)^\ell$, i.e.,

$$(12.1.9) \quad a_0(u,v;\lambda) = \sum_{|\alpha|=m} \int_\Omega \left(Q_\alpha(\omega, \partial_\omega, \lambda)u\,,\, Q_\alpha(\omega, \partial_\omega, 2m-n-\overline{\lambda})v \right)_{\mathbb{C}^\ell} d\omega$$

$$= \frac{1}{2\log 2} \int_{\substack{\mathcal{K} \\ 1/2<|x|<2}} \sum_{|\alpha|=m} \left(\partial_x^\alpha U, \partial_x^\alpha V \right)_{\mathbb{C}^\ell} dx,$$

where $U(x) = r^\lambda u(\omega)$, $V(x) = r^{2m-n-\overline{\lambda}} v(\omega)$.

LEMMA 12.1.2. *There exists a positive constant c independent of u and λ such that*

$$a_0(u,u;\lambda) \ge c \left(\|u\|^2_{W_2^m(\Omega)^\ell} + |\lambda|^{2m} \|u\|^2_{L_2(\Omega)^\ell} \right)$$

for all $u \in W_2^m(\Omega)^\ell$ and sufficiently large $|\lambda|$, $\operatorname{Re}\lambda = m - n/2$.

Proof: Let $u \in W_2^m(\Omega)^\ell$ and $\lambda \in \mathbb{C}$, $\operatorname{Re}\lambda = m - n/2$, be given. We set $U(x) = r^\lambda u(\omega)$. Using the representation

$$(r\partial_r)^m = \sum_{|\alpha|\le m} c_\alpha(\omega)\, r^\alpha\, \partial_x^\alpha,$$

we obtain

$$|\lambda|^{2m} \|u\|^2_{L_2(\Omega)^\ell} = c \int_{\substack{\mathcal{K} \\ 1/2<|x|<2}} \left| (r\partial_r)^m U \right|^2 dx$$

$$\le c \sum_{|\alpha|\le m} \int_{\substack{\mathcal{K} \\ 1/2<|x|<2}} r^{2|\alpha|} \left| \partial_x^\alpha U \right|^2 dx$$

$$\le c \left(a_0(u,u;\lambda) + \sum_{j=0}^{m-1} |\lambda|^{2j} \|u\|^2_{W_2^{m-1-j}(\Omega)^\ell} \right).$$

Furthermore,

$$
\begin{aligned}
\|u\|^2_{W_2^m(\Omega)^\ell} &= c \sum_{|\alpha| \le m} \int_{\substack{K \\ 1/2 < |x| < 2}} \left| \partial_x^\alpha (r^{-\lambda} U) \right|^2 dx \\
&\le c \left(a_0(u, u; \lambda) + \sum_{j=1}^m |\lambda|^{2j} \|u\|^2_{W_2^{m-j}(\Omega)^\ell} \right)
\end{aligned}
$$

The last two inequalities together with the inequality

$$
|\lambda|^j \|u\|_{W_2^{m-j}(\Omega)^\ell} \le \varepsilon \|u\|_{W_2^m(\Omega)^\ell} + c\,\varepsilon^{(j-m)/j} |\lambda|^m \|u\|_{L_2(\Omega)^\ell},
$$

which follows from the interpolation inequality

$$
\|u\|_{W_2^k(\Omega)} \le c \|u\|^{(m-k)/m}_{L_2(\Omega)} \|u\|^{k/m}_{W_2^m(\Omega)},
$$

imply

$$
|\lambda|^{2m} \|u\|^2_{L_2(\Omega)^\ell} + \|u\|^2_{W_2^m(\Omega)^\ell} \le c \left(a_0(u, u; \lambda) + \sum_{j=0}^{m-1} |\lambda|^{2j} \|u\|^2_{W_2^{m-1-j}(\Omega)^\ell} \right).
$$

Hence for large $|\lambda|$ we get the desired inequality. ∎

Using the preceding lemma, we establish the following properties of the quadratic form $a(u, u; \lambda)$.

LEMMA 12.1.3. *Suppose* Condition 1 *is satisfied. Then the following assertions are true.*

1) *There exist a positive constant c_0 and a real-valued function $c_1 = c_1(\lambda)$ such that*

(12.1.10) $|a(u, u; \lambda)| \ge c_0 \|u\|^2_{W_2^m(\Omega)^\ell} - c_1(\lambda) \|u\|^2_{L_2(\Omega)^\ell}$

for all $u \in W_2^m(\Omega)^\ell$, $\lambda \in \mathbb{C}$.

2) *The quadratic form $a(u, u; m - n/2 + it)$ is real-valued for $t \in \mathbb{R}$. Furthermore,*

$$
a(u, u, m - n/2 + it) > 0
$$

for large $|t|$, $t \in \mathbb{R}$, $u \in W_2^m(\Omega)^\ell \backslash \{0\}$.

Proof: Let $u \in W_2^m(\Omega)^\ell$, $U(x) = r^\lambda u(\omega)$, $\operatorname{Re} \lambda = m - n/2$. Then, by (12.1.5), we have

$$
a(u, u; \lambda) = \frac{1}{2 |\log \varepsilon|} \int_{\substack{K \\ \varepsilon < |x| < 1/\varepsilon}} \sum_{|\alpha| = |\beta| = m} \left(A_{\alpha, \beta} \partial_x^\beta U, \partial_x^\alpha U \right)_{\mathbb{C}^\ell} dx.
$$

Since $A_{\alpha, \beta} = A^*_{\beta, \alpha}$, we conclude that $a(u, u; \lambda)$ is real for $\operatorname{Re} \lambda = m - n/2$.

We denote by $\{\eta_j\}$ a finite collection of smooth real-valued functions on S^{n-1} such that the functions η_j^2 form a partition of unity subordinate to a sufficiently fine open covering of the sphere S^{n-1}. Furthermore, let ζ be a smooth real function of the variable r with support in a neighborhood of the point $r = 1$ such that $\zeta(1) > 0$. It can be easily shown that the quantities

$$
\left| b(\zeta U, \zeta U) - a(u, u; \lambda) \int_0^\infty (\zeta(r))^2 r^{-1} dr \right| \quad \text{and} \quad \left| b(\zeta U, \zeta U) - \sum_j b(\zeta \eta_j U, \zeta \eta_j U) \right|
$$

do not exceed the expression

$$c \sum_{\substack{\mu+\nu\leq 2m-1 \\ \mu,\nu\leq m}} |\lambda|^{2m-1-\mu-\nu} \|u\|_{W_2^\mu(\Omega)^\ell} \|u\|_{W_2^\nu(\Omega)^\ell}$$

if $\operatorname{Re}\lambda = m - n/2$. Consequently,

$$(12.1.11) \qquad \left| \sum_j b(\zeta\eta_j U, \zeta\eta_j U) - a(u,u;\lambda) \int_0^\infty (\zeta(r))^2 \, r^{-1} \, dr \right|$$

$$\leq c \sum_{\substack{\mu+\nu\leq 2m-1 \\ \mu,\nu\leq m}} |\lambda|^{2m-1-\mu-\nu} \|u\|_{W_2^\mu(\Omega)^\ell} \|u\|_{W_2^\nu(\Omega)^\ell}$$

with a certain positive constant c. Analogously,

$$(12.1.12) \qquad \left| \sum_j b_0(\zeta\eta_j U, \zeta\eta_j U) - a_0(u,u;\lambda) \int_0^\infty (\zeta(r))^2 \, r^{-1} \, dr \right|$$

$$\leq c \sum_{\substack{\mu+\nu\leq 2m-1 \\ \mu,\nu\leq m}} |\lambda|^{2m-1-\mu-\nu} \|u\|_{W_2^\mu(\Omega)^\ell} \|u\|_{W_2^\nu(\Omega)^\ell} \, .$$

Using (12.1.11), (12.1.12), and the inequality

$$b(\zeta\eta_j U, \zeta\eta_j U) \geq c\, b_0(\zeta\eta_j U, \zeta\eta_j U)\,,$$

which follows from Condition 1, we get the estimate

$$a(u,u;\lambda) \geq c_1 \, a_0(u,u;\lambda) - c_2 \sum_{\substack{\mu+\nu\leq 2m-1 \\ \mu,\nu\leq m}} |\lambda|^{2m-1-\mu-\nu} \|u\|_{W_2^\mu(\Omega)^\ell} \|u\|_{W_2^\nu(\Omega)^\ell} \, .$$

From this and from Lemma 12.1.2 we conclude that

$$(12.1.13) \qquad a(u,u;\lambda) \geq c \left(\|u\|_{W_2^m(\Omega)^\ell}^2 + |\lambda|^{2m} \|u\|_{L_2(\Omega)^\ell}^2 \right)$$

for $u \in W_2^m(\Omega)^\ell$ and for sufficiently large $|\lambda|$, $\operatorname{Re}\lambda = m - n/2$. This proves the second assertion of the lemma. Using the fact that the order of the differential operator $Q_\beta(\omega, \partial_\omega, \lambda) - Q_\beta(\omega, \partial_\omega, \mu)$ is less than m for $|\beta| = m$, we derive the first assertion from (12.1.13). ∎

As a consequence of Theorems 1.2.1, 1.2.2 and Lemma 12.1.3, the following results hold.

THEOREM 12.1.1. *Let* Condition 1 *be satisfied. Then the following assertions are valid.*

1) *The operator* (12.1.6) *is Fredholm for all* $\lambda \in \mathbb{C}$.

2) *The spectrum of the pencil* $\mathfrak{A}$ *consists of isolated eigenvalues with finite algebraic multiplicities.*

3) *For all* $u, v \in W_2^m(\Omega)^\ell$ *there is the equality*

$$a(u,v;\lambda) = \overline{a(v,u;2m-n-\overline{\lambda})}.$$

Therefore, we have

$$\mathfrak{A}(\lambda)^* = \mathfrak{A}(2m-n-\overline{\lambda}).$$

4) *If λ_0 is an eigenvalue of the pencil $\mathfrak{A}$, then $2m - n - \overline{\lambda}_0$ is also an eigenvalue. The geometric, partial and algebraic multiplicities of these eigenvalues coincide.*

12.2. The energy line

12.2.1. Eigenvalues on the line $\mathrm{Re}\,\lambda = m - n/2$. We denote the space
of homogeneous scalar polynomials of degree k in the variables $x_1, \ldots, x_n$ by $\Pi_k^{(n)}$. The dimension of this space is equal to $\binom{n+k-1}{n-1}$.

THEOREM 12.2.1. *Suppose that* Condition 2 *is satisfied.*

1) *If $2m < n$ or $2m \geq n$ and n is odd, then the operator $\mathfrak{A}(\lambda)$ is positive definite for $\mathrm{Re}\,\lambda = m - n/2$. Consequently, in these cases the line $\mathrm{Re}\,\lambda = m - n/2$ does not contain eigenvalues of the pencil $\mathfrak{A}$.*

2) *If $2m \geq n$ and n is even, then the operator $\mathfrak{A}(\lambda)$ is positive definite for $\mathrm{Re}\,\lambda = m - n/2$, $\lambda \neq m - n/2$. The operator $\mathfrak{A}(m - n/2)$ is nonnegative and the number $\lambda_0 = m - n/2$ is the only eigenvalue of the pencil $\mathfrak{A}$ on the line $\mathrm{Re}\,\lambda = m - n/2$. The corresponding eigenfunctions are the restrictions of the homogeneous vector-polynomials of degree $m - n/2$ to the domain Ω on the sphere. For every eigenfunction there exists exactly one generalized eigenfunction.*

Proof: Let ζ be a positive real-valued function on the interval $(0, \infty)$, $\zeta = 1$ on $[1, 2]$, $\zeta = 0$ outside the interval $[1/2, 4]$. Substituting the function

$$U(x) = \varepsilon^{-1/2} r^\lambda \zeta(\varepsilon \log r) u(\omega)$$

into (12.0.3) and passing to the limit as $\varepsilon \to 0$, we obtain

$$(12.2.1) \qquad a(u, u; \lambda) \geq c\, a_0(u, u; \lambda) \geq 0 \qquad \text{for } \mathrm{Re}\,\lambda = m - n/2,$$

where a_0 is the sesquilinear form (12.1.9). In the same way, by means of the estimate $b(U, U) \leq C\, b_0(U, U)$, where $b_0(\cdot, \cdot)$ is the sesquilinear form (12.1.8), we get the inequality

$$(12.2.2) \qquad a(u, u; \lambda) \leq C\, a_0(u, u; \lambda).$$

The quadratic form $a_0(u, u; \lambda)$ is given by the right-hand side of (12.1.9), where $U = V = r^\lambda u$, if $\mathrm{Re}\,\lambda = m - n/2$. If the vector-valued function U is not a polynomial, then this expression is positive. This proves that the line $\mathrm{Re}\,\lambda = m - n/2$ does not contain eigenvalues if $m - n/2 \notin \{0, 1, \ldots\}$ and that in the case of nonnegative integer $m - n/2$ this number is the only eigenvalue of the pencil $\mathfrak{A}$ with real part $m - n/2$. Furthermore, the assertion of the theorem concerning the eigenfunctions holds.

Let $m - n/2$ be a nonnegative integer and let u_0 be the trace on Ω of a homogeneous vector-polynomial of degree k. By Lemma 1.2.2, there exists a generalized eigenfunction associated to the eigenfunction u_0. We prove the absence of second generalized eigenfunctions.

Suppose that u_0, u_1, u_2 is a Jordan chain corresponding to the eigenvalue $\lambda_0 = m - n/2$. Then

$$(12.2.3) \qquad a(u_0, v; \lambda_0) = 0,$$

$$(12.2.4) \qquad a(u_1, v; \lambda_0) + a^{(1)}(u_0, v; \lambda_0) = 0,$$

$$(12.2.5) \qquad a(u_2, v; \lambda_0) + a^{(1)}(u_1, v; \lambda_0) + \frac{1}{2} a^{(2)}(u_0, v, \lambda_0) = 0$$

for all $v \in W_2^m(\Omega)^\ell$, where $a^{(j)}(u,v;\lambda) = d^k a(u,v;\lambda)/d\lambda^k$. By what has been shown above, the function $r^{m-n/2} u_0(\omega)$ is a polynomial in x. Consequently, we have $\partial_x^\alpha (r^{m-n/2} u_0(\omega)) = 0$ for $|\alpha| = m$ or, what is the same,

$$(12.2.6) \qquad Q_\alpha(\omega, \partial_\omega, \lambda_0) u_0 = 0,$$

where Q_α is the differential operator (12.1.3). We set $Q_\alpha^{(k)}(\lambda) = d^k Q_\alpha(\omega, \partial_\omega, \lambda)/d\lambda^k$. By (12.1.4), (12.2.4), and (12.2.6), we have

$$(12.2.7) \quad 0 = a(u_1, u_1; \lambda_0) + a^{(1)}(u_0, u_1; \lambda_0)$$

$$= \sum_{|\alpha|=|\beta|=m} \int_\Omega \left(A_{\alpha,\beta} (Q_\beta(\lambda_0) u_1 + Q_\beta^{(1)}(\lambda_0) u_0) , Q_\alpha(\lambda_0) u_1 \right)_{\mathbb{C}^\ell} d\omega.$$

Furthermore, from (12.2.5) and (12.2.6) it follows that

$$(12.2.8) \; 0 = a(u_2, u_0; \lambda_0) + a^{(1)}(u_1, u_0; \lambda_0) + \frac{1}{2} a^{(2)}(u_0, u_0, \lambda_0)$$

$$= - \sum_{|\alpha|=|\beta|=m} \int_\Omega \left(A_{\alpha,\beta}(Q_\beta(\lambda_0) u_1 + Q_\beta^{(1)}(\lambda_0) u_0) , Q_\alpha^{(1)}(\lambda_0) u_0 \right)_{\mathbb{C}^\ell} d\omega.$$

Subtracting (12.2.8) from (12.2.7), we get

$$\sum_{|\alpha|=|\beta|=m} \int_\Omega \left(A_{\alpha,\beta}(Q_\beta(\lambda_0) u_1 + Q_\beta^{(1)}(\lambda_0) u_0) , Q_\alpha(\lambda_0) u_1 + Q_\alpha^{(1)}(\lambda_0) u_0 \right)_{\mathbb{C}^\ell} d\omega = 0.$$

We substitute the function

$$U(x) = \varepsilon^{-1/2} \zeta(\varepsilon \log r) \, r^{\lambda_0} \left(u_1(\omega) + u_0(\omega) \log r \right)$$

into (12.0.3). Passing to the limit as $\varepsilon \to 0$, analogously to (12.2.1), we obtain

$$0 = \sum_{|\alpha|=|\beta|=m} \int_\Omega \left(A_{\alpha,\beta}(Q_\beta(\lambda_0) u_1 + Q_\beta^{(1)}(\lambda_0) u_0) , Q_\alpha(\lambda_0) u_1 + Q_\alpha^{(1)}(\lambda_0) u_0 \right)_{\mathbb{C}^\ell} d\omega$$

$$\geq c \sum_{|\alpha|=m} \int_\Omega \left| Q_\alpha(\lambda_0) u_1 + Q_\alpha^{(1)}(\lambda_0) u_0 \right|_{\mathbb{C}^\ell}^2 d\omega.$$

From this we conclude that $Q_\alpha(\lambda_0) u_1 + Q_\alpha^{(1)}(\lambda_0) u_0 = 0$ or, equivalently,

$$\partial_x^\alpha \left(r^{\lambda_0} u_1(\omega) + r^{\lambda_0} u_0(\omega) \log r \right) = 0$$

for $|\alpha| = m$. Hence $r^{\lambda_0} (u_1(\omega) + u_0(\omega) \log r)$ is a polynomial in x. This is only possible if $u_0 = 0$ what contradicts our assumption on u_0. The proof is complete. ∎

12.2.2. An example. The following example shows that Condition 2 in the formulation of Theorem 12.2.1 cannot be replaced by the weaker Condition 1.

Let $\sigma < 1$ and $n \geq 3$. Suppose that the boundary $\partial\Omega$ is a smooth manifold. We consider the sesquilinear form

$$b_\sigma(U, V) \overset{def}{=} \int_\mathcal{K} \sum_{j,k=1}^n \frac{\partial^2 U}{\partial x_j \, \partial x_k} \cdot \frac{\partial^2 \overline{V}}{\partial x_j \, \partial x_k} \, dx - \sigma \int_\mathcal{K} \Delta U \cdot \Delta \overline{V} \, dx$$

which generates one of the Neumann problems for the biharmonic operator Δ^2. The corresponding parameter-depending sesquilinear form on $W_2^m(\Omega) \times W_2^m(\Omega)$ is denoted by $a_\sigma(\cdot, \cdot; \lambda)$.

We verify the ellipticity of the corresponding boundary value problem. Since the form $b_\sigma(\cdot,\cdot)$ is invariant with respect to rotations, we may restrict ourselves to the case $\mathcal{K} = \mathbb{R}^n_+$. Then the boundary conditions induced by the sesquilinear form b_σ on the boundary $\partial\mathcal{K}$ have the form

$$(\partial^2_{x_n} - \sigma\Delta)\,U = \big(\partial^2_{x_n} + (\sigma - 2)\Delta\big)\,\partial_{x_n} U = 0.$$

Standard calculations show that this problem is elliptic if and only if $\sigma \neq 3/4$.

Clearly, Conditions 1 and 2 are equivalent in the case of a half-space, since inequality (12.0.3) is invariant with respect to translation and dilation. Furthermore, it can easily verified that the inequality $\sigma < 3/4$ is necessary and sufficient for the local coerciveness of the form $b_\sigma(\cdot,\cdot)$ defined on $W^2_2(\mathbb{R}^n_+) \times W^2_2(\mathbb{R}^n_+)$. Consequently, Condition 1 for the form b_σ defined on $W^2_2(\mathcal{K})$ is also equivalent to the inequality $\sigma < 3/4$.

We prove that for $\sigma \in (1/n, 3/4)$ there exists a cone $\mathcal{K}$ for which the form $b_\sigma(\cdot,\cdot)$ has the following property: the pencil $\mathfrak{A}_\sigma(\lambda)$ corresponding to the form b_σ has eigenvalues with nonzero imaginary part on the line $\operatorname{Re}\lambda = 2 - n/2$. In particular, this implies that the condition in the formulation of Theorem 12.2.1 can not be replaced by Condition 1. Let $\mathcal{K} = \{x \in \mathbb{R}^n : x_n > |x'|/\varepsilon\}$, where ε is a small positive number. We introduce the coordinates $\omega = (\omega_1, \ldots, \omega_n) = x/|x|$ and set

$$u_0(\omega) = \omega_n^{-n/2+it_0}\,(\omega_n^2 + c\,t_0^2\,|\omega'|^2), \qquad t_0 \in \mathbb{R},$$

where $\omega' = (\omega_1, \ldots, \omega_{n-1})$. Then we have

(12.2.9)
$$a_\sigma(u_0, u_0; 2 - n/2 + it_0)$$
$$= \frac{1}{2\log 2} \int\limits_{\substack{\mathcal{K} \\ 1/2 < |x| < 2}} \Big(\sum_{j,k=1}^n \Big|\frac{\partial^2 U}{\partial x_j\,\partial x_k}\Big|^2 - \sigma\,|\Delta U|^2 \Big)\,dx,$$

where

$$U(x) = r^{2-n/2+it_0}\,u_0(\omega) = x_n^{-n/2+it_0}\,(x_n^2 + c\,t_0^2\,|x'|^2).$$

If $|t_0| \geq c(n)$, where $c(n)$ is a sufficiently large positive constant depending only on n, then the term in the integral on the right of (12.2.9) is equal to

$$\Big(\big(4c^2(n-1) + 1 - \sigma(2c(n-1) - 1)^2\big)\,t_0^4 + O(|t_0|^3) + O(\varepsilon^2 t_0^6) \Big)\,x_n^{-n}.$$

Since the infimum with respect to c of the expression in the big brackets is equal to $(1 - \sigma n)/(1 - \sigma(n-1))$ for $\sigma \in (1/n, 1/(n-1))$ and to $-\infty$ for $\sigma \geq 1/(n-1)$, one can choose c, t_0, ε, depending only on σ and n, in such a way that the integral on the right in (12.2.9) is negative. As it was shown in the proof of Lemma 1.2.1, the inequality

$$a_\sigma(u_0, u_0, 2 - n/2 + it_0) < 0$$

implies the existence of an eigenvalue of the pencil $\mathfrak{A}_\sigma$ which has the form $\lambda = 2 - n/2 + it$, where $t \in \mathbb{R}$, $|t| > |t_0|$.

12.3. The energy strip

Additionally to the assumptions of Section 12.1, we suppose now that the cone $\mathcal{K}$ is given by the inequality $x_n > \phi(x')$, $x' \in \mathbb{R}^{n-1}$, in some Cartesian coordinate system, where the function ϕ is positively homogeneous of degree 1 and satisfies the Lipschitz condition

(12.3.1) $$|\phi(x') - \phi(y')| \leq M\,|x' - y'|$$

for all $x', y' \in \mathbb{R}^{n-1}$. We describe the spectrum of the pencil $\mathfrak{A}$ in the strip $|\operatorname{Re}\lambda - m + n/2| \leq 1/2$.

12.3.1. The operators T_δ and S_δ. We introduce the following operator T_δ defined on the set of all functions $\psi \in L_1(S^{n-1})$:

$$(T_\delta\psi)(x) = \int\limits_{S^{n-1}} f\left(\frac{\omega - x}{|\omega - x|}\right) |\omega - x|^{\delta - n}\,\psi(\omega)\,d\omega,$$

where $\delta \in (0,1)$, $f \in C^\infty(S^{n-1})$. The function $T_\delta\psi$ is defined almost everywhere on $\mathbb{R}^n$, or, more precisely, on the set $\mathbb{R}^n \backslash S^{n-1}$, where it is smooth. For large values of $|x|$ there is the inequality

$$|D_x^\alpha (T_\delta\psi)(x)| \leq c_\alpha\,|x|^{\delta - n - |\alpha|} \int\limits_{S^{n-1}} |\psi(\omega)|\,d\omega,$$

where $|\alpha|$ is an arbitrary multi-index and $c_\alpha =$ const.

The following lemma gives an estimate for the W_2^l-norm of the function $T_\delta\psi = (T_\delta\psi)(x)$ with respect to the coordinates $x/|x|$ on the sphere.

LEMMA 12.3.1. *If l is a nonnegative integer and $\psi \in W_2^l(S^{n-1})$, then*

$$\|(T_\delta\psi)(|x|\cdot)\|_{W_2^l(S^{n-1})} \leq c\,\big|1 - |x|\,\big|^{\delta - 1}\,\|\psi\|_{W_2^l(S^{n-1})}$$

for small values of $|1 - |x||$.

Proof: We assume that the support of the function ψ is contained in a small coordinate neighborhood $\mathcal{U}$ and estimate the norm of $(T_\delta\psi)(|x|\cdot)$ in $W_2^l(\mathcal{U})$. Let z', y' be local coordinates of the points $x/|x|$ and ω in $\mathcal{U}$ and let $z_n = |x| - 1$. In these coordinates we can write $T_\delta\psi$ in the form

$$(T_\delta\psi)(z) = \int\limits_{\mathbb{R}^{n-1}} F(y' - z', z_n, y')\,|y' - z'|^{\delta - n}\,\psi(y')\,dy',$$

where F admits the estimate

$$|\partial_{y'}^\alpha F(z, y')| \leq \text{const}, \qquad |\alpha| \leq l,$$

on any compact subset of $\mathbb{R}^n \times \mathbb{R}^{n-1}$. Clearly,

$$
\begin{aligned}
|\partial_{z'}^\alpha (T_\delta\psi)(z)| &\leq c\sum_{\beta \leq \alpha} \int\limits_{\mathbb{R}^{n-1}} |\partial_{z'}^\beta F(y', z_n, z' + y')|\,|\partial_{z'}^{\alpha - \beta}\psi(z' + y')|\,|y'|^{\delta - n}\,dy' \\
&\leq c\sum_{\beta \leq \alpha} \int\limits_{\mathbb{R}^{n-1}} |y' - z'|^{\delta - n}\,|\partial_{y'}^\beta\psi(y')|\,dy'.
\end{aligned}
$$

This implies the inequality

$$\|(T_\delta\psi)(\cdot, z_n)\|_{W_2^l(Q)} \leq c\,|z_n|^{\delta - 1}\,\|\psi\|_{W_2^l(\mathbb{R}^{n-1})}\,,$$

where Q is an arbitrary compact set in $\mathbb{R}^{n-1}$. This proves the lemma for functions ψ with $\operatorname{supp}\psi \subset \mathcal{U}$. If the function ψ has arbitrary support, then the assertion holds by means of a sufficiently fine partition of unity on S^{n-1}. ∎

Let $\mathcal{K}_0$ be the cone $\{x \in \mathbb{R}^n : x_n > 2M |x'|\}$, where M is the Lipschitz constant in (12.3.1). Then the relation $x \in \mathcal{K}$ implies $x + \mathcal{K}_0 \in \mathcal{K}$. Furthermore, let δ be a real number in the interval $(0,1)$ and f a smooth nonnegative function on the unit sphere with support in $\mathcal{K}_0 \cap S^{n-1}$ which does not vanish identically. We consider the integral operator with weak singularity

$$(S_\delta U)(x) = \int\limits_{\mathbb{R}^n} f\left(\frac{y-x}{|y-x|}\right) |y-x|^{\delta-n} U(y) \, dy,$$

i.e., $S_\delta U$ is the convolution of U with the function $F_\delta(x) = f(\omega) \, r^{\delta-n}$. By the assumption on the support of f, it suffices to integrate only on $\mathcal{K}$ in order to obtain the value of $S_\delta U$ at a point $x \in \mathcal{K}$. Therefore, we have

$$(S_\delta U)(x) = \int\limits_{\mathcal{K}} f\left(\frac{y-x}{|y-x|}\right) |y-x|^{\delta-n} U(y) \, dy, \qquad x \in \mathcal{K}.$$

LEMMA 12.3.2. *There exists a positive constant c such that*

(12.3.2) $$\operatorname{Re} \int\limits_{\mathbb{R}^n} S_\delta U \cdot \overline{U} \, dx \geq c \int\limits_{\mathbb{R}^n} |\xi|^{-\delta} |\hat{U}|^2 \, d\xi,$$

where

$$\hat{U}(\xi) = \int\limits_{\mathbb{R}^n} e^{ix\cdot\xi} \, U(x) \, dx$$

denotes the Fourier transform of U.

Proof: Let (ρ, θ) be spherical coordinates of the point ξ. For the Fourier transform of the function $F_\delta(x) = f(\omega) \, r^{\delta-n}$ the following formula (see Gel'fand and Shilov [66]) holds.

$$\hat{F}_\delta(\xi) = \Gamma(\delta) \, \rho^{-\delta} \int\limits_{S^{n-1}} \left(e^{i\delta\pi/2} (\theta \cdot \omega)_+^{-\delta} + e^{-i\delta\pi/2} (\theta \cdot \omega)_-^{-\delta} \right) f(\omega) \, d\omega.$$

Consequently,

$$\operatorname{Re} \hat{F}_\delta(\xi) = \Gamma(\delta) \, \cos(\delta\pi/2) \, \rho^{-\delta} \int\limits_{S^{n-1}} |\theta \cdot \omega|^{-\delta} \, f(\omega) \, d\omega$$

and, therefore, $\operatorname{Re} \hat{F}_\delta > 0$. Using the Parseval equality, we get (12.3.2). ∎

12.3.2. Absence of eigenvalues in $\{\lambda : 0 < |\operatorname{Re}\lambda - m + n/2| < 1/2\}$. The following lemma will be used in the proof of Theorem 12.3.1.

LEMMA 12.3.3. *Let $V = r^{-n/2-\delta/2+it} \, v(\omega)$, $W = r^{-n/2-\delta/2+it} \, w(\omega)$, where $\delta \in (0,1)$, $t \in \mathbb{R}$, and v, w are functions from $L_2(\Omega)$. Furthermore, let χ_ε be the characteristic function of the spherical shell $\{x : \varepsilon < |x| < 1/\varepsilon\}$, $0 < \varepsilon < 1$. Then the integral*

$$I_\varepsilon = \int\limits_{\mathcal{K}} (1 - \chi_\varepsilon) \, V \, S_\delta \, \chi_\varepsilon \overline{W} \, dx$$

is bounded as $\varepsilon \to 0$.

Proof: Obviously,

$$(12.3.3) \qquad I_\varepsilon = \int\limits_{\substack{\mathcal{K} \\ |x| \le \varepsilon}} V\, S_\delta \chi_\varepsilon \overline{W}\, dx + \int\limits_{\substack{\mathcal{K} \\ |x| \ge 1/\varepsilon}} V\, S_\delta \chi_\varepsilon \overline{W}\, dx.$$

We readily check the estimates

$$|(S_\delta \chi_\varepsilon \overline{W})(x)| \le \begin{cases} c\,\varepsilon^{-n/2+\delta/2} & \text{for } |x| \le \varepsilon, \\[2mm] c\,\varepsilon^{-n/2+\delta/2}\, |x|^{\delta-n} & \text{for } |x| \ge 1/\varepsilon. \end{cases}$$

This and (12.3.3) imply the uniform boundedness of I_ε for small ε. ∎

THEOREM 12.3.1. *Let the cone $\mathcal{K}$ be given by the inequality $x_n > \phi(x')$, where ϕ is positively homogeneous of degree 1 and satisfies the Lipschitz condition (12.3.1). Furthermore, let* Condition 3 *be satisfied. Then there are no points of the spectrum of the pencil $\mathfrak{A}$ in the set $\{\lambda \in \mathbb{C} : 0 < |\mathrm{Re}\,\lambda - m + n/2| < 1/2\}$.*

Proof: By Theorem 12.1.1, it suffices to establish the absence of eigenvalues in the strip $m - (n+1)/2 < \mathrm{Re}\,\lambda < m - n/2$. Suppose there exists a solution

$$U(x) = r^{m-n/2-\delta/2+it}\, u_0(\omega), \qquad t, \delta \in \mathbb{R},\ 0 < \delta < 1,$$

of the Neumann problem (12.1.2). Then

$$\sum_{|\alpha|=|\beta|=m} \int\limits_{\mathcal{K}} \Big(A_{\alpha,\beta}\, \partial_x^\beta U,\ \partial_x^\alpha (S_\delta \chi_\varepsilon U) \Big)_{\mathbb{C}^\ell} dx = 0.$$

Since the convolution and the differentiation operators commute and $\chi_\varepsilon^2 = \chi_\varepsilon$, we obtain

$$(12.3.4) \qquad 0 = \sum_{|\alpha|=|\beta|=m} \int\limits_{\mathcal{K}} \Big(A_{\alpha,\beta}\, \partial_x^\beta U,\ S_\delta \partial_x^\alpha (\chi_\varepsilon U) \Big)_{\mathbb{C}^\ell} dx$$

$$= \sum_{|\alpha|=|\beta|=m} \int\limits_{\mathcal{K}} \Big(A_{\alpha,\beta}\, \chi_\varepsilon \partial_x^\beta U,\ S_\delta \chi_\varepsilon \partial_x^\alpha U \Big)_{\mathbb{C}^\ell} dx$$

$$+ \sum_{|\alpha|=|\beta|=m} \int\limits_{\mathcal{K}} \Big(A_{\alpha,\beta}\, \partial_x^\beta U,\ S_\delta\, [\partial_x^\alpha, \chi_\varepsilon] U \Big)_{\mathbb{C}^\ell} dx$$

$$+ \sum_{|\alpha|=|\beta|=m} \int\limits_{\mathcal{K}} \Big(A_{\alpha,\beta}\, \partial_x^\beta U,\ [S_\delta, \chi_\varepsilon]\, \chi_\varepsilon \partial_x^\alpha U \Big)_{\mathbb{C}^\ell} dx,$$

where $[A, B] = AB - BA$. Since $[S_\delta, \chi_\varepsilon]\, \chi_\varepsilon \partial_x^\alpha U = (1 - \chi_\varepsilon)\, S_\delta\, (\chi_\varepsilon \partial_x^\alpha U)$, it follows from Lemma 12.3.3 that the last sum on the right in (12.3.4) is uniformly bounded.

We consider the second integral on the right in (12.3.4). Let e_j be the multi-index with all components equal to zero except for the j-th which is equal to unity. Then

$$(12.3.5) \quad S_\delta [\partial_x^\alpha, \chi_\varepsilon] U = \sum_{\gamma + e_j \le \alpha} c_{j,\gamma}^{(\alpha)}\, \partial_x^\gamma S_\delta \partial_{x_j} \chi_\varepsilon \partial_x^{\alpha - \gamma - e_j} U$$

$$= \sum_{\gamma + e_j \le \alpha} c_{j,\gamma}^{(\alpha)}\, \partial_x^\gamma S_\delta \Big(\omega_j \big(\delta(r - \varepsilon) - \delta(r - \varepsilon^{-1}) \big) \partial_x^{\alpha - \gamma - e_j} U \Big),$$

We introduce the functional (for the time being formally)

$$G_a = \int_{\mathcal{K}} V \cdot \partial_x^\gamma S_\delta \big(\delta(r - a) \overline{W} \big) \, dx, \qquad a > 0,$$

where $V(x) = r^{-n/2 - \delta/2 + it} \, v(\omega)$, $W(x) = r^{m-s-n/2-\delta/2+it} \, w(\omega)$, $v \in L_2(\Omega)$, $w \in W_2^{m-s}(\Omega)$, $|\gamma| + s = m - 1$. Using the homogeneity of the functions V and W, we obtain that $G_a = G_1$. This and (12.3.5) imply the equality

$$(12.3.6) \qquad \sum_{|\alpha|=|\beta|=m} \int_{\mathcal{K}} \Big(A_{\alpha,\beta} \, \partial_x^\beta U \, , \, S_\delta [\partial_x^\alpha, \chi_\varepsilon] U \Big)_{\mathbb{C}^\ell} \, dx = 0.$$

It remains to give a meaning to the functional G_1. Let $\zeta = \zeta(r)$ be a smooth function equal to 1 in a neighborhood of the point $r = 1$ and to zero for $|r - 1| > 1/2$. By Lemma 12.3.1, it suffices to regularize the functional

$$(12.3.7) \qquad \int_{\mathcal{K}} V \, \partial_x^\gamma \Big(\zeta S_\delta \big(\delta(r - 1) \, \overline{W} \big) \Big) \, dx = \int_{\mathcal{K}} V \, \partial_x^\gamma \big(\zeta T_\delta \overline{w} \big) \, dx.$$

We represent the operator ∂_x^γ in the form

$$\partial_x^\gamma = \sum_{k=0}^{|\gamma|} r^{-|\gamma|} \, (r \partial_r)^k \, q_k(\omega, \partial_\omega),$$

where q_k are differential operators of order $|\gamma| - k$ with smooth coefficients on the sphere. Then by the functional (12.3.7) we mean the sum

$$\sum_{k=0}^{|\gamma|} \Big(\frac{n}{2} - |\gamma| - \frac{\delta}{2} + it \Big)^k \int_0^\infty \int_\Omega r^{n/2 - 1 - |\gamma| - \delta/2 + it} \, v(\omega) \, q_k(\omega, \partial_\omega) \big(\zeta T_\delta \overline{w} \big) \, d\omega \, dr,$$

which is obtained from (12.3.7) by integration by parts. The integrals in this sum are absolutely convergent by Lemma 12.3.1. Thus, the relation (12.3.6) is completely proved.

From (12.3.4), (12.3.6) and from the previously proved boundedness of the last sum in the right-hand side of (12.3.4) it follows that the integral

$$J_\varepsilon = \sum_{|\alpha|=|\beta|=m} \int_{\mathcal{K}} \Big(A_{\alpha,\beta} \chi_\varepsilon \partial_x^\beta U \, , \, S_\delta \chi_\varepsilon \partial_x^\alpha U \Big)_{\mathbb{C}^\ell} \, dx$$

is uniformly bounded. Applying Parseval's equality and (12.3.2), we find that

$$\operatorname{Re} J_\varepsilon \geq c \int_{\mathbb{R}^n} |\xi|^{-\delta} \sum_{|\alpha|=|\beta|=m} \big(A_{\alpha,\beta} \hat{U}_{\varepsilon,\beta} \, , \, \hat{U}_{\varepsilon,\alpha} \big)_{\mathbb{C}^\ell} \, d\xi,$$

where $U_{\varepsilon,\alpha}$ is the zero extension of the function $\chi_\varepsilon \partial_x^\alpha U$ given on $\mathcal{K}$. By Condition 3, we obtain

$$(12.3.8) \qquad \operatorname{Re} J_\varepsilon \geq c \sum_{|\alpha|=m} \int_{\mathbb{R}^n} |\xi|^{-\delta} \, |\hat{U}_{\varepsilon,\alpha}(\xi)|^2 \, d\xi.$$

We introduce the integral operator

$$(R_{\delta/2} G)(x) = \int_{\mathbb{R}^n} G(y) \, |x - y|^{-n + \delta/2} \, dy.$$

Then from (12.3.8) we get

$$\operatorname{Re} J_\varepsilon \geq c \sum_{|\alpha|=m} \int_{\mathbb{R}^n} |(R_{\delta/2} U_{\varepsilon,\alpha})(x)|^2 \, dx.$$

Let τ be a positive number and

$$(12.3.9) \qquad J_\varepsilon^{(\tau)} = \int_{\mathbb{R}^n} \chi_\tau \, |(R_{\delta/2} U_{\varepsilon,\alpha})(x)|^2.$$

It is easily seen that $J_\varepsilon^{(\tau)}$ tends to the limit

$$J^{(\tau)} = \int_{\mathbb{R}^n} \chi_\tau \, |(R_{\delta/2} U_{0,\alpha})(x)|^2 \, dx$$

as $\varepsilon \to 0$, where $U_{0,\alpha}$ is the zero extension to $\mathbb{R}^n$ of the function $\partial_x^\alpha U$. Since $U_{0,\alpha}$ is positively homogeneous of degree $-n/2 - \delta/2 + it$, we have

$$|(R_{\delta/2} U_{0,\alpha})(x)|^2 = |x|^{-n} \, \psi_\alpha(\omega).$$

If $\partial_x^\alpha U \neq 0$ for a certain multi-index α, then $\psi_\alpha \neq 0$ and the integral $J^{(\tau)}$ tends to infinity as $\tau \to 0$. Consequently, if some of the functions $\partial_x^\alpha U$ with $|\alpha| = m$ are not zero, then for any positive c one can choose ε in such a way that $\operatorname{Re} J_\varepsilon > c$. The last inequality contradicts the uniform boundedness of J_ε with respect to ε. Thus, $\partial_x^\alpha U = 0$ for all α with $|\alpha| = m$. Hence U is a vector-valued polynomial. Since U is positively homogeneous of degree $m - n/2 - \delta/2 + it$, where $\delta \in (0,1)$, $t \in \mathbb{R}$, we have $U = 0$. This proves the theorem. ∎

We prove now that Condition 3 in Theorem 12.3.1 can be replaced by Condition 1 and estimate (12.0.4). Then together with Theorem 12.2.1 the following result holds.

THEOREM 12.3.2. *Let the cone $\mathcal{K}$ be given by the inequality $x_n > \phi(x')$, where ϕ is positively homogeneous of degree 1 and satisfies the Lipschitz condition (12.3.1). Furthermore, let* Condition 1 *and estimate (12.0.4) be satisfied.*

1) If $2m < n$ or $2m \geq n$ and n is odd, then the strip $|\operatorname{Re} \lambda - m + n/2| < 1/2$ contains no points of the spectrum of the pencil $\mathfrak{A}$.

2) If $2m \geq n$ and n is even, then the strip $|\operatorname{Re} \lambda - m + n/2| < 1/2$ contains the single eigenvalue $\lambda_0 = m - n/2$. The vector-valued function $u = u(\omega)$ is an eigenfunction corresponding to the eigenvalue $\lambda_0 = m - n/2$ if and only if $u = p|_\Omega$, where p is a homogeneous vector-valued polynomial of degree $m - n/2$ in $\mathbb{R}^n$. To each eigenfunction there corresponds precisely one generalized eigenfunction.

Proof: We consider the sesquilinear form $\tilde{b}_\varepsilon(U, V) = b(U, V) + \varepsilon \, b_0(U, V)$, where b, b_0 are defined by (12.0.2) and (12.1.8). By $\tilde{a}_\varepsilon(\cdot, \cdot; \lambda)$ we denote the parameter-depending form generated by $b_\varepsilon(\cdot, \cdot)$:

$$\tilde{a}_\varepsilon(u, v; \lambda) = a(u, v; \lambda) + \varepsilon \, a_0(u, v; \lambda)$$

(see (12.1.4), (12.1.9)). The operator pencil $\tilde{\mathfrak{A}}_\varepsilon$ corresponding to the form $\tilde{a}_\varepsilon$ is equal to $\mathfrak{A} + \varepsilon \mathfrak{A}_0$, where $\mathfrak{A}_0$ is the operator pencil generated by the form a_0. Obviously, the coefficients of the form $\tilde{b}_\varepsilon$ satisfy Condition 3 for $\varepsilon > 0$. Therefore, the set $\{\lambda : 0 < |\operatorname{Re} \lambda - m + n/2| < 1/2\}$ is free of eigenvalues of the pencil $\tilde{\mathfrak{A}}_\varepsilon$. The statement of the theorem for $\tilde{\mathfrak{A}}_\varepsilon$, $\varepsilon > 0$, relative to the line $\operatorname{Re}\lambda = m - n/2$ follows immediately from Theorem 12.2.1.

Since Condition 1 holds for the form $\tilde{a}_\varepsilon$ uniformly with respect to $\varepsilon \geq 0$, it follows that for some σ, independent of ε, the set

$$\{\lambda \in \mathbb{C} \; : \; |\operatorname{Re}\lambda| < \sigma\,|\operatorname{Im}\lambda|, \; |\operatorname{Im}\lambda| > 1/\sigma\}$$

contains no points of the spectrum of the pencil $\tilde{\mathfrak{A}}_\varepsilon$, $\varepsilon \geq 0$. Using Theorem 1.1.4, we conclude that within the set

$$\{\lambda \in \mathbb{C} \; : \; |\operatorname{Re}\lambda - m + n/2| < 1/2\,, \; |\operatorname{Im}\lambda| \leq 1/\sigma\}$$

eigenvalues of the pencil $\tilde{\mathfrak{A}}_0 = \mathfrak{A}$ can lie only on the line $\operatorname{Re}\lambda = m - n/2$ and that the sum of the algebraic multiplicities of these eigenvalues is the same as for the pencil $\tilde{\mathfrak{A}}_\varepsilon$, $\varepsilon \geq 0$. Thus, the theorem is proved for $m < n/2$ and for $m \geq n/2$, n odd.

Let $m - n/2$ be a nonnegative integer. Then $\lambda = m - n/2$ is an eigenvalue of the pencil $\mathfrak{A}$, and the restrictions of homogeneous vector-polynomials of degree $m - n/2$ are eigenfunctions. From (12.0.4) and from the representation (12.1.5) it follows that the quadratic form $a(u, u; \lambda)$ is nonnegative for $\operatorname{Re}\lambda = m - n/2$. Consequently, by Lemma 1.2.2, to each eigenfunction there corresponds at least one generalized eigenfunction. Since the sum of the algebraic multiplicities of eigenvalues of $\mathfrak{A}$ in the strip $|\operatorname{Re}\lambda - m + n/2| < 1/2$ is the same as for the pencil $\tilde{\mathfrak{A}}_\varepsilon$, $\varepsilon > 0$, we conclude that the eigenvalue $\lambda_0 = m - n/2$ has no other eigenfunctions and generalized eigenfunctions and that there are no other eigenvalues of the pencil $\mathfrak{A}$ in the strip $|\operatorname{Re}\lambda - m + n/2| < 1/2$. The proof is complete. ∎

12.3.3. On the validity of the generalized Korn inequality. The example in Section 12.2 shows that Condition 1 does not follow from Condition 2. However, by means of the last theorem, it can be shown that Condition 1 and inequality (12.0.4) imply Condition 2. For the proof of this fact we need the following lemma.

LEMMA 12.3.4. *Let the conditions of* Theorem 12.3.2 *be satisfied and let $m - n/2$ be a nonnegative integer. Then there exist positive constants c_1 and c_2 such that*

$$(12.3.10) \qquad c_1\, a(u, u; m - n/2 + it) \leq t^2\, \|u^{(0)}\|^2_{L_2(\Omega)^\ell} + \|u^{(1)}\|^2_{W_2^m(\Omega)^\ell}$$
$$\leq c_2\, a(u, u; m - n/2 + it)$$

for all $u \in W_2^m(\Omega)^\ell$ and for small $|t|$, $t \in \mathbb{R}$, where $u^{(0)}$ is the orthogonal projection of u onto the subspace of traces on Ω of homogeneous vector-polynomials of degree $m - n/2$ and $u^{(1)} = u - u^{(0)}$.

Proof: From (12.0.4) it follows that the form $b(U, U)$ admits the representation

$$(12.3.11) \qquad b(U, U) = \int_{\mathcal{K}} \sum_{j=1}^{N} \left|P_j(\partial_x)\,U\right|^2_{\mathbb{C}^\ell}\, dx,$$

where P_j are homogeneous matrix-valued differential operators with constant coefficients. Consequently, for the corresponding form a we have

$$(12.3.12) \qquad a(u, u; \lambda) = \sum_{j=1}^{N} \int_{\Omega} \left(\mathcal{P}_j(\omega, \partial_\omega, \lambda)u\,,\, \mathcal{P}_j(\omega, \partial_\omega, 2m - n - \overline{\lambda})\,u\right)_{\mathbb{C}^\ell}\, d\omega,$$

where $\mathcal{P}_j(\omega, \partial_\omega, \lambda)u = r^{m-\lambda} P_j(\partial_x) (r^\lambda u(\omega))$. Let $a^{(\nu)}(u, v; \lambda) = d^\nu a(u, v; \lambda)/d\lambda^\nu$ and $\mathcal{P}_j^{(\nu)}(\lambda) = d^\nu \mathcal{P}_j(\omega, \partial_\omega, \lambda)/d\lambda^\nu$. Furthermore, let $\mathcal{H}_0$ denote the set of eigenvectors of the pencil $\mathfrak{A}$ corresponding to the eigenvalue $m - n/2$, i.e., the space of traces on Ω of homogeneous vector-polynomials of degree $k = m - n/2$. Since $\mathcal{P}_j(k)u = 0$ for $u \in \mathcal{H}_0$ and $j = 1, \dots, N$, we get

$$(12.3.13) \qquad -a^{(2)}(u, v; k) = 2 \sum_{j=1}^{N} \left(\mathcal{P}_j^{(1)}(k)u \,, \mathcal{P}_j^{(1)}(k)v \right)_{L_2(\Omega)^\ell}$$

for arbitrary $u, v \in \mathcal{H}_0$.

We show that the quadratic form $-a^{(2)}(u, u; k)$ is positive definite on $\mathcal{H}_0$. If $a^{(2)}(u_0, u_0; k) = 0$ for some eigenfunction $u_0 \in \mathcal{H}_0 \backslash \{0\}$, then $\mathcal{P}_j^{(1)}(k) u_0 = 0$ for $j = 1, \dots, m$ and, therefore, $a^{(2)}(u_0, v; k) = 0$ for all $v \in \mathcal{H}_0$. Since

$$\left| a^{(1)}(u_0, v; k) \right|^2 \leq \left(\sum_{j=1}^{N} \left| \left(\mathcal{P}_j^{(1)}(k)u_0 \,, \mathcal{P}_j(k)v \right)_{L_2(\Omega)^\ell} \right| \right)^2$$

$$\leq \sum_{j=1}^{N} \| \mathcal{P}_j^{(1)}(k)u_0 \|^2_{L_2(\Omega)^\ell} \sum_{j=1}^{N} \| \mathcal{P}_j(k)v \|^2_{L_2(\Omega)^\ell} = -\frac{1}{2} a^{(2)}(u_0, u_0; k) \cdot a(v, v; k),$$

we get $a^{(1)}(u_0, v; k) = 0$ for all $v \in W_2^m(\Omega)$. Consequently, if u_1 is a generalized eigenvector associated to u_0, then, by (12.2.4), we have $a(u_1, v; k) = 0$ for all $v \in W_2^m(\Omega)^\ell$. Thus, $u_1 \in \mathcal{H}_0$ and $a^{(1)}(u_1, v; k) = 0$ for all $v \in \mathcal{H}_0$. From this we conclude that the equation

$$(12.3.14) \qquad a(u_2, v; k) + a^{(1)}(u_1, v; k) + \frac{1}{2} a^{(2)}(u_0, v; k) = 0, \qquad v \in W_2^m(\Omega)^\ell,$$

for the second generalized eigenvector u_2 is solvable. Since this contradicts Theorem 12.3.2, we have proved that the form $-a^{(2)}(u, u; \gamma/2)$ is positive definite on $\mathcal{H}_0$.

We pass to the proof of (12.3.10). Using the equalities $a(u^{(0)}, u^{(0)}; k) = 0$, $a(u^{(0)}, u^{(1)}; k) = 0$ and the fact that the space $\mathcal{H}_0$ has finite dimension, we can deduce the left estimate in (12.3.10) from the equality

$$a(u, u; k + it) = a(u^{(0)}, u^{(0)}; k + it) + 2 \operatorname{Re} a(u^{(0)}, u^{(1)}; k + it) + a(u^{(1)}, u^{(1)}; k + it).$$

We prove the right estimate in (12.3.10). Since

$$a(u^{(0)}, u^{(0)}; k) = a(u^{(0)}, u^{(1)}; k) = a^{(1)}(u^{(0)}, u^{(0)}; k) = 0,$$

it suffices to establish the inequality

$$(12.3.15) \qquad -\frac{t^2}{2} a^{(2)}(u^{(0)}, u^{(0)}; k) + 2 \operatorname{Re} it \, a^{(1)}(u^{(0)}, u^{(1)}; k) + a(u^{(1)}, u^{(1)}; k)$$

$$\geq c \left(t^2 \| u^{(0)} \|^2_{L_2(\Omega)^\ell} + \| u^{(1)} \|^2_{W_2^m(\Omega)^\ell} \right).$$

If $u^{(1)} = 0$, then this inequality follows from the positive definiteness of the form $-a^{(2)}(u, u; k)$ on $\mathcal{H}_0$ and from the fact that $\mathcal{H}_0$ is finite-dimensional. Using inequality (12.1.10), the compactness of the imbedding $W_2^m(\Omega) \subset L_2(\Omega)$ and the positive definiteness of the form $a(u, u; k)$ on the orthogonal complement of $\mathcal{H}_0$ in $W_2^m(\Omega)^\ell$, we obtain the validity of (12.3.15) in the case $u^{(0)} = 0$. Therefore, in order to prove

(12.3.15), we only have to verify that

$$(12.3.16) \qquad \left| \operatorname{Im} a^{(1)}(u^{(0)}, u^{(1)}; k) \right|^2 \neq -\frac{1}{2} a^{(2)}(u^{(0)}, u^{(0)}; k) \cdot a(u^{(1)}, u^{(1)}; \gamma/2)$$

for all nonvanishing $u^{(0)}$, $u^{(1)}$. By the equality

$$a^{(1)}(u^{(0)}, u^{(1)}; k) = \sum_{j=1}^{m} \left(\mathcal{P}_j^{(1)}(k) u^{(0)} \,,\, \mathcal{P}_j(k) u^{(1)} \right)_{L_2(\Omega)^\ell}$$

and (12.3.13), the inequality (12.3.16) is violated only if

$$(12.3.17) \qquad \mathcal{P}_j(k)\, u^{(1)} = c\, \mathcal{P}_j^{(1)}(k)\, u^{(0)}, \qquad j = 1, \dots, m.$$

By renorming $u^{(1)}$ one may assume that $c = -1$. Then it follows from (12.3.17) that

$$a(u^{(1)}, v; k) + a^{(1)}(u^{(0)}, v; k) = 0 \qquad \text{for all } v \in W_2^m(\Omega)^\ell.$$

Consequently, $u^{(1)}$ is a generalized eigenvector associated to $u^{(0)}$. Using (12.3.17), we get

$$\sum_{j=1}^{m} \left(\left(\mathcal{P}_j(k) u^{(1)} \,,\, \mathcal{P}_j^{(1)}(k) v \right)_{L_2(\Omega)^\ell} + \left(\mathcal{P}_j^{(1)}(k) u^{(0)} \,,\, \mathcal{P}_j^{(1)}(k) v \right)_{\mathcal{H}} \right) = 0.$$

This is equivalent to

$$(12.3.18) \qquad a^{(1)}(u^{(1)}, v; k) + \frac{1}{2} a^{(2)}(u^{(0)}, v; k) = 0 \qquad \text{for } v \in \mathcal{H}_0.$$

Therefore, equation (12.3.14) for the second generalized eigenfunction u_2 is solvable. Since this contradicts Theorem 12.3.2, inequality (12.3.16) must be valid. The lemma is proved. ∎

COROLLARY 12.3.1. *Suppose Condition 1 and inequality (12.0.4) are satisfied and the cone $\mathcal{K}$ is given by the inequality $x_n > \phi(x')$, where ϕ is positively homogeneous of degree one in $\mathbb{R}^{n-1}$ and satisfies the Lipschitz condition (12.3.1). Then the inequality (12.0.3) is valid for all $U \in W_2^m(\mathcal{K})^\ell$.*

Proof: First we prove (12.0.3) for $U \in C_0^\infty(\overline{\mathcal{K}} \backslash \{0\})$. Let $2m \geq n$ and n be even. Then, by Lemma 12.3.4, we have the equivalence relation

$$(12.3.19) \qquad a(u, u, m - n/2 + it) \sim t^2 \, \|u^{(0)}\|_{L_2(\Omega)^\ell}^2 + \|u^{(1)}\|_{W_2^m(\Omega)^\ell}^2$$

for small $|t|$, $t \in \mathbb{R}$, where $u^{(0)}$ is the orthogonal projection of u onto the subspace of traces on Ω of homogeneous vector-polynomials of degree $m - n/2$ and $u^{(1)} = u - u^{(0)}$. For sufficiently large $|t|$ we have

$$(12.3.20) \qquad a(u, u, m - n/2 + it) \sim t^{2m} \, \|u\|_{L_2(\Omega)^\ell}^2 + \|u\|_{W_2^m(\Omega)^\ell}^2$$

(see inequality (12.1.13)). Since, by Theorem 12.3.2 the form $a(u, u; m - n/2 + it)$ is positive for $t \neq 0$, the relation (12.3.20) can be extended over the set of all real t satisfying the condition $0 < c_1 < |t| < c_2$.

Replacing b by b_0 and a by a_0 (see (12.1.8), (12.1.9)) in the above arguments, we arrive at the relations (12.3.19), (12.3.20) for the form $a_0(u, u; m - n/2 + it)$. Therefore,

$$(12.3.21) \qquad a(u, u; \lambda) \sim a_0(u, u; \lambda) \qquad \text{for } \operatorname{Re} \lambda = m - n/2.$$

Setting $u = \tilde{U}(\lambda, \omega)$, where

$$\tilde{U}(\lambda, \omega) = \int\limits_0^\infty r^{-\lambda-1} U(r, \omega)\, dr$$

denotes the Mellin transform with respect to r of the function U, integrating (12.3.21) over the line $\operatorname{Re}\lambda = m - n/2$ and applying the Parseval equality, we get the inequality (12.0.3) for the case $2m \geq n$, n even, $U \in C_0^\infty(\overline{\mathcal{K}}\backslash\{0\})^\ell$.

If $2m < n$ or $2m \geq n$ and n is odd, then the proof of this inequality for $U \in C_0^\infty(\overline{\mathcal{K}}\backslash\{0\})^\ell$ is much simpler, since, by the first part of Theorem 12.3.2, the line $\operatorname{Re}\lambda = m - n/2$ contains no eigenvalues of the pencil $\mathfrak{A}$ and, consequently, (12.3.20) holds for all real t. This again yields (12.3.21), and the application of the Mellin transformation leads to (12.0.3) for $U \in C_0^\infty(\overline{\mathcal{K}}\backslash\{0\})^\ell$. Since $C_0^\infty(\overline{\mathcal{K}}\backslash\{0\})$ is dense in $W_2^m(\mathcal{K})$ for $2m \leq n$, the proof is complete for such m and n.

It remains to get rid of the restriction $U \in C_0^\infty(\overline{\mathcal{K}}\backslash\{0\})^\ell$ for $2m > n$. Since $C^\infty(\overline{\mathcal{K}}) \cap W_2^m(\mathcal{K})$ is dense in $W_2^m(\mathcal{K})$, it suffices to prove (12.0.3) only for smooth functions U. Let

$$p(x) = \sum_{|\alpha| < m - n/2} \frac{1}{\alpha!} (\partial_x^\alpha U)(0)\, x^\alpha$$

and let $\eta \in C_0^\infty(\mathbb{R}^n)$, $\eta(x) = 1$ for $|x| \leq 1$. We set $\eta_N(x) = \eta(x/N)$ for $N = 1, 2, \ldots$. It can be easily seen that the difference $V_n = U - \eta_N p$ can be approximated in $W_2^m(\mathcal{K})^\ell$ by functions from $C_0^\infty(\overline{\mathcal{K}}\backslash\{0\})^\ell$. Therefore, $b(V_N, V_N) \geq c\, b_0(V_N, V_N)$. Finally, we observe that

$$b(\eta_N p, \eta_N p) + b_0(\eta_N p, \eta_N p) \to 0$$

as $N \to \infty$. This proves Corollary 12.3.1. ∎

12.3.4. The spectrum on the lines $\operatorname{Re}\lambda = m - n/2 \pm 1/2$. Let the cone $\mathcal{K}$ be given by the inequality $x_n > \phi(x')$. Now we suppose that the function ϕ is positively homogeneous of degree 1 and smooth on $\mathbb{R}^{n-1}\backslash\{0\}$.

THEOREM 12.3.3. *Let* Condition 3 *be satisfied. Then the following assertions are valid:*

1) If n is even or n is odd and $2m < n - 1$, then the lines $\operatorname{Re}\lambda = m - n/2 \pm 1/2$ do not contain eigenvalues of the pencil $\mathfrak{A}$.

2) If n is an odd number and $2m \geq n - 1$, then every of the lines $\operatorname{Re}\lambda = m - n/2 \pm 1/2$ contains the single eigenvalue $\lambda_\pm = m - n/2 \pm 1/2$. The geometric, partial and algebraic multiplicities of the eigenvalues λ_+ and λ_- coincide. The set of the eigenfunctions corresponding to the eigenvalue λ_+ is exhausted by the traces on Ω of homogeneous vector-polynomials of degree λ_+. The algebraic multiplicity of the eigenvalue λ_+ is equal to

$$\ell\left(\binom{n + \lambda_+ - 1}{n - 1} + \binom{n + \lambda_+ - 2}{n - 1} \right)$$

(in the case $\lambda_+ = 0$ the second binomial coefficient is equal to $\binom{n-2}{n-1} = 0$). To every eigenfunction there corresponds at most one generalized eigenfunction.

Proof: By assertion 4) of Theorem 12.1.1, it suffices to consider the line $\operatorname{Re}\lambda = m - n/2 + 1/2$. Let λ_0 be an eigenvalue of the pencil $\mathfrak{A}$ on this line and let u_0 be an eigenfunction corresponding to this eigenvalue. We set $U(x) = r^{\lambda_0} u_0(\omega)$.

Since $\lambda_0 - 1 = 2m - n - \overline{\lambda}_0$, it follows from the definition of the operator $\mathfrak{A}(\lambda)$ (cf. (12.1.5)) that

$$\int_{\mathcal{K}} \chi_\varepsilon \sum_{|\alpha|=|\beta|=m} \left(A_{\alpha,\beta}\, \partial_x^\beta U\,,\, \partial_x^\alpha\, \partial_{x_n} U\right)_{\mathbb{C}^\ell} dx = 0,$$

where χ_ε is the characteristic function of the spherical shell $\varepsilon < |x| < 1/\varepsilon, 0 < \varepsilon < 1$. Using the equalities $A_{\alpha,\beta} = A_{\beta,\alpha}^*$ for $|\alpha| = |\beta| = m$ and the fact that the integral on the left is real, we get

$$\frac{1}{2} \int_{\mathcal{K}} \chi_\varepsilon\, \partial_{x_n} \sum_{|\alpha|=|\beta|=m} \left(A_{\alpha,\beta}\, \partial_x^\beta U\,,\, \partial_x^\alpha U\right)_{\mathbb{C}^\ell} dx = 0.$$

Because of the cancelling of the corresponding integrals over the spheres $|x| = \varepsilon$ and $|x| = 1/\varepsilon$, integration by parts yields

$$\int_{\partial\mathcal{K}} \chi_\varepsilon\, \nu_n \sum_{|\alpha|=|\beta|=m} \left(A_{\alpha,\beta}\, \partial_x^\beta U\,,\, \partial_x^\alpha U\right)_{\mathbb{C}^\ell} dx = 0.$$

Here the n-th component ν_n of the exterior normal ν is negative on $\partial\mathcal{K}\backslash\{0\}$. Hence it follows from Condition 3 that

(12.3.22) $\partial_x^\alpha U = 0$ on $\partial\mathcal{K}\backslash\{0\}$ for $|\alpha| = m$.

First let $n \geq 3$. Then $\partial\mathcal{K}\backslash\{0\}$ is a connected manifold and, by (12.3.22), there exists a vector-polynomial $p(x)$, $\deg p \leq m - 1$, such that

$$\partial_x^\alpha U = \partial_x^\alpha p \quad \text{on } \partial\mathcal{K}\backslash\{0\} \quad \text{for } |\alpha| \leq m - 1.$$

Since the function U is positively homogeneous of degree λ_0, this implies $p = 0$ if $\lambda_0 \notin \{0, 1, 2, \dots\}$. If λ_0 is a nonnegative integer (i.e., $\lambda_0 = \lambda_+ = m - n/2 + 1/2$, $2m \geq n - 1$, n odd), then we conclude that the polynomial p is homogeneous of degree λ_0. In both cases $U - p$ is a positively homogeneous function of degree λ_0 which satisfies the Dirichlet boundary conditions on $\partial\mathcal{K}\backslash\{0\}$. Moreover,

$$\sum_{|\alpha|=|\beta|=m} \partial_x^\alpha \left(A_{\alpha,\beta}\, \partial_x^\beta (U - p)\right) = 0 \quad \text{in } \mathcal{K}.$$

Using Theorem 11.1.1, we conclude that $U - p = 0$.

In the case $n = 2$ the set $\partial\mathcal{K}\backslash\{0\}$ consists of two half-lines Γ_1 and Γ_2. Consequently, by (12.3.22), there exist polynomials p_1, p_2 of degree $\leq m - 1$ such that

$$\partial_x^\alpha U = \partial_x^\alpha p_1 \quad \text{on } \Gamma_1, \qquad \partial_x^\alpha U = \partial_x^\alpha p_2 \quad \text{on } \Gamma_2, \ |\alpha| \leq m - 1.$$

Since λ_0 is noninteger in this case, we get $p_1 = p_2 = 0$. By the same arguments as in the case $n \geq 3$, this implies $U = 0$.

Thus, we have proved that only in the case $2m \geq n - 1$, n odd, there is an eigenvalue on the line $\operatorname{Re}\lambda = m - n/2 + 1/2$, that this eigenvalue is equal to $m - n/2 + 1/2$, and that the eigenfunctions corresponding to this eigenvalue are traces on Ω of homogeneous polynomials.

We show that for integer $k = m - n/2 + 1/2 \geq 0$ every eigenfunction of the pencil $\mathfrak{A}$ corresponding to the eigenvalue $\lambda_+ = k$ has at most one generalized eigenfunction. Let u_0, u_1, u_2 be a Jordan chain corresponding to the eigenvalue

$\lambda_+ = k$. Then, as it was shown above, the function $U_0 = r^k u_0(\omega)$ belongs to the set $(\Pi_k^{(n)})^\ell$ of homogeneous vector-polynomials of degree k. Furthermore,

$$(12.3.23) \qquad a^{(1)}(u_0, v; k) + a(u_1, v; k) = 0,$$

$$(12.3.24) \qquad \frac{1}{2} a^{(2)}(u_0, v; k) + a^{(1)}(u_1, v; k) + a(u_2, v; k) = 0$$

for all $v \in W_2^m(\Omega)^\ell$. We introduce the operator $T_n(\lambda)$ by the equality

$$T_n(\lambda)\, u = T_n(\omega, \partial_\omega, \lambda)\, u(\omega) = r^{-\lambda+1} \partial_{x_n} (r^\lambda u(\omega)).$$

Then we have

$$Q_\alpha(k-1)\, T_n(k) = r^{-k+m+1} \partial_x^\alpha \partial_{x_n} (r^k u_0) = 0 \qquad \text{for } |\alpha| = m,$$

where Q_α is the operator defined in (12.1.3). Inserting $v = T_n(k)\, u_0$ into (12.3.24), we obtain

$$(12.3.25) \qquad \frac{1}{2} a^{(2)}(u_0, T_n(k)u_0; k) + a^{(1)}(u_1, T_n(k)u_0; k) = 0.$$

If we set $v = T_n(k)\, u_1$ in (12.3.23), then we get

$$(12.3.26) \qquad a^{(1)}(u_0, T_n(k)u_1; k) = -a(u_1, T_n(k)u_1; k).$$

One can readily verify that

$$(12.3.27) \quad |\log \varepsilon|\, a^{(2)}(u_0, T_n(k)u_0; k)$$
$$= -\int_{\mathcal{K}} \chi_\varepsilon \sum_{|\alpha|=|\beta|=m} \left(A_{\alpha,\beta} \partial_x^\beta (\log r\, U_0),\ \partial_x^\alpha (\log r\, \partial_{x_n} U_0) \right)_{\mathbb{C}^\ell} dx$$
$$= -\frac{1}{2} \int_{\partial\mathcal{K}} \chi_\varepsilon \nu_n \sum_{|\alpha|=|\beta|=m} \left(A_{\alpha,\beta} \partial_x^\beta (\log r\, U_0),\ \partial_x^\alpha (\log r\, U_0) \right)_{\mathbb{C}^\ell} dx$$
$$+ \int_{\mathcal{K}} \chi_\varepsilon \sum_{|\alpha|=|\beta|=m} \left(A_{\alpha,\beta} \partial_x^\beta (\log r\, U_0),\ \partial_x^\alpha (\omega_n r^{-1} U_0) \right)_{\mathbb{C}^\ell} dx.$$

Setting $U_1 = r^k u_1$, in a similar way, we obtain

$$(12.3.28) \qquad 2\, |\log \varepsilon|\, a^{(1)}(u_1, T_n(k)u_0; k)$$
$$= -\int_{\mathcal{K}} \chi_\varepsilon \sum_{|\alpha|=|\beta|=m} \left(A_{\alpha,\beta} \partial_x^\beta U_1,\ \partial_x^\alpha (\log r\, \partial_{x_n} U_0) \right)_{\mathbb{C}^\ell} dx$$
$$= \int_{\mathcal{K}} \chi_\varepsilon \sum_{|\alpha|=|\beta|=m} \left(A_{\alpha,\beta} \partial_x^\beta \partial_{x_n} U_1,\ \partial_x^\alpha (\log r\, U_0) \right)_{\mathbb{C}^\ell} dx$$
$$- \int_{\partial\mathcal{K}} \chi_\varepsilon \nu_n \sum_{|\alpha|=|\beta|=m} \left(A_{\alpha,\beta} \partial_x^\beta U_1,\ \partial_x^\alpha (\log r\, U_0) \right)_{\mathbb{C}^\ell} dx$$
$$+ \int_{\mathcal{K}} \chi_\varepsilon \sum_{|\alpha|=|\beta|=m} \left(A_{\alpha,\beta} \partial_x^\beta U_1,\ \partial_x^\alpha (\omega_n r^{-1} U_0) \right)_{\mathbb{C}^\ell} dx.$$

Substituting $v = \omega_n u_0$ in (12.3.23), we get

$$(12.3.29) \qquad a^{(1)}(u_0, \omega_n u_0; k) = -a(u_1, \omega_n u_0; k).$$

The left-hand side of (12.3.29) multiplied by $2\,|\log\varepsilon|$ is equal to the last integral in (12.3.27), while the right-hand side is equal to the last integral in (12.3.28). Therefore, by (12.3.27) and (12.3.28), we can rewrite (12.3.25) in the form

$$(12.3.30) \quad \frac{1}{2}\int_{\partial\mathcal{K}}\chi_\varepsilon\,\nu_n \sum_{|\alpha|=|\beta|=m}\Big(A_{\alpha,\beta}\partial_x^\beta(\log r\,U_0)\,,\,\partial_x^\alpha(\log r\,U_0)\Big)_{\mathbb{C}^\ell}dx$$

$$+\int_{\partial\mathcal{K}}\chi_\varepsilon\,\nu_n \sum_{|\alpha|=|\beta|=m}\Big(A_{\alpha,\beta}\partial_x^\beta U_1\,,\,\partial_x^\alpha(\log r\,U_0)\Big)_{\mathbb{C}^\ell}dx$$

$$-\int_{\mathcal{K}}\chi_\varepsilon \sum_{|\alpha|=|\beta|=m}\Big(A_{\alpha,\beta}\partial_x^\beta\partial_{x_n}U_1\,,\,\partial_x^\alpha(\log r\,U_0)\Big)_{\mathbb{C}^\ell}dx=0.$$

Furthermore, the right-hand side in (12.3.26) is equal to

$$-\frac{1}{2\,|\log\varepsilon|}\int_{\mathcal{K}}\chi_\varepsilon \sum_{|\alpha|=|\beta|=m}\Big(A_{\alpha,\beta}\partial_x^\beta U_1\,,\,\partial_x^\alpha\partial_{x_n}U_1\Big)_{\mathbb{C}^\ell}dx$$

$$=-\frac{1}{4\,|\log\varepsilon|}\int_{\partial\mathcal{K}}\chi_\varepsilon\,\nu_n \sum_{|\alpha|=|\beta|=m}\Big(A_{\alpha,\beta}\partial_x^\beta U_1\,,\,\partial_x^\alpha U_1\Big)_{\mathbb{C}^\ell}dx,$$

and the left-hand side of (12.3.26) and the last integral in (12.3.30) are complex conjugates. Therefore, taking the real part in (12.3.30), we get

$$(12.3.31) \quad \int_{\partial\mathcal{K}}\chi_\varepsilon\,\nu_n \sum_{|\alpha|=|\beta|=m}\Big(A_{\alpha,\beta}\,\partial_x^\beta\,(U_0\log r+U_1)\,,\,\partial_x^\alpha(U_0\log r+U_1)\Big)_{\mathbb{C}^\ell}dx$$
$$=0.$$

This implies the existence of a vector-polynomial p of degree not greater than $m-1$ such that $\partial_x^\gamma(U_0\log r+U_1)=\partial_x^\gamma p$ on $\partial\mathcal{K}$ for $|\gamma|\leq m$. From this it follows that $U_0=0$. This contradicts the assumption that u_0 is an eigenfunction. Thus, we have proved that every eigenfunction has at most one generalized eigenfunction.

Now we give an upper estimate for the algebraic multiplicity of the eigenvalue $\lambda_+=k=m-n/2+1/2$. Let u_0 be an eigenfunction corresponding to this eigenvalue. We assume that the vector-polynomial $U_0=r^k u_0$ does not depend on the variable x_n and show that there are no generalized eigenfunctions associated to the eigenfunction u_0. For this end, we suppose the contrary, i.e., there exists a generalized eigenfunction u_1. Then $U_0\log r+r^k u_1$ is a solution of the problem (12.1.2). If we set $V=\chi_\varepsilon\,\partial_{x_n}(U_0\log r+U_1)$ with $U_1=r^k u_1$ in (12.1.2), we get

$$(12.3.32)$$

$$0=\int_{\mathcal{K}}\sum_{|\alpha|=|\beta|=m}\Big(A_{\alpha,\beta}\,\partial_x^\beta(U_0\log r+U_1),\partial_x^\alpha\big(\chi_\varepsilon\partial_{x_n}(U_0\log r+U_1)\big)\Big)_{\mathbb{C}^\ell}dx$$

$$=\int_{\mathcal{K}}\chi_\varepsilon \sum_{|\alpha|=|\beta|=m}\Big(A_{\alpha,\beta}\,\partial_x^\beta(U_0\log r+U_1),\partial_{x_n}\partial_x^\alpha(U_0\log r+U_1)\Big)_{\mathbb{C}^\ell}dx,$$

since the polynomial U_0 is independent of x_n. Integrating by parts in the right-hand side of (12.3.32) we arrive at (12.3.31), where $U_1=r^k u_1$. Here we have used the fact that the integrals over the spheres $|x|=\varepsilon$ and $|x|=1/\varepsilon$ mutually cancel by the homogeneity of the functions $\partial_x^\beta(U_0\log r)$, $\partial_{x_n}(U_0\log r)$, and U_1. Consequently, $U_0=0$. This implies that the dimension of the subspace of eigenfunctions having a generalized eigenfunctions does not exceed the dimension of the space of the vector-polynomials $x_n\, p(x)$, $p \in (\Pi_{k-1}^{(n)})^\ell$. Thus, the algebraic multiplicity of the eigenvalue λ_+ does not exceed $\ell((\,{}^{n+k-1}_{\;\;n-1}\,) + (\,{}^{n+k-2}_{\;\;n-1}\,))$.

For $k = 0$ this estimate, together with the assertion on the eigenfunctions, implies the statement of the theorem. Let $k \geq 1$. By Theorem 12.1.1, the dimension of the space of eigenfunctions to the eigenvalue λ_+ having a generalized eigenfunction is the same as for the eigenvalues $\lambda_- = k - 1$. The eigenspace of the eigenvalue λ_- contains the traces of the vector-valued polynomials from $(\Pi_{k-1}^{(n)})^\ell$ on Ω. We show that for all of them generalized eigenfunctions exist. In fact, if $u_0 = U_0|_\Omega$, $U_0 \in (\Pi_{k-1}^{(n)})^\ell$, then

$$a^{(1)}(u_0, v; k - 1) = 0 \qquad \text{for all } v = V|_\Omega, \ V \in (\Pi_k^{(n)})^\ell.$$

This is equivalent to the existence of a generalized eigenfunction associated to u_0. Hence the algebraic multiplicity of λ_+ is at least $\ell((\,{}^{n+k-1}_{\;\;n-1}\,) + (\,{}^{n+k-2}_{\;\;n-1}\,))$. The proof of the theorem is complete. ∎

REMARK 12.3.1. In Section 9.2 we got a transcendental equation for eigenvalues of the pencil generated by general elliptic boundary value problems for equations of order $2m$ in a plane angle of size $\alpha \in (0, 2\pi]$. For $\alpha = 2\pi$ this transcendental equation has the form

$$\det(e^{-4\pi\lambda i}\, I - B^{-1} C A^{-1} D) = 0,$$

where A, B, C, D are $m \times m$-matrices depending on the operators of the problem. In the case of the Neumann problem we have $C = A$ and $B = D$. Therefore, every half-integer number is an eigenvalue. Since the roots of the above-mentioned transcendental equation depend continuously on α, for any angle size close to 2π there exist eigenvalues of the pencil close to $m - 1/2$. This implies that the statement of Theorems 12.3.2, 12.3.3 can not be refined. Namely, there exists no strip wider than the strip $m - 1 < \operatorname{Re} \lambda \leq m - 1/2$ which is free of eigenvalue of the pencil $\mathfrak{A}$ without additional assumptions on the operator and the angle size.

The following example shows that Theorem 12.3.3 fails if Condition 3 is replaced by the weaker Condition 2.

Example. We consider the Neumann problem in the example of Section 12.2 in the half-space $\mathcal{K} = \mathbb{R}^3_+$ for $\sigma = 1/2$. Then Condition 2 is satisfied for the corresponding quadratic form. On the other hand, the form

$$\sum_{i,j=1}^{3} |f_{i,j}|^2 - \frac{1}{2} \left| \sum_{i=1}^{3} f_{i,i} \right|^2$$

is not positive, i.e., Condition 3 does not hold. Theorem 12.3.3 fails for this problem, since it has four linearly independent positively homogeneous solutions of degree 1: x_1, x_2, x_3, and $|x|$.

12.4. Applications to the Neumann problem in a bounded domain

Let $\mathcal{G}$ be a bounded domain in $\mathbb{R}^n$ and $0 \in \partial\mathcal{G}$. For simplicity we assume that the set $\{x \in \mathcal{G} : |x| < 1\}$ coincides with $\{x \in \mathcal{K} : |x| < 1\}$, where $\mathcal{K}$ is the same cone as in the preceding section.

Furthermore, let L be the operator

$$L(\partial_x) = (-1)^m \sum_{|\alpha|=|\beta|=m} A_{\alpha,\beta}\, \partial_x^{\alpha+\beta}\,,$$

where $A_{\alpha,\beta}$ are constant $\ell \times \ell$-matrices, $A_{\alpha,\beta} = A_{\beta,\alpha}^*$. We assume the coefficients $A_{\alpha,\beta}$ are subject to Condition 3. The Neumann problem will be considered in two formulations.

PROBLEM I. *Let f be a vector-valued function from $L_2(\mathcal{G})^\ell$ orthogonal to all vector-polynomials of degree not greater than $m-1$. By a solution of Problem I we mean a function $U \in W_2^m(\mathcal{G})^\ell$ satisfying the integral identity*

$$(12.4.1) \qquad \int_{\mathcal{G}} \sum_{|\alpha|=|\beta|=m} \left(A_{\alpha,\beta}\, \partial_x^\beta U\,,\, \partial_x^\alpha V\right)_{\mathbb{C}^\ell} dx = \int_{\mathcal{G}} (f,V)_{\mathbb{C}^\ell}\, dx$$

for all $V \in W_2^m(\mathcal{G})^\ell$.

Condition 3 ensures the solvability of Problem I. The solutions are uniquely determined up to additive polynomial terms of degree not greater than $m-1$.

To formulate Problem II we use the function space $W_2^m(\mathcal{G},0)$, which is defined as the completion of $C_0^\infty(\overline{\mathcal{G}}\backslash\{0\})$ in the norm of the space $W_2^m(\mathcal{G})$. This space differs from $W_2^m(\mathcal{G})$ only in the case $2m > n$.

Furthermore, we denote by $\overset{\circ}{\Pi}$ the set of the polynomials of degree not greater than $m-1$ vanishing at the conic point $x = 0$ together with the derivatives of order less than $m - n/2$.

PROBLEM II. *Let $2m > n$ and let $f \in L_2(\mathcal{G})^\ell$ be a vector-valued function orthogonal to $\overset{\circ}{\Pi}{}^\ell$. By a solution of Problem II we mean a vector-valued function $U \in W_2^m(\mathcal{G},0)^\ell$ satisfying the integral identity (12.4.1) for all $V \in W_2^m(\mathcal{G},0)^\ell$.*

Clearly, this problem is also solvable and its solution is uniquely determined up to polynomials from $\overset{\circ}{\Pi}{}^\ell$.

As an application of Theorems 12.2.1–12.3.3 we find the principal terms of the asymptotics near the point $x = 0$ for the solutions of the two boundary value problems just stated. We limit consideration to a smooth vector function f. Using asymptotic formulas for solutions of general elliptic boundary value problems near conic points (see Section 1.4) and Theorems 12.2.1–12.3.3, we obtain the following asymptotics for the solution U_{II} of Problem II:

$$(12.4.2) \qquad U_{II}(x) = \sum_{|\alpha|=m-n/2} c_\alpha\, x^\alpha + O(r^{m-(n-1)/2+\varepsilon})$$

if n is even and

$$(12.4.3)\quad U_{II}(x) = \sum_{|\beta|=m-(n-1)/2} c_\beta\, x^\beta$$

$$+ \sum_{|\gamma|=m-(n+1)/2} c_\gamma \left(p_\gamma(x)\log r + h_\gamma(x)\right) + O(r^{m-(n-1)/2+\varepsilon})$$

if n is odd. Here c_α, c_β are constant vectors, c_γ are scalars, p_γ are vector-valued polynomials of degree $|\gamma|+1$, h_γ are vector-valued functions positively homogeneous of degree $|\gamma|+1$, and ε is a positive number.

THEOREM 12.4.1. *The solution U_I of* Problem I *has the asymptotics*

$$(12.4.4) \qquad U_I(x) = \sum_{|\alpha| \leq m-(n-1)/2} c_\alpha \, x^\alpha + O(r^{m-(n-1)/2+\varepsilon})$$

near the point $x = 0$, where c_α are constant vectors. (In the case $2m < n-1$ the sum in (12.4.4) is absent.)

Proof: For $2m \leq n$ as well as for $2m > n$ and even n the asymptotics (12.4.4) immediately follows from Theorems 12.2.1–12.3.3. In the case $2m > n$, n odd, these theorems yield the cruder asymptotic representation

$$(12.4.5) \quad U_I(x) = \sum_{|\beta| \leq m-(n-1)/2} c_\beta \, x^\beta$$

$$+ \sum_{|\gamma|=m-(n+1)/2} c_\gamma \left(p_\gamma(x) \log r + h_\gamma(x) \right) + O(r^{m-(n-1)/2+\varepsilon}).$$

We prove that the coefficients c_γ in (12.4.5) are zero. Let $U(x) = p_\gamma(x)\log r + h_\gamma(x)$ and let η be a function of the class $C_0^\infty(\mathbb{R}^n)$ equal to one in a neighborhood of the origin. Obviously, it suffices to prove that the equality

$$(12.4.6) \qquad\qquad b(U, \eta V) = 0$$

is not valid for all homogeneous vector-valued polynomials V of degree $\lambda_+ - 1$, where $\lambda_+ = k = m - (n-1)/2$ and $\eta \in C_0^\infty(\overline{\mathcal{K}})$.

Suppose that (12.4.6) is satisfied for all $V \in (\Pi_{k-1}^{(n)})^\ell$. Since the left-hand side of (12.4.6) does not depend on the cut-off function η, (12.4.6) implies

$$b(U, \chi V) = 0 \qquad \text{for all } V = r^{k-1} v(\omega) \in (\Pi_{k-1}^{(n)})^\ell,$$

where $\chi = \chi(r)$ is the characteristic function of the unit ball. One can readily check the relation

$$\partial_x^\alpha (\chi V) = \sum_{k=1}^m r^{k-m-1} (r\partial_r)^k \chi \cdot q_\alpha^{(k)}(\omega, \partial_\omega) \, v(\omega),$$

where $q_\alpha^{(k)}$ are differential operators on the unit sphere,

$$q_\alpha^{(1)}(\omega, \partial_\omega) = \partial_\lambda Q_\alpha(\omega, \partial_\omega, \lambda)|_{\lambda=k-1}.$$

Since $\partial_x^\beta U(x)$ has the form $r^{k-m} g_\beta(\omega)$ for $|\beta| = m$, we obtain

$$b(U, \chi V) = \int_{\mathcal{K}} \sum_{|\alpha|=|\beta|=m} \left(A_{\alpha,\beta} \partial_x^\beta U, \, r^{k-m} \chi' \, q_\alpha^{(1)}(\omega, \partial_\omega) v \right)_{\mathbb{C}^\ell} dx$$

$$= -\int_{\Omega} \sum_{|\alpha|=|\beta|=m} \left(A_{\alpha,\beta} \left(Q_\beta'(\omega, \partial_\omega, k) u_0 + Q_\beta(\omega, \partial_\omega, k) u_1 \right), Q_\alpha'(\omega, \partial_\omega, k-1) v \right)_{\mathbb{C}^\ell} d\omega,$$

where $u_0 = p_\gamma|_\Omega$, $u_1 = h_\gamma|_\Omega$. Taking into account that u_0 and v are traces on Ω of polynomials of degree k and $k-1$, we get

$$b(U, \chi V) = \frac{1}{2} a^{(2)}(u_0, v; k) + a^{(1)}(u_1, v; k).$$

Thus, we arrive at the equalities

$$a^{(1)}(u_0, w; \lambda_+) + a(u_0, w; \lambda_+) = 0 \qquad \text{for all } w \in W_2^m(\Omega)^\ell,$$

$$\frac{1}{2} a^{(2)}(u_0, v; \lambda_+) + a^{(1)}(u_1, v; \lambda_+) = 0 \qquad \text{for all } v \text{ such that } r^{\lambda_+ - 1} v \in (\Pi_{\lambda_+ - 1}^{(n)})^\ell.$$

Using the same arguments as in the proof of Theorem 12.3.3, from these relations we deduce (12.3.25)–(12.3.31), where U_0 and U_1 are replaced by p_γ and h_γ, respectively. From the equality (12.3.31) modified in this way we obtain

$$\partial_x^\alpha (p_\gamma \log r + h_\gamma) = p \qquad \text{on } \partial\mathcal{K} \quad \text{for } |\alpha| \leq m,$$

where p is a vector-valued polynomial of degree not greater than $m - 1$. Consequently, $p_\gamma = 0$. This proves the representation (12.4.4). ∎

Comparing the asymptotics of the solutions U_I and U_{II} for $2m > n$ and odd n, we see that $U_I \in W_2^{m+1/2+\delta}(\mathcal{G})^\ell$ for some $\delta > 0$ and $U_{II} \in W_2^{m+1/2-\delta}(\mathcal{G})^\ell$ for all $\delta > 0$. For other dimensions U_I and U_{II} belong to the space $W_2^{m+1/2+\delta}(\mathcal{G})^\ell$.

12.5. The Neumann problem for anisotropic elasticity in an angle

Let us consider the Neumann problem for the Lamé system of the plane anisotropic elasticity in the angle $\mathcal{K}$ with opening $\alpha \in (0, 2\pi)$. In other words, we deal with problem (9.2.13)–(9.2.15). We use the notation introduced in Section 9.2.5. By the second assertion of Theorem 12.3.2, the strip $|\text{Re}\,\lambda| < 1/2$ contains only the eigenvalue $\lambda = 0$ of the pencil $\mathfrak{A}$ corresponding to the boundary value problem for the displacement vector. Moreover, this eigenvalue generates two constant eigenvectors, and each of them has precisely one generalized eigenvector.

Here we enhance this information by describing spectral properties of $\mathfrak{A}$ for $1/2 \leq |\text{Re}\,\lambda| \leq 1$. We use the reduction (9.2.17) of the elasticity problem to the fourth order elliptic equation (9.2.18) with boundary conditions (9.2.19), (9.2.20). By assertion 4 of Theorem 12.1.1, it is sufficient to consider the strip $1/2 \leq \text{Re}\,\lambda \leq 1$.

Integrating (9.2.19) and (9.2.20), we obtain that U is linear and the normal derivative of U is constant on each side of $\mathcal{K}$. Since we are interested in power-logarithmic solutions U with the power $\lambda + 1$, the Dirichlet data for U just mentioned should be zero. We note that the power-logarithmic displacement solutions satisfying problem (9.2.13)–(9.2.15) can be recovered from U modulo a rigid body displacement term. Clearly, this term is zero if $\lambda \neq 1$. For $\lambda = 1$ the rigid displacement term is the vector $c\,(x_2, -x_1)$.

Referring to Theorem 8.4.1, we see that for $\alpha < \pi$ the strip $1/2 \leq \text{Re}\,\lambda \leq 1$ contains one simple eigenvalue $\lambda = 1$ with the eigenvector $(\sin\varphi, -\cos\varphi)$.

If $\alpha \in [\pi, 2\pi)$, then, by Theorem 8.5.1, all eigenvalues of $\mathfrak{A}$ in the strip $1/2 \leq \text{Re}\,\lambda \leq 1$ are real and situated in $(1/2, 1]$. Furthermore, all eigenvalues of $\mathfrak{A}$ in the interval $(1/2, 1)$ are simple and strictly decreasing with respect to α. By the same theorem, there exists an angle $\alpha^* \in (\pi, 2\pi)$ such that there is one eigenvalue in the interval $(1/2, 1)$ if $\alpha \in (\pi, \alpha^*]$ and there are two eigenvalues in the same interval if $\alpha \in (\alpha^*, 2\pi)$. If α is close to 2π, then the interval $(1/2, 1)$ contains exactly two eigenvalues subject to the asymptotic formula

$$\lambda_k(\alpha) = \frac{1}{2} + c_k(2\pi - \alpha)^{2k-1} + O(|2\pi - \alpha|^{2k}),$$

where $k = 1, 2$ and $c_k = \text{const} > 0$ (see Theorem 8.3.1). If α is close to π and $\alpha > \pi$, then the pencil has exactly one simple eigenvalue $\lambda_2(\alpha)$ in $(1/2, 1)$ which satisfies

$$\lambda_2(\alpha) = 1 - c\,(\alpha - \pi) + O((\alpha - \pi)^2)\,,$$

where c is a positive constant (see Theorem 8.2.1).

Theorem 8.1.3 and the above information about the eigenvalue $\lambda = 1$ show that, in the case $\alpha = \pi$, the strip $1/2 \leq \operatorname{Re}\lambda \leq 1$ contains the single eigenvalue $\lambda = 1$ of multiplicity 2. A corresponding eigenvector is $(\sin\varphi, -\cos\varphi)$ and another one can be found from (9.2.17), where $U = x_2^2$.

It remains to consider the case of a cut, i.e. $\alpha = 2\pi$. Condition 2 (Korn's inequality) holds because it is valid for the half-plane. Hence and by the second assertion of Theorem 12.2.1, there is only one eigenvalue $\lambda = 0$ of $\mathfrak{A}$ on the line $\operatorname{Re}\lambda = 0$. The corresponding eigenvectors are $(1, 0)$ and $(0, 1)$, and each of them has one generalized eigenvector. Theorem 8.1.3 guarantees that the spectrum of $\mathfrak{A}$ in the strip $0 < \operatorname{Re}\lambda \leq 1$ consists of the eigenvalues $1/2$ and 1. Moreover, the eigenvalue $\lambda = 1$ has the same two eigenvectors as in the case $\alpha = \pi$ and there are no generalized eigenvectors. The remaining eigenvalue $\lambda = 1/2$ has geometric and algebraic multiplicities 2.

12.6. Notes

The results of this chapter, except for the previous section, are taken from the paper [**129**] by Kozlov and Maz'ya.

Bibliography

[1] Adolfsson, V., Jerison, D., L^p-integrability of the second order derivatives for the Neumann problem in convex domains, Indiana Univ. Math. J. **43** (1994) 4, 1123-1138.

[2] Agmon, S., Douglis, A., Nirenberg, L., Estimates near the boundary for solutions of elliptic partial differential equations satisfying general boundary conditions II, Comm. Pure Appl. Math. **17** (1964) 35-92.

[3] Agmon, S., Nirenberg, L., Properties of solutions of ordinary elliptic equations in Banach space, Comm. Pure Appl. Math. **16** (1963) 121-163.

[4] Agranovich, M. S., Vishik, M. I., Elliptic problems with parameter and parabolic problems of general type, Uspekhi Mat. Nauk **19** (1964) 3, 53-161 (in Russian).

[5] Andersson, B., Falk, U., Babuška, I., von Petersdorff, T., Reliable stress and fracture mechanics analysis of complex components using a h-p version of FEM, Internat. J. Numer. Methods Engrg. **38** (1995) 13, 2135-2163.

[6] Apel, T., Nicaise, S., Elliptic problems in domains with edges: Anisotropic regularity and anisotropic finite element meshes, Partial Differential Equations and Functional Analysis. In memory of Pierre Grisvard. Proceedings of a conference held in November 1994, Birkhäuser, Boston, Prog. Nonlinear Differ. Equ. Appl. 22 (1996) 18-34.

[7] Aronszajn, N., On coercive integro-differential quadratic forms, Conference on Partial Differential Equations, Report 14, Univ. of Kansas, pp. 94-106.

[8] Ashbaugh, M. S., Benguria, R. D., Sharp upper bound to the first nonzero Neumann eigenvalue for bounded domains in spaces of constant curvature, J. London Math. Soc. **52** (1995) 2, 402-416.

[9] Aziz, A. K., Kellogg, R. B., On homeomorphisms for an elliptic equation in domains with corners, Differential Integral Equations **8** (1995) 2, 333-352.

[10] Babuška, I., von Petersdorff, T., Andersson, B., Numerical treatment of vertex singularities and intensity factors for mixed boundary value problems for the Laplace equation in $\mathbb{R}^3$, SIAM J. Numer. Anal. **31** (1994) 5, 1265-1288.

[11] Bateman, H., Erdélyi, A. et al. Higher transcendental functions I,II, Mc Graw Hill, New York 1953.

[12] Bažant, Z. P., Keer, M., Singularities of elastic stresses and of harmonic functions at conical notches or inclusions, Internat. J. Solids and Structures **10** (1974), 957-964.

[13] Bažant, Z. P., Estenssoro, L. F., Surface singularity and crack propagation, Internat. J. Solids and Structures **15** (1979) 405-426.

[14] Beagles, A. E., Sändig, A.-M., Singularities of rotationally symmetric symmetric solutions of boundary value problems for the Lamé equations, Z. Angew. Math. Mech. **71** (1991) 11, 423-431.

[15] Beagles, A. E., Whiteman, J. R., Treatment of a re-entrant vertex in a three-dimensional Poisson problem, Singularties and Constructive Methods for Their Treatment, Proceedings Conf. Oberwolfach 1983, Lecture Notes in Math. 1121 (1985) 19-27.

[16] Beagles, A. E., Whiteman, J. R., General conical singularities in three-dimensional Poisson problems, Math. Methods Appl. Sci. **11** (1989) 215-235.

[17] Ben M'Barek, Merigot, M., Régularité de la solution d'un problème de transmission, C. R. Acad. Sci. Paris Sér. I Math. **280** (1975) 1591-1593.

[18] Benthem, J. P., State of stress at the vertex of a quarter-infinite crack in a half-space, Internat. J. Solids and Structures **13** (1977) 479-492.

[19] Benthem, J. P., The quarter-infinite crack in a half-space; alternative and additional solutions, Internat. J. Solids and Structures **16** (1980) 119-130.

[20] Birman, M. Sh., Skvortsov, G. E., *On the quadratic summability of the highest derivatives of the solution of the Dirichlet problem in a domain with piecewise smooth boundary*, Izv. Vyssh. Uchebn. Zaved. Mat. **5** (1962) 12-21.

[21] Blum, H., Rannacher, R., *On the boundary value problem of the biharmonic operator on domains with angular corners*, Math. Methods Appl. Sci. **2** (1980) 556-581.

[22] Boersma, J., *Singularity exponents for complementary sectors*, Electron. Lett. **27** (1991) 1484-1485.

[23] Boussinesq, M. J., *Application des Potentiels a l'étude de l'équilibre et du mouvement des solides élastiques*, Gauthier - Villars, Paris (1885).

[24] Brown, S. N., Stewartson, K., *Flow near the apex of a plane delta wing*, J. Inst. Math. Appl. **5** (1969) 206-216.

[25] I. Chavel, *Lowest eigenvalue inequalities*, Geometry of the Laplace operator, Proc. Symp. Pure Math. **36**, Amer. Math. Soc., Providence, R.I., 1980, 79-89.

[26] I. Chavel, *Eigenvalues in Riemannian geometry*, Academic Press, New York 1984.

[27] Cherepanov, G. P., *Fracture: a topical encyclopedia of current knowledge,* Krieger Pub. Co.1996.

[28] Cherepanov, G. P., *Methods of fracture mechanics: solid matter physics,* Kluwer, Dordrecht 1997.

[29] Costabel, M., Dauge, M., *Construction of corner singularities for Agmon-Douglis-Nirenberg elliptic systems*, Math. Nachr. **162** (1993) 209-237.

[30] Costabel, M., Dauge, M., *General edge asymptotics of solutions of second order elliptic boundary value problems I, II,* Proc. Roy. Soc. Edinburgh **123A** (1993) 109-155, 157-184.

[31] Costabel, M., Dauge, M., *Computation of corner singularities in linear elasticity*, Boundary Value Problems and Integral Equations in Nonsmooth Domains, Proceedings of the conference, held at the CIRM, Luminy, France, May 3-7, 1993. New York, NY: Marcel Dekker, Lect. Notes Pure Appl. Math. 167, 59-68 (1995).

[32] Courant, R., Hilbert, D., *Methoden der mathematischen Physik, I, II,* Springer, Berlin-Heidelberg-New York 1968.

[33] Dahlberg, B. E. J., *L^q estimates for Green potentials in Lipschitz domains*, Math. Scand. **44** (1979) 149-170.

[34] Dahlberg, B. E. J., Kenig, C. E., *Hardy spaces and the Neumann problem in L^p for Laplace's equation in Lipschitz domains*, Ann. of Math. (2) **125** 437-465 (1987).

[35] Dahlberg, B. E. J., Kenig, C. E., *L^p estimates for the 3-dimensional system of elastostatics on Lipschitz domains*, Lecture Notes in Pure and Appl. Math. **122** (1990) 621-634.

[36] Dahlberg, B. E. J., Kenig, C. E., Verchota, G. C., *The Dirichlet problem for the biharmonic equation in a Lipschitz domain*, Ann. Inst. Fourier (Grenoble) **36** (1986) 3, 109-135.

[37] Dahlberg, B. E. J., Kenig, C. E., Verchota, G. C., *Boundary value problems for the systems of elastostatics in Lipschitz domains*, Duke Math. J. **57** (1988) 3, 795-818.

[38] Dauge, M., *Opérateur de Stokes dans des espaces de Sobolev à poids sur des domaines anguleux,* Canad. J. Math. **34** (1982) 4, 853-882.

[39] Dauge, M., *Probleme de Dirichlet pour le laplacien dans un polygone curviligne*, J. Differential Equations **70** (1987) 93-113.

[40] Dauge, M., *Problemes de Neumann et de Dirichlet sur un polyedre dans $\mathbb{R}^3$: regularite dans des espaces de Sobolev L_p*, C. R. Acad. Sci. Paris Sér. I Math. **307** (1988) 1, 27-32.

[41] Dauge, M., *Elliptic boundary value problems in corner domains - smoothness and asymptotics of solutions,* Lecture Notes in Mathematics, Vol. 1341, Springer-Verlag Berlin 1988.

[42] Dauge, M., *Stationary Stokes and Navier-Stokes systems on two- and three-dimensional domains with corners. Part I: Linearized equations*, SIAM J. Math. Anal. **20** (1989), 74-97.

[43] Dauge, M., *Problemes mixtes pour le laplacien dans des domaines polyedraux courbes*, C. R. Acad. Sci. Paris Sér. I Math. **309**(1989) 8, 553-558.

[44] Dauge, M., *Higher order oblique derivatives problems on polyhedral domains*, Comm. Partial Differential Equations **14** (1989) 8/9, 1193-1227.

[45] Dauge, M., Nicaise, S., *Oblique derivative and interface problems on polygonal domains and networks*, Comm. Partial Differential Equations **14** (1989) 8/9, 1147-1192.

[46] Deny, J, Lions, J.-L., *Les espaces du type de Beppo Levi*, Ann. Inst. Fourier (Grenoble) **5** (1955) 305-370.

[47] Deuring, P., *L_p-theory for the Stokes system in 3d domains with conical boundary points*, Indiana Univ. Math. J. **47** (1998) 1, 11-47.

[48] Dimitrov, A., Andrä, H., Schnack, E., *Efficient computation of corner singularties in 3D-elasticity*, to appear in Internat. J. Numer. Methods Engng.

[49] Douglis, A., Nirenberg, L., *Interior estimates for elliptic systems of partial differential equations*, Comm. Pure Appl. Math. **8** (1955) 503-538.

[50] Eck, C., Nazarov, S. A., Wendland, W. L., *Asymptotic analysis for the mixed boundary value problem associated to a contact problem*, Preprint 97-9, Math. Inst., Univ. of Stuttgart, 1997.

[51] Eskin, G. I., *General boundary value problems for equations of principal type in a plane domain with angular points*, Uspekhi Mat. Nauk **18** (1963) 3, 241-242 (in Russian).

[52] Eskin, G. I., *The conjunction problem for equations of principal type with two independent variables*, Tr. Mosk. Mat. Obs. **21** (1970) 245-292; Engl. transl. in: Trans. Moscow Math. Soc. **21** (1970).

[53] Eskin, G. I., *Boundary value problems for second order elliptic equations in domains with corners*, Pseudodifferential Operators and Applications, Proc. Symp., Notre Dame/Indiana 1984, Proc. Sympos. in Pure Math. **43** (1985) 105-131.

[54] Eskin, G. I., *Index formulas for elliptic boundary value problems in plane domains with corners*, Trans. Amer. Math. Soc. **314** (1989) 1, 283-348.

[55] Faber, G., *Beweis, dass unter allen homogenen Membranen von gleicher Fläche und gleicher Spannung die kreisförmige den tiefsten Grundton gibt*, Sitzungsberichte der Math.-Phys. Klasse der Bayr. Akad. der Wiss. München 1923, 169-172.

[56] Fabes, E. B., *Boundary value problems of linear elastostatics and hydrostatics on Lipschitz domains*, Proc. Center for Math. Anal., A.N.U. Miniconference on Linear Analysis and Function Spaces, Vol. 9 (1985) 27-45.

[57] Fabes, E. B., Kenig, C. E., Verchota, G. C., *The Dirichlet problem for the Stokes system on Lipschitz domains*, Duke Math. J. **57** (1988) 3, 769-793.

[58] Fichera, G., *Existence theorems in elasticity*, Handbuch der Physik, Springer-Verlag 1973.

[59] Fichera, G., *Asymptotic behavior of the electric field and density of the electric charge near singular points of a conducting surface*, Russ. Math. Surveys **30** (1975) 3, 107-127.

[60] Fichera, G., *Asymptotic behavior of the electric field near the singular points of the conductor surface*, Atti Accad. Naz. Lincei Cl. Sci. Fis. Mat. Natur. Rend. Lincei (8) **60** (1976) 13-20.

[61] Fichera, G., *Asymptotic behavior of the electric field near singular points of the conducting surface*, Partial Differential Equations, Proc. All-Union Conf., Moscow 1976, dedic. I. G. Petrovskiĭ, 230-235 (1978).

[62] Fichera, G., *Singularities of 3-dimensional potential functions at the vertices and at the edges of the boundary*, Ordinary and Partial Differential Equations, Proc. 5th Conf., Dundee 1978, Lecture Notes in Math. **827** (1980) 107-114.

[63] Friedland, S., Hayman, W. K., *Eigenvalue inequalities for the Dirichlet problem on spheres and the growth of subharmonic functions*, Comment. Math. Helv. **51** (1976) 133-161.

[64] Fufaev, V. V., *On conformal mappings of domains with angles*, Dokl. Akad. Nauk SSSR **152** (1963), 4, 838-840 (in Russian).

[65] Gantmakher, F., *Theory of matrices*, Chelsea, New York 1959.

[66] Gel'fand, I. M., Shilov, G. E., *Generalized functions, Vol. 1: Properties and operations*, Fizmatgiz, Moscow 1958; English transl. in: Academic Press, New York 1968.

[67] Gel'fond, O. A., *The calculus of finite differences*, Nauka, Moscow 1967 (in Russian).

[68] Ghahremani, F., *A numerical variational method for extracting 3D singularities*, Internat. J. Solids and Structures **27** (1991) 1371-1386.

[69] Girault, V., Raviart, P.-A., *Finite elements methods for Navier-Stokes equations*, Springer-Verlag, Berlin - Heidelberg - New York - Tokyo 1986.

[70] Gobert, J., *Une inéquation fondamentale de la théorie de l'élasticité*, Bull. Soc. Roy. Sci. Liège No. 3 and 4 (1962).

[71] Gohberg, I., Goldberg, S., Kaashoek, M. A., *Classes of linear operators, Vol. 1*, Oper. Theory: Adv. Appl. **49** Birkhäuser-Verlag, Basel-Boston-Berlin 1990.

[72] Gohberg, I., Lancaster, P., Rodman, L., *Matrix polynomials*, Academic Press, New York - London - Paris 1982.

[73] Gohberg, I., Sigal, E. I., *Operator generalization of the theorem on the logarithmic residue and Rouché's theorem.*, Mat. Sb.**84** (1971) 4, 607-629; English transl. in: Math. USSR Sb. **13** (1971).

[74] Grisvard, P., *Probleme de Dirichlet dans un cone*, Ricerche Mat. **20** (1971) 175-192.

[75] Grisvard, P., *Alternative de Fredholm relative au probleme de Dirichlet dans un polygone ou un polyedre*, Boll. Un. Mat. Ital., IV. Ser. 5 (1972) 132-164.

[76] Grisvard, P., *Alternative de Fredholm relative au probleme de Dirichlet dans un polyedre*, Ann. Scuola Norm. Sup. Pisa Cl. Sci., IV, Ser. 2 (1975) 359-388.

[77] Grisvard, P., *Singular solutions of elliptic boundary value problems in polyhedra*, Portugal. Math. **41** (1982) 367-382.

[78] Grisvard, P., *Elliptic problems in nonsmooth domains*, Monographs and Studies in Mathematics **21** Pitman, Boston 1985.

[79] Grisvard, P., *Le probleme de Dirichlet dans l'espace W_p^1*, Portugal. Math. **43** (1986) 393-398.

[80] Grisvard, P., *Problemes aux limites dans les polygones. Mode d'emploi*, Bull. Dir. Etud. Rech., Ser. C, No. 1 (1986) 21-59.

[81] Grisvard, P., *Singularities in elasticity theory*, Applications of Multiple Scaling in Mechanics, Proc. Int. Conf. Paris 1986, Rech. Math. Appl. **4** (1987) 134-150.

[82] Grisvard, P., *Le probleme de Dirichlet pour les equations de Lame*, C. R. Acad. Sci. Paris Sér. I Math. **304** (1987) 71-73.

[83] Grisvard, P., *Edge behavior of the solution of an elliptic equation*, Math. Nachr. **132** (1987) 281-299.

[84] Grisvard, P., *Singularities in elliptic boundary value problems*, Delft Prog. Rep. **13** (1989) 349-373.

[85] Grisvard, P., *Singularites en elasticite*, Arch. Rational Mech. Anal. **107** (1989) 2, 157-180.

[86] Grisvard, P., *Singularities in boundary value problems*, Recherches en Mathematiques Appliquees. 22. Paris: Masson. Berlin: Springer-Verlag 1992.

[87] Gromov, M., Shubin, M. A., *The Riemann-Roch theorem for general elliptic operators*, C. R. Acad. Sci. Paris Sér. I Math. **314** (1992) 5, 363-367.

[88] Gromov, M., Shubin, M. A., *The Riemann-Roch theorem for elliptic operators and solvability of elliptic equations with additional conditions on compact subsets*, Invent. Math. **117** (1994) 1, 165-180.

[89] Hanna, M. S., Smith, K. T., *Some remarks on the Dirichlet problem in piecewise smooth domains*, Comm. Pure Appl. Math. **20** (1967) 575-593.

[90] Hinder, R., Meister, E., *Regarding some problems of the Kutta-Joukowski condition in lifting surface theory*, Math. Nachr. **184** (1997) 191-228.

[91] Hsiao, G., Stephan, E. P., Wendland, W. L., *On the Dirichlet problem in elasticity for a domain exterior to an arc*, J. Comput. Appl. Math. **34** (1991) 1, 1-19.

[92] Jerison, D., Kenig, C. E., *The functional calculus for the Laplacian on Lipschitz domains*, Journ. "Equations Deriv. Partielles", St. Jean-De-Monts 1989, Exp. No.4, 10 p. (1989).

[93] Jerison, D., Kenig, C. E., *The inhomogeneous Dirichlet problem in Lipschitz domains*, J. Funct. Anal. 130, No.1, 161-219 (1995).

[94] John, F., *Plane waves and spherical means applied to partial differential equations*, Interscience, New York 1955.

[95] Kalex, H.-U., *On the flow of a temperature-dependent Bingham fluid in non-smooth bounded two-dimensional domains*, Z. Anal. Anwendungen **11** (1992) 4, 509-530.

[96] Kato, T., *Perturbation Theory for Linear Operators*, Springer Verlag, Berlin-Heidelberg-New York 1966.

[97] Keldysh, M. V., *On the eigenvalues and eigenfunctions of certain classes of nonselfadjoint equations*, Dokl. Akad. Nauk SSSR **77** (1951) 1, 11-14 (in Russian).

[98] Keldysh, M. V., *On the completeness of eigenfunctions of certain classes of nonselfadjoint operators*, Uspekhi Mat. Nauk **26** (1971) 4, 15-41 (in Russian).

[99] Keller, J. B., *Singularities at the tip of a plane angular sector*, J. Math. Phys. **40** (1999) 2, 1087-1092.

[100] Kellogg, R. B., *Singularities in interface problems*, Symposium on Numerical Solutions of Partial Differential Equations, II, Academic Press, 351-400 (1971)

[101] Kellogg, R. B., Osborn, J., *A regularity result for the Stokes Problem in a convex polygon*, J. Funct. Anal. **21** (1976) 397-431.

[102] Kenig, C. E., *Boundary value problems for linear elastostatics on Lipschitz domains*, Semin. Goulaouic-Meyer-Schwarz 1983-84, Exp. No. 21 (1984) 1-12, École Polytechnique, Palaiseau.

[103] Kenig, C. E., *Boundary value problems for elliptic equations*, Publ. Mat. (Barcelona) **35** (1991) 1, 281-289.

[104] Kenig, C. E., *Harmonic analysis techniques for second order elliptic boundary value problems: dedicated to the memory of Professor Antoni Zygmund*. Regional Conference Series in Mathematics. 83. Providence, RI: American Mathematical Society, xii, 146 p. (1994).

[105] Kenig, C. E., Pipher, J., *Hardy spaces and the Dirichlet problem on Lipschitz domains,* Rev. Mat. Iberoamericana **3** (1987) 2, 191-247.

[106] Kenig, C. E., Pipher, J., *The oblique derivative problem on Lipschitz domains with L^p data,* Amer. J. Math. **110** (1988) 4, 715-737.

[107] Komech, A. I., *Elliptic boundary value problems on manifolds with piecewise smooth boundary,* Mat. Sb. **92** (1973) (in Russian).

[108] Komech, A. I., Merzon, A. E., *General boundary value problems in regions with corners,* Oper. Theory Adv. Appl. **57** (1992) 171-183.

[109] Kondrat'ev, V. A., *Boundary value problems for elliptic equations in domains with conical or angular points,* Tr. Mosk. Mat. Obs. **16** (1967) 209-292; English transl. in: Trans. Moscow Math. Soc. **16** (1967) 227-313.

[110] Kondrat'ev, V. A., *The smoothness of solution of Dirichlet's problem for second order elliptic equations in a region with piecewise smooth boundary,* Differ. Uravn. **6** (1970) 1831-1843; English transl. in: Differential Equations **6** (1970) 1392-1401.

[111] Kondrat'ev, V. A., *Singularities of a solution of Dirichlet's problem for a second order elliptic equation in the neighborhood of an edge,* Differ. Uravn. **13** (1977) 2026-2032; English transl. in: Differential Equations **13** (1977) 1411-1415.

[112] Kondrat'ev, V. A., Oleĭnik, O. A., *Boundary value problems for partial differential elliptic equations in nonsmooth domains,* Uspekhi Mat. Nauk **38** (1983) 2, 3-76; English transl. in: Russian Mathematical Surveys **38** (1983) 2, 1-86.

[113] Kondrat'ev, V. A., Kopachek, I., Oleĭnik, O. A., *Estimates of solutions of elliptic second order equations and of the system of elasticity theory in a neighborhood of a boundary point,* Uspekhi Mat. Nauk **36** (1981) 1, 211-212 (in Russian).

[114] Kondrat'ev, V. A., Kopachek, I., Oleĭnik, O. A., *Estimates of the modulus of a solution of the biharmonic equation and its derivatives in a neighborhood a nonregular boundary point,* Uspekhi Mat. Nauk **36** (1981) 4, 230-231 (in Russian).

[115] Kondrat'ev, V. A., Kopachek, I., Oleĭnik, O. A., *On the behavior of generalized solutions of elliptic second order equations and of the system of elasticity theory in a neighborhood of a boundary point,* Trudy Sem. Petrovsk. **8** (1982) 135-152.

[116] Kopachek, I., Oleĭnik, O. A., *On asymptotic properties of solutions of the system in elasticity theory,* Uspekhi Mat. Nauk **33** (1978) 5, 189-191 (in Russian).

[117] Kopachek, I., Oleĭnik, O. A., *On the behavior of solutions of the system in elasticity theory in a neighborhood of nonregular boundary points and at infinity,* Tr. Mosk. Mat. Obs. **43** (1981) 260-274 (in Russian).

[118] Kozlov, V. A., *Asymptotics of the spectrum of operator pencils generated by elliptic boundary value problems in an angle,* Some Appl. of Funct. Anal. to Problems of Mathem. Physics, Novosibirsk (1988) 82-96 (in Russian).

[119] Kozlov, V. A., *On the spectrum of operator pencils arising in Dirichlet problems for elliptic equations in an angle,* Mat. Zametki **45** (1989), 117-118 (in Russian).

[120] Kozlov, V. A., *On singularities of solutions of the Dirichlet problem for elliptic equations in the neighborhood of corner points,* Algebra i Analiz **1** (1989) 4, 161-177; English transl. in: Leningrad Math. J. **1** (1990) 4, 967-982.

[121] Kozlov, V. A., *The strong zero theorem for an elliptic boundary value problem in an angle,* Mat. Sb. **180** (1989) 6, 831-849; English transl. in: Math. USSR Sb. **67** (1990) 1, 283-302.

[122] Kozlov, V, A., *On the Dirichlet problem for elliptic equations on domains with conical points,* Differ. Uravn. **26** (1990) 6, 1014-1023; English transl. in: Differential Equations **26** (1990), 739-747.

[123] Kozlov, V. A., *On the spectrum of the pencil generated by the Dirichlet problem for an elliptic equations in an angle,* Sibirsk. Mat. Zh. **32** (1991) 2, 74-87; English transl. in: Siberian Math. J. **32** (1991) 2, 238-251.

[124] Kozlov, V. A., *On the behavior of solutions of elliptic boundary value problems in an angle,* Mat. Zametki **49** (1991) 1, 56-62; English transl. in: Math. Notes **49** (1991) 1-2, 40-45.

[125] Kozlov, V. A., *On regularity of solutions of the Dirichlet problem for the polyharmonic operator,* J. Korean Math. Soc. **37** (2000) 5.

[126] Kozlov, V. A., Kondrat′ev, V. A., Maz′ya, V. G, *On sign variation and the absence of "strong" zeros of solutions of elliptic equations*, Izv. Akad. Nauk SSSR Ser. Mat. **53** (1989) 2, 328-344; English transl. in: Math. USSR-Izv. **34** (1990) 2, 337-353.

[127] Kozlov, V. A., Maz′ya, V. G, *Spectral properties of operator bundles generated by elliptic boundary value problems in a cone*, Funktsional. Anal. i Prilozhen. **22** (1988), 38-46; English transl. in: Funct. Anal. Appl. **22** (1988) 2, 114-121.

[128] Kozlov, V. A., Maz′ya, V. G., *On stress singularities near the boundary of a polygonal crack*, Proc. Roy. Soc. Edinburgh **117A** (1991) 31-37.

[129] Kozlov, V. A., Maz′ya, V. G., *On the spectrum of an operator pencil generated by the Neumann problem in a cone*, Algebra i Analiz **3** (1991), 111-131; English transl. in: St. Petersburg Math. J. **3** (1992) 2, 333-353.

[130] Kozlov, V. A., Maz′ya, V. G., *On the spectrum of an operator pencil generated by the Dirichlet problem in a cone*, Mat. Sb. **182** (1991) 5, 638-660; English transl. in: Math. USSR Sb. **73** (1992) 1, 27-48.

[131] Kozlov, V. A., Maz′ya, V. G., *On eigenvalues of the operator pencil generated by the Dirichlet problem for a strongly elliptic system in a polyhedral angle*, Preprint Lith-Math-R-91-14, Univ. Linköping 1991.

[132] Kozlov, V. A., Maz′ya, V. G, *Singularities of solutions to problems of mathematical physics in nonsmooth domains*, Partial Differential Equations and Functional Analysis, In Memory of Pierre Grisvard, Birkhäuser, Boston-Basel-Berlin 1996

[133] Kozlov, V. A., Maz′ya, V. G, *On "power-logarithmic" solutions of the Dirichlet problem for elliptic systems in $K_d \times \mathbb{R}^{n-d}$, where K_d is a d-dimensional cone*, Atti Accad. Naz. Lincei Cl. Sci. Fis. Mat. Natur. Rend. Lincei (9) Mat. Appl. **7** (1996) 1, 17-30.

[134] Kozlov, V. A., Maz′ya, V. G, *On power-logarithmic solutions to the Dirichlet problem for the Stokes system in a dihedral angle*, Math. Methods Appl. Sci. **20** (1997) 315-346.

[135] Kozlov, V. A., Maz′ya, V. G, *Differential equations with operator coefficients* (with applications to boundary value problems for partial differential equations), Monographs in Mathematics, Springer-Verlag 1999.

[136] Kozlov, V. A., Maz′ya, V. G., Roßmann, J., *Elliptic boundary value problems in domains with point singularities*, Mathematical Surveys and Monographs **52**, Amer. Math. Soc., Providence, Rhode Island 1997.

[137] Kozlov, V. A., Maz′ya, V. G., Roßmann, J., *Spectral properties of operator pencils generated by elliptic boundary value problems for the Lamé system*, Rostock. Math. Kolloq. **51** (1997) 5-24.

[138] Kozlov, V. A., Maz′ya, V. G., Roßmann, J., *Conic singularities of solutions to problems in hydrodynamics of a viscous fluid with a free surface*, Math. Scand. **83** (1998) 103-141.

[139] Kozlov, V. A., Maz′ya, V. G., Schwab, C., *On singularities of solutions to boundary value problems near the vertex of a rotational cone*, Preprint LiTH-MAT-R-91-24, University of Linköping, 1991.

[140] Kozlov, V. A., Maz′ya, V. G., Schwab, C., *On singularities of solutions of the displacement problem of linear elasticity near the vertex of a cone*, Arch. Rational Mech. Anal. **119** (1992) 197-227.

[141] Kozlov, V. A., Maz′ya, V. G., Schwab, C., *On singularities of solutions to the Dirichlet problem of hydrodynamics near the vertex of a cone*, J. Reine Angew. Math. **456** (1994) 65-97.

[142] Kozlov, V. A., Roßmann, J., *On the behaviour of a parameter-dependent operator under small variation of the domain and an application to operator pencils generated by elliptic boundary value problems in a cone*, Math. Nachr. **153** (1991) 123-129.

[143] Kozlov, V. A., Roßmann, J., *Singularities of solutions of elliptic boundary value problems near conical points*, Math. Nachr. **170** (1994) 161-181.

[144] Krahn, E., *Über eine von Rayleigh formulierte Minimaleigenschaft des Kreises*, Math. Ann. **94** (1925) 97-100.

[145] Krahn, E., *Über Minimaleigenschaften der Kugel in drei und mehr Dimensionen*, Acta Comm. Univ. Tartu (Dorpat) **A9** (1926) 1-44.

[146] Krasnosel′skiĭ, M. A., *Positive solutions of operator equations*, Noordhoff, Groningen 1964.

[147] Kufner, A., Sändig, A.-M., *Some applications of weighted Sobolev spaces*, Teubner, Leipzig 1987.

[148] Ladyzhenskaya, O. A., *The mathematical theory of viscous incompressible flow*, Gordon and Breach, New York 1963.

[149] Landkof, N. S., *Foundations of modern potential theory*, Grundlehren der mathematischen Wissenschaften **180**, Springer-Verlag, Berlin-Heidelberg-New York 1972.

[150] Leguillon, D., *Computation of 3d singularities in elasticity*, Proc. of the Conf. held at the CIRM, Luminy, France, 1993, New York, Marcel Dekker, Lecture Notes in Pure and Appl. Math. **167** (1995) 161-170.

[151] Leguillon, D., Sanchez-Palencia, E., *Computation of singular solutions in elliptic problems and elasticity*, Masson, Paris, J. Wiley, New York 1987.

[152] Lemrabet, K., *Régularité de la solution d'un problème de transmission*, J. Math. Pures Appl. **56** (1977) 1-38.

[153] Levin, B. J., *Distribution of the zeros of entire functions*, Translations of Math. Monographs **5**, Amer. Math. Soc., Providence, Rhode Islands 1972.

[154] Liebowitz, H. (ed.), *Fracture, An advance treatise, Vol. I-VII*, Academic Press, New York and London, 1968

[155] Lions, J.-L., Magenes, E., *Problemes aux limites non homogenes et applications, Vol. 1*, Dunod, Paris 1968.

[156] Lopatinskiĭ, Ya. B., *On one type of singular integral equations*, Teoret. i Prikl. Mat. (Lwow) **2** (1963) 53-57 (in Russian).

[157] Lozi, R., *Résultats numériques de régularité du problème de Stokes et du laplacien itéré dans un polygone*, RAIRO Anal. Numér. **12** (1978) 3, 267-282.

[158] Magnus, W., Oberhettinger, F., Soni, R. P., *Formulas and theorems for the special functions of mathematical physics*, Grundlehren der mathematischen Wissenschaften **52**, Springer-Verlag, Berlin-Heidelberg-New York 1966.

[159] Malvern, L. E., *Introduction to the mechanics of a continuous medium*, Prentice-Hall Inc., Englewood Cliffs, New Jersey 1969.

[160] Markus, A. S., *Introduction to the Spectral Theory of Polynomial Operator Pencils*, Translations of Mathematical Monographs **71**, Amer. Math. Soc., Providence, Rhode Island 1980.

[161] Markus, A. S., Sigal, E. I., *The multiplicity of the characteristic number of an analytic operator function*, Mat. Issled. **5** (1970) 3, 129-147 (in Russian).

[162] Maslovskaya, L. V., *Behavior of solutions of boundary value problems for the biharmonic equation in domains with angular points*, Differ. Uravn. **19** (1983) 12, 2172-2175 (in Russian).

[163] Maz'ya, V. G., *Behavior of solutions to the Dirichlet problem for the biharmonic operator at a boundary point*, Equadiff IV (Proc. Czechoslovak. Conference Differential Equations and Applications, Prague 1977), Lecture Notes in Math. **703** (1979) 250-262.

[164] Maz'ya, V. G., *Sobolev spaces*, Izdat. Leningrad Univ., Leningrad 1985; English transl.: Springer-Verlag, Berlin-New York 1985.

[165] Maz'ya, V. G., *On the Wiener-type regularity of a boundary point for the polyharmonic operator*, Appl. Anal. **71** (1999) 149-165.

[166] Maz'ya, V. G., Donchev, T., *Regularity in the sense of Wiener of a boundary point for a polyharmonic operator*, C. R. Acad. Bulgare Sci. **36** (1983) 177–179; English transl. in: Amer. Math. Soc. Transl. **137** (1987) 53-55.

[167] Maz'ya, V. G., Levin, A. V., *On the asymptotics of density of harmonic potentials near the vertex of a cone*, Z. Anal. Anwendungen **8** (1989) 6, 501-514 (in Russian).

[168] Maz'ya, V. G., Morozov, N. F., Plamenevskiĭ, B. A., *On linear bending of a plate with a crack*, Differential and Integral Equations, Boundary Value Problems, I. M. Vekua Memorial Collection, Izdat. Tbilissi Univ. 1979, 145-163; English transl. in: Amer. Math. Soc. Transl. **123** (1984) 125-139.

[169] Maz'ya, V. G., Nazarov, S. A., *The apex of a cone can be irregular in Wiener's sense for a fourth order elliptic equation*, Mat. Zametki **39** (1986) 1, 24-28 (in Russian).

[170] Maz'ya, V. G., Nazarov, S. A., *On the singularities of the solutions of the Neumann problem in a conical point*, Sibirsk. Mat. Zh. **30** (1989) 3, 52-63 (in Russian).

[171] Maz'ya, V. G., Nazarov, S. A., Plamenevskiĭ, B. A., *On the singularities of solutions of the Dirichlet problem in the exterior of a slender cone*, Mat. Sb. **122** (1983) 4, 435-457; English transl. in: Math. USSR Sb. **50** (1985) 2, 415-437.

[172] Maz'ya, V. G., Nazarov, S. A., Plamenevskiĭ, B. A., *On homogeneous solutions of the Dirichlet problem in the exterior of a slender cone*, Dokl. Akad. Nauk SSSR **266** (1982) 2; English transl. in: Soviet Math. Dokl. **26** (1982) 2, 320-323.

[173] Maz'ya, V. G., Nazarov, S. A., Plamenevskiĭ, B. A., *On the singularities of solutions of the Dirichlet problem in the exterior of a thin cone*, Mat. Sb. **122** (1983) 4, 435-457 (in Russian).

[174] Maz'ya, V. G., Nazarov, S. A., Plamenevskiĭ, B. A., *Asymptotische Theorie elliptischer Randwertaufgaben in singulär gestörten Gebieten I*, Akademie-Verlag Berlin 1991, English transl.: *Asymptotic theory of elliptic boundary value problems in singularly perturbed domains, Vol. 1*, Birkhäuser, Basel 2000.

[175] Maz'ya, V. G., Plamenevskiĭ, B. A., *Problem with oblique derivative in a domain with piecewise smooth boundary*, Funktsional. Anal. i Prilozhen. **5** (1971) 3, 102-103; English transl. in: Funct. Anal. Appl. **5** (1971).

[176] Maz'ya, V. G., Plamenevskiĭ, B. A., *On the asymptotics of the Navier-Stokes equations near edges*, Dokl. Akad. Nauk SSSR **210** (1973) 4, 803-806.

[177] Maz'ya, V. G., Plamenevskiĭ, B. A., *On elliptic boundary value problems in domains with piecewise smooth boundary*, Trudy Simp. Mekh. Splosnoĭ Sredy 1971, Metsniereba, Tbilissi 1973, 171-182 (in Russian).

[178] Maz'ya, V. G., Plamenevskiĭ, B. A., *On boundary value problems for a second order elliptic equation in a domain with edges*, Vestnik Leningrad. Univ. Mat. Mekh. Astron. **1** (1975) 102-108; English transl. in: Vestnik Leningrad Univ. Math. **8** (1980) 99-106.

[179] Maz'ya, V. G., Plamenevskiĭ, B. A., *Weighted spaces with nonhomogeneous norms and boundary value problems in domains with conical points*, Elliptische Differentialgleichungen (Meeting in Rostock, 1977), Univ. Rostock, 1978, pp. 161-189; English transl. in: Amer. Math. Soc. Transl. (2) **123** (1984).

[180] Maz'ya, V. G., Plamenevskiĭ, B. A., *On the coefficients in the asymptotics of solutions of elliptic boundary value problems in domains with conical points*, Math. Nachr. **76** (1977) 29-66; English transl. in: Amer. Math. Soc. Transl. (2) **123** (1984) 57-88.

[181] Maz'ya, V. G., Plamenevskiĭ, B. A., *Elliptic boundary value problems on manifolds with singularities*, Probl. Mat. Anal. **6** (1977) 85-142.

[182] Maz'ya, V. G., Plamenevskiĭ, B. A., *Estimates in L_p and Hölder classes and the Miranda-Agmon maximum principle for solutions of elliptic boundary value problems in domains with singular points on the boundary*, Math. Nachr. **81** (1978) 25-82; English transl. in: Amer. Math. Soc. Transl., Vol. 123 (1984) 1-56.

[183] Maz'ya, V. G., Plamenevskiĭ, B. A., *L_p estimates for solutions of elliptic boundary value problems in domains with edges*, Tr. Moskov. Mat. Obs. **37** (1978) 49-93; Engl. transl. in: Trans. Moscow. Math. Soc. **37** (1980) 49-97.

[184] Maz'ya, V. G., Plamenevskiĭ, B. A., *Estimates of the Green function and Schauder estimates for solutions of elliptic boundary value problems in a dihedral angle*, Sibirsk. Mat. Zh. **19** (1978) 5, 1065-1082 (in Russian).

[185] Maz'ya, V. G., Plamenevskiĭ, B. A., *Schauder estimates of solutions of elliptic boundary value problems in domains with edges on the boundary*, Partial Differential Equations (Proc. Sem. S. L. Sobolev, 1978, No. 2) Inst. Mat. Sibirsk. Otdel. Akad. Nauk SSSR, Novosibirsk, 1978, 69-102; English transl. in: Amer. Math. Soc. Transl. **123** (1984) 141-169.

[186] Maz'ya, V. G., Plamenevskiĭ, B. A., *On the asymptotics of the fundamental solution of elliptic boundary value problems in regions with conical points*, Probl. Mat. Anal. **7** (1979) 100-145; English transl. in: Sel. Math. Sov. **4** (1985) 363-397.

[187] Maz'ya, V. G., Plamenevskiĭ, B. A., *On the maximum principle for the biharmonic equation in domains with conical points*, Izv. Vyssh. Uchebn. Zaved. Mat. **2** (1981) 52-59 (in Russian).

[188] Maz'ya, V. G., Plamenevskiĭ, B. A., *On the properties of solutions of three-dimensional problems of elasticity theory and hydrodynamics in domains with isolated singular points* (in Russian), Dinamika Sploshn. Sredy **50** (1981) 99-121; English transl. in: Amer. Math. Soc. Transl. **123** (1984) 109-123.

[189] Maz'ya, V. G., Plamenevskiĭ, B. A., *The first boundary value problem for the classical equations of mathematical physics on piecewise smooth domains*, Part I: Z. Anal. Anwendungen **2** (1983) 335-359, Part 2: Z. Anal. Anwendungen **2** (1983) 523-551 (in Russian).

[190] Maz'ya, V. G., Plamenevskiĭ, B. A., Stupyalis, L. I., *The three-dimensional problem of steady-state motion of a fluid with a free surface*, Diff. Uravn. i Primenen.–Trudy Sem.

Protsessy Optimal. Upravl. **23** (1979); English transl. in: Amer. Math. Soc. Transl. **123** (1984) 171-268).

[191] Maz'ya, V. G., Roßmann, J., *Über die Lösbarkeit und die Asymptotik der Lösungen elliptischer Randwertaufgaben in Gebieten mit Kanten I–III,* Preprint P-MATH 07/84, 30/84, 31/84 Institut für Math., Akademie der Wissenschaften der DDR, Berlin 1984.

[192] Maz'ya, V. G., Roßmann, J., *Über die Asymptotik der Lösungen elliptischer Randwertaufgaben in der Umgebung von Kanten,* Math. Nachr. **138** (1988) 27-53.

[193] Maz'ya, V. G., Roßmann, J., *On the Agmon-Miranda maximum principle for solutions of elliptic equations in polyhedral and polygonal domains,* Ann. Global Anal. Geom. **9** (1991) 3, 253-303.

[194] Maz'ya, V. G., Roßmann, J., *On the Agmon-Miranda maximum principle for solutions of strongly elliptic equations in domains of $\mathbb{R}^n$ with conical points,* Ann. Global Anal. Geom. **10** (1992) 125-150.

[195] Maz'ya, V. G., Roßmann, J., *On the behaviour of solutions to the Dirichlet problem for second order elliptic equations near edges and polyhedral vertices with critical angles,* Z. Anal. Anwendungen **13** (1994) 1, 19-47.

[196] Maz'ya, V. G., Shaposhnikova, T. O., *Theory of multipliers in spaces of differentiable functions,* Monographs and Studies in Mathematics **23**, Pitman, Boston, Masson, London 1985.

[197] Meister, E., Penzel, F., Speck, F.-O., Teixeira, F. S., *Two canonical wedge problems for the Helmholtz equation,* Math. Methods Appl. Sci. **17** (1994) 11, 877-899.

[198] Melrose, R. B., *Pseudodifferential operators, corners and singular limits,* Proc. Int. Congr. Math., Kyoto (Japan), Vol I (1991) 217-234.

[199] Melzer, H., Rannacher, R., *Spannungskonzentrationen in Eckpunkten der Kirchhoffschen Platte,* Preprint 270, SFB 72, Univ. Bonn 1979 (see also Bauingenieur **55** (1980) 181-184)

[200] Mennicken, R., Möller, M., *Root functions, eigenvectors, associated vectors and the inverse of a holomorphic operator function,* Arch. Math. **42** (1984) 455-463.

[201] Mikhlin, S. G., Partielle Differentialgleichungen in der mathematischen Physik, H. Deutsch, Frankfurt/Main 1978.

[202] Nadirashvili, N. S., *Multiple eigenvalues of the Laplace operator,* Mat. Sb. **133** (1987) 2, 223-237 (in Russian).

[203] Nazarov, S. A., *Weight functions and invariant integrals,* Vychisl. Mekh. Deform. Tverd. Tela **1** (1990) 17-31 (in Russian).

[204] Nazarov, S. A., *Crack on a junction of anisotropic bodies, Singularities of strains and invariant integrals,* Prikl. Mat. Mekh. **62** (1998) 3, 489-502 (in Russian).

[205] Nazarov, S. A., *Singularities at angular points of a trailing edge under the Joukowski-Kutta condition,* Bach, M. (ed.) et al, Analysis Numerics and Applications of Differential and Integral Equations, Harlow: Addison Wesley Longman. Pitman Res. Notes Math., Ser. 379 (1998) 153-157.

[206] Nazarov, S. A., *The polynomial property of self-adjoint elliptic boundary value problems and an algebraic description of their attributes,* Uspekhi Mat. Nauk **54** (1999) 5, 77-142; English transl. in: Russian Mat. Surveys **54** (1999) 5, 947-1014.

[207] Nazarov, S. A., Plamenevskiĭ, B. A., *Elliptic problems in domains with piecewise smooth boundaries,* De Gruyter Expositions in Mathematics **13**, Berlin - New York 1994.

[208] Nicaise, S., *About the Lamé system in a polygonal or polyhedral domain and a coupled problem between the Lamé system and the plate equation. I: Regularity of the solutions,* Ann. Scuola Norm. Sup. Pisa Cl. Sci. (4) **19** (1992) 3, 327-361.

[209] Nicaise, S., *About the Lamé system in a polygonal or polyhedral domain and a coupled problem between the Lamé system and the plate equation. II: Exact controllability,* Ann. Scuola Norm. Sup. Pisa Cl. Sci. (4) **20** (1993) 2, 163-191.

[210] Nicaise, S., *Regularity of the weak solution of the Lamé system in nonsmooth domains,* Bandle, C. (ed.) et al., Progress in Partial Differential Equations: Calculus of Variations, Applications. 1st European conference on elliptic and parabolic problems, Pont-a-Mousson, France, June 1991. Harlow, Essex: Longman Scientific; Technical, Pitman Res. Notes Math. Ser. 267 (1993) 272-284.

[211] Nicaise, S., *Polygonal interface problems,* Verlag Peter Lang GmbH, Frankfurt/Main 1993.

[212] Nicaise, S., Sändig, A.-M., *General interface problems,* Part 1: Math. Methods Appl. Sci. **17** (1994) 395-430, Part 2: Math. Methods Appl. Sci. **17** (1994) 431-450.

[213] Nikishkin, V. A., *Singularities of the solutions to the Dirichlet problem for a second order equation in a neighborhood of an edge*, Vestnik Moskov. Univ Ser. I Mat. Mekh. **2** (1979) 51-62; English transl. in: Moscow Univ. Math. Bull. **34** (1979) 2, 53-64.

[214] Nikol'skiĭ, S. M., *The Dirichlet problem in domains with corners*, Dokl. Akad. Nauk SSSR **109** (1956) 1, 33-35 (in Russian).

[215] Nikol'skiĭ, S. M., *Boundary properties of functions defined on domains with angular points I-III*, Mat. Sb. **40** (1956) 3, 303-318, **44** (1957) 1, 127-144, **45** (1958) 2, 181-194 (in Russian).

[216] Noble, B., Hussain, M. A., Pu, S. L., *Apex singularities for corner cracks under opening, sliding and tearing modes*, in Fracture Mechanics in Engineering Application, Bangalore, India, 1979 (Sijthoff & Noordhoff, 1979).

[217] Oleĭnik, O. A., Yosifyan, G. A., *On singularities at the boundary points and uniqueness theorems for solutions of the first boundary value problem of elasticity*, Comm. Partial Differential Equations **12** (1977) 9, 937-969.

[218] Oleĭnik, O. A., Yosifyan, G. A., *A priori estimates for solutions of the first boundary value problem for the system of elasticity theory and applications*, Uspekhi Mat. Nauk **32** (1977) 5, 191-192.

[219] Oleĭnik, O. A., Yosifyan, G. A., Tavkhelidze, I. N., *Estimates of solutions of the biharmonic equation in a neighborhood of nonregular boundary points and at infinity*, Uspekhi Mat. Nauk **33** (1978) 181-182 (in Russian).

[220] Oleĭnik, O. A., Yosifyan, G. A., Tavkhelidze, I. N., *On the asymptotics of solutions of the biharmonic equation in a neighborhood of nonregular boundary points and at infinity*, Tr. Mosk. Mat. Obs. **42** (1981) 160-175 (in Russian).

[221] Orlt, M., Sändig, A.-M., *Regularity of viscous Navier-Stokes flows in nonsmooth domains*, Boundary Value Problems and Integral Equations in Nonsmooth Domains, Proceedings of the Conference in Luminy, France, 1993, Lecture Notes in Pure and Appl. Math. **167** (1995) 185-201.

[222] Osborn, J. E., *Regularity of solutions of the Stokes problem in a polygonal domain.* in "Numerical Solution of PDE III" (SYNSPADE proceedings), B. Hubbard Editor, pp. 393 - 411, Academic Press, New York 1975.

[223] Parton, V. Z., Perlin, P. I., Mathematical methods of the theory of elasticity, Mir Publishers, Moscow 1984 (Russian edition 1981).

[224] Pazy, A., *Asymptotic expansion of solution of ordinary differential equation in Hilbert space*, Arch. Rational Mech. Anal. **24** (1967) 193-218.

[225] Pipher, J., Verchota, G., *The Dirichlet problem in L^p for the biharmonic equation on Lipschitz domains*, Amer. J. Math. **114** (1992) 5, 923-972.

[226] Pipher, J., Verchota, G., *A maximum principle for biharmonic functions in Lipschitz and C^1 domains*, Comment. Math. Helv. **68** (1993) 3, 385-414.

[227] Pipher, J., Verchota, G., *Maximum principles for the polyharmonic equation on Lipschitz domains*, Potential Anal. **4** (1995) 6, 615-636.

[228] Plamenevskiĭ, B. A., Algebras of pseudodifferential operators, Nauka, Moscow 1986; English transl. in: Mathematics and Its Applications, Soviet Series, 43, Kluwer Academic Publishers, Dordrecht etc. 1989.

[229] Pólya, G., Szegö, G., *Isoperimetric inequalities in mathematical physics,* Princeton University Press 1951.

[230] Protter, M. H., Weinberger, H. F., *Maximum principles in differential equations,* Prentice Hall Partial Differential Equations Series, Englewood Cliffs, New Jersey 1967.

[231] Roßmann, J., *The Asymptotics of the solutions of linear elliptic variational problems in domains with edges*, Z. Anal. Anwendungen **9** (1990) 6, 565-578.

[232] Samarskiĭ, A. A., *On the influence of constraints on the fundamental frequencies of closed volumes,* Uspekhi Mat. Nauk **5** (1950) 3, 133-134 (in Russian).

[233] Sändig, A.-M., Richter, U., Sändig, R., *The regularity of boundary value problems for the Lamé equations in a polygonal domain*, Rostock. Math. Kolloq. **36** (1989) 21-50.

[234] Sändig, A.-M., Sändig, R., *Singularities of non-rotationally symmetric solutions of boundary value problems for the Lamé equation in a 3-dimensional domain with conical points*, Symp. "Analysis on Manifolds with Singularities" Breitenbrunn 1990, Teubner-Texte zur Mathematik **131** (1992) 181-193,

[235] Schmitz, H., Volk, K., Wendland, W. L., *Three-dimensional singularities of elastic fields near vertices*, Numer. Methods Partial Differential Equations **9** (1993) 3, 323-337.

[236] Schulze, B.-W., *Corner Mellin operators and reduction of orders with parameters*, Ann. Scuola Norm. Sup. Pisa Cl. Sci. (4) **16** (1989) 1, 1-81.

[237] Schulze, B.-W., *Pseudo-differential operators on manifolds with singularities*, Studies in Mathematics and Its Applications. 24. North-Holland, Amsterdam etc., 1991.

[238] Schulze, B.-W., *The Mellin pseudo-differential calculus on manifolds with corners*, Contributions from the international workshop "Analysis on manifolds with singularities", Breitenbrunn (Germany) 1990, Teubner, Stuttgart, Teubner-Texte zur Mathematik **131** (1992) 208-289.

[239] Schulze, B.-W., *Pseudo-differential boundary value problems, conical singularities and asymptotics*, Mathematical Topics. 4. Akademie-Verlag, Berlin 1994.

[240] Schulze, B.-W., *Boundary value problems in Boutet de Monvel's algebra for manifolds with conical singularities. I*, Pseudo-differential Calculus and Mathematical Physics, Mathematical Topics. 5. Akademie-Verlag, Berlin (1994) 97-209 .

[241] Schulze, B.-W., *Boundary value problems in Boutet de Monvel's algebra for manifolds with conical singularities. II*, Boundary Value Problems, Schrödinger Operators, Deformation Quantization. Mathematical Topics. 8. Akademie-Verlag (1995) 70-205 .

[242] Schulze, B.-W., *Anisotropic edge pseudo-differential operators with discrete asymptotics*, Math. Nachr. **184** (1997) 73-125.

[243] Seif, J. B., *On the Green's function for the biharmonic equation on an infinite wedge*, Trans. Amer. Math. Soc. **182** (1973) 241-260.

[244] Shen, C. L., Shieh, C. T., *Some properties of the first eigenvalue of the Laplace operator on the spherical band in S^2*, SIAM J. Math. Anal. **23** (1992) 5, 1305-1308.

[245] Shubin, M. A., *L^2 Riemann-Roch theorem for elliptic operators*, Geom. Funct. Anal. **5** (1995) 2, 482-527.

[246] Solonnikov, V. A., *On Green's matrices for elliptic boundary value problems I,II*, Tr. Mat. Inst. Steklova **110** (1970), **116** (1971); English transl. in: Proc. Steklov Inst. Math. **110** (1970) 123-170, **116** (1971) 187-226.

[247] Solonnikov, V. A., *Solvability of a three-dimensional boundary value problem with a free surface for the stationary Navier-Stokes system*, Banach Center Publ. **10** (1983) 361-403.

[248] Sperner, E., *Zur Symmetrisierung von Funktionen auf Sphären*, Math. Z. **134** (1973) 317-327.

[249] Stein, M. E., Weiss, G., *Introduction to Fourier analysis on Euclidean spaces*, Princeton Univ. Press 1971.

[250] Stephan, E. P., Whiteman, J. R., *Singularities of the Laplacian at corners and edges of three-dimensional domains and their treatment with finite element methods*, Technical report BICOM 81/1, Brunel University, Institute of Computational Mathematics, 1981, also: Math. Methods Appl. Sci. **10** (1988) 339-350.

[251] Szabo, B. A., Babuška, I., *Computation of the amplitude of stress singular terms for cracks and reentrant corners*, Fracture Mechanics, 19th Symposium, ASTM STP 969, American Society for Testing and Materials, Philadelphia (1988) 101-124.

[252] Szegö, G, *Inequalities for certain eigenvalues of a membrane of given area*, J. Rational Mech. Anal. **3** (1954) 343-356.

[253] Temam, R., *Navier-Stokes equations*, North-Holland 1978.

[254] Vainberg, M. M., Trenogin, V. A., *The theory of branching of solutions of nonlinear equations*, Noordhoff, Leyden, 1974.

[255] Verzhbinskiĭ, G. M., Maz'ya, V. G., *On the asymptotics of solutions of the Dirichlet problem near a nonregular boundary*, Dokl. Akad. Nauk SSSR **176** (1967) 3, 498-501 (in Russian).

[256] Verzhbinskiĭ, G. M., Maz'ya, V. G., *Asymptotic behaviour of solutions of elliptic equations of second order near the boundary I, II*, Sibirsk. Mat. Zh. **12** (1971) 6, 1217-1249, **13** (1972) 6, 1239-1271.

[257] Verzhbinskiĭ, G. M., Maz'ya, V. G., *Closure in L_p of the Dirichlet problem operator in a region with conical points*, Izv. Vyssh. Uchebn. Zaved. Mat. **18** (1974) 6, 8-19.

[258] Volkov, E. A., *On the solution of boundary value problems for the Poisson equation in a rectangle*, Dokl. Akad. Nauk SSSR **147** (1962) 13-16.

[259] Volkov, E. A., *On the differential properties of boundary value problems for the Laplace and Poisson equations on a rectangle*, Tr. Mat. Inst. Steklova **77** (1965) 89-112; Engl. transl. in: Proc. Steklov Inst. Math. (1967) 101-126.

[260] Volkov, E. A., *On the differential properties of boundary value problems for the Laplace equation on a polygon*, Tr. Mat. Inst. Steklova **77** (1965) 113-142; Engl. transl. in: Proc. Steklov Inst. Math. (1967) 127-159.

[261] Vorovich, I. I., *Establishment of boundary value problems of in theory of elasticity with infinite energy integral and basic properties of homogeneous solutions*, in the book "Mechanics of deformating bodies", Moscow, Mashinostroenie 1975, pp. 112-128 (in Russian).

[262] Vorovich, I. I., Kopasenko, V. V., *Some problems of elasticity theory for a half-strip*, Prikl. Mat. Mekh. **30** (1966) 1, 109-115 (in Russian).

[263] Walden, H., Kellogg, R. B., *Numerical determination of the fundamental eigenvalue for the Laplace operator on a spherical domain*, J. Engrg. Math. **11** (1977) 299-318.

[264] Weinberger, H. F., *An isoperimetric inequality for the n-dimensional free membrane problem*, J. Rational Mech. Anal. **5** (1956) 633-636.

[265] Wendland, W., *Bemerkungen über die Fredholmschen Sätze*, In: Methoden u. Verfahren d. Mathematischen Physik, B. I. Verlag Mannheim **3** (1970) 141-176.

[266] Wendland, W. L., Schmitz, H., Bumb, H., *A boundary element method for three-dimensional elastic fields near reentrant corners*, The Mathematics of Finite Elements and Applications VI, MAFELAP 1987, Proc. 6th Conference Uxbridge/UK 1987 (1988) 313-322.

[267] Whittaker, E. T., Watson, G. N., *A course of modern analysis*, Vol. 2, Cambridge University Press, 1927.

[268] Williams, M. L., *Stress singularities resulting from various boundary conditions in angular corners of plates in extension*, J. Appl. Mech. **19** (1952) 4, 526-528.

[269] Wigley, N. M., *Asymptotic expansions at a corner of solutions of mixed boundary value problems*, J. Math. Mech. **13** (1964) 549-576.

[270] Wigley, N. M., *Mixed boundary value problems in domains with corners*, Math. Z. **115** (1970) 33-52.

[271] Wigley, N. M., *Differentiability at a corner for a solution of Laplace's equation*, Amer. Math. Monthly **79** (1972) 1107.

[272] Wigley, N. M., *Schauder estimates in domains with corners*, Arch. Rational Mech. Anal. **104** (1988) 3, 271-276.

[273] Zajaczkowski, W., Solonnikov, V. A., *Neumann problem for second order elliptic equations in domains with edges on the boundary*, Zap. Nauchn. Semin. Leningrad. Otdel. Mat. Inst. Steklov. **127** (1983) 7-48; English transl. in: J. Sov. Math. **27** (1984) 2561-2586.

[274] Zeidler, E., *Nonlinear functional analysis and its applications*, Vol. 1, Springer, New York-Berlin-Heidelberg-Tokyo 1986.

Index

adjoint operator pencil 15
admissible operator 27
Adolfsson, V. 58, 417
Agmon, S. 31, 255, 320, 344, 417
Agranovich, M. S. 31, 50, 255, 260, 417
algebraic multiplicity 11
Andersson, B. 59, 138, 417
Andrä, H. 138, 419
anisotropic elasticity 133, 303, 414
Apel, T. 219, 417
Aronszajn, N. 389, 417
Ashbaugh, M. S. 49, 57, 417
Aziz, A. K. 58, 417

Babuška, I. 59, 138, 248, 417, 427
Bateman, H. 97, 330, 417
Bažant, Z. P. 105, 132, 138, 417
Beagles, A. E. 59, 105, 417
Beltrami operator 40
Benguria, R. D. 49, 57, 417
Ben M'Barek 33, 417
Benthem, J. P. 138, 417
biharmonic equation 227ff
Birman, M. Sh. 58, 418
Blum, H. 248, 418
Boersma, J. 58, 418
Boussinesq, M. J. 97, 176, 418
Brown, S. N. 53, 58, 418
Bumb, H. 138, 428

canonical system 11
capacity 43, 90
Chavel, I. 57, 418
Cherepanov, G. P. 105, 123, 418
Costabel, M. 105, 138, 291, 418
Courant, R. 42, 45, 418

Dahlberg, B. E. J. 59, 106, 249, 418
Dauge 29, 32, 58, 105, 138, 196f, 290f, 345, 418
Deny, J. 91, 418
Deuring, P. 197, 418
Dimitrov, A. 138, 419
Dirichlet conditions 298
Donchev, T. 249, 423

Douglis, A. 27, 320, 344, 417, 419

Eck, C. 138, 419
eigenvalue 10
eigenvector 10
energy line 17
energy strip 17
Erdélyi, A. 97, 330, 417
Eskin, G. I. 31, 32, 291, 419
Estenssoro, L. F. 138, 417

Faber, G. 57, 419
Fabes, E. B. 106, 197, 419
Falk, U. 138, 417
Fichera, G. 58, 59, 419
Fourier transform 323, 335
Fredholm operator 10
Friedland, S. 44, 419
Fufaev, V. V. 58, 419

Gantmakher, F. 265, 270, 419
Gårding, L. 372
Gel'fand, I. M. 322, 328, 382, 400, 419
Gel'fond, O. A. 269, 419
generalized eigenvector 11
generating function 66
geometric multiplicity 10
Ghahremani, F. 138, 419
Girault, V. 192, 419
Gobert, J. 111, 419
Gohberg, I. 12, 13, 32, 67, 419
Goldberg, S. 12, 13,32, 419
Green formula 377
Green function 104, 194, 313, 358
Grisvard 29, 58, 105f, 138, 197, 248, 291, 419f
Gromov, M. 32, 420

Hanna, M. S. 58, 420
Hayman, W. K. 44, 419
Hilbert, D. 42, 45, 418
Hinder, R. 58, 420
holomorphic operator function 12
Hooke law 107, 110, 133
Hsiao, G. 138, 420

Hussain, M. A. 138, 426

index of an eigenvalue 11
index of an operator 29
indicator 301
inverse Fourier transform 339
isoperimetric property 44, 49, 57

Jerison, D. 58, 417, 420
John, F. 313, 317, 344, 420
Jordan chain 11

Kaashoek, M. A. 12, 13, 32, 419
Kalex, H.-U. 225, 420
κ-regular boundary conditions 297
Kato, T. 22, 238, 420
Keer, M. 105, 132, 417
Keldysh, M. V. 32, 420
Keller, J. B. 58, 420
Kellogg, R. B. 33, 58, 196f, 420, 428
Kelvin transform 40, 62, 83, 139, 154
Kenig, C. E. 59, 106, 197, 249, 418f, 420f
Komech, A. I. 58, 291, 421
Kondrat'ev, V. A. 29, 31f, 58, 106, 248f,
    291, 387, 421f
Kopachek, I. 106, 249, 421
Kopasenko, V. V. 138, 428
Korn's inequality 390
Kozlov, V. A. 13, 29, 32, 58, 105, 138, 178,
    197, 225f, 248, 291, 305, 344, 387, 417,
    421f
Krahn, E. 57, 422
Krasnosel'skiĭ, M. A. 42, 422
Kufner, A. 58, 422

Ladyzhenskaya, O. A. 192, 423
Lamé system 61ff, 385
Lancaster, P. 67, 419
Landkof, N. S. 43, 91, 423
Laplace operator 36, 40
leading part of a differential operator 27
Leguillon, D. 33, 138, 423
Lemrabet, K. 33, 423
Levin, A. V. 57, 423
Levin, B. J. 301, 423
Liebowitz, H. 105, 423
Lions, J.-L. 35, 91, 225, 418, 423
local coerciveness 389
Lopatinskiĭ, Ya. B. 31, 32, 423
Lozi, R. 196, 423

Magenes, E. 35, 225, 423
Magnus, W. 44, 423
Malvern, L. E. 78, 110, 423
Markus, A. S. 32, 423
Maslovskaya, L. V. 248, 423
maximum principle 104

Maz'ya, V. G. 13, 29, 32, 43, 45, 52, 57ff,
    105, 138, 142, 178, 194, 197, 199, 225f,
    247ff, 289ff, 344, 384, 387, 389, 417,
    422ff, 427
Meister, E. 33, 58, 420, 425
Mellin transform 9, 356
Melrose, R. B. 32, 425
Melzer, H. 248, 425
Mennicken, R. 32, 425
Merigot, M. 33, 417
Merzon, A. E. 291, 421
Mikhlin, S. G. 40, 425
Miranda-Agmon maximum principle 247,
    249
Möller, M. 32, 425
Morozov, N. F. 248, 423

Nadirashvili, N. S. 45, 57, 425
Nazarov, S. A. 29, 32, 52, 58, 138, 290f, 344,
    384, 387, 419, 423ff
Neumann conditions 298
Nicaise, S. 33, 58, 105f, 138, 291, 417f, 425
Nikishkin, V. A. 291, 426
Nikol'skiĭ, S. M. 58, 426
Nirenberg, L. 27, 31, 255, 320, 344, 417, 419
Noble, B. 138, 426

Oleĭnik, O. A. 32, 106, 249, 421, 426
operator pencil 10
Orlt, M. 225, 426
Osborn, J. E. 196f, 420, 426

Parseval equality 356
partial multiplicity 11
Parton, V. Z. 105, 426
Pazy, A. 344, 426
Penzel, F. 33, 425
Perlin, P. I. 106, 426
von Petersdorff, T. 59, 138, 417
Pipher, J. 59, 249, 421, 426
Plamenevskiĭ, B. A. 29, 32, 52, 58, 105, 142,
    194, 197, 199, 247ff, 289ff, 344, 384, 387,
    423ff
Poisson kernel 307, 320
Pólya, G. 57, 426
polyharmonic operator 228, 239ff, 329, 384
Pu, S. L. 138, 426

Rannacher, R. 248, 418, 425
Raviart, P.-A. 192, 419
regular boundary conditions 297
regular point 10
Richter, U. 105, 138, 426
Rodman, L. 67, 419
Roßmann, J. 29, 32, 58, 138, 225, 248f, 290f,
    387, 422, 425f
Rouché's theorem 13

Samarskiĭ, A. A. 52, 58, 426
Sanchez-Palencia, E. 33, 138
Sändig, A.-M. 33, 58, 105, 138, 225, 417, 425f
Schmitz, H. 59, 138, 426, 428
Schnack, E. 138, 419
Schulze, B.-W. 32, 291, 427
Schwab, C. 32, 105, 138, 178, 197, 422
Seif, J. B. 248, 427
Shapiro-Lopatinskiĭ condition 297
Shaposhnikova, T. O. 59, 425
Shen, C. L. 57, 427
Shieh, C. T. 57, 427
Shilov, G. E. 322, 328, 382, 400, 419
Shubin, M. A. 32, 420, 427
Sigal, E. I. 32, 419, 423
simple eigenvalue 38
Skvortsov, G. E. 58, 418
Smith, K. T. 58, 420
Solonnikov, V. A. 58, 199, 225, 320, 344, 427f
Soni, R. P. 44, 423
Speck, F.-O. 33, 425
spectrum 10
Sperner, E. 44, 57, 427
spherical components 75
spherical divergence 75
spherical gradient 75
Stein, M. E. 330, 427
Stephan, E. P. 59, 138, 420, 427
Stewartson, K. 53, 58, 418
Stokes system 3, 139ff, 385
strain tensor 107
stress tensor 107
strongly elliptic operator 309
Stupyalis, L. I. 199, 225, 424
subspace of first kind 212
subspace of second kind 212
Szabo, B. A. 248, 427
Szegö, G. 57, 426, 427

Tavkhelidze, I. N. 249, 426
Teixeira, F. S. 33, 425
Temam, R. 149, 160, 427
Trenogin, V. A. 124, 270, 282, 427

Vainberg, M. M. 124, 270, 282, 427
variational principle 42, 91f, 119, 168ff, 221
Verchota, G. C. 106, 197, 249, 418f, 426
Verzhbinskiĭ, G. M. 58, 427
Vishik, M. I. 31, 50, 255, 260, 417
Volk, K. 59, 138, 426
Volkov, E. A. 58, 427, 428
Vorovich, I. I. 105, 138, 428

Walden, H. 59, 428
Watson, G. N. 187, 428
Weinberger, H. F. 50, 57, 426, 428

Weiss, G. 330, 427
Wendland, W. L. 32, 59, 138, 419f, 426, 428
Whiteman, J. R. 59, 417, 427
Whittaker, E. T. 187, 428
Wigley, N. M. 58, 428
Williams, M. L. 248, 428

Yosifyan, G. A. 106, 249, 426

Zajaczkowski, W. 58, 428
Zeidler, E. 42, 428

List of Symbols

Chapter 1

$L(\mathcal{X}, \mathcal{Y})$ space of linear and continuous operators $\mathcal{X} \to \mathcal{Y}$, 10

$\mathcal{R}(A)$ range of the operator A, 10

$\Phi(\mathcal{X}, \mathcal{Y})$ space of Fredholm operators, 10

$\mathbb{C}$ set of complex numbers

$m(\varphi_0)$ multiplicity of the eigenvector φ_0, 11

$\kappa(\mathfrak{A}, G)$ sum of algebraic multiplicities of eigenvalues in G, 12

$\partial_r = d/dr$ derivative, 13

$\mathcal{N}(\mathfrak{A}, \lambda_0)$ set of power-logarithmic solutions, 13

$\mathfrak{A}^*(\lambda)$ adjoint operator pencil, 15

$\mathcal{G}$ domain in $\mathbb{R}^n$, 26

$V^l_{p,\vec{\beta}}(\mathcal{G})$ weighted Sobolev space, 28

$N^{l,\sigma}_{\vec{\beta}}(\mathcal{G})$ weighted Hölder space, 31

$V^{l-1/p}_{p,\vec{\beta}}(\partial\mathcal{G})$, $N^{l,\sigma}_{\vec{\beta}}(\partial\mathcal{G})$ trace spaces, 28, 31

Chapter 2

$\mathbb{R}^n$ Euclidean space

$\mathcal{K}$ angle or cone, 36, 40

r, φ polar coordinates in $\mathbb{R}^2$, 36

Δ Laplace operator, 36, 40

$W^l_2((0,\alpha))$, $\overset{\circ}{W}{}^l_2((0,\alpha))$ Sobolev spaces, 37

$W^{-1}_2((0,\alpha))$ dual space to $\overset{\circ}{W}{}^1_2((0,\alpha))$, 37

S^{n-1} unit sphere

Ω domain on the unit sphere, 40

$\partial\Omega$ boundary of Ω

$r, \theta_1, \ldots, \theta_{n-1}$ spherical coordinates in $\mathbb{R}^n$, 40

$q_j = (\sin\theta_1 \cdots \sin\theta_{j-1})^2$, 40

δ Beltrami operator, 40

$d\omega$ measure on Ω, 41

$C^\infty_0(\Omega)$ set of infinitely differentiable functions with support in Ω, 41

$W^1_2(\Omega)$, $\overset{\circ}{W}{}^1_2(\Omega)$ Sobolev spaces, 41

$W^{-1}_2(\Omega)$ dual space to $\overset{\circ}{W}{}^1_2(\Omega)$, 41

Λ_j eigenvalues of the operator $-\delta$ with Dirichlet boundary condition, 41

$\mathbb{R}^n_+$ half-space, 42

S^{n-1}_+ half-sphere, 42

$\text{cap}\, K$ capacity of K, 43

ν exterior normal to $\partial\mathcal{K}\backslash\{0\}$, 46

∇U gradient of U

$C^\infty_0(\overline{\mathcal{K}}\backslash\{0\})$ set of infinitely differentiable functions with compact support in $\overline{\mathcal{K}}\backslash\{0\}$, 46

N_j eigenvalues of $-\delta$ with Neumann boundary condition, 46

Chapter 3

U displacement vector, 61

$\nabla \cdot U$ divergence of U

ν Poisson ratio, 61

$\mathcal{K}$ angle or cone, 64, 74

U_r, U_φ polar components of U, 64

$\mathfrak{A}$ operator pencil, 65

$d_\pm(\lambda)$ functions determining the eigenvalues of $\mathfrak{A}$, 68

Ω domain on the sphere, 74

u vector function on Ω,

r, θ, φ spherical coordinates, 75

u_r, u_θ, u_φ spherical components of u, 75

$J(\theta, \varphi)$ matrix, 75

$\omega = (\theta, \varphi)$, 75

u_ω vector with components u_θ, u_φ, 75

433

$\nabla_\omega \cdot u_\omega$ spherical divergence, 75

$\nabla_\omega v$ spherical gradient, 75

$Q(\cdot, \cdot)$ sesquilinear form, 75

h_2^1, $\overset{\circ}{h}{}_2^1$ function spaces, 75, 76

$\mathcal{L}$ operator pencil, 78

H matrix, 79

$\mathcal{T}(\lambda)$ matrix, 81

Γ, F_ν set functions, 83

$q(u, u; \lambda)$ quadratic form, 84

$\Omega_0 \subset \Omega_1$ relation between sets, 88

$A(\lambda)$ operator pencil, 91

$\tilde{a}(\cdot, \cdot; \lambda)$ sesquilinear form, 91

M set function, 93

$V_{\vec{\beta}, \vec{\delta}}^{l,p}(\mathcal{G})$ weighted Sobolev space, 102

$N_{\vec{\beta}, \vec{\delta}}^{l,\sigma}(\mathcal{G})$ weighted Hölder space, 102

M_τ positive number, 103

Chapter 4

U displacement vector, 107

n exterior normal to the boundary, 107

$\{\varepsilon_{i,j}(U)\}$ strain tensor, 107

$\{\sigma_{i,j}(U)\}$ stress tensor, 107

μ shear modulus, 107

ν Poisson ratio, 107

$\mathcal{K}$ angle or cone, 108, 120, 126, 129, 133

Γ_k part of the boundary $\partial\mathcal{K}$, 108

Ω domain on S^2, 108

γ_k part of the boundary $\partial\Omega$, 108

I_0, I_n, I_τ sets of indices, 108

$\mathcal{H}$ subspace of $W_2^1(\Omega)^3$, 109

$[\cdot, \cdot]$ sesquilinear form on $h_2^1(\Omega)$, 110

$a(\cdot, \cdot; \lambda)$ sesquilinear form, 110, 126

$\mathcal{H}_s$ subspace of $W_2^1(\Omega) \times h_2^1(\Omega)$, 112

$\mathfrak{A}$ operator pencil, 112, 121, 126

$\mathcal{H}_s^r$ subspace of $W_2^1(\Omega)$, 114

$\mathcal{H}_s^\omega$ subspace of $h_2^1(\Omega)$, 114

$q(\cdot, \cdot; \lambda)$ sesquilinear form, 116

α_* solution of the equation $\tan\alpha = \alpha$ in $(0, 2\pi)$, 123

α_{**} second positive root of $\tan\alpha = \alpha$, 123

α_0, α_1 certain positive numbers, 123

$M_m^\pm$ matrices, 129

$m_r^\pm$, $m_\theta^\pm$, $m_\varphi^\pm$ rows of $M_m^\pm$, 129

$N_m^\pm$ matrices, 130

H plane angle, 133

$h = H \cap S^2$ arc on the sphere, 133

Chapter 5

U velocity

P pressure

$\mathcal{K}$ angle or cone, 142, 148, 178

U_r, U_φ polar components of U, 142

$\mathfrak{A}$ operator pencil, 143

$d_\pm(\lambda)$ functions determining the eigenvalues of $\mathfrak{A}$, 144

$\mathcal{L}$ operator pencil, 148

$Q(\cdot, \cdot)$ sesquilinear form, 149

$a(\cdot, \cdot; \lambda)$ sesquilinear form, 149

$\mathcal{S}(\lambda)$ matrix, 153

J_λ matrix, 155

$F(\Omega)$, $\Gamma(\Omega)$ set functions, 155, 156

$\mathcal{X}(\Omega)$ function space, 163

$\mathcal{N}_j$ certain eigenvalues, 163

Ω_α spherical domain, 164

b, b_ω sesquilinear forms, 169

$\mathcal{Y}(\Omega)$ function space, 169

Chapter 6

U velocity

n exterior normal to the boundary, 199

U_n normal component of U, 199

U_τ tangential component of U, 199

$\varepsilon(U)$ strain tensor, 199

$\varepsilon_n(U) = \varepsilon(U) \cdot n$, 199

$\varepsilon_{n,n}(U)$ normal component of $\varepsilon_n(U)$, 199

$\varepsilon_{n,\tau}(U)$ tangential component of $\varepsilon_n(U)$, 199

$\mathcal{K}$ cone, 200, 223

Γ_k part of the boundary $\partial\mathcal{K}$, 200

$\gamma_k = \Gamma_k \cap S^2$, 200

I_0, I_n, I_τ sets of indices, 200

$\mathcal{H}$ subspace of $W_2^1(\Omega)^3$, 201

$\mathcal{H}_s$ subspace of $W_2^1(\Omega) \times h_2^1(\Omega)$, 202

$[\cdot, \cdot]$ sesquilinear form, 202

$a(\cdot, \cdot; \lambda)$ sesquilinear form, 202, 224

$\mathfrak{A}$ operator pencil, 202, 224

$\mathcal{H}_s^{(\lambda)}$ subspace of $\mathcal{H}_s$, 204

$\mathcal{H}_s^r$ subspace of $W_2^1(\Omega)$, 205

$\mathcal{H}_s^\omega$ subspace of $h_2^1(\Omega)$, 205

J_λ matrix, 207

$s(\cdot,\cdot;\lambda)$ sesquilinear form, 207

$\mathcal{H}_c$ set of constant vectors in $\mathcal{H}$, 210

Problem P, 214

Chapter 7

$\mathcal{K}$ angle or cone, 229, 233

α opening of the angle, 229

$\mathfrak{A}$ operator pencil, 230

$d_\pm(\lambda)$ functions determining the eigenvalues of $\mathfrak{A}$, 230

$u^\pm(\lambda,\cdot)$ eigenfunctions of $\mathfrak{A}$, 232

Ω domain on the sphere, 233

$\overset{\circ}{W}{}_2^2(\Omega)$ Sobolev space, 233

$W_2^{-2}(\Omega)$ dual space of $\overset{\circ}{W}{}_2^2(\Omega)$, 233

$\mathcal{L}$ operator pencil, 233, 239

$a(\cdot,\cdot;\lambda)$ sesquilinear form, 234

$\mu_0(\Omega)$ first eigenvalue of $-\delta$ on Ω, 234

$\sigma_0(\Omega)$ set function, 234

$\sigma_1(\Omega)$ set function, 236

M positive number such that $M(M+n-2)=\mu_0$, 237

$\overset{\circ}{W}{}_2^m(\Omega)$ Sobolev space, 239

$\mu_j(\lambda)$ eigenvalue of the operator $\mathcal{L}(\lambda)$, 241

Chapter 8

$\mathcal{K}$ plane angle, 254

α opening of the angle, 254

$L(\partial_{x_1},\partial_{x_2})$ differential operator, 255, 283

$\mathcal{L}$ operator pencil, 255, 283

$\mathcal{B}_\alpha^{(k)}(\varphi,\partial_\varphi,\lambda)$ ordinary differential operator of order k, 256

$\mathfrak{A}$ operator pencil, 256

$v_l(\lambda,\varphi)$ solutions of a Cauchy problem, 256

$\Psi(\lambda)$ quadratic matrix, 256

$z_1,\dots,z_d$ roots of $L(1,z)=0$, 257

$\nu_1,\dots,\nu_d$ multiplicities of the roots $z_1,\dots,z_d$, 257

$(\mu)_l=\mu(\mu+1)\cdots(\mu+l-1)$, 257

$w_{q,j}$ special solutions of $\mathcal{L}(\varphi,\partial_\varphi,\lambda)=0$, 257

T matrix, 258

$\sigma_k(z,\lambda)$ complex function, 258

$\mathcal{D}(\lambda)$ matrix, 258

$\sigma(z)=\cos\alpha+z\sin\alpha$, 259

$D(\lambda)$ matrix, 259

T_0 matrix, 260

U_k, V_k vectors, 262

$\tilde{\sigma}(z,\varepsilon)=\cos\varepsilon-z\sin\varepsilon$, 264

ω complex number, 264, 266

$\mathcal{N}(\omega,\tau)$ matrix, 266

Chapter 9

$\mathcal{K}$ plane angle, 293

α opening of the angle, 293

$L, B_{0,k}, B_{1,k}$ differential operators, 293

$\mathcal{L}, \mathcal{B}_{0,k}, \mathcal{B}_{1,k}$ parameter-depending differential operators, 294

$\mathfrak{A}$ operator pencil, 294

$z_1,\dots,z_d$ roots of $L(1,z)=0$, 294

$\nu_1,\dots,\nu_d$ multiplicities of the roots $z_1,\dots,z_d$, 294

$W_{q,j}, w_{q,j}$ functions, 294

$U_k, V_k(\lambda)$ vectors, 295

$\mathcal{D}(\lambda)$ matrix, 295

Λ_m set of vectors, 295

$\mathcal{P}_m$ polygon, 295

Γ_m set of corners of $\mathcal{P}_m$, 295

Λ_m' set of vectors, 297

Chapter 10

$D_x=-i(\partial_{x_1},\dots,\partial_{x_n})$, 308

$W_2^m(\mathcal{G}), \overset{\circ}{W}{}_2^m(\mathcal{G})$ Sobolev spaces, 308

$C_0^\infty(\mathcal{G})$ set of infinitely differentiable functions with compact support in $\mathcal{G}$, 308

$W_2^{-m}(\mathcal{G})$ dual space of $\overset{\circ}{W}{}_2^m(\mathcal{G})$, 308

S^{n-1} $(n-1)$-dimensional unit sphere

Ω subset of S^{n-1}, 308

$\omega=x/|x|$

$\mathcal{K}$ n-dimensional cone, 308, 332

$\overset{\circ}{W}{}_{2,loc}^m(\mathcal{K},0)$ function space, 308

$W_{2,loc}^{-m}(\mathcal{K},0)$ space of distributions, 308

$L_2(\Omega)$ space of square summable functions on Ω, 308

$W_2^m(\Omega), \overset{\circ}{W}{}_2^m(\Omega)$ Sobolev spaces, 309

L differential operator, 309
L^+ formally adjoint operator to L, 309
$\mathcal{L}, \mathcal{L}^+$ operator pencils, 310
$\mathcal{M}$ pencil generated by the poly-
     harmonic operator, 311
I_ℓ $\ell \times \ell$ identity matrix, 313
G Green matrix of L, 313
$\Pi_k^{(n)}$ space of homogeneous polynomi-
     als of degree k with n variables, 314
$S'(\mathbb{R}^n)$ space of tempered distributions
$\mathbb{R}^n_+$ half-space, 319
G_j Poisson kernel, 320
$\mathbf{R}^1$ line in $\mathbb{R}^n$, 321
$\mathfrak{R}$ ray in $\mathbb{R}^n$, 321
$x_{n+}^{\lambda-1}$ distribution, 322
$\hat{U}$ Fourier transform of U, 323

$\mathcal{K}$ cone, 389
Ω domain on the sphere, 389
$b(\cdot,\cdot)$ sesquilinear form, 389
ν exterior normal, 389
Q_α differential operator on the sphere,
     391
$a(\cdot,\cdot;\lambda)$ sesquilinear form, 391
$\mathfrak{A}$ operator pencil, 391
$b_0(\cdot,\cdot)$ sesquilinear form, 393
$a_0(\cdot,\cdot;\lambda)$ sesquilinear form, 393
$\Pi_k^{(n)}$ set of homogeneous polynomials
     of degree k, 396
ϕ function, 398
T_δ integral operator, 399
S_δ integral operator, 400
$\overset{\circ}{\Pi}$ set of polynomials, 412

Chapter 11

$\mathcal{K}$ cone $x_n > \phi(x')$, 346, 365
Ω domain on the sphere, 346
L differential operator, 346, 353, 376
$\mathcal{L}$ operator pencil, 346, 368, 376
$\nu = (\nu_1, \dots, \nu_n)$ exterior normal, 346
$a(\cdot,\cdot;\lambda)$ sesquilinear form, 352, 354
$\omega = (\omega_1, \dots, \omega_n) = x/|x|$ coordinates
     on the unit sphere
∂_{ω_j} differential operator on the sphere,
     354
G Green function, 358
α_j interior angle between plane parts
     of $\partial\mathcal{K}$, 365
$\mathfrak{R}$ ray, 368
N north pole of the unit sphere, 368
G_β vector function on S^{n-1}, 368
Ω_ε domain on the sphere, 372, 376,
     383
$\mathcal{L}_\varepsilon$ operator pencil, 372, 377, 383
$\mathcal{K}_\varepsilon$ cone, 376, 383
G_{m-1} function on $\mathbb{R}^n_+$, 377

Chapter 12

L differential operator, 389
$A_{\alpha,\beta}$ $\ell \times \ell$ matrices, coefficients of L,
     389

Selected Titles in This Series

(*Continued from the front of this publication*)

53 **Andreas Kriegl and Peter W. Michor,** The convenient setting of global analysis, 1997

52 **V. A. Kozlov, V. G. Maz'ya, and J. Rossmann,** Elliptic boundary value problems in domains with point singularities, 1997

51 **Jan Malý and William P. Ziemer,** Fine regularity of solutions of elliptic partial differential equations, 1997

50 **Jon Aaronson,** An introduction to infinite ergodic theory, 1997

49 **R. E. Showalter,** Monotone operators in Banach space and nonlinear partial differential equations, 1997

48 **Paul-Jean Cahen and Jean-Luc Chabert,** Integer-valued polynomials, 1997

47 **A. D. Elmendorf, I. Kriz, M. A. Mandell, and J. P. May (with an appendix by M. Cole),** Rings, modules, and algebras in stable homotopy theory, 1997

46 **Stephen Lipscomb,** Symmetric inverse semigroups, 1996

45 **George M. Bergman and Adam O. Hausknecht,** Cogroups and co-rings in categories of associative rings, 1996

44 **J. Amorós, M. Burger, K. Corlette, D. Kotschick, and D. Toledo,** Fundamental groups of compact Kähler manifolds, 1996

43 **James E. Humphreys,** Conjugacy classes in semisimple algebraic groups, 1995

42 **Ralph Freese, Jaroslav Ježek, and J. B. Nation,** Free lattices, 1995

41 **Hal L. Smith,** Monotone dynamical systems: an introduction to the theory of competitive and cooperative systems, 1995

40.4 **Daniel Gorenstein, Richard Lyons, and Ronald Solomon,** The classification of the finite simple groups, number 4, 1999

40.3 **Daniel Gorenstein, Richard Lyons, and Ronald Solomon,** The classification of the finite simple groups, number 3, 1998

40.2 **Daniel Gorenstein, Richard Lyons, and Ronald Solomon,** The classification of the finite simple groups, number 2, 1995

40.1 **Daniel Gorenstein, Richard Lyons, and Ronald Solomon,** The classification of the finite simple groups, number 1, 1994

39 **Sigurdur Helgason,** Geometric analysis on symmetric spaces, 1994

38 **Guy David and Stephen Semmes,** Analysis of and on uniformly rectifiable sets, 1993

37 **Leonard Lewin, Editor,** Structural properties of polylogarithms, 1991

36 **John B. Conway,** The theory of subnormal operators, 1991

35 **Shreeram S. Abhyankar,** Algebraic geometry for scientists and engineers, 1990

34 **Victor Isakov,** Inverse source problems, 1990

33 **Vladimir G. Berkovich,** Spectral theory and analytic geometry over non-Archimedean fields, 1990

32 **Howard Jacobowitz,** An introduction to CR structures, 1990

31 **Paul J. Sally, Jr. and David A. Vogan, Jr., Editors,** Representation theory and harmonic analysis on semisimple Lie groups, 1989

30 **Thomas W. Cusick and Mary E. Flahive,** The Markoff and Lagrange spectra, 1989

29 **Alan L. T. Paterson,** Amenability, 1988

28 **Richard Beals, Percy Deift, and Carlos Tomei,** Direct and inverse scattering on the line, 1988

27 **Nathan J. Fine,** Basic hypergeometric series and applications, 1988

26 **Hari Bercovici,** Operator theory and arithmetic in H^∞, 1988

For a complete list of titles in this series, visit the
AMS Bookstore at **www.ams.org/bookstore/**.